14th International Conference "Experimental Fluid Mechanics" (EFM 2019)

EPJ Web of Conferences Volume 269 (2022)

Frantiskovy Lazne, Czech Republic
19 – 22 November 2019

Editors:

J. Pulec
J. Novosad
P. Dančová

ISBN: 978-1-7138-6268-0

Printed from e-media with permission by:

Curran Associates, Inc.
57 Morehouse Lane
Red Hook, NY 12571

Some format issues inherent in the e-media version may also appear in this print version.

For additional information, please contact EDP Sciences – Web of Conferences at the address below.

EDP Sciences – Web of Conferences
17, Avenue du Hoggar
Parc d'Activité de Courtabœuf
BP 112
F-91944 Les Ulis Cedex A
France

Phone: +33 (0) 1 69 18 75 75
Fax: +33 (0) 1 69 28 84 91

contact-edps@webofconferences.org

Additional copies of this publication are available from:

Curran Associates, Inc.
57 Morehouse Lane
Red Hook, NY 12571 USA
Phone: 845-758-0400
Fax: 845-758-2633
Email: curran@proceedings.com
Web: www.proceedings.com

TABLE OF CONTENTS

Experimental Fluid Mechanics 2019

Františkovy Lázně, Czech Republic, November 19th-22nd 2019
efm.kez.tul.cz
efm.conf@gmail.com

14th International Conference "Experimental Fluid Mechanics 2019" was held in the INGO Casino Františkovy Lázně (Czech Republic), November 19th–22nd, 2021.

The EFM conference is focused on obtaining overviews of projects under investigation and experimental methods used in the field of fluid mechanics and thermodynamics. Its aim is to organize a network of research workers who are interested in experimental and also theoretical work in the field of fluid mechanics and thermodynamics. The conference should contribute to the creation of closer contacts and sharing experience in the application of various experimental methods, preparation and implementation of experiments, processing of results, and numerical simulations of experiments.

Keynote lectures for 2019 were:

Dr. Marco La Mantia from The Charles University: *Visualization of superfluid helium flows.*

Dr. Sergeev Daniil from The Russian Academy of Sciences: *Experimental investigation of multiphase hydrodynamics of the ocean-atmosphere boundary layer within laboratory modeling.*

We would like to thank our sponsors (in alphabetical order)
DANTEC DYNAMICS GmbH (www.dantecdynamics.com),
LAVISION (http://www.lavision.de/en/),
MIT s.r.o. (www.mit-laser.cz),
TSI GmbH (www.tsi.com),
4JTECH (http://4jtech.cz/en/247-2/).

Františkovy Lázně

INGO Casino

We would like to invite you to the EFM 2022 held in Safari Park in Dvůr Králové nad Labem, November 29 - December 2, 2022. Please visit our web pages efm.kez.tul.cz, or for any info, contact us via efm.conf@gmail.com or efm@tul.cz.

Safari Park Dvůr Králové nad Labem (conference venue for EFM 2022)

Numerical investigation on NACA0012 airfoil with tubercular structure

Halil Hakan Açıkel[1,*], *Melike* Tosun[1], *Mustafa Serdar* Genç[1], *Kemal* Koca[1]

[1]Wind Engineering and Aerodynamic Research Center, Department of Energy Systems Engineering, Erciyes University, KAYSERİ, TURKEY

Abstract. Humpback whales have very special fins, which reduce the friction force by the projections on the fins and enhance the lift force with increasing angle of attack. Thus, it has opportunity to move with less energy. In this study, the tubular structure was formed and adapted to the NACA0012 airfoil profile. Numerical modelling of tubercle wing was performed in ANSYS FLUENT. In numerical modelling, k-ω SST and Transition SST turbulence models were used to investigate tubercle wing performance. The tubercle structure on the wing provided to canalize the flow in the tubercle cavities, thus flow separation decreased. Consequently, it was observed that the aerodynamic performance of the wing increased when the sinusoidal protrusions known as tubercle structure were applied to the leading edge of the wing.

1 Introduction

All living creatures in nature have optimized their systems in close familiarity to perfection, depending on their living conditions and ecosystems. People inspired these systems tried to find solutions to their problems by imitating nature and living things. Recently, many different wing structures have been obtained in the leading edge region of wing inspired by the fin structure of the humpback whales. These structures, which are affected by the morphology of humpback whales, are referred to as "tubercle structure". The aerodynamic properties of the modified tubercle wing were investigated in the wind tunnel and compared with the rigid NACA 0020 [1]. The lift and drag coefficients of the modified wing profile were measured experimentally in a water tunnel and compared with the NACA 63(4)421 [2]. Pedro and Kobayashi [3] were studied two different aerofoils and modified tubular structure wing at low Reynolds number both experimentally and numerically. Yao et al. [4] were examined aerodynamic performance of aerofoil by using numerical methods. They investigated the effects of angle of attack on aerodynamic coefficients and surface pressure. Rostamzadeh et al. [5] were done a numerical study to investigate flow over leading edge tubercle airfoil at high and low Reynolds (Re) numbers. The numerical analyzes were performed by using computational fluid dynamics at Re=120000 and 1500000. In these analyzes, it was aimed to see the effects of Re number on tubercle wing. Şahin and Acır [6] examined NACA0015 airfoil performance both experimentally and numerically. In numerical simulation, Spalart Almaras and k-ε models was used to compare airfoil performance with experimental results. The experimental results had a good agreement with Spalart Almaras turbulence model. A numerical study of leading edge tubercles at low and high Reynolds numbers was done by Gawad [7]. Another experimental analysis of the tubercle wings with an aspect ratio of 4 with the Re number range between 700,000 to 3,000,000 was performed by Rocha et.al. [8].

In this study, tubercle structure design inspired by the fin of the humpback whale was made to the leading edge by using the wing profile of NACA0012.

2 Numerical Modelling

In the design, which is shown in Figure 1a and Figure 1b, the chord length is 0.1 m based on the NACA-0012 profile. The new design of aerofoil's span is divided into five tubercle structures. The tubercles amplitude is 0.1 chord length (c_t). The half wave length (λ) is 0.011 m. Operating Reynolds number is 1.36×10^5 in this study.

a)

* Corresponding author: hhacikel@erciyes.edu.tr

b)

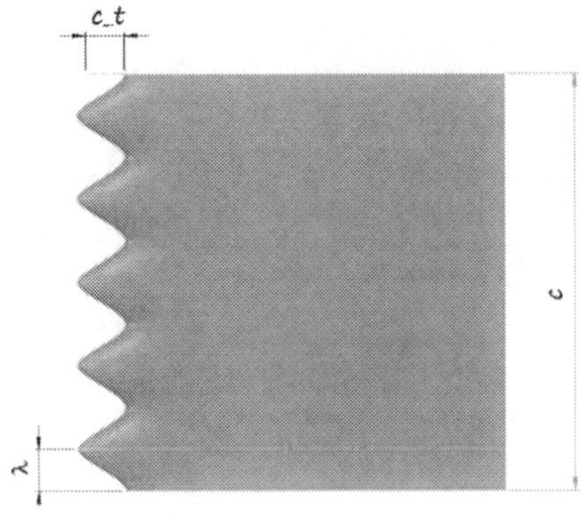

Fig. 1. The design of aerofoil.

Figure 2 represents the mesh structure of the flow domain. The flow domain was extended to 20c towards the trailing edge of the airfoil and 10c in other directions. The y^+ value is lower than 1 in these numerical simulations. The k-ω SST and Transition SST turbulence models were utilized for numerical calculations. The solution algorithm was selected SIMPLE, and momentum and turbulence kinetic energy solution methods were selected as Second Order Upwind method.

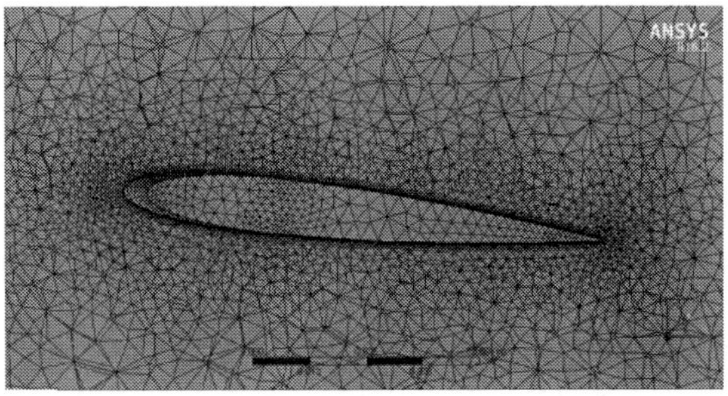

Fig. 2. The mesh structure of the wing.

2.1 Grid independence study

The grid independence study was conducted to determine mesh independence for numerical simulations which was shown in Table 1. The table gives information about mesh number and aerodynamic force coefficients. According to the results obtained from numerical analysis, 4.6 million meshes were selected for numerical solutions.

Table 1. Grid independence study

Mesh number	C_L	C_D
2006584	1.97399e-01	2.78839e-02
3232490	1.97678e-01	2.80084e-02
4673533	1.97678e-01	2.80084e-02

3 Results

3.1. Aerodynamic Coefficients

Non-dimensional aerodynamic coefficients such as lift and drag coefficients were used to investigate aerodynamic performance of the wing. These coefficients are expressed in follow respectively.

$$C_L=[(2L))/((\rho V^2A)] \quad (1)$$

$$C_D=[(2D))/((\rho V^2A)] \quad (2)$$

In these equations, ρ is air density, V is free stream velocity, A is area and L and D are forces.

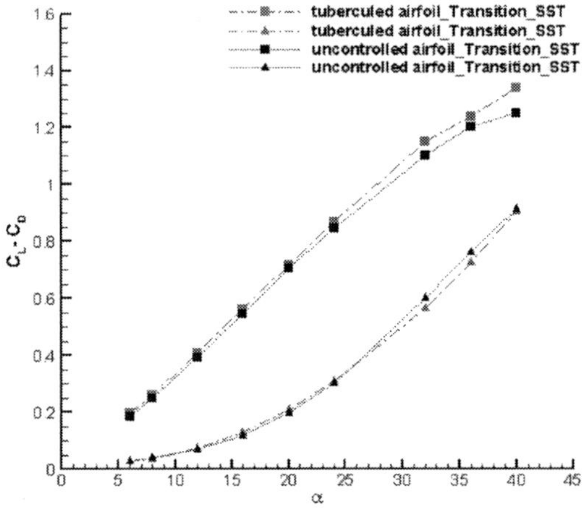

Fig. 3. Aerodynamic for coefficients for the Transition SST turbulence model.

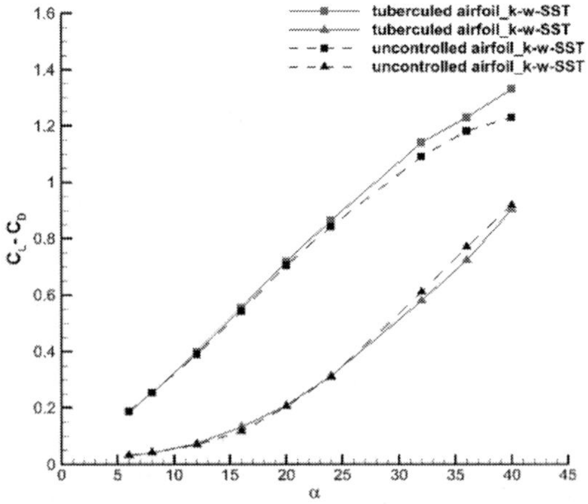

Fig. 4. Aerodynamic for coefficients for the k-ω SST turbulence model.

Fig. 3 shows aerodynamic force coefficients for tubercle and rigid wings at different angles of attack. In lower angles of attack, tubercle and uncontrolled wings presented similar performance as seen from two figures. As the angle of attack increased, the tubercled wings had a good aerodynamic performance, which meant the lift coefficient increased and the drag coefficient decreased. At angles of attack between 25°-40°, the lift coefficient was increased by using tubercle structures, and the drag coefficient was decreased slightly.

Due to the fact that both turbulence models presented similar performance, flow characteristics results on the tubercled and clean wing were given for Transition SST turbulence model. Transition SST turbulence model is a transition model which is a successful model on low Re number flows [9-13].

3.2 Velocity Distributions

The velocity distributions give information about flow characteristics over airfoil. The velocity contours on the clean and tubercled wings airfoils at α=24° and α=32° in Transition SST model was presented in Figure 5. At these angles of attack, the tubercle gaps canalized the flow and using the tubercle structure helped the flow control.

a) α=24°

b) α=32°

c) α=24°

d) α=32°

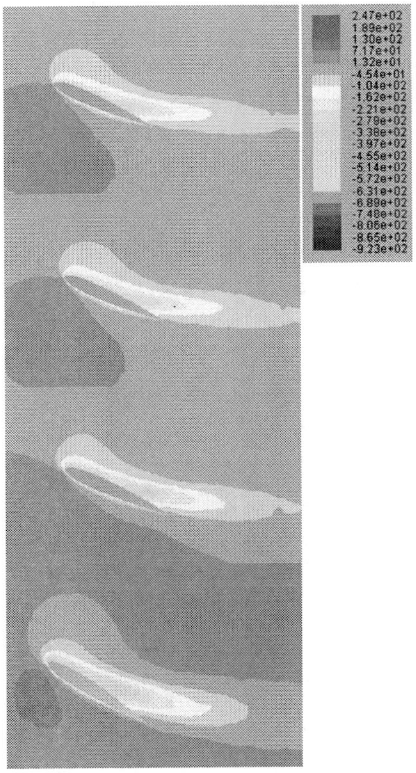

Fig. 5. The velocity contours over the wings in the middle plane at α=24° and α=32° with Transition SST model for rigid wing (a-b) and for tubercle wing (c-d).

The velocity contours at cross plane of the wings presented in Figure 6. The separated flow and growing with the increasing of the angle of attack over the clean wing was showed in Figure 6 as the blue region at the rear of the wing. The tubercle structure on the wing provided to canalize the flow in the tubercle cavities, thus flow separation decreased.

a) α=24°

b) α=32°

c) α=24°

d) α=32°

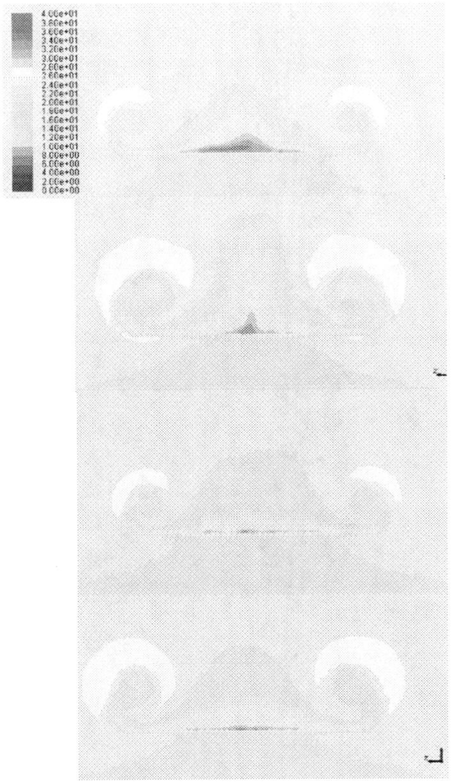

Fig. 6. The velocity contours at cross plane of the wings at α=24° and α=32° with Transition SST model for rigid wing (a-b) and for tubercle wing (c-d).

3.3 Turbulence Kinetic Energy

The turbulence kinetic energy shows the transition from laminar to turbulence and flow separation over the airfoil/wing. The turbulence kinetic energy distributions on the clean and tubercled wings airfoils at α=24° and α=32° in Transition SST model was given in Figure 7. The flow on the clean wing separated from the trailing edge of the wing and the flow separation enlarged with the increasing of the angle of attack. Utilizing the tubercle structure on the wing provided to trigger the transition. The flow over the wing become the turbulent flow and gained a resistance to flow separation. The canalized flow with the tubercled structure supplied the flow control.

https://doi.org/10.1051/epjconf/202226901001

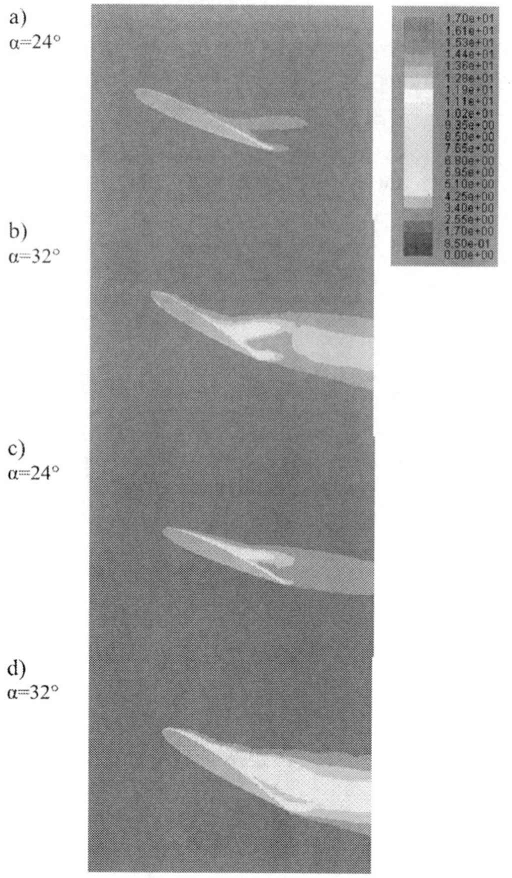

Fig. 7. The turbulence kinetic energy contours over the wings in the middle plane at α=24° and α=32° with Transition SST model for rigid wing (a-b) and for tubercle wing (c-d).

4 Conclusion

In this study, the tubercle structure adapted to the NACA0012 airfoil profile was investigated numerically. Numerical modelling of tubercle wing was carried out in ANSYS FLUENT using k-ω SST and Transition SST turbulence models. The tubercle structure on the wing provided to canalize the flow in the tubercle cavities, thus flow separation decreased. Utilizing the tubercle structure on the wing provided to trigger the transition. The flow over the wing become the turbulent flow and gained a resistance to flow separation. The canalized flow with the tubercled structure supplied the flow control. Consequently, the aerodynamic performance of the wing increased when the tubercle sinusoidal protrusions were applied to the wing.

References

1. Miklosovic, D. S., Murray, M. M., Howle, L. E., & Fish, F. E. Physics of fluids 16.5 (2004): L39-L42.
2. Johari, H., Henoch, C. W., Custodio, D., & Levshin, A. AIAA journal 45.11 (2007): 2634-2642.
3. Carreira Pedro, Hugo, and Marcelo Kobayashi. 46th AIAA aerospace sciences meeting and exhibit. 2008.
4. Yao, Ji, et al. Procedia Engineering 31 (2012): 80-86.
5. Rostamzadeh, Nikan, Richard M. Kelso, and Bassam Dally. Theoretical and Computational Fluid Dynamics 31.1 (2017): 1-32.
6. Şahin, İzzet, and Adem Acir. International Journal of Materials, Mechanics and Manufacturing 3.1 (2015): 22-25.
7. Gawad, Ahmed Farouk Abdel. International Mechnaical Engineering Congresss & Exposition. 2012.
8. Rocha, F. A., de Paula, A. A., Sousa, M. D., Cavalieri, A. V., & Kleine, V. G. (2018). 2018 Applied Aerodynamics Conference (p. 3816).
9. M.S. Genç, U. Kaynak, H. Yapıcı, European Journal of Mechanics B/Fluids, 30(2) 218-235 (2011).
10. M.S. Genç, Proc IMechE, Part C- Journal of Mechanical Engineering Science, Vol 224 (10), pp. 2155 – 2164, 2010.
11. I. Karasu, M. Özden, M.S. Genç, Journal of Fluids Engineering-Transactions of The ASME, 140(12) 121102-1-121102-15 (2018).
12. H. Demir, M. Özden, M.S. Genç, M. Çağdaş. In EPJ Web of Conferences (Vol. 114, p. 02016). EDP Sciences, (2016)
13. M.S. Genç, K. Koca, H.H. Açıkel. Energy, 176, 320-334 (2019)

Experimental and numerical simulation of the wall-pressure fluctuation on the isolated helicopter fuselage

Andrei Batrakov[1,*]

[1]Researcher, Aerodynamics Department, 420111 K. Marx street 10, Kazan, Russia

Abstract. This work is devoted to wall-pressure fluctuation analysis. The object of investigation was an isolated helicopter fuselage. Investigation was carried out by experimental and numerical methods. Numerical simulation was based on RANS and DES approaches. The experiment was carried out in a low-speed wind tunnel with an open test section. Wall-pressure fluctuation was measurement by high-frequency pressure probes ENDEVCO 8510B-2. The experimental results were compared with both DES data and results by the semi-empirical model based on the RANS simulation. It was shown that DES modelling provides a wall-pressure spectrum for low and middle-frequency parts. For simulation high-frequency part of the spectrum, the semi-empirical model is preferable.

1 Introduction

In recent years the helicopter aeroacoustics is of great interest to researchers. The actuality of this research field is due to increased requirements for the helicopters. The main trend of requirements is the decreasing of the noise emission and improvement of flight comfort. These tasks have especially relevant to the development of the advanced high-speed aircraft.

Helicopter has different noise sources that are characterized by amplitude, frequency range and radiation direction. The main helicopter noise sources are the main and the tail rotors, the transmission (gearbox), the engine and the turbulent boundary layer. According to the radiation direction of the acoustic field can be separated into the interior and the exterior parts. For interior noise, the most important frequencies are related to the 500 Hz – 4 kHz range due to the human ear sensitivity. The typical acoustic signature inside the helicopter cabin presented in Figure 1 consists of tonal noise from the main gearbox and the engines and the broadband noise from aerodynamic excitation.

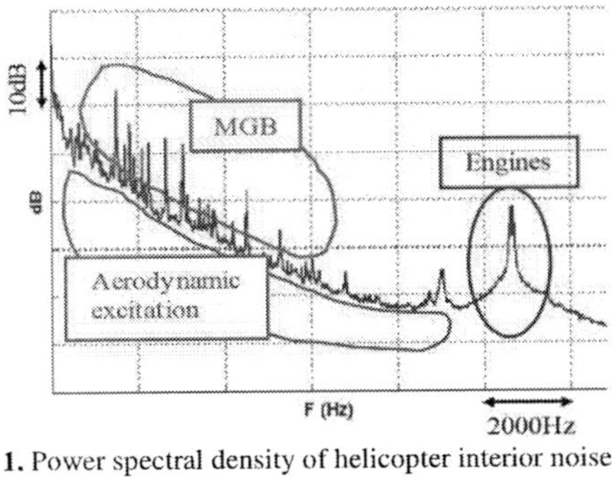

Fig. 1. Power spectral density of helicopter interior noise [1].

The tonal noise from gearbox and engines has a structure-borne path and can be reduced by active or passive vibration dumping. The aerodynamic excitation noise reaches into the helicopter cabin crossing the fuselage skin and windows. Recent studies proposal a new helicopter trim panel with high acoustic transmission loss [2]. Development material with high acoustic transmission loss for helicopter windows is very difficult task. Anyway, the accurate prediction of the interior noise is a very actual problem. For this goal, it is necessary to evaluate the parameters of the aeroacoustics sources, in particular, the wall-pressure fluctuation due to the turbulent boundary layer.

Evaluation of the turbulent wall-pressure fluctuation can be performed by different approaches like direct numerical simulation of pressure fluctuation or using models based on auto-spectral/cross-spectral analysis.

The present paper focuses on the prediction of the wall-pressure fluctuation on the isolated helicopter fuselage. Investigation is carried outed by both experimental and numerical approaches.

2 Experimental approach

The experimental investigation is carried out by the open test section closed-circuit low-speed T-1K wind tunnel of Kazan National Research Technical University (KNRTU). The diameter of the test section is 2.25m. The wind tunnel can be operated at wind speeds up to 50 m/s and it has freestream turbulence intensity below 0.5% in the jet core. Science 2014 year the wind tunnel has equipped by anechoic chamber as it presented in Figure 2.

The object of investigation is the model of helicopter fuselage that is presented in Figure 3. The considered model of the fuselage is the prototype of the multi-

* Corresponding author: asbatrakov@kai.ru

proposal helicopter ANSAT produced by Kazan Helicopter Plant.

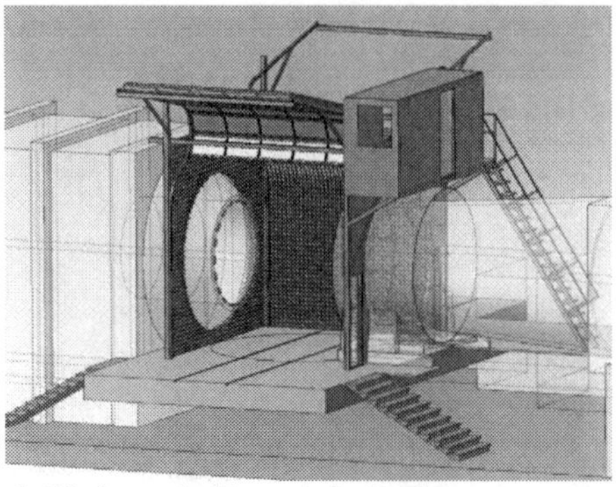

Fig. 2. Wind tunnel with anechoic chamber.

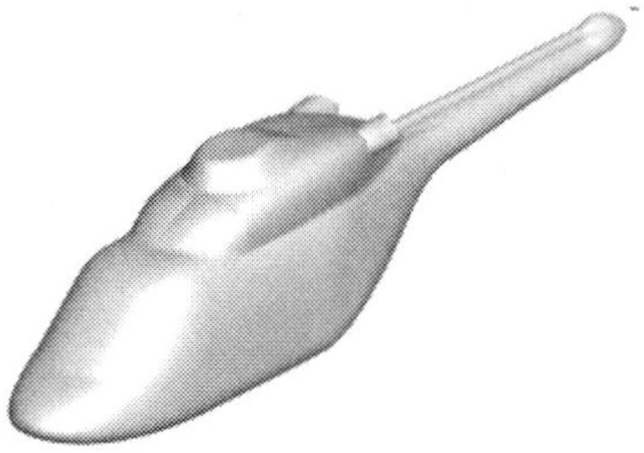

Fig. 3. The isolated fuselage model.

The wind tunnel model has a fuselage length of 1.8 m, and a reference area of S_F= 0.106 m^2. At the previous work [3] the integral aerodynamic characteristics of the model were analysed by experiment and numerical simulation.

Investigation of the pressure fluctuation was carried out at the following flow conditions: Mach number is M=0.1 and Reynolds number is Re=$4.4 \cdot 10^6$. For the wall-pressure measurement, the fuselage model was equipped by high-frequency pressure probes ENDEVCO 8510B-2. Due to probes are differential types it is necessary to have a reference pressure tube. To get more information during the experiment the model was equipped by accelerometer. The equipment layout is presented in Figure 4.

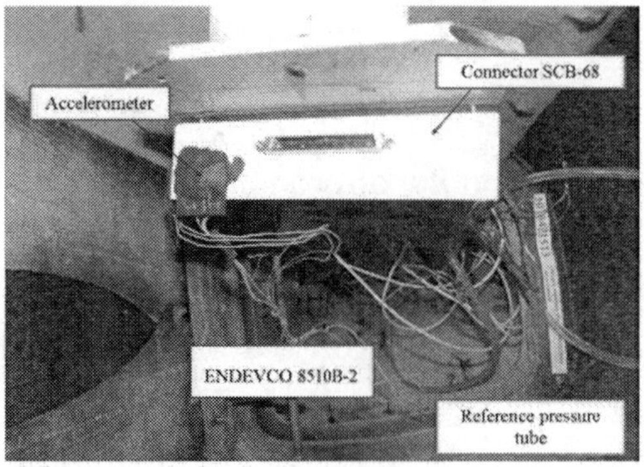

Fig. 4. Equipment inside the fuselage.

The power supply and data wires with the reference pressure tube were routed through the hole near the rotor head. The fuselage was mounted by six-component Prandtl-type balance as it is presented in Figure 5.

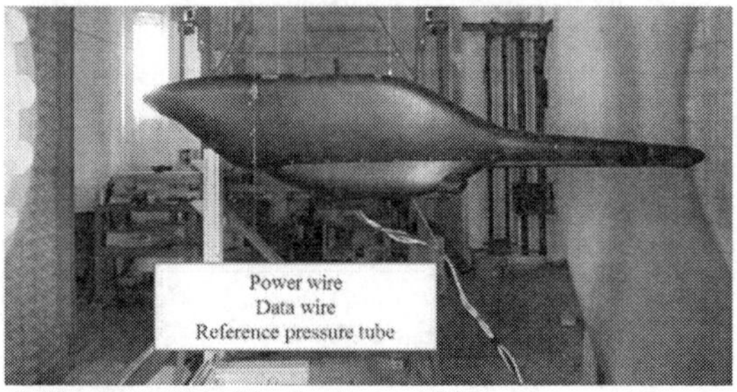

Fig. 5. The mounting of the fuselage model.

The high-frequency pressure probes were distributed on the fuselage surface as it is illustrated in Figure 6 to focus on the front and rear parts of the fuselage.

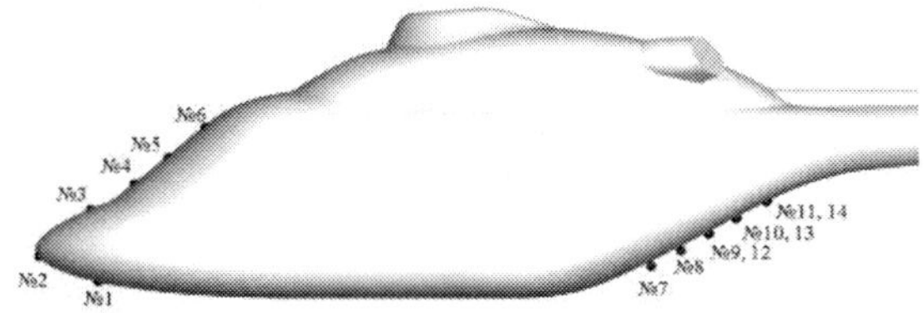

Fig. 6. High-frequency pressure probes location.

3 Numerical approach

The numerical simulation was carried out using the in-house CFD code Helicopter Multi-Block (HMB). Some details of this CFD code are presented in [4]. The solver provides a Reynolds averaged Navier – Stokes simulation (RANS/URANS) with different turbulence models as well as the hybrid approaches like a scale adaptive simulation (SAS), detached eddy simulation (DES) and large eddy simulation (LES). Investigation of the averaged flow characteristics was based on the RANS simulation with the k-ω SST turbulence model. Unsteady pressure was evaluated by DES approach.

For the spatial discretization of Navier Stokes equations, the ANSYS ICEM Hexa-mesh generation tool has been used. The computational grid has a multi-block topology. The O-grid type of topology with the high resolution is used near the fuselage to resolve surface boundary layers (Figure 7). The full grid consists of 964 blocks. The number of cells and parameters of cells distribution were determined by the study of grid independence [3, 5]. As a result, the spacing of the near-wall grid in the normal to surface direction was $1 \cdot 10^{-5}$ of the fuselage length, the cells size ratio was 1.2, and the total number of cells was $13.5 \cdot 10^6$.

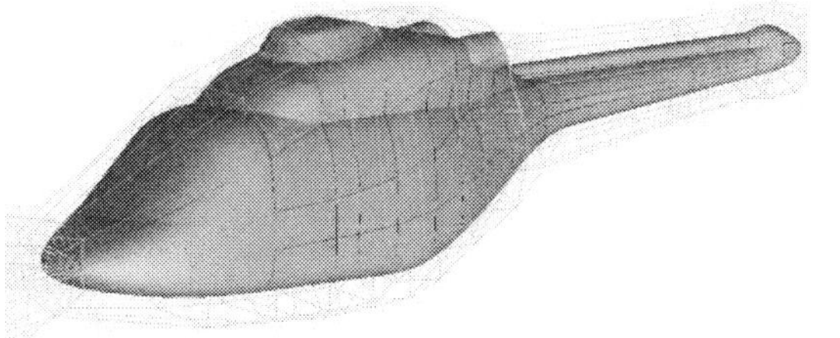

Fig. 7. Mesh topology near the fuselage.

3.1 Wall-pressure fluctuation model

The wall-pressure fluctuation can be evaluated by the semi-empirical model. Recent studies proposal different models, which are trim to the different experimental data. One of famed model was proposed by Goody [6]. This model is used as a start point by many works [7-10] with the aim to extend the range of application including non-zero gradient pressure.

The original Goody model (1) uses aerodynamic quantities like the local boundary layer thickness δ_l, the boundary layer edge velocity U_e, the wall shear stress τ_ω and the friction velocity U_τ

$$\frac{\Phi(\omega)U_e}{\tau_\omega^2\delta_l} = \frac{a\left(\frac{\omega\delta_l}{U_e}\right)^b}{\left[\left(\frac{\omega\delta_l}{U_e}\right)^c+d\right]^e+\left[fR_T^g\left(\frac{\omega\delta_l}{U_e}\right)\right]^h}, \quad (1)$$

where

$$R_T = \frac{U_\tau^2\delta_l}{U_e\nu}, \qquad U_\tau = \sqrt{\frac{\tau_\omega}{\rho}}. \quad (2)$$

Parameters $a - h$ in the original model are constant coefficients presented in Table 1.

Table 1. Parameters of Goody model.

a	b	c	d	e	f	g	h
3.0	2.0	0.75	0.5	3.7	1.1	-0.57	7

Other models keep the main form of equation 1 but propose to use different aerodynamic parameters like the boundary layer displacement thickness δ_*, the boundary layer momentum thickness θ, the boundary layer shape factor H, velocity and dynamic pressure of the freestream. To increase the model flexibility some parameters from a to h are represented as a function of Reynolds number. To take into account non-zero pressure gradient conditions Cole's wake parameter [11] with Clauser's equilibrium parameter [12] can be introduced to the model.

The combination of wall-pressure fluctuation model with numerical simulation based on RANS seems very promising. This approach allows analysing pressure spectrum with a relatively low computational cost. However, some parameters of the model are difficult to determine with high accuracy by CFD results, especially for the complex objects. For example, the boundary layer thickness defined as the distance from surface to location at 99% of the freestream velocity is difficult to use.

On the other hand, the main reason for the wall-pressure fluctuation is the turbulence inside the boundary layer. RANS simulation allows evaluating averaged parameters of the turbulence like as kinetic energy (k) and specified kinetic energy dissipation rate (ω_t). For example, Klabes et. al. [13] proposes to use kinetic energy instead of wall shear stress.

In present work is proposed to use the characteristic source frequency ω_0 as the frequency scale. The characteristic source frequency can be evaluated by the Strouhal relation for the turbulent flow [14]:

$$\frac{\omega_0 l_0}{u_o} \approx 1.7. \quad (3)$$

Characteristic length l_0 and velocity scale u_o for the turbulence can be determined at the point of the boundary layer with a maximum of kinetic energy:

$$u_o = \max\left(\sqrt{k(y)}\right) = \sqrt{k(y_{max})}, \quad 0 \le y \le \delta_l \quad (4)$$

$$l_0 = \frac{u_o}{\omega_t(y_{max})} \quad (5)$$

Taking into account these assumptions it is seen that the characteristic source frequency ω_0 proportional to the specified kinetic energy dissipation rate ω_t.Thus, the wall-pressure spectrum model is rewritten as:

$$\Phi(\omega) = \frac{\rho^2 k(y_{max})^2}{\omega_0}\frac{a\left(\frac{\omega}{\omega_0}\right)^b}{\left[\left(\frac{\omega}{\omega_0}\right)^c+d\right]^e+\left[fRe_t^g\left(\frac{\omega}{\omega_0}\right)\right]^h}, \quad (6)$$

where Re_T is the turbulent Reynolds number.

$$Re_t = \frac{k(y_{max})}{\omega_t(y_{max})\nu}. \quad (7)$$

To determine the model parameters the test case form [8] was considered. The test case represents the flow around the flat plate. The turbulent parameters of the boundary layer were obtained by the RANS simulation. Numerical simulation was carried outed in two-dimensional mode. Computational domain is presented in Figure 8.

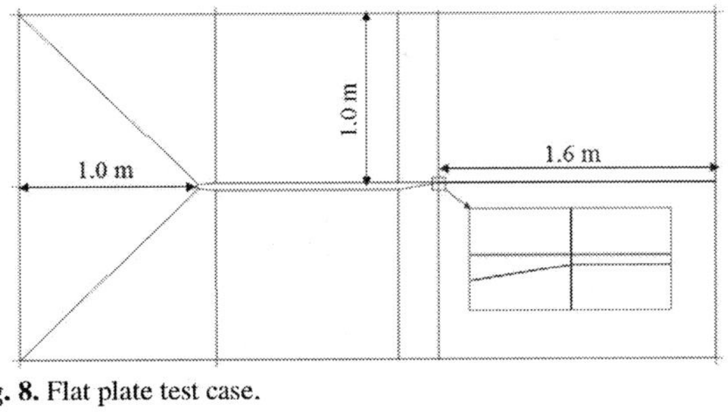

Fig. 8. Flat plate test case.

CFD data were compared with experiment in terms of boundary layer parameters at the control point (x=1210 mm from the leading edge). The results are presented in Table 2.

Table 2. Boundary layer parameters.

	δ_l (mm)	δ_* (mm)	θ (mm)	H
Experiment	19.7	3.51	2.49	1.41
CFD	19.7	3.25	2.42	1.34
	U_τ	Re_x	Re_τ	Re_θ
Experiment	1.125	$2.4\cdot10^6$	1439	4889
CFD	1.123	$2.4\cdot10^6$	1504	4971

According to the data from the Table 2, the numerical results have a good agreement with experiment. Based on these parameters the power spectral density was calculated by the model presented at [8] (Figure 9).

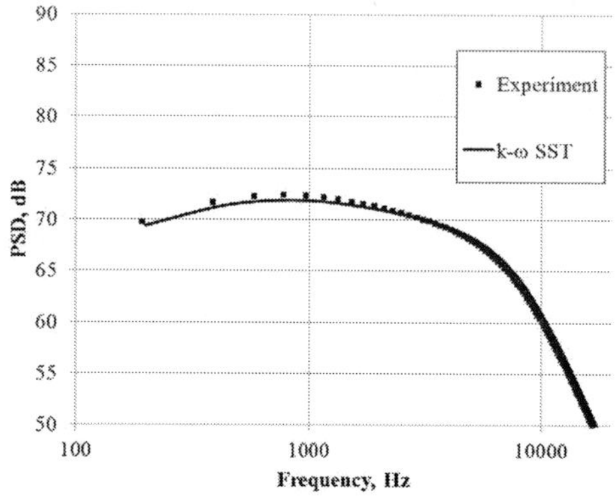

Fig. 9. Power spectral density at the control point.

In order to ensure greater consistency models, four points along the flat plate were considered. The power spectral density at these points was calculated by model [8] based on the boundary layer parameters and by proposal model based on the turbulent parameters of the near-wall flow. Results are presented in Figure10.

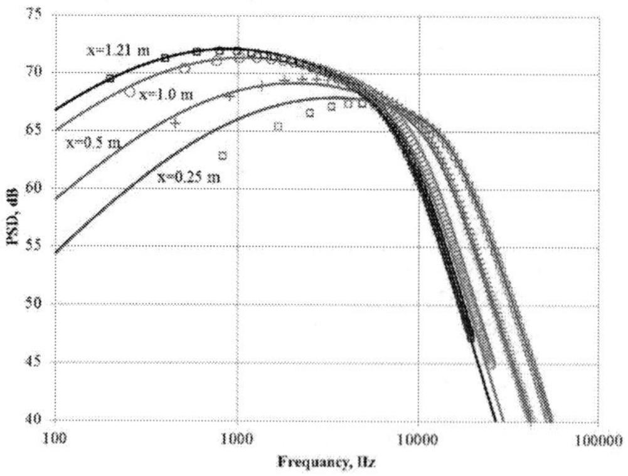

Fig. 10. Power spectral density at the points along the flat plate. Solids lines – current model, symbols – model [8].

Presented in Figure 10 results by the current model were obtained using the coefficients are presented in Table 3.

Table 3. Parameters of the current model.

a	*b*	*c*	*d*	*e*	*f*	*g*	*h*
1.3	*f(Re$_t$)*	0.75	0.2	4	2	-0.18	6.7

Here

$$b = 3.6 - Re_t^{0.21}. \quad (8)$$

4 Results and discussions

At this paper, the analyzing of results focuses on the two pressure probes at the front part (№1, №4) and the two pressure probes at the rear part of the fuselage (№8 and №11). The power spectral density of the pressure based on the experimental data is presented in Figure 11.

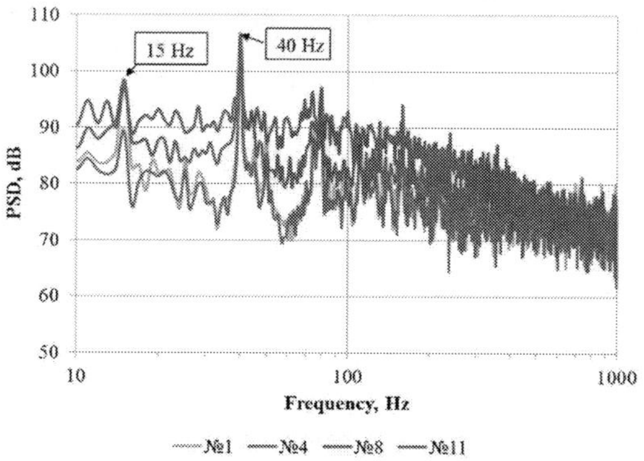

Fig. 11. Experimental PSD of the wall-pressure.

The Figure 11 shows that the position along the fuselage of the probe significant influences on the pressure power spectral density The pressure PSD at the lower frequency range is increased due to increasing of the X position along the fuselage. The spectral analysis shows that there are two peaks at the frequencies 15 Hz and 40 Hz. The first peak at the 15 Hz is observed due to the freestream flow characteristics in the wind tunnel. This assumption was confirmed by analysis of the velocity spectrum. The freestream velocity was measurement by the Constant Temperature Anemometry (CTA) system. The CTA transducer was located in front of the fuselage as it is presented in Figure 12.

Fig. 12. Locations of the CTA transducer.

As a result, it can be concluded that the first peak at the 15 Hz frequency is not related to the turbulent boundary pressure fluctuation. This frequency also is observed in the spectrum of the accelerometer signal.

The second characteristic peak at frequency 40 Hz (and the double frequency 80 Hz) perhaps relate to the natural frequency of the fuselage mounting system. This assumption needs further investigation.

The numerical data based on the DES-simulation presented in Figure 13. The DES simulation was carried out with time step $5 \cdot 10^{-5}$s. At the front point, the pressure spectrum has a very low power that relates to a very thin boundary layer and idealized conditions of numerical simulation.

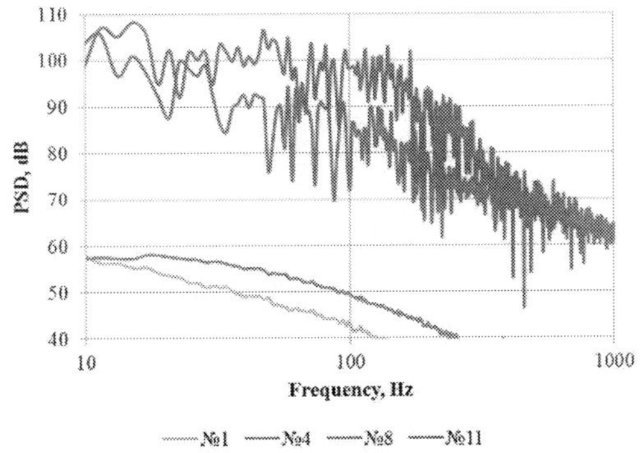

Fig. 13. PSD of the wall-pressure based on the DES model.

At the rear part of the fuselage, the flow is characterized by complex vortex structure [15], that provides enough strong wall-pressure fluctuation.

The PSD of the pressure fluctuation also was calculated by the proposal at the current paper model based on the RANS simulation. The results of the calculation are presented in Figure 14.

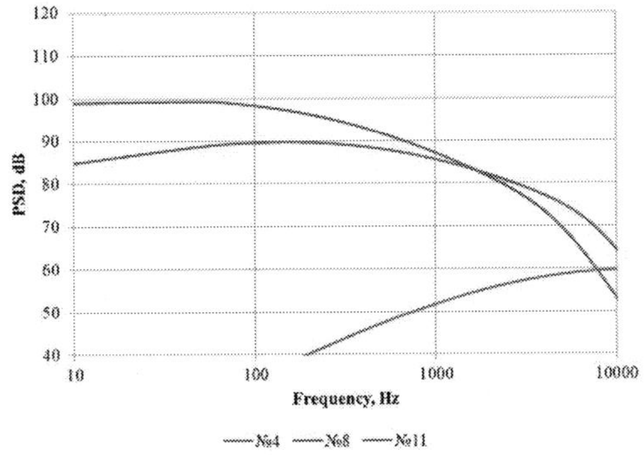

Fig. 14. PSD of the wall-pressure based on the proposed model.

The numerical simulation data for the rear part of the fuselage were compared with experiment. The result of comparison is presented in Figure 15 and 16.

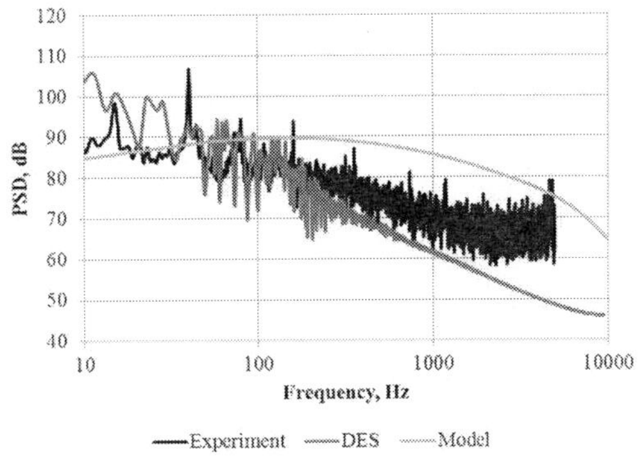

Fig. 15. Comparison of PSD at point №8 based on the different approaches.

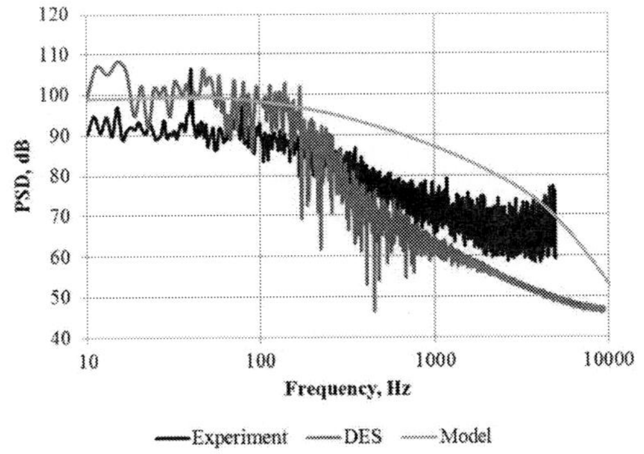

Fig. 16. Comparison of PSD at point №11 based on the different approaches.

As it is shown the DES simulation data have some agreement with experiment at low-frequency range up to 200…500 Hz. At the high-frequency range, the DES simulation data show fast descent of power spectral density. The proposal model overestimates the PSD at the high-frequency range. Both numerical approaches didn't show any peaks at the characteristic frequency in contradistinction to the experiment.

5 Conclusion

This paper presents results of experimental and numerical estimation of the wall-pressure fluctuation on the helicopter fuselage. Analysis of the experimental data shows that parameters of the wind tunnel and mounting system influence on the results. Further research will be devoted to development filter to extract pressure fluctuation due to only the turbulent boundary layer. The DES approach allows estimate wall-pressure fluctuation, but high spectrum resolution requires a high computational cost. The proposed model of the wall-pressure PSD based on the RANS simulation seems attractive to use, but this model needs further investigation to improve accuracy and to expand the operating range.

This work was supported by the grant from the Russian Science Foundation (project № 18-79-00318).

References

1. J. Caillet, F. Marrot, F. Malburet, J-C. Carmona, 31st European Rotorcraft Forum, 13-15, September 2005, Florence, Italy (2005).
2. J. Derr´e, F. Simon, 42nd European Rotorcraft Forum, 5-8, September 2016, Lille, France (2016).
3. A. Batrakov, L. Garipova, A. Kusyumov, S. Mikhailov, G. Barakos, Journal of Aircraft, **52**(5), 1634–1643, (2015).
4. S. Lawson, G. N. Barakos, *Progress in Aerospace Sciences*, **47** (3), 186–216, (2011).
5. A.S. Batrakov, A.N. Kusyumov, S.A. Mikhailov, G.N. Barakos, Aerospace Science and Technology, 77, 704–712, (2018).
6. M. Goody, AIAA Journal, **42** (9), 1788–1794, (2004).
7. Y. Rozenberg, G. Robert, S. Moreau, AIAA Journal, **50** (10), 2168-2179, (2012).
8. N. Hu, M. Herr, Aeroacoustics Conferences, 30 May – 1 June, Lyon, France (2016).
9. M. Kamruzzaman, D. Bekiropoulos, T. Lutz, W. Würz, International journal of aeroacoustics, **14,** 833–882, (2015).
10. M. R. Catlett, J. M. Anderson, J. B. Forest, D. O. Stewart, AIAA Journal, **54** (2), 569-587 (2016).
11. D. Cole, J. Fluid Mech., **1** (2), 191-226, (1956).
12. F. H. Clauser, Journal of the Aeronautical Sciences, **21**(2), 91–108, (1954).
13. A. Klabes, C. Appel, M. Herr, S. Callsen. Inter-Noise, Hamburg, Germany, (2016).
14. G. M. Lilley. Journal of Sound and Vibration, **239**(4), 849-859, (2001).
15. A. S. Batrakov, A. N. Kusyumov, S. A. Mikhailov, L. I. Garipova, G. N. Barakos, 40th European Rotorcraft Forum, 2-5, September 2015, Southampton, UK, (2014).

https://doi.org/10.1051/epjconf/202226901003

Numerical simulation of fuel spraying into an industrial object

Vojtech Betak[1,*]

[1]Czech Aerospace Research Centre, Department of Engines, 19905 Prague, Czech Republic

Abstract. Numerical simulation of fuel spraying into industrial objects is described in this paper. This spatter may be caused by a vehicle or storage tank accident. Prediction of fuel spill is the main aim of simulations that allow predicting possible sources of secondary fires. Numerical simulations are based on the solution of the system of Navier-Stokes equations complemented by the solution of Lagrangian spray cloud and thin fluid film. The simulation results will help to identify better fire prevention and other preventive measures to minimize fire damage.

1 Introduction

In modern times, the industry is focused on various numerical simulations in the projection phase. This allows reducing the production cost, to improve operative cost and safety features. In the case of large industrial buildings, the simulation allows us to better dimension e.g. ventilation. The effect and propagation of fire can now be studied in projection phase due to increased computing power and software availability. It is especially important in the case of fire safety, where it is possible to determine safe escape routes and regulate the amount of smoke.

Results of simulating fuel spraying into an industrial building are presented in the following paper. This can be caused by a vehicle accident when the fuel tank is broken, and fuel is sprayed/spilled into the building. When the fuel spray is fine enough, it may catch fire.

The simulations are based on previous experience with liquid fuel combustion in aircraft engines [1] which are modified for large scale fire.

2 Mathematical model

The mathematical model of reacting fluid flow is based on the finite volume method which is applied to the extended system of Navier-Stokes equations

$$\frac{\partial}{\partial t}\rho + \frac{\partial}{\partial x_j}(\rho u_j) = S_\rho, \tag{1}$$

$$\frac{\partial}{\partial t}\rho u_i + \frac{\partial}{\partial x_j}(\rho u_i u_j) = -\frac{\partial p}{\partial x_i} + \frac{\partial}{\partial x_j}\left[(\mu + \mu_t)\left(\frac{\partial u_i}{\partial x_j} + \frac{\partial u_j}{\partial x_i} - \frac{2}{3}\frac{\partial u_k}{\partial x_k}\delta_{ij}\right)\right] + S_u, \tag{2}$$

$$\frac{\partial}{\partial t}\rho h + \frac{\partial}{\partial x_j}(\rho h u_j) = \frac{\partial}{\partial x_j}\left[\left(\alpha + \frac{\mu_t}{Pr_t}\right)\frac{\partial h}{\partial x_j}\right] + S_h, \tag{3}$$

$$\frac{\partial}{\partial t}\rho Y_i + \frac{\partial}{\partial x_j}(\rho Y_i u_j) = \frac{\partial}{\partial x_j}\left[(\mu + \mu_t)\frac{\partial Y_i}{\partial x_j}\right] + S_Y. \tag{4}$$

where density is represented by ρ, components of velocity vector are represented by u_i, pressure is represented by p, dynamic viscosity is represented by μ. The energy equation is written in the form of enthalpy h where heat diffusion is represented by α. Turbulent quantities are represented by turbulent viscosity μ_t and turbulent Prandtl number Pr_t. The transport equation for fraction of species Y_i must be solved represent volume fraction of species and S_ρ, S_u, S_h, S_Y source term in corresponding equations.

These equations must be completed by the equation of state $p = f(\rho)$. First, we must define the molar fraction as

$$y_i = Y_i \frac{M}{M_i}. \tag{5}$$

where M is an average molar mass and M_i is a molar Mass of fraction i, Then is applied Dalton Law

$$p_i = \rho_i n_i RT. \tag{6}$$

where the partial pressure of fraction i is defined as p_i, partial density is defined as ρ_i and number of moles of fraction i is defined as n_i. Temperature is represented by T and universal gas constant by R. Finally, we obtain density in the following form

$$y_i = \frac{p_i}{p}, \tag{7}$$

$$\rho = \sum y_i \rho_i \tag{8}$$

* Corresponding author: betak@vzlu.cz

and the temperature is defined as

$$T = \frac{h}{\Sigma y_i c_{p_i}}. \tag{9}$$

The system is complemented by Menter [2] two equations $k - \omega SST$ model which is defined as

$$\frac{\partial \rho k}{\partial t} + \frac{\partial}{\partial x_j}(\rho k u_j) = \frac{\partial}{\partial x_j}[\Gamma_k \frac{\partial k}{\partial x_j}] + G_k - Y_k, \tag{10}$$

$$\frac{\partial \rho \omega}{\partial x_j} + \frac{\partial}{\partial x_j}(\rho \omega u_j) = \frac{\partial}{\partial x_j}[\Gamma_\omega \frac{\partial \omega}{\partial x_j}] + G_\omega - Y_\omega + D_\omega, \tag{11}$$

where the kinetic energy of turbulence is defined as k and the specific speed of dissipation as ω. This model is used in DES formulation.

The ODE system is used to describe the movement of the fuel spray particles

$$\frac{du_p}{dt} = F_D(u - u_p) + g\frac{\rho_p - \rho}{\rho_p} + F_x, \tag{12}$$

$$F_D = \frac{18}{24}\frac{\rho}{\rho_p d_p} C_D Re \tag{13}$$

and

$$F_X = \frac{\rho}{\rho_p}\left[0.5\frac{d}{dt}(u - u_p) + u_p\frac{\partial u}{\partial x}\right] \tag{14}$$

where u_p is represented speed vector of spray particles and ρ_p is represented fluid density of spray particles The drag coefficient $C_D = \frac{24}{Re}$ is depended as a function of modified Reynolds number $Re = \rho d_p \frac{|u - u_p|}{\mu}$.

3 Results

Several scenarios of fuel spray are solved in this project. These scenarios are associated with the crash of a small commercial aircraft into a building of critical infrastructure facility (for example power plants, fuel storage, chemical industry, ...). Significant environmental and economic damage is achieved in a wide area in case of accident near to these buildings.

Building geometry (Fig. 1) is selected for the demonstration of numerical simulations. Due to the impact, the sidewall is broken, and the inner space is exposed.

The simulation is initialized by approx. 12000 spray particles with initial velocity of 180 m/s and diameter 0.05 m. These droplets are moving through the computational domain where they hit the walls and break up into smaller droplets. Total pressure and pressure directed velocity boundary conditions are prescribed at the outlet from the computational domain where ambient atmosphere conditions are assumed.

It is assumed that the wall is broken at the moment of the aircraft's impact and debris and airframe disappear. At the same time, the fuel is heated by 200 degrees which initialized evaporation and burning of fuel gases.

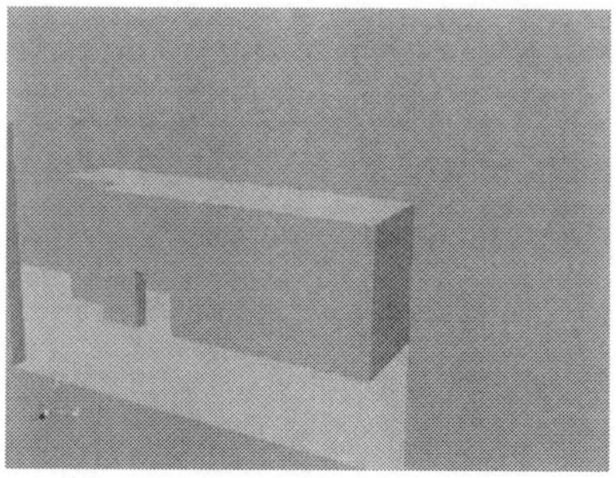

Fig. 1. Example of computational geometry.

Two types of simulations are solved. In the first simulation, the chemistry is switched off and therefore the drops do not evaporate and burn. In the second one, the chemistry is included.

In figures 2 and 3 is shown the difference between reacting and nonreacting simulations. In the case of reacting simulations is formed very fine fuel spray with average droplets diameter in the range from 10^{-5} to 10^{-4} m. Fuel particles are floated due to buoyancy forces caused by fire. In contrast in the case of nonreacting simulations where bigger droplets are formed. The average droplet diameter is in the range from $5 \cdot 10^{-4}$ to 10^{-2} m. These droplets are falling and therefore the fuel film is formatted as is shown in Fig. 4. This film slowly spreads across the floor of the building and flows out of the hole after impact.

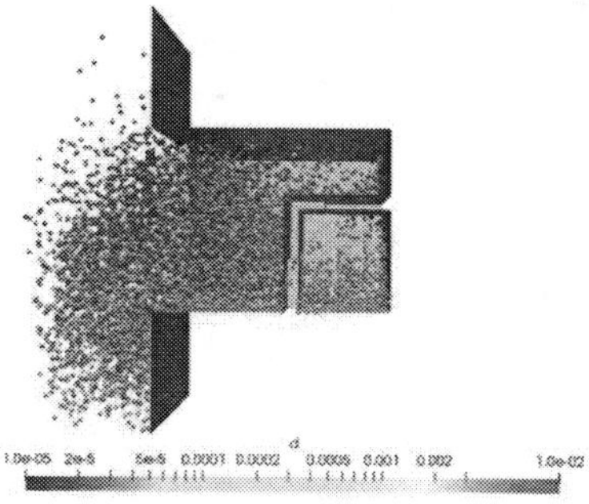

Fig. 2. Position of spray particle for nonreacting flow in time t = 20 s.

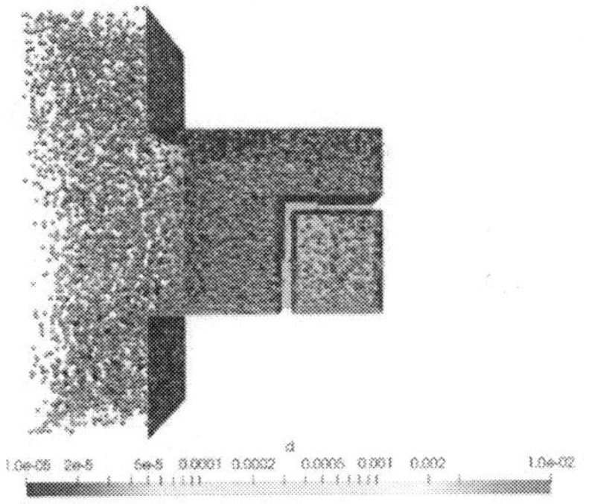

Fig. 3. Position of spray particle for reacting flow in time t = 20 s.

The formation of a fuel film is not observed in the case of the reacting flow field. Instead the "fireball" is predicted by simulation as is possible to see in Fig. 5 where is shown isosurface of temperature at value 800 K. This "fireball" is caused by the burning of all droplets in computational domain. This explains lower diameter of droplets which are observed in Fig. 3.

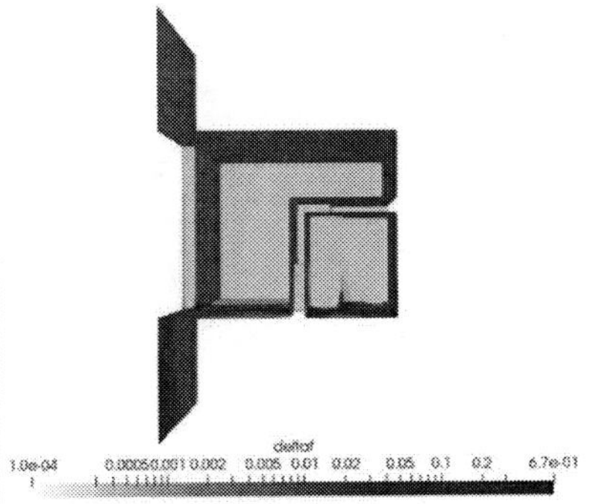

Fig. 4. Formation of fuel film in the case of nonreacting simulation in time t = 60 s.

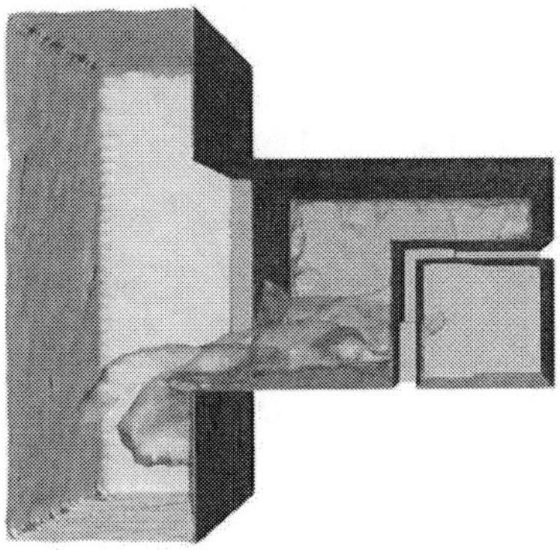

Fig. 5. Isosurface of temperature (800 K) in reacting case in time t = 60 s.

If similar boundary conditions are used on different computational geometry with the large open area, then different results are observed. Changes are visible on droplets diameter where the value between and 5e-3 m and 5e-2 m is observed (Fig. 6.).

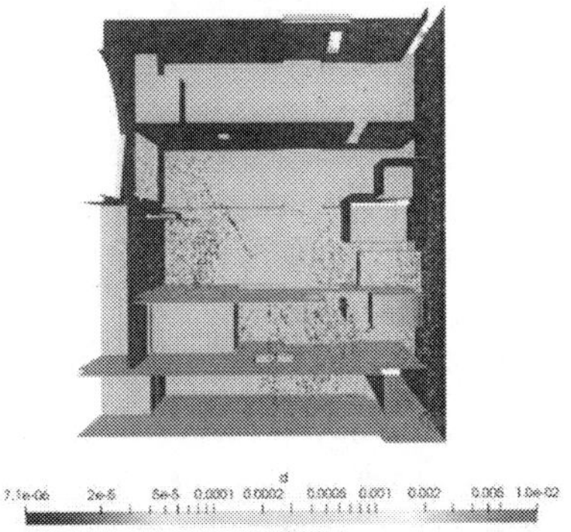

Fig. 6. Position of spray particles for flow time t = 20 s.

Also, the formation of fuel film is observed which flows into the basement (Fig. 7.) where the fuel is accumulated and slowly burns out (Fig. 8.).

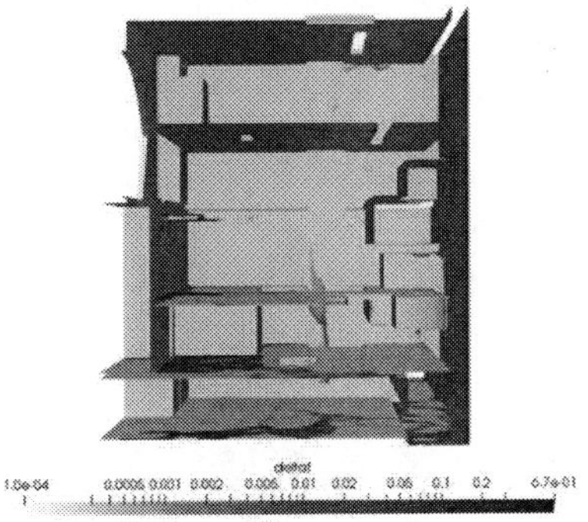

Fig. 7. Formation and position of fuel film in large open area in time t = 60 s.

Thus, due to fire can affect the storage area where flammable liquids can be placed in a building and intensify the fire. At the same time, considerable amounts of smoke will be generated and spread through the building area.

https://doi.org/10.1051/epjconf/202226901003

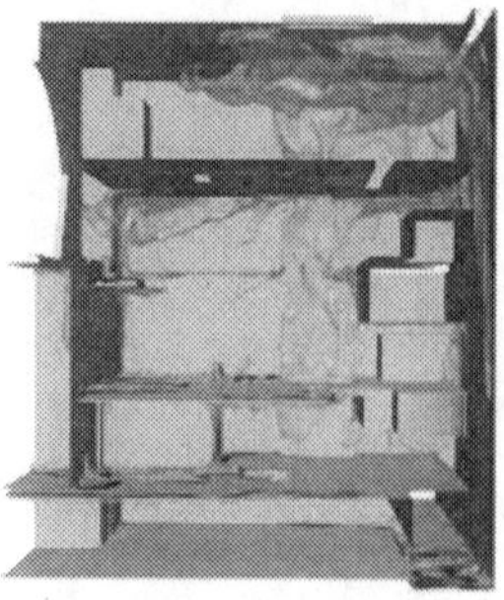

Fig. 8. Isosurface of temparature (800 K)in large open area in time t = 60 s.

4 Conclusion

Results of a fire simulation after an aircraft crashes into a building are shown in this paper. These results are burdened with the assumptions that are applied. These assumptions relate to neglecting the influence of debris, infiltration of fuel and breaking the airplane. If these simplifying assumptions were not applied then the complexity of simulation and hardware and software resources increase significantly.

Two different scenarios of fire propagation are indicated in the results. If the fuel is sprayed into close space, then it is combusted in the form of aerosol and fuel film is not created. Compared to the case, where the fuel is sprayed into open space then the only ¼ of fuel is burned in the form of aerosol and fuel film is formed from the rest of the fuel.

Simulations are mostly limited by using the simplest combustion model. If a more complex combustion model is used, then it is expected improvement results. Especially a soot and CO production is captured. For this purpose, it is a necessary modified combustion model which was used in previous work [1] and it was optimized for combustion of liquid fuels in jet engines.

Acknowledgments

This work has been supported by the grant project of the Ministry of the Interior of the Czech Republic no. VI20172020092 and by The Ministry of Education, Youth and Sports from the Large Infrastructures for Research, Experimental Development and Innovations project „IT4Innovations National Supercomputing Center – LM2015070“.

References

1. V. Betak, J. Kubata, *Proceedings of the International Conference Experimental Fluid Mechanics* (2015)
2. F. R. Menter, F. Kuntz, R. Langhtry, Turb., He. Ma. Trans. (2003)

Direct visualization of hollow fiber membrane fouling distribution

Petr Bílek[1,*], *Filicia Wicaksana*[2]

[1]Department of Nanotechnology and Informatics, Technical University of Liberec, Studentská 1402/2, 461 17, Liberec 1, Czech Republic[1]
[2]Department of Chemical and Materials Engineering, the University of Auckland, 2-6 Park Avenue, Grafton 1023, Auckland, New Zealand[2]

Abstract. Submerged membrane bioreactor system is commonly used for water and wastewater treatment processes. This system is a more economical alternative to conventional activated sludge process due to small footprint and better effluent quality. Nevertheless, membrane fouling is a major issue that requires further studies. This paper covers the use of an optical microscope to observe membrane fouling. Bentonite was used as model particles to simulate activated sludge. In-situ observation of particle deposition along the surface of hollow fiber membranes during filtration was performed at various operating conditions. Results from this studies indicated that the flux distribution was affected by the permeate flux and cross flow velocity. Lower CFV and higher flux caused the flux to be more non-uniformly distributed due to more severe fouling. These findings provided useful information to further improve the membrane module design to achieve more evenly distributed permeate flux.

1 Introduction

Membranes are widely used for various applications (wastewater treatment, drinking water production, food processing and pharmaceutical industry). Compared to flat sheet membranes, hollow fiber membranes offer higher packing density and higher mechanical strength since hollow fiber membranes are self-supporting. Hollow fiber membranes are commonly used in submerged membrane bioreactors for wastewater treatment. The fibers are arranged in parallel and directly immersed in a bioreactor. However, since the membranes are in direct contact with the mixed liquor, membrane fouling is inevitable. The deposition of foulants on the surface of fibers gradually reduces the permeate flow. There are several ways to reduce membrane fouling, such as back-flushing, back washing, the use of air bubbles to scour the membrane surface and ultrasound [1-2].

The small aspect ratio of hollow fiber membranes causes high axial pressure gradient inside the fibers and consequently, the flux gradient. In addition to this, the pressure gradient is also governed by the mode of operation. Submerged hollow fiber membranes are typically operated with the permeate extracted from the upper end of the module, while the other end is sealed. This leads to non-uniformity of permeate flux distribution. It is hypothesized that the highest flux would occur close to the permeate suction point and decrease towards the opposite (sealed) end (Figure 1). If the permeate is extracted from both ends of the fiber, the lowest flux would be in the middle of the fiber. It is expected that the fouling pattern mirrors the flux gradient: the higher the flux, the more severe the fouling.

A model was developed by Carroll, et al. [3] to correlate local permeate flux to fiber length and diameter. Results from their modelling study highlighted the presence of non-uniform axial flux profile that resulted in dynamic fouling patterns.

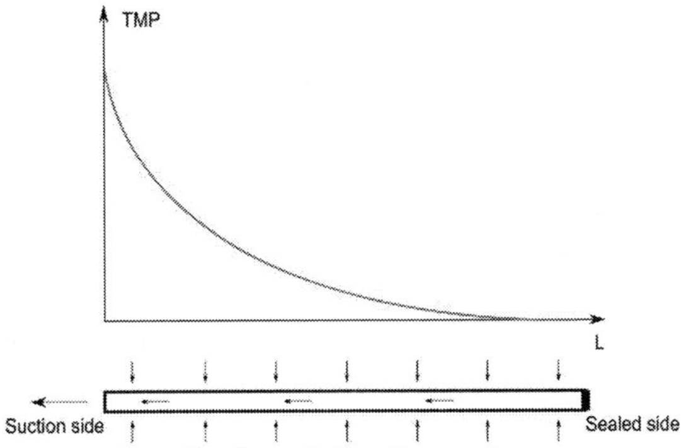

Fig. 1. TMP profile along a hollow fiber membrane.

The non-uniformity of local permeate flow was also reported by former researchers through a simulation study [4] and hollow fiber membrane backflushing [5]. The majority of former studies on fouling distribution focused on modelling work [6]. To the best of authors' knowledge, there is only one experimental study on permeate flux distribution [11]. However, this former study relied on the transmembrane pressure (TMP) measurement to analysis the permeate flux distribution. Membrane fouling can be observed by various methods, such as scanning electron microscopy (SEM) [7]. However this ex situ examination is invasive. The

* petr.bilek@tul.cz

membrane module needs to be destructed for SEM analysis. Hence, the fouling layer could be destroyed during sample preparation and handling. Furthermore, SEM analysis can only be performed after the filtration experiment. Direct microscopic observation method (DO) allows in situ and non-invasive visualization of the membrane surface in real-time. DO technique involves the use of an optical microscope and a camera to observe the development of membrane fouling. This technique had been used in various studies in microfiltration and ultrafiltration processes, such as alginate fouling [7], membrane backwashing [8] and microalgae deposition [9] as well as in forward osmosis process [10]. However, it has not been used to visualize the fouling distribution progress along hollow fiber membranes.

This study involved the use of an optical microscope to observe particle deposition along a hollow fiber membrane with bentonite as model foulant. The study was aimed to provide better understanding of fouling distribution along the hollow fiber membrane which would help to improve the membrane module design and the development of fouling control strategies.

2 Materials and Methods

2.1 Materials

Bentonite particles (ECP ltd) in deionized water were used as model foulants to simulate activated sludge. The size distribution of bentonite particles was measured by Master-sizer 2000. It was reported that bentonite particles could swell in water, therefore, this effect was examined by leaving the bentonite solution overnight. The particle size was then measured and compared with freshly prepared bentonite solution. Results showed that there was no significant difference in bentonite sizes. Furthermore, the filtration experiment was performed for a relatively short period of time (up to 5 hours). Hence, the particle size of bentonite should not change during experiment.

The membranes used in this study were ultrafiltration hollow fiber membranes manufactured by *Mann+Hummel*. The specifications of the membranes can be seen in Table 1.

Table 1. Specifications of hollow fiber membrane.

Material:	polyacrylonitrile (PAN)
Inner diameter:	1 mm
Outer diameter:	2 mm
Nominal pore size:	50 nm

2.2 Experimental setup

The experimental setup used in this study is shown in Figure 2 and the cross flow membrane filtration channel is shown in Figure 3. The rectangular membrane flow channels were made of glass tubes with three different lengths of 155, 305 and 505 mm. The inner dimensions of rectangular cross-section were 30 mm in width × 4 mm in height. One hollow fiber membrane was placed in the middle of the glass tube and fixed with epoxy glue. The permeate was withdrawn from one end of the fiber by using a peristaltic pump (Gilson, Minipuls 3), while the other end of the fiber was sealed. Circulation of a feed solution through the filtration channel was performed by a gear pump (Cole-Palmer). The retentate was recirculated back to the feed tank. The feed tank was equipped with a magnetic stirrer to avoid particle settlement.

Three pressure transducers (Cole-Palmer) were used to measure the transmembrane pressure (two pressure transducers installed on the feed side and one on the permeate side of the membrane). A precision balance (Sartorius) was used to record the permeate flux. The pressure transducers and the balance were connected to LabVIEW data acquisition system.

The development of membrane fouling layer was observed by an optical microscope (Nikon, Eclipse Ni-U) with 10x magnification and a color digital camera (Sony DSC-W630) with an optical zoom set at 1.2, at certain time interval and at three different locations along the hollow fiber membrane: near the suction point, in the middle of the membrane and near the sealed end. Further details of the image recording and analysis are discussed in section 2.4.

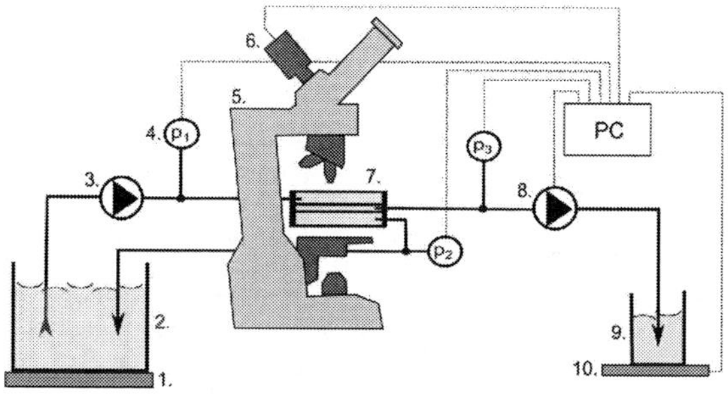

Fig. 2. The diagram of the experimental setup (1. Magnetic stirrer, 2. Beaker for bentonite solution, 3. gear pump, 4. Pressure transducers, 5. Optical microscope, 6. Digital camera, 7. Cross-flow filtration channel equipped with a hollow fiber membrane, 8. Peristaltic pump, 9. Beaker for permeate, 10. Digital balance).

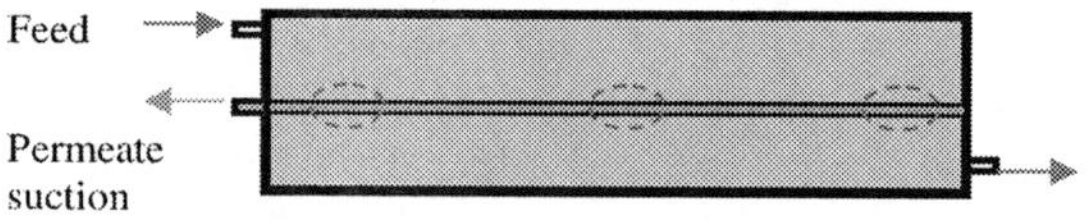

Fig. 3. The cross-flow filtration channel, depicting the feed flow, permeate suction point and the locations of fouling observation

2.3 Filtration experiments and cleaning

Filtration experiments were performed with 300 mg/L bentonite solution as feed at constant permeate flux for up to 5 hours, while the increase in suction pressure was recorded throughout the experiment. The room temperature was approximately 22°C. Various parameters were studied, these included fiber lengths (15 cm to 50 cm), cross flow velocity (18 mm/s and 33 mm/s), permeate flux (35 – 60 L m^{-2} h^{-1}). The transmembrane pressure (TMP) and the thickness of fouling layer during filtration were measured.

After each experiment, the membrane was backwashed for 20 minutes by circulating deionized water in reverse direction (inside-out) at cross flow velocity greater than 50 mm/s. The deionized waterwas then changed to 0.5% HCl and the filtration was run for the next 20 minutes. After that, the system was rinsed by running the filtration for another 20 minutes with deionized water as feed. The membrane was then ready to be re-used for the next experiment.

Transmembrane pressure (TMP) was calculated according to formula:

$$TMP = \frac{p_1+p_2}{2} - P_3\ [kPa]. \quad (1)$$

where P_2 is feed pressure, P_2 is retentate pressure and P_3 is permeate pressure.

The effective hollow fiber membrane area was calculated as below:

$$A = d \cdot \pi \cdot L\ [m^2]. \quad (2)$$

where d is the outer fiber diameter and L is the fiber length.

Cross-flow velocity was calculated from feed flow rate [m^3/s] divided by the cross-sectional area of the flow channel (width x height of the rectangular glass tube).

2.4 Image analysis

The thickness of a fouling layer along the fiber h(x, t) was measured optically with the help of visualization and image analysis. Images were taken at 3 locations along the hollow fiber membrane equally spread from the suction point to the sealed end (x = 0, 0.5L and L). All 3 pictures were taken every 20 minutes throughout the filtration experiment.

The recorded images were processed with ImageJ software and Meazure software [12-13]. All pictures were uploaded onto the software and automatically processed with help of a macro. The macro was written to speed up the image analysis process. Firstly, the color images were converted to 8 bit grey scale images. Then an evaluative rectangle was created in the middle of the picture. The dimensions of the rectangle were determined with default width of 50 pixels and default length of the same as the width of the picture (Figure 4). The mean grey value along the evaluative rectangle was calculated. This was carried out to avoid a statistical error and it resulted in a smoother curve. The curve of the mean grey value along the evaluative rectangle is shown in Figure 5. The red line connecting the gap on the curve in Figure 5 indicated the thickness of the fouling layer.

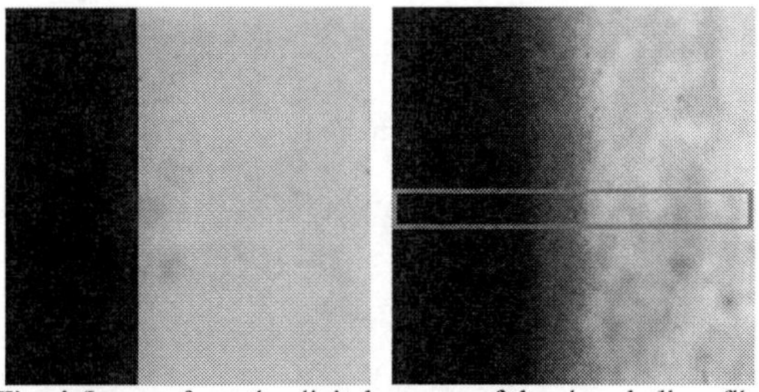

Fig. 4. Images from the digital camera of the clean hollow fiber membrane and after 1 hour of filtration (10x magnification lens).

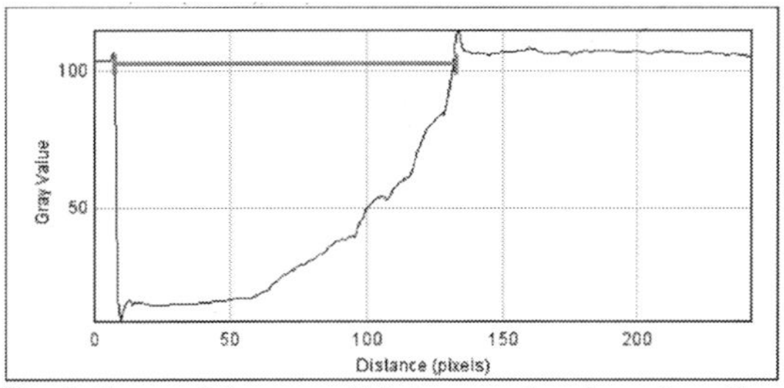

Fig. 5. Curve of the mean digital grey value along the yellow evaluative rectangle.

A scale was used to estimate the actual fouling layer thickness of an object (H_o) in [μm] as follow:

$$H_o = \frac{1}{\frac{S}{h_{i\_px}}}\ [\mu m]. \quad (3)$$

where S [px/μm] is scale of the image and hi_px is height of the object expressed as a number of pixels in the image. S was calculated from number of pixels npx, which represented 1 mm height of a scaled object (micro-ruler with 0.1 mm precision scale was used in this study).

$$S = \frac{n_{px}}{1000}\ [px/\mu m]. \quad (4)$$

Data from ImageJ were uploaded by Matlab. A threshold was set to measure the width of the gap. The same threshold was used for all images. Hence, the contrast of all images was also the same. The width of the gap in pixels was then obtained and transformed to micrometers with the help of scale S according to equation (4). For each picture, one value of thickness was calculated, which means 3 values of thickness along the hollow fiber membrane at certain time t. Since the fouling layer was developed on top of the membrane, the thickness of fouling layer was determined by subtracting the thickness of the membrane from the total thickness. The accuracy of the software in detecting the fouling layer was verified by manual measurements.

The thickness h(x, t) was expressed as cumulative distribution function (CDF) of the deposit CDFd(t) at

time t. The uniformity of the fouling layer thickness was indicated as an area A_{CDF} between CDF of the observed real fouling layer CDFd and CDF of an ideal uniform layer (Figure 6). The smaller the A_{CDF} indicated the more uniform the fouling layer thickness (hence, uniform flux distribution) along the fiber (more efficient filtration). The uniformity coefficient U [%] was calculated by using an expression shown in Figure 6.
The purpose of the measurements was to determine optimum permeate flux J and fiber length L to achieve uniform flux distribution.

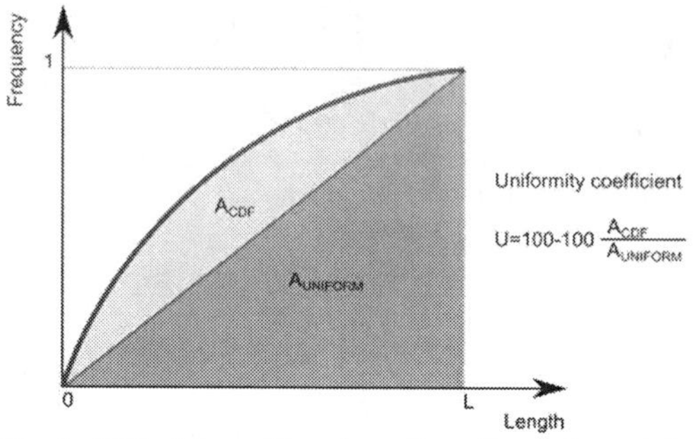

Fig. 6. CDF of the uniform layer (in red) and the real fouling layer (in blue).

3 Results and discussion

3.1 Fouling distribution along hollow fiber membrane

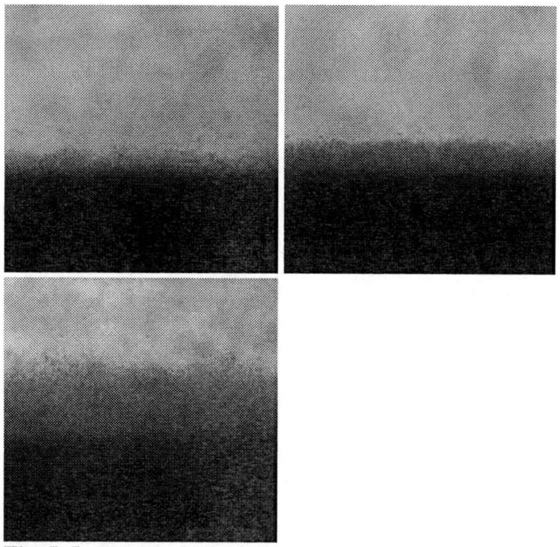

Fig. 7. Images captured at the beginning, in the middle and at the end of the 15 cm hollow fiber membrane after 1 hour of filtration and 33 mm/s CFV.

Figure 7 shows the formation of fouling layer on the membrane surface at three different locations: near the suction point, in the middle of the fiber and near the sealed end of the fiber. The images were captured at the same time after one hour of filtration. It can be seen that the cake deposition along the hollow fiber was not uniform. The fouling layer near the permeate suction point was thinner as compared to the downstream of the fiber. By referring to figure 3, the feed solution inlet was situated close to the permeate suction point (at the beginning of a fiber). Hence, the CFV was the highest at this region. Higher cross-flow shear in this region reduced the cake deposition on the membrane.

The growth of fouling layer over time can be seen in Figure 8. The images were captured near the sealed end of the fiber. It can be seen that fouling layer started to develop within the first 20 minutes of the filtration. The layer became thicker over time. The thickness of fouling layer was measured via image analysis software as described earlier and plotted in figures 9 and 10.

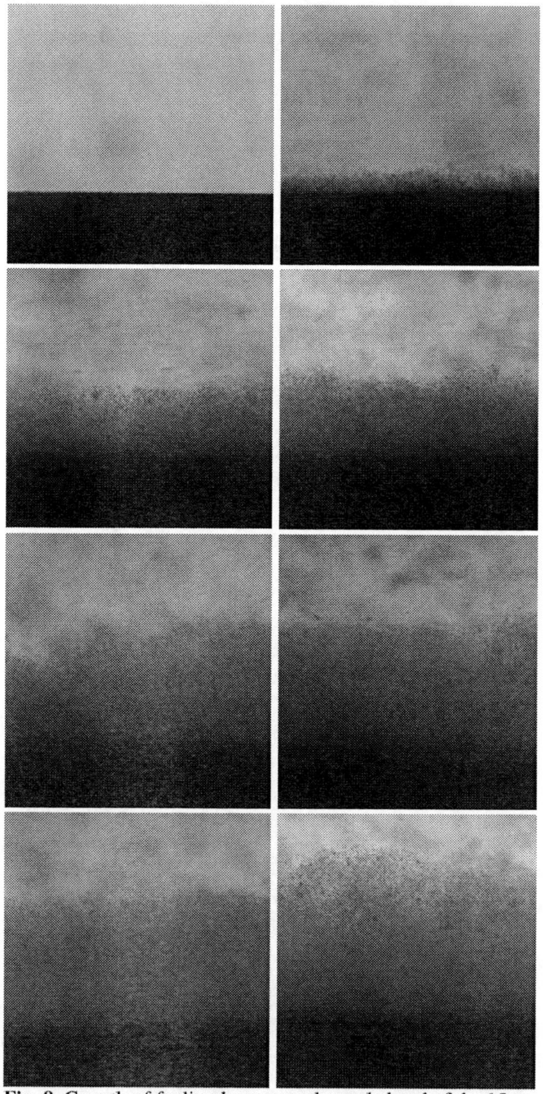

Fig. 8. Growth of fouling layer near the sealed end of the 15 cm hollow fiber membrane (images were taken at 20 minutes interval, the first image was taken at t = 0, while the last image was captured at t = 160 minutes).

At higher CFV (33 mm/sec), the cake at the beginning of a hollow fiber membrane was the thinnest (Figure 9). Interestingly, an opposite trend occurred at low CFV (18 mm/sec), the cake was the thickest near the suction point (at the beginning of the fiber) (Figure 10). As explained earlier, higher CFV near the suction point generated higher shear that helped to reduce the development of fouling layer in this region. When the CFV was decreased to 18 mm/s, fouling became more severe. At the location near the suction point, the cake buildup rapidly. After 3 hours of experiment, the cake buildup reached a plateau (Figure 10). At the beginning of experiment, the hollow fiber membrane was clean. Hence, there was more area for deposition. Once the available filtration area reduced, the rate of deposition decreased. In addition to this, although the permeate was maintained constant, the pump could not cope when the fouling became too severe. The decrease in permeate pump capacity resulted in the decreased in permeate flux, therefore, the fouling rate reduced. The filtration had to be terminated beyond this point as constant permeate flux could no longer be maintained.

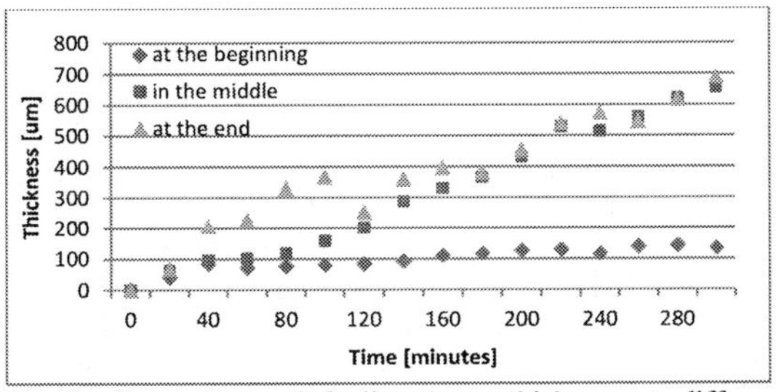

Fig. 9. Comparison of fouling layer thickness at different locations along a hollow fiber membrane vs time (15 cm fiber length, CFV 33 mm/s).

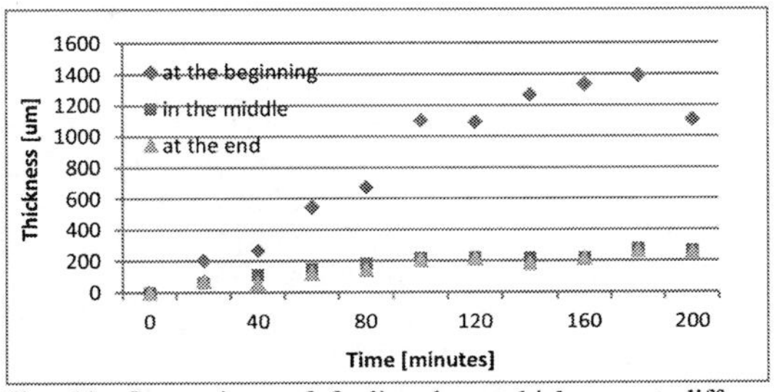

Fig. 10. Comparison of fouling layer thickness at different locations along a hollow fiber membrane vs time (50 cm fiber length, CFV 18 mm/s).

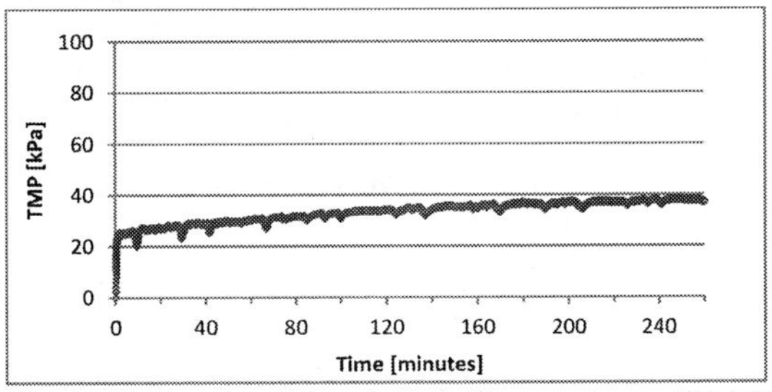

Fig. 11. TMP vs time (15 cm hollow fiber membrane, 60 LMH, CFV 33 mm/s).

The TMP profile vs time is shown in Figure 11. As expected, the TMP increased as fouling progressed. However, the TMP measurement could only reflect the severity of fouling. It could not indicate how the fouling distributed along the hollow fiber membrane. This detailed information could only be achieved by online microscopy observation method.

3.2 Effect of permeate flux

The 15 cm and 50 cm long hollow fiber membranes were tested in different permeate fluxes J = 35 LMH and 60 LMH. The trans-membrane pressure TMP(t) and thickness h(x, t) of the deposition along the hollow fiber membrane were measured during 4 hours of experiment. A uniformity coefficient U [%] of the fouling layer was calculated based on the fouling layer thickness h(x) along the fiber at time t.

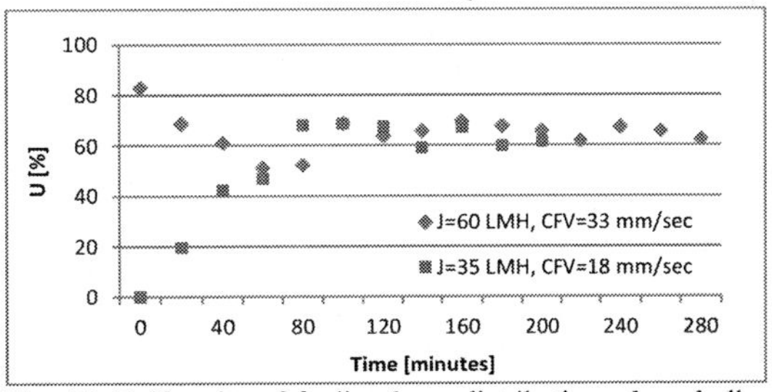

Fig. 12. Uniformity of fouling layer distribution a long hollow fiber membrane vs time at various CFV and permeate flux values (15 fiber length).

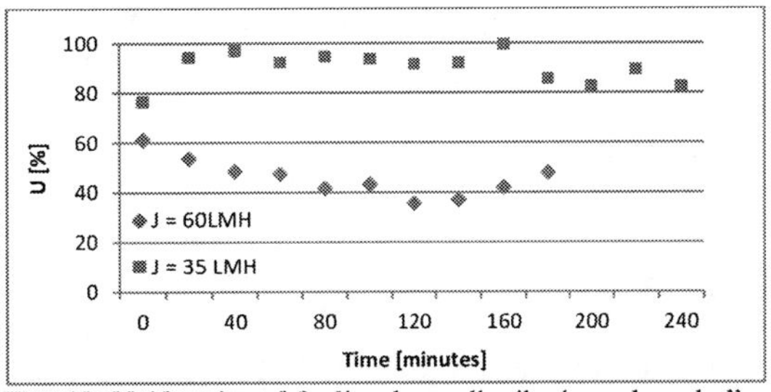

Fig. 13. Uniformity of fouling layer distribution a long hollow fiber membrane vs time at various permeate flux values (50 cm fiber length, 18 mm/s CFV).

Figure 12 represents U vs time t for different fluxes and CFVs. For filtration at higher flux and higher CFV (60 LMH and 33 mm/s, respectively), the uniformity coefficient was high at the start of the experiment which represented uniform flux distribution (the blue curve). The uniformity was decreasing over time because of fouling. Different situation was seen at lower flux and CFV (the red curve). Due to the absence of strong CFV, the non-uniformity of flux distribution developed as soon as the filtration started, with the location near the permeate suction point experienced more severe fouling. However, the uniformity of flux distribution increased over time. Once the upstream of the hollow fiber was fully covered by the cake, the permeability of this membrane section dropped, the permeate was then shifted downstream approaching the cleaner section of the membrane. Hence, the fouling distributed more evenly along the fiber. The uniformity coefficients of both cases (red and blue curves) were relatively stable over the remaining 2 hours of experiments.

Figure 13 depicts uniformity of flux distribution as a function of time for different permeate fluxes. It can be seen that filtration at lower flux resulted in more uniform flux distribution as compared to the one at higher flux. This was due to less severe fouling occurred at lower flux. Hence, the flux distributed more evenly along the fiber. Fouling pattern mirrored the TMP and flux distribution patterns along a hollow fiber membrane.

3.3 Effect of crossflow velocity

Two different CFVs were studied for 50 cm long hollow fiber membrane. Figure 14 shows a significant difference in uniformity of filtration at low and high CFV values. Higher CFV generated higher shear-induced particle removal. Hence, the flux distributed more evenly for operation under high CFV.

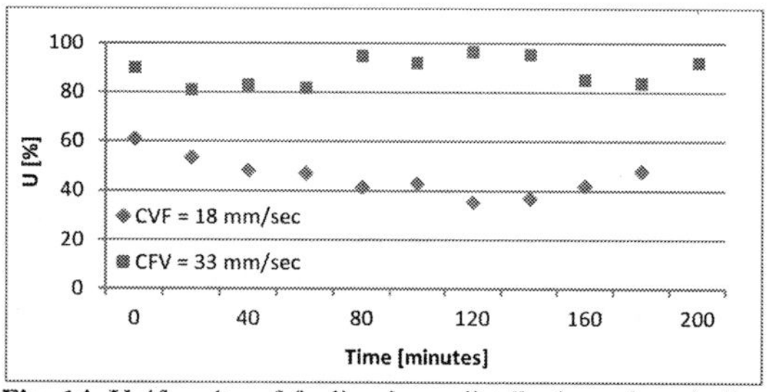

Fig. 14. Uniformity of fouling layer distribution a long hollow fiber membrane vs time at various CFV values (50 cm fiber length, 60 LMH permeate flux).

4 Conclusion

Direct microscopic observation method could provide better understanding of how fouling distributed along a hollow fiber membrane under various conditions. The image analysis method had been successfully developed to accurately measure the thickness of fouling layer. Results from this studies indicated that the flux distribution was affected by the permeate flux and cross flow velocity. Lower CFV and higher flux caused the flux to be more non-uniformly distributed due to more severe fouling. These findings provided useful information to further improve the membrane module design to achieve more evenly distributed permeate flux.

Acknowledgments

This work was supported by the Ministry of Education, Youth and Sports of the Czech Republic and the European Union - European Structural and Investment Funds in the frames of Operational Programme Research, Development and Education - project Hybrid Materials for Hierarchical Structures (HyHi, Reg. No. CZ.02.1.01/0.0/0.0/16_019/0000843).

These studies were carried out in Membrane Research Group, Department of Chemical and Materials Engineering, University of Auckland, New Zealand.

References

1. Sparks T, Chase G, Section 4 - Solid–Liquid Filtration - Filters and Filtration Handbook (Sixth Edition), Butterworth-Heinemann 2016, pp. 199-295, ISBN 9780080993966.
2. Qaisrani, T. M., Samhaber, W. M. Impact of gas bubbling and backflushing on fouling control and membrane cleaning (2011) Desalination, 266 (1-3), pp. 154-161.
3. T. Carroll, N. A. Booker. Axial features in the fouling of hollow-fibre membranes. Journal of Membrane Science 168 (2000) 203 – 212.
4. S. Chang, A. G. Fane. The effect of fibre diameter on filtration and flux distribution – relevance to submerged hollow fibre modules. Journal of Membrane Science 184 (2001) 221 – 231.
5. J. Wang, Z. Cui, H. Jia, H. Zhang. The effect of fibre length on non-uniform and hysteresis phenomenon in hollow fibre membrane backflushing. Desalination 337 (2014) 98 – 108.
6. Wang, Z., Su, K., Shu, T., Wang, W. Numerical simulation of filtration performance in submerged membrane bioreactors: Effect of particle packed structure (2017). Water Science and Technology. 76 (9). pp. 2503-2514.
7. P. Le-Clech, Y. Marselina, Y. Ye, R. M. Stuetz, V. Chen. Visualisation of polysaccharide fouling on microporous membrane using different characterization techniques. Journal of Membrane Science 290 (2007) 36 – 45.
8. Y. Marselina, Lifia, P. Le-Clech, R. Stuetz, V. Chen. Characterisation of membrane fouling deposition and removal by direct observation technique. Journal of Membrane Science 341 (2009) 163 – 171.
9. F. Wicaksana, A. G. Fane, P. Pongpairoj, R. Field. Microfiltration of algae (Chlorella sorokiniana): Critical flux, fouling and transmission. Journal of Membrane Science 387 - 388 (2012) 83 – 92.
10. Y. Wang, F. Wicaksana, C. Y. Tang, A. G. Fane. Direct microscopic observation of forward osmosis membrane fouling. Environmental Science and Technology 44/18 (2010) 7102 – 7109.
11. W. Lee, W. Cheong, K. Yeon, B. Hwang, C. Lee. Correlation between local TMP distribution and bio-cake porosity on the membrane in a submerged MBR. Journal of Membrane Science 332 (2009) 50 – 55.
12. ImageJ. Image analysis software [online: 22.9.2019], URL:<http://rsbweb.nih.gov/ij/index.html>.
13. Meazure. Image analysis software [online: 22.9.2019]. URL: <http://www.cthing.com>.

Numerical modelling of a non-Newtonian fluid in the twin-screw pump

Marian Bojko[1,*], *Sylva* Drábková[1], *Josef* Veselý[2], and *Rostislav* Zapletal[2]
[1] VŠB-TU Ostrava, Faculty of Mechanical Engineering, Department of Hydromechanics and Hydraulic Equipment, 17. listopadu 2172/15, Ostrava-Poruba 70800, Czech Republic
[2] Hydrosystem project a. s., tř. Kosmonautů 1103/6a, Hodolany, 77900 Olomouc, Czech Republic

Abstract. Twin screw pump is a positive displacement pump, which is widely used in numerous type of foodstuff applications. Numerical CFD analysis presented in this paper focuses on the investigation of the influence of the clearances and the non-Newtonian behaviour of the fluid. The pump operates at low rotational speed and low inlet pressure. As the first step 3D CFD simulation of twin-screw pump non-rotating segment has been carried out using static mesh without rotation. Various empirical equations have been applied to represent the viscosity relations for non-Newtonian fluid in simple shear. The results are compared for two different values of clearances and various approaches to non-Newtonian fluid description. The obtained results showed the effect of both tested parameters. The physical experiment is under preparation to validate the calculation results.

1 Introduction

Twin screw pump can be classified as the rotary positive displacement pump that displaces a fixed quantity of fluid for every revolution of the driving shaft. The most common arrangement includes two counter rotating screw rotors, one of them is directly connected to the drive and the second is driven by the means of timing gears. Each of the screws comprises two symmetrically opposed screw parts (double suction) with the flow passing from the ends of the screws to the centre of the pump. The scheme of the twin screw pump is presented in Fig. 1.

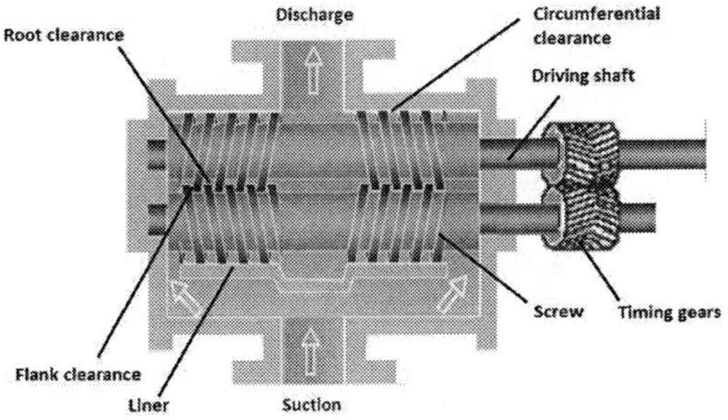

Fig. 1. Scheme of the twin screw pump.

The screws rotate in a liner housing with very small clearances. Figure 1 shows different locations in the pump where clearances occur. We can distinguish the circumferential clearance between screw rotors and liner, the flank clearance between flanks of screw rotors and the root clearance between tip diameter of one screw and root diameter of another screw. These clearances cause an internal leak flow between the discharge higher-pressure and suction lower-pressure sides. The leakage reduces the theoretical (displacement volume) flow rate of the pump and so the volumetric efficiency. In case of highly viscous liquids the leak flow can be assumed laminar and increasing linearly with Δp. The rheological properties of the pumped material have strong influence on screw pump parameters. In case of many food products, the non-Newtonian behaviour must be taken into account.

A lot of research works dealing with the screw pumps have been presented in the field of polymer extrusion. The studies present various approaches to screw extruder design and prediction of its parameters. Differential equations for thermally and hydrodynamically fully developed flow were described in early sixties by Griffith [1]. An analytical model for slightly non-Newtonian biopolymers is presented in [2] and [3]. In recent years the computational fluid dynamics have brought new possibilities to obtain deeper insights in screw pump function and operation. Commercial CFD codes were used first for single screw pump modelling [4]. Development of new grid generation tools have enabled to use the numerical modelling also in case of twin screw pumps with screws intermeshing [5], [6].

The goal of this work was to predict the leakage flow rate first for non-rotating screws. The influence of the clearances and the non-Newtonian behaviour of the fluid was tested on steady screw segments. The laminar flow in internal clearances was investigated for three different geometries of the screws and five rheological functions describing the shear thinning behaviour of the transported material.

* Corresponding author: marian.bojko@vsb.cz

 https://doi.org/10.1051/epjconf/202226901005

2 Model setup

Model studies have been carried out for non-rotating screw segments with four threads. The design of the screws is conventional, based on constant rotor diameter, pitch and profile shape as depicted in Fig. 2.

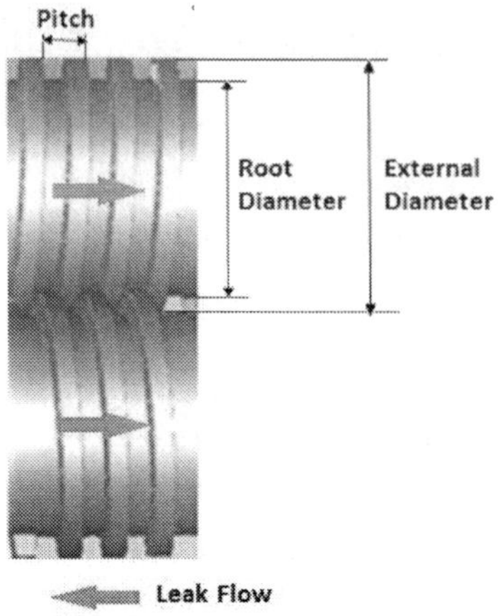

Fig. 2. Screw segments with main dimensions.

The basic geometrical parameters for three variants of screw geometry are defined in Table 1.

Table 1. Geometrical parameters of the screws.

	Geometry version 1:	Geometry version 2:	Geometry version 3:
External diameter (mm)	152	152	159,9
Root diameter (mm)	127	127	119,5
Channel depth (mm)	12,5	12,5	20
Aver. channel width (mm)	12,5	12,5	20
Channel depth (mm)	12,5	12,5	20
Pitch (mm)	25	25	40
Circumferential clearance (mm)	0,7	0,1	0,1
Flank and root clearance (mm)	0,5	0,3	0,3

Inlet and outlet part were connected to the screw segments to enable the definition of boundary conditions as illustrated in Fig. 3.

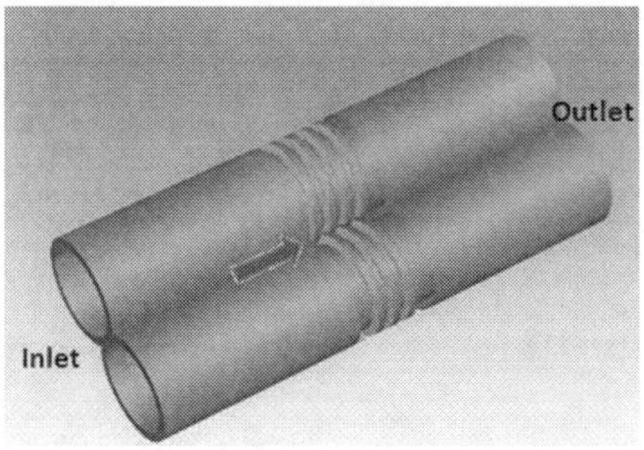

Fig. 3. The modelled geometry with inlet and outlet parts.

3 Non - Newtonian characteristics of the transported material

Tangential shear stress τ is caused by internal friction of the fluid. If the adjacent layers of liquid move at different velocity, their interface shear stress τ arises and it prevents movement. According to Newton's law, this shear stress is linearly proportional to the velocity gradient in the direction perpendicular to the fluid motion.

$$\tau = \eta \frac{du}{dy} \tag{1}$$

where τ is shear stress, du/dy is velocity gradient and η is dynamic viscosity. The ration of tangential shear stress to velocity gradient in Newtonian fluids is constant. For non-Newtonian fluids, this ratio is not constant and different rheological models are defined (for example Power-law model and Herschel-Bulkley model). The Power-law model defines the relationship between the shear stress τ and the shear rate $\gamma = du/dy$ by two parameters, the consistency k and the flow index n in a form:

$$\tau = k\gamma^n \tag{2}$$

The consistency is a constant of proportionality, the flow index defines the degree to which the fluid is shear-thinning. Slightly non-Newtonian materials have a power law index value between 0,7 and 1 and those classified as highly non-Newtonian have values between 0 and 0,7. [3]

The Herschel–Bulkley fluid is generalized, non-linear model of non-Newtonian fluid which combines the Bingham and Power-law fluids in a single relation. Three parameters characterize the relationship between the shear stress τ and the shear rate γ: the consistency k, the flow index n, and the yield shear stress τ_0:

$$\tau = \tau_0 + k\gamma^n \tag{3}$$

The yield stress τ_0 quantifies the amount of stress that the fluid resists before it yields and begins to flow. Applied Power-law functions are presented in Fig. 4 and Herschel–Bulkley characteristics used in simulation are given in Fig. 5.

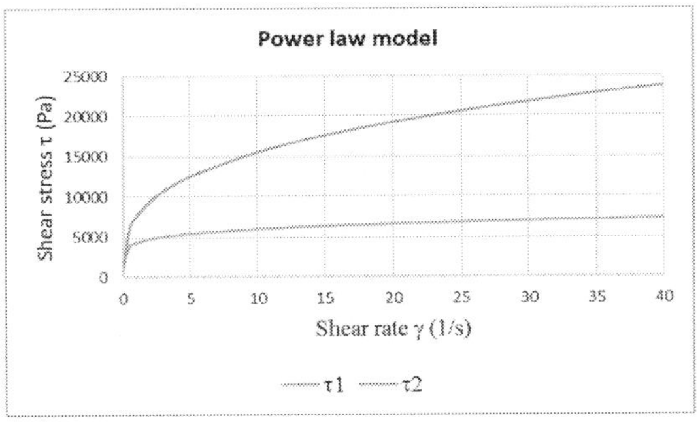

Fig. 4. Power-law characteristics of transported material.

$$\tau_1 = 4295{,}8\gamma^{0{,}1408} \quad (4)$$

$$\tau_2 = 7632{,}6\gamma^{0{,}3081} \quad (5)$$

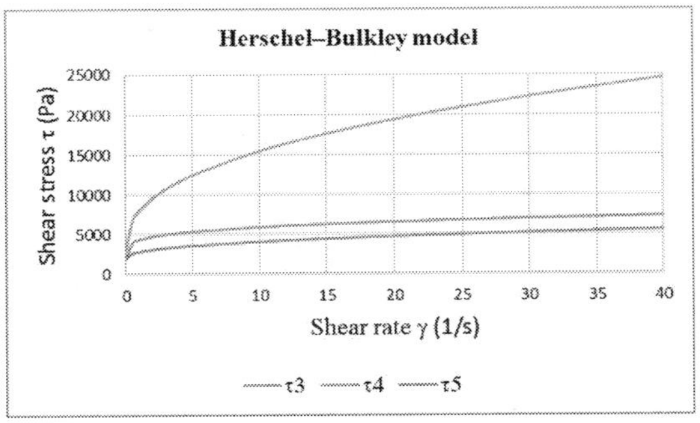

Fig. 5. Herschel–Bulkley characteristics of transported material.

$$\tau_3 = 2050 + 2330\gamma^{0{,}219} \quad (6)$$

$$\tau_4 = 3100 + 4908\gamma^{0{,}401} \quad (7)$$

$$\tau_5 = 1986{,}78 + 823{,}38\gamma^{0{,}3978} \quad (8)$$

4 Mathematical model of fluid flow

The flow of highly viscous substances in the screw pump is characterized as a laminar, stationary flow of incompressible substance in a general three-dimensional region. The resulting mathematical model is defined by basic balance equations such as continuity equations and Navier-Stokes equations for laminar flow [7]:

Mass conservation (continuity equation)

$$\frac{\partial(\rho)}{\partial t} + \nabla \cdot (\rho\vec{u}) = 0 \quad (9)$$

Momentum conservation (Navier- Stokes equation)

$$\frac{\partial(\rho\vec{u})}{\partial t} + \nabla \cdot (\rho\vec{u}\vec{u}) = -\nabla p + [\nabla \cdot \bar{\bar{\tau}}] + \rho\vec{a} + S_m \quad (10)$$

The viscosity of substances is defined by the functional relationships given in chapter 3.

Boundary conditions of the steady isothermal flow computation were the same for each geometry variant and are stated below:

- pressure boundary condition is specified at the inlet ($p = p_i = 3$MPa) and outlet port ($p = p_o = 0.5$MPa),
- no-slip boundary condition is adopted on the stationary walls ($v = 0$).

The geometry of the model and computational grid were prepared in ANSYS DesignModeller [8]. Three variants of geometry were created with different dimensions defined in Table 1. Unstructured computational grid comprises combination of hexagonal, tetrahedral and pyramidal elements. The total number of elements for each geometry variant is presented in Table 2.

Table 2. Number of grid elements

Variant	Number of cells
Geometry variant 1	11 640 331
Geometry variant 2	11 350 618
Geometry variant 3	19 288 383

The grid of the fluid domain is illustrated in Fig. 6.

Fig. 6. Grid example of the fluid domain.

The defined mathematical model was applied to three variants of geometry. The same inlet and outlet boundary conditions were defined for each geometry with the objective to evaluate the mass flow-rate for different non-Newtonian fluids defined in chapter 3. Five numerical calculations were performed in ANSYS Fluent for each geometry variant [9].

5 Evaluation of results

The modelling of the flow of non-Newtonian fluids described with function $\tau_1\tau$,$_{25}\tau$,$_4\tau$,$_3\tau$, was focused on the evaluation of the mass flow- rate for the same pressure drop value of 2.5 MPa. Altogether 15 variants of numerical simulations were realized in ANSYS Fluent environment with the same boundary conditions.

The mass flow-rate predictions from individual CFD analyses are given in Tables 3 and 4 and then compared in the graph (Fig. 7).

Table 3. Calculated mass-flow rate for geometry var.1.

Rheology Description	Geometry variant 1
$\tau_1 = 4295{,}8\gamma^{0{,}1408}$	$Q_{m\text{-}\tau_1}$ = 1760 kg/h
$\tau_2 = 7632{,}6\gamma^{0{,}3081}$	$Q_{m\text{-}\tau_2}$ = 8,87 kg/h
$\tau_3 = 2050 + 2330\gamma^{0{,}219}$	$Q_{m\text{-}\tau_3}$ = 1123 kg/h
$\tau_4 = 3100 + 4908\gamma^{0{,}401}$	$Q_{m\text{-}\tau_4}$ = 10,73 kg/h
$\tau_5 = 1986{,}78 + 823{,}38\gamma^{0{,}3978}$	$Q_{m\text{-}\tau_5}$ =889 kg/h

Table 4. Calculated mass-flow rate for geometry variant 2 and 3.

Geometry variant 2	Geometry variant 3
$Q_{m\text{-}\tau_1}$ = 293 kg/h	$Q_{m\text{-}\tau_1}$ = 733,75 kg/h
$Q_{m\text{-}\tau_2}$ = 1,28 kg/h	$Q_{m\text{-}\tau_2}$ = 3,11 kg/h
$Q_{m\text{-}\tau_3}$ = 178 kg/h	$Q_{m\text{-}\tau_3}$ = 480,24 kg/h
$Q_{m\text{-}\tau_4}$ = 1,59 kg/h	$Q_{m\text{-}\tau_4}$ = 3,49 kg/h
$Q_{m\text{-}\tau_5}$ = 109 kg/h	$Q_{m\text{-}\tau_5}$ =271,8 kg/h

Calculated mass-flow rate for three geometry variants and five rheological functions is compared in Figure 7.

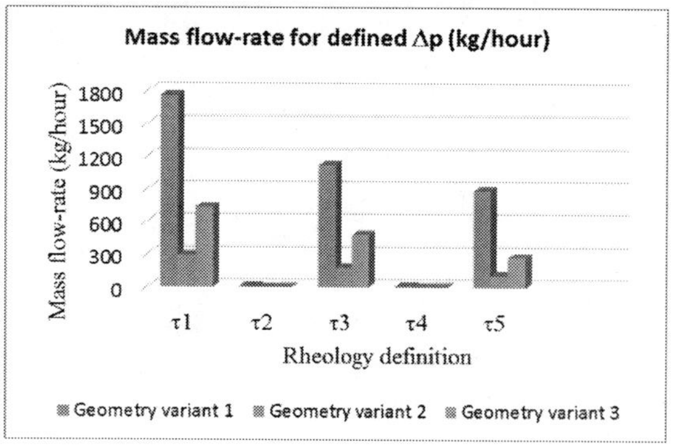

Fig. 7. Comparison of calculated mass flow-rate.

Significantly lower mass flow-rate for non-Newtonian fluid defined by function τ_2 and 4τ can be observed. The reduction of clearances contributes to leak flow decrease, as can be concluded from comparison of geometry variant 1 and 2 which differ only in clearances.

The example of pressure drop dependent on the length l of the modelled geometry variant 1 is evaluated in Fig. 8. The nett pressure drop on the screw segment without inlet and outlet part is given in Table 5.

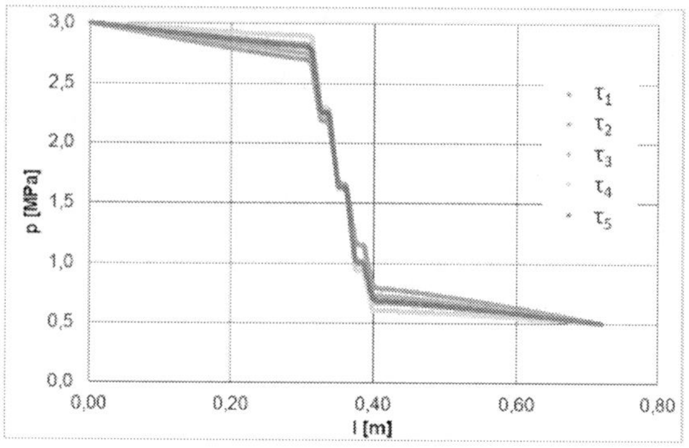

Fig. 8. The pressure drop dependent on the length l.

Table 5. The nett pressure drop on the screw segment without inlet and outlet part.

Geometry version 1	Geometry version 2	Geometry version 3
Δp_{τ_1}=1,84 MPa	Δp_{τ_1}=1,99 MPa	Δp_{τ_1}=2,18MPa
Δp_{τ_2}=2,11 MPa	Δp_{τ_2}=2,28 MPa	Δp_{τ_2}=2,37 MPa
Δp_{τ_3}=1,99 MPa	Δp_{τ_3}=2,02 MPa	Δp_{τ_3}=2,2 MPa
Δp_{τ_4}=2,27 MPa	Δp_{τ_4}=2,45 MPa	Δp_{τ_4}=2,48 MPa
Δp_{τ_5}=2,07 MPa	Δp_{τ_5}=2,21 MPa	Δp_{τ_5}=2,33MPa

The pattern of the non-Newtonian fluid flow is evaluated with pathlines colored with velocity magnitude for geometry variants 1 and 2 and rheological function τ_1 and τ_2. It can be observed that the maximum velocity is reached in the narrow gaps between the intermeshing threads of the screws.

The velocity magnitude is corresponding to the value of mass flow-rates given in table 4. In case of τ_1 the flow index n is 0,1408 and the mass flow-rate $Q_{m\text{-}\tau_1}$ = 1760 kg/hour for the bigger initial clearances.

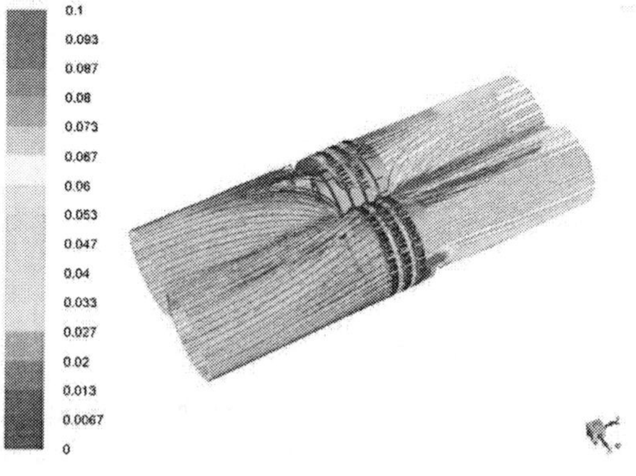

Fig. 9. Pathlines colored by velocity magnitude u (range 0 m·s^{-1} ÷ 0,1 m·s^{-1}) for geometry variant 1 and rheology τ_1.

After reduction of clearances the mass flow-rate decreased to 293 kg/hour and was even more concentrated in the narrow gaps between screw threads.

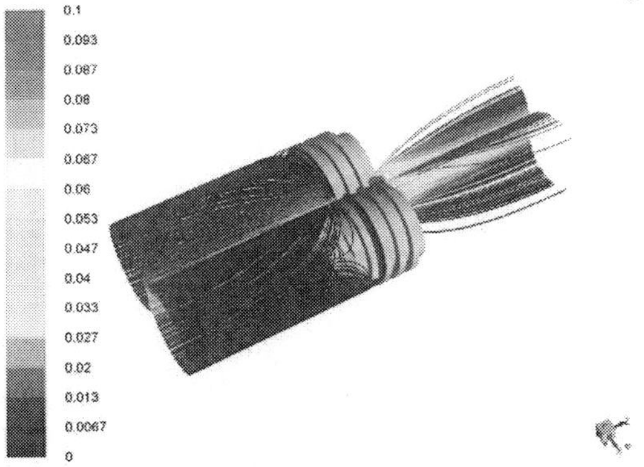

Fig. 10. Pathlines colored by velocity magnitude u (range 0 m·s^{-1} ÷ 0.1 m·s^{-1}) for geometry variant 2 and τ_1.

Similar behaviour can be observed for rheology τ_2 with the flow index n = 0,3081 with mass flow-rate $Q_{m\text{-}\tau_2}$ = 8,87 kg/hour in case of the bigger clearances. After reduction of clearances the mass flow-rate decreased to 1.28 kg/hour and again was concentrated between the screws.

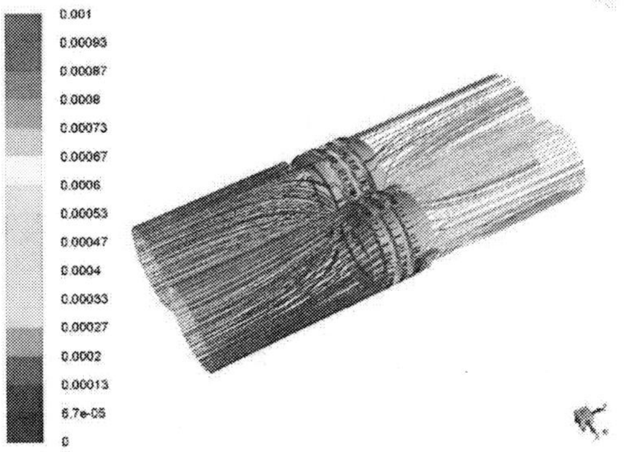

Fig. 11. Pathlines colored by velocity magnitude u (range 0 m·s^{-1} ÷ 0,001 m·s^{-1}) for geometry version 1 and τ_2.

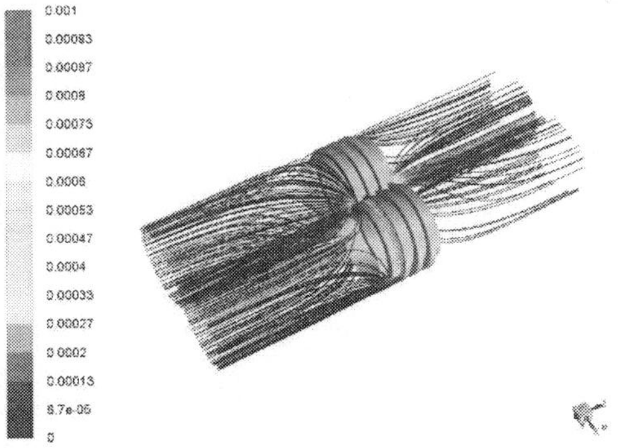

Fig. 12. Pathlines colored by velocity magnitude u (range 0 m·s^{-1} ÷ 0,001 m·s^{-1}) for geometry version 2 and τ_2.

6 Conclusion

This paper deals with the definition of a mathematical model of non-Newtonian fluid flow with different viscosity definitions. Furthermore, the mathematical model is applied to three different geometries of the non-rotating screw segments of the twin screw pump.

The geometry variants 1 and 2 differ only in clearances. The first version of the geometry (Geometry variant 1) has the greatest clearances, and therefore the evaluated mass flow-rate for the same inlet and outlet boundary conditions is significantly higher than for other two geometries in case of all tested rheology functions. The velocity in circumferential clearance between screw rotors and liner is lower than that observed in flank and root clearances. The flow is of similar character as a flow through a straight labyrinth seal [10].

Reduction of clearances leads to decrease of the flow-rate. The geometry variant 3 combines increase of screw channel dimensions with small clearances, which results in higher flow-rates in comparison with variant 2. For comparison of twin screw pump simulation with measurement, the model 1:2 comprising non-rotating screws was prepared and measured. The measured data will be used for validation of results obtained with numerical modelling. The next goal of the work is modelling rotating screws to get more insight of the flow in a twin screw pump.

Acknowledgement

This work was supported by the project Trio FV40105 "Development and production of a prototype of a high-pressure screw pump for pumping highly viscous materials" funded by Ministry of Industry and Trade, Czech Republic.

https://doi.org/10.1051/epjconf/202226901005

References

1. R. M. Griffith. Fully Developed Flow in Screw Extruders. Theoretical and Experimental Study. Industrial & Engineering Chemistry Fundamentals 1962 *1* (3), Pages 180-187
2. J. I.Orisaleye, S. J. Ojolo. Parametric analysis and design of straight screw extruder for solids compaction. Journal of King Saud University. Volume 31, Issue 1, January 2019, Pages 86-96
3. J. I. Orisaleye, O. A. Adefuye, A. A. Ogundare, O.L. Fadipe. Parametric analysis and design of a screw extruder for slightly non-Newtonian (pseudoplastic) materials. Engineering Science and Technology. Volume 21, Issue 2, April 2018, Pages 229-237
4. A. Covas, A. Gaspar-Cunha. A computational investigation on the effect of polymer rheology on the performance of a single screw extruder, e-rheo.pt. 1 (2001) 41–62.
5. K. Wilczynski , A. Nastaj, A. Lewandowski, K. J. Wilczynski. Lewandowski, K. Wilczynski. Multipurpose Computer Model for Screw Processing of Plastics. Polymer-Plastics Technology and Engineering. Volume 51, Issue 6, April 2012, Pages 626-633.
6. A. Lewandowski, K. Wilczynski. Computer modeling for polymer processing co-rotating twin screw extrusion – nonconventional screw configurations. Mechanik NR 4/2017.
7. M. Kozubková, T. Blejchař, M. Bojko, Modelování přenosu tepla, hmoty a hybnosti. VŠB-TUO, pp. 174. 2011
8. ANSYS Fluent User's Guide, Release 18.2, 2017 ANSYS, Inc.
9. ANSYS Fluent Theory Guide, Release 18.2, 2017 ANSYS, Inc.
10. M. Kozdera, M. Bojko, S. Drábková, M. Kozubková. Studying the interaction of liquid and solid in the flow of liquid in annulus by CFD analysis. Applied Mechanics and Materials, Volume 630, p. 220-227, 2014.

Experimental tests of the air flowing out of a swirl diffuser

Marek Borowski[1], *Marek* Jaszczur[2*] *Michal* Karch[1], *Sławosz* Kleszcz[2]

[1] Faculty of Mining and Geoengineering, AGH University of Science and Technology, 30-059 Krakow, Poland
[2] Faculty of Energy and Fuels, AGH University of Science and Technology, 30-059 Krakow, Poland,

Abstract. The geometry of swirl diffusers affects the shape of the air stream injected into the rooms. In this work, the air flow from the swirl diffuser was tested using the Particle Image Velocimetry (PIV) method. The presented laboratory model was a simplification of the real object and was made at a geometric scale of 1:10. The swirl diffuser was tested with two vanes settings. The measured speed fields are shown. Normalized air velocity profiles obtained at different values of volume flow were compared. This will allow better understanding of the air flow from the swirl diffusers and better shape the air flow in the ventilated rooms.

1 Introduction

Ventilation of buildings should ensure adequate air quality in the rooms. The ventilation system is designed to bring the right amount of air and ensure the desired air distribution inside the rooms. Poor flow and too little air exchange in the room causes discomfort, it can also lead to a sick building syndrome. Poor climatic conditions affect the health of the users of the rooms as well as the construction of the building. Knowledge and forecasts of indoor air flow fluxes are therefore important for optimizing indoor climate. This has an impact on energy saving and environmental protection[1,2]. People in the indoors spend nearly 80% of their time during the day. Therefor indoor air quality is an important aspect of the indoor environment.

Swirl diffusers are intended for use in low- and medium-pressure ventilation systems. They are especially recommended for rooms with a height of 3 to 12 meters, where an important parameter is the exact setting of the final air speed to the occupied zone. The diffusers of this type are widely used in supply and exhaust systems in rooms requiring comfort, such as offices, shopping centers, restaurants, cinemas or so-called "Clean rooms" [3].

When choosing this type of diffusers, it is important to determine the coverage area of the air stream being blown. For this purpose, manufacturers of this type of diffusers present tables, nomograms or diagrams that show the range of the air stream depending of air flow. It can be noticed, however, that there are discrepancies between the data of the different producers in determining the range of the supply air stream. These discrepancies result mainly from the measurement methods used by manufacturers. Therefore, it is still a valid task to verify the results on an experimental way, adapting new measurement methods.

The air flow through swirl diffusers was analyzed by CFD methods by other authors [4,5]. Computational fluid dynamics is a very useful technique for predicting the distribution of air in a ventilated space. But the accuracy of CFD forecasts depends on the adopted and delivered boundary conditions. Simplifying the geometric model of the swirl diffuser is often used. This, together with inadequate turbulence modeling, can lead to a result that is far from real. In addition to the few items [6,7], the results of experimental research related to similar issues are missing in the literature. Therefore, an experimental investigation of the air velocity field flowing through the vortex diffusers was selected

In the presented work, Particle Image Velocimetry (PIV) was used. A laboratory model was made at a geometric scale of 1:10. He was a simplification of the real object. The flow conditions were adapted to the scale.

2 Experimental setup

Figure 1 shows a photograph of the measuring section of swirl diffusers. The aim of the research was to obtain velocity fields at the outflow from the diffuser. The measurement was carried out using the PIV Stereo method [6,8,9]. A model of the installation with a horizontal swirl diffuser was developed for the tests. The air flow was forced by a Venture MPB-K 200S radial fan. An inverter was installed, which enabled the fan speed control. The experimental setup consisted of a channel with a diameter of 50 millimeters and a length of 2 meters, to which the air was supplied by the stabilizing system of the air stream which had a length of 1.35 meters and consisted of a diffuser, a confusor and a

* Corresponding author: marek.jaszczur@agh.edu.pl

channel with a diameter of 0.2 meters. It equalized the air velocity distribution at the inlet to the tested diffuser.

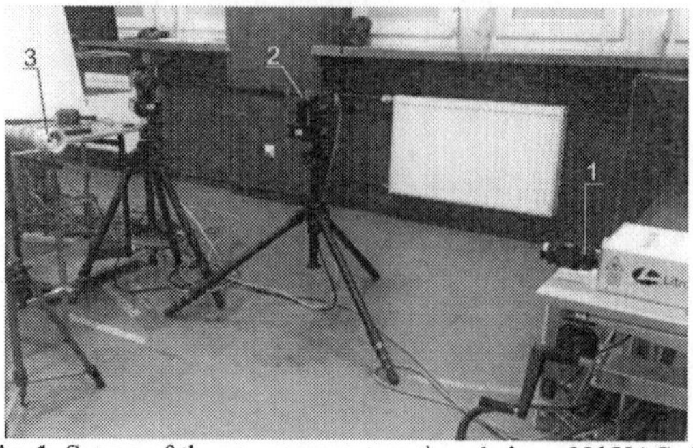

Fig. 1. Set-up of the measurement section: 1- laser Nd:YAG, 2-camera CCD, 3-swirl diffuser.

The air was used as the experimental fluid in the channel, thus the equality of the Pr number was automatic. Having equal Re numbers ensures the averaged velocity fields are similar. The Reynolds numbers were equal to Re1=47000, Re2=85000, Re3=143000, Re4=183000. Particle Image Velocimetry (PIV) method was used to evaluate the velocity vector components. The particles were illuminated with a double-pulse Nd:YAG laser of energy of about 400 mJ per pulse. The digital images were acquired by 4 Mpx monochromatic CCD camera. In each experiment, 1000 double frame images were recorded with the camera recording at a frequency of 3 Hz, which resulted in an overall time of one measure of around 5 minutes. During the calculations, the size of interrogation windows that exhibit satisfying results was set to 32x32 px. The Davis 7 software was used to analyse the images and to find the velocity vector components.

The seeding for the PIV method was DEHS oil sprayed through an atomiser, to obtain a particle diameter of about 1 μm. The seeding particles were introduced to the inlet air stream, immediately behind the flow-stabilising section.

The measurements were performed for the steady and isothermal flow conditions.

The tested swirl diffuser was made in the 1:10 scale, in two variants differing in the arrangement of the vanes. The vanes were set at an angle of 60 and 75 degrees. Analyzed swirl diffusers are shown in Figure 2. The volume air flow was measured using a standardized measuring orifice. The experiment was carried out at four volumetric flow rates of the air:100, 185, 310, 380 m^3/h.

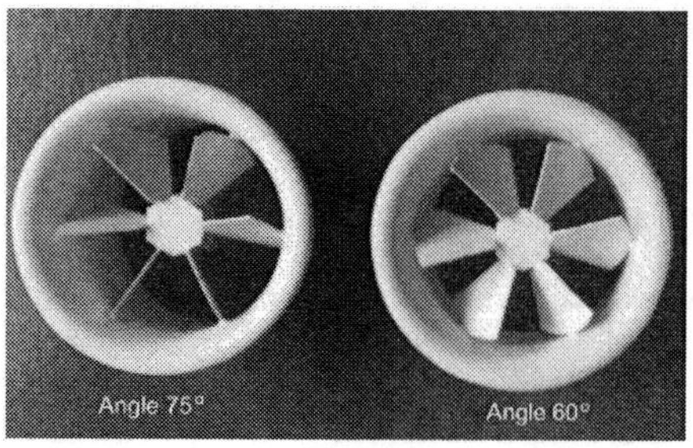

Fig. 2. Setting diffuser vanes: angle of 75 degrees and angle of 60 degrees.

3 Results

The measurement method used allows obtaining of a vector field velocity images throughout the tested area. On this basis, the flow velocity can be analyzed at any point in the area under test. Figures 3 and 4 shows the image of the measured velocity field. The results presented were obtained for 60° and 75° diffusers and for a flow of 380 m^3/h.

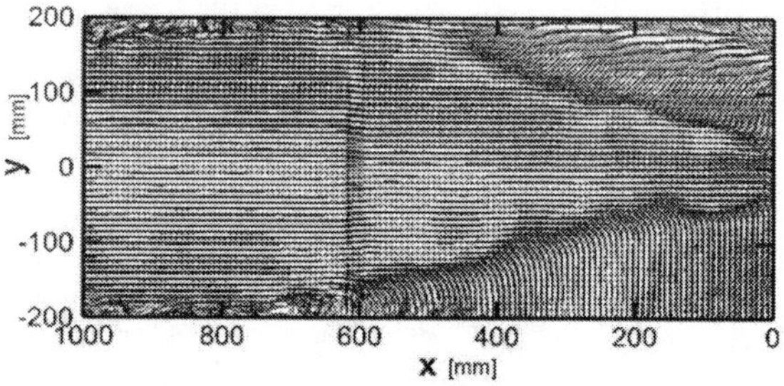

Fig. 3. The image of the measured velocity field. For the case: 60° diffuser and flow rate 380 m^3/h.

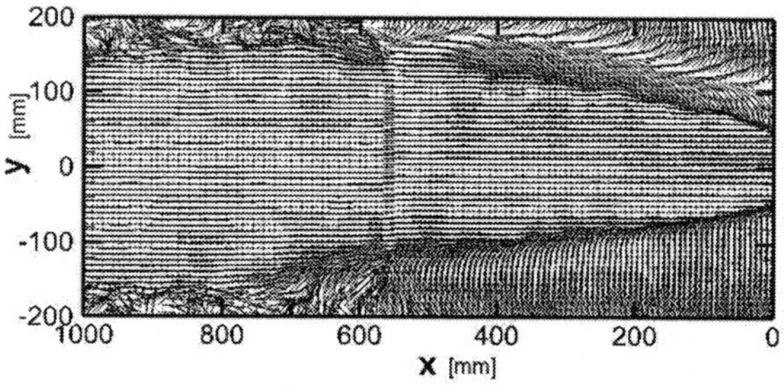

Fig. 4. The image of the measured velocity field. For the case: 75° diffuser and flow rate 380 m^3/h.

On the basis of the measurements made, for each diffuser a series of isolines with the same value of the velocity vector component Vx (axial) was plotted for subsequent values of the Reynolds number. Figure 5 shows the lines for 6 [m/s], 7 [m/s] and 8 [m/s] for the diffuser set with an angle of 60° plotted at Re2= 85000, Re3= 143000 and Re4= 183000. The range of the isoline understood as the coordinate x of the most distant point of the considered isoline from the outlet of the diffuser increases with the number of Re. At the same time, the isoline drawn for higher speeds have shorter and shorter ranges. The images of isolines is similarly presented for the other tested diffusor vanes settings.

Figure 6 shows the lines 9 [m/s], 11 [m/s] and 13 [m/s] for a diffuser with vanes set at an angle of 75° obtained at Re2= 85000, Re3= 143000 and Re4= 183000. Also here is the growing range of individual line with the increase in the value of the number Re.

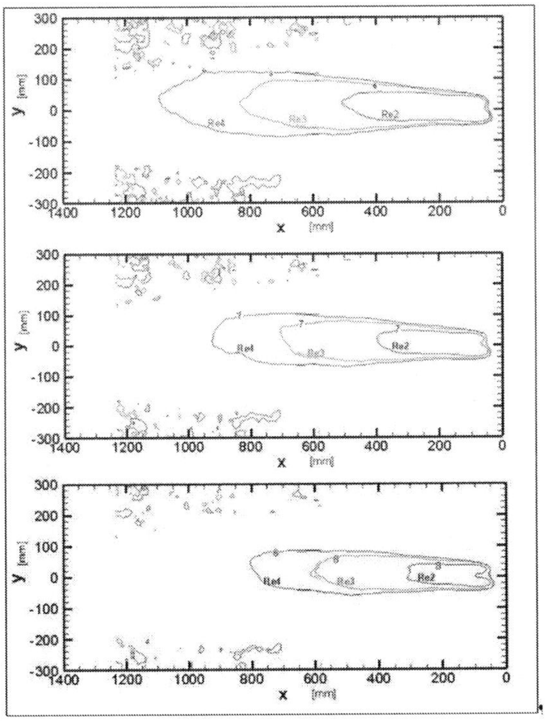

Fig. 5. The lines with the same value velocity vector axial component Vx at the air flow rate 385m3/h and for the 60 angle vanes.

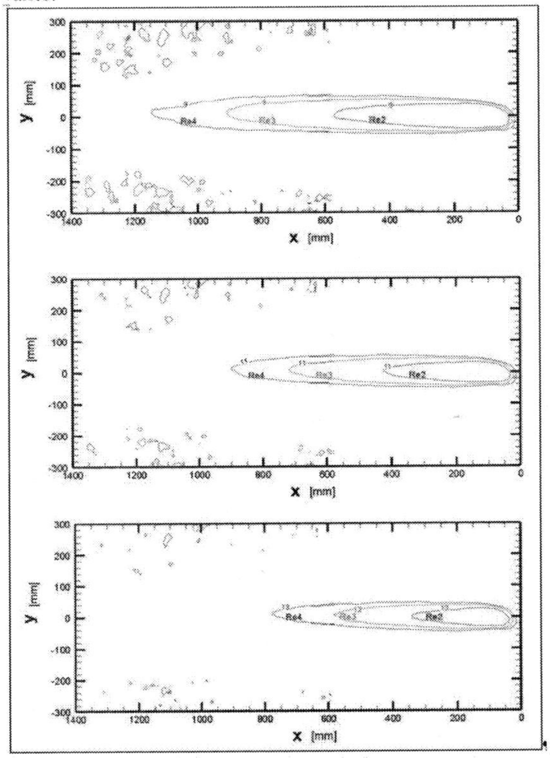

Fig. 6. The lines with the same value velocity vector axial component Vx at the air flow rate 385m3/h and for the 75 angle vanes.

The results of measurements in the form of normalized speed Vnor profiles are presented below:

$$Vnor = \frac{V}{V_{av}}$$

where:
V-the value of the air velocity at the point being analyzed,
Vav- the average speed value in the outlet cross-section of the tested swirl diffuser.

The speed normalization was done by dividing the local velocity values by the average velocity value at the outlet from the diffuser. The average velocity value was determined based on the measuring volumetric flow rate using a orifice. This operation is aimed at comparing speed profiles obtained at different values of the air flow through the diffuser.

Figure 7 shows normalized speed profiles in 100, 200, 300 and 400 mm from the diffuser outlet with vanes set at an angle of 60°, and Figure 8 with vanes set at an angle of 75°. You can see a large similarity of profiles obtained with different values of Re numbers. As the distance from the diffuser increases, the value of the normalized speed decreases. For the swirl diffuser, the speed profiles expand with the increasing distance from the diffuser outlet. The speed profiles for the diffuser with vanes at an angle 75 are thinner than at an angle 60°.

The degree of compliance of distributions of a given size c, for subsequent (i and i+1) numbers of Re, can be expressed in quantitative terms by the mean relative difference:

$$\alpha_w = \frac{\frac{1}{n}\sum_{i=1}^{n}\left|C_{Re_{i+1}} - C_{Re_i}\right|}{\frac{1}{n}\sum_{i=1}^{n}\left|C_{Re_{i+1}}\right|} \cdot 100\%$$

where:
C_{Rei} -value of the c-value measured for the i-th number Re_i: Re1= 47000, Re2= 85,000, Re3= 143000, Re4= 183000

The α_w index was calculated for the velocity vector values. The smaller the relative difference, the greater the degree of matching of distributions.
Table 1 shows the values of α_w calculated for a diffuser with vanes set at an angle of 60°. The measurements were made for the same values of the numbers Re as given above.

Tab. 1 Relative difference of velocity vector value distributions for diffuser with vanes at an angle of 60°.

Distances from the outlet of the diffuser	α_w	
	$Re_2 \rightarrow Re_3$	$Re_3 \rightarrow Re_4$
X=70mm	43,0%	20,6%
X=100mm	35,4%	17,2%
X=150mm	23,2%	15,0%
X=200mm	20,3%	11,8%
Average	30,5%	16,2%

For larger Re values, the index α_w decreases and its average value is 16,2%.

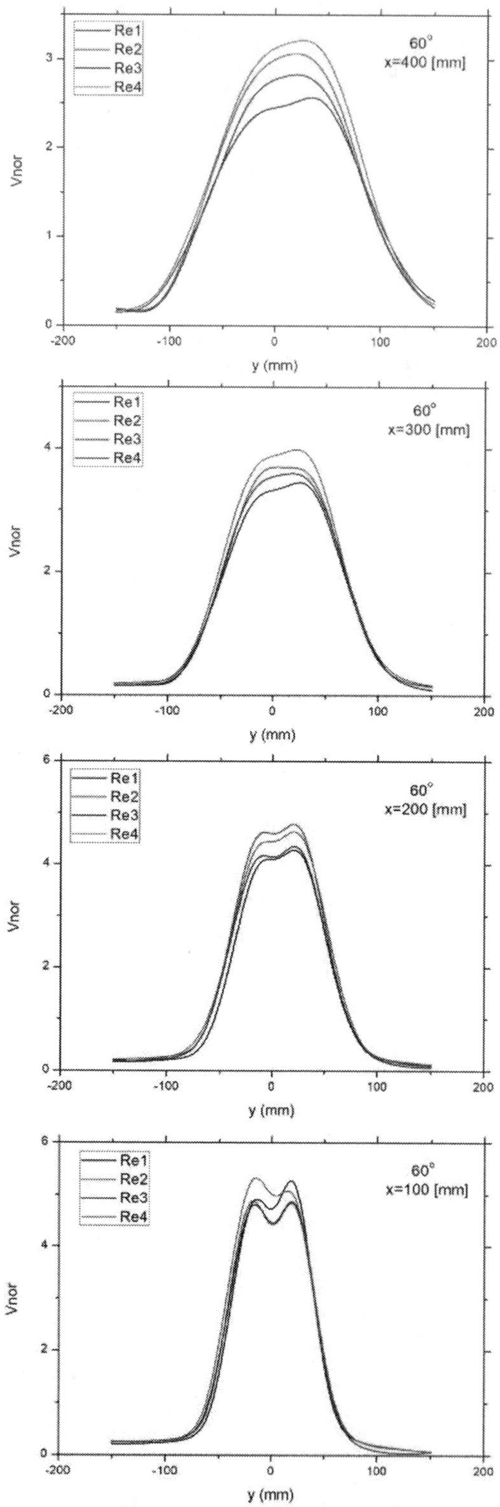

Fig. 7. Standardized profiles of air velocity flowing out of the diffuser with vanes directed at an angle of 60° at different distances (x = 100 ÷ 400mm) from the outlet of the diffuser.

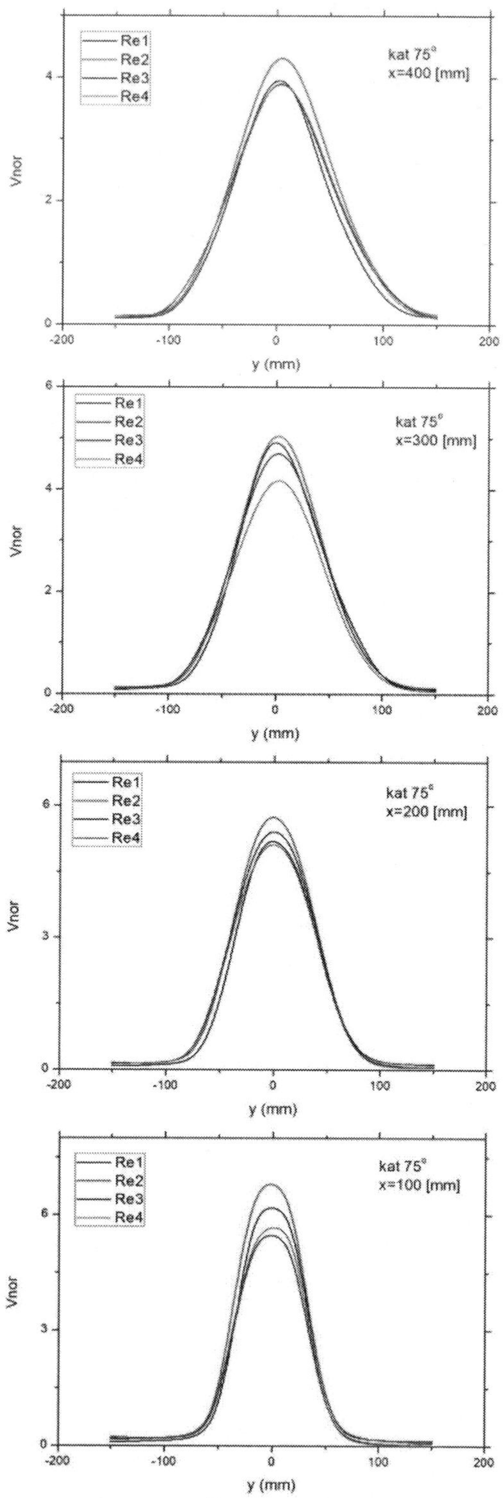

Fig. 8. Standardized profiles of air velocity flowing out of the diffuser with vanes directed at an angle of 75° at different distances (x = 100 ÷ 400mm) from the outlet of the diffuser.

Table 2 shows the values of α_w calculated for a diffuser with vanes set at an angle of 75°. Also in this example, the index α_w decreases for larger Re values, and its average value is 13,2%.

Tab. 2 Relative difference of velocity vector value distributions for diffuser with vanes at an angle of 75°.

Distances from the outlet of the diffuser	α_w	
	$Re_2 \rightarrow Re_3$	$Re_3 \rightarrow Re_4$
X=70mm	28,4%	20,0%
X=100mm	42,2%	13,8%
X=150mm	25,8%	9,9%
X=200mm	18,9%	9,2%
Average	28,8%	13,2%

4 Conclusions

The PIV method was used. This method allows simultaneous measurement in a large area of flow and allows the determination of velocity vectors at any point of this flow. The ability to determine the characteristic features of the diffuser, such as the shape and range of the air stream being formed, demonstrates the suitability of the measurement method used in the issues of ventilation. Information about the formed speed field has a decisive impact on the process of shaping the air flow in the room and thus on the efficiency of ventilation. That is, the rate of removal of gaseous pollutants, the spread of solid particles and microclimate in the zone people occupied.

Many speed profiles normalized in different cross-sections of the stream were compared. For them, a relative difference in the distribution of velocity vector values was calculated. This ratio decreases with the increase in the number of Re. The smallest values were obtained for flows with the Reynolds number above 143000. This means that with the increase in the number of Re, the degree of conformity of normalized distributions of velocity fields increases. In the tested laboratory model of a swirl diffuser, the turbulent flow for Re numbers above 143000 has the ability to self-model.

References

1. H. Awbi, 2003. Ventilation of buidings. London: Taylor and Francis.
2. A. Alfahaid, 2000. Effects of ventilation on human thermal comfort in rooms. Ph.D.Thesis. Norfolk: Old Dominion University.
3. A. Pełech, 2013. Ventilation and air conditioning. Basics. Wrocław: Wrocław University of Technology Publishing House.
4. K.C. Ng, K. Kadirgama, E.Y.K. Ng, 2008. Response surface models for CFD predictions of air diffusion performance index in a displacement ventilated office Energ. Buildings 40 774
5. R. Pandey, A. Kumar Rao, V. Yadav 2015 Comparison between simulation and experimental technique inside an air conditioned room to analyze the performance and flow characteristics of ceiling swirl diffuser under different operating and flow parameters, IJESRT 4 424
6. M. Jaszczur, R. Nowak, J. Szmyd, M. Branny, M. Karch, W. Wodziak 2011. An application of SPIV technique to experimental validation of the turbulence model for the air flow in the intersection of the mining face with the ventilation gallery J Phys Conf Ser. 318 5
7. B. Sajadi, A. Mohebbian, M.H. Saidi, Proc.of the 2nd Int. Conf. on Heating, Ventilation and Air-Conditioning, Tehran, Iran, June 1-3 (2010)
8. J. Westerweel, Digital particle image velocimetry: Theory and applications Delft University Press (1993)
9. LaVision Gmbh, Flow Master Manual for Davis 7.0 (2007)

 https://doi.org/10.1051/epjconf/202226901007

Analysis of velocity profile measurements obtained by different methods in low-speed, small-scale wind tunnel

Ondřej Cejpek[1,*], *Jan* Šip[1], *Milan* Malý[1], *Jan* Jedelský[1], and *Zdeněk* Tomáš[2]

[1]Energy Institute, Faculty of Mechanical Engineering, Brno University of Technology, Technická 2896/2, 616 69 Brno, Czech Republic
[2]Provyko s.r.o., Vinařská 558/3a, Pisárky, 603 00 Brno, Czech Republic

Abstract. A low-speed small-scale wind tunnel was designed for spray investigation under cross and co-flow conditions. Precisely defined profiles of mean velocity and turbulent intensity in the wind tunnel test section are crucial parameters for any rigorous flow study. Different velocity measurement techniques were used to evaluate the velocity and turbulent intensity profiles in the test section. Two non-intrusive techniques, Phase-Doppler Anemometry and Laser-Doppler Anemometry (PDA, LDA), and two intrusive techniques, Constant Temperature Anemometry (CTA) and pitot static tubes of S and L-type were applied. The velocity was measured in 19 equidistantly spaced positions in a centrally placed horizontal plane. The data were obtained for four different mean velocities in the test section (7, 14, 21, 28 m/s). Results of different measurement techniques were mutually compared, and repeatabilities and uncertainties of PDA and CTA measurements were assessed. Turbulent velocity spectra measured by the CTA were analysed. The effect of declination of the pitot static tubes (L-type and S-type) was briefly discussed and compared with an industrial velocity probe QuadraTherm® 640i In-Line Mass Flow Meter with a measuring range of 5–300 m/s. Velocity and declination of pitot static tubes were analysed only in the central point of the test section. The results show that a fully turbulent and uniform flow is developed 15 mm upstream the test section area. Mean velocity and turbulent intensity profiles obtained by different techniques are in a good agreement. Uncertainties of type B of PDA and CTA measurement method are below 5%. Turbulent intensity in the main stream is under 5%. Advantages and drawbacks of the presented measurement techniques were discussed. The PDA was found to be the most suitable measurement technique due to its precision and non–intrusive flow probing.

1 Introduction

Wind tunnels have been used for more than a century. Their size can range from several meters to more than ten meters. Two designs of a wind tunnels used are closed and open loop wind tunnels [1, 2]. The wind tunnels can be also divided according to the velocity in the test section into four groups to subsonic, transonic, supersonic, and hypersonic wind tunnels [3, 4, 5]. Each wind tunnel design is suitable for different types of experiments and required flow characteristics in the test section. Wind tunnels can simulate velocity conditions in combustion chambers, turbines, conditions in the atmosphere and in many other situations related to air movement and air-structure interaction.

Precisely defined velocity profiles and controllable and sufficiently low turbulent intensity in the test section of the wind tunnel are crucial parameters for any rigorous flow study. A new small-scale low-speed wind tunnel was built at Energy Institute, Faculty of Mechanical Engineering, Brno University of Technology, Spray laboratory.

Together with Phase-Doppler Anemometry (PDA), Laser-Doppler Anemometry (LDA), Constant Temperature Anemometry (CTA) and other velocity measurement techniques, wind tunnels create a tool for investigation of high complex, spatially and temporally unstable flows, i.e. flow around the turbine blades, velocity profile around a car, investigation of fluid structure interaction, and spray flows in cross and co-flow.

Laser-Doppler Velocimetry (LDV) or Laser-Doppler Anemometry (LDA) were first implemented and described in 1964 [6]. After this publication, various systems of LDV were proposed. Over a decade, new systems were probed, and improvements of the system were made. The first commercial system was available in 1970. Because of a wide field of application of LDV, many set-ups of the system arise during the LDV commercial use [7].

* Corresponding author: 182752@vutbr.cz

The PDA measurement system is an extension of LDA concept used for velocity measurement. A need to measure the velocity and the size of the particles in investigated flows simultaneously resulted in the development of PDA measurement system. PDA is a very complex system, hence the process of invention, development, and acceptance of laser–based system took a long time. Crucial was the invention of laser in the 1960s, a measurement concept in 1975, and extensive work done by many research groups during 1980s. [8]. The PDA measurement system was not fully accepted at the early stage; the results were verified by other well-known measurement techniques to prove the PDA system accuracy and measurement abilities. After almost 30 years of application of PDA, this measurement technique becomes a standard for the experimental fluid flow studies, and a major part of the flow research is done by the PDA system. Nowadays, PDA is used to validate newly designed techniques, i.e. Particle Image Velocimetry (PIV) [9]. PDA can be also used to validate volumetric measurement methods.

LDA and PDA are optical-based non-intrusive measurement techniques. These techniques identify velocities of the seeding particles introduced into the flow, based on the scattered light by the particles. These techniques can provide results with high spatial and temporal resolution and with high accuracy. LDA measures the velocity of the tracer particles; in addition, PDA is capable of measuring the particle/droplet diameter. The inspected flow had to be seeded with particles which will scatter sufficient light and will follow the flow with high accuracy to identify the finest vortices in the flow. An optical access to the flow is also required [10]. The PDA and LDA system responses are a liner function of the fluid flow velocity in contrast with CTA which is sensitive to the ambient temperature.

Two used intrusive techniques are pitot static tube and CTA. The pitot static tube is a well-known measurement device which has been used in industry for more than two centuries. A simple and robust design, function principle, and accurate results ensure the pitot static tube a wide range of applications in industry (racing cars, aircraft speed measurement, measurement of the flow velocity in duct or pipes, laboratory use). The pitot static tube was initially used only in laboratory for more than a century. After the improvements by Henry P.G. Darcy, the pitot static tube started to be used in industry. [11]

CTA is a measurement technique based on a principle of hot-wire anemometry, where the constant temperature of the measurement element is maintained. A cooling effect of the flow is compensated by the Wheatstone bridge, and the temperature of the element is kept constant. The voltage drop across the circuit is proportional to the inspected fluid flow velocity. The CTA system needs to be calibrated prior to every measurement. CTA is a very useful tool for measurement of the turbulent flow because this system has a very high response to the fluid flow change and can capture the finest vortices in the flow. The probe is usually a thin wire with a diameter of several micrometres; therefore, it has a low thermal inertia.

A fundamental principle of CTA was discovered in 1888, when the relation between the convection and the fluid velocity was experimentally derived. In 1907, cooling of a thin wire in the air stream was observed, and the first suggestion for measuring of the fluid flow velocity appeared. Experimental work continued for the next 30 years; during this period, 2D and 3D probes were designed, and two main principles of hot-wire anemometry emerged (CCA, CTA). Many other improvements of the CTA system arose during this period [12]. Further work was devoted mainly to maximisation of the response of the electrical circuit in the CTA measurement system [13]. The first CTA system was commercially available in 1950s.

2 Experimental setup

Experiments were performed at Spray Laboratory, Faculty of Mechanical Engineering, Brno University of Technology. Mean velocity profiles and turbulent intensities were measured in a newly designed small-scale low-speed wind tunnel test section with cross-section of 200×200 mm. The measurement of mean velocity flow was performed in one plane and in 19 equidistantly spaced positions. Figure 1 shows the schematic layout of the measurement points in the wind tunnel test section.

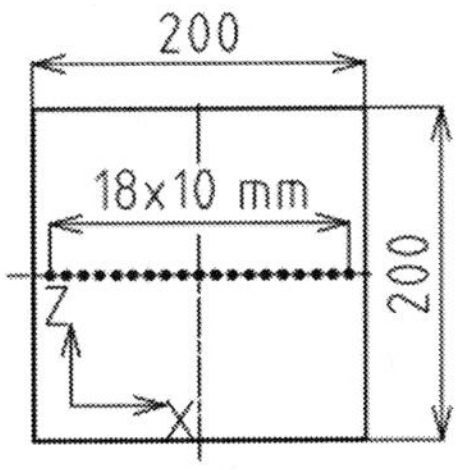

Fig. 1. Measurement points.

2.1 PDA Experimental Setup

Particle sizes and velocities were probed by two-component Phase Doppler anemometer (PDA) Dantec Dynamics A/S, see a schematic layout in Figure 2. In each position, 5 kHz sampling frequency was used, and the measurement took 5 s in each point. Information about the PDA are outlined in Table 1.

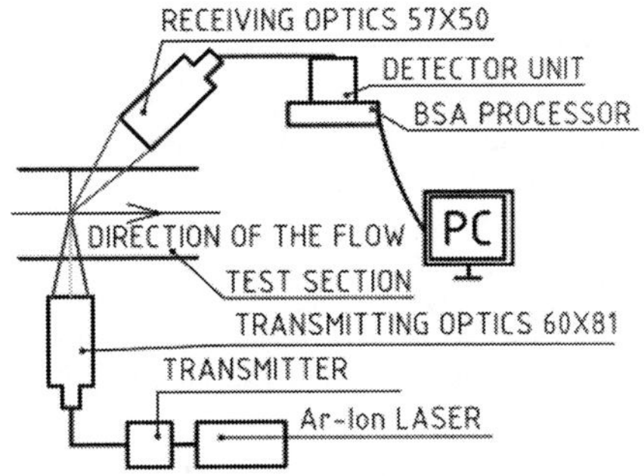

Fig. 2. PDA experimental setup.

Table 1. PDA experimental setup.

Parameter	Value
Laser power output	1 W
Wavelength	514.5 nm
Focal length of transmitting optics	310 mm
Focal length of receiving optics	500 mm
Scattering angle	45 °
Mask	A
Spatial filter	0.1 mm
Sensitivity	1100 V
SNR	0 dB
Signal gain	20 dB
Level validation ratio	6

2.2 CTA Experimental Setup

A thermoanemometry method was applied (HWA), specifically the CTA measurement technique. StreamLine hot-wire anemometry from Dantec Dynamic was used. Velocity measurement was performed with the wired 1D 55R01 probe. The length of the wire was 3 mm, the measuring part of the wire was 1.25 mm long.

Calibration was performed prior to the measurement. The Dantec streamline Pro Automatic calibrator was used. The wire probe was calibrated in the velocity range of 1–45 m/s in fifteen points. A sampling frequency was set to 10 kHz, and the duration of the measurement was set to 5 s in each point.

The Linear traverse system (ISEL AUSTRIA GMBH & CO. KG) was used to traverse the probe. The span of 3 s was set before the measurement to suppress the vibration of the traverse. The CTA experimental setup is illustrated in Figure 3.

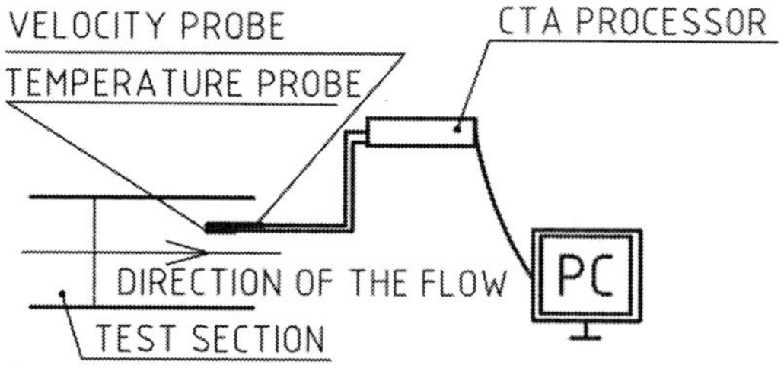

Fig. 3. CTA experimental setup.

2.3 LDA Experimental Setup

FlowExplorer (Dantec Dynamics A/S, Skovlunde, DK) 1D LDA system with Diode Pumped Solid State (DPSS) laser was used. Wavelength of the laser was 660 nm. The system was working in backscatter mode; the receiver and the transmitter were placed in one body. The beam was split into two parallel beams with the power of 30 mW each. One of the beams was shifted by 80 MHz. A converging transmitting/receiving lens with the focal length of 300 mm was used. Dantec BSA P80 signal processor was used to process the measured signal. BSA flow software v5.20 was used to control the data acquisition. A schematic layout of the LDA measurement system is shown in Figure 4.

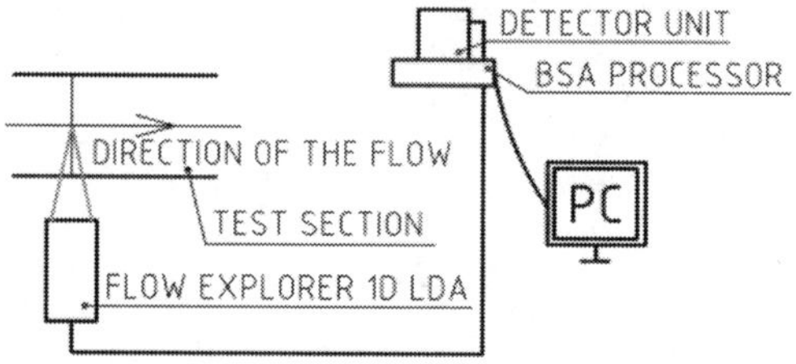

Fig. 4. LDA experimental setup.

2.4 Seeding Particles

A fog generator was used to generate seeding particles. The particles with a diameter D [μm] in the range of 1-3.5 μm were applied on the suction side of the driving fan. The diameter of particles was measured by the PDA measurement system. The particles are small enough to follow the flow reasonably well; the particle Stokes number was below 0.1. A histogram of the particles diameter measured by the PDA is shown in Figure 5.

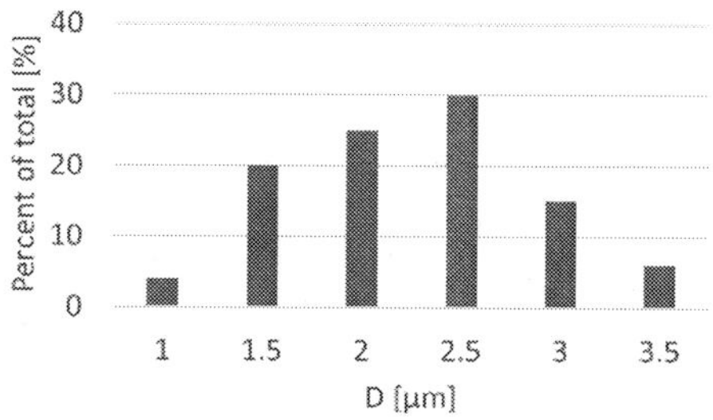

Fig. 5. Histogram of seeding particles diameter measured by PDA system.

2.5 Experimental Setup of Pitot static tube

The L-type pitot static tube was used. The measuring device with accuracy of ±0.1% of measurement value was used. A pressure difference measurement range was 0–1000 Pa. The measurement system does not allow for PC connection or data export; the values were read on the device display. The measurement system can be seen in Figure 6.

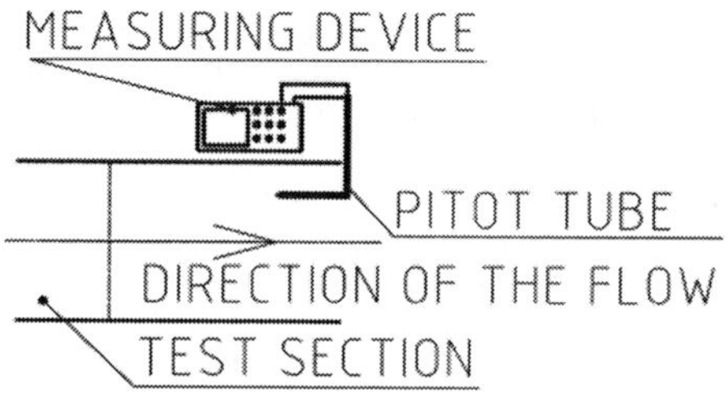

Fig. 6. Pitot static tube experimental setup.

3 RESULTS AND DISCUSION

Important characteristics of every wind tunnel are the velocity profile and turbulent intensity in the tunnel test section. Uniform velocity profile and low turbulence are required in the test section of almost all kinds of wind tunnels. As was mentioned before, velocity and turbulent intensity measurement were performed by different measurement systems, which were described in the foregoing section.

3.1 Velocity Profiles

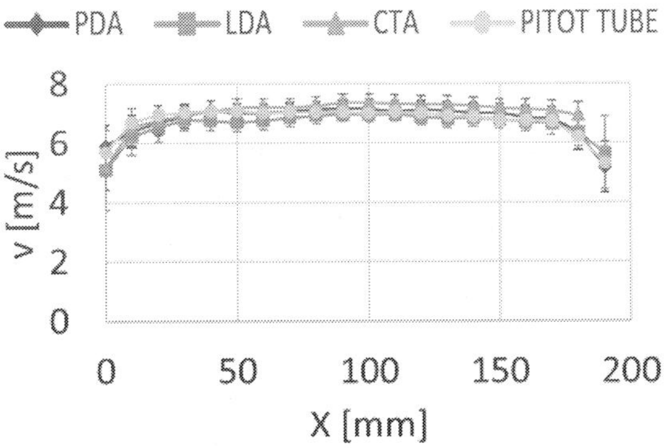

Fig 7. Velocity profile, mean velocity of 7 m/s.

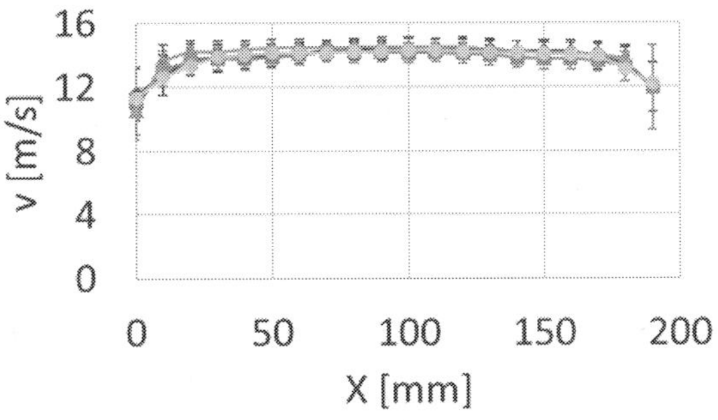

Fig 8. Velocity profile, mean velocity of 14 m/s.

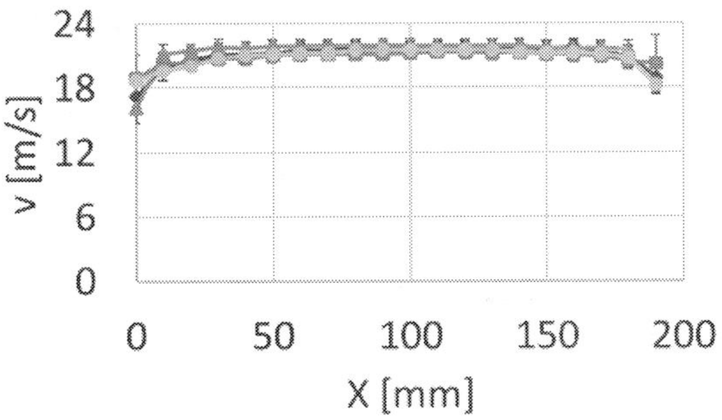

Fig. 9. Velocity profile, mean velocity of 21 m/s.

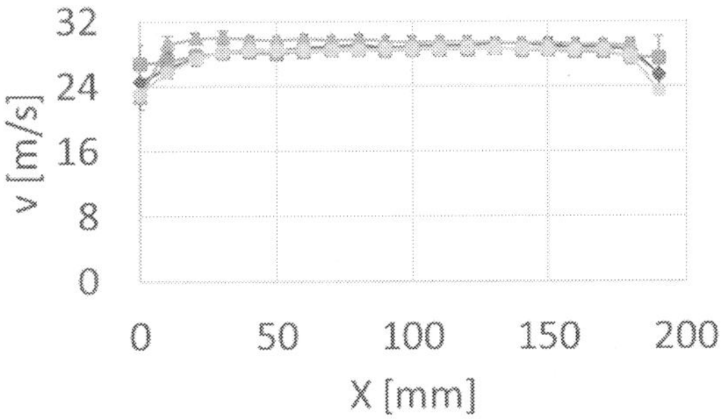

Fig. 10. Velocity profile, mean velocity of 28 m/s.

Figures 7–10 show velocity profiles for different mean velocities in the test section and for different measurement systems. As seen in these images, the results obtained by different measurement methods are in a good agreement. Velocity profiles in Figures 7 to 10 are uniform across the cross-section of the test section. The thickness of boundary layer is defined as the distance from the wall, to the point, where the velocity magnitude reaches value of 99% of free stream velocity. A boundary layer thickness is represented in Table 2. In the boundary layer region, velocity decreases and the largest deviations between the measurement systems are observed.

Table 2. Boundary layer thickness.

Mean velocity [m/s]	Free stream velocity [m/s]	Boundary layer thickness [mm]
7	7,02	15
14	14,31	18
21	21,42	20
28	28,2	23

3.2 Turbulent Intensity

The pitot static tube did not allow to measure the turbulent intensity; a comparison of turbulent intensity is performed only for the PDA, LDA and CTA measurement techniques. Turbulent intensity of four different regimes is illustrated in Figures 11–14. Turbulent intensity in the main flow reaches the values under 5%*. In the boundary layer, turbulent intensity is higher. It is caused by the spatial and temporal instabilities in the boundary layer near the surface. As seen in Figures 11–14, turbulent intensity is slightly different when measured by different techniques, which is most visible in Figure 14. The CTA measurement system measures a lower turbulent intensity than the LDA or PDA systems. According to [14], turbulent intensity measured by the PDA and LDA measurement systems will always be higher (broadening effect).

A lot of improvements of the flow quality can be obtained in the test section. Turbulent intensities in well-designed wind tunnels are under 0.1% [15]. At Tokohu University in Japan, a closed loop wind tunnel with 0.02% turbulent intensity in the free stream was designed [16]. Steps to reduce the turbulent intensity in the test section will be taken in the future.

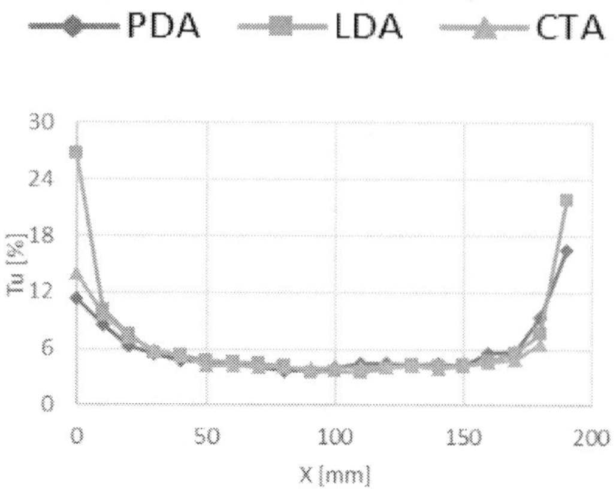

Fig 11. Turbulent intensity, mean velocity of 7 m/s.

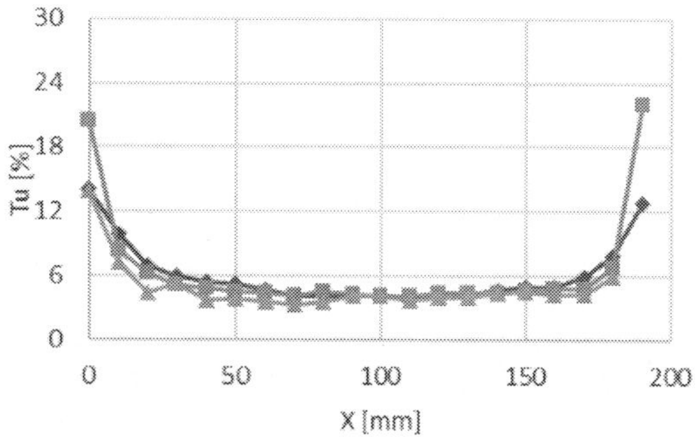

Fig. 12. Turbulent intensity, mean velocity of 14 m/s.

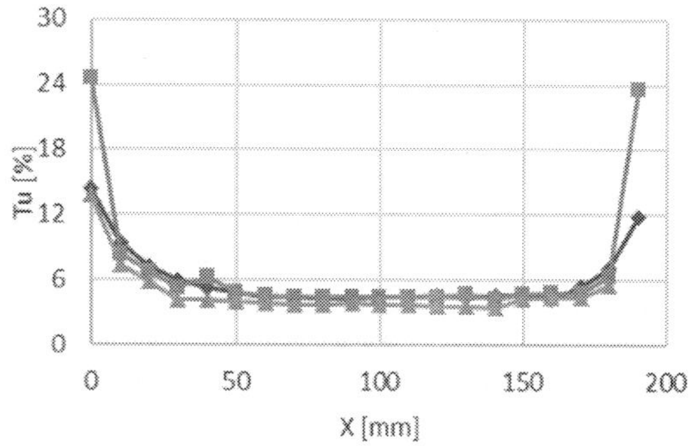

Fig. 13. Turbulent intensity, mean velocity of 21 m/s.

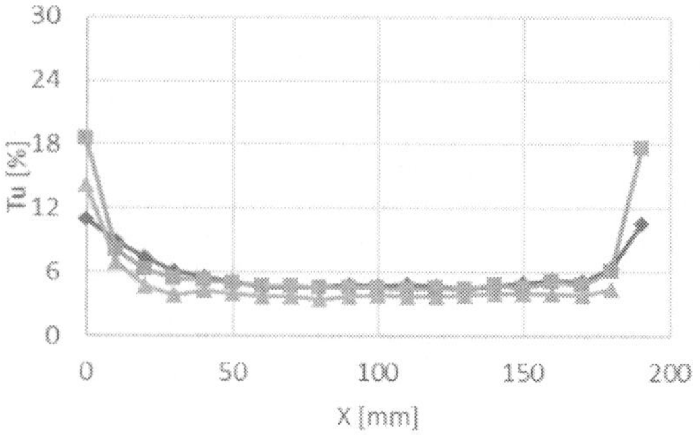

Fig. 14. Turbulent intensity, mean velocity of 28 m/s.

3.3 PDA and CTA Repeatability and Uncertainties

Six measurements were made to determine the PDA and CTA measurement system repeatability. The attempt was made to assess systematic and random errors of measurement. Combined uncertainty for the PDA measurement system was determined according to [17] to be 1% of the measured value for the velocity measurement. Combined uncertainty of the CTA measurement was determined to be 3.5% of the measured value. Uncertainty was determined according to the procedure described in [18]. There are incorporated uncertainties of the calibration (2%), anemometers (0.2%), linearization (0.5%), resolution of A/D converter (0.02%), uncertainty of position of the probe (0.01%). Uncertainty of temperature change of hot wire, ambient temperature change, atmospheric pressure change and humidity of air change are low comparing to aforementioned uncertainty and are not incorporated. In Figures 15 and 16, the uncertainty for one regime of the velocity and turbulent intensity measurement can be seen (mean velocity of 21 m/s). According to the results shown in Figures 15 and 16, the PDA and CTA velocity, turbulent intensity results can be declared to be the same. The uncertainties are higher in the boundary layer region, where the flow has an unstable and unpredictable behaviour. Also, the uncertainties are lower for the PDA measurement system.

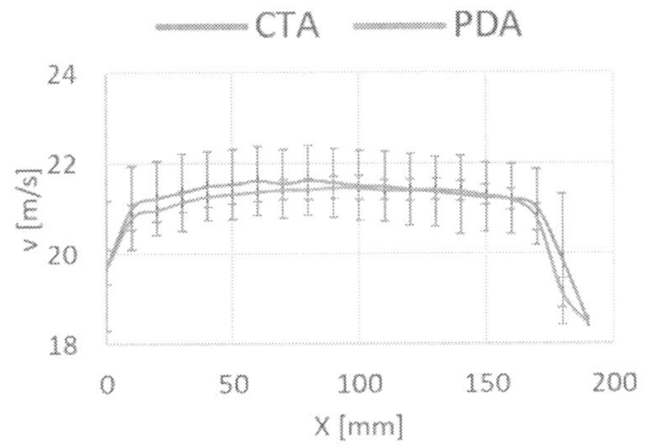

Fig. 15. Uncertainty of velocity measurement for PDA and CTA systems.

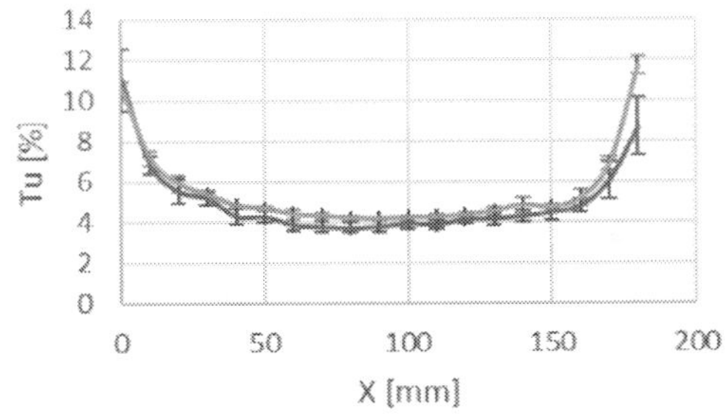

Fig. 16. Uncertainty of turbulent intensity measurement for PDA and CTA systems.

3.4 Turbulent Spectra

Turbulent spectra for the CTA measurement were analysed. The shape of the Fast Fourier transform (FFT) is pictured in Figure 17. Results in Figure 17 contains a dissipation range, the inertial subrange with slope of −5/3, and the energy-containing range. The difference in the dissipation range and the energy-containing range between the regimes is caused by different Reynolds numbers.

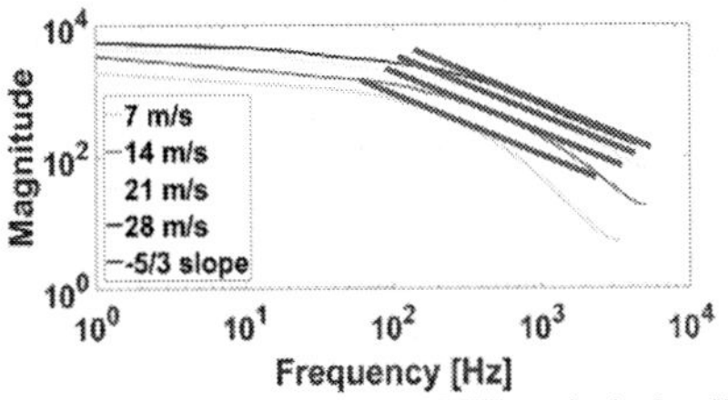

Fig. 17. Results obtained by FFT analysis for different velocity regimes.

3.5 Pitot static tube Declination

Declination α [°] of the pitot static tube from the velocity vector direction and its influence on the flow velocity was measured. The L-type and S-type pitot static tubes were investigated for two regimes (mean velocity of 7 and 28 m/s). A schematic drawing of the pitot static tubes can be seen in Figure 18 and the experimental setup in Figure 19.

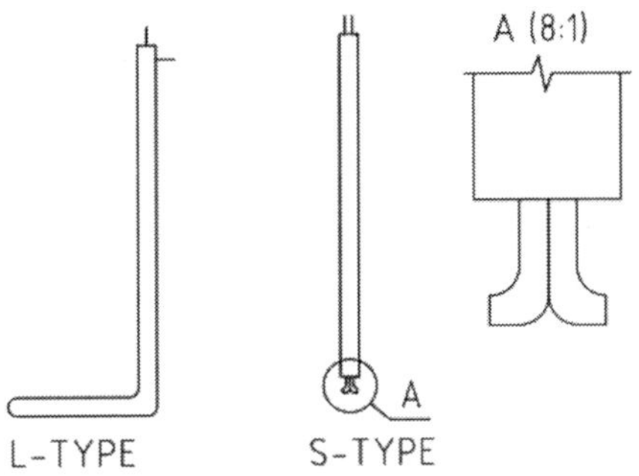

Fig. 18. L-type and S-type pitot static tubes design.

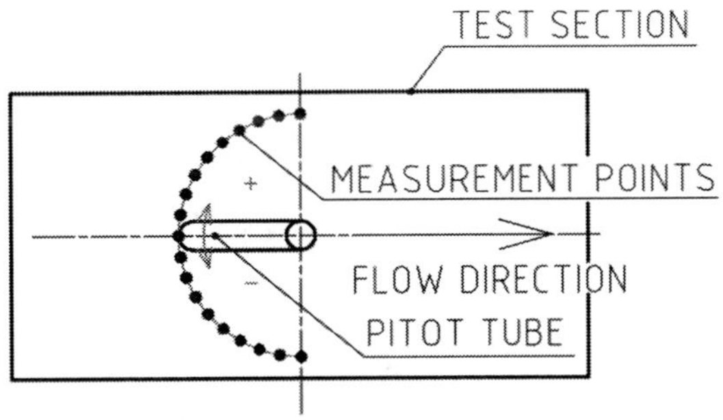

Fig. 19. L-type and S-type pitot static tube experimental setup.

Both types of pitot static tubes are relatively insensible to the probe declination because the first declination derivative of velocity is zero when the declination reaches 0 °. As indicated in Figures 20 and 21, the characteristics are symmetrical. There are small asymmetries on mean velocity of 28 m/s; this asymmetry can be caused by the flow instability or inaccurate measurement. Velocities measured by the S-type Pitot static tube need to be calibrated by the factor of 0.84 according to the standard ČSN ISO 10780. Both S-type and L-type measured the velocity very precisely when there was no declination of the probe. A comparison with the industrial measurement system, QuadraTherm® 640i In-Line Mass Flow Meter, referred here as a "QuadraTherm", can be also seen in Figures 20 and 21. The data obtained by QuadraTherm® 640i In-Line Mass Flow Meter are not symmetrical and the measurement system shows a sensitivity to the probe declination because the first declination derivative of velocity is not zero when the declination is 0 °. This asymmetrical behaviour is caused by non-symmetrical probe location of the measurement system. On the other hand, the measurement range of this industrial system is in the range of 1–300 m/s, and the system is very robust and stable for a wide velocity range. The system can be used in the flow with relatively large polluting particles because of robust probes. The system is able to measure the volume and mass flux.

According to ČSN ISO 10780 S and L-type pitot static tubes accuracy is under 3% of measurement value for declination in range of ±15 ° from velocity vector direction. For higher declination accuracy is under 5% of measurement value. QuadraTherm® 640i has an accuracy in range of ±0.5 of readings plus 0.5% of full scale, when the measured velocity is bellow 50% of full scale. Accuracy of QuadraTherm® 640i measuring the velocity above 50% of full scale is ±0.5 of readings. QuadraTherm® 640i is not suitable for measuring velocities bellow 150 m/s. The results in Figures 20 and 21 show a relatively good agreement with S and L-type Pitot static tubes for zero declination.

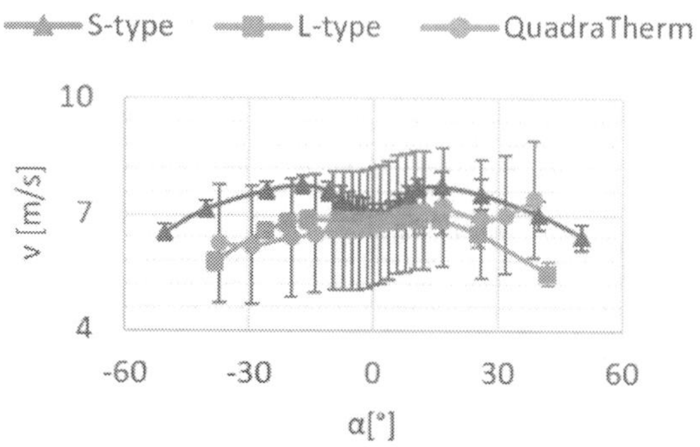

Fig. 20. S, L-type and QuadraTherm velocity probe declination measurement, mean velocity 7 m/s.

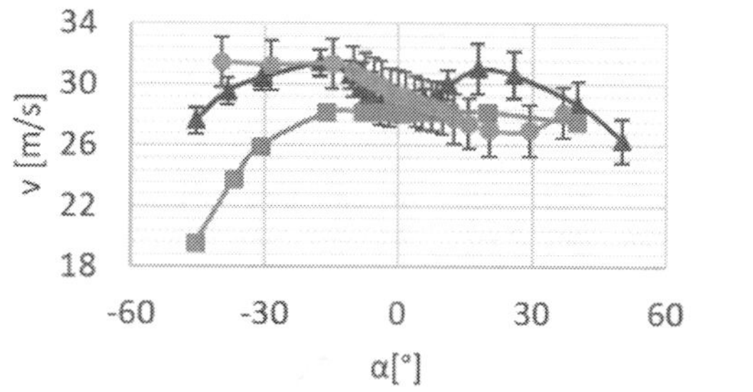

Fig. 21. S, L-type and QuadraTherm velocity probe declination measurement, mean velocity 28 m/s.

4 CONCLUSIONS

The average velocity in the newly designed low-speed small-scale wind tunnel was measured by different velocity measurement techniques (PDA, LDA, CTA, and the pitot static tube.) PDA, LDA, CTA and pitot static tube experimental setups are briefly discussed in the first section. Velocity profiles in the test section are uniform across the cross-section area for all the assumed cases. A uniform flow is developed 15 mm from the test section sides. Turbulent intensity measured by PDA, LDA and CTA in the free stream is under 5% with an increase near the wall where the velocity gradients are present. Turbulent intensities measured by PDA and LDA show expectably higher values than for the CTA measurement system.

The uncertainties of PDA and CTA measurement systems was assessed to be 1% and 4.8%, respectively, and it is worse in the boundary layer region. Both techniques are in a good agreement for both the average velocity and the turbulent intensity.

Declination of different types of pitot static tubes (S and L-type) and its influence on the flow velocity measured was briefly discussed. The experimental setup is briefly described and the comparison with Industrial QuadraTherm® 640i is provided. The results show the independence of S and L-type Pitot static tubes of declination, and the probes have a symmetrical behaviour. The industrial probe shows a dependence on the probe declination and a non-symmetrical dependence on the probe declination. Systems measures the average velocity with relatively high accuracy when declination of the probes is 0 °.

The results obtained by different measurement systems are in a good agreement. For the velocity measurement in the transparent test section of the wind tunnel, the PDA measurement system can be readily used because of the measurement abilities and easy introduction of the measurement volume into the flow. Another advantage of PDA is non-intrusive measurement which does not disturb the flow in the test section.

Acknowledgements

The authors acknowledge the financial support from the project No. 18-15839S funded by the Czech Science Foundation and the project "Computer Simulations for Effective Low-Emission Energy engineering" funded as project No. CZ.02.1.01/0.0/0.0/16_026/0008392 by Operational Programme Research, Development and Education, Priority axis 1: Strengthening capacity for high-quality research and project No. RV9090000335 founded by Faculty of Mechanical Engineering, Brno University of Technology. Authors would like to express their gratitude to Radek Kalík, Provyko s.r.o, Petr Komp and Milan Hladký, KROHNE for their effective support during probes testing.

References

1. S. Mauro, S. Brusca, R. Lanzafame, F. Famoso, A. Galvagno, M. Messina, IJAER, **12**, pp. 13649-13662 (2017)
2. Rahman, Md. Ali, Md. Sidique, F. Pabel, IJERT, **3** (2014)
3. O. Almeida, F. Miranda, O. Neto, F. Saad, JATM, **10** (2018)
4 T. Kumaraswamy, V. V. S. Nikhil Bharadwaj, P. Garre, GJRE-D, **14** (2014)
5. Y. Dwivedi, B. PARVATHAVADHANI, N. Mishra, IJTP, **1**, pp. 14-23 (2012)
6. Y. Yeh, H. Z. Cummis, Appl. Phys. Lett., **4**,176 (1964)
7. A. Boutier, Laser Velocimetry in Fluid Mechanics, **21,** (2012)
8. E. D. Hirleman, Part. Part. Syst. Charact., **13**, pp. 59-67 (1996)
9. B. Husted, P. Peterson, I. Lund, G. Holmstedt, FIRE SAFETY J., **44**, pp. 1030-1045 (2009)
10. F. Lízal, J. Jedelský, K. Morgan, K. Bauer, J. Llop, U. Cossio, S. Kassinos, S. Verbanck, J. Ruiz-Cabello, A. Santos, E. Koch, Ch. Schnabel, Eur. J. Pharm. Science, **113,** pp. 95-131 (2018)

11. Glenn O. Brown, Third National Congress on Civil Engineering History and Heritage, pp. 360-366, (2001)
12. G. Comte-Bellot, ANNU REV FLUID MECH, **8**, pp. 209-231 (2003).
13. Knoblock, Frederick Delbridge (1939) Investigations on the applications of the hot wire anemometer for turbulence measurements. Dissertation (Ph.D.), California Institute of Technology. http://resolver.caltech.edu/CaltechETD:etd-04252008-094935
14. Zh. Zhang, LDA Application Methods - 2010, Laser Doppler Anemometry for Fluid Dynamics, pp. 224-226 (2010)
15. D. Quinn, A. Watts, T. Nagle, D. Lentink, R. Soc. Open sci., **4**, (2017)
16. Y. KOHAMA, R. KOBAYASHI, H. ITO, 28th Joint Propulsion Conference and Exhibit, (1992)
17. Particle Dynamic Analyser (PDA) User´s manual, Skovlunde, Denmark, Dantec Measurement Technology, 1992.
18. F. JØRGENSEN, How to measure turbulence with hot-wire anemometers: a practical guide. 1st edition. Dantec Dynamics A/S, pp. 40-45 (2001)

Numerical simulation of the gas flow with radiation and heat transfer

Jan Česenek[1,*], *Martin* Kyncl[1], and *Jaroslav* Pelant[1]

[1]VZLU - Czech Aerospace Research Center, Beranových 130, 199 05 Praha - Letňany, Czech Republic

Abstract. In this article we focus on the heating of the solid body placed in the vicinity of the intense heat source. We simulate the heat transfer caused by the direct contact with surrounding heated gas. Furthermore it is necessary to take into account the heat transfer caused by the radiation, which radically affects the resulting heat flux. We work with the non-stationary viscous compressible fluid flow with additional heat sources, described by the RANS equations. The heat radiation is simulated using other supplemented equations. The heat transfer equations are solved inside the considered body.

1 Introduction

CFD methods are used in many industrial applications. In some application it is not only important to study influence of the flow around obstacle but it is also necessary to study the heat transfer. In this article we will deal with impact of the heat source to the surroundings in particular to the solid wall. We will simulate heat transfer by the conduction, convection and also by the radiation transfer. For this type of problems we have two regions. First region is occupied by the fluid. This part is described by the compressible Navier-Stokes equation. Second region is considered as a solid wall. In this part heat equation is used for the description of the heat flux through such wall. Between this two regions we have to use suitable boundary conditions for heat transfer. We will present some simulations with two types of heat source. In the first case it will be hot cube and in the second the heat will be released from chemical reaction. The results from the simulations will be shown.

2 Formulation of the $k-\omega$ turbulence model

Compressible flow is described by the system of Navier-Stokes equations:

$$\frac{\partial \mathbf{w}}{\partial t} + \sum_{s=1}^{3} \frac{\partial \mathbf{f}_s(\mathbf{w})}{\partial x_s} = \sum_{s=1}^{3} \frac{\partial \mathbf{R}_s(\mathbf{w}, \nabla \mathbf{w})}{\partial x_s}, \qquad (1)$$

where for $s = 1,...,3$ we have

$$\begin{aligned}
\mathbf{w} &= (w_1, ..., w_5)^T = (\rho, \rho v_1, ..., \rho v_3, E)^T,\\
\mathbf{f}_s(\mathbf{w}) &= (f_{s,1}, ..., f_{s,5})^T\\
&= (\rho v_s, \rho v_1 v_s + \delta_{1s} p, ...,\\
&\quad \rho v_3 v_s + \delta_{3s} p, (E+p) v_s)^T,\\
\mathbf{R}_s(\mathbf{w}, \nabla \mathbf{w}) &= (R_{s,1}, ..., R_{s,5})^T\\
&= \Big(0, \tau^V_{s1}, ..., \tau^V_{s3},\\
&\quad \sum_{r=1}^{3} \tau^V_{sr} v_r + \left(\frac{c_p \mu_L}{Pr}\right) \frac{\partial \theta}{\partial x_s}\Big)^T,
\end{aligned}$$

where

$$\tau^V_{sr} = -\frac{2}{3}\mu_L \,\mathrm{div}\boldsymbol{v}\, \delta_{sr} + 2\mu_L\, d_{sr}(\boldsymbol{v}),$$

$$d_{sr}(\boldsymbol{v}) = \frac{1}{2}\left(\frac{\partial v_s}{\partial x_r} + \frac{\partial v_r}{\partial x_s}\right).$$

We use the following notation: $\boldsymbol{v} = (v_1, ..., v_3)$ - velocity, ρ - density, p - pressure, θ - absolute temperature, E - total energy, γ - Poisson adiabatic constant, κ - heat conduction coefficient, c_v - specific heat at constant volume, c_p - specific heat at constant pressure, where $c_p = \gamma c_v$, Pr is the laminar Prandtl number, which can be express in the form $Pr = \frac{c_p \mu_L}{\kappa}$, μ_L is the dynamic viscosity coefficient dependent on temperature via Sutherland's formula. The above system is completed by the thermodynamical relations

$$\begin{aligned}
p &= (\gamma - 1)\left(E - \frac{1}{2}\rho |\boldsymbol{v}|^2\right),\\
\theta &= \frac{1}{c_v}\left(\frac{E}{\rho} - \frac{1}{2}|\boldsymbol{v}|^2\right)
\end{aligned}$$

*e-mail: cesenek@vzlu.cz

To this system of equations we have to add equation describing radiation model

$$\nabla \cdot (\Gamma \nabla G) - aG \quad = \quad -4\epsilon\sigma T^4 - E$$

where G is the so-called irradiation, which can be expressed as $G := \nabla q_r$, where q_r is heat flux $[qr] = Wm^{-2}$, absorbivity a, emissivity ϵ and E, which charaterize volume radiation source. In the solid region we solve the heat equation.

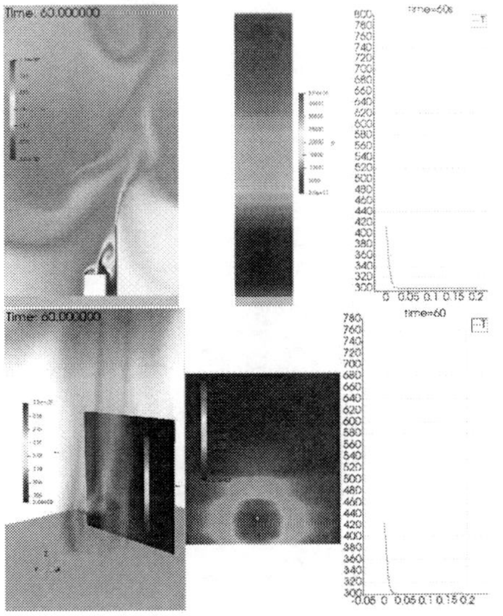

Fig. 1. Heated cube, simulation in 2D and 3D. (From left) The air temperature, the heat flux between the wall and the cube and the distribution of the temperature through the heated wall at time instance $60s$.

3 Numerical experiments

In this section we present computation results. For the simulations we have used *chtMultiRegionFoam* solver from the open source code OpenFoam.

In the first example the source of the heat is presented by very hot solid cube, which has temperature $1200K$. Surrouding air and wall has temperature $300K$ at the beginnig of the simulation. This hot cube radiates heat to the surroundings and warms up the solid wall, which is at distance $1.5m$ from the cube. These simulations were carried out on fine mesh in 2D, and on the coarser mesh in 3D. Figures (1) - (2) show the results of the temperature and the heat flux at two different time instances. The total heat flux between hot cube and wall surface is slowly diminishing as the wall heats up.

In further simulations we used the combustion (and radiation) as the source of the heat. Due to the large complexity of the proper combustion simulation we reduced chemical reactions to the following equation:

$$\begin{aligned} & C_{12}H_{23} + 28.75O_2 + 107.8125N_2 \\ \rightarrow \quad & 23CO_2 + 11.5H_2O + 107.8125N_2, \end{aligned}$$

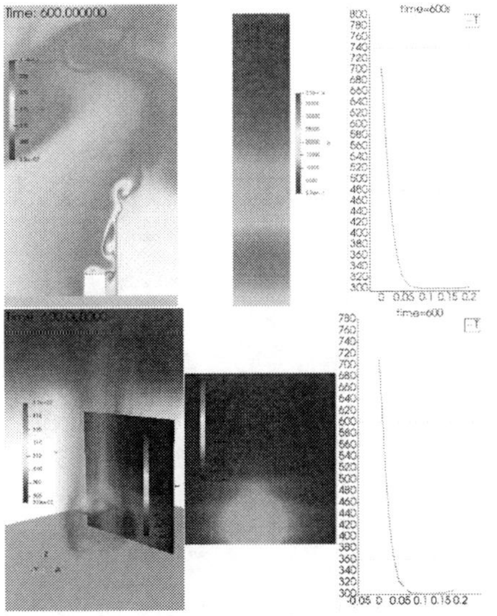

Fig. 2. Heated cube, simulation in 2D and 3D. (From left) The air temperature, the heat flux between the wall and the cube and the distribution of the temperature through the heated wall at time instance $600s$.

where $C_{12}H_{23}$ is the fuel. Figure (3) shows the 2D and 3D simulation of the combustion with prescribed fuel mass flow $10\, g \cdot m^{-2} \cdot s^{-1}$ at the selected area $3.0m \times 3.0m$. Figure (4) shows the simulation with concrete wall (d=$0.2m$, h=$10m$) placed $1.5m$ far from this source. The fire gets attached to the wall in 2D simulation, and the wall is the heated up.

For further simulations we set the concentration of the fuel to 0.5 in the area $2.0m \times 2.0m \times 0.1m$, which is placed 1.5m far from the wall. This presents the pool of the fuel, where the concentration is fixed during the whole simulation. We studied the chemical reaction and its heat and radiation impact to the surround area which is occupied by the air and heat transfer to the solid floor and concrete wall. Figures (5) - (6) show the results of the temperature and heat flux distribution at two different time instances.

Last example is the 3D simulation of the previous case with the fuel in volume $2.0m$ x $2.0m$ x $0.5m$ situated $1.5m$ far from the wall. The concentration fraction of the fuel was set to 0.5. Here we used coarser meshes than in the 2D cases. Presented figures (7) resp. (8) show the distribution of the temperature on the solid wall and floor resp. in the air, figure (9) shows the concentration of the fuel and CO_2. Such approach can be used for the simulation of the poured burned fuel in some real area.

4 Conclusion

In this paper we dealt with the radiation and its impact to the surrounding area. Presented results show, that it is possible to simulate poured burning fuel by a pool filled with the fuel. Here the large amount of heat is transferred by the radiation, and therefore it is necessary to include some radiation model for proper practical application.

Acknowledgements. The results originated with the support of Ministry of the Interior of the Czech Republic, project VI20172020092 - Simulation of fire and smoke dispersion in critical infrastructure area in case of accident or aircraft attack

References

1. M.B. Hermosilla, *Study of heat and mass transfer applications in the field of engineering by using openfoam*, Bachelor's Thesis, (2016)
2. J. Česenek, *EPJ WoC* **67**, 02015 (2014)
3. J. Česenek, *EPJ WoC* **180**, 02016 (2018)
4. J. Česenek, *EPJ WoC* **114**, 02012 (2016)
5. J. Česenek, *EPJ WoC* **143**, 02014 (2017)
6. J. Česenek, *EPJ WoC* **213**, 02011 (2019)
7. M. Kyncl and J. Pelant, *EPJ WoC*, **45**, 01054, (2013)
8. M. Kyncl and J. Pelant, *EPJ WoC*, **180**, 02052, (2018)
9. M. Kyncl and J. Pelant, *EPJ WoC* **213**, 02050 (2019)
10. M. Kyncl and J. Pelant, *EPJ WoC* **213**, 02051 (2019)
11. M. Kyncl and J. Pelant, *EPJ WoC* (to be published) (2020)

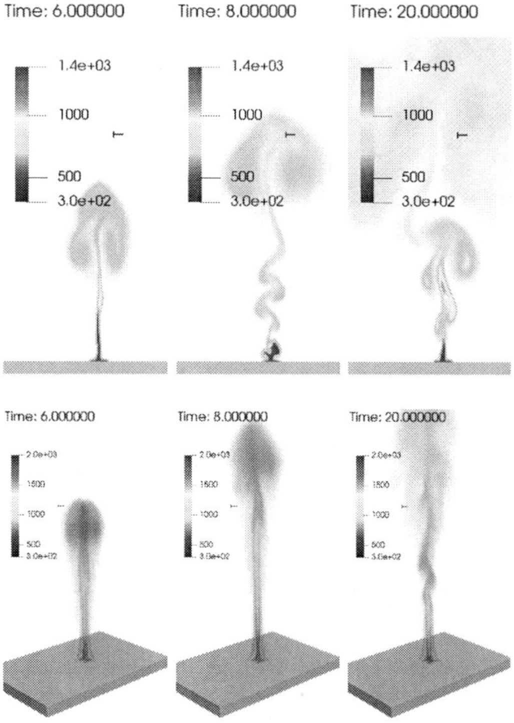

Fig. 3. Simple combustion simulation in 2D (31536 elements) and 3D (834616 elements), temperature field at various time instants. Constant fuel source fixed to 10 $g \cdot m^{-2} \cdot s^{-1}$.

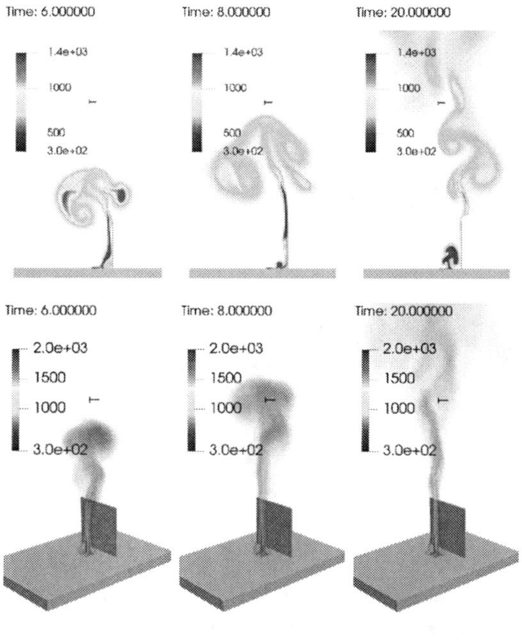

Fig. 4. Combustion simulation in 2D (124223 elements) and 3D (834616 elements), temperature field at various time instants. Constant fuel source fixed to 10 $g \cdot m^{-2} \cdot s^{-1}$. In 2D case the fire becomes attached to wall.

https://doi.org/10.1051/epjconf/202226901008

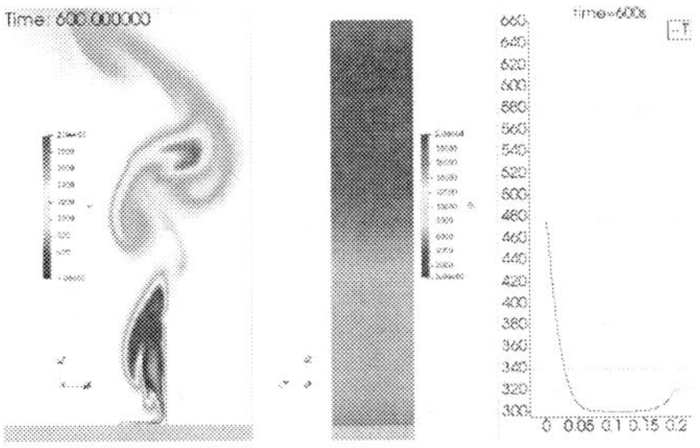

Fig. 5. 2D combustion simulation. (From left) The air temperature, the heat flux through the wall surface, and the distribution of the temperature through the heated wall at h=1.5m, time instance 600s.

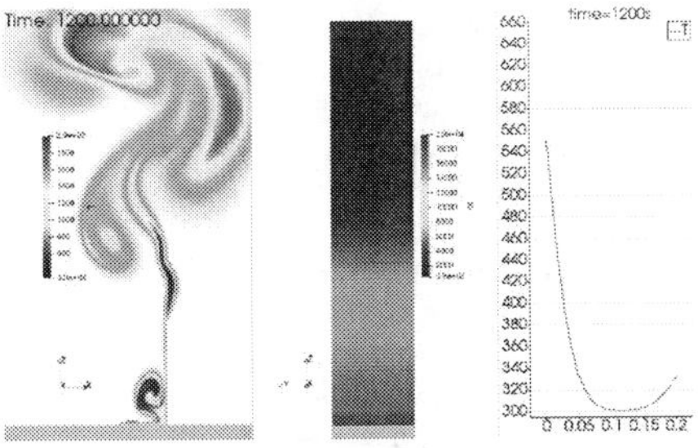

Fig. 6. 2D combustion simulation. (From left) The air temperature, the heat flux through the wall surface, and the distribution of the temperature through the heated wall at h=1.5m, time instance 1200s.

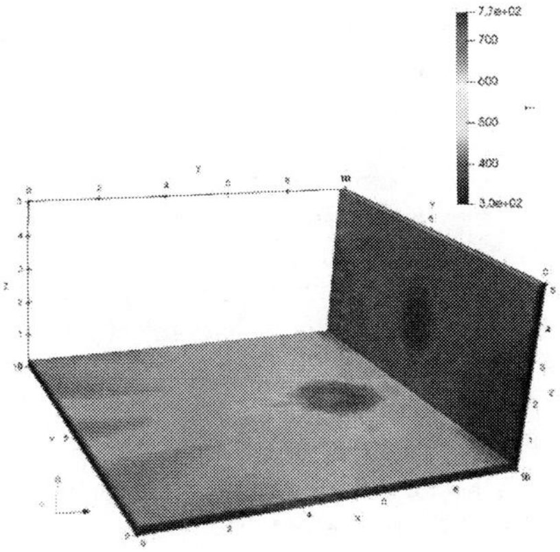

Fig. 7. The distribution of the temperature on the wall and floor.

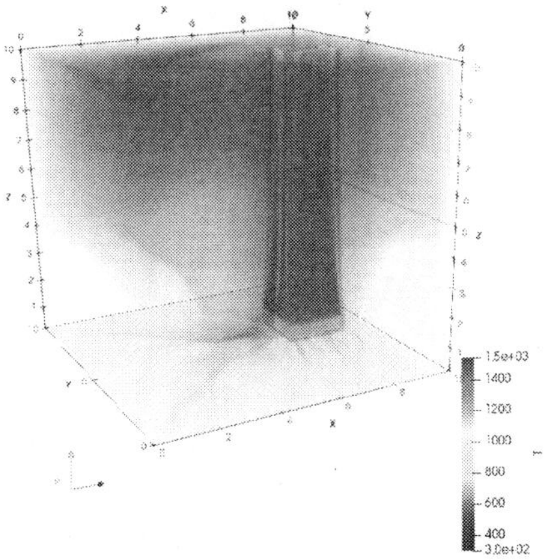

Fig. 8. The distribution of the temperature in the air.

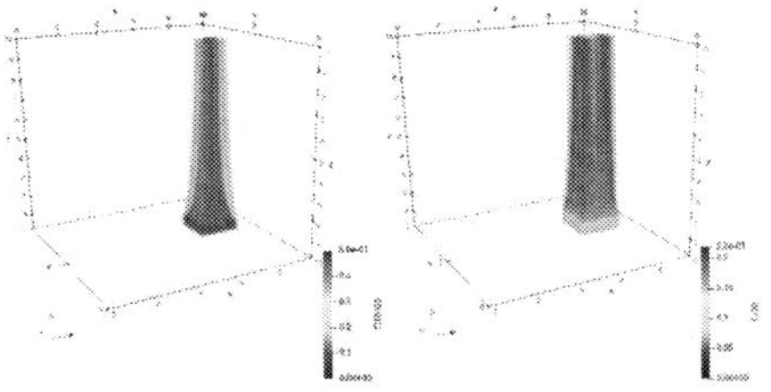

Fig. 9. The distribution of the concentration of the fuel (left) and CO_2 (right).

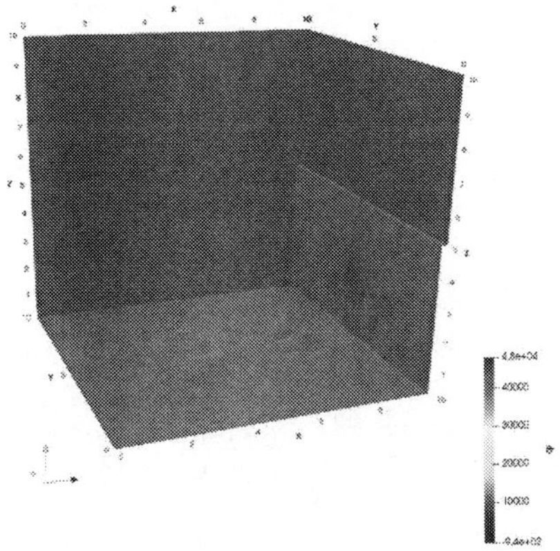

Fig. 10. The distribution of the heat flux.

 https://doi.org/10.1051/epjconf/202226901009

Gnielinski's correlation and a modern temperature-oscillation method for measuring heat transfer coefficients

Martin Dostál[1,*], *Karel* Petera[1], and *Stanislav* Solnař[1]

[1]Czech Technical University in Prague, Faculty of Mechanical Engineering, Department of Process Engineering, Technická 4, 166 07 Prague, Czech Republic

Abstract. The heat transfer coefficient is one of the most important parameters in the design of apparatuses in which convective heat transport takes place. Classical direct methods based on determining basic thermal quantities can be used for measuring heat transfer coefficients. Another option is to measure concentrations, electric current or other quantities that can be transformed to thermal quantities using the analogy between heat and mass transport. The temperature-oscillation method is less frequently used, although the theoretical basis of the method dates back to 1997, and although the method has the major advantage that the heat transfer coefficients can be measured without making any contact with the heat transfer surface. In the temperature-oscillation method, the heat transfer surface is exposed to an oscillating heat flux, and the temperature response on this surface can be measured by a contactless method (e.g. infra-red thermography). The heat transfer coefficients can be determined on the basis of mathematical relations between the oscillating heat flux and the temperature response. However, the method depends on an appropriate method for processing the measured data when it is necessary to correct some conditions that are not included in the mathematical model. This paper evaluates the impact of processing the experimental data on the resulting heat transfer coefficients in one of the basic geometrical configurations – the flow of a liquid in a pipe with a circular cross section. In this paper, we present the results of a comparison of real experiments based on the temperature oscillation method and numerical modeling of the heat transfer in this geometry, using the ANSYS CFD commercial system.

1 Introduction

Experimental methods based on temperature oscillations have been known and used for a long time. They mainly use knowledge of an analytical solution of the temperature field in a solid body that is exposed to a periodic (oscillating) heat flux. For example, the excellent book by Carslaw and Jaeger [1] presented an analytical solution of heat conduction in infinite and semi-infinite bodies, walls, cylinders, spheres, etc., for step changes or oscillating in temperature. Carslaw and Jaeger mentioned the use of these non-stationary solutions for determining the thermophysical properties of materials. Hausen [2] also provided an analysis of determining the heat transmission coefficient (the overall heat transfer coefficient, which describes the transfer of heat through a wall) based on the known temperature field in a regenerator.

Roetzel [3] described an experimental technique and an evaluation of local heat transfer coefficients on the surface of a pipe or other channels based on measuring the wall temperature oscillations caused by a periodic change in the temperature of the liquid flowing in the channel. Good agreement of the experimental results compared with the Hausen correlation were mentioned. Roetzel et al. [4] described a similar technique for measuring the heat transfer coefficient in a plate heat exchanger. The mathematical model of the heat transfer presented in paper [4] included the axial dispersion coefficient, which can also be determined. Roetzel and Luo [5] further improved this technique. They used a computer-controlled electric heating body to generate sine temperature oscillations at the inlet of the heat exchanger. Leblay et al. [6] used the temperature oscillation method to measure the heat transfer in a circular tube and in a multiport-flat tube at Reynolds numbers between 800 and 14000. The temperature oscillations were invoked by Joule's heat, and the temperature response was measured by infrared thermography.

Roetzel et al. [7] described a new experimental technique based on periodic heating of the heat transfer surface by a laser. They measured the heat transfer coefficients on the inner side of an agitated vessel. The temperature response on the outer side of the heat transfer surface was determined by a contactless method. Smaller heat transfer coefficients than the standard correlations were obtained, and better agreement was observed at higher Reynolds numbers (about 15000). They measured approximately 50% values at lower Reynolds numbers (2000).

In 1997, Wandelt and Roetzel [8] published the solution of the temperature field in an infinite plate with oscillating heat flux irradiating one side. The resulting harmonic oscillations in the temperature field are influenced

*e-mail: martin.dostal@fs.cvut.cz

by the heat transfer coefficients on both sides of the plate, and they can be used in determining these coefficients.

Freund and Kabelac [9] tested the method described in [8], known also as Lock-In-Thermography, on a semi-infinite body and on the inlet section of a (copper) pipe, where hydrodynamic and thermal boundary layers were developing. They used a laser diode array with a wave length of 785 nm and power of 15 W to generate temperature oscillations. The temperature responses at the surface were monitored by an infra-red camera with a frame rate of 30 fps. They obtained a 0.1 % deviation from the theoretical phase shift for frequencies of 0.1 Hz, 0.5 Hz, and 0.05 Hz. A laser spot 30 mm in diameter and with sine-modulated intensity with a frequency of 0.1 Hz was used for measuring the heat transfer in the pipe, and the preliminary results agreed with standard engineering correlations. Freund [10] focused on the temperature oscillation method, using contactless measurements of the temperature response (Temperature Oscillation Infra-Red Thermography) and algorithms for processing the measured data. Freund and Kabelac [11] applied the temperature oscillation method for measuring the heat transfer on the plates of a plate heat exchanger. The oscillating heat flux was generated by an array of halogen spotlights with a total of more than 3 kW electric power. The turbulent flow between the plates of the heat exchanger was modeled using the ANSYS® CFX® commercial CFD software package. The Shear-Stress-Transport (SST) and Reynolds-Stress Model (RSM) turbulence models were used. A comparison between the experiments and the numerical simulations showed that the mean heat transfer coefficients based on the SST model are 33 % lower than the experimental values. In the case of RSM, they are 25 % lower.

The works cited here show that the biggest advantage of this method is the contactless determination of the surface temperatures (Temperature Oscillation Infra-Red Thermography = TOIRT). However, few papers have been published on this topic in recent years, although heat transfer coefficients are very important in the design of many industrial apparatuses. Possible reasons may be that experimental investigations of heat transfer have been replaced by numerical simulations, which are, in many cases, cheaper than building an experimental device and measuring the results in the laboratory or on real equipment. However, experimental data still have an important role in verifying the results of numerical models.

Another reason for the less frequent application of the TOIRT method could lie in the contactless measurements of the temperatures. Precise infra-red cameras are nowadays available, but they may be too expensive for some experimental groups. Widely-available cameras, mostly based on the bolometric method, have relatively large thermal sensitivity, but they still have substantial uncertainty when measuring absolute temperature values. For example, the thermal sensitivity (NETD) of our thermoImager TIM 160 camera is 80 mK, but the resolution is 0.1 °C, and the accuracy is ±2 °C or ±2 %, whichever is greater [https://www.micro-epsilon.com]. The Gobi-640-GigE camera has thermal sensitivity of 55 mK at 30 °C [http://www.xenics.com]. The FLIR A325sc infra-red camera has thermal sensitivity lower than 50 mK and accuracy ±2 °C or ±2 % [https://www.termokamery-flir.cz].

Taking into account that oscillations with amplitudes smaller than 1 °C can be measured with the TOIRT method, the accuracy of the temperature measurements might be limiting. An advantage of TOIRT is that it is not necessary to know the absolute value of the temperature, it is sufficient to determine the shape of its response, the phase shift, which, however, can be influenced by the integration type (rolling shutter, snapshot) and by the integration time. The integration times are of the order of μs up to tens of ms. The resolution of the camera, and also the specific lens, affects the size of the monitored region and the corresponding surface on which the heat transfer coefficients can be investigated.

Our tests of the TOIRT method investigated the flow of a Newtonian liquid in a pipe with a circular cross section. In the fully developed turbulent flow regime, Gnielinski's correlation [12] describes the mean Nusselt number ($\mathrm{Nu} = \alpha d/\lambda_f$) in a straight pipe for $\mathrm{Re} \geq 10^4$ ($10^6 \geq \mathrm{Re} \geq 10^4$, $0.1 \leq \mathrm{Pr} \leq 1000$)

$$\mathrm{Nu_m} = \frac{(\xi/8)\,\mathrm{Re\,Pr}}{1 + 12.7\sqrt{\xi/8}\left(\mathrm{Pr}^{2/3} - 1\right)}\left[1 + \left(\frac{d}{L}\right)^{2/3}\right], \qquad (1)$$

where ξ is the Darcy-Weisbach friction coefficient, which can be expressed, as recommended by [12], using the Konakov relation for smooth pipes

$$\xi = (1.8\log\mathrm{Re} - 1.5)^{-2}\,. \qquad (2)$$

The last term in (1) represents the Hausen correction factor, which takes into account the substantial increase in heat transfer for short pipes. This correction factor is for a pipe with $d = L$ equal to 2. The correction factor decreases to 1.2 for the case $d = 2L$, to 1.025 for $d = 5L$, to 1.005 for $d = 10L$, and it approaches a value of 1 with increasing pipe length. A local increase in the heat transfer coefficient is still present at the beginning of a longer pipe, but it has a small impact on the average value along its length. If we measure local values of the heat transfer coefficients, we can compare the results with the derivative of Eq. (1), as stated by [12].

$$\mathrm{Nu_x} = \frac{(\xi/8)\,\mathrm{Re\,Pr}}{1 + 12.7\sqrt{\xi/8}\left(\mathrm{Pr}^{2/3} - 1\right)}\left[1 + \frac{1}{3}\left(\frac{d}{x}\right)^{2/3}\right] \qquad (3)$$

In the case of laminar flow in a pipe with a circular cross section with a constant wall temperature, the Hausen correlation [2] describes the mean value of the Nusselt number

$$\mathrm{Nu_m} = 3.66 + \frac{0.0668\,\mathrm{Gz}}{1 + 0.045\,\mathrm{Gz}^{2/3}}\,, \quad \mathrm{Gz} = \mathrm{Re\,Pr}\,d/L\,. \qquad (4)$$

This relation meets the limits for long pipes and also for short pipes. In the case of short pipes, Léveque derived [12],

$$\mathrm{Nu_m} = 1.615\,\mathrm{Gz}^{1/3} \qquad (5)$$

which can be transformed to

$$\mathrm{Nu_x} = 1.077\,\mathrm{Gz_x}^{1/3}\,, \quad \mathrm{Gz_x} = \mathrm{Re\,Pr}\,d/x \qquad (6)$$

Table 1. Thermophysical properties of widely-used engineering materials at a temperature of 20 °C, see [13] and [https://www.foamglas.com].

Material	λ W m^{-1} K^{-1}	c_p J kg^{-1} K^{-1}	ϱ kg m^{-3}
stainless steel	15.2	501	7900
carbon steel	49.8	465	7840
aluminium	236	902	2710
copper	398	386	8930
FOAMGLASS ®W+F	0.038	1000	100
glass	0.76	840	2710

for local values of the Nusselt numbers at distance x from the beginning of the pipe.

In our experimental work, we had to deal with many problems connected with the application of the TOIRT method. In the following text, we therefore analyze the impact of some parameters on the accuracy of this method.

2 TOIRT basis

As has been mentioned above, Wandelt and Roetzel [8] published an analytical steady periodic solution of the temperature field on the surfaces of an infinite plate with width δ (Figure 7, on the right) described by the one-dimensional Fourier equation

$$\frac{\partial T}{\partial t} = a\frac{\partial^2 T}{\partial x^2} . \tag{7}$$

The plate has zero initial temperature ($T_0 = 0$). It is surrounded by a fluid at zero temperature ($T_f = 0$). Boundary conditions of the third kind are described by the heat transfer coefficients on both sides of the plate (α a α_δ), and a boundary condition of the second kind is represented by the heat flux on one side of the plate ($q|_{x=\delta} = \hat{q}\sin\omega t$), which oscillates around a mean value of zero

$$-\lambda \left.\frac{\partial T}{\partial x}\right|_{x=0} = -\alpha\, T|_{x=0} \ , \ -\lambda \left.\frac{\partial T}{\partial x}\right|_{x=\delta} = -\hat{q}\sin\omega t + \alpha_\delta\, T|_{x=\delta} . \tag{8}$$

The steady periodic temperature solution can be expressed in a general form as a function of amplitude A and phase shift φ

$$T(x,t) = A(x)\sin\left[\omega t - \varphi(x)\right] . \tag{9}$$

Wandelt and Roetzel [8] presented analytical formulas for A and φ, which depend on geometric parameters, thermophysical properties and the heat transfer coefficients α and α_δ on the two sides.

We will now illustrate here the graphical solution of Wandelt and Roetzel for some parameters that can be met in practical situations. The width of the heat exchange surface δ is usually within the order of millimeters, and we will therefore use 0.5, 1, 1.5 and possibly 2 mm. Table 1 describes thermophysical parameters of widely-used engineering materials (stainless steel, carbon steel, aluminium, and copper). All these materials can be met in the design of real apparatuses, but we will place the emphasis here on stainless steel, because it is the material that we used in our experimental work.

Figure 1 describes the dependency of the heat transfer coefficient α on the phase shift $\varphi|_{x=\delta}$ between the heat flux and the surface temperature response for a plate 1.5 mm in width made of stainless steel, for three different frequencies of the oscillating heat flux. It can be clearly seen here that the common frequency 0.1 Hz covers heat transfer coefficients values from hundreds to thousands of W m^{-2} K^{-1}. However, this dependency is quite steep for large heat transfer coefficients values, so a small inaccuracy in the phase shift has a big impact on the evaluated coefficients. The relative inaccuracy of the heat transfer coefficient is depicted by dotted lines. These lines represent the relative change in the heat transfer coefficient with respect to the change in the phase shift. The relative inaccuracy can be expressed as derivative $1/\alpha \cdot \mathrm{d}\alpha/\mathrm{d}\varphi$ (displayed as a percentage, i.e., it has been multiplied by 100 in the Figure). It is clear that the smallest errors in the heat transfer coefficients can be achieved when the phase shift is around 45°. In this region, a change of ±1° in the phase shift causes a relative error of approximately ±5 %. Similar relationships can be illustrated for other considered plate widths, 0.5 and 1.0 mm.

Figures 2 illustrate the Wandelt-Roetzel solution for various widths (left) and materials (right) of the plate. It is interesting that the plate material (i.e. the heat transfer surface) does not have a large impact on the shape and the position of the dependency. We can therefore state that the plate can be made from stainless steel, and also from copper (of course, taking other parameters into account, e.g. the plate width of 1 mm). However, the width of the plate has a greater influence, see Figure 2, on the left. If we want to be in a region with reasonable accuracy, we should therefore have in mind that a thicker plate substantially reduces the region with corresponding phase shifts. Interestingly, the relative error of the heat transfer coefficient is same as for thinner plates.

The graphical dependencies presented here clearly show that the phase shift should lie within 20 – 70° for

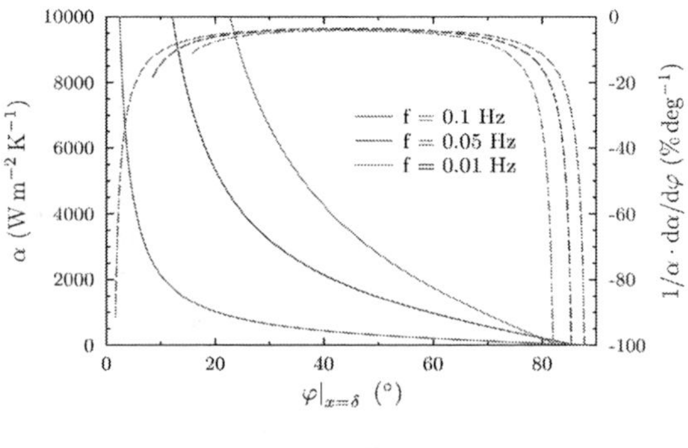

Fig. 1. The dependency of heat transfer coefficient α on phase shift $\varphi|_{x=\delta}$ for different frequencies of the oscillating heat flux (full lines). The dotted lines represent the relative sensitivity of the heat transfer coefficient with respect to the phase shift, which is the derivative $1/\alpha \cdot \mathrm{d}\alpha/\mathrm{d}\varphi$ (see the right axis). The plate width is 1.5 mm, stainless steel, see properties in Table 1. The heat transfer on the irradiated side is 3 W m^{-2} K^{-1}.

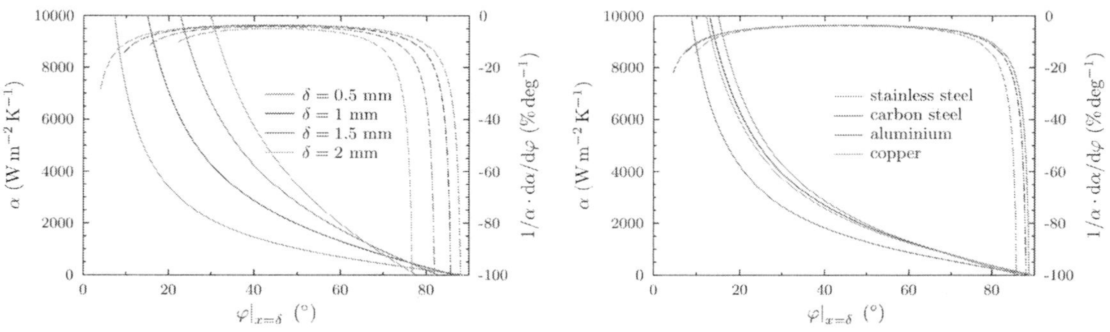

Fig. 2. The dependency of heat transfer coefficient α on phase shift $\varphi|_{x=\delta}$ for different wall widths (left) and for different materials (right). The dotted lines represent the relative sensitivity of the heat transfer coefficient with respect to the phase shift, which is the derivative $1/\alpha \cdot d\alpha/d\varphi$ (see the right axis). The dependency is illustrated for 0.1 Hz frequency of the oscillating heat flux. The wall material is stainless steel, see properties in Table 1. The heat transfer on the irradiated side is 3 W m^{-2} K^{-1}.

the frequency of the oscillating heat flux 0.1 Hz and for plate width up to 1.5 mm. In this case, the relative error is within an acceptable range $\mp 5 - 10$ % for inaccuracy $\pm 1°$ of the phase shift.

The figures illustrate Wandelt and Roetzel's solution in graphical form. In cases where we are able to measure the surface temperature on the side irradiated by the sine-oscillating heat flux and where we can determine the phase shift, we can read the corresponding heat transfer coefficient value. The basis of this method appears simple, but it is based on assumptions that cannot be precisely met in a real experiment:

- Zero ambient fluid temperature conditions and zero initial temperature condition, or, in other words, conditions with the same initial temperature of the plate and of the surrounding fluid.
- A steady-state periodic regime that does not take into account the transitional change at the beginning of the temperature oscillations (at the beginning of the measurements).
- Temperature oscillations with a zero mean value, i.e. they are positive (heating) in the first part of the sine period and negative (cooling) in the second part. These oscillations around a zero value cannot be achieved in a real experiment. Due to the continuous temperature increase of the wall irradiated by the heat flux, the measured temperature response will oscillate around a non-oscillating component which changes with time.

3 Transitional change and dynamic equilibrium

To better analyse the assumptions of the Wandelt-Roetzel solution and their impact on the evaluated heat transfer coefficient, we implemented a numerical solution of the heat conduction equation (7), along with the initial and boundary conditions (8), using the **pdepe** function in MATLAB® software. Some details of the solution can be found in the Appendix. Only graphical results will be presented and discussed in this section.

Figure 3 (left) compares the numerical and Wandelt-Roetzel solutions for the case with zero mean value of the heat flux. It is clear that the transitional change at the beginning disappears approximately after the second period, and the numerical solution then copies the analytical solution of Wandelt and Roetzel. This figure represents the case of heat transfer coefficient $\alpha = 1000$ W m^{-2} K^{-1}, and the impact of the transient change at the beginning is reduced with higher heat transfer coefficients. The impact of the transitional change is illustrated here by the green dashed line, as the absolute difference between the numerical solution and the analytical solution. It is interesting that this difference is very similar to the exponential function, which represents the response of the first-order system to the step change of the boundary condition, which is characterized by some time constant (for corresponding parameters, of course, it is not correct in general).

For technical reasons, it is not possible to have the heat flux oscillating around a mean value of zero. In practice, the plate is heated by an oscillating heat flux with non-zero mean value $\bar{q}$, and the oscillations of the temperature response are also around a non-zero mean value, and, in addition, that non-zero temperature mean value is time-dependent. Figure 3 (right) presents the solution of the problem with a non-zero mean value of the heat flux. In this case, the mean value is $\bar{q} = 2000$ W m^{-2} and the amplitude is $\hat{q} = 1000$ W m^{-2}. A transient change can be observed at the beginning. It briefly approaches the steady harmonic oscillations of the temperature corresponding to the Wandelt-Roetzel solution, but now around non-zero value $T|_{x=\delta,\mathrm{eq}}$. The steady mean temperature is due to the stationary solution of the heat conduction in the plate (7) with boundary conditions (8). In this case, for the boundary condition on the side of the heat flux, which is in the position $x = \delta$, the harmonic function $\hat{q} \sin \omega t$, is replaced by constant value $\bar{q}$. The solution of this equation for the steady-state case, i.e. $\partial T/\partial t = 0$, and zero surrounding

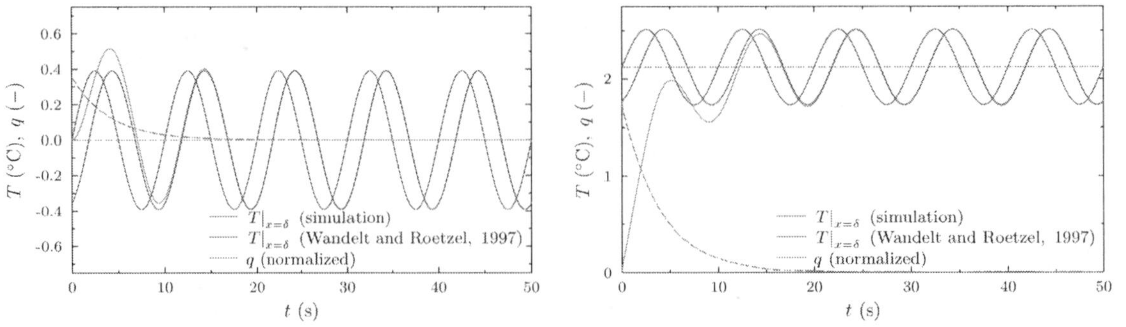

Fig. 3. Temperature response on the surface of the plate (red line) to the harmonic oscillating heat flux (black line), and a comparison with oscillations corresponding to the Wandelt-Roetzel solution (blue line, and the green dashed line represents the difference between them) for heat transfer coefficient 1000 W m^{-2} K^{-1}. The plate 1 mm in width is made from stainless steel. Heat transfer coefficient α_δ is 3 W m^{-2} K^{-1}. The amplitude of the heat flux oscillations is 1000 W m^{-2}, with a zero mean value (left), in comparison with a dependency with a mean value of 2000 W m^{-2} (right).

temperature $T_\mathrm{f} = 0$ is

$$T|_{x=\delta,\,\mathrm{eq}} = \bar{q}\frac{\delta + \dfrac{\lambda}{\alpha}}{\alpha_\delta\delta + \dfrac{\lambda}{\alpha} + \lambda}\,. \tag{10}$$

Similarly as with the zero mean value of the heat flux, the length of the transitional initial period decreases with increasing heat transfer coefficients. For a 1 mm plate, it is relatively short. To illustrate its impact, see Figure 4 corresponding to the solution with a plate width of 10 mm. It is clear that the transitional period is very long and we do not approach constant mean temperature even after 10 oscillations.

Making measurements with plates of greater width means that we would have to wait a long time to reach the steady-state mean value, especially for smaller heat transfer intensities. In practise, we can encounter this situation

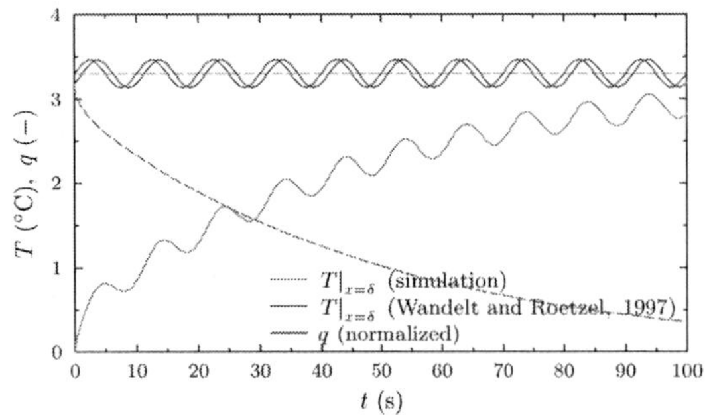

Fig. 4. Temperature response on the surface of the plate (red line) to the harmonic oscillating heat flux (black line), and a comparison with oscillations corresponding to the Wandelt-Roetzel solution (blue line, and the green dashed line represents the difference between them) for heat transfer coefficient 1000 W m^{-2} K^{-1}. The plate 10 mm in width is made from stainless steel. Heat transfer coefficient α_δ is 3 W m^{-2} K^{-1}. The amplitude of the heat flux oscillations is 1000 W m^{-2}, with a mean value of 2000 W m^{-2}.

even with a plate width of 1 or 1.5 mm. The question is whether we can use the measured response even in cases when we do not reach these steady state oscillations.

Another question is whether we could make an estimate of the length of this transitional region. As has already been mentioned above, the exponential character of the time dependency would correspond to the response of a 1st order system induced by a unit step change, which is the response of the system characterized by the heat capacity of the plate. Assuming the simplest case, with a constant temperature inside the plate, we could determine the time constant of the transition process as

$$\tau = \frac{M\,c_p}{(\alpha + \alpha_\delta)\,S} = \frac{S\,\delta\varrho c_p}{(\alpha + \alpha_\delta)\,S} = \frac{\delta\varrho c_p}{(\alpha + \alpha_\delta)} = \frac{\delta\lambda}{a\,(\alpha + \alpha_\delta)}\,, \tag{11}$$

where M is the mass of a plate of width δ, surface S and density ϱ. Assuming that the whole mass of the plate is concentrated into a single mass point of constant temperature, i.e. for the Biot number $\mathrm{Bi} = \alpha\delta/\lambda \ll 1$,we could simply express the temperature oscillations around the temperature given by the transition process as

$$T|_{x=\delta} = T|_{x=\delta,\,\mathrm{eq}} - \left(T|_{x=\delta,\,\mathrm{eq}} - A|_{x=\delta}\sin\varphi|_{x=\delta}\right)\mathrm{e}^{-t/\tau} + \\ + A|_{x=\delta}\sin\left(\omega t - \varphi|_{x=\delta}\right)\,, \tag{12}$$

where the amplitude and the phase shift of the harmonic oscillations are given by relation (9), and the steady-state temperature and the time constant are given by (10, 11).

4 Eliminating a non-oscillating component

The crucial step in the TOIRT method, which affects its accuracy, is to eliminate the non-oscillating component of the measured temperature response. The aim is for the resulting signal to oscillate around the zero mean value after this elimination, as depicted in Figure 3 (left).

One option could lie in a theoretical solution of the non-oscillating component. In the simplest case, it can be replaced by an exponential function (especially when the heat transfer surface is made of a material with high heat conductivity – e.g. aluminum or copper). In a more general case, when this kind of simplification cannot be used, we can use an empirical approach and find a suitable algorithm, and then verify it. Solnař [14] describes an algorithm based on several steps: (a) the measured temperature signal is divided into whole periods, (b) the mean temperature is calculated in each period, (c) this temperature moves to the beginning of the period and creates a piecewise linear correction function, (d) this correction function is subtracted from the measured signal. The whole procedure is repeated several times, until the correction function is smaller than a chosen number.

Figure 5 at the top illustrates a practical example of the time dependency of the surface temperature when measuring heat transfer coefficients in a pipe (see the last section of this paper for specific parameters). The bottom part of this figure displays the results of this procedure (red line with black dots).

These transformed data can then be used to find the corresponding phase shift. This can be determined by

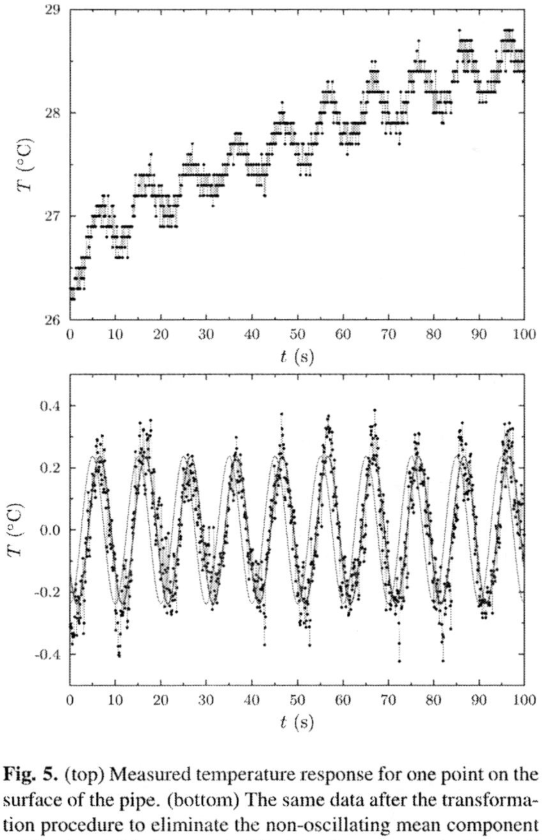

Fig. 5. (top) Measured temperature response for one point on the surface of the pipe. (bottom) The same data after the transformation procedure to eliminate the non-oscillating mean component of the signal (black points). The blue line represents the result of nonlinear regression using sine function, see Eq. (9), which should correspond to the oscillating heat flux irradiating the heat transfer surface (see the gray line, which here offers just an approximate illustration of the temperature response phase shift).

the Fourier transform, or by using the nonlinear regression procedure (the **nlinreg** function in MATLAB®). It is based on minimizing the sum of squares, in our case

$$\sum_{i=1}^{N} [T_i - A\sin(\omega t_i + \varphi)]^2 = \min\,, \qquad (13)$$

where A and φ are the amplitude and the phase shift of the sine temperature response, N is the number of experimental points, i.e. couples of temperatures T_i and times t_i, and ω is the angular frequency of the oscillating heat flux. We could evaluate the phase shift 303.678°, or −56.322°, for the experimental data in Figure 5.

It is obvious that the algorithm for eliminating the non-oscillating component of the measured temperature response can have a huge impact on the evaluated experimental results. Let us test its influence on model data prepared with a function of given parameters, so that we can make an estimate of the inaccuracies when using this procedure. The model data were based on the following function

$$T = \underbrace{a + b\left(1 - \mathrm{e}^{-t/\tau}\right)}_{\text{steady part}} + \underbrace{A\sin(\omega t + \varphi)}_{\text{harmonic part}} + \underbrace{\mathrm{randn}\,(\sigma)}_{\text{noise}}\,, \qquad (14)$$

where a and b are additional parameters affecting the shape of the signal, and σ is the standard deviation of random data with a normal distribution around a mean value of zero.

The algorithm was tested for a signal with frequency 0.1 Hz, sampled frequency 10 Hz, and various shapes (a pure sine signal, a sine signal superposed on an exponential function with a shift, a signal with superposed noise, and a quantum signal with a specific minimal temperature resolution size, for example 0.1 °C).

Table 2 shows the results of eliminating the non-oscillating component in the case of a model harmonic signal with two amplitudes. The algorithm that is used clearly provides adequate results for the phase shift in the whole range – the maximum absolute deviation is 0.31°. These deviations decrease with increasing amplitude of the oscillations.

In practice, the measured signal is noisy and, according to the infra-red camera, the measured temperatures are resolved with quantum steps of 0.1 °C. In this case, the deviations are substantially larger, see Table 3. The tables show the results for only one run and one shape. When processing many randomly generated noisy data, we get a maximum deviation of ±2.45°, and for the case of a quantum signal a maximum deviation of ±2.88°.

5 Phase shift in real measurements

In the previous section, algorithms for evaluating the heat transfer coefficient with the TOIRT method were described, under the assumption that only the temperature response suffers from measurement inaccuracies, but that the heat flux and time are measured and set precisely. This is quite important, because the resulting phase of the temperature response is used in the evaluation. In reality, however, the situation is different.

Table 2. Deviations in processing a harmonic signal with frequency 0.1 Hz ($\omega = 2\pi 0.1\ \mathrm{s}^{-1}$) superposed on the exponential function for amplitudes $A = 0.5$ °C and 1 °C. The sampled frequency is 10 Hz, and the signal length is 100 s.

$T = 25 + 4\left(1 - \mathrm{e}^{-t/100}\right) + A \sin(\omega t + \varphi_{\mathrm{set}})$

	$A = 0.5$ °C		$A = 1$ °C	
φ_{set} (°)	φ (°)	$\Delta\varphi$ (°)	φ (°)	$\Delta\varphi$ (°)
0	-0.15	-0.15	-0.08	-0.08
5	4.88	-0.12	4.94	-0.06
10	9.90	-0.10	9.95	-0.05
15	14.93	-0.07	14.97	-0.03
20	19.96	-0.04	19.98	-0.02
25	24.99	-0.01	25.00	-0.00
30	30.02	0.02	30.01	0.01
35	35.05	0.05	35.03	0.03
40	40.08	0.08	40.04	0.04
45	45.11	0.11	45.06	0.06
50	50.14	0.14	50.07	0.07
55	55.17	0.17	55.08	0.08
60	60.19	0.19	60.10	0.10
65	65.22	0.22	65.11	0.11
70	70.24	0.24	70.12	0.12
75	75.26	0.26	75.13	0.13
80	80.28	0.28	80.14	0.14
85	85.30	0.30	85.15	0.15
90	90.31	0.31	90.16	0.16

Table 3. Deviations in processing a harmonic signal with frequency 0.1 Hz ($\omega = 2\pi 0.1\ \mathrm{s}^{-1}$) superposed on an exponential function with standard deviation of 0.2 °C for amplitude $A = 0.5$ °C. Sample frequency 10 Hz, signal length 100 s. The results for a non-rounded signal are shown in the columns on the left. The results for a signal with quantum resolution of 0.1 °C are shown in the columns on the right.

$T = 25 + 3\left(1 - \mathrm{e}^{-t/100}\right) + A \sin(\omega t + \varphi_{\mathrm{set}}) + \mathrm{randn}\,(0.2)$

	noise		noise & quant	
φ_{set} (°)	φ (°)	$\Delta\varphi$ (°)	φ (°)	$\Delta\varphi$ (°)
0	1.21	1.21	1.12	1.12
5	6.20	1.20	6.23	1.23
10	11.18	1.18	11.20	1.20
15	16.16	1.16	16.30	1.30
20	21.12	1.12	21.04	1.04
25	26.08	1.08	26.10	1.10
30	31.03	1.03	31.10	1.10
35	35.97	0.97	36.04	1.04
40	40.91	0.91	41.08	1.08
45	45.83	0.83	45.80	0.80
50	50.76	0.76	50.76	0.76
55	55.67	0.67	55.94	0.94
60	60.58	0.58	60.32	0.32
65	65.49	0.49	65.49	0.49
70	70.39	0.39	70.54	0.54
75	75.29	0.29	75.28	0.28
80	80.19	0.19	80.28	0.28
85	85.09	0.09	84.88	-0.12
90	89.99	-0.01	90.12	0.12

The oscillating heat flux is generated by a set of one or more halogen lamps, each with power of 500 W in our experiments. The power supply of direct voltage 0 – 230 V with a maximum of 1600 W [https://www.bke.cz/cs/produkty/js-2k0-2k0nv] is controlled by the BK Precision 4052 precise two-channel signal generator [http://www.bkprecision.com/products/signal-generators/4052-5-mhz-dual-channel-function-arbitrary-waveform-generator.html]. One channel of this generator creates the sine signal that controls the power supply of the halogen lamps. The second channel generates trigger square pulses for the thermoIMAGER TIM 160 infra-red camera [https://www.micro-epsilon.com/temperature-sensors/thermoIMAGER/]. The timings of the trigger pulses and the generated signal for the power supply are depicted in Figure 6. It is clear that the first trigger pulse for the camera measuring the temperature is delayed behind the signal controlling the heat flux. This delay does not cause any problem in the evaluation, because it is precisely defined, and can be taken into account. The problem lies especially in the dynamic characteristics of the power supply and the halogen lamps, which generate the heat flux. Assuming that the system, the power supply – the halogen lamps, behaves as a system of the first order, we can describe it by time constant τ (which is of course not same as the time constant of the temperature response in the previous section). In the case of harmonic oscillations, this time constant will represent the delay of the generated heat flux with respect to the control signal in the generator, which ensured the synchronization of the measurement. This time constant, or its corresponding phase shift, must be used to decrease the evaluated phase shift from the experimental data.

The most practical way to determine the time constant would be by measuring the heat transfer coefficient on a geometry where it is precisely known. Using the methods mentioned above, we could determine this additional compensation of the phase shift in such a way that the determined value would be equal to the known value. The problem is the precision of the known value of the heat transfer coefficient. Most verified correlations (a semi-infinite

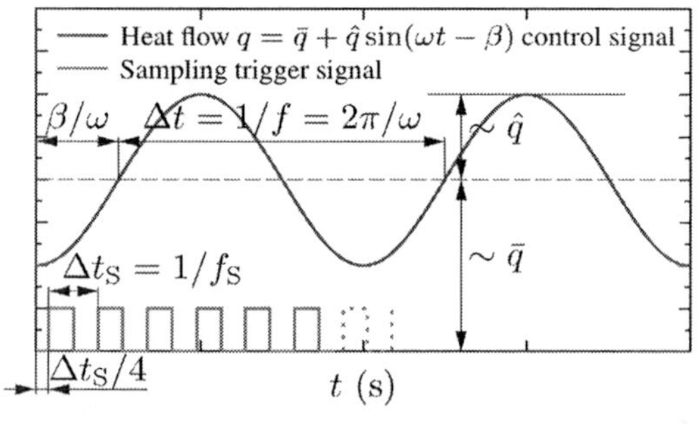

Fig. 6. Illustration of the time dependencies of the signal controlling the generated heat flux and the sampling signal for measuring the temperature.

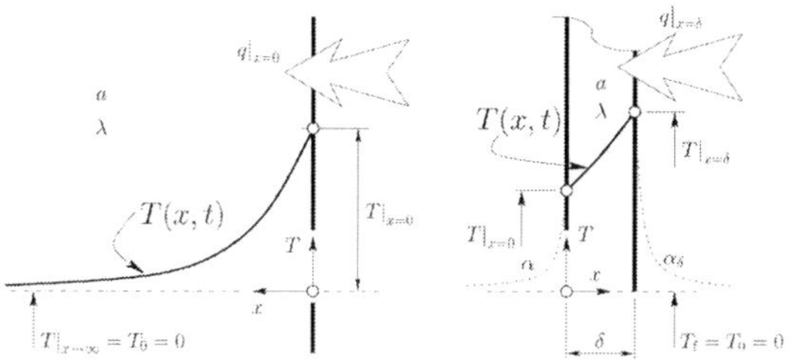

Fig. 7. A semi-infinite body as a calibration tool of the TOIRT method (left), and the heat transfer surface of an infinite plate (right). The boundary condition of the harmonic oscillating heat flux $q|_{x=\delta} = \bar{q} + \hat{q}\sin(\omega t - \beta)$ is applied to the heat transfer surface (wall) with the TOIRT method (the Wandelt-Roetzel solution assumes $\bar{q} = 0$ and $\beta = 0$).

body, a plate of finite width, an impinging jet, the flow in a pipe of circular cross section) already have some non-zero deviation from the "theoretical" value, although they have been in use for many years.

Another possible approach would be to make measurements on a geometry where the value of the heat transfer coefficient has no impact or a very small impact on the results. A semi-infinite body is one of the options, see Figure 7 (left).

Carslaw and Jaeger [1] describe the solution of the temperature field in a semi-infinite body for the case of harmonic oscillations of the surface temperature (page 64), which is the boundary condition $T|_{x=0} = A\cos\omega t$ (including the phase shift β which is not mentioned here) and for the zero initial temperature $T_0 = 0$.

$$T(x,t) = Ae^{-\sqrt{\frac{\omega}{2a}}x}\cos\left(\omega t - \sqrt{\frac{\omega}{2a}}x\right) + \\ - \frac{2A}{\sqrt{\pi}}\int_0^{x/\sqrt{4at}} \cos\left\{\omega\left(t - \frac{x^2}{4a\mu^2}\right)\right\} e^{-\mu}\mathrm{d}\mu\,. \quad (15)$$

The solution consists of the steady periodic part (first term/row) and the transient part (second term/row). The transient part expresses the disturbance caused by starting the oscillations, which disappears with increasing time. The term $x\sqrt{\omega/2a}$ represents the dimensionless coordinate here, and we can define the characteristic length using the penetration depth $\delta \sim \sqrt{a/\omega}$.

Expressing the heat flux on the surface of the semi-infinite body using the Fourier law and only the steady-periodic part, we get

$$q|_{x=0} = -\lambda \left.\frac{\partial T}{\partial x}\right|_{x=0} = -\lambda \cdot A\sqrt{\frac{\omega}{2a}}e^{-\sqrt{\frac{\omega}{2a}}x} \cdot \\ \cdot \sin\left(\omega t - \sqrt{\frac{\omega}{2a}}x\right) - \cos\left(\omega t - \sqrt{\frac{\omega}{2a}}x\right)\Bigg|_{x=0} = \\ = \underbrace{\lambda A\sqrt{\frac{\omega}{a}}}_{\hat{q}}\cos\left(\omega t + \frac{\pi}{4}\right). \quad (16)$$

It can be seen here that the heat flux on the surface of the semi-infinite body is phase-shifted by $\pi/4$, and the leading factor of the goniometric function represents the amplitude. Assuming periodic heat oscillations on the surface, which is the boundary condition $q|_{x=0} = \hat{q}\cos\omega t$, the corresponding temperature response on the surface of the semi-infinite body (steady periodic part) can be described as

$$T|_{x=0} = \frac{\hat{q}}{\lambda\sqrt{\frac{\omega}{a}}}\cos\left(\omega t - \frac{\pi}{4}\right). \quad (17)$$

We can see here that the semi-infinite body could be advantageously used in calibrating the whole system, because the phase shift between the heat flux and the temperature signal is exactly $\pi/4$.

But what about the transition process, which complicates the measurements even on thin plates? In real cases, where the temperature response oscillates around a non-zero mean value, the impact of the transition process and the influence of the heat accumulation in the semi-infinite body can be assessed by a numerical simulation of the temperature field. This was performed by the **pdepe** function in MATLAB® again. Figure 8 shows the temperature dependency of a semi-infinite body heated by the heat flux oscillating around a non-zero mean value.

The figure also shows the steady periodic solution oscillating around a non-zero mean value of the heat flux and the difference between the steady periodic solution and the numerical solution. It is clear that the difference does not approach zero so quickly as with a plate of finite width.

Carslaw and Jaeger [1] also described the solution of the temperature field in a semi-infinite body heated by the constant heat flux on the surface (page 75) and for zero

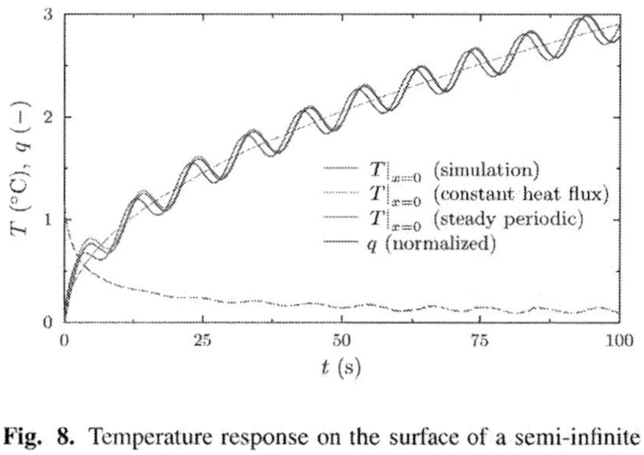

Fig. 8. Temperature response on the surface of a semi-infinite body made from stainless steel. The amplitude of the temperature oscillations is 1000 W m^{-2}, and the mean value is 2000 W m^{-2}. The heat transfer coefficient at the interface is zero. The red line represents the numerical simulation, in comparison with the steady periodic oscillations (blue). The green line slowly approaching a zero value represents the difference between them, magnified by a factor of 10. The gray line represents the heat flux oscillating around the constant mean value.

initial temperature $T_0 = 0$.

$$T = \frac{2\bar{q}}{\lambda}\left\{\sqrt{\frac{at}{\pi}}\,e^{-\frac{x^2}{4at}} - \frac{x}{2}\operatorname{erfc}\frac{x}{\sqrt{4at}}\right\} \tag{18}$$

which can be used simply to express the dependency of the surface temperature

$$T|_{x=0} = \frac{2\bar{q}}{\lambda}\sqrt{\frac{a\,t}{\pi}}\,. \tag{19}$$

Carslaw and Jaeger [1] also described the solution for the case of oscillating heat flux $q|_{x=0} = \sin\omega t$ on the surface (page 76)

$$T = \frac{1}{\lambda\sqrt{\frac{\omega}{a}}}e^{-x\sqrt{\frac{\omega}{2a}}}\sin\left(\omega t - x\sqrt{\frac{\omega}{2a}} - \frac{\pi}{4}\right) + \frac{2a}{\pi\lambda}\int_0^{\infty}\frac{\omega\cos ux}{\omega^2 + a^2u^4}e^{-au^2t}\mathrm{d}u\,. \tag{20}$$

The last term in this equation, when compared to the steady periodic solution on the surface of a semi-infinite body, describes the impact of the transition process caused by the oscillating heat flux at the beginning.

We performed a numerical simulation of the heat transfer in a semi-infinite body, see Figure 8, and the phase shift evaluated by the algorithm was 48.61°. This is close to the theoretical value of 45°, but due to the impact of its deviation on the accuracy of the heat transfer coefficient, a value of 48.61° is unacceptable.

The last question concerns the practical realization of the semi-infinite body. In practice, we would like to use a finite body that would provide results not too different from the results for the semi-infinite body. What should the width of such a body be? Using the dimensionless coordinate $\eta = x/\sqrt{4at}$, we can transform solution (18) to

$$T = \bar{q}\frac{\sqrt{4at}}{\lambda}\left\{\sqrt{\frac{1}{\pi}}e^{-\eta^2} - \eta\operatorname{erfc}\eta\right\}. \tag{21}$$

Let us find a value of dimensionless coordinate η such that, for the given time, the value of the term in brackets (containing real coordinate x) was equal to, for example, 0.01, 0.005 or 0.001 of the value on the surface ($\eta = 0$ for $x = 0$, and the term is $1/\sqrt{\pi}$). For the chosen values, we get $\eta_{0.01} = 1.6056$, $\eta_{0.005} = 1.7688$ and $\eta_{0.001} = 2.1127$. The temperature field is practically unaffected (0.1 %) at time 100 s at a distance of 82.8 mm for a stainless steel, and 24.4 mm for glass. This value does not reflect the penetration depth of the oscillating solution, However, looking at the damping term in (15) and evaluating its value for stainless steel, for example, criterion 0.1 %, with the same time and frequency of the oscillating heat flux 0.1 Hz, we get $\exp\left(-x\sqrt{\omega/2a}\right) = \exp\left(-\eta\sqrt{2\omega t}\right) = \exp(-2.1127\sqrt{4\pi 0.1\cdot 100} \sim 10^{-11}$. We can see that the impact of the penetration depth is much smaller with an oscillating heat flux than with a constant heat flux.

6 Gnielinski

Heat transfer coefficients were measured on the surface of a horizontal pipe with inner diameter $d = 32$ mm and wall width $\delta = 1.5$ mm. A pipe with length $L = 2000$ mm was made of stainless steel (see Table 1 of the corresponding thermophysical properties). The pipe section 200 mm in length at a distance of 1700 mm from the pipe inlet was irradiated by the heat oscillations generated by two 500 W halogen lamps.

Water at a temperature of 25.16 °C ± 0.2 °C and with a flow rate of 27.37 l min^{-1} ran in the pipe from a closed hydraulic circuit. The thermophysical properties of water at this temperature [13] are: density $\varrho = 997.0$ kg m^{-3}, specific heat capacity $c_p = 4179.2$ J kg^{-1} K^{-1}, thermal conductivity $\lambda_f = 0.6114$ W m^{-1} K^{-1}, and kinematic viscosity 8.8998×10^{-7} m^2 s^{-1}.

The corresponding Reynolds number (based on a mean volumetric velocity of 0.5672 m s^{-1}) is 20394 in this case, so we are in the region of fully developed turbulent flow [12], i.e. in the region described by Gnielinski's correlation (1). Using this correlation in the calculation of the Nusselt number, without the Hausen correction for now, we get $\mathrm{Nu_m} = \alpha_m d/\lambda_f = 147.97$ and the corresponding mean heat transfer coefficient $\alpha_m = 2827$ W m^{-2} K^{-1}.

The turbulent flow in this geometry was modeled using ANSYS® CFD® software. A structured mesh with 40000 elements and the ANSYS® Fluent™ solver were used. Rotational symmetry and the RANS approach (Reynolds Averaged Navier-Stokes), along with the $k-\omega$ SST (Shear Stress Transport) turbulence model were used in the solution of the momentum and energy transport equations. The SIMPLE algorithm for pressure-velocity coupling and second order spatial discretizations for all quantities were used. Iterations were performed until the residuals fell below 10^{-10}. A constant velocity profile was specified at the pipe inlet. A stable turbulent velocity profile developed in the pipe (entrance length 1700 mm). The inlet temperature was 300 K, and the wall temperature of the measuring section 200 mm in length was set to 400 K. The constant thermophysical properties mentioned above were assumed. For the evaluation of the heat transfer coefficient on the inner surface, the reference temperature corresponding to the arithmetic mean of the mixed cup temperatures at the inlet and the outlet of the measuring section was evaluated as (300.035 + 302.639)/2 = 301.337 K. The maximum dimensionless size of the mesh element at the pipe wall was $Y^+ = 0.152$.

Table 4 compares the results of our numerical simulation with values based on the Gnielinski correlation. Figure 9 compares local values of the heat transfer coefficients along the measurement section based on Eq. (3) with values obtained in the numerical simulation. It is clear that the mean value based on the numerical simulation is approximately 10 % lower than the Gnielinski correlation (the accuracy can be expected to be 6 – 10 %). The dependency of the local values shows that this difference is mainly affected by the increase in heat transfer at the beginning of the measurement section. After the thermal boundary layer is developed, this difference is much

Table 4. A comparison between the results based on numerical simulations and the analytical results based on engineering correlations. The pressure drop can be expressed from the Darcy-Weisbach equation $\Delta p = \xi \cdot L/d \cdot \varrho \bar{u}^2/2$, and the wall shear stress from $\tau_w = \Delta p\, d/4L$.

Quantity	Analytical	Numerical	
Δp	25.602	26.982	Pa
ξ	0.0255	0.0269	
τ_w	1.0241	1.0789	Pa
α_m	3660	3300	$\mathrm{W\,m^{-2}\,K^{-1}}$

smaller. Figure 9 illustrates that the substantial increase in the heat transfer in the development of the thermal boundary layer must be taken into account when using any experimental method focused on local values. We should therefore take into account the location where the measurement is made, and also the way in which the heat transfer surface is heated.

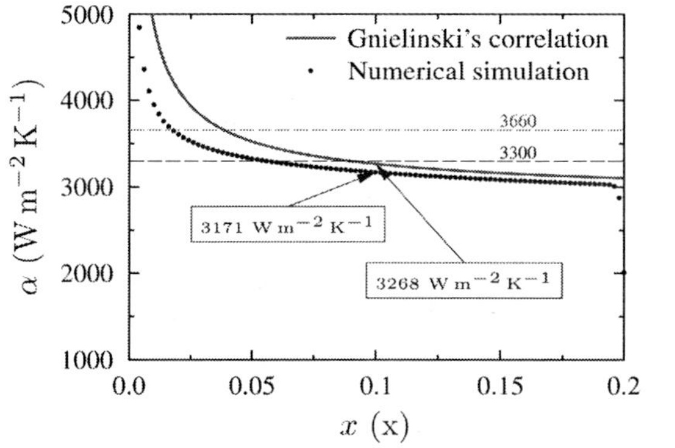

Fig. 9. A comparison of local values of the heat transfer coefficients along the measurement section. Blue color corresponds to (3), black color represents values based on the numerical simulation.

Using the numerical model, we can solve the case of measuring the heat transfer coefficient using the TOIRT method. Because the TOIRT method is a transient method, we have to solve the time dependency of the temperature field and monitor the temperature on the surface of the measuring section. The temperature oscillations are generated by the oscillating heat flux on the surface of the measurement section. The following user-defined function (UDF) was used in the ANSYS® Fluent™ solver as the corresponding boundary condition.

```
DEFINE_PROFILE(hf_t_alpha, thread, position)
{
  real flow_time = RP_Get_Real("flow-time");
  real wTemp, q, q_ave, q_ampl, omega, freq,
  face_t f; real alpha_delta, T_amb;

  freq = 0.1; omega = 2*PI*freq;
  alpha_delta = 3; T_amb = 300;  q_ampl = 2000;
  q_ave = 2*q_ampl;
  begin_f_loop(f, thread)
  {
    wTemp = F_T(f,thread);
```

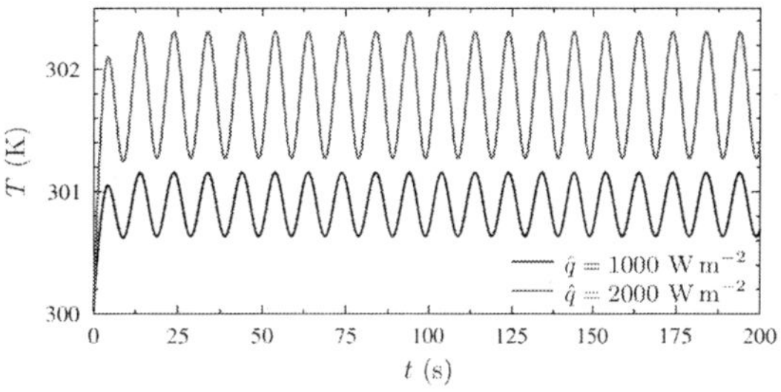

Fig. 10. Time dependency of the surface temperature at a distance of 100 mm from the beginning of the measuring section for two cases of heat flux amplitudes $\hat{q}$ and the heat transfer coefficient between the pipe surface and surroundings 3 $\mathrm{W\,m^{-2}\,K^{-1}}$.

```
    q = q_ave + q_ampl*sin(omega*flow_time)
        + alpha_delta*(T_amb-wTemp);
    F_PROFILE(f, thread, position) = q;
  }
  end_f_loop(f, thread)
}
```

The mean value of the oscillating heat flux was set equal to twice the amplitude of the heat flux, as is done in real measurements. The amplitude was set in such a way that the resulting temperature response amplitude was 0.5 °C, again corresponding to real measurements. Figure 10 shows the time dependency of the surface temperature obtained in the numerical simulation of the heat transfer in the middle of the measurement section.

Processing these data using the algorithm for eliminating the non-oscillating component, we can get the phase shift and the corresponding value of the heat transfer coefficient based on the theoretical Wandelt-Roetzel solution. Assuming zero heat transfer coefficient on the irradiated side ($\alpha_\delta = 0$), the resulting phase shift is 49.575° (ignoring the first period of the temperature response, and if the first period is included, the phase shift is 52.085°). For the heat transfer coefficient on the irradiated side $\alpha_\delta = 3\ \mathrm{W\,m^{-2}\,K^{-1}}$, the phase shift is 49.541° (52.048° if the first period is included). Both results are for amplitude of the heat flux oscillations 2000 $\mathrm{W\,m^{-2}}$ and time 100 s. The corresponding heat transfer coefficient values, based on the Wandelt-Roetzel solution, are 2844 $\mathrm{W\,m^{-2}\,K^{-1}}$ in both cases. The local heat transfer coefficient value evaluated in the numerical simulation is 3171 $\mathrm{W\,m^{-2}\,K^{-1}}$, see Figure 9, which is 10 % higher than the Wandelt-Roetzel solution. Gnielinski's correlation (3) predicts a value of 3268 $\mathrm{W\,m^{-2}\,K^{-1}}$ at the monitored point.

Let us recall that the evaluation of the experimental data depicted in Figure 5, which represent the same geometry as in the numerical simulations, resulted in a phase shift of 56.322°. The corresponding heat transfer coefficient is then 2101 $\mathrm{W\,m^{-2}\,K^{-1}}$. This value is not correct, because it does not take into account the time constant of the whole measuring system, which was discussed in section 5. If we want to fit the phase shift to the known heat transfer coefficient value of 3268 $\mathrm{W\,m^{-2}\,K^{-1}}$, based on Gnielinski's correlation (3), we will have to determine a value of 46.253°. We can see that the real phase shift is greater by 56.322 − 46.253 = 10.069°, and this value corresponds to the time constant 10.069/360 × 10000 = 279.7 ms for a period of 10 s, i.e. for 0.1 Hz frequency of the temperature oscillations.

We performed repeated measurements of the heat transfer coefficients along the whole measuring section as a calibration experiment. Using these measurements, we employed a non-linear regression technique to find the time constant of 254 ms, which corresponds to the mean value, according to Gnielinski (see Table 4), or 185 ms for the numerical solution.

7 Conclusions

We have presented some aspects of the TOIRT experimental methods in this paper. It is obvious that further investigations are needed, but we can summarize the following findings.

The measured phase shift should be in the range between 20 and 70° when using the TOIRT method for heat transfer measurements on the surface of a stainless steel plate up to 1.5 mm in width and with 0.1 Hz frequency of the oscillating heat flux. We can expect a relative measurement error of around 10 % for a sufficient large heat flux, creating temperature oscillations with amplitude of 1 °C.

The time constant of the transient process corresponds to ~ $\delta\lambda/(a\alpha)$ for the standard application of the method (~ 1 mm, ~ 1000 $\mathrm{W\,m^{-2}\,K^{-1}}$, ~ 10 $\mathrm{W\,m^{-1}\,K^{-1}}$). The corresponding time of the transient process is at least 5 times longer.

A numerical analysis of the algorithm for eliminating the non-oscillating component of the measured temperature response showed that its error increases with decreasing amplitudes of the temperature oscillations, and with an increased noise-signal ratio and increased deformation of the signal. To reduce the corresponding error, temperature oscillations with an amplitude of 1 °C or larger are recommended.

When testing the algorithm for eliminating the non-oscillating component for the case of a semi-infinite body, a phase shift of 48.61° was determined, whereas the theoretical value is 45°.

The local heat transfer coefficient in a pipe with turbulent flow was monitored in the middle of a heated section. The corresponding value based on Gnielinski's correlation is 3268 $\mathrm{W\,m^{-2}\,K^{-1}}$, and the value based the numerical simulation is 3171 $\mathrm{W\,m^{-2}\,K^{-1}}$. The value based on applying the Wandelt-Roetzel method to the numerical simulation results is 2844 $\mathrm{W\,m^{-2}\,K^{-1}}$. Is the neglected lateral heat conduction the main reason for this difference, or are there some other reasons (the type of boundary condition, insufficient accuracy of the numerical solution, etc.)?

A comparison was made between the numerical simulation results, the engineering correlations and the experimental data for the case of turbulent flow in the pipe. A phase shift of 254 or 185 ms corresponding to the time constant of the whole measurement process was evaluated, depending on the "correct" mean value that was used for the heat transfer coefficient (Gnielinski's correlation or the numerical solution). Let us recall that a phase shift of one degree at a frequency of 0.1 Hz corresponds to a time delay of 30 ms with a 10 % relative error of the method.

It is clear that further investigations are necessary, especially into various transformation algorithms, the impact of the lateral heat conduction, and the application of the TOIRT method to the laminar flow of liquids. It is obvious that the algorithm for eliminating the non-oscillating component of the temperature response has a large impact on the accuracy of the results. It is necessary to devise a better test, using polynomials of higher orders, using the theoretical solution, etc. The dynamic behavior of the whole measurement system also has a huge impact on the measurement results. There are a various potential approaches. Our future research will involve directly monitoring the light generated by halogen lamps, instead of the calibration experiments mentioned here. The light intensity can then be used for directly controlling the heat flux, or as an input parameter for the TOIRT mathematical model that accounts for the general time dependency of the heat flux.

This work was supported by the Ministry of Education, Youth and Sports of the Czech Republic under OP RDE grant number CZ.02.1.01/0.0/0.0/16_019/0000753 "Research centre for low-carbon energy technologies". Authors also acknowledge support from the Grant Agency of the Czech Technical University in Prague, grant number SGS18/129/OHK2/2T/12.

Finally, the authors express their thanks to Robin Healey, from the Czech Technical University in Prague, for his huge efforts to convert our "English text" into an English text that can be offered to our readers.

Nomenclature

a	thermal diffusivity, $a = \lambda/\rho c_\mathrm{P}$ ($\mathrm{m^2\,s^{-1}}$)
$a,\ b$	general parameters (°C, K)
A	amplitude of temperature oscillations (°C, K)
c_p	specific heat capacity ($\mathrm{J\,kg^{-1}\,K^{-1}}$)
d	diameter (m)
f	frequency (Hz, $\mathrm{s^{-1}}$)
f_S	sampling frequency (Hz, $\mathrm{s^{-1}}$)
Gz	Graetz number, $\mathrm{Gz} = \mathrm{Re\,Pr}\,d/L$ (−)
$\mathrm{Gz_x}$	local Graetz number, $\mathrm{Gz} = \mathrm{Re\,Pr}\,d/x$ (−)
L	length (m)
M	mass (kg)
N	number of experimental points (−)
Nu	Nusselt number, $\mathrm{Nu} = \alpha d/\lambda_\mathrm{f}$ (−)
$\mathrm{Nu_m}$	mean Nusselt number (−)
$\mathrm{Nu_x}$	local Nusselt number (−)
Pr	Prandtl number (−)
q	heat flux ($\mathrm{W\,m^{-2}}$)
$\bar{q}$	mean value of heat flux ($\mathrm{W\,m^{-2}}$)
$\hat{q}$	amplitude of oscillating heat flux ($\mathrm{W\,m^{-2}}$)
Re	Reynolds number, $\mathrm{Re} = \bar{u}d/\nu$ (−)
S	surface, heat transfer surface ($\mathrm{m^2}$)
t	time (s)
t_i	time (s)
T	temperature (°C, K)
T_i	measured temperature at time t_i (°C, K)

T_0	initial temperature (°C, K)
T_{eq}	equilibrium fluid (liquid) temperature (°C, K)
T_f	fluid (liquid) temperature (°C, K)
$\bar{u}$	mean velocity ($m\,s^{-1}$)
x, y, z	spatial coordinates of the Cartesian coordinate system (m)
Y^+	dimensionless distance from the wall, dimensionless wall spacing (–)
α	heat transfer coefficient (at $x = 0$) ($W\,m^{-2}\,K^{-1}$)
α_δ	heat transfer coefficient at $x = \delta$ ($W\,m^{-2}K^{-1}$)
β	phase shift (excitation) (–)
δ	width of impinged plate (m)
Δt	time difference, time step (s)
η	dimensionless coordinate, $\eta = x/\sqrt{4at}$ (–)
σ	standard deviation (°C, K)
λ	thermal conductivity, wall thermal conductivity ($W\,m^{-1}\,K^{-1}$)
λ_f	fluid thermal conductivity ($W\,m^{-1}\,K^{-1}$)
ν	kinematic viscosity ($m^2\,s^{-1}$)
ξ	Darcy-Weisbach friction factor (–)
ρ	density ($kg\,m^{-3}$)
τ	time constant (s)
τ_W	wall shear stress (Pa)
φ	phase shift (response) (–)
ω	angular frequency of oscillating heat flux, $\omega = 2\pi f$ (s^{-1})

References

1. H. Carslaw, J. Jaeger, eds., *Conduction of Heat in Solids* (Oxford Univesrity Press, London, 1959)
2. H. Hausen, *Heat Transfer: In Counterflow, Parallel Flow and Cross Flow* (McGraw-Hill, 1983), https://ia801303.us.archive.org/26/items/in.ernet.dli.2015.133989/2015.133989.Heat-Transfer-In-Counterflow-Parallel-Flow-And-Cross-Flow.pdf
3. W. Roetzel, Chem. Eng. Technol. **12**, 379 (1989)
4. W. Roetzel, S.K. Das, X. Luo, Int. J. Heat Mass Transfer **37**, 325 (1994)
5. W. Roetzel, X. Luo, Rev Gén Therm **37**, 277 (1998)
6. P. Leblay, J. Henry, D. Caron, D. Leducq, A. Bontemps, L. Fournaison, *Infrared Thermography applied to measurement of Heat transfer coefficient of water in a pipe heated by Joule effect*, in *QIRT - 11th Quantitative InfraRed Thermography* (2012), pp. 1–10
7. W. Roetzel, S. Prinzen, M. Wandelt, Chem. Eng. Technol. **16**, 89 (1993)
8. M. Wandelt, W. Roetzel, *Lockin thermography as a measurement technique in heat transfer*, in *QIRT 96 - Eurotherm Series 50* (1997), pp. 189–194
9. S. Freund, S. Kabelac, *Measurement of local convective heat transfer coefficients with temperature oscilation IR thermography and radiant heating*, in *Proceedings of HT2005, 2005 ASME Summer Heat Transfer Conference* (2005), pp. 1–7
10. S. Freund, Ph.D. thesis, Der Fakultät für Maschinenbau der Helmut-Schmidt-Universität / Universität der Bundeswehr Hamburg (2007)
11. S. Freund, S. Kabelac, Int. J. Heat Mass Transfer **53**, 3764 (2010)
12. P. Stephan, ed., *VDI Heat Atlas* (Springer, Berlin, Heidelberg, 2010)
13. J. Šesták, J. Bukovský, M. Houška, *Heat processes. Transport a thermodynamic data. (in Czech)* (Publishing Czech Technical University in Prague, Prague, 1996)
14. S. Solnař, Master's thesis, Czech Technical University in Prague (2016)

Appendix

A numerical solution of the heat conduction equation with the given boundary and initial conditions with the **pdepe** function in MATLAB® was made [https://www.mathworks.com]. The function solves an initial-boundary value problem of the parabolic partial differential equation with one independent spatial coordinate. The detailed usage of this function can be found at https://www.mathworks.com/help/matlab/ref/pdepe.html, and our shortly commented source code is presented below.

```
function youcantryme()
% Global variables (parameters)
global OMEGA A LAMBDA ALPHA ALPHAD QAVE QAMPL QFI

% Given parameters
OMEGA = 2*pi()*0.1;
LAMBDA = 15.2; A = 15.2/(7900*501);
ALPHA = 1000; ALPHAD = 3;
QAVE = 2000; QAMPL = 1000; QFI = 0;

t = linspace(0,100,2000); % Time space

% Plate thickness and spatial distribution
DELTA = 0.001; x = linspace(0,DELTA,100);

% PDE solution (0 means Cartesian coordinate system = slab)
[sol] = pdepe(0,@pdefun,@icfun,@bcfun,x,t);

% Results presentation
figure(1); surf(x,t,sol);

end % End of main script

%  PDE, IC and BC definitions
function [c,f,s] = pdefun(x,t,u,dudx) % PDE function
global A % Thermal diffusivity
c = 1/A; f = dudx; s = 0;
end

function u = icfun(x)  % Initial condition
u = 0;
end

function [pl,ql,pr,qr] = bcfun(xl,ul,xr,ur,t)  % BC
global OMEGA A LAMBDA ALPHA ALPHAD QAVE QAMPL QFI
pl = ALPHA*ul; ql = - LAMBDA;
pr = - ALPHAD*ur + QAVE + QAMPL*sin(OMEGA*t + QFI);
qr = - LAMBDA;
end
```

Grid turbulence studied by Particle Image Velocimetry

Daniel Duda[1,*], *Jindřich* Bém[1], *Vitalii* Yanovych[1], and *Václav* Uruba[1,2]

[1]Faculty of Mechanical Engineering, University of West Bohemia, Pilsen, Czech Republic
[2]Institute of Thermomechanics, Czech Academy of Sciences, Prague, Czech Republic

Abstract. We studied the grid-generated turbulence by using Particle Image Velocimetry (PIV) technique. We test on this already well studied flow the new ways of analyzing spatially resolved PIV data, such as the spatial spectra and structure functions. We compare some of the turbulence characteristics with results of Laser Doppler Anemometry (LDA).

1 Introduction

The decay of turbulence produced by flow passing a regular grid is a standard turbulence problem [1–6]. The grid turbulence is interesting from the theoretical point of view as it is a prototype of an „ideal“ Kolmogorov-type turbulence, which is homogeneous and isotropic (of course, it is not fully true) [3, 7]. It decays in space, but from the point of view of the fluid, it decays in time. The decay realizes via the Richardson cascade, which transports the energy into both: smaller and larger scales. The energy at smallest scale, called Kolmogorov scale, dissipates due to viscosity and, as there is no other energy input than the transfer from larger scales, the turbulence decays from bottom: growing the Kolmogorov scale.

This flow has been studied experimentally many times by using mainly the *Hot Wire Anemometry* (HWA) [8], which is the method best suitable for turbulence research. Its result is a perfect time-series of velocities in *single point*.

In this study we use another widely used experimental method: *Particle Image Velocimetry* (PIV) [8, 9]. It is based on direct optical observation of small particles (or droplets) carried by the fluid in the entire field of view of a camera. This method has lower precision than the HWA, but it offers the spatially resolved instantaneous velocity fields, which is the aspect, we found to be crucial to distinguish the main building blocks of turbulence – vortices [10–12].

This investigation of already a well-studied flow focuses to testing new ways of analysis of the spatially resolved PIV data. We used the LDA (Laser Doppler Anemometry) technique as well, which offers higher accuracy, but only in one point. Therefore we compare turbulence characteristics such as turbulence intensity obtained by both techniques. More details can be found in our previous publication [13].

2 Experimental setup

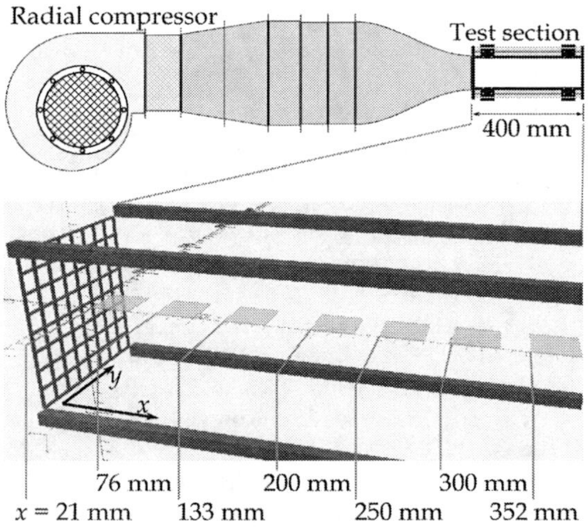

Fig. 1. Sketch of the wind tunnel (top) and the position of the studied areas in respect to the grid, which is located at the inlet into the transparent test section.

The flow has been generated by WT-tech wind tunnel of maximum achieved velocity 25 m/s (this experiment) with the transparent test-section of size $125 \times 125 \times 400$ mm. At the beginning of the test section, there is the grid with grid spacing $M = 125\,\text{mm}/8 = 15{,}625\,\text{mm}$, thickness of square rods is $d = 2$ mm, therefore the porosity $\beta = (M - d)^2/M^2 = 76\%$.

PIV measurement is performed by using a commercially available Dantec PIV system. Illumination is realized by a double-pulse Nd:YAG solid-state laser

*e-mail: dudad@kke.zcu.cz

NewWave Solo which produces a pair of 500 mJ shots lasting 5 ns. Double-frame CCD cameras FlowSense MkII can distinguish two expositions separated by only $1,5\,\mu s$, but the repetition frequency does not exceed 7, 4 Hz, thus the obtained instantaneous velocity fields can be considered as statistically independent. As the tracer particles we use droplets of $\sim\,\mu$m diameter produced by commercially available fog generator SAFEX.

We have measured the flow in 7 square areas of size 32×32 mm, i.e. $\sim 2M$, located 21, 76, 133, 200, 250, 300 and 352 mm from the grid (this is the location of the leading edge of these areas). Their z-coordinate is 7, 5 mm under the axis of the test section, which is approximately close to the middle of the first grid hole. We have measured at velocities u ranging from 1, 2 to 25 m/s. M-based Reynolds number $Re = u \cdot M/\nu$, where ν is the kinematic viscosity of air. Re ranges from $1,3 \cdot 10^3$ to $2,7 \cdot 10^4$.

3 Results

General description of the grid turbulence is that, the turbulence is first produced by the interacting jets behind the holes, or by interacting the wakes behind the rods, as is apparent in figure 2(a). Wakes are widening as described e.g. by [14] – note the double maxima of TKE at small distance (Fig. 3(a)), at later stage (Fig. 3(b)), the pair of maxima is separated more than the maxima produced by neighboring rods. Even later, it starts to be influenced by the horizontal rods up and down the measuring plane.

3.1 Turbulent kinetic energy

Turbulent kinetic energy (TKE, K) is average kinetic energy associated with vortices and other coherent flow structures, which the turbulence consists of. It is calculated as *variance* of velocity, i.e.

$$K(\vec{x}) = \frac{1}{2}\left(\left\langle u^2(\vec{x}) + v^2(\vec{x})\right\rangle_T - \langle u(\vec{x})\rangle_T^2 - \langle v(\vec{x})\rangle_T^2\right) \quad (1)$$

Where u and v denote the x- and y-component of the velocity vector, $\vec{x}$ is location and $\langle\cdot\rangle_T$ denote the averaging over time, i.e the ensemble averaging in this case of relatively „slow" PIV.

The flow is naturally three-dimensional, while we have measured only the planar cut, therefore we lack any information about the third velocity component w. Some authors solve this issue by assuming that the third component fluctuations are statistically equal to the second component (v), thus they multiply that term by two. We prefer to keep the measured data and not to introduce more artificial assumptions, than it is needed.

Equation (1) can be rewritten by using the energies of instantaneous velocity field E and of averaged velocity field E_T as $K = E - E_T$, where $E = \frac{1}{2}\left\langle \vec{u}^2\right\rangle_T$ and $E_T = \frac{1}{2}\langle\vec{u}\rangle_T^2$.

The spatial distribution of TKE at first three positions is plotted in figure 3, the rest is no more interesting, because the spatial distribution is flat in span-wise direction (y). The change of the amount of TKE in stream-wise direction (x) is better recognizable in figure 4, where the average over y of TKE is displayed. When we take into account only the data more far away from the grid, we find the decay with power -1.95. Note that this value can differ, when we would use an *virtual origin*, i.e. shifting of x by some x_0, which is used by some authors [2]. In article [13], we have shown, that the decay exponent depends on the length-scale as well. TKE of small scale decays faster at small Reynolds numbers, while at large Re, the decay rate is same for all scales (Figure 8 in [13]).

The value of TKE is underestimated when measured by using the PIV method, because this method assumes a homogeneous motion inside the so-called *interrogation area*, whose size is 0.51 mm in our case. Thus all flow structures smaller than this size are neglected and does not counted in our TKE. On the other hand, the majority of energy occurs at larger scales due to the quite steep decay of power spectral density, which is (in the homogeneous and isotropic turbulence) $-5/3$. Thus we need to test our result by another method more sensitive even to the small scale, which is the Laser Doppler Anemometry (LDA).

3.2 Comparison with Laser Doppler Anemometry

Laser Doppler Anemometry (LDA) measures the velocity of small particles caried by the fluid (similarly as PIV does) in a one small volume, where two coherent laser beams cross. The size of this measurement volume is around $xx\,\mu m^3$, thus it can be considered as a point. LDA is generally though to be a precise trustworthy absolute method, which is limited only by the ability of used tracer particles to follow the flow.

We measured the stream-wise velocity u and compared the results obtained by both techniques, as shown in figure 5. The value of stream-wise velocity strongly depends on the position, specially in the vicinity to the grid. Generally, speaking, the flow is accelerated in the jets and decelerated in the wakes behind the rods. The measurement point of LDA has been located into the wake. Therefore for comparison with PIV, we have to select the cut from the wake, as it is done in figure 5. There are plotted also PIV data averaged over the span-wise direction, thus containing both, jets and wakes. The data plotted in figure 5 suggest that the LDA measuring point has not been exactly in the wake, as its position has been selected manually before the measurement, while the track of the wake in PIV data has been selected after the measurement from the data. This difference vanishes in higher distances from the grid.

The plot of standard deviation $\sigma_T[u]$

$$\sigma_T[u] = \sqrt{\left\langle u^2 - \langle u\rangle_T^2\right\rangle_T} \quad (2)$$

shows better agreement between LDA and PIV data from the rod wake. The standard deviation in LDA data is systematically higher by about 13%. It is too much to have only physical reasons discussed above, it can partly be caused by non-identical coordinate systems in LDA and PIV data. This issue we will fix in future by using an motorized traverser for the LDA measuring head.

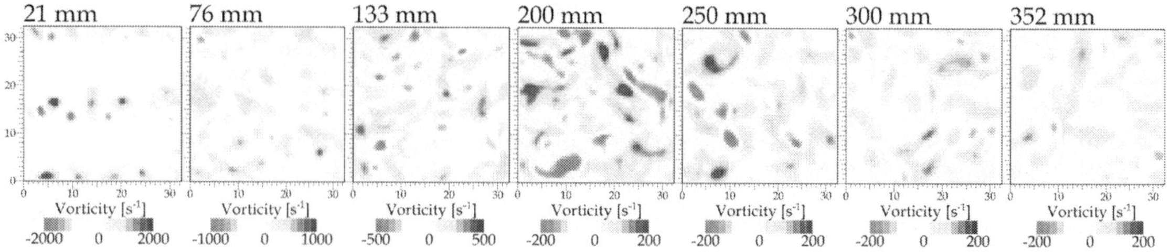

Fig. 2. Examples of spatial distribution of instantaneous vorticity ω in 7 studied areas (see figure 1). Stream-wise velocity (from left to right) is 3.0 m/s corresponding to Reynolds number $Re = 3.1 \cdot 10^3$. Note the different color-scale of first three images.

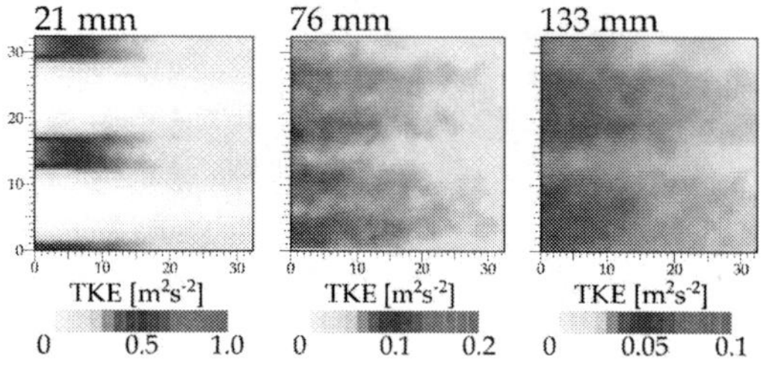

Fig. 3. Spatial distribution of the turbulent kinetic energy. Stream-wise velocity is 3.0 m/s, corresponding $Re = 3.1 \cdot 10^3$.

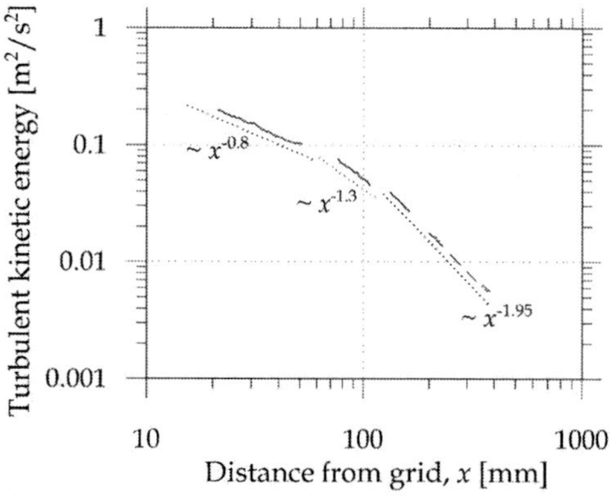

Fig. 4. Turbulent kinetic energy as a function of the distance from the grid for Reynolds number $3.1 \cdot 10^3$. The plotted value is averaged over the span-wise coordinate (y). There are displayed powers of the dotted lines just under the data (solid lines). Note the log-log plot.

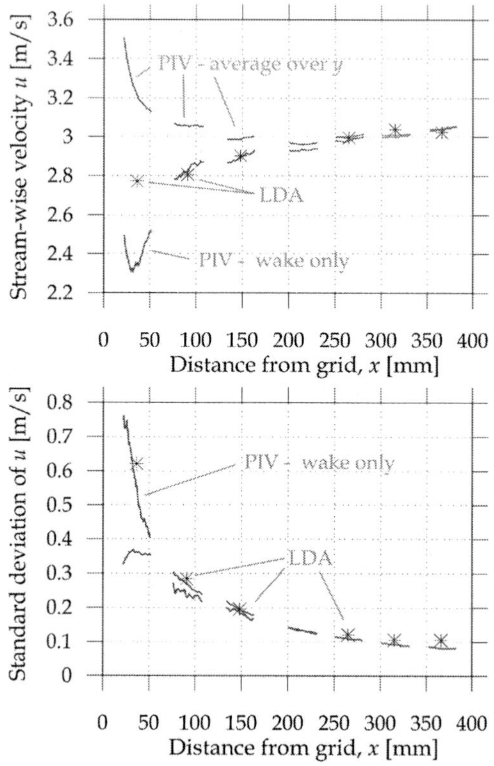

Fig. 5. Top: Stream-wise velocity u measured by using PIV (solid lines) and LDA (stars). The PIV data is processed by 2 ways: it is averaged over the span-wise direction (y) or there is selected only a single profile in the wake of the grid rod. **Bottom**: Standard deviation of the stream-wise velocity. This is plotted instead of TKE, which needs all three velocity component, while our LDA system measures only one component. Again, the PIV data are selected to the averaged ones and those from the wake only, which gives highest difference at locations close to the grid.

3.3 Intensity of turbulence

The turbulence produces at the grid and later on it dissipates according to a power law, which is shown in figure 4 for one velocity. The PIV offers two ways of defining the turbulence intensity, which is a quantity useful for describing the decay processes: the classical definition based on the temporal averaging we denote I_T and we calculate it as

$$I_T = \frac{(E - E_T)^{1/2}}{E_T^{1/2}} \quad (3)$$

, where $E = \frac{1}{2}\vec{u}^2$ is the average energy of instantaneous velocities, while E_T is the energy of temporal average velocity, $E_T = \frac{1}{2}\langle\vec{u}\rangle_T^2$. This turbulence intensity does not contain the spatial changes inside the average velocity field, such as the wakes or jet structures. Therefore we can use an alternative definition of spatial intensity of turbulence

I_P calculated similarly as

$$I_P = \frac{(E - E_P)^{1/2}}{E_P^{1/2}} \tag{4}$$

, where the subscript P plays for spatial averaging, i.e. averaging over the field of view. In principle, this quantity can be calculated by using a single instantaneous velocity field only. The effect is evident in figure 6, where close to the grid, the spatial intensity is larger due to the structure even in the averaged velocity field (the mentioned jets and wakes), while more far from the grid, the temporal one is grater due to the fluctuations larger than the investigated field of view as the integral length-scale grows.

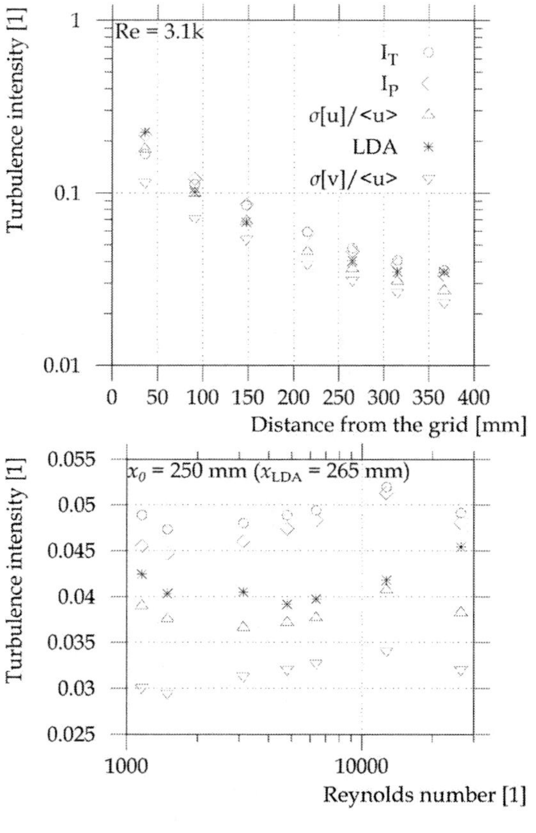

Fig. 6. Top: intensity of turbulence obtained by using different approaches as a function of distance from the grid. The Reynolds number is $3.1 \cdot 10^3$. **Bottom**: intensities of turbulence as a function of Reynolds number at constant mean distance from the grid $x_0 = 250\,\text{mm} = 16M$, $x_{\text{center}} = 265\,\text{mm} = 17M$. I_T and I_P reflects the temporal and spatial intensity calculated according to equations (3) and (4) respectively, The data obtained from LDA, which measures x-component only, can be compared only with the stream-wise turbulence intensity $\sigma[u]/\langle u\rangle$. For complexity, there is also plotted the span-wise component $\sigma[v]/\langle u\rangle$, which is normalized by stream-wise velocity, as the mean value of span-wise velocity is close to zero.

The comparison with LDA data can be done only for the corresponding quantity, which is the ratio of standard deviation of stream-wise velocity and its average value, as this is the component, which is measured. This quantity obtained from PIV data is displayed in figure 6 as well. As this contains only fluctuations in one direction, its value is naturally lower, than that containing both measured components (and if we would have measured the third component as well, the value might be even larger). On the other hand, it is evident, that, again as discussed above, the PIV slightly underestimates the fluctuations.

3.4 Length-scale dependent fluctuations

The idea of averaging over space instead of over time can be broaden by using different length scales of averaging than just the entire field of view. We can focus on spatial fluctuations in respect to different vicinity of the probed point. This can be done by convoluting the velocity field $\vec{u}(\vec{x})$ with band-pass filter [15], which consists of two Gaussian functions:

$$\vec{u}_{lh}(\vec{x}) = \vec{u}(\vec{x}) * \left(\frac{1}{2\pi\sigma_l^2} e^{-\frac{x^2}{2\sigma_l^2}} - \frac{1}{2\pi\sigma_h^2} e^{-\frac{x^2}{2\sigma_h^2}} \right) \tag{5}$$

The resulting velocity field $\vec{u}_{lh}(\vec{x})$ is smoother than the lower bound σ_l and rougher than the upper bound σ_h, thus it contains only fluctuations of specific length-size interval. The process is illustrated in figure 7.

When we have the velocity fields with highlighted different scale of fluctuations [16], we calculate the TKE of each ensemble getting the spatial distribution of each interval of length-scales [17]. Then we select three of this intervals and combine them into a single image. The red channel comes from the small length-scale, i.e. $\sigma_l^R = 0.5\,\text{mm}$ and $\sigma_h^R = 0.7\,\text{mm}$, the green channel does from the middle scales, i.e. $\sigma_l^G = 1.5\,\text{mm}$, $\sigma_h^G = 2.0\,\text{mm}$ and the largest scales are coded into blue channel, i.e. $\sigma_l^B = 3.9\,\text{mm}$ and $\sigma_h^B = 5.9\,\text{mm}$. The relative magnitude of colors is weighted according to the *thickness* of the interval and to the estimated Kolmogorov scaling (thus that the ideal homogeneous and isotropic turbulence would be displayed in gray-scale).

3.5 Spatial spectrum

By using this approach we can obtain the spatial spectrum as a plot of standard deviation of energies of a band-pass filtered velocity field in dependence on the effective wave vector of the band. Figure 8 shows the resulting spectra, first in dependence on the distance from the grid for constant velocity (i.e. *Re*), second, in dependence on velocity for constant distance from the grid.

The spectrum is already smoothed, which is in consistency with the general uncertainty principle: the particular localization in direct space is redeemed by delocalization in the Fourier space. Additionally, the available range is quite short: we are limited from one side by the size of the measuring grid, which is 0.5 mm, from the other side by the size of the largest σ_h, which is 5.9 mm. One can ask, why not by the size of field of view, which is around 5× larger. This is due to the fact, that the real size of the

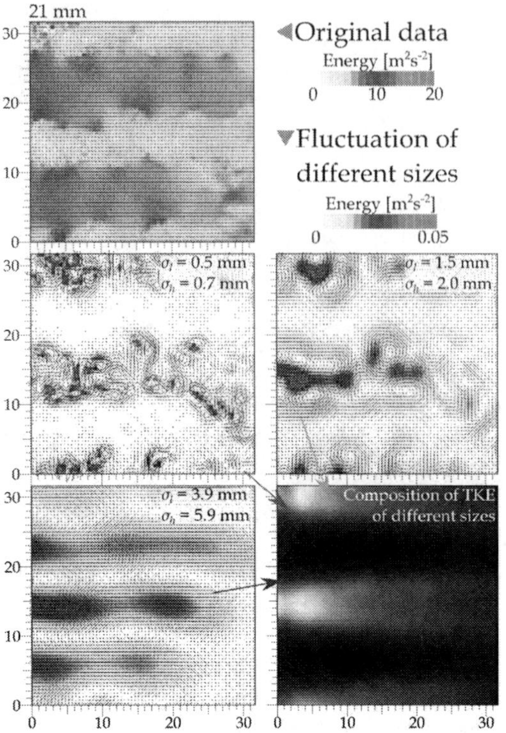

Fig. 7. Top left: example of the instantaneous velocity field colored by the energy (i.e. half of square of velocity magnitude). **Middle left**: the same data with highlighted fluctuations of size interval from $\sigma_l = 0.5$ to $\sigma_h = 0.7$ mm. **Middle right**: from $\sigma_l = 1.5$ to $\sigma_h = 2.0$ mm. **Bottom left**: from $\sigma_l = 3.9$ to $\sigma_h = 5.9$ mm. **Bottom right**: Length-scale dependent fluctuations, red colors correspond to smallest length-scales, green to middle scales and blue to largest scales. The Reynolds number is $3.1 \cdot 10^3$, the left side of the panels is 21 mm from the grid.

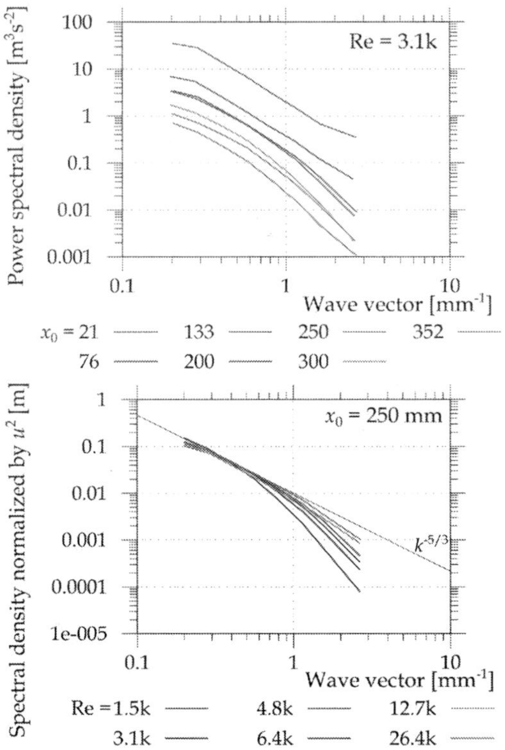

Fig. 8. Top: The power spectral density as a function of the distance from the grid (the x_0 means the leading side of the probed areas) at constant Reynolds number $Re = 3.1 \cdot 10^3$. **Bottom**: the power spectral density divided by square of velocity as a function of Reynolds number at constant distance from the gird $x_0 = 250\,\text{mm} = 16M$. For comparison there is plotted the famous Kolmogorov $-5/3$ law.

effective support of the band pass is much larger than its half-width, therefore the value in such a point is already influenced by too many points lying outside of the measured field of view. We want to minimize the boundary effect of the points outside, whose value is extrapolated by using the „Extend Edge Handling" approach (i.e. that the nearest border points are conceptually extended as far as necessary to provide values for the convolution. Corner points are extended in 90° wedges. Other edge points are extended in lines).

As the spectrum is spatial, it is questionable, if it makes sense close to the grid, where the flow contains strong structures even in the average, as are jets and wakes produced by the grid rods. The spatial spectrum naturally mix the data from both very different regions and thus its physical interpretation is not easy. On the other hand, the fact, that the spectrum close to grid is despite the just mentioned reason very similar to the others, suggests that the famous Kolmogorov law developed for homogeneous and isotropic turbulence has more mathematical than physical reasons, because it should *not* be valid in that case.

The dependence on Reynolds number (Figure 8 bottom) shows that at lower velocities, the spectrum falls down faster than the Kolmogorov scaling suggesting the vicinity of the Kolmogorov scale in that distance.

The power spectral density shown in figure 8 bottom is normalized by the square of the mean stream-wise velocity. Note, that the lines meet at the larger scales, which again suggest, that the energy normalization follows these larger scales, while the energy deficit (towards the ideal Kolmogorov profile) does not affect the overall mean velocity.

3.6 Structure functions of second order

The structure function [18] of second order of the stream-wise velocity component u is defined as

$$S^2_{uu}\left(\vec{x}, \vec{\xi}\right) = \frac{\left\langle \left(u\left(\vec{x}\right) - u\left(\vec{\xi}\right)\right)^2 \right\rangle_T}{\sigma_T\left[u\left(\vec{x}\right)\right] \cdot \sigma_T\left[u\left(\vec{\xi}\right)\right]} \tag{6}$$

where $\vec{x}$ denote the probed point, where the structure function is calculated, while $\vec{\xi}$ is the reference point; σ_T means

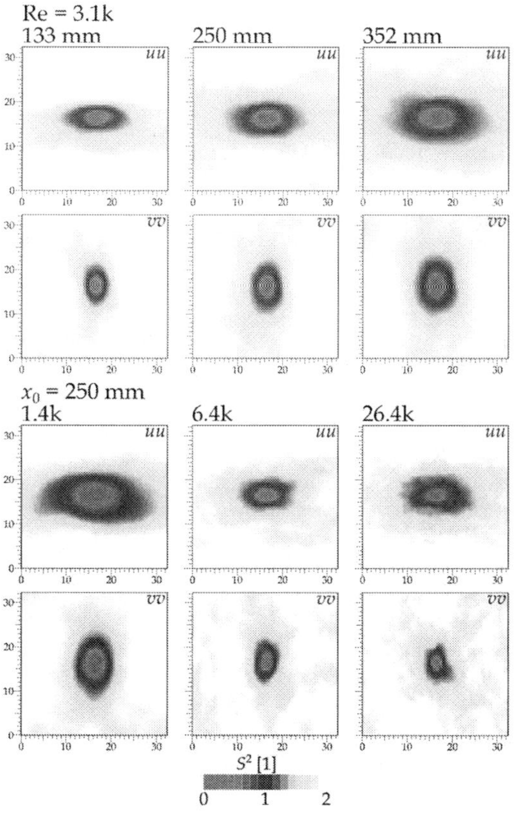

Fig. 9. Top: spatial distribution of the structure function around the point $\vec{\xi}$ in the center of the field of view in different distances from the grid. The first line shows the structure function of u, i.e. x-component of the velocity, second line does the v, which is the span-wise component. The velocity is constant corresponding to $Re = 3.1 \cdot 10^3$. **Bottom**: the distance from the grid is constant and equals to $x_0 = 250$ mm, different panels show different Reynolds numbers.

the standard deviation (2) in respect to time, i.e. ensemble in this case of relatively slow PIV. S^2_{uu} thus says, how much is the value of u at the point $\vec{x}$ statistically related to the value of u at point $\vec{\xi}$. When the values are same (which is identically fulfilled when $\vec{x} = \vec{\xi}$), the value of $S^2 = 0$, when the values are not correlated, the value of $S^2 = 2$.

As has been proven in [18], the S^2 is in close relation to the autocorrelation function C_{uu} as $S^2_{uu} = 2 - 2C_{uu}$, when the sample size is infinite. But, in respect to autocorrelation function, the structure function is not sensitive to the mean value and it converges better.

Analogically to the equation (6) we can define S^2_{vv} for the span-wise velocity component v:

$$S^2_{vv}\left(\vec{x}, \vec{\xi}\right) = \frac{\left\langle \left(v(\vec{x}) - v\left(\vec{\xi}\right)\right)^2 \right\rangle_T}{\sigma_T\left[v(\vec{x})\right] \cdot \sigma_T\left[v\left(\vec{\xi}\right)\right]} \tag{7}$$

These functions are autostructure functions as they probe the same quantity only in different locations. There can be defined also a crossstructure function S_{uv}, which we do not discuss in this article.

Spatial distribution of both functions is plotted in figure 9, first for different distances from the grid at constant velocity, second for different velocities at constant position. In all cases is the reference point $\vec{\xi}$ chosen in the center of the field of view.

In all panels of figure 9 we can see that the distribution is not axially symmetric, e.g. the difference between values of stream-wise velocity grows faster in span-wise direction than in the stream-wise one, which suggests that the turbulence is not isotropic even in this quite large distances from the grid. In addition, this anisotropy does not seems do decrease with distance from the grid.

The range of the similar points grows with increasing distance as can be estimated from the fact, that the turbulence decays from bottom (i.e. the Kolmogorov dissipative scale grows). This range decreases with increasing Reynolds number, as can be seen in the bottom half of figure 9.

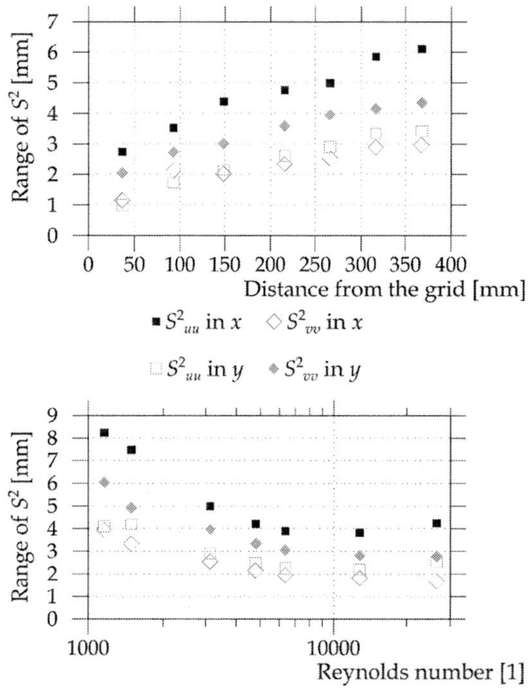

Fig. 10. Top: Range of structure function S^2. The velocity is constant corresponding to $Re = 3.1 \cdot 10^3$. **Bottom:** the distance from the grid is constant and equals to $x_0 = 250$ mm.

We try to quantify the range R of structure function as a distance of the reference point $\vec{\xi}$ to point, where $S^2 = 1$, in some important direction, which is the stream-wise or the span-wise direction. The result is plotted in figure 10.

Again, we see, that the range of structure function increases with distance from the grid and decreases with velocity. In the first dependence we can see an interesting *slowing of increase* around distance 200 mm; this effect is probably connected with the lost of averaged large-scale structure of alternating jets and wakes. On the other hand, the decrease with velocity is not fully uniform, as can be seen at the highest velocities in the bottom part of figure 10. At this moment, we doubt that this effect was physical; we prefer the hypothesis, that it is an error of PIV measurement technique connected with higher velocities. In the future we plan to perform a similar comparison with LDA data, as it has been done above for the turbulence intensity, but we are not yet familiar with the LDA processing software.

4 Conclusion

The grid-generated turbulence has been studied by using both, PIV and LDA techniques. We observe that both methods give similar results, mainly at lower velocities. At higher velocities the PIV technique underestimate the fluctuations, probably due to the finite size of the *interrogative area*, inside which the velocity is averaged.

The turbulent kinetic energy decays with distance as $x^{-1.95}$.

We used a new method of spatial turbulence spectra [17] and we have found that they follow the Kolmogorov law even in the vicinity of the grid, where the turbulence is not homogeneous neither isotropic. At smaller velocities, we observed a deviation from this law caused probably by the vicinity of the dissipative length-scale.

Structure functions show that the turbulence is surprisingly not isotropic as the shape of the structure function is deformed in the direction of the flow (it is prolate in the axis of the investigated velocity component). The range of structure function increases with distance and decreases with velocity, although there is an inexplicable increase at largest velocity, which might be an PIV artifact, but we lack the comparison with LDA to proof that at this moment.

Acknowledgements We thank to Václav Koudelka and Karsten Dörner from Dantec Dynamics for valuable technical help with the LDA system. This work was supported by the project of Technology Agency of the Czech Republic TAČR No TH02020057 „Program Epsilon".

References

[1] T. Kurian, J.H.M. Fransson, Fluid Dynamics Research **41**, 021403 (2009)

[2] M. Mohamed, J. LaRue, J. Fluid Mech. **219**, 195 (1990)

[3] P. Roach, International Journal of Heat and Fluid Flow **8**, 82 (1987)

[4] G. Comte-Bellot, S. Corrsin, J. Fluid Mech. **25**, 657 (1966)

[5] Z. Warhaft, J.L. Lumley, J. Fluid Mech. **88**, 659 (1978)

[6] W.Z. Grzelak J., TRANSACTIONS OF THE INSTITUTE OF FLUID-FLOW MACHINERY **130**, 93 (2015)

[7] S.B. Pope, *Turbulent flows* (Cambridge University Press, Cambridge, UK, 2000)

[8] C. Tropea, A. Yarin, J.F. Foss, *Springer Handbook of Experimental Fluid Mechanics* (Springer, Heidelberg, DE, 2007)

[9] D. Jašíková, M. Kotek, R. Horálek, J. Horčička, V. Kopecký, ILASS – Europe 2010, 23rd Annual Conference on Liquid Atomization and Spray Systems, Brno, Czech Republic, September 2010 **23** (2010)

[10] D. Duda, J. Bém, V. Yanovych, P. Pavlíček, V. Uruba, European Journal of Mechanics, B/Fluids **79**, 444 (2020)

[11] D. Duda, in *Vortex Dynamics – From Physical to Mathematical Aspects [Working Title]*, edited by I. Bakırtaş, N. Antar (IntechOpen, 2021)

[12] D. Duda, *Individual vortex searching algorithm*, in *In Proceedings Topical Problems of Fluid Mechanics 2020*, edited by D. Šimurda, T. Bodnár (2020), pp. 56–63

[13] D. Duda, V. Yanovych, V. Uruba, Processes **8**, 1355 (2020)

[14] D. Duda, V. Yanovych, V. Uruba, Energies **14**, 4712 (2021)

[15] A. Agrawal, Experiments in Fluids **39**, 836 (2005)

[16] D. Duda, V. Uruba, AIP conference proceedings **2000**, 020005 (2018)

[17] D. Duda, V. Uruba, ASME J of Nuclear Rad Sci. **5** (2019)

[18] E.O. Schulz-DuBois, I. Rehberg, Applied physics **24**, 323 (1981)

Wake flow past an ALUTEC profile

Daniel Duda[1,*], *Yasar* Dokak[1], *Vitalii* Yanovych[1], and *Václav* Uruba[1,2]

[1]Faculty of Mechanical Engineering, University of West Bohemia, Pilsen, Czech Republic
[2]Institute of Thermomechanics, Czech Academy of Sciences, Prague, Czech Republic

Abstract. Aluminum construction profiles become a standard construction material for short-term stands in science and in industry as well. It has the shape of square prism with an groove in each side to easy connecting the nut or bolt in any location. We studied the wake past the profile of 45 mm side. We studied the topology of the wake in the dependence on the angle of attack by using the Particle Image Velocimetry technique.

1 Introduction

Due to its properties and the development of mechanical, thermal and chemical processing technologies, the aluminum profile is increasingly used in various fields: in the aircraft industry, shipbuilding, car building, in the construction of subways, in the electronic industry, in the food industry, in mechanical engineering, in the automotive industry, in the electrical industry and instrument making, household appliances, etc.

Aluminum alloys possess a number of advantages, and that led to their wide application in various industries. In particular, this high mechanical strength at low density, resistance to weathering, durability, significant compared with competing materials, turnaround time, good machinability pressure and cutting, flexibility, high cold resistance, the possibility of recycling with little energy consumption etc.

Aluminum systems offer almost unlimited possibilities for architectural design of buildings and structures. In addition, the use of this metal in architecture and construction is advantageous in the process of operation, and due to its exceptional corrosion resistance, aluminum is valued as one of the most convenient raw materials for recycling. The wide use aluminum due to its high strength, durability and workability.

It is these characteristics that determine the broad prospects for the development of the use of aluminum profiles for 15-20 years in advance. The market for the consumption of aluminum and aluminum alloy profiles is currently one of the fastest growing in the world.

The raw materials for the production of profiles are technical aluminum and wrought aluminum alloys obtained by alloying primary aluminum or aluminum scrap with alloying additives. ALUTEC KK Structural Aluminum System (ALS) was introduced to the market 25 years ago, and currently includes more than 100 versions of aluminum profiles and more than 1500 accessories. A key component of the entire structural system is an aluminum profile with a longitudinal track.

The structural aluminum system offers an operational and functional solution for various industries. It is characterized by simplicity and speed of installation, as well as high adaptability to other structural systems. Any part of it can be connected to the rest without the use of special tools or the need for a strong (one-piece) connection, which is welding.

The aluminum constructional system allows the production of frames, single-purpose or serial machines, ensuring the safety of workstations and machines, workbenches, trolleys, transport systems, assembly lines, conveyors, manipulators and entire workstations from standard parts. The system can be used in industrial premises, provided compatibility with ESD. The profiles are made of Al Mg Si 0.5 F 25 material in accordance with DIN 3.3206.72, and processed by natural eloxification (thickness 15 μm). This type of surface treatment improves the functional properties of the base alloy, increases the corrosion resistance, hardness and wear resistance of the profile [1].

The advantages of the ALUTEC KK system are as follows: dismantling (reconfiguration), reuse of system elements; low system weight (1/3 of the weight compared to steel); high accuracy, durability and service life of parts; simplicity, fast and intuitive work with elements; application does not require expensive equipment; application in interiors and exteriors; adaptability to other solutions and new requirements.

2 Experimental setup

2.1 Wind tunnel

We used a new wind tunnel developed by the company WTtech. The radial compressor power is controlled automatically to keep a constant velocity, which is measured as

*e-mail: dudad@kke.zcu.cz

a pressure drop on a confuser just before the test-section. The projected maximum wind speed is 90 m/s, but it still a subject of testing and optimization.

2.2 The obstacle and its position

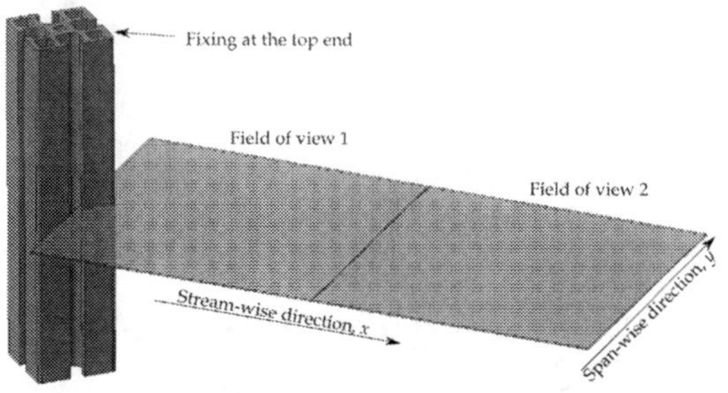

Fig. 1. Sketch of the positions of both fields of view relative to the obstacle. Obstacle dimensions: 45 × 45 × 200 mm, field of view dimensions: 164 × 164 mm, distance of FoVs from the axis of the obstacle: 17 mm and 180 mm.

The aerodynamic tunnel at which the study was conducted has a closed test section. The height and width of this camera are 305 × 203 mm respectively, and the length of the test section is 745 mm. The side walls and bottom of the test section are made in the form of a monoblock of extruded polycarbonate. Characterized by high optical purity. Whereas the upper wall is modular and allows to combine its parts according to the necessity of the experiment.

The ALUTEK KK profile, measuring 45 × 45 mm in length and 200 mm in length, was used for the experimental studies. A through mounting hole was used to mount this profile in the middle of the test section. In which the threaded connection M8 was made. A similar hole was made in the modular section to cover the test section. Then test specimen and part of the coating section were fastened by the M8 screw. This type of connection is convenient. Because when you change the angle of the profile in the middle of the section, you only need to turn the screw with the profile to a given angle. It should also be noted that as a result of mounting the profile, there is a remained a distance of 3 mm between the profile and the bottom of the test section in order to avoid damaging the bottom desk.

2.3 Particle Image Velocimetry technique

Particle Image Velocimetry technique (PIV) [2] is in its principle a very simple method: it is based on observing the motion of small particles carried by a transparent fluid. Then we use the assumption, that small particles follow the motion of the fluid as the viscous drag dominates their own inertia, buoyancy and other hydrodynamical forces connected with local flow accelerations. In this measurement, we used droplets of oil-glycerol solution as a tracer particles. This fog produced by commercially available fog generator Safex and the particle size is in order of microns. The use of this „theater fog" has been proven by the world-wide experience e.g. in [3–5] and many others.

These particles are illuminated in one plane of small thickness less than 1 mm. This plane of light is in our case produced by a laser beam defocused in one direction by using a cylindrical diverging lens. We use the solid state laser New Wave Solo of dominant wave-length 532 nm with pulse energy up to 500 mJ. The duration of a single pulse is 5 ns. In fact, there are two lasers in single box in order to produce a *pair* of laser shots, whose interval is dependent on velocity and we used 200 μs for the velocity mainly discussed in this article 5 m/s.

The illuminated particles are observed by using a camera and the figure is processed in computer by using an appropriate software. In the current measurement we used CCD camera MkII with spatial resolution 2048 pixels. This camera is able to distinguish both frames illuminated by the two laser shots. The objectives focus length 60 mm lead to the resolution of the grid of the velocity vectors of 2.6 mm at size of 32 pixels. We used the Dantec Dynamic Studio software for data analysis.

3 Observation

3.1 Averaged fields

The time-averaged data shows the pair of recirculation bubbles in the near-wake region as visible in the spatial distribution of the span-wise velocity shown in figure 2.

The back flow in the near-wake region is visible also in the stream-wise profile of the average stream-wise velocity displayed in figure 3 as a region with stream-wise velocity smaller than zero.

The fact, that the angle of attack changes, causes different crosssectional size of the obstacle: at $\alpha = 0°$, i.e. when the sides are perpendicular to the incoming flow, its cross-sectional size is just the side of the profile, i.e. $a = 45$ mm, at $\alpha = 45°$, the crosssectional size is $\sqrt{2} \cdot a = 63.6$ mm. This change can have stronger effect on the values of turbulent kinetic energy (displayed in figure 4) and on the recovery of the stream-wise velocity in the wake (displayed in figure 3bottom), and to the position of saddle point in the wake.

Similarly, we can discuss, if the definition of Reynolds number is correct, when it is based on the side of the obstacle, not on the crosssectional size. In that case, the imposed velocity should change in order to keep the Reynolds number constant. In the present study we keep the standards used in the literature and we use the size of obstacle side as the characteristic dimension in the Reynolds number definition, which causes, that the constant imposed velocity signifies constant Reynolds number. Physically, the most important length is the wake width [6], but that quantity is a kind of result, not control parameter.

3.2 Turbulent kinetic energy

Turbulent kinetic energy (TKE, K) is average kinetic energy associated with vortices and other coherent flow

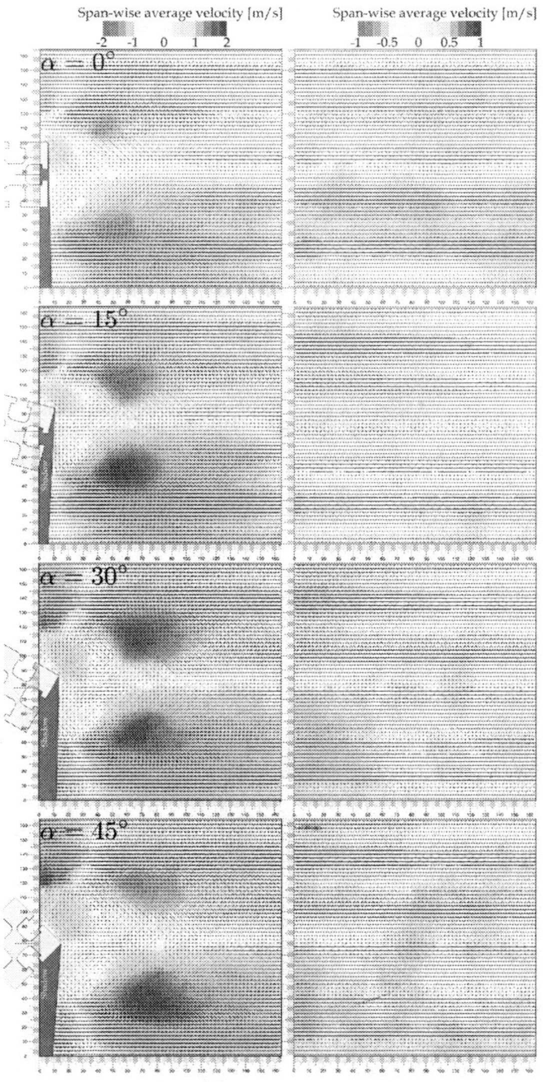

Fig. 2. Spatial distribution of the of the averaged span-wise velocity. Stream-wise velocity is 5.0 m/s, corresponding $Re = 1.5 \cdot 10^4$. The angle of attack α is from top 0°, 15°, 30° and 45° respectively. Note that the legend limits are different in left and right column.

structures, which the turbulence consists of. It is calculated as *variance* of velocity, i.e.

$$K(\vec{x}) = \frac{1}{2}\left(\left\langle u^2(\vec{x}) + v^2(\vec{x})\right\rangle_T - \langle u(\vec{x})\rangle_T^2 - \langle v(\vec{x})\rangle_T^2\right) \quad (1)$$

Where u and v denote the x- and y-component of the velocity vector, $\vec{x}$ is location and $\langle\cdot\rangle_T$ denote the averaging over time, i.e the ensemble averaging in this case of relatively „slow" PIV.

The flow is naturally three-dimensional, while we have measured only the planar cut, therefore we lack any information about the third velocity component w.

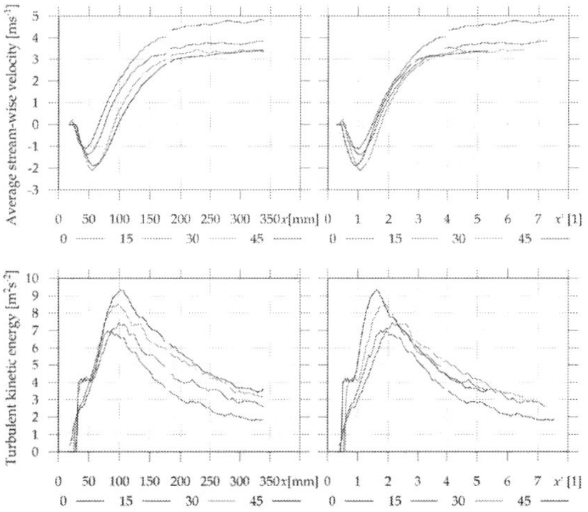

Fig. 3. Top: the development of of the average stream-wise velocity on the wake axis. The drop at $x = 180$ mm is caused by gluing the data from the two cameras. **Bottom**: the development of the turbulent kinetic energy along the wake axis. Color depicts the angle of attack: red for 0°, yellow for 15o, green for 30° and blue for 45°. **Left**: the distance x is in [mm] counted from the axis of the obstacle. **Right**: the stream-wise distance x' is normalized by the crosssectional size of the obstacle, i.e. $x' = (x/a) \cdot \cos\alpha$, where $a = 45$ mm is the side of the studied aluminum profile.

Equation (1) can be rewritten by using the energies of instantaneous velocity field E and of averaged velocity field E_T as $K = E - E_T$, where $E = \frac{1}{2}\left\langle \vec{u}^2\right\rangle_T$ and $E_T = \frac{1}{2}\langle\vec{u}\rangle_T^2$.

The turbulent kinetic energy is dominantly produced by the fluctuations of span-wise velocity v. The stream-wise one is dominant only in the narrow shear layer just on sides of the obstacle, while the fluctuations of the span-wise component v are massive in the wake axis, where the structure similar to „von-Kármán vortex street" splits the direction of span-wise velocity. This effect is displayed in figure 5 for single angle of attack $\alpha = 15°$, but for other angles this behavior is similar. This behavior is observed in the wake past airfoil [7–9] and even in grid turbulence [10].

3.3 Distribution of span-wise velocity

The swapping direction of the span-wise velocity component in the wake can be visible from the instantaneous velocity fields and from the probability density function of this velocity component in the wake (both is displayed in figure 6).

Probability density function of span-wise velocity component v is plotted in figure 6 (b) and (c) as a function of non-dimensionalized span-wise velocity $v' = (v - \langle v\rangle)/\sigma[v]$, where the averaging $\langle\cdot\rangle$ and standard deviation $\sigma[\cdot]$ is performed in respect to the data points counted in the histogram. The double peak observable in the entire

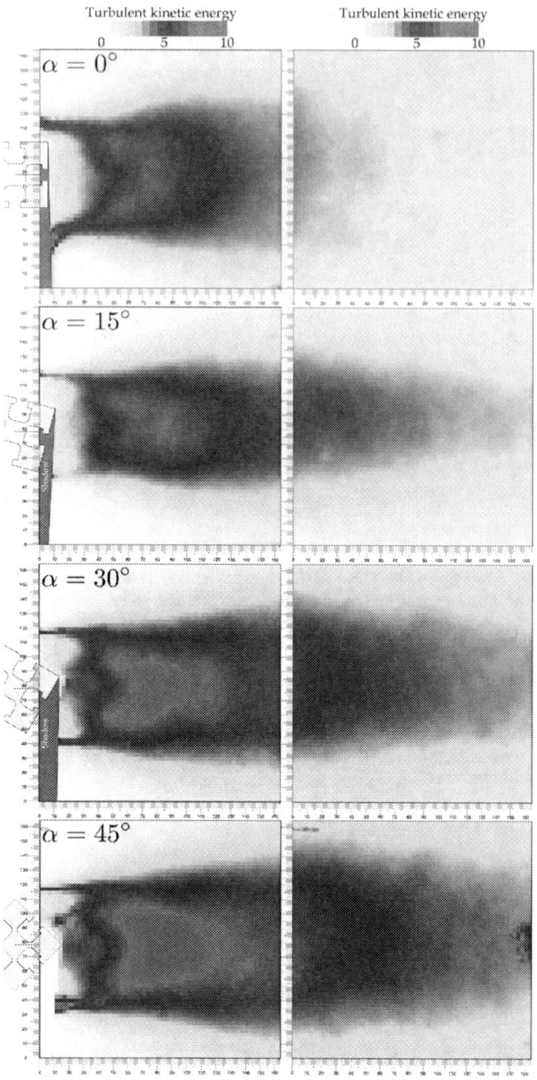

Fig. 4. Spatial distribution of the of the turbulent kinetic energy. Stream-wise velocity is 5.0 m/s, corresponding $Re = 1.5 \cdot 10^4$. The angle of attack α is from top 0°, 15°, 30° and 45° respectively. The data at 45° contain more noise due to measuring during the sunny days.

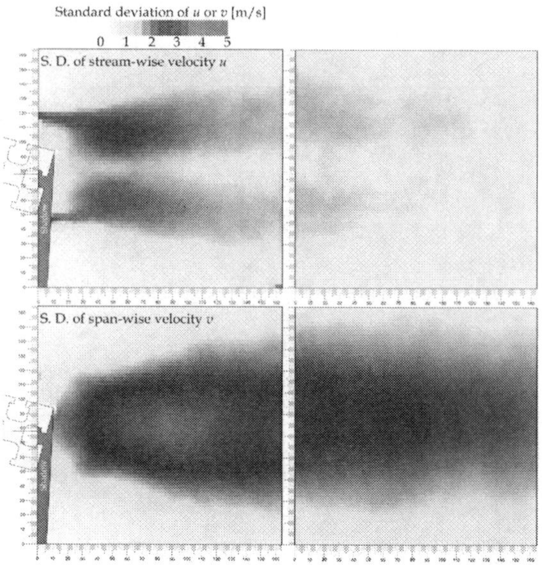

Fig. 5. Comparison of the standard deviations of stream-wise velocity u (**top**) and span-wise velocity v (**bottom**). Here are displayed only data at angle $\alpha = 15°$, other data have similar ratio of these two components.

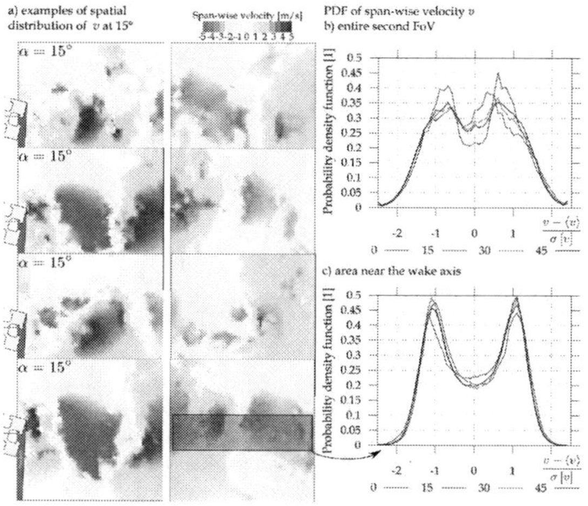

Fig. 6. Left: example of few instantaneous fields of span-wise velocity v at angle $\alpha = 15°$ and stream-wise velocity $u = 5$ m/s. **Right**: probability density functions of span-wise velocity in the entire second field of view (**b**) and in the limited area (**c**) near the wake axis. This area is depicted in the bottom image of the instantaneous velocity as a gray rectangle.

field of view clearly suggest the two dominant velocity directions. The fact, that this signal comes from the wake is even supported by selecting only that area for the processing, which is done in figure 6(c), and where a pair of symmetrical maxims is dominating the distribution.

Note that the maximum frequency of the PIV system is 7.4 Hz, which is very slow in comparison with the typical processes in the flow at the probed velocity (5 m/s), because during the time between the consecutive frame pairs (the frames are photographed in pairs delayed by few microseconds) the fluid moves by around 675 mm and thus it leaves not only the field of view, but also the tunnel test section. Therefore we cannot use the *slow* PIV data for determining the *Strouhal number* from the characteristic frequency of large vortices in the wake, as it is done e.g. in [11]

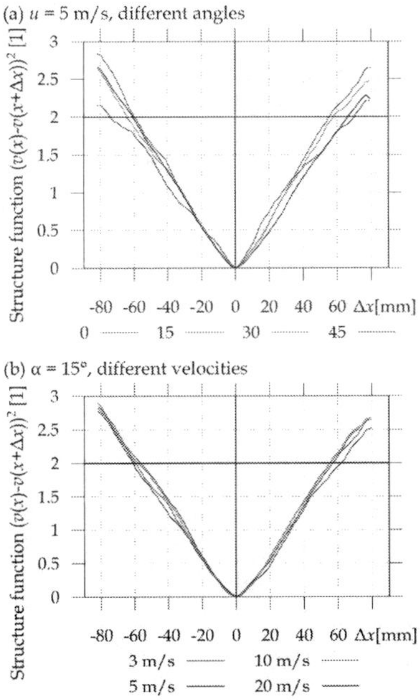

Fig. 7. Top: structure function S^2_{vv} of the span-wise velocity as a function of Δx, stream-wise velocity $u = 5\,\mathrm{m/s}$, different colors represent different angles of attack. **Bottom**: structure function S^2_{vv} for different velocities at constant angle of attack $\alpha = 15°$.

3.4 Structure function of span-wise velocity

The structure function [12] of second order of the span-wise velocity component v is defined as

$$S^2_{vv}(\vec{x}, \Delta\vec{x}) = \frac{\left\langle (v(\vec{x}) - v(\vec{x}+\Delta\vec{x}))^2 \right\rangle_T}{\sigma_T[v(\vec{x})] \cdot \sigma_T[v(\vec{x}+\Delta\vec{x})]} \quad (2)$$

where $\vec{x}$ denote the reference point, which has been chosen at the wake axis in the center of the second field of view, i.e. 270 mm i.e. $6 \cdot a$ from the obstacle axis. $\Delta\vec{x}$ is the relative vector from the reference point $\vec{x}$ to the probed point, where the structure function is calculated. σ_T means the standard deviation in respect to time, i.e. ensemble in this case of relatively slow PIV. S^2_{vv} thus says, how much is the value of v at the point $\vec{x}$ statistically different from the value of v at point $\vec{x} + \Delta\vec{x}$. When the values are same (which is identically fulfilled when $\Delta\vec{x} = 0$), the value of $S^2 = 0$, when the values are *not correlated*, the value of $S^2 = 2$.

This function is autostructure function as it probes the same quantity only in different locations. There can be defined also a crossstructure function S_{uv}, which relates the span-wise and the stream-wise velocity component; we do not discuss it in this article.

As has been proven in [12], the S^2 is in close relation to the autocorrelation function C_{vv} as $S^2_{vv} = 2 - 2C_{vv}$, when the sample size is infinite. But, in respect to autocorrelation function, the structure function is not sensitive to the mean value and it converges better.

Figure 7 shows the structure function S^2_{vv} in respect to the stream-wise displacement Δx (without vector, just displacement along the wake axis). We can see, that in all cases S^2_{vv} reaches the value of 2 (where it is considered to be not correlated) straightly continuing even up. This observation suggests, that the values are in phase shift by 90° instead of not correlated, and, if we would have larger field of view, we could probably observe a maximum at double the distance of reaching 2 and then drop down and another minimum at 4× this distance. But this are just speculations not supported by measured data. But, what is clearly observable is, that the structure function of span-wise velocity does not depend on the imposed stream-wise velocity – it is surprisingly similar. On the other hand, it changes little bit with angle of attack, but not in a systematic way.

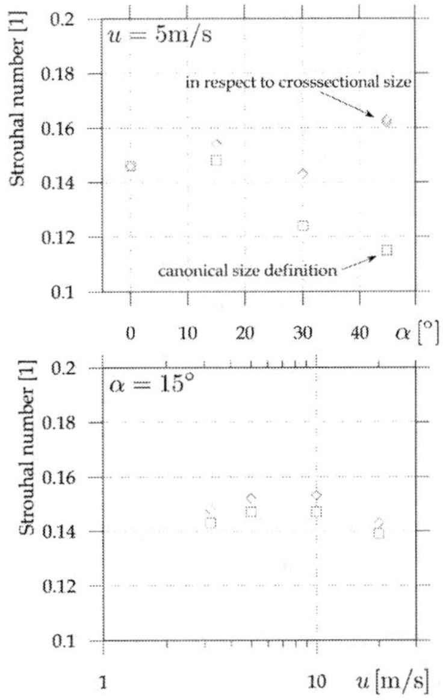

Fig. 8. Top: dependence of the Strouhal number Sr on the angle of attack α at constant imposed stream-wise velocity $u = 5\,\mathrm{m/s}$. **Bottom**: Strouhal number as a function of imposed stream-wise velocity u at constant angle of attack $\alpha = 15°$. Red squares represent the „standard“ definition with the obstacle size as the characteristic length-scale, while blue diamonds use the crosssectional size of the obstacle, i.e. $a' = a/\cos\alpha$, where $a = 45\,\mathrm{mm}$ is the side of the studied aluminum profile.

The distance, where the autostructure function reaches the value of 2, which we can consider to be one quarter of the spatial period of the „von-Kármán-like-structure“, is $R_+ = 58.5\,\mathrm{mm}$ and $R_- = 55.7\,\mathrm{mm}$, where the subscript + and − means the stream-wise direction or the counter-stream-wise one; it is not surprise that the range in the stream-wise direction is larger, as the convective average velocity grows along the wake axis. We can recalculate the spatial period to the time-period by using the stream-wise velocity in the wake axis and calculate the *Strouhal number* Sr

$$\mathrm{Sr} = f\frac{L}{U} = \frac{u_{\mathrm{wake}}}{2\,(R_+ + R_-)} \cdot \frac{a}{u} \qquad (3)$$

where u_{wake} is the convective stream-wise average velocity at the wake axis in the distance of the probed point, u is the imposed stream-wise velocity, a denotes the size of the aluminum profile side and R_+ and R_- are the just discussed distances, where autostructure function of span-wise velocity reaches the value 2, which is estimated to be in one quarter of the spatial period. Similarly to the discussion above, we can use a *crosssectional size* of the obstacle $a' = a/\cos\alpha$, where $a = 45$ mm is the side of the studied aluminum profile.

Figure 8 shows, that this estimation of Strouhal number reaches the classical values for cylinder of square crosssection only approximately, e.g. Dutta et al. [11] report the value around 0.14 with similar development with angle of attack – first an increase with maximum at 22.5° and then faster decrease in a such way, that $\mathrm{Sr}\,(\alpha = 30°) < \mathrm{Sr}\,(\alpha = 0°)$ and $\mathrm{Sr}\,(\alpha = 45°)$ is smallest. Chauhan et al. [13] report the value of 0.143 for $\alpha = 0°$.

4 Conclusion

By using Particle Image Velocimetry (PIV) technique we observed wake behind a aluminum construction profile of 45 mm side. We investigated the dependence of the wake on the angle of attack for four different angles $\alpha = 0°$, 15°, 30° and 45° at velocity 5 m/s corresponding to side-based Reynolds number $Re = 1.5 \cdot 10^4$.

We found that there is a pair of recirculation vortices in the averaged field of view. The average stream-wise velocity in the near wake region is negative and it returns towards the imposed value along the wake axis. We observed, that this return collapses better, when the distance is calculated in crosssectional sizes of obstacle. The turbulent kinetic energy profile creates a plateau just behind the obstacle for angles of attack 30° and 45° but not for 0° and 15°; we do not have theoretical explanation at this moment. We confirmed, that the turbulent kinetic energy is caused mainly by the fluctuations in span-wise velocity, which is caused by the periodic structure similar to the von Kármán vortex street in the wake of circular cylinders. Its existence causes the double peak shape of the probability density function (PDF) of the span-wise velocity in the wake area. The second order structure function grows almost linearly over the value of 2, which corresponds to not correlated values, but due to the periodic character we believe, that it means 90° phase shift rather than loss of correlation. Based on this assumption, we estimated the value of Strouhal number, which is a nondimensionalised frequency of a such pattern, and we approximately fit the values reported in literature for square cylinders.

In the future we plan to check our result based on artificial assumptions (i.e. that the value of structure function of 2 corresponds to one quarter of the spatial period, that the convective velocity is the velocity at the point probed by the structure function) by using Laser Doppler Anemometry technique (LDA), which is a method with very good time-resolution and thus it is suitable to, among others, direct measurement of Strouhal number.

Acknowledgements We thank to Ing. Buhumil Laštovka for valuable technical help. This work was supported by the project of Technology Agency of the Czech Republic TAČR No TH02020057 „Program Epsilon".

References

[1] V. Yanovych, D. Duda, V. Uruba, European Journal of Mechanics, B/Fluids **85**, 110 (2021)
[2] C. Tropea, A. Yarin, J.F. Foss, *Springer Handbook of Experimental Fluid Mechanics* (Springer, Heidelberg, DE, 2007)
[3] D. Jašíková, M. Kotek, R. Horálek, J. Horčička, V. Kopecký, ILASS – Europe 2010, 23rd Annual Conference on Liquid Atomization and Spray Systems, Brno, Czech Republic, September 2010 **23** (2010)
[4] V. Koschatzky, P.D. Moore, J. Westerweel, F. Scarano, B.J. Boersma, Experiments in Fluids **50**, 863 (2011)
[5] M. Hirota, H. Fujita, H. Yokosawa, H. Nakai, H. Itoh, International Journal of Heat and Fluid Flow **18**, 170 (1997)
[6] D. Duda, V. Yanovych, V. Uruba, Energies **14**, 4712 (2021)
[7] D. Duda, V. Yanovych, V. Tsymbalyuk, V. Uruba, Energies **15**, 1227 (2022)
[8] I. Solís-Gallego, A. Meana-Femández, J.M.F. Femández Oro, K.M.A. Argüelles Díaz, S. Velarde-Suárez, Journal of Applied Fluid Mechanics **10**, 1013 (2017)
[9] V. Yanovych, D. Duda, V. Uruba, P. Antoš, SN Appl. Sci. **3**, 885 (2021)
[10] D. Duda, V. Yanovych, V. Uruba, Processes **8**, 1355 (2020)
[11] S. Dutta, P.K. Panigrahi, K. Muralidhar, Journal of Engineering Mechanics **134**, 788 (2008)
[12] E.O. Schulz-DuBois, I. Rehberg, Applied physics **24**, 323 (1981)
[13] M.K. Chauhan, S. Dutta, B.K. Gandhi, Journal of Wind Engineering and Industrial Aerodynamics **184**, 342 (2019)

Advanced simulation of Stirling engine

Peter Ďurčanský[1,*], Richard Lenhard[1] and Jozef Jandačka[1]
[1]University of Žilina, Faculty of mechanical engineering, Department of energy technology, Univerzitna 1, 01026 Žilina, Slovak Republic

Abstract. Energy demand reaches high values with its negative impact on the environment, it presents a considerable danger. For environmental and economic reasons, the main strategic tasks in this area are associated with reducing and minimizing energy consumption. The possible solution is preceded in the development phase by the application of optimization of thermal processes using computer simulation of thermal processes of whole systems. The result is numerical model with sufficiently accurate information that help to verify the accuracy of the proposed procedures and decisions and thus contribute to the efficiency and rationalization of production. New technologies are needed to utilize and use the low potential heat, or other fuel than fossil fuels. Stirling engine, as engine with outer combustion is a possible solution.

1 Unconventional micro cogeneration unit

Today´s research priority is focused on more efficient transformation process of primary energy from renewable energy sources. There are many possibilities in low potential heat utilization from renewable energy sources and the absence of effective ways of transforming heat into electricity, thus efficient heat cycles in general. Currently there is no cost-effective energy system for converting heat of medium or low potential. Thermal cycle quality significantly affects the final energy conversion efficiency, it is currently at Organic Rankin Cycle for example about 17%, which is about 35% compared to the efficiency in the conventional Rankine cycle and a 44% efficiency in a supercritical cycle in a large facility. [1], [2] The theoretical thermodynamic cycle of Stirling engine varies considerably from the real one. This is due to imperfect isothermal conditions, heat leakage to the environment and other influences. For this reason, the efficiency of the ideal cycle is higher than the real cycle. They work on the principle of external combustion, so fuel combustion does not take place in the working cylinder. This allows, in contrast to conventional internal combustion engines, control the course of the combustion process, the related quality, which is reflected in the composition of pollutants released into the atmosphere. The air is compressed in the compressor, it passes through the heat exchanger under constant pressure and takes heat. Then the air expands adiabatically in the cylinder and makes a work. Some of this work is used to drive the compressor and the other part is converted to mechanical work with the help of the generator into electrical energy. On generating of heat energy may be used a wide range of fuel, because this is the external combustion engine and for this is very common the use of renewable sources. The fuel is combusted in a separate combustion chamber, and thermal energy is transformed with the aid of the heat exchanger to the working fluid. [3]

The proposed micro-cogeneration unit with Stirling engine consists of two heat sources, one serves as a cooler and the other as the heater. Both of them have different requirements for operation. The proposal of measurement system is in Fig. n. 1.

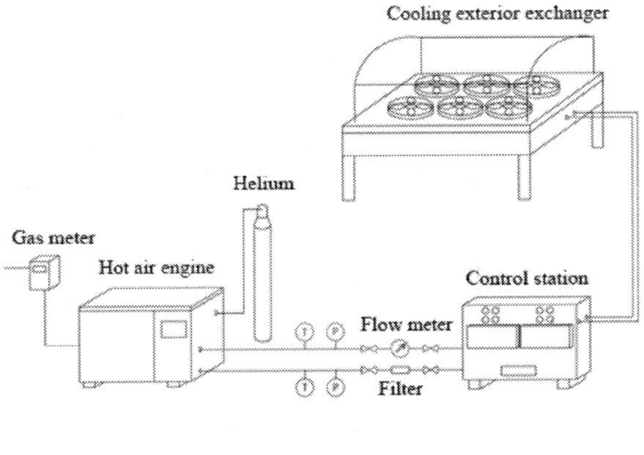

Fig. 1 Measurement setup

2 Numerical simulations

The thermal efficiency of the ideal Stirling cycle is just the same as the thermal efficiency of the Carnot cycle with the same inlet and outlet temperatures. For an ideal

* Corresponding author: peter.durcansky@fstroj.uniza.sk

regenerator, the heat removed during cooling period is absorbed by this regenerator and transferred to the working medium within compression – heating period. Computation of Stirling engine cycle was later generalized by Reitlinger, where he replaced the isotherms with polytropes. Significant clarification of the Stirling engine's analytical solution is the regenerative Rallis Circulation. It consists of two isotherms and two regenerative processes, which are carried out in part at constant pressure, in part at a constant volume. In the case of the expansion of gas compression in the working space of a heat engine, which are considered to be isotherms in the Stirling Circulation, they are in fact polytropes with an exponent approaching, in particular, the higher frequencies of the pistons of the motor of the exponent adiabatic engine. Rallis therefore replaced the isotherms with adiabatic changes in their final form. For the assessment of the potential electrical power of the cogeneration unit, a performance characteristic can be constructed from the inlet temperature on the heater and the average working pressure of the medium in the engine. Theoretical electrical power can also be calculated for temperature differences of heater and cooler. These simple methods are based on an ideal lossless model with experimentally determined efficiency and power correction factors applied, which represent generalized effects of various energy losses. These less accurate methods serve primarily as a starting point for estimating engine performance, suitability for use in a given application, and the relationship between engine size and power. However, they do not give a more specific view and their results are relatively optimistic. The first-order method is the frequently used Schmidt analysis. A dimensionless Beale or West number, as in Fig. 2, is most often used for preliminary performance determination.

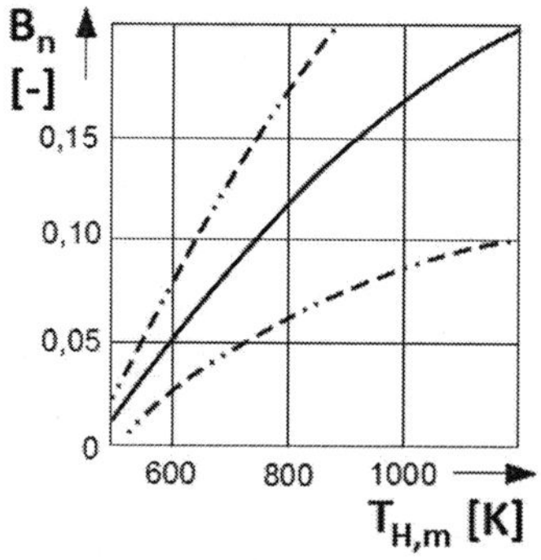

Fig. 2 Beale number according to mean temperature

The Beale number is determined as a function of the mean working gas temperature of the hot side of the engine. The upper dotted line corresponds to the engines with a smaller self-capacity and higher power and the lower ones with a larger dead volume and lower power. The assumptions under which the relationships characterizing the circulation of Stirling, Reitlinger and Rallis were derived are a great idealization of reality. Their practical use is only possible in the case of elementary preliminary calculations. A much more accurate expression of the events actually taking place is the Schmidt cycle, whose relations were derived in 1871. Modern Stirling engine is a clean device operating with considerable efficiency, which is related to the fact that heat, as an energy source, is extracted from the outside of the engine and transferred to the working medium of the engine via a heat exchanger. External combustion, which takes place under almost ideal conditions (sufficient oxygen, low time, theoretically any fuel, etc.) is not burdened with the production of undesirable emissions, especially nitrogen oxides. In connection with ecology, other sustainable sources for a modern engine, such as solar or geothermal energy, should be mentioned. Other aspects such as extra silent engine operation, vibration-free operation, or reverse operation capability make it an interesting engine for modern society.

At present, the Stirling engine is undergoing breakthroughs in several areas, particularly as a submarine drive. Last but not least, as a heat source in households, energy sources in space and solar power plants. Even the production of passenger cars is not completely closed for the Stirling engine. Other applications may be power sources for generators, on-board heaters and aircraft propulsion. Powered by a Stirling engine, the yacht's electricity generator is a practical combination of its smooth and silent operation and the disadvantage of cooling, which should certainly not be a problem on a boat surrounded by a lot of water. Also, driving a low-vibration silent engine is not a bad idea, moreover, any aircraft engine used up to now loses power to an increasing altitude. The sealed Stirling engine is not dependent on ambient air density, moreover, as the temperature decreases, the temperature difference between the hot and cold parts, and thus its performance, would increase. The main purpose of this publication is to show possible accuracy of numerical simulation in regard to real measurements.

3 Basic setup

Simple analysis of an ideal Stirling engine model, where the compression and expansion spaces are maintained at some designed temperatures led to the paradoxical situation that neither the heater nor the cooler produce any net heat transfer during the cycle and it is therefore redundant. All the required heat transfer took place across the boundaries of the isothermal work spaces, which is not correct because the cylinder walls are not designed to transfer heat. In real engines, the workspaces will act as adiabatic rather than isothermal, which means that the heat transferred during the cycle must be provided by heat exchangers.

An alternative adiabatic model for Stirling cycle engine, based on the previous design (Urieli, 1984) assumes the following conditions:

- volume variations are sinusoidal,
- working fluid has ideal gas behaviour,
- the temperatures on the wall of the heater and cooler remain constant,

- compression and expansion spaces are adiabatic,
- cyclic steady state equilibrium is reached,
- the pressure drop along the engine is initially neglected,
- heat and mechanical losses are initially neglected,
- no gas leakage; therefore, the total mass of the gas in the system is constant.

Depending on the model, the engine is theoretically divided into five characteristic control volumes: expansion chamber (E), heater (H), compression space (K), regenerator (R), and cooler (C). These are represented by control volumes, shown in the Fig. 3 along with the appropriate variables describing the system - volume, weight, temperature, pressure and mass flow.

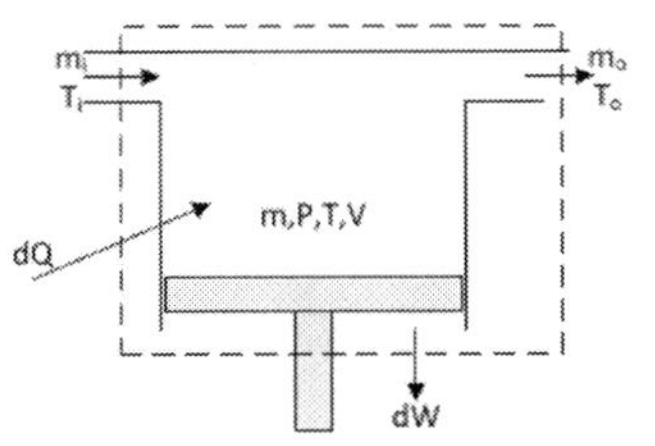

Fig. 3 Cylinder control volume

4 Energy balance

The advanced adiabatic model, worked out according to [3], is based on the basic equations of mass and energy equilibrium, with a state equation applied to each control volume. Control volume represents every part of the engine, the expansion chamber (E), heater (H), compression space (K), regenerator (R), and cooler (C).
By applying the mass balance equation to the control volume, the mass change according to the piston crank position (φ) can be expressed as in [3]:

$$\dot{m}_i - \dot{m}_o = \frac{dm}{d\varphi} \tag{1}$$

By including mass flow into the energy equilibrium balance with neglected kinetic components, whole system energy can be expressed as:

$$\frac{dQ}{d\varphi} + c_{pi}T_i\dot{m}_i - c_{po}T_o\dot{m}_o = \frac{dW}{d\varphi} + c_v\frac{d(mT)}{d\varphi} \tag{2}$$

The evaluation of the pressure change along the working space of the engine shall be calculated taking into account the mass balance applied to the whole engine, where the whole mass is not changing:

$$M = m_c + m_k + m_r + m_h + m_e \tag{3}$$

The working gas mass is defined by ideal isothermal Schmidt analysis. [3] This analysis is taken into account as first step, where these assumptions make it possible to obtain the following analytical expressions:

$$M = \int_0^{2\pi} \frac{p}{r}\left(\frac{V_c}{T_k} + \frac{V_k}{T_k} + \frac{V_r}{T_r} + \frac{V_h}{T_h} + \frac{V_e}{T_h}\right) d\varphi \tag{4}$$

$$M = \frac{p_m(s\sqrt{1-b^2})}{r} \tag{5}$$

With parameters s and b defined as:

$$b = \frac{c}{s} \tag{6}$$

$$c = \frac{1}{2}\sqrt{\left(\frac{V_{swe}}{T_h}\right)^2 + 2\frac{V_{swe}}{T_h}\frac{V_{swc}}{T_k}\cos(\alpha_f) + \left(\frac{V_{swc}}{T_k}\right)^2} \tag{7}$$

$$s = \frac{V_{swc}}{2T_k} + \frac{V_{clc}}{T_k} + \frac{V_k}{T_k} + \frac{V_r \ln\frac{T_h}{T_k}}{T_h - T_k} + \frac{V_h}{T_h} + \frac{V_{swe}}{2T_h} + \frac{V_{cle}}{T_e} \tag{8}$$

The non-isothermal mass volumes for the compression and expansion cylinder should be described in differential form:

$$\frac{dm_c}{d\varphi} = \frac{p\left(\frac{\partial V_c}{\partial\varphi}\right) + \frac{V_c\left(\frac{\partial p}{\partial\varphi}\right)}{\kappa}}{rT_{ck}} \tag{9}$$

$$\frac{dm_e}{d\varphi} = \frac{p\left(\frac{\partial V_e}{\partial\varphi}\right) + \frac{V_e\left(\frac{\partial p}{\partial\varphi}\right)}{\kappa}}{rT_{he}} \tag{10}$$

The actual values of the temperatures in the compression and expansion cylinder during the cycle are calculated using the differential form of the working gas state equation:

$$\frac{dT_c}{d\varphi} = T_c\left(\frac{\frac{\partial p}{\partial\varphi}}{p} + \frac{\frac{\partial V_c}{\partial\varphi}}{V_c} - \frac{\frac{\partial m_c}{\partial\varphi}}{m_c}\right) \tag{11}$$

$$\frac{dT_e}{d\varphi} = T_e\left(\frac{\frac{\partial p}{\partial\varphi}}{p} + \frac{\frac{\partial V_e}{\partial\varphi}}{V_e} - \frac{\frac{\partial m_e}{\partial\varphi}}{m_e}\right) \tag{12}$$

By applying the energy equilibrium equation to each heat exchanging volume, considering that it is an isothermal space, we can according to [3], write the following expressions:

$$\frac{dQ_k}{d\varphi} = \frac{V_k\left(\frac{\partial p}{\partial\varphi}\right)c_v}{r} - c_p(T_{ck}m_{ck} - T_{kr}m_{kr}) \tag{13}$$

$$\frac{dQ_r}{d\varphi} = \frac{V_r\left(\frac{\partial p}{\partial\varphi}\right)c_v}{r} - c_p(T_{kr}m_{kr} - T_{rh}m_{rh}) \tag{14}$$

$$\frac{dQ_h}{d\varphi} = \frac{V_h\left(\frac{\partial p}{\partial\varphi}\right)c_v}{r} - c_p(T_{rh}m_{rh} - T_{he}m_{he}) \tag{15}$$

Although the ideal adiabatic model is independent of the operating frequency, in order to evaluate power and other time-dependent effects, such as conduction heat loss in the regenerator, it must be defined. The system is designed as a quasi-static flow, i.e. mass flows are constant for each integration interval. For the calculation it is necessary to know at least the initial values of working temperatures, which are the result of adiabatic processes and enthalpy flow. The only clue to their correct selection is that their end-of-cycle values should be equal to their respective values at the start of the cycle. [3], [4] Due to the cyclical nature of the system, the problem is solved by entering the initial conditions and integrating them over several cycles. [4], [6] The best measure of the convergence to cyclic equilibrium is the residual amount of regenerative heat or at the end of the cycle, which should be zero. [7], [8] The fourth order Runge-Kutta method is used to solve the system of differential equations with initial conditions.

5 Results and conclusions

The measurement was carried out at different engine operating pressure settings using natural gas as fuel. The measured values were the inlet and outlet temperature from the heat exchanger, which is a part of the boiler. To measure them, wells were created on the inlet and outlet pipes, in which the sensors were placed. Temperature measurements were made with resistance thermometers, magnetic flowmeter and pressure gauges. The internal engine parameters such as the temperature of the radiator, heater, combustion chamber, and helium working pressure are recorded by the engine control panel. These values are recorded to the computer. The cogeneration unit was connected to a heat exchanger control station, which cooled the excess heat energy through an external cooler of 50 kW. The results are in Fig. 4.

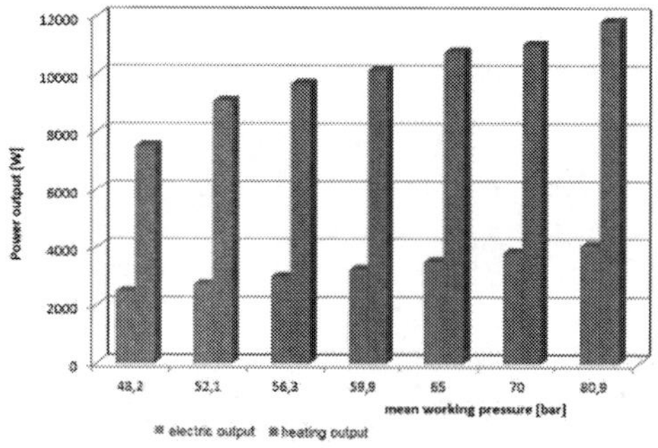

Fig. 4 Real power output

A graph was constructed to compare the measured values of electrical power with the values obtained by the calculation model. The results are shown in Fig. 5. It can be observed that model power values differ only slightly from those measured, with the largest deviation of 425 W recorded at an operating pressure of 80.9 bar. The graph shows that model power at 56.3 bar is less than measured. This is due to a decrease in the instantaneous temperature of the heater head when the highest measured power is reached. The instantaneous head temperature at the measured power is set into the model, which translates the momentary drop into the total power

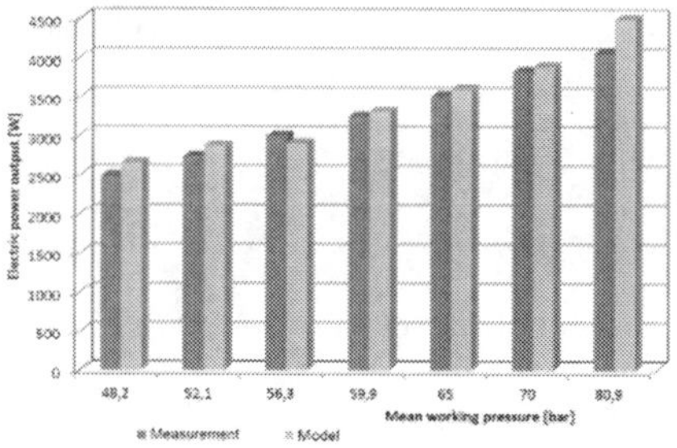

Fig. 5 Comparison between model and measurement

The experimental equipment for determining the operational and performance parameters of the individual parts of the system was designed and implemented under laboratory conditions. Comparison of the measured parameters with the corresponding values obtained from the model for the same operating conditions showed small deviations and correctness of the design.

Acknowledgments
Work on article has been financially supported by the project VEGA-1/0738 / 18 „Optimization of energy inputs for the rapid generation of natural gas and biomethane hydrates for the accumulation of high potential primary energy ", and the project VEGA-1/0479/19 „Influence of combustion conditions on production of solid pollutants in small heat sources " .The project is co-financed by EU funds, Project title: "Research on new methods of heat conversion from RES to electricity using new progressive thermal cycles" ITMS 26220220117.

References

1 Creyx M., "Energetic optimization of the performances of a hot air engine for micro-CHP systems working with a Joule or an Ericsson cycle" in Elsevier, France, 2012.

2 Darlington, N. – Strong, K. 2005. Stirling and hot air engines. Ramsbury: The Crowood Press Ltd, 2005. 236 s. ISBN 978-1-86126-688-0

3 Urieli, I. - Berchowitz, D. M. 1984. Stirling cycle engine analysis. Bristol: Hilger

4 NEMEC, P., HUŽVÁR, J. Proposal of heat exchanger in micro cogeneration unit, configuration with biomass combustion, Materials science and technology,Žilina, Slovakia, 2011

5 R. Lenhard, M. Malcho, "Numerical simulation device for the transport of geothermal heat with forced circulation of media" in Mathematical and Computer Modelling, 2013, vol. 57, iss. 1-2, p. 111-125.

https://doi.org/10.1051/epjconf/202226901012

6. Organ, A. J. Stirling Cycle Engines. London: John Wiley & Sons, Ltd., 2014. 272 s. ISBN 978-1-118-81843-5
7. Invernizzi, C. M. 2013. Closed Power Cycles. London: Springer Verlag, 2013. 264 s. ISBN 978-1-4471-5140-1
8. Karabulut, H. A. Torque and power characteristics of a helium charged Stirling engine with a lever controlled displacer driving mechanism. In: Renewable Energy. 2010, 35(1)

https://doi.org/10.1051/epjconf/202226901013

A Preliminary Study On The Inclination And Shape Of Jets Issuing From An Orifice In The Lateral Wall Of A Pipe

Simone Ferrari[1,*], *Luca* Salvadori[1], *Michela* Garau[1], and *Giorgio* Querzoli[1]

[1]DICAAR – Dip. Ingegneria Civile, Ambientale e Architettura, University of Cagliari, via Marengo 2, 09123 Cagliari, Italy

Abstract. The topic of the behaviour of jets issuing from an orifice in the lateral wall of a pipe is of interest in many engineering fields, such as sea discharges of pollutants via submerged pipes, diffusers to mix different fluids in tanks and reactors, leakage in industrial and civil pipelines, etc. Most of the previous researches focuses on the pressure loss across the orifice, without taking into account the modification of the internal flow and, most important, the direction and shape of the jet issuing from the outlet: as a matter of fact, this is usually considered perpendicular to the axis of the pipe and axisymmetric. We have investigated these issues in the laboratory via Digital Image Analysis techniques, namely LIF (Laser Induced Fluorescence) to measure concentration fields and FTV (Feature Tracking Velocimetry) to measure velocity fields, varying the ratio of the flow rate of the jet to the flow rate in the pipe before the orifice. Here we present some preliminary results, highlighting how the jet is not always perpendicular to the axis of the pipe nor axisymmetric: the inclination increases and the axisymmetry decreases when the ratio of the flow rates above described decreases.

1 Introduction

A jet issuing from the lateral wall of a pipe where a fluid is flowing can be used to discharge a pollutant in a fluid receiving body, to mix a fluid into another fluid in a tank, a reactor or a reservoir, or can be the unwanted consequence of a failure of the wall material, with a consequent fracture and leak.

The first case is the one, for instance, of marine outfall via submerged pipes, where a common configuration for the diffusers is the use of orifices in the lateral wall of a submerged outfall pipe (see, for instance, Missimer et al. 2015 [1], Ferrari and Querzoli, 2010 [2], Avanzini 2006 [3], Wood 1993 [4]). The topic of the behavior of the jet issuing from an orifice in the lateral wall of the outfall pipe is even more relevant when multi-port diffusers are employed, as the jets are usually considered as perpendicular to the pipe axis and axisymmetric (Wue et al., 2014 [5], Al-Ghamdi 2010 [6]). The second case is the one, for instance, of the diffusers adopted to mix different fluids in a tanks or reactors. One example is can be find in Photobioreactors (PBRs) and chemical reactors (in order to drive completely developed chemical reactions or to realize industrial products, Anderson et al., 2014 [7]). Another example is the diffuser optimization to reduce the thermal stratification in the tanks used for thermal energy storage (TES), in order to reduce costs and energy consumption in the air conditioning systems of large buildings (Osman et al. 2008 [8]).The third case, the accidental leak jets generating from a fracture in the wall of a pipeline (as a leak from a fracture in a pipeline can be considered as a leak from an orifice, van Zyl, 2014 [9]), has received a particular attention during the last years. In fact, the new scenarios related to the availability and cost of water, oil and gas and of their transportation have forced to pay more attention to the efficiency of the distribution systems and, in particular, to the leakage control and detection (Soleimani nia et al., 2019 [10], Ferrante et al., 2013) [11]). Moreover, a high-pressure non-controlled jet can be dangerous for the safety of people working or living in the vicinity of industrial plants and pipelines (Mohsin et al., 2015 [12]).

Most of the previous researches focus on the pressure drop across the orifice or fracture in the flow direction, via numerical simulations (Cassa et al, 2010 [13]), laboratory experiments (van Zyl and Clayton, 2007 [14]) or theoretical studies (van Zyl, 2014 [9]). Other Authors investigated the jet behaviour outside the pipe of reacting or flammable fluids, such as the release of hydrogen or helium into the air, with the buoyancy (i.e. the difference in density) and the chemical reactions playing a relevant role in the behaviour of the jet (Soleimani nia et al., 2018 [15] and 2019 [10]). Anyway, they concluded that the conventional round jet assumptions are not adequate to predict the behavior of turbulent buoyant jets issuing from a slot on the lateral wall of a pipe, in particular regarding the direction and axisymmetry of the jets.

Here we focus on the behavior (in particular, shape and direction) of non-buoyant or simple jets, issuing from a sharp-edged orifice on the lateral wall of pipe, released into a fluid with the same density. The employment of a sharp-edged orifice derives from the fact that previous investigations on simple jets have measured that the mixing efficiency is higher in sharp-edged orifice jets than

* Corresponding author: ferraris@unica.it

in jets released in other ways, such from contoured nozzles or at the end of a pipe (Hashiehbaf and Romano (2013) [16], Mi et al. 2001 [17]). Moreover, as stated in Mi et al., 2007 [18], this configuration is widely used in many engineering field.

We have performed some experimental campaigns in the Hydraulics Laboratory of the DICAAR – University f Cagliari (Italy), measuring via Digital Image Analysis techniques both concentration and velocity fields, inside and outside the pipe in the region close to the orifice. In the present work, we will show some preliminary results regarding the concentration measurements.

2 Physical parameters

A simple jet is the release, with a non-negligible initial momentum, of a fluid into a fluid with the same density. A buoyant jet is a jet released into a fluid with a different density. Therefore, if a fluid jet is released into the air, like in the case of oil or water pipelines, a buoyant jet will develop, whilst if a water jet is released into water, a simple jet will develop, like in the present case.

Regarding the shape, a jet can be usually classified as round or axisymmetric, when the orifice it is released from is circular; as two-dimensional, when the orifice is a line; as three-dimensional, when the orifice has a non-circular shape (Hashiehbaf and Romano, 2013 [16]). In the present case, the jet is released from a sharp-edged round orifice.

The key non-dimensional number governing the flow regime is the ratio of the inertial force to the viscous force, the Reynolds number *Re*: if *Re* is larger than a critical value, the flow is turbulent. In the present case, there are three relevant *Re*: a jet Reynolds number Re_J, a pipe Reynolds number Re_P and a backward step Reynolds number Re_h.

The jet Reynolds number is defined as:

$$Re_J = \frac{U_J D_J}{\upsilon} \quad (1)$$

where U_J is the jet mean velocity at the orifice, D_J the diameter of the orifice (0.08 m) and υ the kinematic viscosity of the released fluid (tap water). The critical value of Re_J has been measured as 500.

The pipe Reynolds number is defined as:

$$Re_P = \frac{U_P D_P}{\upsilon} \quad (2)$$

where U_P is the mean velocity in the pipe before the orifice, D_P the internal diameter of the pipe (0.064 m).

The backward step Reynolds number is defined as:

$$Re_h = \frac{U_h h}{\upsilon} \quad (3)$$

where U_h is the mean velocity in the pipe before the divergent reach and h is the step height (0.019 m).

The distance of flow reattachment after the backward step is X_d.

In the case of buoyant jets, another relevant non-dimensional number is the Richardson number *Ri*, that is the ratio of gravitational force to inertial force (the opposite is called the densimetric Froude number). In the present case, *Ri* is null, as the jet and the receiving body have the same density.

The flow rate in the pipe before the orifice (Figure 1) is Q_1, the flow rate after the orifice Q_2, the flow rate of the jet through the orifice is Q_3, and the ratio of the flow rate of the jet to the flow rate before the orifice is r:

$$r = \frac{Q_3}{Q_1} \quad (4)$$

The angle between the jet axis and the line perpendicular to the pipe axis is θ.

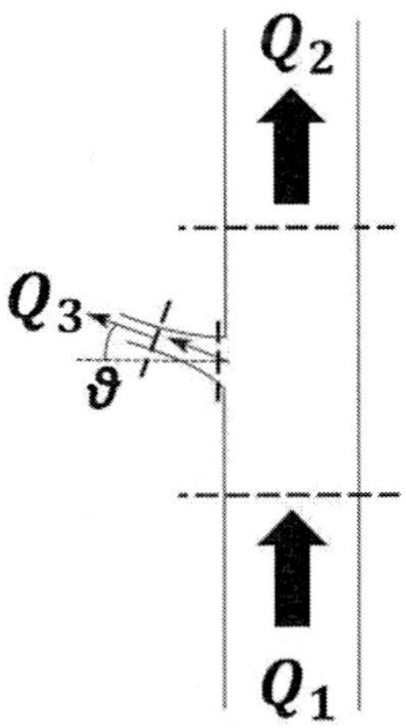

Fig. 1. Definition of the three flow rates in the pipe: Q_1 is the flow rate in the pipe before the orifice, Q_2 the flow rate in the pipe after the orifice, Q_3 the flow rate of the jet.

3 Experimental set-up and data acquisition and elaboration

The jet is released from a sharp-edged round orifice (with a diameter D_J of 8.0 mm) on the lateral wall of a pipe of transparent Perspex (with an internal diameter D_P of 64.0 mm) with a length of 600.0 mm and with a divergent reach at its beginning and a convergent reach at its end (Figure 2).

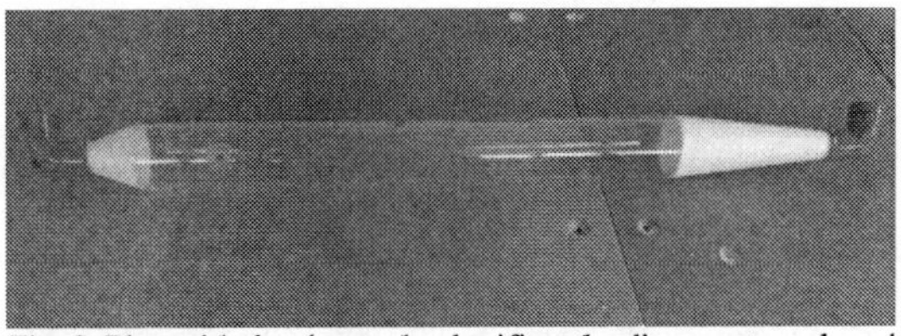

Fig. 2. Pipe with the sharp edged orifice, the divergent reach and the convergent reach; flow from right to left.

Divergent and convergent reaches are needed to connect the Perspex pipe to a rubber pipe (with a diameter of 25.4 mm) that, in turns, are connected to a constant-head closed-loop hydraulic circuit. The divergent has a length of 185.0 mm and its wall has an inclination of around 6°, in order to minimize the flow separation. As a matter of fact, following Westphal et al (1984) [19], the divergent reach is long enough to allow the flow reattachment with Re_h higher than $1\cdot10^3$, that is with velocity in the pipe U_h higher than 0.05 ms^{-1} (Figure 3). In the experiments presented hereafter, U_h was around 0.1 ms^{-1} and U_P was around 0.1 ms^{-1}, corresponding to a $Re_h \approx 12\cdot10^4$ and to a $Re_p \approx 6.4\cdot10^3$. Moreover, to maximize the flow stability before the orifice, this is placed close (around 100.0 mm) to the end of the pipe in the flow direction.

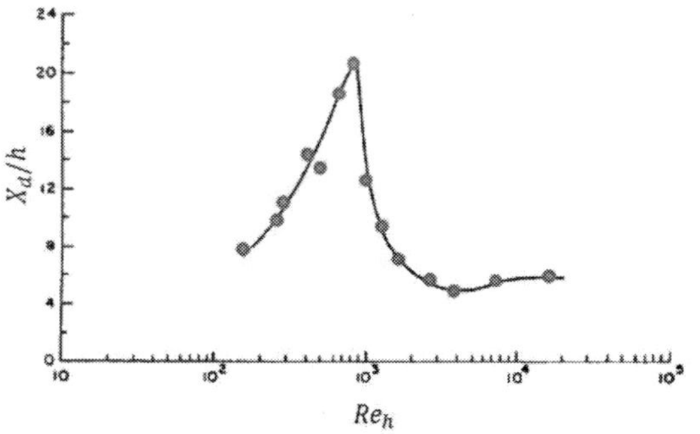

Fig. 3. Non-dimensional distance of reattachment versus Reynolds number (modified from Westphal et al, 1984 [19]).

The convergent reach is 50.0 mm long and with a wall inclination of around 20°, in order to minimize the flow separation also in this case.

The pipe of Figure 1 is immersed in a constant head tank with a volume of 2.0 m^3 (2.0 m long and with a section of 1.0 m × 1.0 m), with transparent Perspex walls (Figure 4). The transparent Perspex walls are needed in order to have an optical access to the flow and to perform non-intrusive and quasi-continuous-in-space measurements via Digital Image Analysis (DIA) techniques (a brief review on DIA techniques can be found in Ferrari, 2017 [20]). Consequently, the experimental set-up includes a high-frequency video acquisition system (with a camera placed vertically above the pipe) and a laser with a cylindrical lens, in order to obtain a horizontal light sheet. As during the experimental campaign both velocity (via Feature Tracking Velocimetry, FTV, see Besalduch et al., 2013 [21] and 2014 [22] for details) and concentration (via Laser induced Fluorescence, LIF) measurements were performed, a green laser was employed in the first case and a blue laser in the second case. In the case of FTV, the flow was seeded with non-buoyant pine pollen particles; in the case of LIF, the flow was seeded with fluorescein dye (the quantity was in the range allowing a linear relation between the dye concentration and the light intensity, Sutton et al., 2008 [23]). An example of an instantaneous LIF image is shown on Figure 5, where light green means a high concentration, pale green a low concentration and black zero concentration.

The two DIA techniques employed have been designed and developed by the Hydraulics Group of the DICAAR - University of Cagliari, and have been applied in the recent past in laboratory investigations to measure tracer concentrations (Ferrari et al., 2015 [24], Ferrari et al., 2018a [25] and 2018b [26], Badas et al. (2018) [27], Ferrari et al (2019) [28]), wave position (Ferrari et al., 2016 [29], Gallerano et al (2019) [30]), flow velocity (Querzoli, 1996 [31], Badas and Querzoli, 2011 [32], Simone et al., 2013 [33], Garau et al., 2018 [34], Badas et al. (2019) [35]) and acceleration (Ferrari et al, 2007 [36], Ferrari and Rossi, 2008 [37], Ferrari et al, 2008 [38], Lardeau et al., 2008 [39], Rossi et al, 2009 [40]). Among the engineering fields of interest, the DIA techniques have been employed in the civil and industrial field (such as in the present paper), in the environmental field (Ferrari et al., 2016 [41], Badas et al., 2017 [42], or in Garau et al., 2017 [43], Di Bernardino et al. (2017) [44], Ferrari et al., 2017 [45], Garau et al., 2018 [46], Badas et al., 2019 [47]) and biomedical fields (Badas et al., 2015 [48], Espa et al, 2012 [49], Querzoli et al (2019) [50]).

During the velocity measurements, the flow rates were computed through the velocity measurements FTV, while during the concentration measurements a flowmeter (Transonic 400-series coupled with a TS410 tubing module was employed. The flowmeter was placed downstream the convergent reach, so Q_1 was measured closing the orifice on the lateral wall of the pipe, then the orifice was opened and Q_2 was measured while the jet was flowing, and Q_3 was computed as the difference between Q_1 and Q_2. The ratio between the flow rates can be adjusted via two valves, one upstream the divergent reach and one downstream the convergent reach.

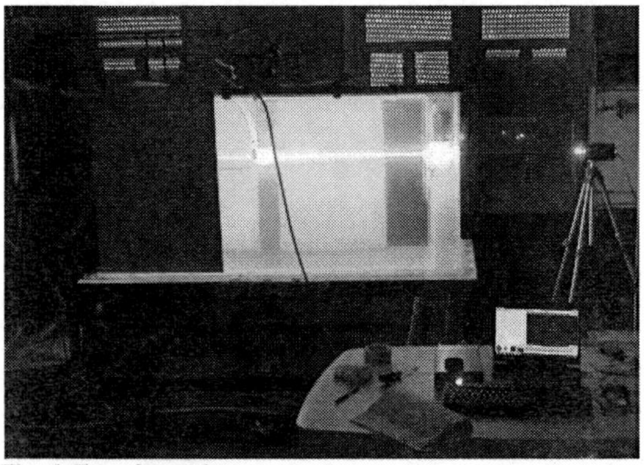

Fig. 4. Experimental setup.

As stated before, on the acquired LIF images the concentration field is proportional to the light intensity, so a quantitative measure of the concentration is allowed. Under the hypothesis of ergodicity, the mean and the variance on each pixel were obtained by time averaging over the 6000 samples recorded for each run. Then the mean concentration values were normalised with the mean value at the outlet C_0 (initial jet concentration), in order to obtain the fields of the reduction of the concentration C/C_0 (e.g., Figure 6). The fields of the variance of the concentration were normalised by the square of C_0 (Figure 10).

In order to measure the inclination of the jet, the axis of the jet itself has to be defined. As here the focus is on the jet region close to the orifice, the jet axis was defined as the straight line passing through the jet origin that best fits, in a least-square sense, the concentration maxima on the vertical sections close to the jet origin (e.g., see Figure 6). As previously stated, the jet inclination angle θ was consequently defined as the one between the above defined jet axis and a line perpendicular to the pipe axis. On Figures from 6 to 9, the concentration maxima are highlighted with white asterisks, the jet axis with a green line and the line perpendicular to the pipe axis with a blue line.

Fig. 5. An example of a LIF instantaneous image.

4 Results

In this section, some preliminary results concerning the concentration measurements via LIF will be shown. In particular, two of the measured quantities will be shown (mean and variance of the concentration), so this section is divided into two subsections.

In the first one, some of the non-dimensional mean concentration fields will be shown, in order to show the modifications induced on the jet inclination and shape by the different flow rate ratios *r*.

In the second subsection, the map of the non-dimensional variance of the concentration will be shown, in order to further highlight the lack of axisymmetry of this kind of jet when r increases.

On all the Figures from 6 to 10, the jet flows from the right to the left, the x- and y-axis are non-dimensionalized by the orifice diameter D_J and the origin of the x-axis and x-axis is placed on the centre of the orifice (jet origin).

4.1 Mean Concentration fields

The fields of the mean non-dimensional concentration C/C_0, where C_0 is the mean concentration measured at the outlet (that is the initial jet concentration), are shown in false colours from Figure 6 to Figure 9: dark red implies high concentration values (or a low dilution), dark blue implies low concentration values (or a low dilution), with increasing values of the flow rate ratio *r*, form very low values ($r = 0.015$, Figure 6) to higher values ($r = 0.482$, Figure 9).

On Figure 6, it is possible to see that when *r* is low (that is when most of the flow rate is going beyond the orifice withut passing throut it), the jet is not perpendicular to the pipe axis but, instead, there is a relevant angle θ between the jet axis and the line perpendicular to the pipe axis ($\theta \approx 27°$). Moreover, at a certain distance from the orifice, the upstream jet region (i.e. towards positive values of y/D) tends to widens more than the downstream jet region (i.e. towards negative values of y/D): this implies that the entrainment in the upstream region is higher than in the downstream region, so the upstream region experiences slowest velocities and, as a consequence, the jet tends to bend towards the positive y/D values. From another point of view, we can state that the jet has a low rigidity that tends suddenly to lose. So we can say that, for low values of *r*, the jet is not pependicular to the pipe axis and not even axysimmetric.

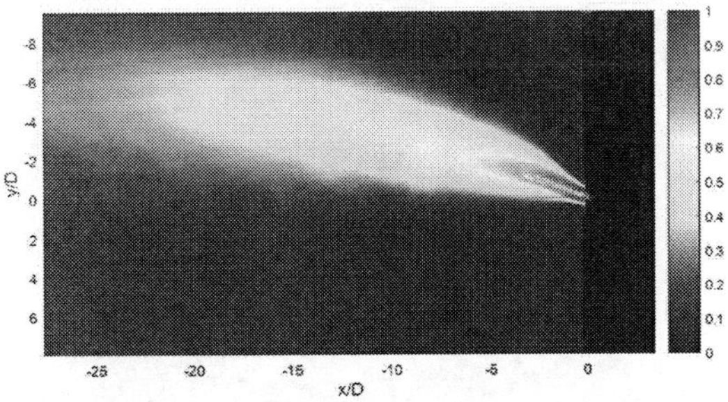

Fig. 6. Mean non-dimensional concentration C/C_0 field (C_0 is the mean concentration measured at the outlet, i.e. the initial jet concentration) for a jet with $r = 0.015$ ($\theta \approx 27°$); the white asterisks are the concentration maxima in the initial region of the jet, the green line is the jet axis, the blue line is the line perpendicular to the pipe axis.

Looking at Figure 7 ($r = 0.053$) and Figure 8 ($r = 0.103$), it is possible to not how, when *r* increases i.e. when more flow rate is passing through the orifice and less is going beyond it, the angle θ tends to decrease, the differencies between the upstream and downstream regions of the jet tend to reduce and the curvature is less accentuated.

On Figure 8, a jet with a higher value of *r* is shown ($r = 0.482$, meaning that almost half of the flow rate before the orifice becomes jet flow rate through the orifice), the initial region of the jet is almost perpendicular to the pipe axis ($\theta \approx 2°$) and almost axisymmetric.

As a consequence, we can state that, when a jet issues from an orifice in the lateral wall of pipe where a flow is moving, the jet is not perpendicular to the pipe axis but, instead, there is an angle of inclination in the downstream direction that is inversely proportional to the ratio of the jet flow rate to the pipe flow rate before the orifice *r*. The same consideration applies to the jet axisymmetry, as this tends to decrease when *r* decreases.

4.2 Concentration Variance

The field of non-dimensional variance of the concentration (normalized with the square of C_0) for a jet with $r = 0.011$ is shown, in false colours, on Figure 10. As the concentration variance is proportional to the turbulence intensity, high variance values (dark red) highlight regions of intense turbulent fluctuations, low variance values (dark blue) regions of very low turbulent fluctuations.

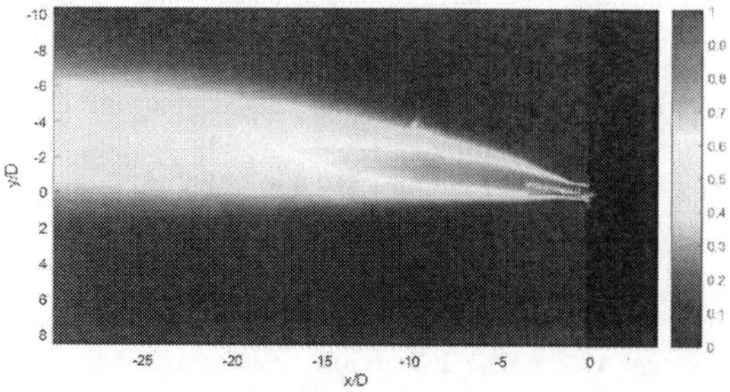

Fig. 7. Mean non-dimensional concentration C/C_0 field (C_0 is the mean concentration measured at the outlet, i.e. the initial jet concentration) for a jet with r = 0.053 ($\theta \approx 12°$); the white asterisks are the concentration maxima in the initial region of the jet, the green line is the jet axis, the blue line is the line perpendicular to the pipe axis.

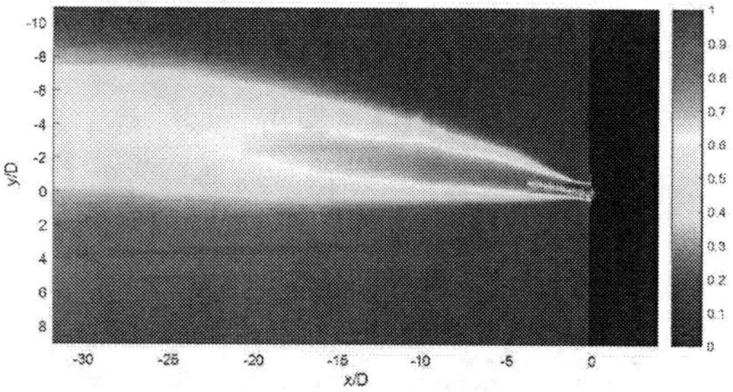

Fig. 8. Mean non-dimensional concentration C/C_0 field (C_0 is the mean concentration measured at the outlet, i.e. the initial jet concentration) for a jet with r = 0.103 ($\theta \approx 8°$); the white asterisks are the concentration maxima in the initial region of the jet, the green line is the jet axis, the blue line is the line perpendicular to the pipe axis.

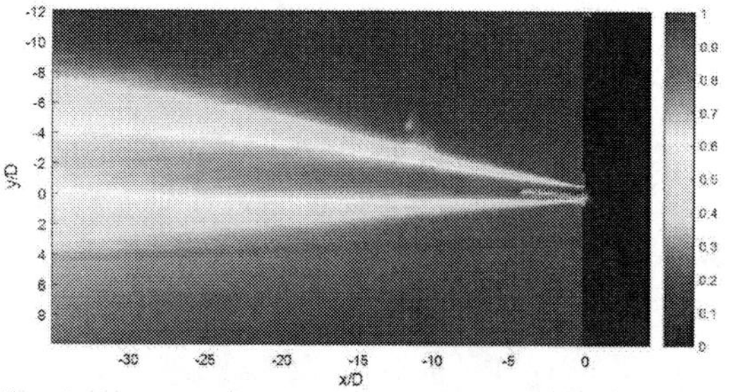

Fig. 9. Mean non-dimensional concentration C/C_0 field (C_0 is the mean concentration measured at the outlet, i.e. the initial jet concentration) for a jet with r = 0.482 ($\theta \approx 2°$); the white asterisks are the concentration maxima in the initial region of the jet, the green line is the jet axis, the blue line is the line perpendicular to the pipe axis.

This Figure highlights what stated above about the lack of axisymmetry of this kind of jets. First of all, the upstream and downstream region are different. The highest values are found in the downstream region, close to the jet origin, so the entrainment is larger on the downstream region: this could be the reason for the fact that the jet tends to initially deflect to the downstream direction. Alternatively, considering that the jet inclination is also a consequence of the flow pattern inside the pipe near the orifice, the largest variance in the upstream jet region close to the outlet could also be the consequence of this. This is currently under investigation, as we are performing the analysis on the velocity measurements also in the region inside the pipe and close to the outlet.

Even if the highest variance values are found on the downstream region, the area of high variance values on the upstream region is longer than the downstream one, giving an explanation to the reason of the second deflection of the jet, in an opposite direction.

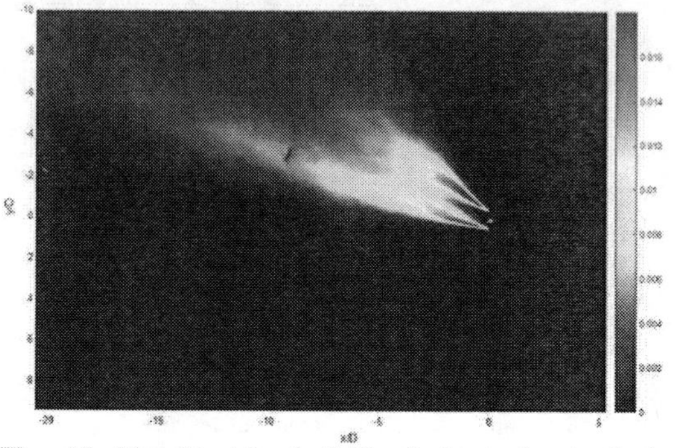

Fig. 10. Non-dimensional field of the variance of the concentration, non-dimensionalized with the square of the mean concentration measured at the outlet (highlighted by a yellow point), i.e. the initial jet concentration C_0, for a jet with r = 0.011.

5 Conclusions

In this work, we have shown some preliminary results regarding the behaviour of fluid jets released, in a receiving body of the same fluid, from a sharp-edged orifice in the lateral wall of a pipe where a certain flow rate is flowing. The concentration fields for various ratios of the jet flow rate to the pipe flow rate (r) have been measured in the Hydraulics Laboratory of the DICAAR – University of Cagliari, via a LIF technique, a non-intrusive and quasi-continuous-in-space Digital Image Analysis technique.

The preliminary results showed above highlight that the jets are not perpendicular to the pipe flow nor axisymmetric. In particular, the inclination of the jet tends to increase when r decreases. This result can have a relevant practical application, as it can be used to forecast where the jet will move once outside the orifice.

A similar consideration can be drawn regarding the jet axisymmetry, as this tends to decrease when r decreases. Moreover, the variance field seems to suggest that the different entrainments experienced from the two regions of the jet makes it bend twice, on the downstream direction soon after the release and on the upstream direction at a certain distance from the orifice.

We are currently performing and analysing velocity measurements, in order to investigate, among the others, the flow inside the pipe in the region close to the orifice, the velocity fields and the Reynolds stresses in the jet and the relation of r versus θ.

The Authors would like to acknowledge Mr. Antonio Mascia.

References

1. T. M. Missimer, B. Jones, and R. G. Maliva, *Intakes and Outfalls for Seawater Reverse-Osmosis Desalination Facilities* (Springer Berlin Heidelberg, 2015).
2. S. Ferrari and G. Querzoli, J. Hydraul. Res. **48**, 632 (2010).
3. C. Avanzini, in *Proc 4th Intl Conf Mar. Waste Water Dispos. Mar. Environ. 2nd Intl Exhib. Mater. Equip. Serv. Coast. Waste Water Treat. Plants Outfalls Sealines Antalya Turk.* (Mem Ajans, Istambul, 2006).
4. J. R. Wood, in *Ocean Dispos. Wastewater* (World Sci. Publ., Singapore, 1993), pp. 71–84.
5. W. Xue, W. Huai, Z. Qian, Z. Yang, and Y. Zeng, Eng. Comput. **31**, 1379 (2014).
6. A. S. Al-Ghamdi, J. Coast. Conserv. **14**, 63 (2010).
7. G. A. Anderson, A. Kommareddy, T. Suess, and S. P. Gent, in *ASME 2014 8th Int. Conf. Energy Sustain. ES 2014 Collocated ASME 2014 12th Int. Conf. Fuel Cell Sci. Eng. Technol.* (ASME, 2014).
8. K. Osman, S. M. N. Al Khaireed, M. K. Ariffin, and M. Y. Senawi, Energy Convers. Manag. **49**, 3270 (2008).
9. J. E. van Zyl, Procedia Eng. **89**, 273 (2014).
10. M. Soleimani nia, P. Oshkai, and N. Djilai, Int. J. Hydrog. Energy **44**, 15262 (2019).
11. M. Ferrante, C. Massari, E. Todini, B. Brunone, and S. Meniconi, J. Hydroinformatics **15**, 666 (2013).
12. R. Mohsin, Z. A. Majid, and F. L. Tan, Eng. Fail. Anal. **48**, 30 (2015).
13. A. M. Cassa, J. E. van Zyl, and R. F. Laubsher, Urban Water J. **7**, 109 (2010).
14. J. E. van Zyl and C. R. I. Clayton, Proc. Inst. Civ. Eng. Water Manag. **160**, 109 (2007).
15. M. Soleimani nia, B. Maxwell, P. Oshkai, and N. Djilali, Int. J. Hydrog. Energy **43**, 9379 (2018).
16. A. Hashiehbaf and G. P. Romano, Int. J. Heat Fluid Flow **44**, 208 (2013).
17. J. Mi, G. J. Nathan, and D. S. Nobes, J. Fluids Eng. **123**, 878 (2001).
18. J. Mi, P. Kalt, G. J. Nathan, and C. Y. Wong, Exp. Fluids **42**, 625 (2007).
19. R. V. Westphal, J. P. Johnston, and J. K. Eaton, 284 (n.d.).
20. S. Ferrari, EPJ Web Conf. **143**, 01001 (2017).
21. L. A. Besalduch, M. G. Badas, S. Ferrari, and G. Querzoli, EPJ Web Conf. **45**, 01012 (2013).
22. L. A. Besalduch, M. G. Badas, S. Ferrari, and G. Querzoli, EPJ Web Conf. **67**, 02007 (2014).
23. J. A. Sutton, B. T. Fisher, and J. W. Fleming, Exp. Fluids **45**, 869 (2008).
24. S. Ferrari and G. Querzoli, EPJ Web Conf. **92**, 02018 (2015).
25. S. Ferrari, M. Badas, and G. Querzoli, Water **10**, 726 (2018).
26. S. Ferrari, M. G. Badas, and G. Querzoli, EPJ Web Conf. **180**, 02025 (2018).
27. M. G. Badas, M. Garau, A. Seoni, S. Ferrari, and G. Querzoli, J. Phys. Conf. Ser. **1110**, 012003 (2018).
28. S. Ferrari, M. G. Badas, M. Garau, L. Salvadori, A. Seoni, and G. Querzoli, EPJ Web Conf. **213**, 02017 (2019).
29. S. Ferrari, M. G. Badas, and G. Querzoli, EPJ Web Conf. **114**, 02022 (2016).
30. F. Gallerano, G. Cannata, M. Tamburrino, S. Ferrari, M. G. Badas, and G. Querzoli, WSEAS Trans. FLUID Mech. **14**, 84 (2019).
31. G. Querzoli, Atmos. Environ. **30**, 2821 (1996).
32. M. G. Badas and G. Querzoli, Exp. Fluids **50**, 1093 (2011).
33. F. Simone, M. G. Badas, L. A. Besalduch, and Q. Giorgio, Int. Symp. Turbul. Shear Flow Phenom. TSFP 2013 (2013).
34. M. Garau, M. G. Badas, S. Ferrari, A. Seoni, and G. Querzoli, Int. J. Environ. Pollut. **65**, 103 (2019).
35. M. G. Badas, S. Ferrari, M. Garau, S. Seoni, and G. Querzoli, Bound.-Layer Meteorol. **in press**, (2019).
36. S. Ferrari, L. Rossi, and J. C. Vassilicos, in *Adv. Turbul. XI - Proc. 11th EUROMECH Eur. Turbul. Conf.* (2007), pp. 485–487.
37. S. Ferrari and L. Rossi, Exp. Fluids **44**, 873 (2008).
38. S. Ferrari, P. Kewcharoenwong, L. Rossi, and J. C. Vassilicos, in *IUTAM Symp. Flow Control MEMS*, edited by J. F. Morrison, D. M. Birch, and P. Lavoie (Springer Netherlands, Dordrecht, 2008), pp. 267–272.
39. S. Lardeau, S. Ferrari, and L. Rossi, Phys. Fluids **20**, 127101 (2008).
40. L. Rossi, S. Bocquet, S. Ferrari, J. M. Garcia de la Cruz, and S. Lardeau, Int. J. Heat Fluid Flow **30**, 505 (2009).
41. S. Ferrari, M. G. Badas, M. Garau, A. Seoni, and G. Querzoli, HARMO 2016 - 17th Int. Conf. Harmon. Atmospheric Dispers. Model. Regul. Purp. Proc. 351 (2016).
42. M. G. Badas, S. Ferrari, M. Garau, and G. Querzoli, J. Wind Eng. Ind. Aerodyn. **162**, 24 (2017).
43. M. Garau, M. G. Badas, S. Ferrari, A. Seoni, and G. Querzoli, HARMO 2017 - 18th Int. Conf. Harmon. Atmospheric Dispers. Model. Regul. Purp. Proc. 470 (2017).
44. A. D. Bernardino, P. Monti, G. Leuzzi, F. Sammartino, and S. Ferrari, in *HARMO 2017 - 18th Int. Conf. Harmon. Atmospheric Dispers. Model. Regul. Purp. Proc.* (Hungarian Meteorological Service, 2017), pp. 460–464.
45. S. Ferrari, M. G. Badas, M. Garau, A. Seoni, and G. Querzoli, Int. J. Environ. Pollut. **62**, 22 (2017).
46. M. Garau, M. G. Badas, S. Ferrari, A. Seoni, and G. Querzoli, Bound.-Layer Meteorol. **167**, 123 (2018).
47. M. G. Badas, M. Garau, G. Querzoli, L. Salvadori, and S. Ferrari, Int. J. Environ. Pollut. **in press**, (2019).
48. M. G. Badas, S. Espa, S. Fortini, and G. Querzoli, EPJ Web Conf. **92**, 02004 (2015).
49. S. Espa, M. G. Badas, S. Fortini, G. Querzoli, and A. Cenedese, Eur. J. Mech. - BFluids **35**, 9 (2012).
50. G. Querzoli, V. Satta, G. Matta, S. Ferrari, M. G. Badas, and G. Bitti, J. Phys. Conf. Ser. **1249**, 012006 (2019).

Practical aspects of the wettability

Simona Fialová[1], *František* Pochylý[1], *Alexander* V. Volkov[2] and *Jiří* Šoukal[3]

[1] Faculty of Mechanical Engineering, Brno University of Technology, Technická 2896/0, 61669, Brno, Czech Republic, corresponding author email: fialova@fme.vutbr.cz
[2] Moscow Power Engineering Institute (National Research University), Krasnokazarmennaja 14, Moscow, 111250 Russia
[3] Hydraulic Research Center Sigma, Lutín, Czech Republic

Abstract. A new knowledge of fluid flow on hydrophobic surface will be presented. The boundary condition will be refined considering the effects of air adhering to the rough, hydrophilic surface. Possibilities of turbulent viscosity effects as a result of liquid slippage on the hydrophobic surface will also be considered. These effects will be demonstrated on the basis of performed experiments. The experimental study will be focused on the determination of the pressure difference in the tube, whose surface is provided with an ultrahydrophobic coating with a contact angle higher than 140 °. Results will be reported based on Reynolds number. In addition to the results of the experiment, the thesis will also include a theoretical study aimed at refining the boundary condition, expressing the interaction of the liquid with the hydrophobic surface. In conclusion, the dissipation function will be derived, expressing the effects of liquid slippage on the hydrophobic surface and compared with its value taking into account the hydrophilic properties of the surface.

1 Introduction

Various mathematical models of fluid flow on a hydrophobic surface have been demonstrated in previous works [1], [7]. In these models, Navier's hypothesis of the dependence of shear stress vs. slip velocity was taken as the basis. This condition has been enhanced for the liquid motion on a general curvilinear surface [4]. Another upgrade of the mathematical model was made by Tesca [8], who assumed that the movement of the liquid on the hydrophobic surface would not occur until a certain value of shear stress was overcome. These models put the size of the contact angle in correlation with the angle of the inclined plane, when the droplet detaches from the surface and moves. The adhesion coefficient k [7] was then determined according to the droplet velocity. Since the value of k is not dimensionless, a more accurate coefficient of adhesion κ is introduced as a function of the velocity of the droplet movement on the inclined plane v_t.

Using this defined mathematical model, computational modelling of both laminar and turbulent flow was performed. As a result of the analysis it was obtained, that the hydrophilic surfaces characterized by a coefficient $\kappa >$ 10 behave as hydrophilic [2]. Furthermore, it has been shown that the fluid slip increases the cavitation zone width and reduces the number of Taylor vortices. Hydraulic losses are also reduced at $\kappa <$ 10. These mathematical models were verified by experiments in both laminar and turbulent flow. Experiments have shown that the hydrophobic surface has the ability to bind air molecules to one another. The result is a thin air layer adjacent to the surface. Its stability is dependent on the value of the adhesion coefficient κ. The thickness of the air layer decreases as the adhesion coefficient increases. Experiments carried out on tubes of the size close to capillaries have shown that in the laminar flow, as the pressure drop increases, the air layer is compressed. While the pressure drop is lowered, the air layer increases again, so that the air layer restores.

In the turbulent flow, part of the air layer is washed away because the thickness of the air layer is not restored when the pressure drop is reduced back. These statements will be demonstrated in details using examples.

2 Analysis of mathematical model - velocity profiles

The flow in a rigid tube of circular cross-section is considered. Turbulent flow is solved using the power model for incompressible liquid.

Hydrophilic surface - laminar flow

$$v = \Delta p \frac{R^2}{4L\eta}\left[1 - \left(\frac{r}{R}\right)^2\right]; \quad v_s = 0 \tag{1}$$

Hydrophobic surface - laminar flow

$$v = \Delta p \frac{R^2}{4L\eta}\left\{\frac{2\eta}{R\kappa} + \left[1 - \left(\frac{r}{R}\right)^2\right]\right\}; \quad v_s = \Delta p \frac{R}{2L}\frac{1}{\kappa} \tag{2}$$

Hydrophilic surface - turbulent flow

$$v = \Delta p \frac{R^2}{2L\eta}\frac{n}{n+1}\left[1 - \left(\frac{r}{R}\right)^{\frac{n+1}{n}}\right]; \quad v_s = 0 \tag{3}$$

Hydrophobic surface - turbulent flow

$$v = \Delta p \frac{R^2}{2L\eta_S(R)}\frac{n}{n+1}\left\{\frac{\eta_S(R)}{R\kappa}\frac{n+1}{n} + \left[1 - \left(\frac{r}{R}\right)^{\frac{n+1}{n}}\right]\right\}; \tag{4}$$

$$v_s = \Delta p \frac{R}{2L}\frac{1}{\kappa} \tag{5}$$

From these terms it is clear that they do not consider the effect of the air layer near the surface. In the case of turbulent flow, it is assumed that near the surface due to the slip of the liquid on the surface, a turbulent viscosity develops. This contains the expression (4). However, it is clear from (5) that the turbulent viscosity value does not affect the slip velocity in this model. However, this does not correspond to the experiment, as it will be shown in the next chapter. In [4] this model is refined by the effect of turbulent viscosity on κ. This stems from the fact that after the air layer is washed away, the liquid must slip on the surface. However, slippage will cause turbulent viscosity, which will affect liquid slippage due to turbulent fluctuations. This may reduce the slip of the drop across the surface. This, however, contradicts the equation (5).

3 New mathematical model

In a mathematical model describing the flow between two parallel walls, a layer of air near the hydrophobic surface [1], [8] was considered. Both laminar and turbulent models were derived.

3.1 Laminar flow

The boundary condition was defined by [4]:

$$\tau = \eta_B \frac{\partial u}{\partial y} = \kappa u_s \quad , \quad \text{where} \quad \eta_B = \frac{\eta}{1 + c\eta u_0} \quad \text{and} \quad \kappa = \rho v_t \alpha \tag{6}$$

From this it is obvious that at $\kappa = 0$ is $\tau = 0$ which indicates the correct physical sense, since in the intact air layer the air adheres to the hydrophobic surface, so it holds that $u_s = 0$. The slip velocity [1], [4] corresponds to the value of the viscosity η_B:

$$v_s = \Delta p \frac{(4Hc\eta u_0^3 - c\eta u_0^3 + 6H^2)}{12LH(c\eta u_0^2 + 1)\kappa + 12L\eta} \tag{7}$$

For $\kappa = 0$ the velocity $v_s \neq 0$, which is contrary to the physical reality. Therefore, we propose a new mathematical model that is in accordance with the definition of κ. Since dimension κ is equal to the dimension of the viscosity gradient, we propose the following mathematical model to formulate the boundary conditions:

$$\tau = \eta_B \frac{\partial \mathbf{u}}{\partial y} = \left(\kappa + \frac{\partial \eta_B}{\partial y}\right)\mathbf{v} \tag{8}$$

This model is generic enough to describe different flow variants considering the small air layer near the surface. The following viscosity model was proposed in [1] for the air layer:

$$\eta_B = \frac{\eta}{1 + c\eta(u_0 - u)^2} \tag{9}$$

This model well describes the behavior of the fluid in the air layer.

$$\left.\frac{\partial \eta_B}{\partial y}\right]_{y=0} = 2\frac{\eta}{u_0}\frac{\alpha}{(1+\beta)^2} \ , \quad \beta = c\eta u_0^2 \tag{10}$$

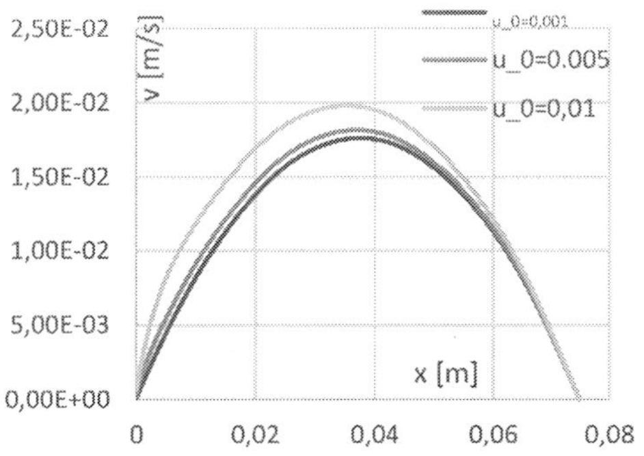

Fig. 1 Examples of the velocity profiles

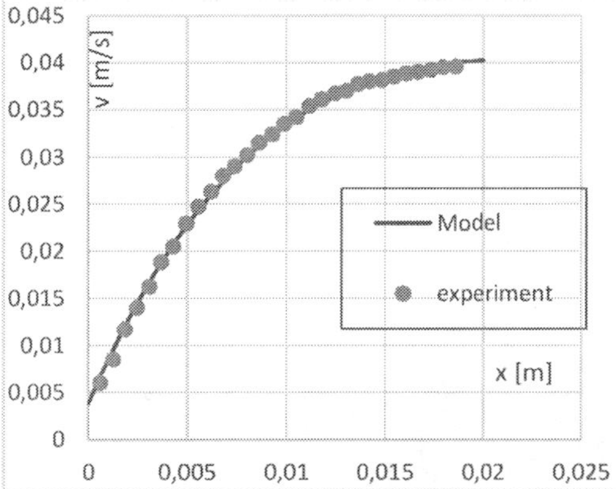

Fig. 2 Model comparison with the experiment

3.2 Turbulent flow

In the case of turbulent flow, two variants can be considered:

- Assuming that the air layer is washed away, it can be assumed that turbulent fluctuations in velocity and pressure will occur on and near the surface, causing turbulent viscosity on the surface. At the same time, due to the turbulent viscosity, the slip of the liquid on the hydrophobic surface decreases. In this case, the boundary condition will depend on the turbulent viscosity and its gradient so that the shear stress can be assumed to depend on the slip rate in the form:

$$\boldsymbol{\tau} = \underbrace{\hat{\eta}}_{\hat{\eta}=\eta+\eta_t} \frac{\partial \mathbf{v}}{\partial r} = \frac{\partial \eta_t}{\partial r}\mathbf{v} \qquad (11)$$

- Assuming that the hydrophilic surface is not completely smooth, it is either coarse due to corrosion or artificially created with the surface uneven with the holes or grooves. In such a case, the surface is partially hydrophobic and partly formed by holes filled with the air. In this case, the boundary condition for shear stress can be written in the form:

$$\boldsymbol{\tau} = \eta_B \frac{\partial \mathbf{v}}{\partial r} = \left(\kappa + \frac{\partial \eta_t}{\partial r}\right)\mathbf{v} \qquad (12)$$

These mathematical models are based on the following experimental results.

4 Experiments

Figures 3, 4 show the flow in a minichannel with parallel walls. There is an apparent thickness of the air layer in laminar flow compared to the liquid layer. The following figure 5 shows the change in air layer thickness as a function of the Reynolds number and the degree of hydrophobia expressed by the contact angle Θ [3].

From this view, the influence of the air layer and its compression up to the value of Re 5000 is evident. In the mathematical model (11) this effect can be easily understood.

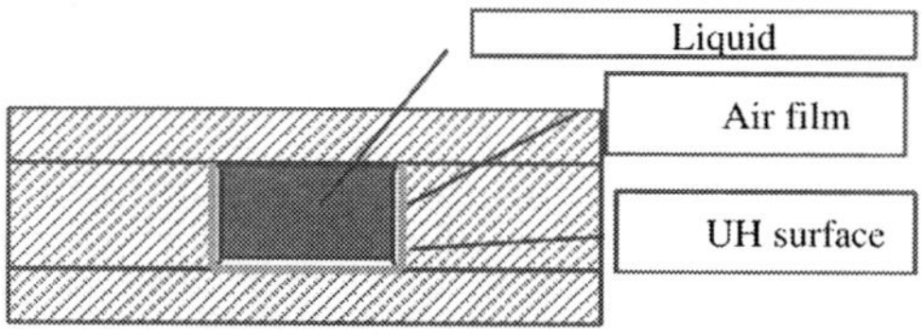

Fig 3. The geometry of the air film formation in the mini-channel. [3]

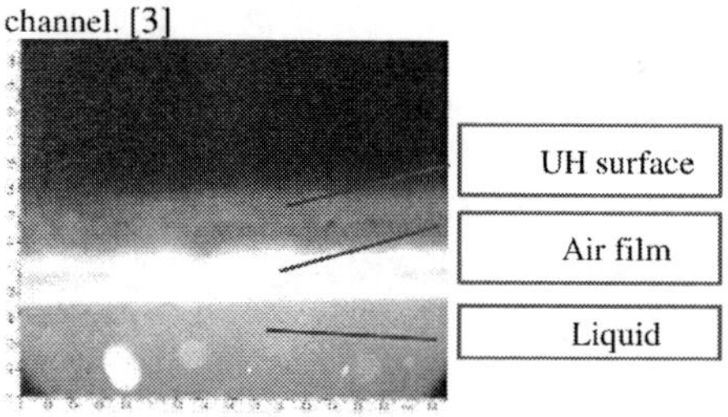

Fig.4 The visualization of the air film with the microscope magnification 10x. [3]

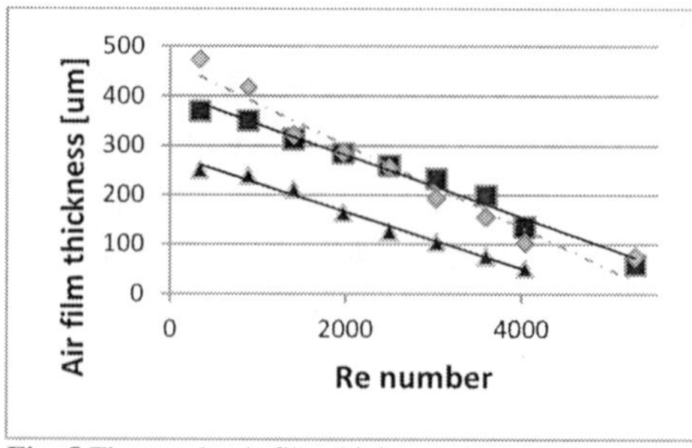

Fig. 5 The graph: air film thickness dependant on Re number. [3]

4.1 Smooth hydrophobic surface

Further experiments were carried out again in a channel with parallel walls, one of which was provided with a hydrophobic surface. The profile dimensions were 75x80mm [2]. Figures 6, 7, 8 show the measured velocity profiles for Re 2000, Re 9000, Re 13000.

The wall provided with a hydrophobic surface had a contact angle of 150 °. It is apparent from Figures 6,7 that the velocity profile near the surface is influenced by the air layer bound to the hydrophobic surface, which results in a relatively high velocity near the surface.

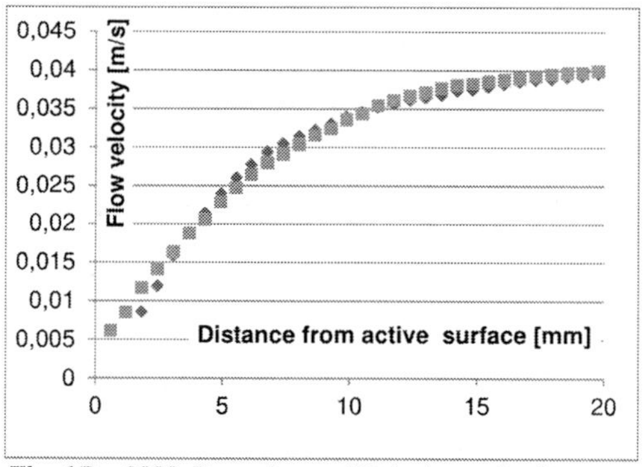

Fig. 6 Re=2000; Dependence of Velocity on the distance from active surface, which is on the position 0

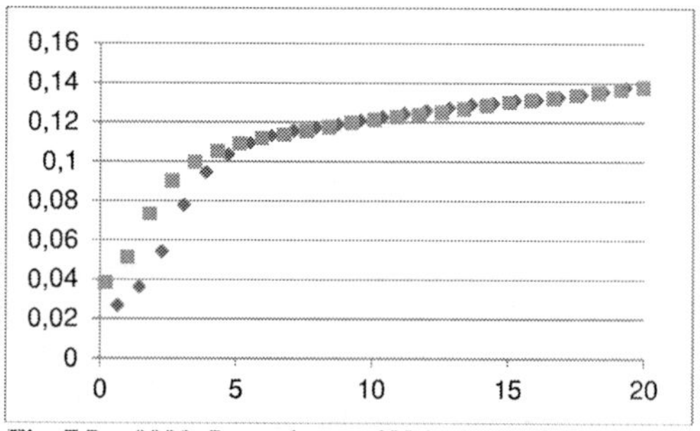

Fig. 7 Re=9000; Dependence of Velocity on the distance from active surface, which is on the position 0

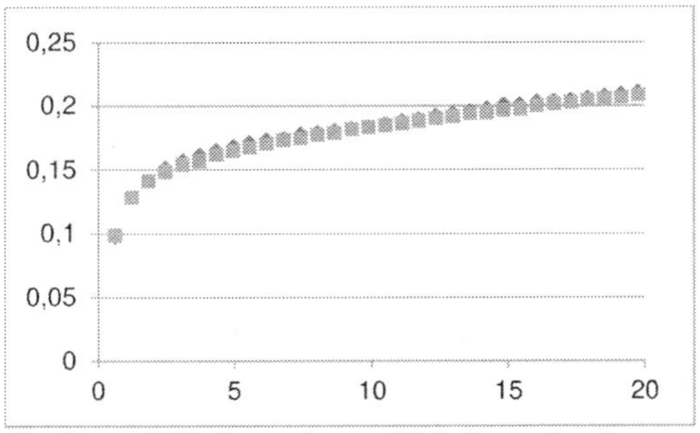

Fig. 8 Re=13000; Dependence of Velocity on the distance from active surface, which is on the position 0

Another situation occurs at higher Reynolds numbers. Even at Re 13000, some of the air layer residues are partially washed away and compressed, so that the wall behaves as wettable.

To verify this, another experiment is performed, this time with a tube of circular cross-section whose inner surface has been treated with three types of hydrophobic coatings with the following parameters:

$$D = 0{,}042m \text{ a } L = 6m$$

a) Stainless steel $\Theta = 80°$
b) Stainless steel + UED $\Theta = 158°$
c) Ceramics + UED $\Theta = 157°$

The measured section is shown in Fig. 9.

Fig. 9 Experimental stand

The dependence of specific energy on Reynolds number is shown in Fig. 7. Stainless steel has a contact angle of only 80 °, which is at the edge of hydrophilia. It can therefore be assumed that liquid adheres to this surface.

The other two surfaces are ultrahydrophobic, so they can be assumed to be aerophillic. However, it is apparent from the value of the measured pressure difference that the air layer has already been washed away or so compressed when the turbulent flow is developed that the surface already behaves as wettable as the liquid adheres to the surface.

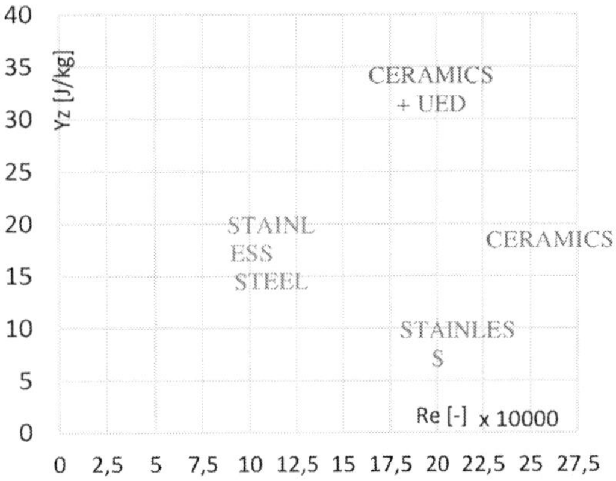

Fig. 10 The dependence of specific energy on Reynolds number

If the air layer is washed away and the liquid slips, the conditions for turbulent viscosity are created directly on the surface and the pressure difference may increase. From the measured values shown in Fig. 10, it is evident that the ultrahydrophobic smooth surface is characterized by a higher pressure difference than the smooth hydrophilic surface corresponding to the stainless steel material. This result is very surprising and will be verified by further experiments.

4.2 Coarse-textured surfaces

A surface with a rough structure is shown schematically in Figure 11. It is characterized by the air filled in the holes (grooves) of the hydrophobic coarse textured surface. This type of surface is characteristic for surfaces affected by corrosion. Experiments carried out on this type of surfaces draw a very important conclusion on hydraulic losses [5].

Fig. 11 Surface with a rough structure [5]

Figure 12 shows the pressure difference for a surface affected by corrosion. Its profile is shown in Fig. 13.

Different types of hydrophobia were considered in the experiments depending on the contact angle Θ and the angle of the inclined plane at which the drop motion occurs.

The experiments show that the pressure difference is lower for hydrophobic surfaces, compared to the original surface variant affected by corrosion. The cause is an air layer that adheres to the superhydrophobic surface and is not washed away by the liquid stream. For this type of surfaces, boundary conditions (12) can be used in computational modeling.

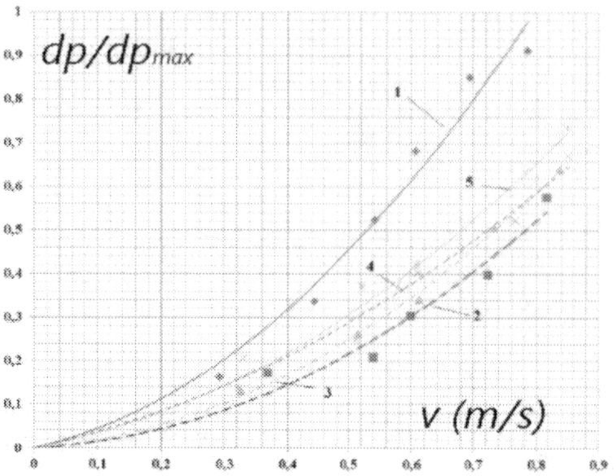

Fig. 12 Pressure difference ratio for different types of contact angles (1-78°, 2-110°, 3-128°, 4-130°, 5-133°)

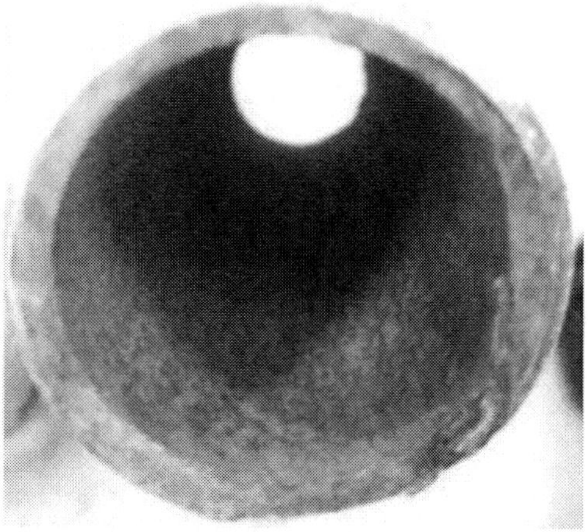

Fig. 13 Coarse-textured surface [5]

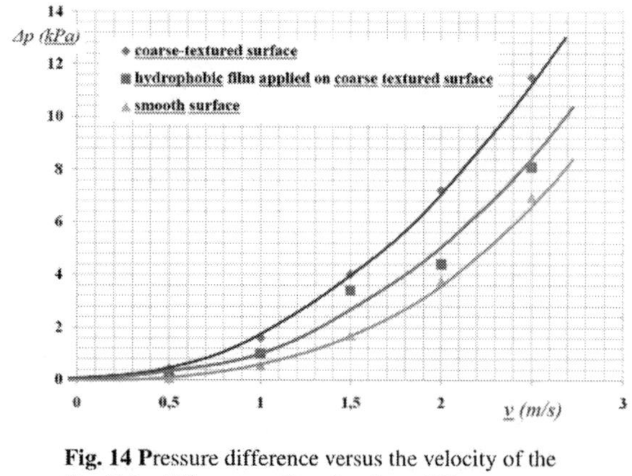

Fig. 14 Pressure difference versus the velocity of the liquid for different quality of surfaces [6]

In addition to these experiments, another [6], taking into account the flow of oil, was also performed in a circular-section tube with a rough structure. The results are analogous to the previous study. Figure 14 shows three curves of the pressure difference versus the velocity of the liquid.

Based on these experiments, it can be concluded that hydrophobic technology applied to surfaces with high roughness reduces hydraulic losses when fluid flows through a pipe.

The energy dissipation characterizing the power input to cover hydraulic losses in a constant cross section tube is determined from:

$$p = \int_V \hat{\eta} v_{ij} v_{ij} dV = \Delta p Q - |\boldsymbol{\sigma}||\mathbf{v}_s|\Gamma \quad (13)$$

$$\text{ale } \boldsymbol{Q} = \hat{\mathbf{v}}S, |\boldsymbol{\sigma}|\Gamma = \Delta p\mathbf{S}, \text{ takže} \quad (14)$$

$$p = \Delta p\mathbf{S}(\hat{\mathbf{v}} - |\mathbf{v}_s|) \quad (15)$$

$\hat{\mathbf{v}}$ mean velocity of liquid in section S

$\mathbf{v}_s$ slip speed, or velocity at the boundary of the air layer.

5 Conclusion

The detailed analysis of the mathematical model of the boundary condition for the liquid in contact with the hydrophobic surface was performed. This analysis was based on a new knowledge gained from experiments of incompressible fluid flow in tubes of constant cross-section. The experiments showed a strong dependence of the velocity of the liquid near the hydrophobic surface on the thickness of the air layer during both laminar and turbulent flow. Based on these findings from the experiment, a mathematical model of the boundary condition for shear stresses on the wall of the hydrophobic surface was formulated. This was included in the $k - \varepsilon$ turbulence model.

Acknowledgement This paper was supported by the project "Computer Simulations for Effective Low-Emission Energy Engineering", funded as project No. CZ.02.1.01/0.0/0.0/16_026/0008392 by the Operational Program Research, Development and Education, Priority axis 1: Strengthening capacity for high quality research.

References

1. Pochylý F., Fialová S., Havlásek M.: New boundary conditions for fluid interaction with hydrophobic surface, EPJ Web Conf. 180 02084 (2018), DOI:10.1051/epjconf/201817002084
2. Jašíková, D.; Kotek, M.; Fialová, S.; Kopecký, V.: A liquid interaction with ultrahydrophobic surfaces. *EPJ Web of Conferences,* 2016, vol. 114, no. 02052 (2016), p. 1-6. ISSN: 2100-014X.
3. Jasiková D., Gašič M., Fialová S., Kotek M. and Kopecký V.: The interaction between fluid flow and ultra-hydrophobic surface in mini channel, EPJ Web Conf., 143 (2017) 02045, DOI: DOI:10.1051/epjconf/201714302045

4. Pochylý F., Fialová S., Havlásek M., Malík J.: Influence of boundary conditions on fluid flow on hydrophobic surfaces, September 2017, AIP Conference Proceedings 1889(1):020008, DOI: 10.1063/1.5004342
5. Michail A. Morozov: Numerical-experimental research of the hydraulic characteristics of the termal pipeline systems with the hydrophobic effect of the surface (in Russian), Ph.D. thesis, Moscow, 2016
6. Artem V. Ryzhenkov: Methodology of the effectivity increasing of the termal pipeline systems of oil transport by using the hydrophobic surfaces (in Russian), Ph.D. thesis, Moscow, 2017
7. Fialová S., Pochylý F.: Identification and Experimental Verification of the Adhesive Coefficient of Hydrophobic Materials, 2015, Wasserwirtschaft Extra (13): 125-129, DOI: 10.1007/s35147-015-0529-8
8. Kučera, R., Šátek, V., Haslinger, J., Fialová, S., and Pochylý, F. (October 10, 2016). "Modeling of Hydrophobic Surfaces by the Stokes Problem With the Stick–Slip Boundary Conditions." ASME. *J. Fluids Eng*. January 2017; 139(1): 011202. https://doi.org/10.1115/1.4034199

https://doi.org/10.1051/epjconf/202226901015

Dispersion of immiscible liquid-liquid system in a vessel agitated by a Sawtooth impeller: drop size time evolution

Roman Formánek[1,*], *Radek* Šulc[1] and *Zdeněk* Chára[2]

[1]Czech Technical University in Prague, Faculty of Mechanical Engineering, Department of Process Engineering, Technická 4, 160 00 Prague, Czech Republic
[2]Czech Academy of Science, Institute of Hydrodynamics, Pod Paťankou 30/5, 166 12, Prague, Czech Republic

Abstract. Mixing of immiscible liquid-liquid system is the crucial process for many branches of industry. The main purpose of this process is to increase mass transfer efficiency in processes occurring in chemical and metallurgical industry. The knowledge of drop size and surface area of drops is important for the determination of mass transfer time. The aim of this contribution is to study drop size time evolution in immiscible liquid-liquid system agitated by high-shear Sawtooth impeller in baffled cylindrical vessel. The drop size evaluation was investigated in three different regions of interest. The time evolution of drop size was described by the kinetic model proposed by Hong and Lee (1983) for each region of interest. Simultaneously, the drop size distribution (DSD) was determined and compared for each regions of interest. The time evolution of drop sizes was investigated by non-invasive measurement directly in control volume. The images of illuminated region of interest were captured by a high-speed camera. The captured images were processed by the Image Analysis method.

1 Introduction

The knowledge of drop size kinetics is vital for many processes in wide branches of industry. The information about Sauter mean diameter d_{32} in the final stage of the process has been mostly presented in the literature. The drop size distribution (DSD) in a mixing tank has been simulated by means of the population balance approach. Unfortunately, there is a lack of kinetic experimental data to verify the simulation predictions of DSD kinetics. The different experimental methods for the measurement of drop size exist which may be affected by many factors.

The kinetics of Sauter mean diameter d_{32} and drop size distribution in an agitated vessel has been investigated by many researchers. Hong and Lee [1] investigated unsteady-state in the agitated vessel of a liquid-liquid dispersion system. They proposed a simple model of drop size describing d_{32} time evolution with model parameters depending on the mixing system characteristics. The time evolution of the DSD was investigated by Šulc et al. [2].

In our previous work [3] the time evolution of Sauter mean diameter and DSD and the homogeneity of liquid-liquid dispersion were investigated in a cylindrical baffled vessel agitated by a Rushton turbine. We found that the differences between the three different regions of interest exist. The highest differences were found between the highest placed region and both other lower placed regions of interest. The results obtained in both lower placed regions of interests were practically the same.

The aims of this contribution are 1) to measure the time evolution of DSD and Sauter mean diameter and 2) to investigate the homogeneity level of liquid-liquid dispersion in baffled vessel agitated by a high-shear Sawtooth impeller. The homogeneity was assessed comparing numbers of evaluated drops and Sauter mean diameters and DSD in three investigation areas placed in different off-bottom distance. The drop size was evaluated by the in-situ measurement with the following Image Analysis (IA).

2 Experimental

The experiments were carried out in a cylindrical baffled vessel of inner diameter T = 300 mm with a flat bottom and the vessel was filled with water-oil emulsion. The mixing of the emulsion was provided by a high-shear Sawtooth impeller with diameter $D = T/3$. The impeller was placed with off-bottom clearance $C = T/4$ (see Figure 1). The impeller revolution of 600 rpm was used through all of the performed experiments. The speed of the impeller was set to obtain the turbulent regime and also to prevent the oil settling on the bottom of the vessel due to the higher oil density comparing with water. The impeller Reynolds number was equal to Re = 107 262.

The continuous phase was distilled water and the dispersed phase was silicone oil WACKER AP 200. The physical properties of used silicon oil (dispersed phase) and distilled water (continuous phase) are given in Table

* Corresponding author: roman.formanek@fs.cvut.cz

1 for measured temperature T = 23 °C. The dispersed phase volume fraction was 0.00047 v/v (dilute system).

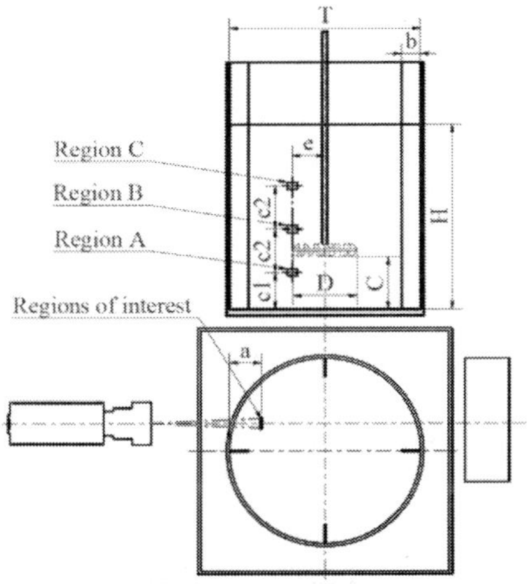

Figure 1. Scheme of the image acquisition system.

Table 1. Physical properties of continuous and dispersed phase.

Phase	Dynamic viscosity μ [mPa.s]	Density ρ [kg.m^{-3}]	Surface tension σ [mN.m^{-1}]
Continuous	0.93	997.54	71.92
Dispersed	228	1075.18	26.42

The image sequences were captured by a high-speed camera SpeedSence MK III with frame rate 30 fps and resolution 1280 x 1024 pixels which was equipped with objective Laowa 60 mm f 2.8 Ultra-Macro 2:1. This combination of camera and objective has no influence on the distortion of the images (Fig. 3a). The one-point high-power light source of own design was used for illumination of all regions. The image distortion was eliminated also using the optical box filled distilled water in which the agitated vessel was placed in a center of optical box (Fig. 1).

The homogeneity of the water-oil emulsion in a volume of the agitated vessel was investigated in three different regions of interest (A, B, C). The images were captured in each region of interest. The regions of interest were placed in different off-bottom clearance in control volume (Fig. 1). The geometrical parameters of regions of interest are following: $c1$ = 50 mm; $c2$ = 70 mm; e = 55 mm; a = 25 mm.

The area of regions was approximately 7.8 x 6.3 mm. The image resolution was 0.0061 mm/pixel for these configurations used. The image resolution was evaluated from 1x1 mm grid captured on the image (Fig. 2a).

2.1 Experimental procedure

The experimental procedure proposed in our previous work [3] was used. The ten sets of 1000 images were captured in all regions of interest (Fig. 2c). The sets were captured after 5 min with a shutter speed of 3 μs and frame rate of 30 fps. At the time t = 0 min silicone oil was added and the first measurement was performed at the time t = 5 min. Thus the total measurement time was 50 min.

The Image Analysis procedure used was calibrated by means of balls of precise diameter (d_p = 1.19 mm) captured in the set of images (Fig. 2b).

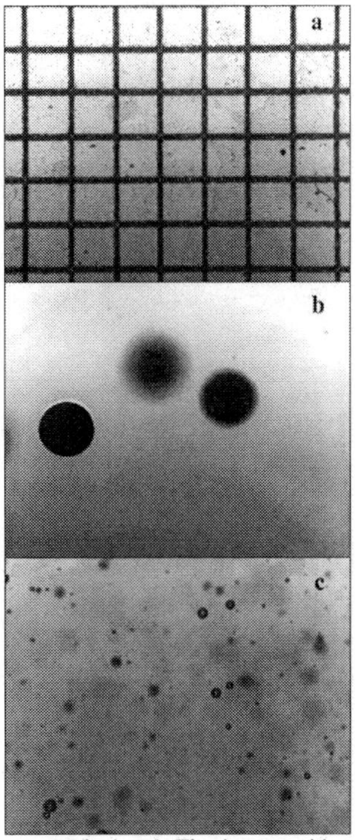

Figure 2. Image analysis: a) The image with captured scale (1x1 mm grid); b) Image with captured precise balls (d_p = 1.19 mm); c) The image with captured drops of dispersed phase.

3 Results

3.1. Number of evaluated drops

As it follows from the theory of drop break-up in a dilute system, the number of evaluated drops should be increasing with increasing time due to disruption of large drops into smaller drops. This prediction has been assumed for the whole agitated system, i.e. regardless of point of location in an agitated vessel. In our previous work [3] we found that time evolution of a number of drops was different from region to region in the system agitated by a Rushton turbine. The question arises, whether this heterogeneity exists also in the system agitated by high-shear Sawtooth impeller.

The number of evaluated drops in dependence on dimensionless time in each time steps and each regions of interest is shown in Figure 3. The dimensionless time was defined as it follows:

$$t^* = N\,t \qquad (1)$$

where N [rpm] is a rotation of impeller and t [min] is an agitation time.

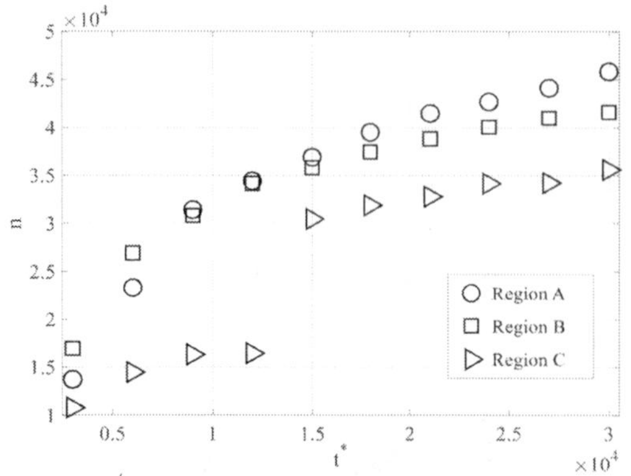

Figure 3. The evolution of the number of evaluated drops.

It was found that the number of evaluated drops increases with increasing dimensionless time in each regions of interest as it was expected. The dimensionless time evolution of a number of evaluated drops is practically the same for region A and B. Unlike this, the number of drops evaluated in region C is significantly lower.

3.2 Sauter mean diameter kinetics

The evolution of drop sizes on dimensionless time was described using by Sauter mean diameter d_{32}. The Sauter mean diameter was calculated using the following formula:

$$d_{32} = \frac{\sum_{i=1}^{m} n_i d_i^3}{\sum_{i=1}^{m} n_i d_i^2} \quad (2)$$

where d_i [mm] is drop diameter in ith class and n_i [-] is a number of the drops in ith class. The d_{32} was evaluated for each time step of measurement and for every region of interest.

Hong and Lee [1] proposed a simple model describing d_{32} kinetics in the following form:

$$d_{32}/d_{32}^{*} = 1 + \alpha e^{-\beta t^{*}} \quad (3)$$

where d_{32} [μm] is the value of Sauter mean diameter at time t^*, d_{32}^* [μm] is the value of Sauter mean diameter at steady state, α and β [-] are model parameters.

The model was tested on the measured data. The evaluated parameters of this model for the measured data are presented in Table 2. The comparison of model and experimental data is presented in Figure 4. The correlation is very good.

The differences between d_{32} kinetics and predicted drop diameter at steady state in each regions are evidently significant. The smallest drop diameter at steady-state d_{32}^* was determined in region C. Comparing with this region the drop diameter is twice and three times higher in region A and region B respectively. The parameters α and β vary similarly.

Table 2. Sauter mean diameter kinetics: evaluated parameters of Hong and Lee [1] model for measured data.

Region	d_{32}^* [μm]	α [-]	$\beta \cdot 10^{-5}$ [-]
A	31.639	1.386	-1.469
B	45.076	0.682	-2.918
C	16.874	3.435	-0.726

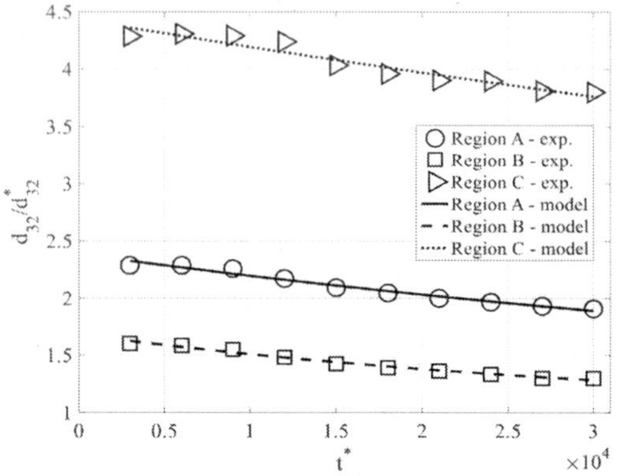

Figure 4. Sauter mean diameter kinetics: comparison of Hong and Lee [1] model and measured data.

3.3 Time evolution of DSD

The drop size distribution was described by Log-normal distribution function plotted as a function of the logarithm of the drop diameter. The cumulative volume fraction was determined for each time step and for each region of interest. For illustration, the time evolution of the cumulative volume fraction of region A is shown in Figure 5.

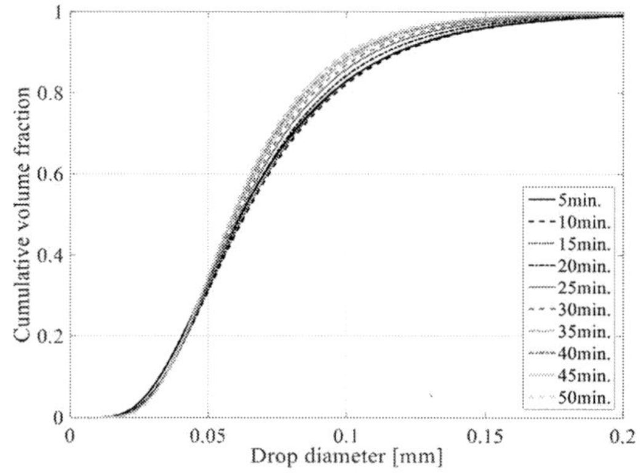

Figure 5. Time evolution of cumulative distribution for measured region A.

It was found that the DSD develops unstably during the first 10 minutes after the oil injection (oil injection at the $t = 0$ min). In this period d_{32} diameter increases and then again decreases. This volatility is probably due to low pumping effect of high-shear Sawtooth impeller. Thus, some "stabilization" time is necessary to overcome this transient state.

The progress of dispersion can be visibly demonstrated comparing the beginning and last run of measurement. At the beginning of drop breaking (5 min) 60 % drops had a diameter lower than 0.75 mm. At the last measurement run (50 min) the percentage of drops having a smaller diameter than 0.75 mm was approximately 70 %.

For comparison the DSD at the beginning and last time step are presented in Figure 6 for each region of interest. The DSD at the beginning was practically the same in all investigated regions of interest. The difference between beginning states was only several percents and it can be negligible from the viewpoint of the number of evaluated drops. In regions A and B the DSDs at last time steps were practically similar. Unlike those, the DSD changes with time in region C were not significantly visible. It signalizes that the drop size distribution was not homogeneous in the agitated vessel for our experimental conditions. In our previous work [3] the same results were obtained for dilute oil-in-water system agitated by a Rushton turbine. It seems that the location of the measurement region influences the taken results in dilute systems.

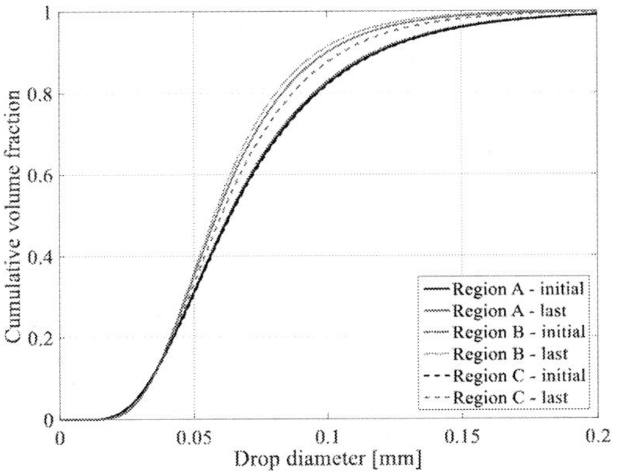

Figure 6. Cumulative distribution at the initial and last time step for each regions of interest.

4 Conclusions

The drop size time evolution and the homogeneity degree of liquid-liquid dispersion were investigated in a vessel agitated by a high-shear Sawtooth impeller. The experiments were executed in a fully baffled vessel with a diameter 300 mm. The tests have been carried out with silicone oil - distilled water dispersion (oil in water) of dispersed phase volume fraction 0.00047.

The in-situ measurement based on image analysis was used for the determination of drop size distribution and Sauter mean diameter. The drop sizes were measured in three regions of interest with different off-bottom clearance. The impeller revolution of 600 rpm was used for all provided experiments.

The number of evaluated drops was found to be increasing with increasing time of measurement in each region of interest as it was expected on the basis of the theory of drop break-up in a dilute system. The number of evaluated particles was in range from 1 to 1.75 x 10^4 at the beginning and from 3.5 to 4.5 x 10^4 at the last time step of measurement.

The time evolution of Sauter mean diameter was found to be different in each regions of interest. The experimental data were treated using the model of d_{32} kinetics proposed by Hong and Lee [1]. The correlation is very good. The model predicts different values of drop diameter at steady-state d_{32}^* in each region of interest. The smallest drop diameter at steady-state d_{32}^* was determined in the region C. Comparing with this region the drop diameter is twice and three times higher in the region A and region B respectively.

The cumulative volume fraction was determined for each time step and for each region of interest. The increasing proportion of small drops with increasing time was significantly evident in the dispersed system in both regions of interest A and B. Unlike those, the change of smaller drops was not so visible in region C.

It was found that the DSD develops unstably during first 10 minutes after the oil injection (oil injection at the t = 0 min). In this period d_{32} diameter increases and then again decreases. Thus, some "stabilization" time is necessary to overcome this transient state. This time period is two times higher comparing with the system agitated by a Rushton turbine [3]. This longer period is caused probably due to the low pumping effect of high-shear Sawtooth impeller.

It was found that the DSD profiles obtained at the beginning (10 min) and at the last time step were different for each region of interest. It signalizes that the drop size distribution was not homogeneous in the investigated regions. The effect of the sampling point is known for strongly coalescing agitated liquid-liquid systems [4]. The possible effect of location on DSD will be tested for dilute liquid-liquid dispersion and various process conditions in future experiments.

It was found that time that is necessary to achieve steady-state is longer for system agitated by high-shear Sawtooth impeller comparing with the system agitated by Rushton turbine probably due to lower pumping effect. Following it the measuring time period must be prolonged for system agitated by Sawtooth impeller.

This work was supported by GA CTU SGS project number SGS18/129/OHK2/2T/12 and by the Ministry of Education, Youth and Sports of the Czech Republic under OP RDE grant number CZ.02.1.01/0.0/0.0/16_019/0000753 "Research center for low-carbon energy technologies".

References

1. P. O. Hong, J. M. Lee, Industrial and Engineering Chemistry, Process Design and Development, **22**, 130–135 (1983)
2. R. Šulc, B. Kysela, P. Ditl, Chemical Papers, **72**, 543-553 (2018)
3. R. Formánek, B. Kysela, R. Šulc, Chemical Engineering Transaction, **74**, 1039-1044 (2019)
4. Sprow, AIChE J., **13**, 995-998 (1967)

Unsteady aerodynamics over surface of a chambered airfoil at stall angle and low Reynolds number

Gamze Genç[1,*], *Kemal* Koca[1], *Mustafa Serdar* Genç[1]

[1]Wind Engineering and Aerodynamic Research Center, Department of Energy Systems Engineering, Erciyes University, Kayseri, TURKEY

Abstract. The proposal of this paper is clear to reveal and understand the unsteady flow characteristics over the surface of chambered airfoil operating especially at low Reynolds numbers by detecting boundary layer separation and laminar separation bubble (LSB). Experiments including smoke-wire technique for flow visualization, velocity measurement via hot-wire sensor and quasi-wall shear stress measurement by means of the hot-film sensor have been performed over the suction surface of NACA 4412 airfoil at Reynolds number of $5x10^4$ and angle of attack of 14°. Experiments indicate that the airfoil at a stall angle exhibits the bistable characteristics over the suction surface because of the unsteady bubble bursting at low Reynolds numbers. That is, either laminar flow separates from the leading-edge of the airfoil owing to a dominant adverse pressure gradient (APG) or it does not separate along downstream. It is also noted that unsteady boundary layer formation may affect the flow characteristics by changing airfoil's stability appreciably, causing the periodic fluctuations with the occurring bubble bursting and the leading edge and trailing edge vortex merging. This may conclude undesirable problems such as noise, vibration and flutter for the flight vehicles and wind turbines.

1 Introduction

Understanding of flow characteristics around airfoils is crucial in engineering systems including micro air vehicles, small-scale wind turbines and unmanned aerial vehicles operating at low Reynolds numbers regime ($1x10^4<Re<5x10^5$) [1], because these directly influence the aerodynamic performance of wind turbines and aircrafts' energy consumption. Therefore, they have been investigated in detail recently. Since the high drag and low lift generated in these flow regimes, the aerodynamic performance of the aforementioned engineering applications has been frequently gained attentions by researchers [2].

The essential impediment for aerial vehicles operating at low Reynolds numbers is laminar boundary layer separation [3]. The flow behaviour at low Reynolds numbers regimes is different from flow at higher Reynolds numbers because APG plays a dominant role and opposes the incoming flow, causing the boundary layer to separate from the surface [4]. After a while, separated flow in the transition region may reattach to the surface by gaining momentum, leading to the occurrence of a laminar separation bubble (LSB) over the surface. The laminar separated shear layer at low Reynolds numbers and higher angles of attack are highly unstable. Therefore, aerodynamic researchers have made a contribution to the understanding of flow development and unstable flow phenomena in the separated region and, especially, transition region [5-7]. Mechanism of boundary layer separation and flow transition in a separated shear layer over a NACA0012 airfoil at $Rec=5x10^4$ was numerically investigated with Direct Numerical Simulation (DNS) by Rodriguez et al. [8]. They found that airfoil exhibited a composition of training / leading edges stall, bringing about massive flow separation. Simoni et al. [9] carried out an experimental study to investigate laminar boundary layer separation, shear layer transition and reattachment phenomena at $Re=7x10^4$ and $Re=1x10^5$. Their conclusion revealed that the essential mechanism of transition is the inviscid Kelvin-Helmholtz instability. They also concluded that lower frequency oscillations must be amplified and reach satiation for emerging of transition. LSB on a NACA0012 airfoil at $Re_c=5x10^4$ was investigated by Jones et al. [10]. They also discussed that the effect of volume forcing is important for the promotion of transition to turbulence. Koca et al.

* Corresponding author: gamzegenc@erciyes.edu.tr

[5], Genç et al. [11-12], Karasu et al. [13], Ducoin et al. [14], Lambert and Yarusevych [15] investigated the vortex dynamics in LSB (leading edge vortex) and wake region of airfoils (trailing edge vortex) with a chance of angle of attack and Reynolds number. Their results showed that the size of LSB decreased and vortex shedding characteristics changed when the angle of attack increased, causing the shedding frequency to increase and the streamwise wavelength to reduce.

The purpose of this letter is to reveal the unsteady flow characteristics, LSB and transition to turbulence formation on NACA 4412 airfoil with experiments including velocity measurement, time-dependent flow visualization and quasi-wall shear stress measurement by means of hot-wire, smoke-wire and hot-film sensor, respectively. All experiments have been conducted at Rec=5x10^4 and angle of attack of (α) of 14°.

2 Experimental Rigs

The experiments have been performed in a low-speed wind tunnel having a square test section of 500mm x 500mm as shown in Figure 1. The detailed information about wind tunnel can be found in the authors' previous studies [5, 16, 17]. NACA 4412 airfoil as the experimental test model has been fabricated by means of 3D printer [11].Two transparent plexiglass end-plates have been attached at each tip of the airfoil to be protected the effect of tip-vortices while allowing the researchers to observe flow patterns over surface easily.

Fig. 1. Experimental set-up.

2.1 Flow visualization system

Regarding smoke-wire and hot-wire experiments, a sketch has been ensured as illustrated in Figure 2. The smoke-wire consists of the electrically resistive heating to burn oil, a wire with 0.03 mm diameter, three halogen lamps, a compact camera and oil. The wire has been vertically positioned at a distance of 1c (chord length) in front of leading-edge of the airfoil. The electrically resistive heating has been used to heat when oil has been manually dropped at the top of the test section. The streamlines visualized by smoke sheets have been, then, captured by means of a compact camera.

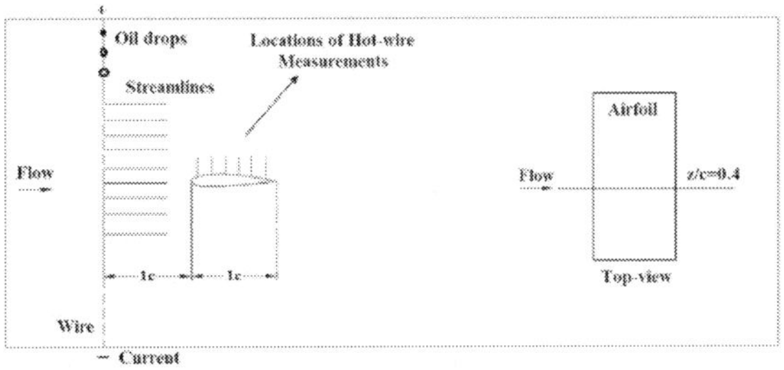

Fig. 2. A sketch of smoke-wire and hot-wire experiments

2.2 Velocity Distribution and Turbulence Statistics

Dual sensor probe has been utilized to obtain velocity distribution and U_{rms} values over suction surface airfoil. After calibration process, hot-wire probe has been positioned and run by means of traverse system and velocity measurements have been performed by starting from x/c=0.1 to x/c=1.0 (x refers to the distance of leading-edge) with 0.1c intervals. During the experiment, 2 kHz has been selected as the sample rate and 20480 data has been collected.

Regarding equations, averages of u velocity component and the velocity standard deviation have been calculated with Equation (1), Equation (2), respectively as follows:

$$U_{mean} = \frac{1}{N}\sum_{i}^{N} U_i \quad (1)$$

$$U_{rms} = \left(\frac{1}{N-1}\sum_{i}^{N}\left(U_i - U_{mean}\right)^2\right)^{0.5} \quad (2)$$

2.3 Shear Stress Measurement

As the researchers found in their study [5, 16, 17], the exploration boundary layer separation, transition and LSB phenomena based on wall shear stress measurement via hot-film sensors have been performed in this letter. Since the total heat transfer from the hot-film sensor may be negatively affected because of undesired factors such as model material and flow properties [18], the calibration process of the hot-film sensor is too complicated. Therefore, the quasi-wall shear stress measurement in this letter has been recognized as carried out by Zhang et al. in their study [19]. They identified:

$$\tau = \left(\frac{E^2 - E_0^2}{E_0^2}\right)^3 \quad (3)$$

where E means output voltage while E_0 indicates the air temperature's voltage with no flow condition in the test section. During the experiment, data has been recorded via multi-channel Constant Temperature Anemometer (CTA) at the line of z/c=0.4 and along the chordwise from x/c=0.1 to x/c=0.8.

3 Results

3.1 Smoke-wire results

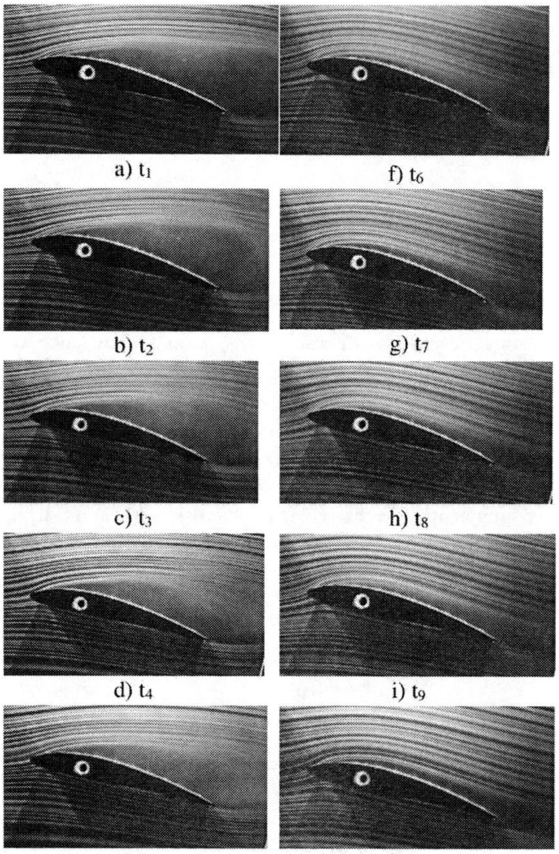

Fig. 3. Results of time-dependent smoke-wire experiment $Re=5x10^4$, $\alpha=14°$.

From the results of time-dependent smoke-wire experiments at $Re=5x10^4$, $\alpha=14°$ in Figure 3, two different flow phenomena have observed over the airfoil owing to unsteady flow at low Reynolds number. In the experiment, duration t_1-t_5, it has been observed that flow separation has occurred at the distance of 15% chord length from the leading edge (x/c=0.15), due to the fact that long bubble burst as a result of the presence of favourable role APG (dP/dx>0). During t_1-t_3 (Figure 3(a)-(c)) the leading-edge separation between x/c=0.15 and x/c=0.70, located between x/c=0.15 and x/c=0.35 at t_4 and t_5 (Figure 3(d)-(e)). This means the shorten bubble, which leading edge vortex becomes small (inherently have bigger frequencies). It causes more vibrations and noise especially at trailing edge of airfoil (another trace of unsteadiness). On the other hand, the condition at which reattachment to the surface occurs has been observed in conjunction with the image sequences indicated from Figure 3(f) to Figure 3(j). Unlike in Figure 3(a)-3(e), APG could not play a dominant role on flow, result in emerging the condition, at which boundary layer separation does not form even though they have been performed with same Reynolds number and angle of attack. As seen the authors' previous study [5], LSB forming at leading-edge of the airfoil especially at α=10° and α=12° might burst based on Reynolds number, resulting in leading-edge separation and abrupt stall. Figure 3 shows the unsteady leading-edge separation based on the interaction of APG and laminar flow.

3.2 Velocity Distribution and U_{rms} Results

u/U velocity profile over the airfoil obtained from hot-wire measurement results have been illustrated as seen in Figure 4. In this figure, the boundary layer (dashed line) and separations (solid line) were sketched based on velocity profile over the airfoil. It is clearly shown that the flow at x/c=0 and x/c=0.1 is laminar. As observed at results of the smoke-wire experiment where flow separation has occurred, the velocity at x/c=0.2 has abruptly decreased in the boundary layer, indicating that flow has started to separate from the solid surface because of APG. Moreover, it has been noticed that larger vortices trajectory has occurred near the boundary layer. These statements have caused the flow, which has different frequencies (a trace of unsteadiness), to occur to the surface of the airfoil. Furthermore, moving downstream the velocity fluctuations owing to the boundary layer instability have been amplified. These fluctuations having large momentum cause the unstable flow in the boundary layer to be had. Regarding the more detailed information of different physical mechanisms as observed at a study performed by Deng et al. [20], it has been revealed that surface (or leading-edge vortex) vortex (SV) between at x/c=0.6 and x/c=0.8 occurs near the boundary layer. After x/c=0.8, larger vortex called as trailing-edge vortex (TEV) having more momentum are formed due to the wake effects. These two flow characteristics near the boundary layer cause the alternating downstream shedding to occur, resulting in different periodic characteristics in the wake flow [5].

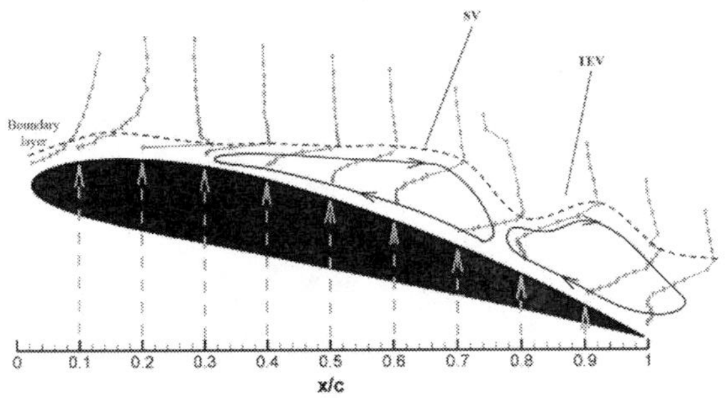

Fig. 4. u/U velocity profile in the boundary layer of NACA 4412 airfoil.

Figure 5 shows the U_{rms} velocity profile, turbulent kinetic energy contours, separations (solid), the boundary layer (dashed) and maximum RMS (dash-dotted) lines sketched by authors based on velocity measurements over the airfoil. By comparison with Figure 4, the flow is laminar before x/c=0.2. After x/c=0.2, leading-edge flow separation has occurred. The maximum RMS line corresponds to the middle part of the SV and TEV separations. At the same time, it has been also pointed out that the highest values of maximum RMS line correspond with core region of the turbulent kinetic energy contour. Besides, the fluctuations in the flow have simultaneously increased along the downstream, promoting complex and chaotic flow behaviour at the trailing edge mostly due to the fact that the flow separations over the suction surface of the airfoil encounter with those coming from pressure surface. This situation is understood from zigzags in U_{rms} profile near the trailing edge.

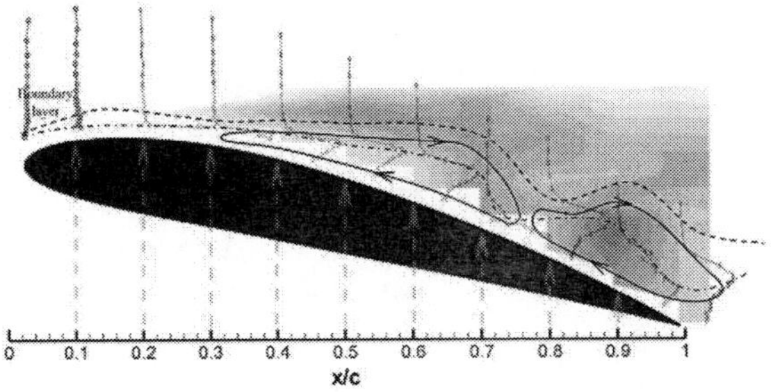

Fig. 5. U_{rms} profile and turbulent kinetic energy contours over the NACA 4412 airfoil.

3.3 Shear Layer Measurement Result

The different pathway of understanding of the physical characteristic of flow by performing the definite near-surface flow measurements has been shown with the results of hot-film sensors as denoted in Figure 6. y-axis at graphs illustrates the time-dependent voltage variance, whilst x-axis indicates the time (data has been recorded at 10 seconds during the hot-film experiment). As the flow is under the laminar regime before at x/c=0.2, therefore the voltage fluctuation is low. At x/c=0.2, surges in the voltages have started to increase, indicating the occurrence of boundary layer flow separation. Moreover, fluctuations in the voltage variances have apparently increased the downstream and have reached the peak levels especially x/c=0.6, x/c=0.7 and x/c=0.8. These means that more energized SVs and TEVs have been occurred at the trailing edge. Moreover; voltage undulations changes based on the time and occasionally the voltage peak occurs intermittently, which points out unsteady flow characteristic over the airfoil at the stall angle.

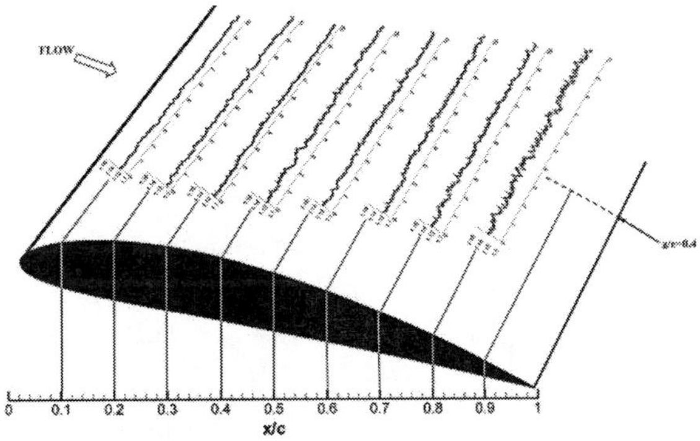

Fig. 6. Time-dependent voltage variance results of hot-film sensor.

4 Conclusion

As a conclusion, the experimental results have proven that the flow condition under low Reynolds number exhibits different regime over the surface of NACA 4412 airfoil: (*i*) boundary layer separation without reattachment due to the long bubble bursting, (*ii*) attached flow with a bubble. Besides, the frequency bands of fluctuations under the flow separation have exponentially amplified along downstream of airfoil and they have reached the top level probably with saturation of energized flow, resulting in emerging of unsteady and chaotic flow characteristics. Experimental results have also revealed that they exhibit a subordinate role with each other. Unsteady flow condition over the surface of the airfoil has inevitably emerged with these two flow phenomena, affecting flow regimes or aerodynamic performances of flight vehicles operating at low Reynolds number negatively.

References

1. B. H. Carmichael, NASA CR. (1981).
2. T. J. Mueller, J. D. DeLaurier, Annu. Rev. Fluid Mech. 35, 89-111, (2003)
3. E. J. Fitzgerald, T. J. Mueller. AIAA J, **28(4)**, 584-592, (1990).
4. M. O'meara, T. J. Mueller. AIAA J, **25(8)**, 1033-1041, (1987).
5. K. Koca, M. S. Genç, H. H. Açıkel, M. Çağdaş, T. M. Bodur. Energy, **144**, 750-764, (2018).
6. M. S. Genç, İ. Karasu, H. H. Açıkel. Exp. Therm. Fluid Sci., **39**, 252-264, (2012).
7. K. Koca, M.S. Genç, H.H. Açıkel. Çukurova Üniversitesi Mühendislik-Mimarlık Fakültesi Dergisi, **31(ÖS2)**, 127-134 (2016)
8. I. Rodríguez, O. Lehmkuhl, R. Borrell, A. Oliva, Int. J. Heat Fluid Flow, **43**, 194-203, (2013).
1. D. Simoni, M. Ubaldi, P. Zunino, D. Lengani, F. Bertini. Flow Turbul. Combust. **88**, 45-62, (2012).
9. L. E. Jones, R. D. Sandberg, N. D. Sandham. J. Fluid Mech., **602**, 175-207, (2008).
10. M. S. Genç, K. Koca, H. H. Açıkel, G. Özkan, M. S. Kırış, R. Yıldız. In EPJ Web Conf., **114**, 2029, (2016).
11. M.S. Genç, G. Özkan, M. Özden, M.S. Kırış, R. Yıldız. Proc IMechE, Part C: Journal of Mechanical Engineering Science, **232(22)**, 4019-4037 (2018)
12. I. Karasu, M. Özden, M.S. Genç, Journal of Fluids Engineering-Transactions of The ASME, 140(12) 121102-1-121102-15 (2018).
13. A. Ducoin, J. C. Loiseau, J. C. Robinet. Eur. J. Mech: B/Fluids, **57**, 231-248, (2016).
14. A. Lambert, S. Yarusevych. Phys. Fluids, 31(6), 064105, (2019).
15. M. S. Genç, K. Koca, H. H. Açıkel. Energy, **176**, 320-334, (2019).
16. H. H. Açıkel, M. S. Genç. Energy, **165**, 176-190, (2018).
17. C. Ghouila-Houri, J. Claudel, J. C. Gerbedoen, Q. Gallas, E. Garnier, A. Merlen, R. Viard, A.
18. Talbi, P. Pernod. Appl. Phys. Lett., **109(24)**, 241905, (2016).
19. H. H. Bruun. Meas. Sci. Technol.,7, 10, (1996).
20. X. Zhang, A. Mahallati, S. Sjolander. AIAA Conf., 3643, (2002).
21. J. Deng, L. Sun, X. Shao. Phys. Fluids, 31(2), (2019).

Experimental investigation on effect of partial flexibility at low aspect ratio airfoil – Part I: Installation on suction surface

Mustafa Serdar Genç[1,*], *Halil Hakan* Açıkel[1], *Kemal* Koca[1]

[1]Wind Engineering and Aerodynamic Research Center, Department of Energy Systems Engineering, Erciyes University, Kayseri, TURKEY

Abstract. Effects of flexible membrane mounted over suction surface of NACA 4412 airfoil were experimentally investigated at Reynolds number of $5x10^4$ and low aspect ratio (AR=1) in this paper. The smoke-wire visualization method has been performed for flow visualization to demonstrate flow phenomena as laminar separation bubble (LSB), leading edge separation at z/c=0.4 and tip vortices at z/c=0.1. Values of velocity, Reynolds stress and turbulence statistics were measured by means of a constant temperature anemometer (CTA) system. Results of smoke-wire experiment revealed that size and height of LSB formed along z/c=0.4 at lower angles of attack such as α=8° was mitigated. Moreover, stall phenomenon as a result of boundary layer separation was apparently postponed at higher angles of attack. Velocity value was increased, whereas values of Reynold stress and turbulent kinetic energy was decreased with reduction of amount of fluctuations in flow. Consequently, using flexible membrane over suction surface of airfoil allowed the LSB to be mitigated or extinguished, resulting in exhibition of more stable flow characteristics.

1 Introduction

Micro Air Vehicles (MAVs) mostly run under Reynolds (Re) number of $1x10^5$ [1-4]. In this flow regime, accurately determination of LSB and prediction of transition phenomenon inside LSB are obligatory. Because, these two flow phenomena can occur the suction surface of airfoils, resulting in the abrupt reduction at aerodynamic performance [4-9] because of low Reynolds number airfoil drag generally depends upon position of transition (essentially in a short LSB) as a function of angle of attack. Until now, aerodynamic researchers have carried out many experimental and numerical methods to investigate and observe the location of LSB and transition. Moreover, they have applied few control techniques such as acoustic excitation [10-11], leading-edge slats [12], vortex generators [13], roughness material [14-15] since they want to eliminate LSB and hinder the deterioration of aerodynamic efficiency.

In addition to these control techniques, the bio-inspired design such as flexible membrane for MAVs have been started to gain interests and reputations recently. Maximum stall angle and lift coefficient as well as vehicle agility can be ensured with the use of flexible membrane wings for MAVs [16-32].

In this study, the experiments containing smoke-wire and hot-wire measurements are performed for partial flexible NACA 4412 wing, which has aspect ratio (AR) of 1, at Reynolds number of $5x10^5$ and angle of attack (α) of 8°, 12°, 20°. In addition, the results of partial flexible airfoil are compared with the results of our former study [19] to see how flexible membrane affects the aerodynamic performance.

2 Experimental Rigs

The wing airfoil has 18 cm chord and 18 cm span length. The experiments are carried out wind tunnel where the free stream turbulence intensity is quite low. The detailed information with regards to wind tunnel has been ensured in Ref. [9, 14-15]. 3D printer has been utilized for manufacturing process of partial flexible wing. As shown in the Figure 1, membrane is mounted between x/c=0.2 and x/c=0.7 since LSB and separation, which affects the aerodynamic performance negatively, occurs between these points. As the technical features of flexible membrane, latex rubber sheet which has Young's modulus (E) of 2.2 MPa and density (ρ_m) of 1 g/cm^3 [19] is 0.2 mm. It is adhered onto rigid skeleton of airfoil with double side tape.

* Corresponding author: musgenc@erciyes.edu.tr

As demonstrated in the Figure 2, measurements are conducted along z/c=0.1 and z/c=0.4. The experiments along z/c=0.1 ensures tip vortices and their effects very well, whilst LSB and flow separation are especially obtained from experiments along z/c= 0.4.

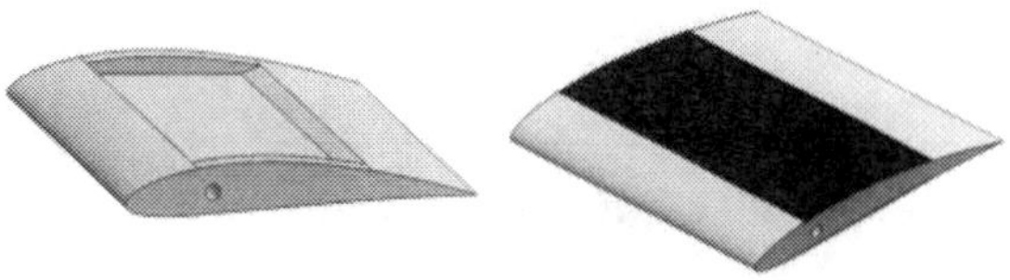

Fig. 1. The fabricating process of partial flexible airfoil [19].

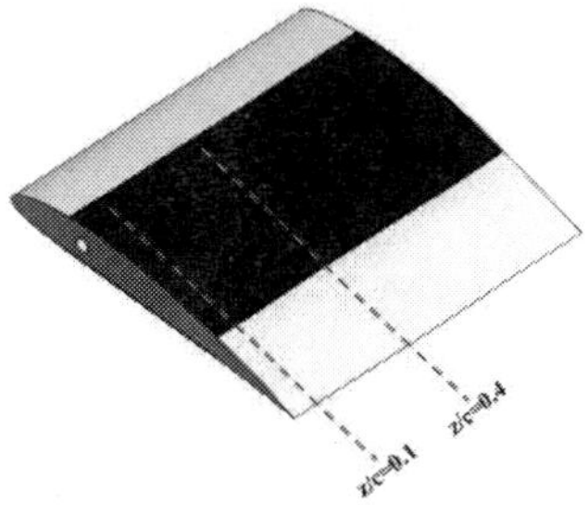

Fig. 2. Sketch of the measurements line.

2.1 Flow visualization system

Tip vortices occurred at tip of NACA 4412 and flow separation, LSB formed over suction surface have been visualized by means of smoke-wire experiment. The essential components of this technique are electrical resistive heating, 0.03 mm diameter steel wire, an oil mixture, a compact camera (the Canon EOS-D1100 type), and two or three halogen lambs. The arrangement of smoke-wire technique, the most prevalent of which has been performed by Koca et al. [9] and Genç et al. [14-15], can be found in mentioned References.

2.2 Velocity Distribution and Turbulence Statistics

In addition to the smoke-wire experiment, hot-wire experiment has been performed to investigate flow velocity, turbulence intensity and Reynolds stresses. Two-dimensional wire probe has been functioned with utilizing of a constant temperature anemometer (CTA) at the over and wake of airfoil (at the distance of 1c and 2c). The probe has been run with the intervals of x/c=0.1 along the chord-wise via traverse system. 20480 data have been obtained with a sample rate of 2 kHz at selected x/c. Furthermore, the reference velocity probe has been used for probe calibration and the calibration error has been measured about ±1%.

Regarding equations, averages of u and v velocity components, the velocity standard deviation, the turbulence intensity and Reynolds stresses have been calculated with Equation (1), Equation (2), Equation (3), Equation (4), respectively as follows:

$$U_{mean} = \frac{1}{N}\sum_{i}^{N} U_i \tag{1}$$

$$U_{rms} = \left(\frac{1}{N-1}\sum_{i}^{N}\left(U_i - U_{mean}\right)^2\right)^{0.5} \tag{2}$$

$$Tu = \frac{U_{rms}}{U_{mean}} \tag{3}$$

$$\overline{u'v'} = \frac{1}{N}\sum_{i}^{N}(U_i - U_{mean})(V_i - V_{mean}) \tag{4}$$

3 Results

3.1 Smoke-wire results

It is observed from Figure 3(a) that the flow separates (referred as **S** with green arrow) from x/c=0.2 and almost reattaches (referred as **R** with green arrow) to x/c=0.8, resulting in formation of long separation bubble. On the other hand, it is clearly seen that size of LSB is reduced with the using of partial flexible membrane. Because LSB is occurred between x/c=0.2 and x/c=0.5. Flexible membrane causes the flow to gain momentum and this momentum transfer ensures the separated flow to reattaches at x/c=0.5. Therefore, the reduction of 50% (0.3c) at size of LSB is observed.

a)
b)

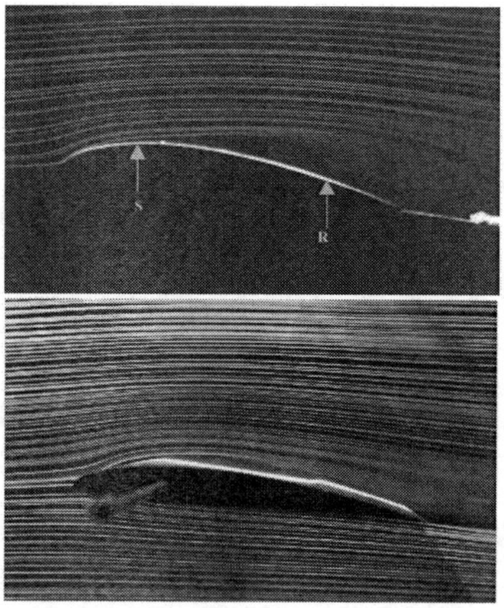

Fig. 3. Smoke-wire results for Re=5x10^4, α=8° and z/s=0.4 a) uncontrolled case [19], b) partial flexible airfoil.

Figure 4 and Figure 5 have indicated tip-vortices formation at α=12° and at α=20° for both uncontrolled case and partially flexible airfoil, respectively. It has been clearly noticed from the figures that size of tip-vortices has decreased with using of membrane material. Moreover, the wake region of partially flexible skinned airfoil has narrower than the wake region of uncontrolled case, meaning that airfoil drag force and coefficient decreased. Thereby, utilizing the membrane material over suction surface of the airfoil causes aerodynamic performance to increase.

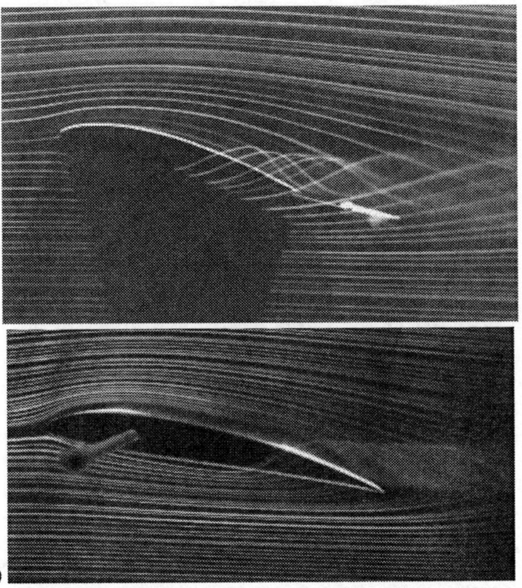

Fig. 4. Smoke wire results for Re=5x10^4, α=12° and z/s=0.1 a) uncontrolled case, b) partial flexible airfoil.

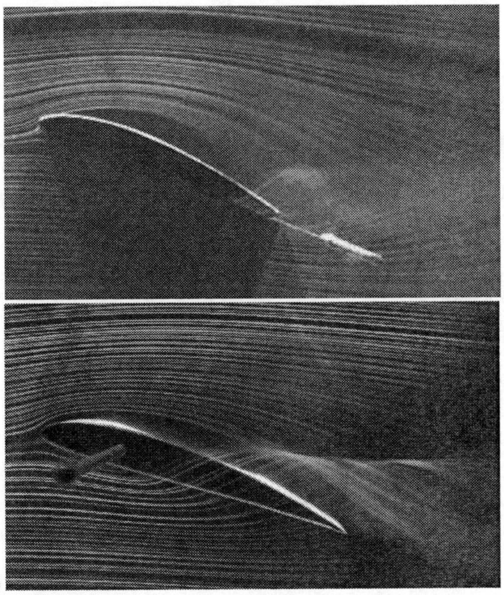

Fig. 5. Smoke wire results for Re=5x10^4, α=20° and z/s=0.1 a) uncontrolled case, b) partial flexible airfoil.

3.2 Velocity Distribution and Turbulence Statistics

Velocity measurement and turbulence statistics including turbulence intensity and Reynolds stresses at Re=5x10^4 and α=8° are investigated for better understanding of using of partial flexible membrane over suction surface of airfoil. Uncontrolled case of velocity measurement (in Figure 6(a)) reveal that flow separation which is caused by adverse pressure gradient (APG) occurs between x/c=0.4 and x/c=0.5 and lower velocity region at leading-edge is observed. But, as seen in Figure 6(b), amount of velocity over suction surface is apparently increased via flexible membrane, causing the increase of lift and aerodynamic performance. Furthermore, separated flow does not nearly occur over suction surface and energized flow by using flexible membrane causes the separated shear layer to approach the surface of airfoil.

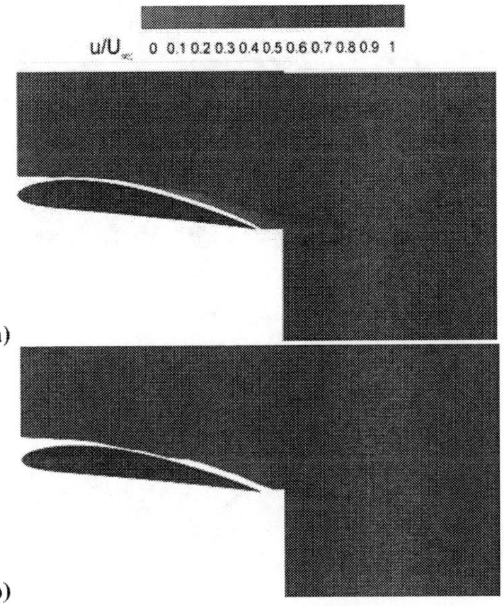

Fig. 6. Velocity measurements results for Re=5x10^4, α=8° and z/s=0.4 a) uncontrolled case, b) partial flexible airfoil.

Results of Reynolds stresses and turbulent intensity for both uncontrolled case and partially flexible airfoil are shown in Figure 7 and Figure 8, respectively. Different contour colours at both Reynolds stress and turbulent intensity indicate amount of fluctuations in the shear layer and flow-field. As seen in Figure 7(a) and Figure 8(a), fluctuations have increased at x/c=0.42, showing the transition to turbulence. As remarked at study of Koca et al. [9], vortex shedding driven by both LSB and trailing edge have especially played a dominant role between x/c=1.1 and x/c=1.25 (near wake of airfoil). This means that larger vortices having lower frequency

 https://doi.org/10.1051/epjconf/202226901017

have occurred in this region, causing the drag forces to increase. On the other hand, transition point is x/c=0.39 when membrane material has been used over suction surface as a flow controller. That is, transition point has moved towards to leading edge. Unlike the uncontrolled case, fluctuations because of vortex sheds at near wake are less, meaning less drag force and steadier flow in terms of aerodynamic performance.

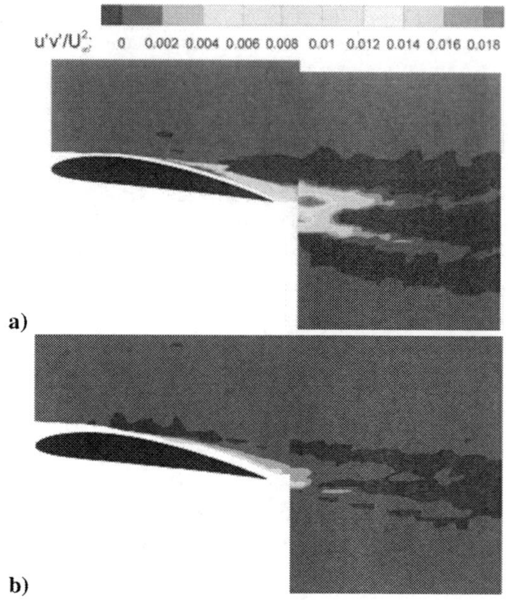

Fig. 7. Reynolds Stress results for Re=5x10^4, α=8° and z/s=0.4 a) uncontrolled case, b) partially flexible airfoil.

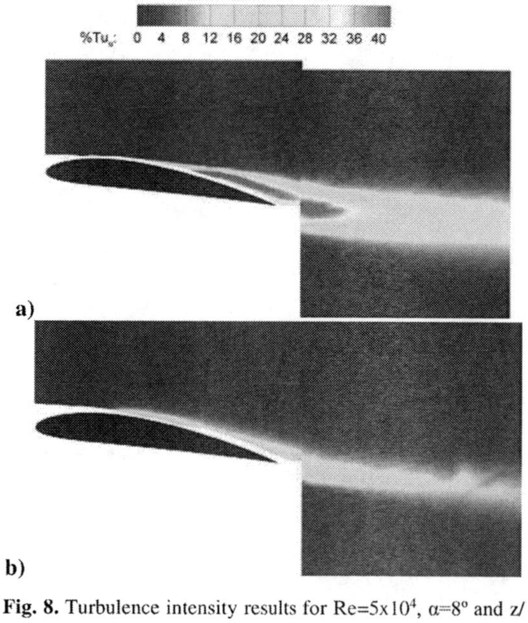

Fig. 8. Turbulence intensity results for Re=5x10^4, α=8° and z/s=0.4 a) uncontrolled case, b) partially flexible airfoil.

4 Conclusion

The objective of this experimental based study is to investigate the flow phenomena such as flow separation and LSB over suction surface of NACA 4412 airfoil with AR=1 at Re= 5x10^4 and different angles of attack. After flow investigation, flexible membrane material has been partially mounted between x/c=0.2 and x/c=0.7 to understand how membrane material has affected the flow characteristics under same flow condition. Smoke-wire experiment has been performed for z/c=0.1 and z/c=0.4 to observe LSB formation, flow separation and tip vortices at α=8°, α=12° α=20°. Additionally, hot-wire experiment have been employed to obtain velocity distributions, Turbulent intensity and averaged Reynolds stresses at α=8°. Results showed as follows:

- Smoke-wire results along z/c=0.4 indicated that along LSB have formed over suction surface of uncontrolled case between x/c=0.2 and x/c=0.8 at α=8°. Utilizing flexible membrane material over surface have obviously affected the flow characteristics especially by decreasing the size of LSB.

- Smoke-wire results along z/c=0.1 showed that increasing of angles of attack from α=12° to α=20° have led to improve the formation of tip-vortices throughout mid-span and wake region of uncontrolled case. For both angles of attack, tip-vortices have gradually been narrower especially at near wake region as partially flexible membrane material have been used, causing effect of drag forces to mitigate.

- Velocity distribution from hot-wire results demonstrated that LSB and flow separation have formed near at mid-span of uncontrolled case and α=8° because of APG, while these flow phenomena have suppressed by making the flow to reattaches to surface with use of flexible material.

- Finally, vortex sheds and amount of fluctuations obtained from turbulent statics including turbulent intensity and Reynolds stresses have illustrated that onset of transition have occurred at x/c=0.42 and α=8° and it has moved towards to leading-edge as flexible membrane has been used. Moreover, vortex sheds coming from both LSB and trailing-edge have played prevailing role on near wake region for the uncontrolled case, whilst mounting flexible material have restrained the power of fluctuations by doing damping effect, resulting in presence of narrower wake region (inherently meaning of lower drag forces).

References

1. Tani I., Prog Aerosp Sci, 5: 70–103, (1964).
2. Gaster, M. Aeronautical Research Council Reports and Memoranda, No: 3595, London, (1967).
3. T.J. Mueller, J.D. DeLaurier. Annual review of fluid mechanics, **35(1)**, 89-111 (2003)

4. M.S., Genç, I. Karasu, H.H. Açıkel, M.T., Akpolat. In Low Reynolds number aerodynamics and transition. InTech. (2012)
5. H. Demir, M. Özden, M.S. Genç, M. Çağdaş. In EPJ Web of Conferences (Vol. **114**, p. 02016). EDP Sciences, (2016)
6. M.S. Genç, K. Koca, H.H. Açıkel, G. Özkan, M.S. Kırış, R. Yıldız. In EPJ Web of Conferences (Vol. **114**, 2029 (2016)
7. M.S. Genç, G. Özkan, M. Özden, M.S. Kırış, R. Yıldız. Proc IMechE, Part C: Journal of Mechanical Engineering Science, **232(22)**, 4019-4037 (2018)
8. I. Karasu, M. Özden, M.S. Genç, Journal of Fluids Engineering-Transactions of The ASME, 140(12) 121102-1-121102-15 (2018).
9. K. Koca, M.S. Genç, H.H. Açıkel, M. Çağdaş, T.M. Bodur. Energy, **144**, 750-764 (2018)
10. H.H. Açıkel, M.S. Genç. Proceedings of the Institution of Mechanical Engineers, Part G: Journal of Aerospace Engineering, **230(13)**, 2447-2462 (2016).
11. M.S. Genç. H.H. Açıkel, M.T. Akpolat, G. Özkan, İ. Karasu, Journal of Aerospace Engineering-ASCE, 29 (6), 04016045, 2016.
12. M.S. Genç, Ü. Kaynak, G.D. Lock. Proc IMechE, Part G: Journal of Aerospace Engineering, **223(3)**, 217-231 (2009)
13. H. Wang, B. Zhang, Q. Qiu, X. Xu. Energy, **118**, 1210-1221 (2017)
14. K. Koca, M.S. Genç, H.H. Açıkel. Çukurova Üniversitesi Mühendislik-Mimarlık Fakültesi Dergisi, **31(ÖS2)**, 127-134 (2016)
15. M.S. Genç, K. Koca, H.H. Açıkel. Energy, **176**, 320-334 (2019)
16. W. Shvy, F. Klevebring, M. Nilsson, J. Sloan, B. Carroll, C. Fuentes. Journal of Aircraft, **36(3)**, 523-529 (1999)
17. Y. Lian, W. Shvy, D. Viieru, B. Zhang. Progress in Aerospace Sciences, **39(6-7)**, 425-465 (2003)
18. P. Rojratsirikul, M.S. Genç, Z. Wang, I. Gursul. Journal of Fluids and Structures, **27(8)**, 1296-1309 (2011)
19. H.H. Açıkel, M.S. Genç. Energy, **165**, 176-190 (2018)
20. Genç, M. S., Açikel, H. H., Demir, H., Özden, M., Çağdaş, M. and Isabekov, EPJ Web of Conferences. 114 02028 (2016).
21. Genç, M. S., Özden, M., Açikel, H. H., Demir, H. and Isabekov, EPJ Web of Conferences. 114, 02030 (2016).
22. Demir, H. and Genç, M. S. European Journal of Mechanics-B/Fluids, 65, 326-338 (2017).
23. Shyy, W., Berg, M. and Ljungqvist, D. Progress in aerospace sciences, 35(5), 455-505 (1999)..
24. Timpe, A., Zhang, Z., Hubner, J. and Ukeiley, L. Experiments in Fluids, 54(3), 1-23 (2013).
25. Genç, M. S. Experimental Thermal and Fluid Science, 44, 749-759 (2013).
26. Tregidgo L., Wang Z., Gursul I., Aerospace Science and Technology, Vol. 28, pp. 79-90, (2013).
27. Beguin B., Breitsamter C., Aerospace Science and Technology, 37, 138-150, (2014).
28. Alioli M., Masarati P., Morandini M., Albertani R., Carpenter T., Aerospace Science and Technology, 69, 419-431, (2017).
29. Nguyen A.T., Han J.H., Aerospace Science and Technology, 79, 468-481, (2018).
30. Ryu Y.G., Chang J.W., Chung J., Aerospace Science and Technology, 86, 558-571, (2019).
31. Bleischwitz, R., De Kat, R., & Ganapathisubramani, B. In 45th AIAA Fluid Dynamics Conf. 2764, (2015).
32. Hu, H., Kumar, A. G., Abate, G., & Albertani, R. Aerospace Science and Technology, 14(8), 575-586 (2010).

Planar laser-induced fluorescence measurement of the angular pattern of the cone-shaped spray

Oleg Gobyzov*, *Mikhail* Ryabov, *Konstantin* Inzhevatkin, *Artur* Bilsky, *Dmitriy* Markovich

[1]Kutateladze Institute of Thermophysics, Siberian Branch of Russian Academy of Sciences, 630090, Novosibirsk, Russia

Abstract. In the present work, a PLIF-based approach for a large cone-shaped spray patternation is presented and experimental results of its application are discussed. The patternation approach is based on simple time-averaged PLIF imaging. Special attention is paid to the image and data processing in order to simplify a patternation results interpretation. Experimental testing was performed on a water spray, formed by centrifugal nozzle at atmospheric ambient pressure. Circumferential relative concentration profiles in spray crosswise plane are presented in the paper and their validity is analysed.

1 Introduction

Spray testing is one of the important problems for many different practical applications. Despite the many years of development and a large number of different techniques developed for the task, spray testing still remains challenging. One of the major tasks in spray testing is a spray patternation, in other words – acquisition of spatial structure of a spray. In industrial measurements, patternation is usually performed with mechanical patternators that collect portions of the spray into special coleector bins. Mechanical patternators measure time-averaged mass-flow rate within the section of the spray.

This technique is straightforward and robust, but have certain inherent disadvantages. For example, mechanical patternators lack flexibility and often a specific configuration of collectors is required for different nozzles. Besides that, their spatial resolution is restricted by the construction and size of the spray collector bins, which often hinders the interpretation of test results. Considering this, an optical patternation is seen as a promising alternative. Optical setups are nonintrusive, more flexible, and can provide data with higher spatial resolution.

A large number of optical methods are available for spray testing: elastic (Mie) scattering, planar laser-induced fluorescence (PLIF), phase-doppler particle analysis (PDPA), time-shift technique, various methods based on laser light extinction or attenuation, and many others. Pointwise techniques, very powerful for in-depth investigation of atomization, in general are not as well suited for the patternation tasks as planar or volumetric techniques. An optical patternation based on the line-of-sight laser light extinction/transmission [1] is one of the specialized optical methods for spray patternation that gained certain recognition. This method employs tomographic reconstruction and yields local liquid surface area distribution in spray crossection.

An elastic (Mie) scattering and PLIF has become wide-spread techniques for spray studies as they allow for detailed spatially-distributed measurements and employ rather multipurpose equipment, such as lasers and digital cameras. Additionally, in PLIF the data acquired is directly associated with the volumetric concentration of liquid in the measurement plane.

A large number of examples of spray diagnostics with PLIF and elastic scattering, as well as modifications of these techniques, is presented in papers, yet only a small number of them is focused on the patternation task, i.e. measurements in crosswise plain with quantification of spray structure. For example in [2] a PLIF and Mie scattering patternation of a fuel injector in gas-turbine combustor was performed. Authors employed several optical methods to describe the spray structure and had investigated an evolution of the spray with increase of a distance from a nozzle exit. A circumferential (angular) patternation of a spray generated by an aero-engine atomizer at elevated pressure was presented in [3], though this work did not provide details about experimental technique and data processing. Another example is the work [4], in which authors used dual light-sheet lighting to reduce the laser light attenuation effect and validated the optical patternation results with mechanical patternator measurements.

In the present work, a PLIF-based approach for a large cone-shaped spray patternation is presented and experimental results of its application are discussed. Sprays of such kind are quite common in fire extinguishing systems, irrigation systems and in some of the continuous fuel injection nozzles.

* Corresponding author: oleg.a.g.post@gmail.com

2 Experimental setup and measurement system

PLIF system used in experiments included a continuous DPSS laser (532 nm emission wavelength) with light-sheet forming optics and a CCD-camera with a lens and a colored glass filter used to suppress light at emission wavelngth. A spray was formed by a centrifugal nozzle, a diameter of the spray in the patternation plane was about 350 mm. A nozzle was connected to a supply line through a straight swivel fitting which allowed to revolve a nozzle around its axis of symmetry. A distilled water doped with a small amount of Rhodamine 6G luminophore (30 μg per liter) was used as a working liquid.

Patternation by means of PLIF assumes that spray is illuminated with a light sheet aligned in a horizontal plane, orthogonal to the spray nozzle axis. Usually, a collimated light sheet is used to avoid the energy variation along the light-sheet, but forming a collimated light-sheet requires optical elements of a size, comparable to the size of the measurement area and thus technically challenging in case of large spray. For this reason, a divergent light sheet was employed.

A camera was placed above the spray nozzle at an angle to the measurement area. In this case, as the size of the spray is comparable to the distance from the measurement plane to the lens, different light collection angles for different parts of the spray, as well as the different light path length through the spray volume can affect a PLIF image. On the whole, the selected PLIF system configuration was simple in realization but had many potential sources of error that had to be corrected through the imaging and data processing procedures. A scheme of the PLIF experiment is shown in figure 1.

During measurements, a nozzle was rotated around its axis of symmetry with a step of 15°, and for each position (24 positions in total) a 16-bit image with a long exposure of about 1 second was captured. To perform a perspective distortion correction (geometric reconstruction) an image of a calibration target placed in a light-sheet plane was captured before measurements. Set of images captured at different nozzle rotation angles θ were used in correction procedures and for estimation of the measurement's uncertainty.

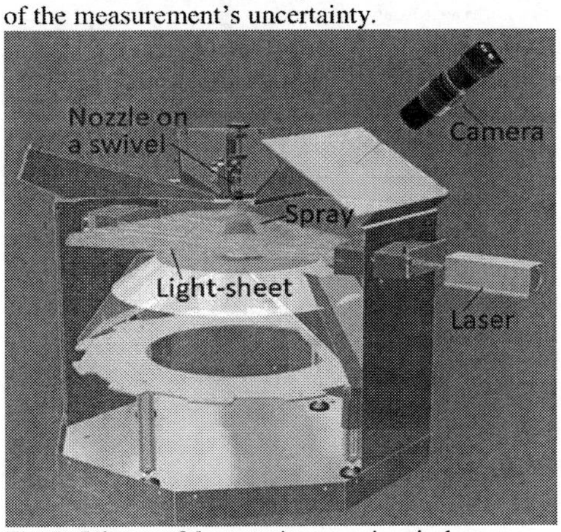

Fig. 1. Scheme of the experiment and optical setup.

3 Data processing

The data processing consisted of several steps: image geometric reconstruction, identification of the spray pattern center, coordinate system transformation and integration of the signal, correction and normalization of the spray angular profile.

Geometric reconstruction was performed using a calibration target image and a back-projection algorithm with a second-order polynomial model [5].

The geometric reconstruction procedure was followed by a spray pattern center search. There can be various definitions of the spray center and, therefore, different methods for estimation of its position. The spray center can be fixed beforehand, as in circumferential mechanical patternators, assuming that the nozzle and measurement area are perfectly aligned and the spray is not tilted. In other cases, a center can be set manually by the operator. Sometimes a centroid is considered as a spray center, as in [3], but in a spray with strong local nonuniformities, centroid position will be biased from the 'geometric' center of the spray pattern. Assuming that an 'expected' spray shape is known in advance and that its variations are relatively small a shape approximation can be employed. For example, for hollow or filled cone sprays an approximation by a circle or ellipse can be feasible. In the present work, a right circular hollow cone spray was an 'expected' shape in experiments and a circle approximation in the measurement plane was employed. To perform an approximation, two image brightness maxima positions (one maximum in each half of the image) over each of the image rows and columns were detected, and then positions of the maxima were approximated by a circle. Obviously, outliers can appear in the list of image brightness maxima due to the spray shape deformation, stray light in the image or hot pixels in the sensor. Strong outliers bias the simple least-squares fitting, thus a more robust RANSAC [6] circle parameters estimation was used to prevent bias. To define an outer boundary of the spray pattern, a circle radius was increased until the average intensity over its perimeter was smaller than a threshold value (which was set slightly above the background brightness).

After the center and a bounding radius of a spray were evaluated, a transformation of the image to the polar coordinates with a coordinate system center at a found circle center, and numerical integration of the image intensity over the sectors with defined central angle was performed.

Preliminary tests of the processing procedure on synthetic images was performed to ensure its viability. Synthetic images were ring-shaped patterns with radial intensity gradient, with introduced Gaussian local shape and brightness distortions, random brightness noise up to 5% of the maximum brightness, and nonuniform stray light distribution with brightness from 0 up to 10 % of the maximum signal level. An example of the synthetic image and corresponding circumferential profile are shown in fig 2.

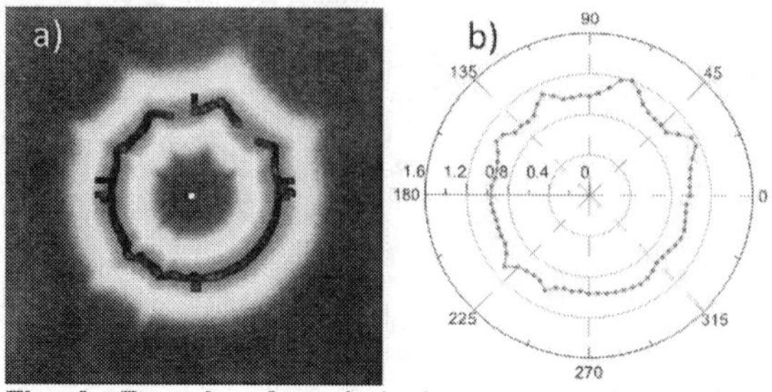

Fig. 2. Example of synthetic image (a) with identified valid maxima, center and outer boundary, and corresponding circumferential profile (b).

The processing procedure proved to be robust towards introduced distortions and yielded the center position with the precision better than 2 pix (0,2 % of the synthetic image size and about 0,4 % of the characteristic ring diameter). From the working principle of the RANSAC algorithm, it is expected to remain stable until large-scale structured spray shape distortions take place, or, in other words, until a concurrent fitting model appears.

4 Experimental results and discussion

The described processing procedure was applied to the set of experimental images. To suppress brightness variations produced by an excitation light intensity variations (figure 3) a light-sheet correction procedure was employed beforehand. To perform the correction, a test section was filled with a fine mist of water doped with luminophore using a Laskin nozzle aerosol generator. After the mist got settled in a test section, a set of PLIF images of the mist were captured and a standard light-sheet correction of spray images was performed as follows:

$$I_c(x,y) = I_0(x,y)\frac{\overline{I_{ls}}}{I_{ls}} \quad (1)$$

where I_0 is the initial image brightness, I_{ls} is the light-sheet brightness at the same pixel and $\bar{I}_{ls}$ is the light-sheet brightness averaged over the measurement area.
After a light-sheet correction an image is assumed to represent a distribution of relative concentration of the liquid.

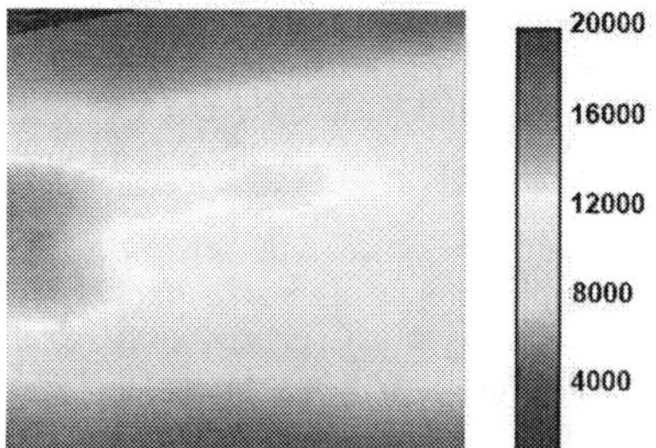

Fig. 3. Light-sheet brightness distribution.

An example of the raw PLIF image of the spray crossection and the result of the geometric reconstruction and light-sheet correction are shown in figure 4. From the image, it can be seen that a spray pattern shape is close to circular in general with a large number of local intensity variations.

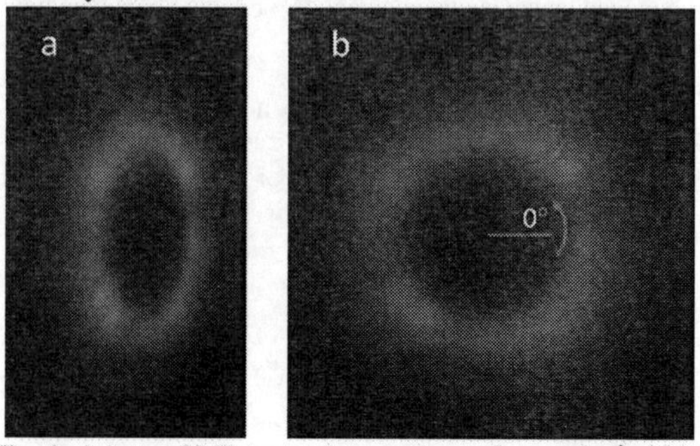

Fig. 4. a) A raw PLIF image for nozzle rotation angle θ=0°, b) PLIF image after geometric reconstruction and light-sheet correction.

The center and radius search procedures applied to the images yielded qualitatively good results, yet there was no proper technique to quantify a precision of center identification in the experimental image.

Besides the signal variations seen in the images, the acquired circumferential spray pattern also contained strong large-scale distortions, which remained stationary with the rotation of the nozzle on a swivel (see fig. 5). They were identified as the artifacts of the optical arrangement: as it was noted in ‘experimental setup’ section, there is a number of effects that can lead to such distortions. These distortions, which remained after the light-sheet correction, were most likely related to the variation of the light path length through the spray volume for different parts of the measurement area and light-sheet attenuation in the bulk of the spray.

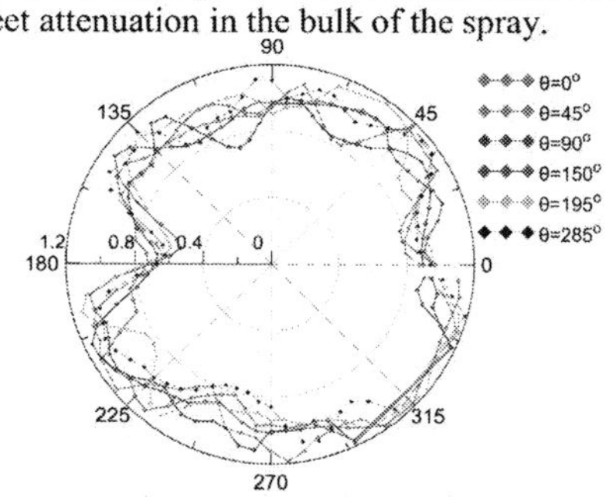

Fig. 5. Spray angular profiles with spurious distortions for different nozzle rotation angles.

Assuming that flow conditions inside the feed line do not change with nozzle rotation, spray pattern features will rotate accordingly while spurious brightness variations should remain stationary. This allows to use the circumferential profile, averaged over the different nozzle rotation angles, as a reference to normalize profiles for separate rotation angles in order to eliminate the rest of the spurious brightness variations as follows:

$$Q(\varphi,\theta_k) = \frac{q(\varphi,\theta_k)}{N_\theta}\sum_i q(\varphi,\theta_i) \quad (2)$$

where Q is the normalized relative concentration within the sector φ, q is not normalized relative concentration, θ is the nozzle rotation angle and N_θ is the total number of rotation angles.

To test this approach profiles averaged over 24, 12, 6 and 3 rotation angles θ were calculated (see fig. 6). Profiles averaged over 24 and 12 rotation angles match each other quite well, which indicates a sufficient number of samples in the statistic. These profiles feature only large-scale variations. Profiles averaged over a smaller number of rotation angles show local variations and worse match to each other and to the profiles with larger statistics. Apparently, a sufficient number of samples (nozzle rotation angles) may vary depending on the optical setup and a spray structure.

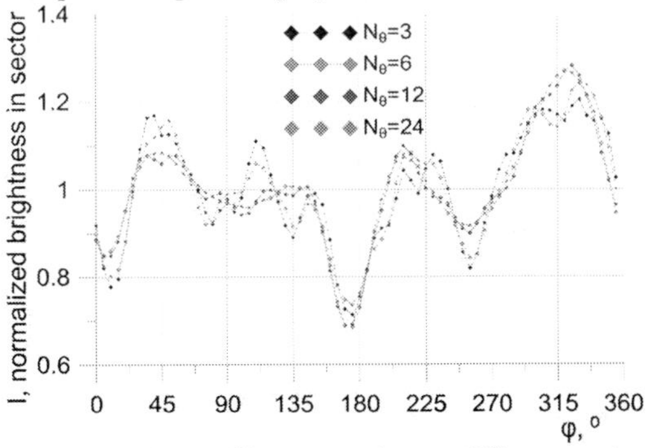

Fig. 6. Angular profiles averaged over different number of samples.

In figure 7b a normalized angular relative concentration profiles for 0° and 90° nozzle rotation angles are shown. Profiles match closely and demonstrate common features of the spray pattern, which can be also noticed in raw images (see fig 3). Therefore, it can be concluded that the proposed correction procedures allowed to eliminate most of the error sources and yield circumferential profiles that clearly reveal the angular nonuniformity of the spray.

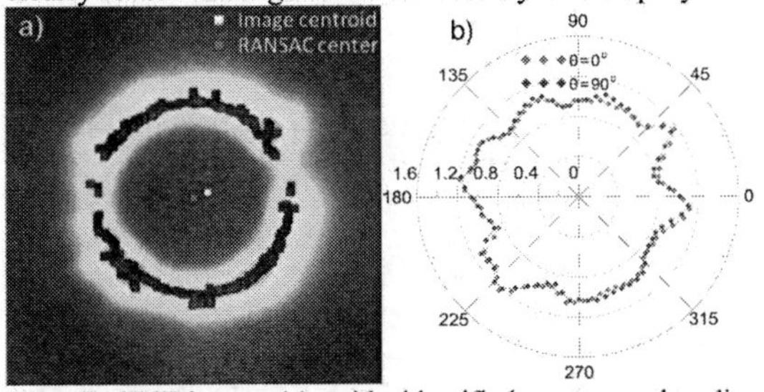

Fig. 7. PLIF-image (a) with identified center and radius and angular relative concentration profiles for 0° and 90° nozzle rotation angles (b). A 90° profile is revolved to match the orientation of 0°-profile for better comprehension.

To quantify a measurement error an RMS of the relative concentration was evaluated from profiles for 24 separate nozzle rotation angles. An RMS of the relative concentration was less than 10% for the profile with sectors division $\Delta\varphi$=5° (see figure 8). Note that error could partially come from the flow conditions changes and error in setting the nozzle rotation angle, so this value should be considered as an upper boundary of the measurement error.

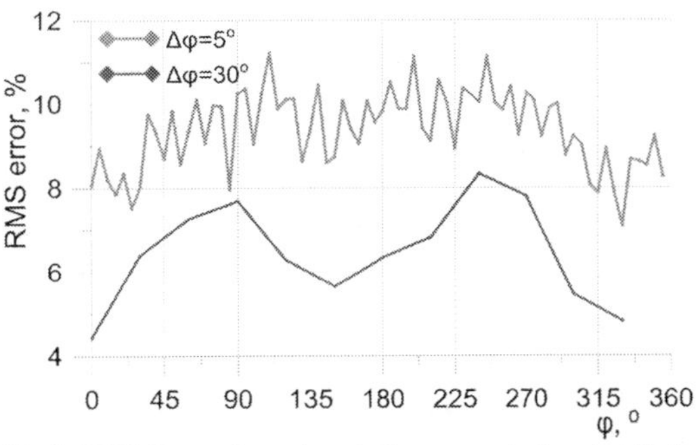

Fig. 8. RMS Error of angular relative concentration distribution for sectors division $\Delta\varphi$=5° and $\Delta\varphi$=30°.

5 Conclusion

A method for PLIF circumferential patternation of the large cone-shaped spray was developed. The method benefits from low equipment requirements and relatively free camera positioning. In current realization it is not meant for application as a sole patternation instrument, as some of the error sources are yet to be considered. Nevertheless, it can be used as a supplementary method for more informative spray testing.

In the future work, the method can be substantially improved by introducing such techniques as a correction of planar laser-induced fluorescence distributions for local nonuniform laser attenuation and suppression of secondary reflections.

The main disadvantage of the procedure is the need for a large number of PLIF-images captured at different nozzle rotation angles. This part can be also improved, for example by rotating the measurement system on a rotational stage instead of rotating the nozzle.

Acknowledgements: The research was supported by the Russian Science Foundation (Grant No 19-79-30075).

References

1. J. Lim, Y. Sivathanu, V. Narayanan, S. Chang, Atomization and Sprays, **13**, 27 (2003)
2. R. J. Locke , Y. R. Hicks , R. C. Anderson, M. M. Zaller, Combustion Science and Technology, **138**, 297 (1998)
3. M.K. Lai, W.G. Freeman, P.R. Yankowich, J.D. Bryant, P. Walterscheid, Proceedings of the ASME Turbo Expo 2004, **665** (2004).
4. K. Jung, H. Koh, Y. Yoon, Meas .Sci. Technol. **14**, 1387 (2003)
5. S.M. Soloff, R.J. Adrain, Z-C. Liu, Meas. Sci. Technol., **8**, 1441 (1997)
6. M. Fischler, R. Bolles, Communications of the ACM, **24**, 381 (1981)

Correction of self-illumination for pressure and temperature sensitive paints

Oleg Gobyzov[1*], *Artyom* Rabetskiy[2], and *Artur* Bilskiy[1]

[1]Kutateladze Institute of Thermophysics, Siberian Branch of Russian Academy of Sciences, 630090, Novosibirsk, Russia
[2]Sigma-Pro LLC, 630090, Novosibirsk, Russia

Abstract. The present work reports self-illumination correction procedure and results of its experimental testing on a simple model. In the procedure reflectivity value is assessed directly from the set of model images captured at wind-off conditions with different reflecting surfaces covered or removed. The paper covers the theoretical part of the problem and describes exact implementation of a processing sequence, which included image mapping on a 3-D surface, building a mesh of surface elements, evaluation of radiative exchange between elements and calculation of the corrected brightness map.

1 Introduction

Pressure and Temperature sensitive paints (PSP and TSP) [1] are optical techniques for measurement of the spatial distributions of pressure and temperature on surfaces. Both techniques are mostly employed in aerodynamics. They are based on imaging of the luminescence of special layers, applied on the surfaces. Luminescent layers are sensitive to temperature and pressure variations because of the quenching mechanisms, which allows evaluating pressure and temperature distributions on the surfaces from the image brightness through the calibration procedures.

PSP and TSP techniques have different sources of error, related to experimental conditions, equipment characteristics, imaging and data processing. One of such error sources is so-called secondary reflection or self-luminescence. It occurs at concave surfaces, surface joints and internal angles due to the radiative exchange between surfaces. For example, in two-surface configuration, a reflected excitation light and emitted luminescent light propagate from the one surface to another and then, being reflected or absorbed (with subsequent emission) at the second surface, form brightness variations in the image that are not related to the pressure or temperature distribution. An illustration of self-luminescence in configuration with two plane surfaces is shown in fig 1a.

There are several methods for the elimination of this effect. Elimination of the part, related to the excitation light reflections can be performed by the so-called 'wind-off' correction procedure, but it does not solve the problem in general. The effects of self-illumination in PSP and procedures of its correction were studied by Ruyten and Fischer [2], and Le Sant [3]. In order to perform a correction, one must account for the 3D-shape of the surfaces and reflectivity, or, in more general case - bidirectional reflectance distribution function (BRDF) over the surfaces. Correction procedures usually employ a diffusive surface model, but a reflectance value, required for the correction, poses a separate problem.

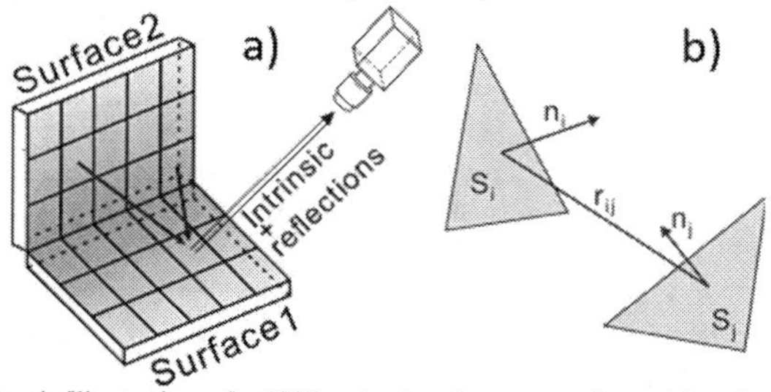

Fig. 1. Illustration of self-illumination for two surfaces (a) and spatial configuration of surface elements (b).

In the present work, a realization of a self-illumination correction procedure is reported and results of its experimental testing are shown. Instead of direct radiometric measurement of the reflectivity, its value is assessed from the set of images captured at wind-off conditions with different reflecting surfaces removed or covered. Other steps of the correction procedure include 2-D brightness field (image) reprojection on the 3-D surface, building a mesh of surface elements, evaluation of radiative exchange between elements and calculation of the corrected brightness map.

2 General description of the problem

To correct PSP/TSP data for self-luminescence several steps of the data processing and preparation should be performed. Particularly, a mapping of the image to the 3-D surface of the model is required, which means that model geometry should be known in advance.

Today methods of surface geometry reconstruction (3D scanning) are very versatile and common in many

* Corresponding author: oleg.a.g.post@gmail.com

applications. A good survey of such methods is provided by Daneshmand et al in [4]. For the PSP/TSP methods it is preferred that the mapping is available with information acquired directly from the luminescence image. This is usually performed using reference points (markers), position of which on a 3D-surface is known [5]. Ideally, every pixel of the image should be bijectively mapped to the 3D-surface.

The self-luminescence correction is based on the idea that the total brightness of each element i consists of its own 'intrinsic' luminescence and 'external' incident light from other elements j, reflected from element i. In a general case of reflecting surface, for each point of the surface, there is a function that connects an angular distribution of the reflected light intensity to the incidence angle of incoming light (so-called spatially varying bidirectional reflectance distribution function, SV-BRDF). In case, if all of the surface points have similar reflection properties, the reflection properties of the whole surface are described by the BRDF that does not account for the position of the point. Calibration for the BRDF is still quite complicated and often a diffusive reflection model, in which luminance is equal for all directions of observation, is employed. In this case, reflection properties of the surface are described by a single reflectivity coefficient R. A diffusive model is justified for at least one type of PSP in [2]. Moreover, any model with nonuniform BRDF drastically complicates calculations, which makes diffusive model more suitable in practice. With a diffusive reflection model a total brightness luminance of the element i is:

$$I_i = I_{lum,i} + I_{refl,i} = I_{lum,i} + R_i \sum_j^N I_j K_{ij} \tag{1}$$

where R_i is the surface element reflectivity, N is the number of other elements involved in a radiative exchange with element indexed i, and K_{ij} is a form factor or 'geometric factor', that takes into account size, position and orientation of the elements. Note, that we still hold the index for the reflectivity, assuming that it's value is a property of a considered element. K_{ij} depends on the position and orientation of elements i and j and on the surface area of element i:

$$K_{ij} = \frac{(r_{ij}, n_i)(r_{ij}, n_j)}{\pi |r_{ij}|^4} dS_i \tag{2}$$

where n_i and n_j are unit normals to the surface elements and r_{ij} is the distance vector between elements (see also figure 1b). Mentioned considerations can be also found, for example, in [2] and [3].

If a 2-D brightness distribution (image) is mapped to the 3-D model surface, all of the values in eq (1) can be evaluated, except for the reflectivity. The reflectivity value, though, should be acquired from separate measurements. In [2] and [3] radiometric measurements were suggested to acquire reflectance value. In the current work, an alternative approach of reflectivity estimation directly from the luminescence images is proposed. A very similar approach was suggested in [6], though in this work a separate unit consisting of two rectangular plates, connected at one side through a regulated joint, was proposed for calibration and the average reflectivity value for the paint was assessed without 3D-mapping of the image.

For the proposed approach, first, an image of the surface, for which a reflectivity value is required, without outer sources of light that may be reflected is taken. After that, another image with one or more external light sources is captured. To avoid errors, both images should be taken at the same conditions, such as temperature, pressure, excitation and ambient light. In this case, the total brightness in the second image can be separated into intrinsic and reflected parts using first image and the reflectivity value can be evaluated. Let us consider the case with two surfaces: first is the surface, for which R is evaluated, and the second is the source of external light. In this case Ri for each element i of the first surface is evaluated as follows:

$$\begin{aligned}
I_{1,i} &= I_{lum,i} + R_i \sum_j^{N1} I_j K_{ij} \\
I_{2,i} &= I_{lum,i} + R_i \left(\sum_j^{N1} I_j K_{ij} + \sum_j^{N2} I_j K_{ij} \right) \\
I_{1,i} - R \sum_j^{N1} I_j K_{ij} &= I_{2,i} - R_i \left(\sum_j^{N1} I_j K_{ij} + \sum_j^{N2} I_j K_{ij} \right) \\
R_i &= \frac{(I_{2,i} - I_{1,i})}{\sum_j^{N2} I_j K_{ij}}
\end{aligned} \tag{3}$$

R_i can be used as a spatially varying reflectivity function if reflectivity exhibits noticeable variations over the surface. Alternatively, R_i can be averaged over the surface to acquire a single reflectivity coefficient. The rest of the correction procedure implies solving the equation (1) for each element of emitting/reflecting surfaces that take part in the radiative exchange. In the next sections, a setup for experimental testing of the correction procedure and realization of the processing scheme are considered.

3 Experimental setup

To test the procedure a simple model of 3 small rectangular plates assembled into the 90° internal angle with 3 walls was manufactured (see figure 2). A PtTFPP pressure-sensitive paint was applied on the surfaces of the plates.

For the mapping of the image brightness onto the 3D-model a set of 9 evenly spread markers was manually applied on each of the angle sides. A number of markers was excessive, but additional markers could improve the precision of mapping and could be used in verification of the mapping procedure.

A high-power LED with a maximum emission wavelength at 390 nm and emission spectra half-width at half maximum of about 20 nm was used as the excitation light source. LED was operated in pulse mode with pulse length of about 100 μs and was synchronized with a camera. Images were captured with a scientific-grade CCD-camera with 1 Mpix image resolution in 14-bit mode and saved as 16-bit images. The image resolution was 70 μm per pixel. The camera lens was supplied with a band-pass filter to suppress excitation light and stray background light. To reduce random error in image brightness for each configuration statistics of 100 images was captured and series-averaged images were used for further processing.

Fig. 2. Test model with 3 luminescent walls. Numbering is according to the one employed in the paper.

To exclude luminescence of separate surfaces from images they were covered with matte black sheets. The model images were captured in three configurations: with surfaces 2 and 3 covered, so that only intrinsic luminescence of surface 1 was visible, with surface 3 covered and with all three surfaces visible. An example of images is shown in figure 3.

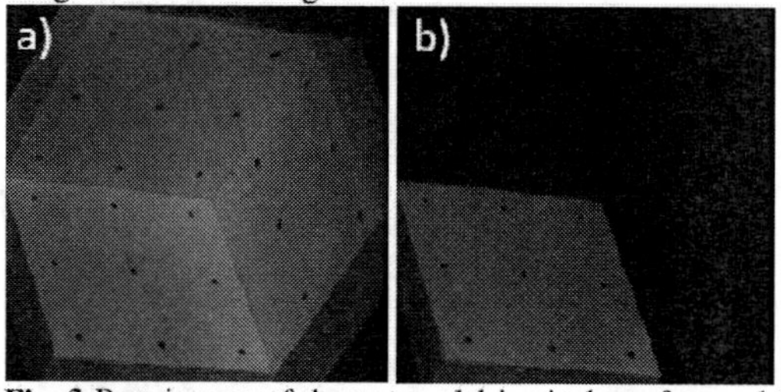

Fig. 3 Raw images of the test model in single-surface and 3-surfaces configuration.

4 Processing sequence

4.1 Data preparation

All of the processing, except for the initial 3D surface modeling, was performed using Python 3.5 and open-source libraries for this programming language. An initial 3D surface model of the plates assembly was created in CAD software and stored in the STL format, which describes surface as a set of adjacent triangles (unstructured triangular mesh) and unit normals to the triangles. As it was mentioned before, ideally, a mesh resolution should be enough to provide mapping of each pixel to a separate triangle vertex. To fulfill this requirement more than $3 \cdot 10^5$ vertices for one plate and about $2 \cdot 10^{11}$ rays in ray tracing must be evaluated. To avoid heavy calculations a 3D-surface mesh resolution was reduced to 0.5 mm, which was still much smaller than the characteristic size of brightness variations. Note, that STL format was used merely as a tool for mesh building and navigating over vertices. Brightness values were projected onto the vertices, and for the future evaluation a surface area dS, equal to the triangle area, and a unit normal, calculated form the normals of adjacent triangles, were assigned to each vertex.

The next step was a search of markers position in the image and calculation of camera model parameters for image mapping. A markers search was performed using the 'SimpleBlobDetector' method from OpenCV library.

The projection parameters were acquired from the camera transform equations, that connect 3-D spatial coordinates of an object and the corresponding 2-D spatial coordinates in an image [5]. Internal camera orientation parameters were measured separately using a calibration target to achieve better parameters fitting. Parameters of projection were fitted using the Levenberg-Marquardt algorithm. In fact, solving a series of equations requires 6 points with known coordinates in 3-D space and in the image, thus, larger number of markers served mainly for random error reduction.

The final 3D model contained about $2 \cdot 10^5$ vertices (about $9 \cdot 10^4$ visible), for each of those 3 sets of data were stored: a brightness value, a reference pixel coordinates, and visibility information, indicating if the vertex is seen in the image.

4.2. Self-luminescence correction

A self-luminescence correction procedure was based on the ray-tracing method. To perform required calculations for each vertex on the considered surface a set of rays, connecting the vertex with all other vertices in the model was built. In general, each ray should be checked for occlusion, i.e. it should be checked whether there are any obstacles that block the light path. Although the tested model does not require such check due to relatively simple configuration, it was still implemented in a processing sequence. The occlusion check was based on the BVH algorithm [7] with an STL model as an input data.

The reflectivity of the surface was estimated as described in the 'general description' section. For this purpose, images with only one visible surface and with two visible surfaces were used. The first step for reflectivity calculation is the evaluation of form factor in two-surface system for the elements of the surface 1, for which the correction was to be performed.

A diffusive emission model, in which the light, emitted from the surface is evenly distributed in a half-space in front of the surface element, was employed. An assumption that all the light from the vertex is emitted directly from the vertex point was also used, although each vertex has an assigned surface area value. The assumption is valid if the distance between emitting and reflecting elements is much larger than the size of any of the elements.

With these assumptions, a form factor for each element can be evaluated from the eq. (2). Knowing the intrinsic luminescence brightness for each element $I_{lum\ 1,i}$ of the surface 1 (from the images with only this surface visible), brightness of each element $I_{2,j}$ of the surface 2, and all required form-factors, it is possible to evaluate $R_{i,1}$ of each element of the surface 1 and average R for the whole surface.

Finally, after the evaluation of R, a self-luminescence correction can be performed. To test the validity of the correction, an image of the model with three luminescent surfaces was used. In this case, the surface 1 is illuminated by a light emitted from the surfaces 2 and 3. Using the same approach, the image was mapped to the 3-D model and the amount of light incident from surfaces 2 and 3 onto each of the elements of the surface 1 was calculated and then subtracted from the total brightness of those elements. Brightness distributions in corrected and initial brightness maps on the 3-D surface are shown in figure 4. Acquired distributions show, that corrected brightness qualitatively match the luminescence of single surface, but the correction with distributed reflectivity produce more noise in the data.

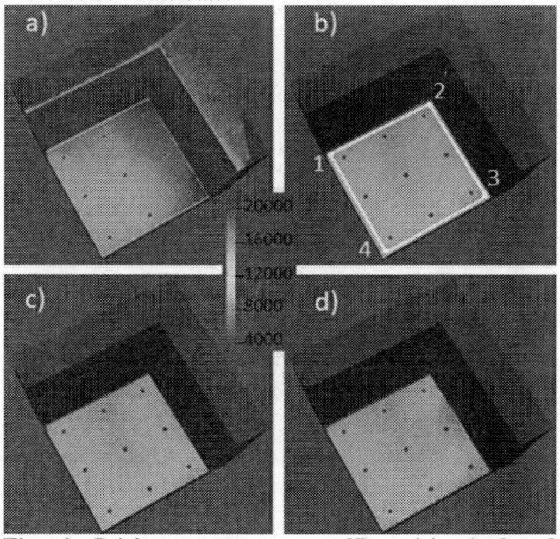

Fig. 4. Brightness maps on a 3D-model. a) 3-surface configuration, not corrected, b) single-surface configuration, c) 3-surface configuration, corrected with average R, d) 3-surface configuration, corrected with distributed R_i.

The luminescence profiles for corrected and uncorrected images with three surfaces and corrected luminescence versus luminescence with only one surface visible are shown in figure 5. The profiles were built along the closed line, shown in fig. 4b. Each section of the line was parallel to the border of the plate and was placed at a distance of 2.5 mm from the border. The normalized difference between corrected brightness and single surface brightness is presented in figure 6.

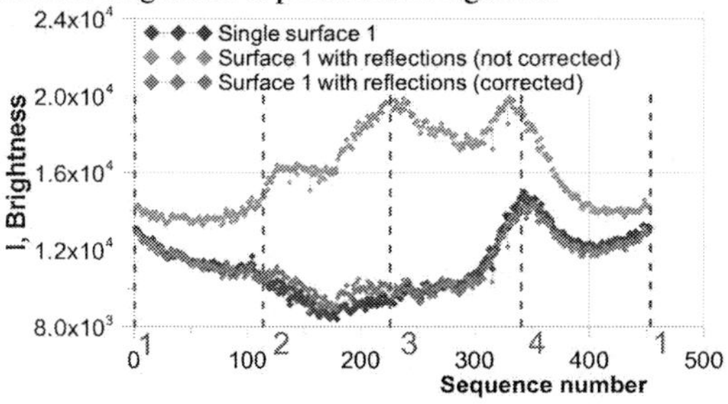

Fig. 5. Brightness profiles in corrected and initial brightness maps on the 3-D model.

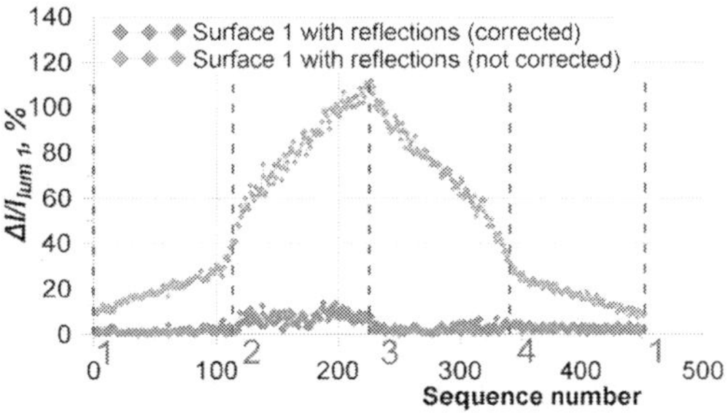

Fig. 6. Relative error before and after correction.

These profiles show, that self-luminescence was successfully corrected over the whole surface. Small discrepancy (within 10% of single plate image brightness) can be seen in the vicinity of the joint of three surfaces. There can be several sources of this residual difference: a misalignment in projection and back-projection of the brightness map, effect of the reflections of higher orders, and non-validity of the element infinitesimality assumption near the wall. More precise mapping and an increase in 3D-model resolution, as well as reflectivity calculation from several model configurations could further improve the result.

5 Conclusion

A variation of the method for the correction of self-luminescence in PSP/TSP experiment has been presented. The method was experimentally tested and allowed to reduce the error, related to the self-luminescence from more than 100% to 10%. The proposed approach for reflectivity estimation can be applied directly to the model in situ. It should be noted, that at least in current implementation the processing sequence is quite computationally heavy, so for more complex models and higher spatial resolution it requires further optimization in order to provide reasonable processing time for experimental data.

Acknowledgments: The work was carried out within the state assignment of Ministry of Science and Higher Education of the Russian Federation.

REFERENCES

1. T. Liu, J.P. Sullivan, *Pressure and temperature sensitive paints* (Springer-Verlag, 2005)
2. W. M. Ruyten, C. Fisher, AIAA J., **39 (8)**, 1587 (2001)
3. Y. Le Sant, ICIASF '2001 Record, 19th International Congress on Instrumentation in Aerospace Simulation Facilities', 5.2, (2001)
4. M. Daneshmand, A. Helmi, E. Avots, F. Noroozi, F. Alisinanoglu, H. Arslan, J. Gorbova, R. Haamer, C. Ozcinar, G. Anbarjafari, arXiv:1801.08863 (2018)
5. A.W. Burner, T. Liu, J. of aircraft, **38 (4)**, 745 (2001)
6. W. M. Ruyten, Rev. Sci. Instr., 68, 3079 (1997)
7. J. Goldsmith, J. Salmon, IEEE Computer Graphics and Applications, 7 (5), 14 (1987)

Study on incompressible fluid analysis by three-dimensional particle method with finite volume techniques

Shigeyuki Hibi[1,*], *Kazuki* Yabushita[1], and *Takayuki* Tsutsui[1]

[1]The National Defense Academy, Department of Mechanical Systems Engineering, 1-10-20 Hashirimizu, Yokosuka, Kanagawa 239-8686, Japan

Abstract. Simulating violent free-surface flows by particle methods is effective. In recent years, the SPH (Smoothed Particle Hydrodynamics) method and the MPS(Moving Particle Simulation method) method are considered representative especially in the fields of ship engineering and civil engineering but these methods need various artificial parameters in calculation and may have unnatural large numerical oscillation in the calculated pressure values. In order to avoid these problems, the authors have proposed another particle method based on the finite volume techniques. The usefulness of the proposed method in two-dimensional problems has been verified by comparing the pressure fluctuation when a water tank of the two-dimensional model is forcibly oscillated with the corresponding experiment [1]. Therefore, in this research, the authors extended the algorithm so that the calculation formula can be applied to three-dimensional problems. The effectiveness of the calculation is demonstrated by applying to the dam break problem.

1 Introduction

Incompressible fluid analysis using particle methods is effective for flows with large fluid deformations, such as when a liquid such as ballast water or LNG/LPG ships strikes a wall of tank in a ship, or when a wave breaks up after a wave rides on a ship. In recent years, the SPH[2] method (Smoothed Particle Hydrodynamics) and the MPS[3] method (Moving Particle Simulation method) have attracted attention in the fields of ship engineering and civil engineering, but these methods need various artificial parameters in calculation and may have unnatural large numerical oscillation in the calculated pressure values. In order to avoid these problems, the authors have proposed another particle method [4] based on the finite volume techniques. In this particle method, the equation of continuity and the N-S equation are transformed into forms that are easy to adapt to the particle method. Based on the finite volume techniques the boundary condition is uniformly treated so that particles can freely move not only on the free surface but also in the air, whereas SPH [5] and MPS [6] methods need some special treatments. Through the research up to last year, the usefulness of the proposed method in two-dimensional problems has been verified by comparing the pressure fluctuation when a water tank of the two-dimensional model is forcibly oscillated with the corresponding experiment. Therefore, in this research, the authors extended the algorithm so that the calculation formula can be applied to three-dimensional problems. The effectiveness of the proposed method is validated by comparison with the dam break experiment executed by the authors refering to the experiment by Zhou [7]. The dam break problem is usually used for a benchmark problem. By comparing the time history of pressure values occured on the wall with the experimental results, the effectiveness of the proposed particle method was assured.

2 Numerical simulation by three-dimensional particle method with finite volume techniques

In particle methods, fluid is modeled as a collection of particles and discretized by considering the interaction between particles without a calculation lattice. Therefore, particle methods are suitable for intense fluid analysis with strong nonlinearity.

2.1 Discretization with finite volume techniques

The fundamental equations described by a fluid particle coordinates $\hat{r}$ are represented by the Navier–Stokes equations and the continuity equation.

$$\left.\frac{\partial \boldsymbol{v}}{\partial t}\right|_{\hat{r}} = -\frac{1}{\rho}\hat{\nabla} p + \boldsymbol{K} + \nu \hat{\nabla} \cdot \hat{\nabla} \boldsymbol{v} \tag{1}$$

$$\hat{\nabla} \cdot \hat{\nabla} p = \rho \lim_{\Delta \to 0} \frac{\hat{\nabla}(t, \hat{r}) \cdot \boldsymbol{v}(t - \Delta t, \hat{r} - \Delta \hat{r})}{\Delta t} \tag{2}$$

Here $\boldsymbol{v}$ is the speed of particles, p is the pressure of particles, ρ is the density of fluid, $\boldsymbol{K}$ is outer forces and ν is the kinematic coefficient of viscosity of fluid. From

* Corresponding author: hibiteru@nda.ac.jp

https://doi.org/10.1051/epjconf/202226901020

Gauss's theorem equations (1) and (2) are transformed into flow equations of integral shape.

$$\left.\frac{\partial \boldsymbol{v}}{\partial t}\right|_{\hat{r}} = -\lim_{V\to 0}\frac{1}{\rho V}\int_S p\boldsymbol{n}dS + \boldsymbol{K} + \lim_{V\to 0}\frac{\nu}{V}\int_S \frac{\partial \boldsymbol{v}}{\partial n}dS \quad (3)$$

$$\lim_{V\to 0}\frac{1}{V}\int_S \frac{\partial p}{\partial n}dS = \lim_{\substack{\Delta t\to 0\\ V\to 0}}\frac{\rho}{V\Delta t}\int_S \boldsymbol{v}(t-\Delta t)\cdot\boldsymbol{n}dS \quad (4)$$

Here $\boldsymbol{n}$ is the normal vector of a particle and $\boldsymbol{V}$ is the volume of a particle.

Eventually, the discritized equations are derived from equations (3) and (4) in the proposed method.

$$\boldsymbol{v}_i - \frac{\Delta t\nu}{\frac{4}{3}\pi R^3}\sum_{j=1,j\neq i}^{N}\frac{\boldsymbol{v}_j-\boldsymbol{v}_i}{r_{ij}}\Delta S_{ij}$$
$$= \boldsymbol{v}_i^{\,n-1} - \frac{\Delta t}{\frac{4}{3}\pi R^3}\sum_{j=1,j\neq i}^{N}\left(p_i + (p_j - p_i)\frac{R}{r_{ij}}\right)\boldsymbol{n}_{ij}\Delta S_{ij} + X\Delta t \quad (5)$$

$$\sum_{j=1,j\neq i}^{N}\frac{p_j-p_i}{r_{ij}}\Delta S_{ij} - \frac{p_i}{R}(4\pi R^2 - S_i)$$
$$= \frac{\rho}{\Delta t}\sum_{j=1,j\neq i}^{N}\left(\boldsymbol{v}_i^{\,n-1} + (\boldsymbol{v}_j^{\,n-1} - \boldsymbol{v}_i^{\,n-1})\frac{R}{r_{ij}}\right)\boldsymbol{n}_{ij}\Delta S_{ij} - \frac{\rho}{\Delta t}\boldsymbol{v}_i^{\,n-1}\cdot\boldsymbol{S}_i \quad (6)$$

Here

$$\boldsymbol{r}_i = (x_i, y_i, z_i)$$
$$\boldsymbol{r}_{ij} = \boldsymbol{r}_j - \boldsymbol{r}_i$$
$$r_{ij} = \sqrt{(x_j-x_i)^2 + (y_j-y_i)^2 + (z_j-z_i)^2}$$
$$\boldsymbol{n}_{ij} = \frac{\boldsymbol{r}_{ij}}{r_{ij}}$$
$$R = \frac{D}{\sqrt[3]{\frac{3}{4}\pi}}$$
$$\boldsymbol{S}_i = \sum_{j=1,j\neq i}^{N}\boldsymbol{n}_{ij}\Delta S_{ij},\ S_i = \sum_{j=1,j\neq i}^{N}\Delta S_{ij}$$
$$\boldsymbol{K} = (0, 0, -g)$$

$$\Delta S_{ij} = 4\alpha\frac{r_e - r_{ij}}{r_e - D}R^2 d\theta_{ij}d\phi_{ij}$$
$$\alpha = \frac{4\pi R}{\sum_{j=1,j\neq i}^{N}4\frac{r_e - r_{ij}}{r_e - D}R^2 d\theta_{ij}d\phi_{ij}} = \frac{\pi}{\sum_{j=1,j\neq i}^{N}\frac{r_e - r_{ij}}{r_e - D}R d\theta_{ij}d\phi_{ij}}$$

Subscript i means the particle number of interest and subscript j means the neighboring particle number of the particle i. Superscript n indicates the time step number. $\boldsymbol{r}_i$ means position vector of the i-th particle. The particles are assumed to be spherical (just round in the two dimensional model) and the influence of the neighboring particle j to the particle i is carried out through the micro surface area ΔS_{ij} expressed by equation (7) according to the concept of the Finite Volume Method.

R stands for the radius of a particle and satisfies the following equation.

$$\frac{4}{3}\pi R^3 = D^3 \quad (8)$$

D is the average inter-particle distance and is determined by the spatial resolution of the calculation model. This parameter corresponds to the smoothing length h in the SPH[2] method and the effective radius in the MPS method.

Considering the efficiency and stability of calculation, it is considered that neighboring particles far from the 'effective radius' r_e named in the MPS method, do not have a effect. In this study the following vallue is adopted for the parameter r_e as same as the MPS method.

$$r_e = 3.1\,D \quad (9)$$

g is the gravity acceleration and in this study the outer force is only the gravity. The pressure value of particles at each time step is obtained by implicitely solving the equation (6) using the PCGS method (Preconditioned Conjugate Gradient Squared method). The velocity value of particles is also obtained from the equation (5) implicitly using the PCGS method in case of considering viscosity of fluid and explicitly in case that viscosity is not taken into consideration.

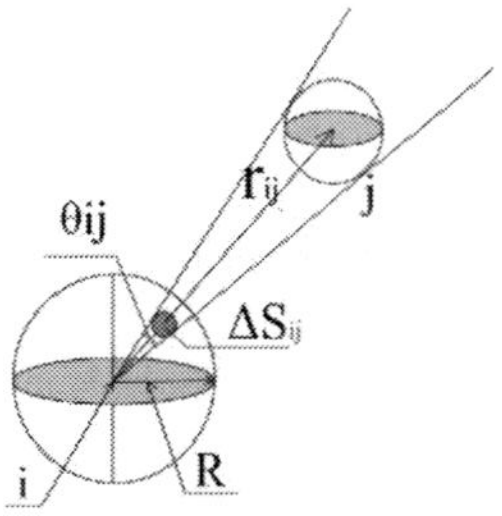

Fig. 1. Micro surface area ΔS_{ij} between particle i and particle j.

3 Dam break experiment

3.1. Specification of the apparatus

The apparatus for the dam break experiment was made by the authors refering to Zhou [4]'s model (See Figs. 2 and 3). The size of Zhou's apparatus is relatively bigger than other researcher's one in order to ignore the effect of the surface tension sufficiently because their study field is in the naval architecture. The authors determined almost half size of Zhou's one considering both the production cost and production time.

The size of tank is 590 × 1620 × 300 (mm) (Height × Width × Depth). At the beginning of a experiment, water is sotred in the left side of the tank devided by the flap and right side of the tank is empty. Water depth is 300 mm.

A pressure sensor is attached on the right side wall of the tank mounted on the apparatus. The position of the sensor is 40 mm or 80 mm from the bottom. The pressure receiver has round area with diameter of 8 mm. When the weight connected to the flap by a tough rope is fallen and the flap is lifted, the water column starts to collapse and it will cause a impact force on the right side wall of the tank. We measured time histories of this pressure on the wall repeatedly.

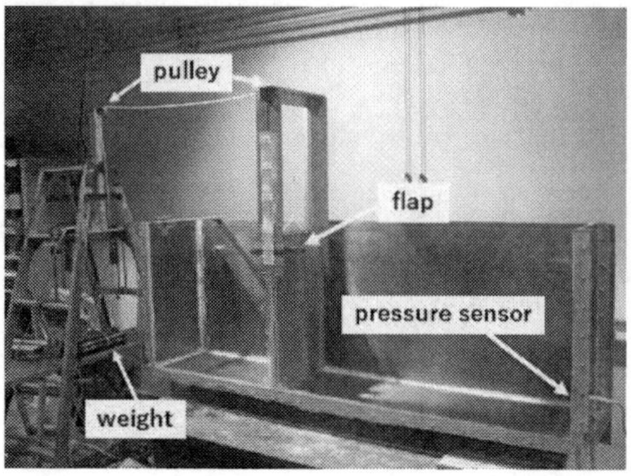

Fig. 2. Dam breaking apparatus overview.

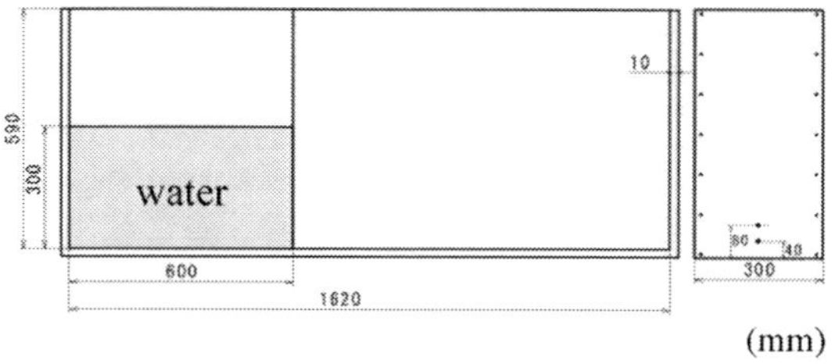

Fig. 3. Dimensions of the dam breaking apparatus.

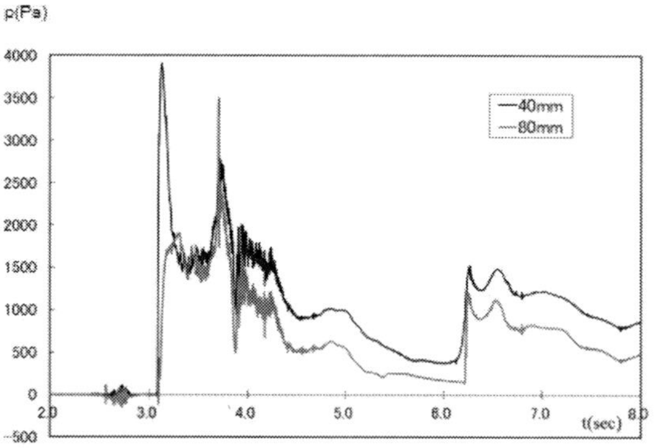

Fig. 4. Time history of pressure on the right side of the tank.

Fig. 4 shows a typical experimental result about time history of pressure on the right side of the tank. The red and black curves represent values of the pressure sensors mounted 40mm and 80mm from the bottom each other. From the figure, it can be seen that there is not much difference in the timing of the first and second impacts and the dam break is a steep and violent phenomenon.

Fig. 5 shows a comparison with time history of pressure between the Zhou's experimental result and ours at the corresponding position, 160mm and 80mm each other. Eventhough the diameter of pressure sensors are so differnt between Zhou and us, 90mm and 8mm each other, the time history of pressure is so similar except the peak values in impact.

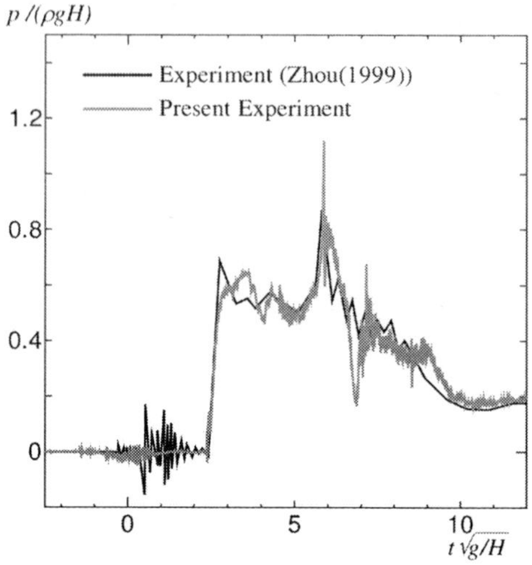

Fig. 5. Comparison with time history of pressure.

3.2 Estimation of the confidence intervals based on the *t*-distribution and uncertainty analysis

Independent random variables subject to the normal distribution $X_1, X_2, ... X_N$ are assumed that mean values are μ and variances are σ^2 (usually unkonwn). In case of a sample of size N the confidence interval at a confidence level z % is given by eq. (10).

$$\bar{x} - t_{N-1}(z\%)\frac{u}{\sqrt{N}} \leq \mu \leq \bar{x} + t_{N-1}(z\%)\frac{u}{\sqrt{N}} \quad (10)$$

Fig. 6 shows the moved distance of the flap in 0.2 second. Error bars mean to the cconfidence interval of 95%. Fig. 7 and 8 show time average of pressure values for 2 seconds by eq. (11) at both 40mm and 80mm positions from the bottom on the right side of the tank.

$$\int_0^2 pdt/2 \quad (11)$$

From these figures the 8th experiment seemes to be different from other experiments, possibly differences in how the weight falls. However, error bars of ttime average of pressure values at both positions tend to decrease. This implies that there is no significant difference in the evaluation of the impact force on the wall, considering the integrated value of time average.

Uncertainty analysis was also carried out according to the GUM (Guide to the Expression of Uncertainty in Measurement) guidance. The expanded uncertainty calculated from Table 1(at the last of the paper) is 31.9 (Pa).

The number of iterations used here is 11 times. From this result, it can be seen that the pressure variation is not so large as compared with the pressure fluctuation on the wall surface.

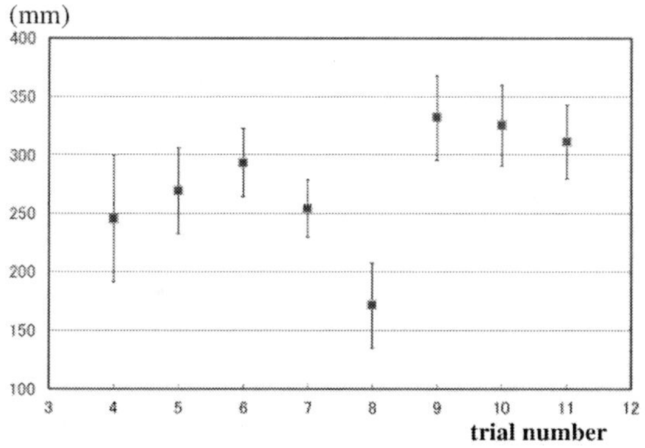

Fig. 6. Moved distance of the flap in 0.2 seconds.

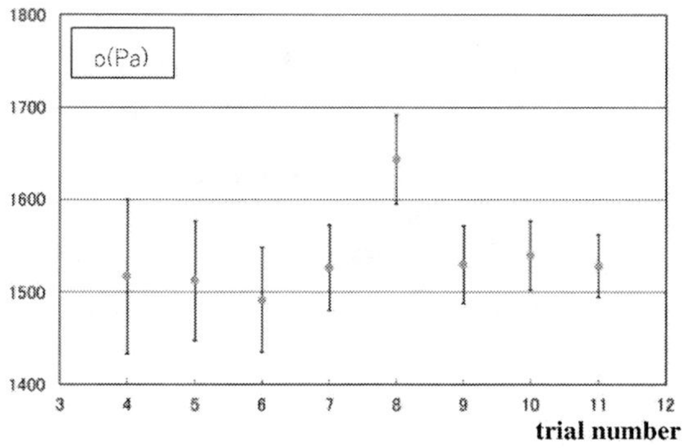

Fig. 7. Time average of pressure values (40mm).

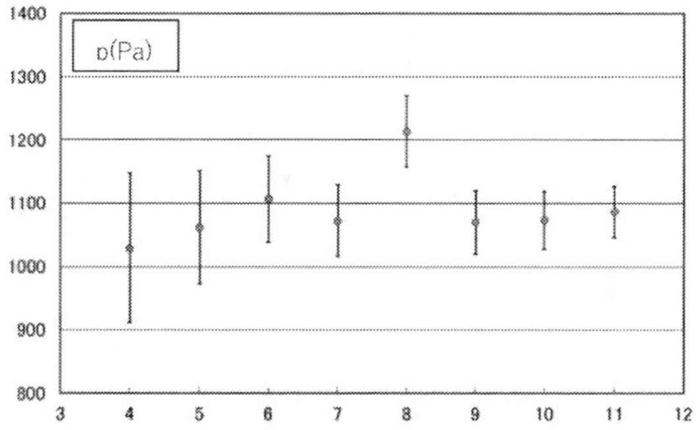

Fig. 8. Time average of pressure values (80mm).

4 Numerical simulation and discussion

4.1 Comparison of calculation accuracy of the particle method with static pressure

At first we confirm accuracy of the proposed particle method by static pressure. In Fig. 9 a tank model for still water is shown. Water depth is 0.3 m as same as the dam break experiment. The total particle number is 7,901.The average inter-particle distance D is 38.10 mm.

Numerical simulation was carried out with the time step of 0.0001 sec. in all cases. The CPU of PC used for the simulation is i7 (2.9 GHz) and the memory of the PC is 3 Gbytes. Fig. 10 shows the time history of static pressure at the bottom of the tank. The pressure value is derived from just one particle at the bottom of the tank. We can see some fluctuations for the first few seconds in simulation time and after that the pressure values become fairly stable. These fluctuations of pressure at the beginning occur when relative position of particles turns stabler and closer from the initial particle arrangement in a tetragonal lattice.

In order to avoid the fluctuation of pressure value, the particle layout of this still water after enough simulation time is used for the initial particle layout of the dam break simulation later.

Fig. 11 shows hydrostatic pressure distribution to whole water particles in depth direction. The red line and blue line respectively represent the hydrostatic linear curve and the analytic pressure value of the bottom of tank. From this figure, we can see that there are some particles with different pressures, but as a result, the hydrostatic pressure at the bottom can be estimated well by the least squares method.

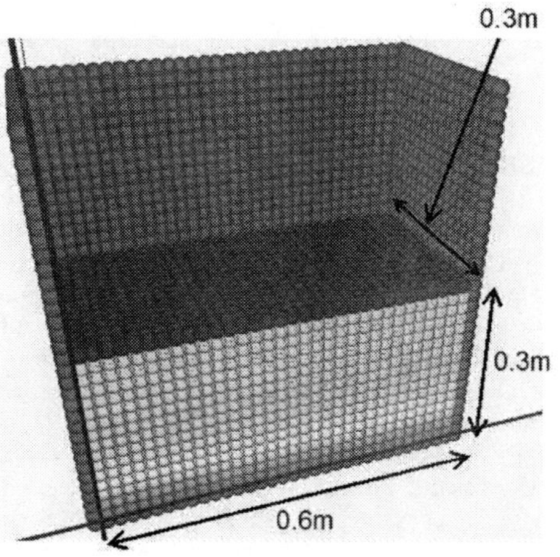

Fig. 9. Initial particle arrangement for still water simulation.

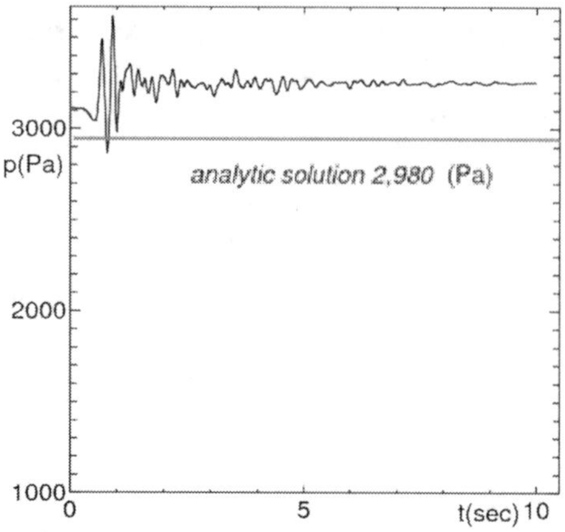

Fig. 10. Time history of static pressure at the bottom of tank.

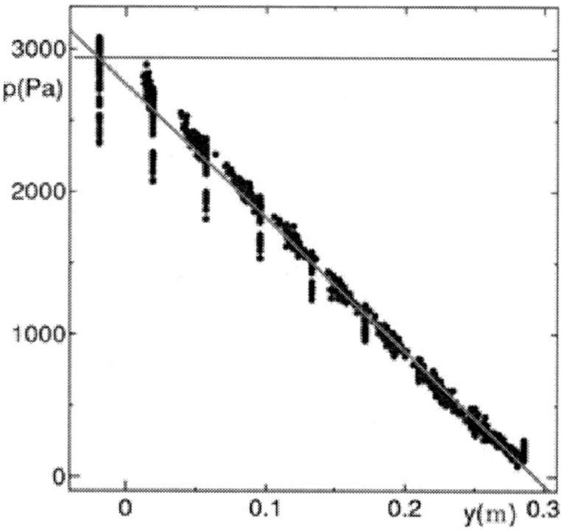

Fig. 11. Hydrostatic pressure distribution in depth direction.

4.2 Numerical Simulation of dam break experiment

We prepared two models, named X2(13,855 particles) and X4(51,209 particles). The average inter-particle distance *D* of the X4 model is half length for the one of the X2 model. The initial particle arrangement for X2 model is shown in Fig. 12. The drawing is made by the OpenGL. The front wall is present but not drawn to observe easily particle movements.

Numerical simulation was carrid out under the same conditions as the still water experiment in the section 4.2. The appearance of the simulation for X4 model is shown from Fig. 13(a) to Fig. 13(e). The gradation of the particle colors represents the magnitude of the pressure value from blue to red.

The pressure values obtained by particle methods generally fluctuate in time and space domains and then in order to obtain a stable pressure values without fluctuation some numerical tecniques are often used. In this study we introduced a numerical filter for pressure value every time step. The filter simply takes an average value of 500 steps before and after the pressure value every moment.

From these figures it can be seen that the water column collapses smoothly and hit the right side wall of the tank and intense impulsive force can occure on the wall.

Fig. 14 shows comparison of the experiment and simulation results for time history of pressure value at the position where the pressure sensor is attached.

From this result, it can be seen that the numerical results for each models have good agreement with the experiment except the peak value. The reason for this is due to the relation between the area of the pressure sensor and the spacial resolution in the simulation. While the diameter of the pressure sensor is 8mm, the average inter-particle distance D is 38.10 mm in the X2 model and 19.05 mm in the X4 model. Further study should be needed for this problem.

However, even though the number of particles is less, it is observed that the accuracy of the analysis is similar. This shows the superiority of the proposed particle method in three-dimensional simulation.

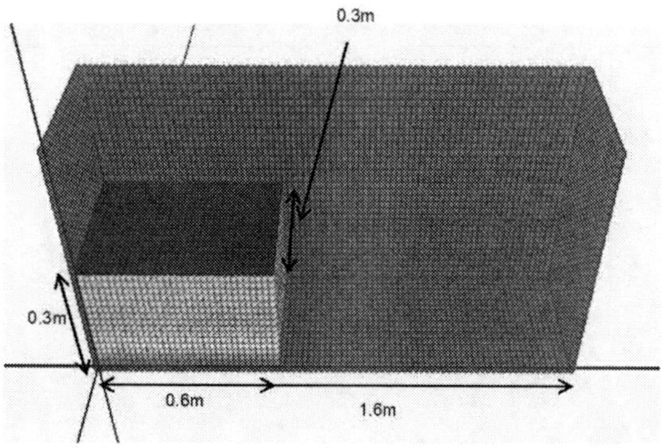

Fig. 12. Initial particle arrangement for dam break simulation (X2 model).

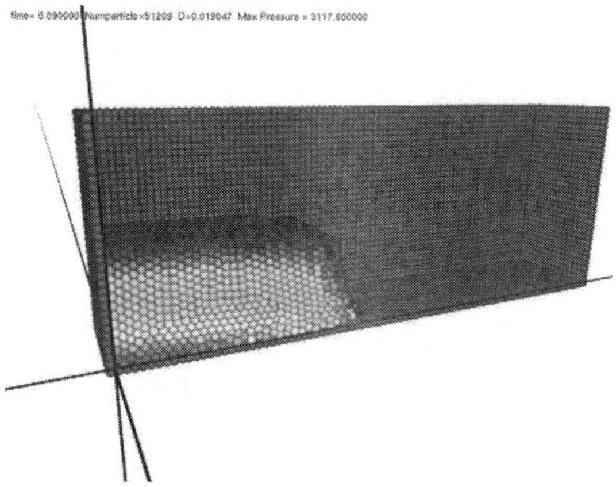

Fig. 13(a). Bird's eye view of dam break simulation (t = 0.09 sec.).

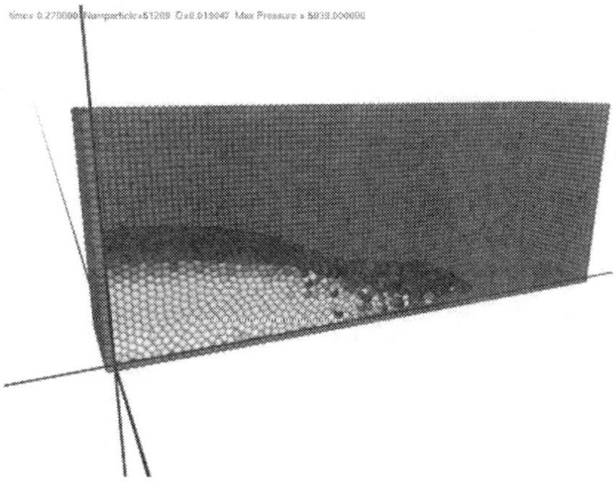

Fig. 13(b). Bird's eye view of dam break simulation (t = 0.27 sec.).

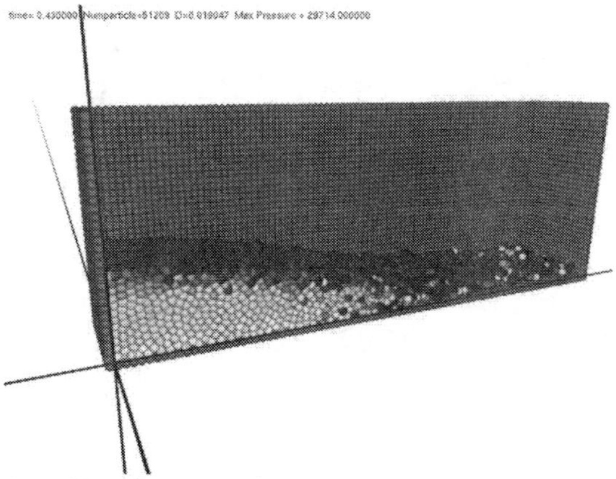

Fig. 13(c). Bird's eye view of dam break simulation (t = 0.43 sec.).

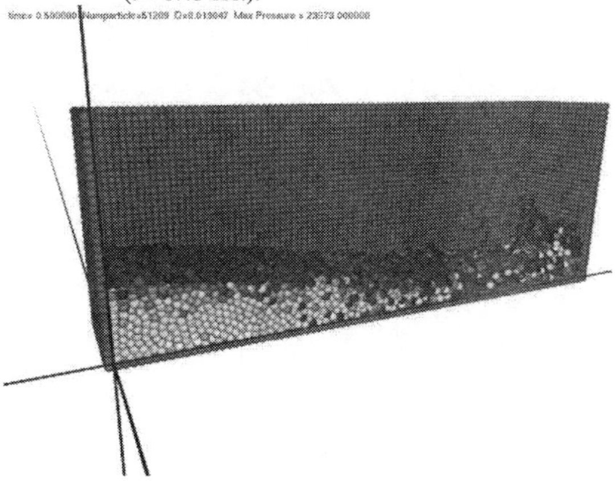

Fig. 13(d). Bird's eye view of dam break simulation (t = 0.50 sec.).

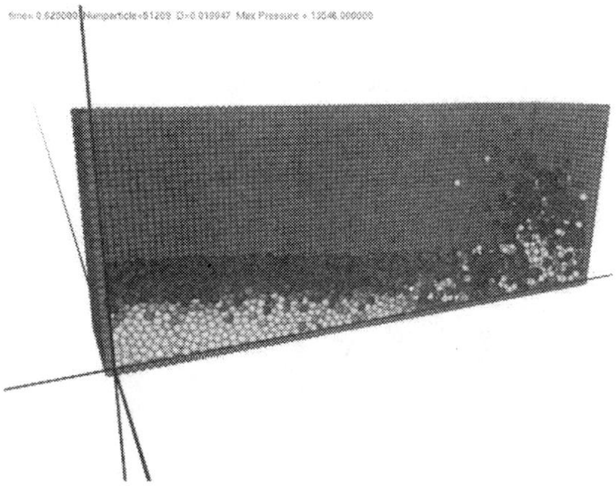

Fig. 13(e). Bird's eye view of dam break simulation (t = 0.62 sec.).

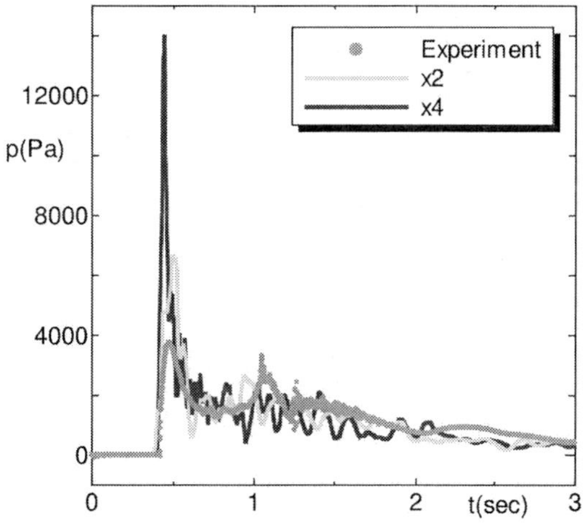

Fig. 14. Time history of pressure at the 40mm position from the bottom on the right side wall.

5 Conclusion

In this study the authors carried out the dam break experiment at first and then validate the proposed numerical method comparing the experimental results. Some consideration were found about it as below.

- Effectiveness of the proposed three-dimensional particle method is assured through the numerical simulation of intense non-linear phenomena.
- Even less number of particles can simulate the impulsive pressure value in the dam break problem by the proposed method.

References

1. Shigeyuki Hibi, EPJ Web Conf. **180**, 02034 (2018)
2. J.J.Monaghan, Journal of Computational Physics **110**, 2, 399(1994)
3. Koshizuka S, Oka Y. Moving-particle semi-implicit method for fragmentation of incompressible fluid. Nucl. Sci. Eng. ,**123**:421–34. (1996)
4. Shigeyuki Hibi and Kazuki Yabushita, ACC J., Vol. **25** Issue 1(2019)
5. Xingye Ni, Jinyu Sheng, Weibing Feng, Journal of Atmospheric and Oceanic Technology, Vol. **33**, 11, 2435 -2460 (2016)
6. Gen Li, Jinchen Gao, Panpan Wen et al., Computer Methods in Applied Mechanics and Engineering, Vol. 367 ,(2020) 113116
7. Z.Q. Zhou, J.O. De Kat and B. Buchner, *Proc. 7th Int. Conf. Num. Ship Hydrod.*, **5.1 1**(1999)

Table 1. Budget table for estimation of uncertainty concerning the pressure value at 40mm position from the bottom.

marks	factors of uncertainty	cause of uncertainty	type of uncertainty	probabilistic distribution	mean value	sensitivity factor $\frac{\partial f}{\partial x_i}$	standard uncertainty		
ρ	density of water	water temperature	A	normal	9.99E-01	871	6.62E-05		
a	Coefficient for conversion	repeated experiment	A	normal	87.96	9.9	0.342		
$V(0)$	Voltage value against atmospheric pressure	repeated experiment	A	normal	1.63	-861	1.53E-02		
$V(100)$	voltage value at the 100mm water depth	repeated experiment	A	normal	2.64	861	2.16E-02		
V_p	calibration of the pressure sensor	calibration error of the pressure sensor	B	uniform		1.0	1.62		
V_{amp}	calibration of the amplifier	calibration error of the amplifier	B	uniform		1.0	4.98		
u_c	combined standard uncertainty							16.0	(Pa)
u	expanded uncertainty							31.9	(Pa)

Development of a two-dimensional mathematical model of flow boiling heat transfer in micro- and minichannels

Sylwia Hożejowska[1,*], *Magdalena* Piasecka[2], and *Artur* Piasecki[3]

[1] Faculty of Management and Computer Modelling, Kielce University of Technology, Al. 1000-lecia P.P. 7, 25-314 Kielce, Poland
[2] Faculty of Mechatronics and Mechanical Engineering, Kielce University of Technology, Al. 1000-lecia P.P. 7, 25-314 Kielce, Poland
[3] Echo Investment S.A., Al. Solidarności 36, 25-323 Kielce, Poland

Abstract. The paper concerns flow boiling heat transfer in micro- and minichannels. In the mathematical model, the steady state heat transfer process in a single asymmetrically heated minichannel was considered. Calculations with the use of Trefftz functions were based on the data from own experiments. The temperature of the heater and the refrigerant were assumed to satisfy the Laplace equations and the energy equation respectively. The problem was solved by the Trefftz method using two sets of Trefftz functions. The known heater and refrigerant temperature distributions were used to determine the heat transfer coefficient at the heater – refrigerant contact. To verify the proposed mathematical model, data from experiments were applied to calculations. The essential part of the experimental stand was the test section which comprises a minichannel heat sink. The heated element for Fluorinert FC-72 flowing along minichannels was a thin foil. The temperature of its outer side was measured using infrared thermography. Thermocouples and pressure transducers installed at the inlet and outlet of the test section monitored fluid temperature and pressure. Mass flow rate, the current supplied to the heater and the voltage drop were also recorded. The resulting graphs presented thermograms of measured temperature on outer surface of the heater, temperature distributions of fluid temperature and local values of the heat transfer coefficient.

1 Introduction

Two-phase microchannel heat sinks are very suitable for resolving thermal problems in electronic components. The boiling process during fluid flow in micro- and minichannels allows obtaining the highest possible heat flux at a low temperature difference between a heated surface and a working fluid on a small heat transfer area.

Main of authors earlier works concerned steady state studies on flow boiling heat transfer in minigaps of different geometry: in an annular minigap [1,2] and in minichannels of rectangular cross sections [2-6]. It was also noticed that the use of enhanced surfaces often increases the intensification of heat transfer both in flow boiling [2-6] and in pool boiling research [7-9]. Maciejewska and Piasecka focused on flow boiling heat transfer under unsteady state conditions in minichannels [10,11].

This paper focuses on building and developing an appropriate mathematical model to describe the heat transfer processes in flow boiling of a refrigerant in micro- and minichannel heat sinks. The acquired experimental data influences the form of energy equations and boundary conditions. The model has been formulated so as to avoid the occurrence of experimentally determined constants. The procedure of calculating the temperature of the fluid flowing through a minichannel is coupled with concurrent determination of temperature distribution in the heating surface. The knowledge of temperature distributions in the heating surface and fluid permits to determine local heat transfer coefficients at the heating surface - fluid interface. The proposed model of the problem consists of energy equations subject to boundary conditions corresponding to the experimental data. The solution involves two kinds of heat transfer problems: a direct (in the fluid) and an inverse problem (in the heater). A problem like that requires an effective and stable solution method. Such requirements are met by the Trefftz method [12] which applies to various types of heat transfer problems [13-15].

2 The mathematical model

In the mathematical model it was assumed that the heat transfer process in the minichannel heat sink was in steady state and the physical parameters of the minichannel heated wall were independent of temperature. Two dimensions were taken into account in the model: one in the direction of the flow (x) and one in the perpendicular direction (y) referred to the thickness of the heater (δ_H) and the depth of the minichannels (δ_M). The temperature of the outer side of the heater was monitored continually. We assumed that the heat transfer process occurring at the side surfaces of the minichannel heat sink did not affect the thermodynamic parameters of its central part. It allows

* Corresponding author: ztpsf@tu.kielce.pl

limiting the considerations to the central channel only (along its length L). Additionally, it was assumed that heat losses to the surroundings are negligible and are not included in the model.
Assumptions for refrigerant Fluorinert FC-72 were as follows:
– fluid flow in the minichannel is laminar (Re < 2100) and steady with a constant mass flux density,
– fluid velocity in the minichannel has one component $w(y)$ parallel to the heated foil surface while the other component equals zero,
– the vapor quality value is low and is not considered in the model,
– fluid temperature at the inlet and outlet of the minichannel is known.

The steady state heat transfer in the heated thin foil and flowing fluid was described using two-dimensional Poisson equation and two-dimensional energy equation, respectively:
a) for heater in the domain
$\Omega_H = \{(x,y) \in R^2 : 0 < x < L,\ \ 0 < y < \delta_H\}$

$$\frac{\partial^2 T_H}{\partial x^2} + \frac{\partial^2 T_H}{\partial y^2} = -\frac{q_V}{\lambda_H} \tag{1}$$

b) for fluid in the domain
$\Omega_f = \{(x,y) \in R^2 : 0 < x < L,\ \ \delta_H < y < \delta_H + \delta_M\}$

$$a\left(\frac{\partial^2 T_f}{\partial x^2} + \frac{\partial^2 T_f}{\partial y^2}\right) = w(y)\frac{\partial T_f}{\partial x} \tag{2}$$

where: q_V – the volumetric heat flux supplied to the heated foil, λ_H – the thermal conductivity of the heater, a – the thermal diffusivity.

For equation (1), the boundary conditions took into account the measurements of the heated foil temperature $T_{H,i}$ conducted in K points $(x_k, 0)$ on the basis of measurements by an infrared camera:

$$T_H(x_k, 0) = T_{H,k} \text{ for } k=1,2,\ldots,K \tag{3}$$

Additionally, it was assumed that the boundaries of the heater temperature for $x = 0$ and $x = L$ were isolated, i.e.

$$\frac{\partial T_H}{\partial x} = 0 \text{ for } 0 < y < \delta_H \text{ and a) } x=0, \text{ b) } x=L \tag{4}$$

For equation (2), the boundary conditions depend, among others, on the assumptions made earlier

$$\text{a) } T_f(0,y) = T_{f,in},\ \text{b) } T_f(L,y) = T_{f,out}$$

$$\text{for } \delta_H < y < \delta_H + \delta_M \tag{5}$$

$$T_H(x,\delta_H) = T_f(x,\delta_H) \tag{6}$$

$$\frac{\partial T_f}{\partial y} = 0 \text{ for } y = \delta_H + \delta_M \text{ and } 0 < x < L \tag{7}$$

In the first step, two-dimensional temperature distribution of the heater is determined by solving the system of equations (1), (3) and (4). Next, the solution of the system of equations (2), (5) - (7) allows us to determine the fluid temperature. Determining the heater temperature is an inverse heat conduction problem (IHCP) while determining the fluid temperature is a direct heat conduction problem (DHCP). Having found both temperature distributions (in the heater and fluid) one can evaluate the heat transfer coefficient from the Robin condition, which gives

$$\alpha(x) = \frac{-\lambda_H \frac{\partial T_H}{\partial y}(x,\delta_H)}{T_H(x,\delta_H) - T_f(x,\delta_H)} \tag{8}$$

3 Methodology of calculations

For the differential equations with suitable boundary conditions written in the mathematical model (1) – (7), approximate solutions were found using the Trefftz method [12]. The choice of the method was based on the fact that determining solutions to equations (1), (3), (4) lead to finding the solution to (IHCP).

The Trefftz method, as reported in the available literature [1, 10-17], gives stable results for the inverse problems. The Trefftz method is based on approximating the unknown solution of a differential equation using the linear combination of Trefftz functions, that is, the functions that satisfy the equation exactly. Two sets of Trefftz functions were used in the numerical calculations: harmonic polynomials $u_n(x,y)$ for two-dimensional Laplace's equation in Cartesian coordinates [12] and the Trefftz functions $h_n(x,y)$ for the energy equation (2) described in detail [14]. In this case the heater temperature T_H and fluid temperature T_f were approximated by linear combinations of adequate Trefftz functions in the following form

$$T_H(x,y) = \sum_{n=1}^{N_H} a_n u_n(x,y) \tag{9}$$

$$T_f(x,y) = \sum_{n=1}^{N_f} b_n h_n(x,y) \tag{10}$$

where N_H and N_f denote the number of adequate Trefftz functions in linear combinations (9) and (10).

The computed functions T_H and T_f satisfy exactly the governing equations (the Laplace equation and the energy equation respectively), and the assumed boundary conditions are satisfied approximately. The numerical procedure included IHCP in the heated foil and DHCP in flowing fluid.

Unknown coefficients a_n, b_n in equations (9) and (10) were determined by minimizing the error functionals that described the mean square error with which functions T_H and T_f satisfied the assumed suitable boundary conditions. For example, error functional for fluid is given in the form

$$J_f = \int_{\delta_H}^{\delta_H+\delta_M} \left(\sum_{n=1}^{N_f} b_n h_n(0,y) - T_{f,in} \right)^2 dy + \int_{\delta_H}^{\delta_H+\delta_M} \left(\sum_{n=1}^{N_f} b_n h_n(L0,y) - T_{f,out} \right)^2 dy + \int_0^L \left(\sum_{n=1}^{N_f} b_n h_n(x,\delta_H) - T_H(x,\delta_H) \right)^2 dx + \int_0^L \left(\sum_{n=1}^{N_f} b_n \frac{\partial h_n(x,\delta_H+\delta_M)}{\partial y} \right)^2 dx \quad (11)$$

Its minimization leads to solution of equations

$$\frac{\partial J_f}{\partial b_n} = 0 \text{ for } n=1,2,\ldots,N_f \quad (12)$$

4 Experimental data

4.1 Experimental stand

The experimental stand consists of several loops and systems. A view of the setup is shown in Fig. 1. The main is a flow loop in which working fluid – Fluorinert FC-72 is recirculated. This loop comprises: a test section with minichannel heat sink as the main element, a mass flow meter with a controller for flow stabilizing, a pump, a filter, a deaerator, a compensating tank (a pressure regulator) and a heat exchanger. The system for data and image acquisition system consists of: an infrared camera, two data acquisition stations and a PC computer with appropriate software. Temperature and pressure of the fluid at the inlet and outlet of the test section are measured by thermocouples (K-type) and pressure transducers, respectively. The supply and control system contains a power supply unit.

4.2 The test section

The test section is illustrated in Fig. 2. It comprises a group of parallel minichannels (each is 32 mm long, 0.5 mm wide and 0.5 deep). The heated element for FC-72 flowing in channels is a foil made of Haynes-230 alloy, 0.1 mm thick. Temperature of the outer side of the foil is measured using FLIR E60 infrared camera.

4.3 Experimental methodology

During experimental series, there is FC-72 flow along minichannels. The experiments are conducted in stationary state. Mass flow rate, temperature and pressure of the fluid at the inlet and outlet of the test sections, current intensity and voltage drop across the heater are monitored by the data acquisitions stations. Thermograms of the outer foil surface are measured due to infrared thermography.

Fig. 1. A view of the experimental stand.

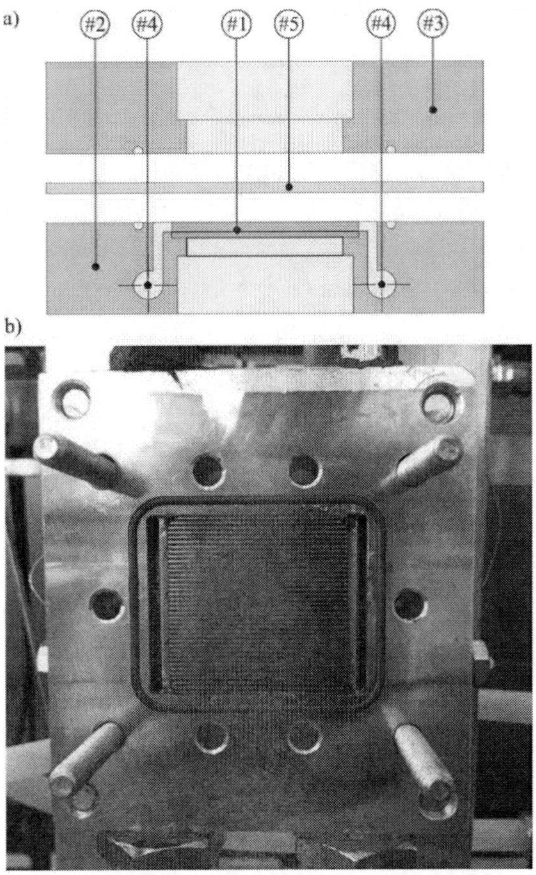

Fig. 2. The test section: a) a schematic diagram, #1 - a group of parallel minichannels, #2 - a channel body, #3 - a front cover, #4 - inlet/outlet chambers,#5 - a heated foil, b) a view.

5 Results and discussion

The calculations were performed using the model described in the previous section, with data collected during experiment conducted on the set up presented above.

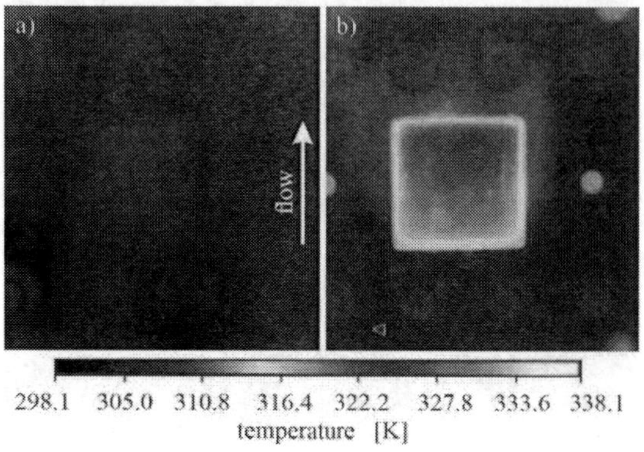

Fig. 3. Thermograms (infrared thermographic images) of the outer surface of the heated foil, for two values of the heat flux: a) q_w= 1.8 kW/m^2, b) q_w = 13.98 kW/m^2.

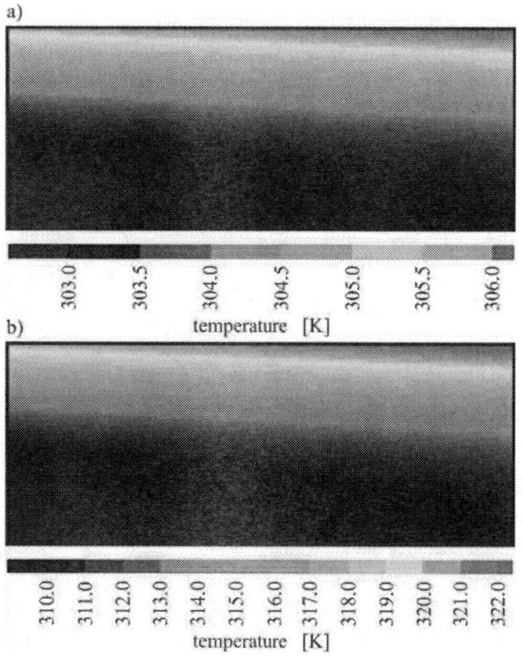

Fig. 4. Two dimensional fluid temperature distribution for two values of the heat flux: a) q_w= 1.8 kW/m^2,b) q_w = 13.98 kW/m^2.

The experimental parameters were as follows: average mass flux of 124 kg/(m^{2}s), average inlet pressure of 134.21 kPa, average inlet liquid subcooling of 31.58 K and 14 values of the heat flux in the range of 0.17 kW/m^2 ÷ 28.30 kW/m^2.

Figure 3 shows the thermograms of the outer foil surface, obtained from infrared thermography. Distribution of fluid temperature along the minichannel width are presented in Fig. 4.The heat transfer coefficient in the function of the distance from the minichannel inlet is illustrated in Fig. 5. It can be noticed that the heat transfer coefficient increased with the distance from the minichannel inlet for each heat flux values. The values of the heat transfer coefficient are in the range 0.53 ÷ 5.22 kW/(m^2 K).

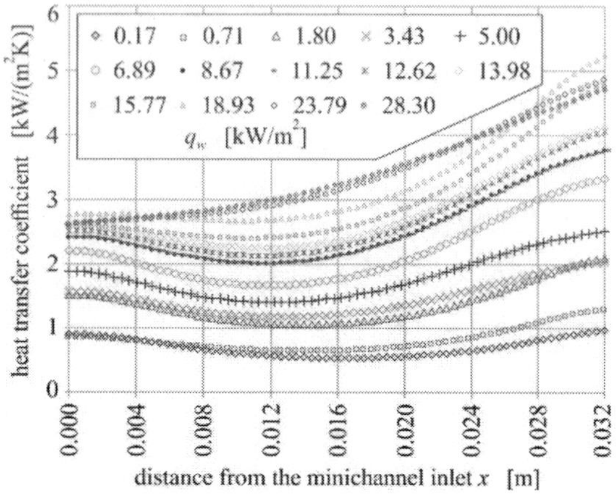

Fig. 5. The heat transfer coefficient vs. the distance from the minichannel inlet.

6 Conclusions

The two-dimensional mathematical model of flow boiling heat transfer in micro- and minichannels was proposed. Calculations with the use of Trefftz functions were based on the data from own experiments. The mathematical model described well the considered phenomenon of heat transfer in flow boiling in a single minichannel.

The Trefftz method allowed to effectively calculate two-dimensional temperature distributions in the channel's heated wall and in the flowing refrigerant. Obtained solutions satisfy the governing partial differential equation exactly.

The proposed numerical method was applied to DHCP and IHCP and the calculations are not overly complicated – the Trefftz functions are, in this case, polynomials. The Trefftz method is not limited by the number or the type of boundary conditions: temperature-related, flow-related, discrete or continuous. To verify the proposed mathematical model data from experiments were applied in calculations. The central minichannel of the group of parallel minichannels in a minichannel heat sink was selected for consideration. The heated element for Fluorinert FC-72 flowing along minichannels was a thin foil. The temperature of its outer side was measured using infrared thermography. The resulting graphs presented thermograms of measured temperature on outer surface of the heater, temperature distributions of fluid temperature and local values of the heat transfer coefficient. It was noticed that the heat transfer coefficient increased with the distance from the minichannel inlet for each heat flux value and achieved values up to 5.22 kW/(m^{2}K).

Acknowledgments

The research reported herein was supported in part by a grant from the National Science Centre, Poland, No. UMO-2018/31/B/ST8/01199.

References

1. S.Hożejowska, M. Piasecka, MATEC Web of Conf. 240, 01012 (2018)
2. M. Piasecka, T. Musiał, A. Piasecki, EPJ Web of Conf. 213, 02066 (2019)
3. K. Strąk, M. Piasecka, B. Maciejewska, Int. J. Heat Mass Transf. **117**, 375-387 (2018)
4. B. Maciejewska, K. Strąk, M. Piasecka, Procedia Eng. **157**, 83-88 (2016)
5. W Depczyński, A. Piasecki, M. Piasecka, K. Strąk, E3S Web of Conf. 19, 03012, 6 (2017)
6. B. Maciejewska, P. Labedzki, A. Piasecki, et al. EPJ Web Conf. **143**, paper No. 02070 (2017)
7. J.T. Cieśliński, T. Z. Kaczmarczyk, Arch. Thermodyn. **35** (2), 3-20 (2014)
8. Ł. J. Orman, N. Radek, J. Brončeк, Mater. Res. Proc. **5**, 189–193 (2018)
9. Ł. J. Orman, N. Radek, A. Kapjor, Mater. Res. Proc. **5**, 216–219 (2018)
10. B. Maciejewska, M. Piasecka, Heat Mass Transf. **53**, 1211-1224 (2017)
11. B. Maciejewska, M. Piasecka, Int. J. Heat Mass Transf. **107**, 925-933 (2017)
12. E. Trefftz, E. Trefftz, Proc. 2nd Int. Con. App. Mech. 131-137(1926)
13. K. Grysa, A. Maciąg, A. Pawińska, Int. J. Heat MassTransf. **55**, 7336–7340 (2012)
14. S. Hożejowska, L. Hożejowski, Eng. Anal. Bound. Elem. **98**, 27-34 (2019)
15. M.J. Ciałkowski, A. Frąckowiak, K. Grysa, Int. J. Heat Mass Transf. **50**, 2170-2181 (2007)
16. E. Kita, N. Kamiya, Adv. Eng. Softw. **24**, 3–12 (1995).
17. I. Herrera, Numer. Methods Part. Differ. Equ. **16**, 561–580 (2000)

Experimental circuit for the generation of cavitation in oil flow

Jana Jablonská [1], *Milada Kozubková* [1] *and Patrik Marcalik* [1,*]

[1] VŠB-Technical University of Ostrava, Faculty of Mechanical Engineering, Department of Hydromechanics and Hydraulic Equipment, 17. listopadu 15, 708 33 Ostrava-Poruba, Czech Republic

Abstract. The paper deals with the flow of oil through an experimental hydraulic circuit with a convergent divergent nozzle with a circular cross-section. Under different physical conditions, i.e. changes in pressure, flow and temperature, the formation and development of cavitation is monitored. As it is a highly dynamic flow in a transparent nozzle, cavitation is monitored using a high-speed camera and the frequency of formation and dissolution of the cavitation bubble and the movement of the cavitation cloud is then determined. The authors deal with the issue of the amount of dissolved or undissolved air in the hydraulic oil. Both variants of air influence the formation, development and size of the cavitation area. This cavitation is in this case called air cavitation. In technical practice, the issue of air cavitation is relevant, especially in the pump suction, where vacuum, leaks and hence air intake may occur.

1 Introduction

Cavitation is a problematic and usually undesirable phenomenon in the technical practice of hydraulics and the transport of liquids. In hydraulic circuits it can cause material wear, degradation of working fluid and degradation of hydraulic components. Cavitation is a phenomenon in which gas filled cavities are formed and destroyed in the liquid. If these cavities implode near the walls (material), the surface may be damaged, this is what is called cavitation damage. If we compare cavitation in two typical fluids used in hydraulics, i.e. in water and oil, it can be said that cavitation in water is more often described in literature [1, 3, 7, 8], and both steam and air cavitation can be achieved. The slow growth of the cavitation bubble caused by air diffusion is called air (gaseous) cavitation. Sudden immediate enlargement of the cavitation bubble, predominantly filled with a steam of the flowing liquid, we call the vapor cavitation.

The steam cavitation phenomenon occurs when the pressure in the liquid is reduced to the saturated vapor pressure of the selected liquid.

The liquid begins to evaporate and the cavitation bubbles are formed, filled with steam evaporating from the liquid. These bubbles are further carried by flowing liquid into the area of higher pressure, where the vapor condensates and the cavitation cavities are formed. The cavity is filled with the surrounding liquid, which penetrates into them at high speed, which creates a large shock that negatively affects the material and also the liquid. Cavitation Bubbles filling a part of the liquid stream form the so-called cavitation area. When the pressure drops on the level of saturated vapor pressure, it is the so-called initial cavitation, where less pressure pulsation of the flow medium is manifested in the unstable cavitation zone. Further reduction of the inlet pressure leads to the enlargement and stabilization of the cavitation area. The cavitation area is described primarily by shape as a capsule (where a certain space is filled with cavitation bubbles).

The reason for the emergence of the volatile pocket cavitation area can be the separation of the current from the wrapped surface. The cause of the formation is the whirling of the liquid between the main current and the wrapped surface, where the instability of this area is expressed as Strouhal's number

$$Sr = \frac{L \cdot f}{v} \qquad (1)$$

Where L is the characteristic dimension of the area, f is the frequency of the loosed vortex, v is the velocity of the flow. [1, 2, 3]

The saturated vapor pressure p_w is dependent on the temperature of the flowing medium, for example for water temperature of 20 °C the saturated vapor pressure is 2367.8 Pa, where for oil of the same temperature saturated pressure is 1 Pa [4]. The dependence of water temperature (under operating conditions) to cavitation does not affect the development of cavitation very much, since the dependence of density and viscosity on the temperature is not so significant.

The simplest form of the cavity is a gas bubble filled with dissolved air. These bubbles form the germs of cavitation, they arise sooner than steam bubbles and suppress steam cavitation. [7] The chapter "Properties of mineral oil" is in more detail devoted to air.

The simplest hydraulic device for examining the hydrodynamic conditions and the influence of the liquid temperature at the beginning of the cavitation phenomenon is convergent divergent nozzle with a circular cross-section (onward referred to as CD nozzle). The advantage of the CD nozzle is simple, you can

* Corresponding author: patrik.marcalik@vsb.cz

approximately determine the pressure by use of the Bernoulli equation and estimate the formation of cavitation. One of the typical parameters by which cavitation is determined is the cavitation number

$$\sigma = \frac{2(p_{in} - p_w)}{\varrho \cdot v_{max}^2} \quad (2)$$

where p_{in} is the absolute pressure at the inlet of the CD nozzle, p_w is the saturated vapor pressure of the liquid at a given temperature, ρ is the density of the liquid at a given temperature, v_{max} is the speed at the narrowest point of the CD nozzle. The critical cavitation number indicates when cavitation in the given physical conditions (density, temperature, media type) starts to occur in the specified geometry. [1, 2]

2 Properties of mineral oil

In hydraulic oils, air exists in two forms, either as air undissolved in a mixture with oil or as dissolved air in oil. If the liquid (oil) is not saturated with air, then the air dissolves by the diffusion theory. [5, 8] The undissolved air in the oil is found in the form of bubbles, adversely affecting its compressibility and therefore the rigidity of the system. Dissolved air represents the molecules of air that are in the oil and its quantity is governed by Henry's law. The release of air from the oil causes a so-called gas cavitation

$$V_D = V \cdot k \cdot \frac{p_{out}}{p_{in}} \quad (3)$$

Where V_D is the volumetric amount of air in the oil in the dissolved state, V the volume of the oil, p_{out} is the atmospheric final pressure, p_{in} is the initial atmospheric pressure, k is the absorption coefficient ($k = 0.093$ to 0.11) [5, 8].

Henry's dependence can be neglected for temperatures ranging from 20 to 80 °C. The amount of air in the oil is approximately 11 % of the oils volume, where the value proportionally increases with the pressure. If, due to the change in pressure and temperature, the equilibrium level is disturbed, at which the oil is saturated with air, the release of the air molecules followed by the formation of bubbles which subsequently increases the amount of undissolved air in the oil and creates a mixture of oil with air, or, conversely, more air in the oil is dissolved. The release of air from the oil takes place much faster than its dissolution, and depends on three main parameters: [5]

1. oil viscosity,
2. the size of the surface area between oil and air,
3. movement state (calm, laminar or turbulent flow).

Literature [6], fig. 1 describes the relationship between the pressure, temperature and amount of dissolved air in the oil, where at 20 °C and atmospheric pressure 10 % of the air is present in the oil. At the same pressure at 80 000 Pa, there is approximately 8% of the dissolved air. Table 1 of the same literature shows the limit values of the volume fraction of air for standard new oils. In the described hydraulic system, the used hydraulic oil is RENOLIN VG 46, whose limit value of dissolved air in the oil is 10%.

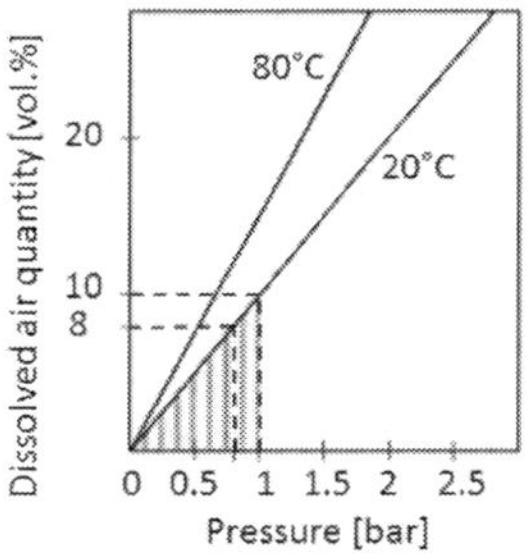

Fig. 1. Dissolved air in oil in volume percentage depending on pressure and temperature. [6]

Table 1. Limit values of air volume fraction for typical standard fresh oil [6]

ISO VG/Type	32	46	68	100
Turbine Oil DIN 51515 ISO 8068	5	5	6	-
Hydraulic Fluid HLP/HM DIN 51524/2 ISO 11158	5	10	13	21st

The literature [1] states that the maximum air solubility in oil is 9.3 % - 11.3 %. The time at which half of the air is released or dissolved is referred to as the half-life of t_p and for the air release in oil is $t_p = (3.6$ to $7.6)$ s and for dissolution $t_p = (6.1$ to $10.2)$ s [1]. The literature [5] states that 25 % to 30 % of the air is dissolved in oil immediately, i.e. over a period of up to 1 s, which is explained by the immediate saturation of the thin layer of oil in contact with air.

For the experiment hydraulic oil marked RENOLIN VG 46 was used with its dependence of density and kinematic viscosity at temperature.

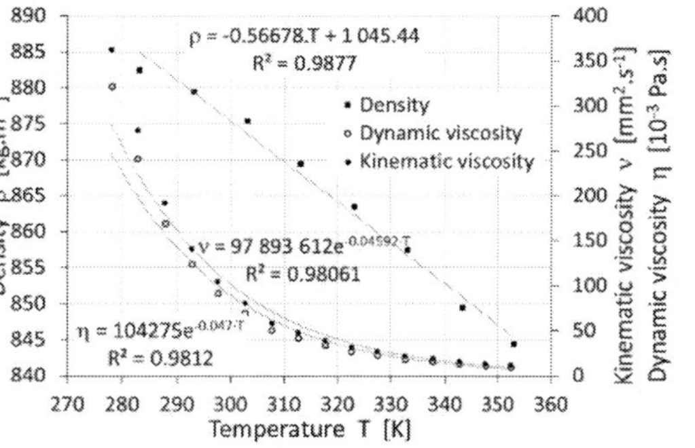

Fig. 2. Physical Properties OF VG46 oil depending on temperature.

3 Methods of diagnosis of cavitation

More often, high-speed cameras are used to identify and measure cavitation. This is a device with a high frame rate, so it is mainly used for short time in terms of storing large data. Using this method, a transparent equipment and suitable illumination should be used. For subsequent analysis, we get a digital record from which you can identify many other data such as bubble size, speed of growth and extinction of bubbles, speed of movement of bubbles and the like.

The cavitation status can also be assessed by means of a laser, using either the Tyndale phenomenon, where the dispersion of the light on the cavitation bubbles is examined in a homogeneous environment of the liquid, or by the method of dispersion of the light on the emerging and disappearing cavitation bubbles. [3]

Cavitation can also be specified using vibroacoustic diagnostics or by examining worn parts. The use of vibroacoustic diagnostics can be a problem in technical practice, as it is necessary to isolate the system from surrounding sources of noise and vibrations, but in laboratory conditions this method is usable. [9]

4 Measuring circuit

From tank 1 the liquid is pumped through the regulation hydrogenerator with tilting plate 3. Heater 2 represents the heating element inside the tank, through which the oil is heated to the given temperature. The circuit is protected against overload by a safety valve 4. The flowmeter 6 is located between the ball valves 5, on which the temperature and flow rate values are obtained. Then the vertical CD nozzle 9 is placed. Before and after CD nozzle pressure sensors 7 are placed, P_1 before the nozzle and P_2 behind the nozzle. From the nozzle the liquid flows through the waste filter 10 to the next tank. From there, it is pumped back into tank 1 by means of a hydrogenerator 11 and this causes the liquid to be mixed.

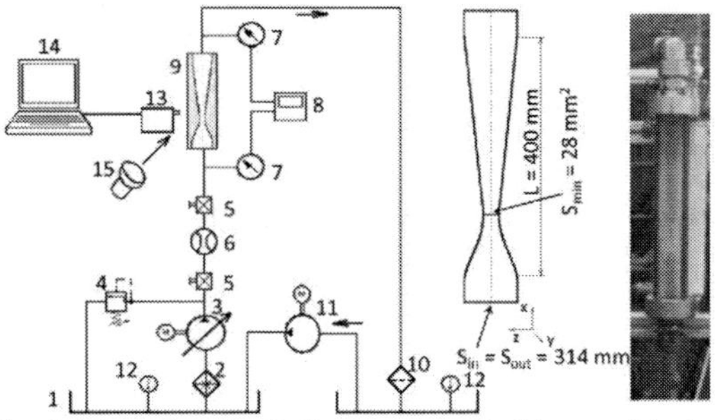

Fig. 3. Measuring circuit diagram – 1 tank, 2 heater, 3 hydrogenerator, 4 safety valve, 5 ball valve, 6 flow meter, 7 pressure sensors, 8 evaluation and recording equipment, 9 convergent divergent nozzle, 10 waste filter, 11 hydrogenator, 12 thermometer, 13 high-speed camera, 14 PC – computer, 15 light.

The measured element is the circular convergent divergent nozzle, a well-known shape with an inlet diameter of Ø 20 mm and diameter of the narrowest section of Ø 6 mm. The distance between the pressure sensors is 400 mm, see fig. 3. The relative pressure sensors of type PR15 with a range (-1 to 6) bar and ± 0.5% accuracy are used. For temperature measurement, the sensor Pt 100 is used with a range (-50 to 200) °C and the accuracy of ± 1 %, which is mounted on the flowmeter. Flow rate is measured by flowmeter GFM-70 with range (0.7 to 70) $dm^3 \cdot min^{-1}$ and accuracy ± 0.5 % [6, 8]. The high-speed Mini UX50 camera was used for visualization, with 8000 fps sampling frequency and 1280 x 296 resolution, aperture 1/10000 s. [7].

5 Evaluation of measurements

A series of measurements were made at the temperature change (T1 - T5) and the change of flow. Temperature T1 = 300 K corresponds to the oil temperature at the first series of measurements when the oil was not heated and the temperature T5 = 325 K corresponds to the last series of measurements that took place after the last oil heating. The oil was gradually heated with step of 5 K. Further heating was not possible due to the risk of oil degradation.

The measured flow rate values are from $1.7 \cdot 10^{-4}$ m^3/s to $5.9 \cdot 10^{-4}$ m^3/s. Fig. 4 presents dependence of pressure drop on volumetric flow for CD nozzle and effect of oil temperature. It is evident that pressure gradient increases with increasing flow. However, with increasing temperature, the pressure loss decreases, which is due to the change in the physical properties of the oil (see Fig. 2), when viscosity decreases with increasing temperature.

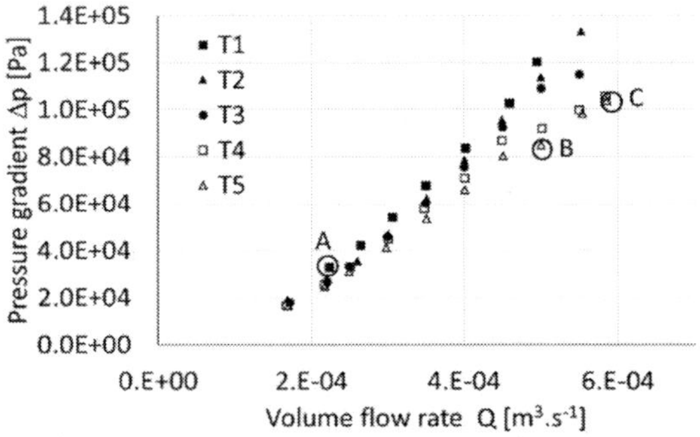

Fig. 4. Dependence of pressure gradient vs. volume flow rate.

In fig. 4, 5 and 6 the points are shown – measurements that will then be evaluated using a high-speed camera. Point A represents the flow without cavitation, point B of the initial cavitation and point C of the fully developed cavitation.

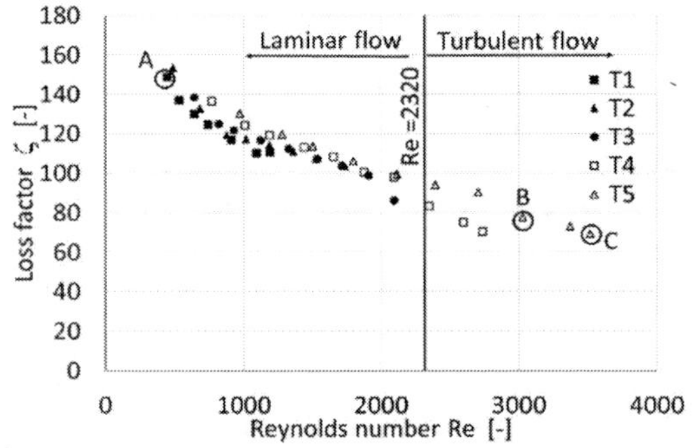

Fig. 5. Dependency of loss factor vs. Reynold number in narrowest section of veneer.

The Reynolds number *Re* is a function of the velocity *v* and the strongly kinematic viscosity *ν*, which depends on the temperature *T*.

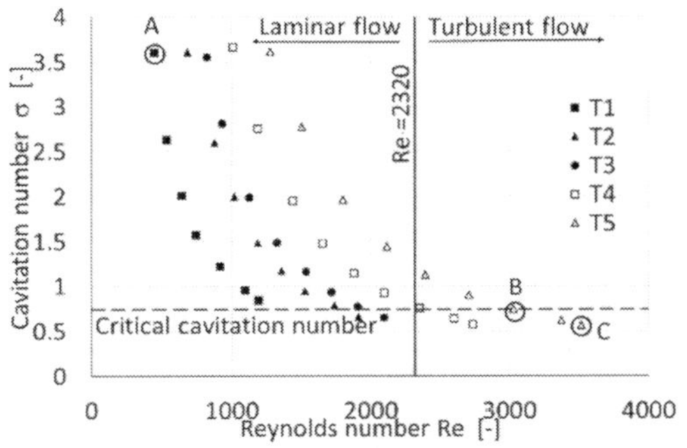

Fig. 6. Dependence of cavitation number on Reynolds number in the narrowest section of CD nozzle for different temperatures.

From fig. 5 it is evident that with the increasing Reynolds number and therefore with increasing flow rate the loss coefficient decreases. However, the flow coefficient will rise. In a turbulent area, the reverse of the trend is not significant, as we still deal with a low-Reynolds numbers, i.e. the transition area between laminar and turbulent flow. The transverse line here shows the critical value of Reynolds number 2320.

The value of Reynolds number at the inlet to the nozzle is between 133 and 1055, so it is a laminar flow. From fig. 6 it is evident that with the growing Reynolds number at the narrowest point of the nozzle the cavitation number decreases, and thus the conditions for the formation of cavitation are better. The first cavitation occurred during the second series of measurements (T2 see fig. 6), when the cavitation number was 0.658. Another measurement found that the critical value of the cavitation number, when the cavitation phenomenon occurred in the nozzle, was 0.747 (line, see fig. 6). Cavitation occurred at any other measurement when the cavitation number lower than 0.747 was reached.

Table 2. Evaluation of measurement *T* – temperature, *Q* – volumetric flow, p_1 - relative pressure at the entrance to the CD nozzle, p_2 - relative pressure on the output from CD nozzle and errors of measurement, σ – cavitation number.

Variant	A	B	C
	Without cavitation	Initial cavitation	Developed cavitation
T [K]	300	325	325
Errors *T* [K]	± 1.7	± 1.9	± 1.9
Q [m^3·s^{-1}]	2.64·10^{-4}	5·10^{-4}	5.86·10^{-4}
Errors *Q* [m^3·s^{-1}]	±7.62·10^{-7}	±1.44·10^{-6}	±1.69·10^{-6}
p_1 [Pa]	41900	84700	106900
Errors p_1 [Pa]	± 121	± 245	± 309
p_2 [Pa]	-2700	-300	3120
Errors p_2 [Pa]	± 8	± 0,9	± 9
σ	2.632	0.747	0.563
Errors σ	± 0,013	± 0,004	± 0,003

In Table 2, the measurement error of individual quantities is determined from the sensor error and assuming a uniform distribution of the error.

From fig. 7 A It is evident that there is no cavitation. Fig. 7 B presents the first cavitation bubbles, it is the initial cavitation phase when, according to table 2, the cavitation number is critical. In fig. 7 C is evident the fully developed cavitation phenomenon, the cavitation number has a value of 0.563.

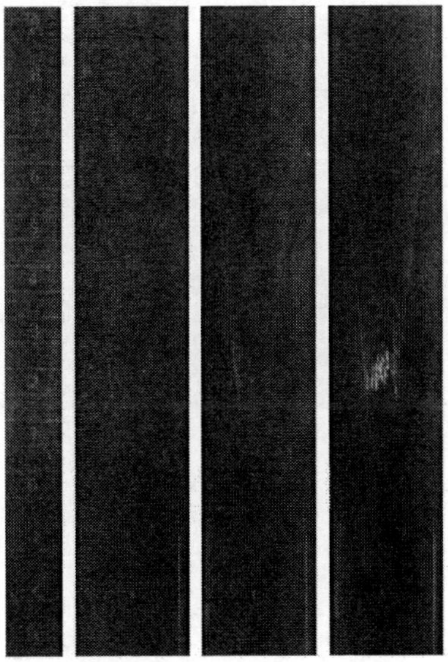

Fig. 7. High-speed Camera images for each cavitation variant A, B, C.

The development of the bubble at the initial cavitation was recorded by a high-speed camera, see fig. 8. The total interval of sensing the initial cavitation is < 253.5 ms, 255.875 ms >, each frame is shifted by the 0.125 ms, and the repeatability of the bubble is on every 7 frames, i.e. every 0.88 ms, corresponding to the frequency of the bubble formation 1142 Hz. As it is a slow flow, there is no significant displacement of the bubble, but the bubbles arise individually and do not affect each other excessively. Remarkably, the bubbles are noticeable only after the narrow section.

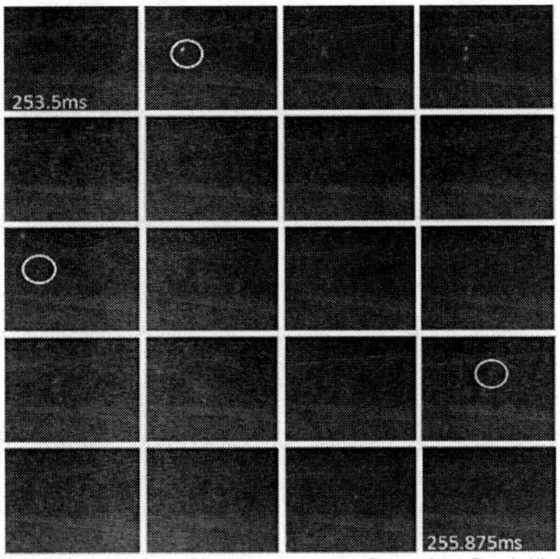

Fig. 8. Images from high speed camera for variant B - initial cavitation.

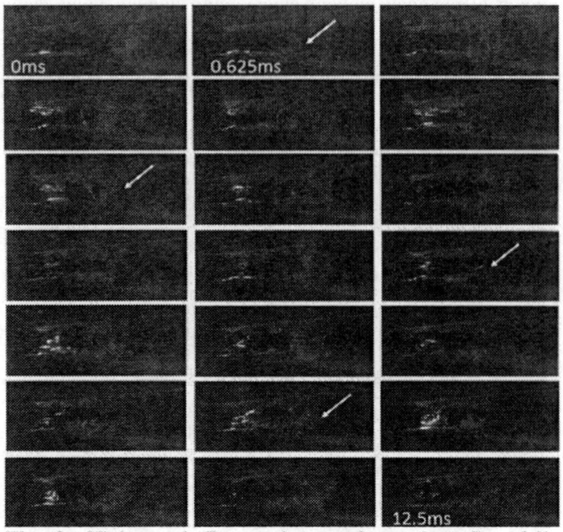

Fig. 9. High-speed Camera images for variant C – fully developed cavitation.

The development of cavitation cloud and bubbles in fully developed cavitation is evident from fig. 9. The first image is at a time of 0 ms, the last frame is at a time of 12.5 ms. Every fifth frame is displayed, so the time interval between frames is 0.625 ms. The repeatability of the bubble, or the disappearance of the cloud, is on every 4 frames (the image with an arrow that tracks the extinction of the cloud), i.e. every 2.5 ms, which corresponds to the frequency of the extinction (creation) of the air cloud 400 Hz. Individual bubbles cannot be observed at this time.

6 Conclusion

In the article, the hydraulic parameters of the formation of cavitation in the flow of oil are examined using the transparent nozzle of the circular cross section and the subsequent evaluation is performed. First of all, attention is paid to the development of cavitation depending on the temperature of the flowing medium. A critical cavitation number is specified for the given geometry.

In variant B, after the extinction of the bubble, a faint fog is observed. The air in the bubble causes suspension and the emergence of a new bubble. If there was no air in the bubble, then the cavity would disappear, and no new bubble in that spot would emerge. This phenomenon is also described in literature [1]. It is therefore clear that this will only be the case for air cavitation. It is likely that neither in the C variant the steam cavitation has been achieved, which will subsequently be verified by the mathematical model.

Cavitation depends on the amount of air in the oil (which cannot be easily determined). This is related to the oil temperature, pressure in the liquid and the dissolution rate. The results of the physical experiment will be used to create a mathematical model of air cavitation in the oil. Measured hydraulic quantities will be used as boundary conditions, the amount of air in the oil will be estimated subsequently, then specified more accurately by numerical calculations.

Acknowledgement

This work has been supported by The Ministry of Education, Youth and Sports of the Czech Republic from the Specific Research Project SP2019/123 (Numerical Modeling of Dynamic Phenomena in Fluid Mechanics with Support of Experimental Research).

References

1. Noskievič, J. Kavitace. Praha: Academia (1969).
2. Brennen, C. E. Cavitation and bubble dynamics. Cambridge University Press, USA (2014).
3. Himr, D., Habán, V., Hudec, M., Pavlík, V. Experimental investigation of the check valve behavior when the flow is reversing. EPJ 114 doi: 10.1051/epjconf/201714302036 (2016).
4. Li, W. Modeling Viscous Oil Cavitating Flow in a Centrifugal Pump. J. Fluids Engineering 138 (1) (2015).
5. Kopáček, J. Vzduch v minerálním oleji a jeho měření. Strojírenství 36 (1986).
6. HYDAC. Fluid Control Contamination Handbook [online]. [CIT. 2019-05-10]. Available from: https://www.hydac.fi/wp-content/uploads/2018/08/EN7603-9-04-18_Fluid-Control-Contamination-Handbook-web.pdf.
7. Jablonská, J., Kozubková, M. Visualization of gas bubbles during cavitation in the nozzle. AIP 1889, 020014 (2017) DOI: 10.1063/1.5004348.
8. Bureček, A., Hružík, L., Vašina, M. Determination of undissolved air content in oil by means of a compression method. Strojniski Vestnik – J. Mechanical Engineering, 61, 7-8 (2015) DOI: 10.5545/SV-JME. 2015.247.
9. Parish, J. Controlling cavitation [Online]. Valve magazine, 2009 [cit. 2019-05-10]. Available from: https://www.pro-quip.com/images/proquip/PDFs/Controlling%20Cavitation.pdf

An experimental analysis of the fluid flow on performance and frost formation in exhaust air energy recovery heat exchanger

Marek Jaszczur[1,*], *Marek* Borowski[2], *Michał* Karch[2], *Sławosz* Kleszcz[1,3]

[1]AGH University of Science and Technology, Faculty of Energy and Fuels, Kraków, Poland
[2]AGH University of Science and Technology, Faculty of Mining and Geoengineering, Kraków, Poland
[3]Frapol Sp. z o.o., Kraków, Poland

Abstract. In the present paper, results for fluid flow, temperature and air humidity experimental measurements for innovative energy recovery counterflow air-to-air heat exchanger have been shown. For the low outdoor air temperatures case, frost formation is also observed and analysed. Base on the results of analysis the temperature, humidity and total efficiency of the exchanger are presented. The analysis showed that fluid flow rate, humidity and system efficiency significantly drop down under unfavourable conditions in the winter period when the outdoor temperature decrease below -5°C. In the present system, the heat exchanger equipped with a set of opposing air dampers can use intake air to absorb the moisture from the exhaust air. This solution has the advantage to prevent the exchanger frosting as well as melt ice layer if any growth on heat exchanger walls.

1 Introduction

In 2010 the Energy Performance of Buildings Directive (EPBD) [1] had been introduced by the European Commission. This document states that from 2019 all public buildings in UE will have to meet the requirements for nearly zero-energy buildings.

Heating, as well as air conditioning systems, are responsible for substantial energy consumption. In many cases, this consumption exceeds 60% of the total energy in operated buildings. Due to new regulations, such systems have to be designed to meet the requirements for the maximum use of non-renewable energy for heating, ventilation, cooling and hot water purpose. On the other hand, heating/cooling and ventilation units are often increasingly applicable in the domestic application as well as in industry [2-5]. Public buildings as also quite often with modern ventilation systems [6,7].

With increasing share of ventilation, the efficient energy recovery appears to be an essential issue and the unique solutions for energy saving [7-9]. But such ventilation type is necessary to keep certain indoor air quality [10-13]. With the implementation of energy recovery exchanger, it is a possibly significant decrease building energy consumption. The typically used for this purpose heat exchanger (HE) are characterised by high (often above 90%) recuperation efficiency [14]. Catalogue sheets of heat exchangers typically provide the temperature efficiency under optimum fluid flow conditions. Such high-efficiency also results from the fact that it is usually determined under laboratory conditions and for very low fluid flow velocity through the heat exchanger. However, these relatively high-efficiency values are usually evaluated without taking into account a dynamic change of fluid flow velocity or flow rate as well as outdoor and indoor air conditions which significantly affecting the reduction of heat recovery efficiency.

Additionally during the winter air temperature in cold and moderate climate zones may drop below minus 5°C which results in danger for HE icing. This phenomenon is caused by water dropping out from the exhaust air and occurs at the contact of the heat exchanger surface with cold air drawn from the outside [15,16]. Under the temperatures below zero Celsius, the condensate deposited in HE change its phase and cause a significant decrease in the heat recovery and finally increase the flow resistance. During low outdoor air temperatures, there is a risk of frost large formation on the heat exchanger surface which can lead to high-pressure drop, flow velocity decrease or even serious unit damage. To prevent heat exchanger from this situation and to keep constant fluid flow velocity, additional energy supply has to be implemented. Such a system is usually energy consuming and significantly reduce system performance. In the literature, various solutions have been proposed to melt ice deposited on HE surface. One of the most popular solutions includes the implementation of an electrical heater inside of HE in the supply air duct. Protection systems warm up the heat exchanger surface and melt the ice layer. The electrical heater significant increase the energy consumption and decrease central unit efficiency. Often a fresh air bypass is also applied. Different manufacturers also apply own solution, in which to warm-up the heat exchanger they turn-off the fresh air fans. However, the fresh air flow prevention

* Corresponding author: jaszczur@agh.edu.pl

fully disturbs the air flow in the entire building not ensuring acceptable comfort [17]. In many situations, the ice layer can also be melted by different the ratio of air supplied to the exhausted application. But such a solution reduces the amount of fresh air delivery to the building as well as create substantial negative pressure in the ventilated space.

In most of the central units, parallel plate counterflow air-to-air energy recovery heat exchanger are used. This type of heat exchanger was analysed in [18] in order to provide a guideline for optimal ventilation unit design. In their study two-dimensional model has was used in order to find analytical expressions. In other work [19] authors examined a new type of geometry for HE used in ventilation systems in cold and moderate climate zones. The methodology for system performance was presented base on the theoretical model and verified with experimental data. Fluid flow in parallel-plate turbulent open channel [20] has been studied successfully using Direct Numerical Simulation methods which provide very accurate results. In [21] the influence of channel geometrical configuration on the counter flow heat exchanger efficiency has been studied.

In this work, flow, temperature and humidity for innovative energy recovery counterflow air-to-air heat exchanger equipped with a set of opposing air dampers have been studied for the selected set of temperatures. Base on the results of analysis the temperature, humidity and total efficiency of the exchanger is evaluated. For the low outdoor air temperatures case, frost formation is also analysed. The present study was aimed to implement this solution to large industrial units.

2 Methodology

The periodic-flow heat exchanger unit under consideration consists of a standard counter-flow exchanger equipped with a set of opposing air dampers with a short opening/closing time, used in order to cyclically modify air flow direction. With this solution, the heat exchanger can use intake air to absorb the moisture from the exhaust air. It has also advantage to prevent the exchanger frosting as well as melt ice layer if any appears. The analysed HE used four identical counterflow exchanger arranged in the right order.

Additionally, the central unit is equipped with four counter-rotating aluminium dampers controlled by the actuators. Switching position allows changing the direction of air in different part of the exchanger. The fourth tightness class (according to EN-1751) prevent fresh and used air mixing. The measurement was done to determine the efficiency of the air handling unit with a new generation of the high-performance periodic counter-flow air-to-air heat exchanger with dampers.

Table 1. Flow Conditions during the study.

Parameter	V_{21}	t_{21}	ϕ_{21}	V_{11}	t_{11}	ϕ_{11}
Unit	m³/h	°C	%	m³/h	°C	%
Value	2000	-18	70	2000	23	46

where V, t, ϕ are volume flow rate, air temperature and humidity respectively.

The experimental measurements were done under real conditions reproduced in the laboratory to generate low air temperature and the possibility of occurrence of the phenomenon of heat exchanger frosting (Table 1).

The test section consisted of the central unit with a control system, which activates the anti-icing mode when the fresh air temperature t_{21} was below zero Celsius. The air handling unit was also equipped with a fresh air bypass, which was closed and deactivated for the current analysis. The time of dampers switching was set to 300s. The central unit was placed in an insulated chamber, inside which the air temperature was about 20+/-1°C. The air was supplied to the air handling unit by two ducts equipped with glycol heaters/coolers. Figure 1 shows the measurement section of the heat exchanger. The temperature of the exhaust air from the room was maintained on a constant level of 23°C, while the temperature of the fresh air was equal to -18°C. JUMO EE650-T2L200 airspeed converters were used to record the volume flow rates (V_{11} ,V_{12}, V_{21}, V_{22}). JUMO EE210-HT3xPBFxB hygroscopic humidity and temperature converters were used to measure the air temperature and humidity (t_{11}, t_{12}, t_{21}, t_{22} and ϕ_{12}, ϕ_{12}, ϕ_{21}, ϕ_{22}). The data acquisition from sensors was carried out using a Mitsubishi Electric FX5U module and a MAPS HMI 750 panel. The measured values were recorded with 1-second intervals during the time equal to 2000s. The data recording was activated at the time of receiving constant temperature of the fresh air.

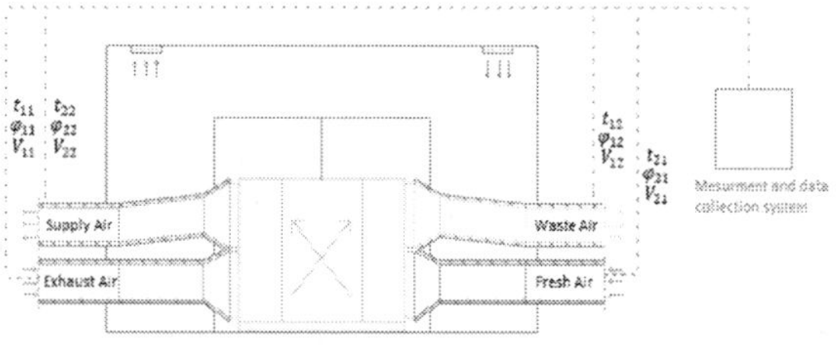

Fig. 1. Diagram of the test section.

Moisture content in the air:

The absolute humidity is defined as a ratio of water vapour mass in the air to the dry air mass. Using Dalton's law and the gas equation of state, it is possible to obtain a formula to evaluate the moisture content x in the air based on temperature and relative humidity as follows:

$$x = 0.622 \cdot \frac{\varphi \cdot p_{gs}}{p_b - \varphi \cdot p_{gs}} \tag{1}$$

where

$$p_{gs} = 6.1121 \cdot e^{\frac{17.502T}{T+240.97}} \tag{2}$$

$$p_b = 1013.25$$

Temperature efficiency:

The temperature η_t efficiency is calculated according to the following equations:

$$\eta_t = \frac{t_{22} - t_{21}}{t_{11} - t_{21}} \tag{3}$$

where t_{22}, t_{21}, are the temperature of supply and outdoor air, while t_{11} is exhaust air temperature.

https://doi.org/10.1051/epjconf/202226901023

3 Results and discussion

Figures 2 to 5 shows the experimental results of the air flow rate of fresh air V_{21}, and exhaust air V_{11}, temperatures t_{21}, t_{11} and relative humidity ϕ_{21}, ϕ_{11} of the air on each side involved in the energy exchange process. The volume flow rate of both flows was set-up to be 2000 m^3/h. On the basis of laboratory and experimental test measurements, the interval of rapid air flow damper position changes was set-up at the level of 300 s. Different the dumpers position results in different the direction of air flow through the air-to-air heat exchanger, enabling the condensate accumulated evaporation from the walls of the exchanger in both operating modes. The deposited at the wall surface water droplets are evaporated and absorbed by the stream of outside air. The results in Figure 2 shows that by switching the air damper position result in the air flow direction change through the heat exchanger. At the same time the resistance of the channel flows has changed which impact on the fluid flow velocity on both sides of the heat exchanger.

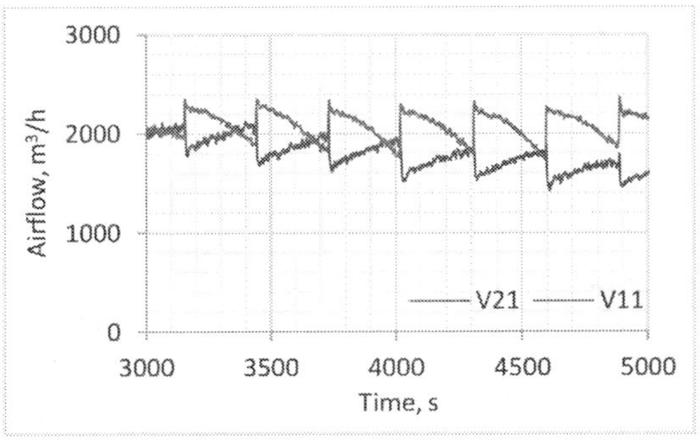

Fig. 2. Fresh V_{21} and exhaust V_{11} air volume flow rates.

The temperature of the supply air t_{22} presented in Figure 3 was about 14°C, with the temperature values for the intake t_{21} and extract air t_{11} stable all equal to -18°C (+/-0.5°C), and 23°C (+/-0.5°C), respectively. The above conditions result in the estimated temperature efficiency of the exchanger with dumpers in the range of 76-78% (see Fig. 6).

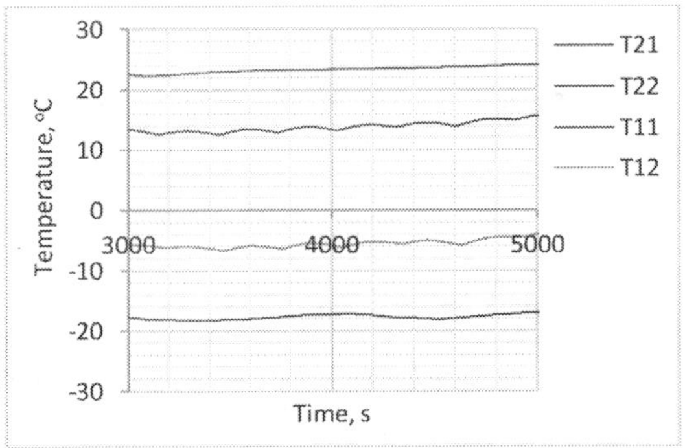

Fig. 3. Fresh t_{21}, supply t_{22}, exhaust t_{11} and waste t_{12} air temperature changes.

It is worth notice that despite critical air conditions due to dumpers implementation and flow direction variation, no frost layer was observed on the heat exchanger surfaces (see Fig. 7). Despite of not full water droplets evaporation (condensate), heat exchanger surface was at good condition. The average amount of condensed water from extract air was about 5.9 g/kg, and the evaporated water in supply air increase was about 4.72 g/kg. The deposited water content was observed as the iced surface of drip tray shown in Figure 8.

The relative humidity is presented in Figure 4. During the measurements, it changes dynamically between 40%–60% on supply air and 80–90% on waste air. During the study, the supplied air had the average humidity level higher than 50%, which effectively prevented the drying of the ventilated rooms, contributing to maintaining the feeling of thermal comfort for the users.

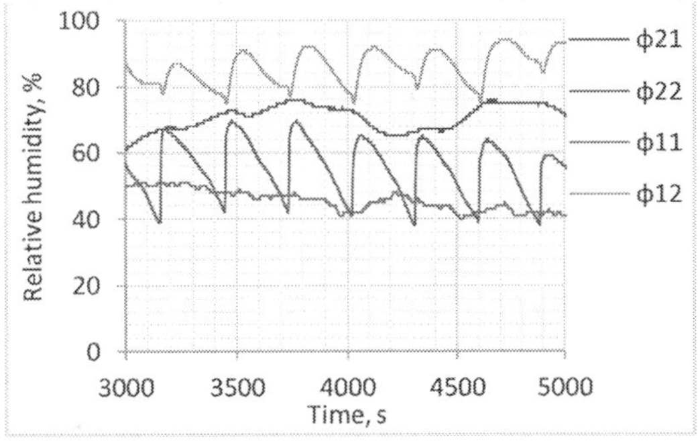

Fig. 4. Fresh φ_{21}, supply φ_{22}, exhaust φ_{11} and waste φ_{12} air relative humidity changes.

On the basis of the measurements of temperature and relative humidity values of the air flowing through the heat exchanger, the moisture content changes in each of the four air flow streams were evaluated. All these values, which contribute to the heat and moisture exchange processes, are presented in Figure 5.

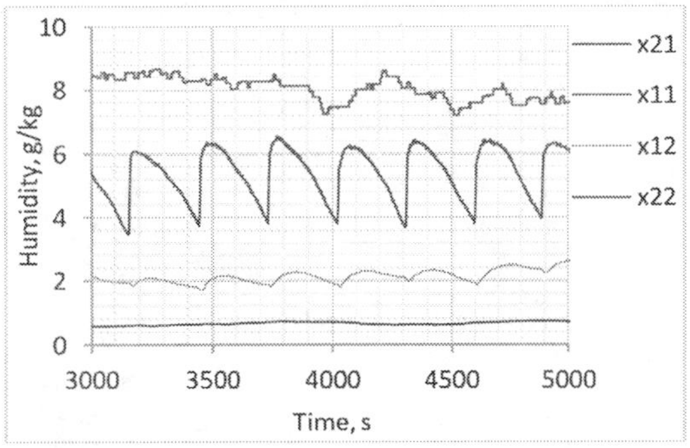

Fig. 5. Fresh x_{21}, supply x_{22}, exhaust x_{11} and waste x_{12} air humidity changes.

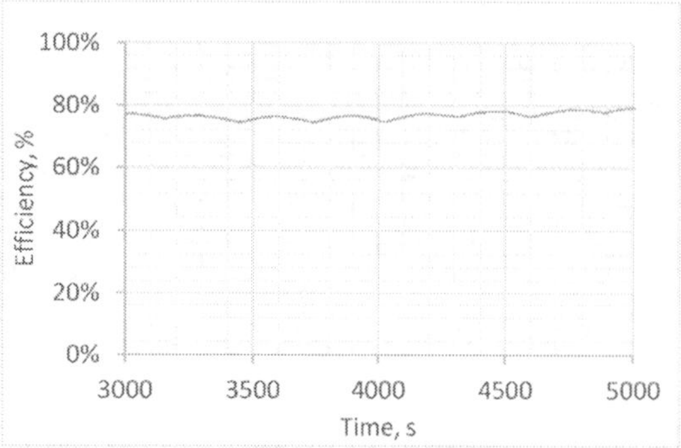

Fig. 6. The temperature efficiency over time.

Based on the experimental measurement the average humidity efficiency was evaluated to be equal to approx. 64%. According to Table 2 when humidity is not exchanged (as the case in traditional standard ventilation units), the process of heat exchange between fresh and exhaust air can be shown on Mollier chart - Figure 9. On these terms, the relative humidity would be only 10%.

Table 2. Average value of temperature, relative humidity and humidity.

t_{21}	−17.8°C	φ_{21}	70.7%	x_{21}	0.66 g/kg
t_{22}	13.8°C	φ_{22}	55.4%	x_{22}	5.38 g/kg
t_{11}	23.3°C	φ_{21}	45.6%,	x_{11}	8.04 g/kg
t_{12}	−5.6°C,	φ_{22}	85.7%	x_{12}	2.13 g/kg

Fig. 7. The heat exchanger surface at the end of measurements.

Fig. 8. Iced drip tray at the end of measurements.

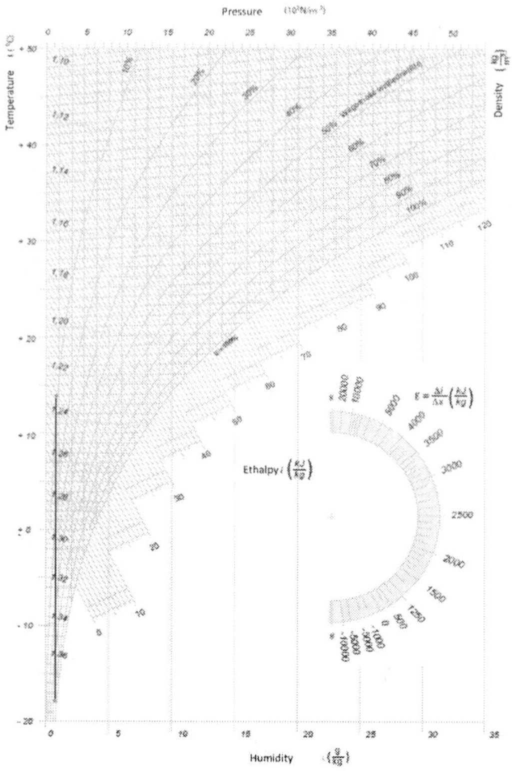

Fig. 9. Air process heating on Mollier chart (x=const).

4 Conclusions

In this research, the results of experimental measurements of an air handling unit equipped with an innovative periodic parallel-plate counterflow heat exchanger with dumpers are presented. Based on the analysis it is seen that traditional "anti-freezing" system which utilises different flow rates or even prevents fresh air flow to melt ice layer on heat exchanger surface should not be used at low-temperature conditions and can't be applied without an additional usually energy consuming protection method. Additionally, such a solution doesn't fulfil its primary role - proper building ventilation. The deficit of fresh air will result in significant growth of CO_2 concentration in the room as well as in the air temperature drop. The implementation of the heat exchanger with a system of air dampers which change air flow direction may solve wholly issue of heat exchanger icing. With dumpers, it is possible to achieve very high-efficiency values - not available by any other commercial unit. The proposed solution, remove anti-freeze systems (usually electrical heaters) completely. For the stable fresh air temperature equal to -18°C the evaluated temperature efficiency of the exchanger with dumpers was in the range of 76-78%.

In addition to that, based on the experimental measurement the average humidity efficiency is equal to approx. 64%. The mean amount of condensed water from extract air was 5.9 g/kg, while the evaporated water in supply air increase was 4.72 g/kg. During the analysis, the supplied air the average humidity was higher than 50%, which entirely prevents the drying of the ventilated space. It is important to notice that low moisture content in air flow stream may lead to health problems, malaise and worsens the thermal comfort. At autumn and winter seasons there is even recommended to humidify the air in dry rooms. However using additional devices people are exposed to unnecessary investment, operational and maintenance costs. The proposed periodic-flow heat exchanger with flow direction switching using dampers may be an excellent solution for all types of buildings (public, private, schools or hospitals), allowing to keep the necessary measure at a comfort level.

It has been found that it is important to precisely determine the time of changing the direction of the fluid flow (the positions of the air dampers) with correlation to the outside and indoor conditions thus enabling a controlled recovery of moisture from the extract air.

The works on the periodic-flow heat exchanger are supported in the Smart Growth Operational Programme 2014-2020 by the Polish National Centre for Research and Development and presentation by the Polish Ministry of Science Grant AGH No. 11.11.210.312.

References

1. Directive 2012/27/EU of the European Parliament and of the Council of 25 Oct. (2012)
2. M. Mijakowski, J. Sowa, P. Narowski, Civil Engineering **4**, 107 (2010)
3. G. P. Vasilyev, A. Tabunshchikov, M. M. Brodach, V. A. Leskov, N. V. Mitrofanova, N. A. Timofeev, V. F. Gornov, G. V. Esaulov, Ener. Build. **112**, 96 (2016)
4. V.Yu. Borodulin, M.I. Nizovtsev, App. Therm. Eng. 130, 1246 (2018)
5. S. Koestera, M. Falkenberga, M. Logemanna, M. Wessling, J. Memb. Sci. **525**, 68 (2017)
6. B. R Hughes, H. N. Chaudhry, J. K. Calautit, Appl. Energy **113**,127 (2014)
7. C. Simonson, Energy Build. **37**, 23 (2005)
8. X. P. Liu, J.L. Niu, Appl. Energy **129**, 364 (2014)
9. El Fouih, Y. Stabat, P. Rivie`re, P. Hoang, P. Archambault, Ener. Build. **54**, 29 (2012)
10. M. Orme, Energy Build. **33**, 199 (2001)
11. J. K. Calautit, B. R. Hughes, H. N. Chaudhry, S. A. Ghani, Appl. Energy **112**, 576 (2013)
12. M Jaszczur, M Branny, M Karch, M Borowski, J. Phys. Conf. Ser. **745**, 032049 (2016)
13. M. Borowski, M. Karch, Chłodnictwo & Klimatyzacja **4**, 65 (2015)
14. Mure, Monitoring of energy efficiency trends and policies in the EU, 36 (2015)
15. M. Besler, M. Skrzycki, Rynek Instalacyjny **3**, 26 (2013)
16. D. Seker, H. Kartas, N. Egrican, Int. J. Refrig. **27**, 367 (2004)
17. L. Huang, Z. Liu, Y. Liu, Exp. Ther. Fluid Sci. **33**, 1049 (2009)
18. M. Vera, A. Linan, Int. J. Heat Mass Tran. **53**, 4885 (2010)
19. J. Kragh, J. Rose, T. R. Nielsen, S. Svendsen, Ener. Build. **39**, 1151 (2007)
20. M. Jaszczur, Arch. of Mech. **63**,77 (2011)
21. M. I. Hasan, A. A. Rageb, M. Yaghoubi, H. Homayoni, Int. J. Therm. Sci. **48**, 1607 (2009)

Pool boiling of ethanol on surfaces with parallel microchannels

Robert Kaniowski[1,*], *Robert* Pastuszko[1]

[1]Kielce University of Technology, Faculty of Mechatronics and Mechanical Engineering, al. 1000-lecia Państwa Polskiego 7, PL-25-314 Kielce, Poland

Abstract. The paper presents investigations into pool boiling heat transfer for open minichannel surfaces. The experiments were carried out with saturated ethanol at atmospheric pressure. Parallel microchannels fabricated by machining were about 0.2, 0.3 and 0.4 wide and 0.2 to 0.5 mm deep. The measurements were performed with increasing heat flux and variable geometric parameters of the minichannels. The image acquisition speed was 493 fps (at resolution 400 x 300 pixels with Photonfocus PHOT MV-D1024-160-CL camera) and an EX-FH20 (*Casio*) camera was used to record the images of the entire surface of the specimen. The analysis of boiling curves for the tested surfaces does not give an unambiguous response to the influence of geometrical parameters, i.e. the height of the microchannels on the heat transfer process.

1 Introduction

Modern technology needs compact heat exchangers, small in size but capable of removing high heat fluxes. High-performance miniaturised electronic systems necessitate finding ways of providing the most powerful cooling at smaller volumes and weight. Pool boiling is an attractive cooling technique that allows dissipating large amounts of heat flux at low superheats. Surface modification with microchannels in various shapes is a relatively simple method of pool boiling heat transfer enhancement.

Authors of [1,2] experimented with copper Ω-shaped samples having round cavities and narrow inlet directed upwards, with water and ethanol as boiling liquids at atmospheric pressure. The surface used highly increases the number of active nucleation sites, which makes it possible to remove large heat flux amounts at low superheats and high CHF. Visualization study showed bubble nucleation, growth and departure behaviours in reentrant cavities, and, at higher heat fluxes, the tendency of the bubbles to coalesce.

Kalani and Kandlikar in [3] used open microchannel structures to investigate the pool boiling performance of various geometric parameters (channel depth 245 – 470 μm, channel width 194 – 406 μm) with ethanol as the working fluid. Microchannels were machined over an area of 10 x 10 mm^2 in the middle of the 20 x 20 mm^2 test section. The authors concluded that the best performing chip had deep (400 μm) and narrow (200 μm) channels.

Studies on pool boiling heat transfer with water, ethanol, FC–72 and Novec-649 from tunnel structures, microfins, microchannels, microcavities [4–7] and boiling heat transfer with FC-72 flowing in narrow channels [8–10] have been conducted in order to determine the most favorable enhanced surfaces parameters in terms of the possibility of obtaining the highest heat transfer coefficients.

2 Experimental setup

The test set up [4-7], Figure 1 and Figure 2, is designed to determine boiling curves and to visually record the vapour bubble generation process. The heat transfer experiments on microfins were conducted for pool boiling in the range of nucleate boiling up to the boiling crisis. The level of liquid above the surface of the specimen was 50 mm. Prior to the measurements, ethanol was degassed by boiling for 15 minutes. Measurement data were recorded after all temperatures stabilized. The tests were performed at increasing heat flux up to the boiling crisis at atmospheric pressure.

The cylinder diameter corresponded to the diagonal of the sample base. A 1000 W electric cartridge heater 19 mm in diameter and 130 mm in length was mounted in a heating bar. Type *K* (NiCr-NiAl) sheathed thermocouples of 0.5 mm in diameter were used for the measurements. The measurement data acquisition system was a product of FLUKE Hydra 2635A. Calibration of the thermocouples was carried out using the Altek calibrator.

The images of the bubbles at the moment of their nucleation, growth and departure were obtained using a PHOT MV-D1024-160-CL (*Photonfocus*) digital monochromatic camera with a resolution of 1024x758 pixels. The camera recorded the photos at a speed of 493 fps at a resolution of 400x300 pixels, mean error 2 fps. An EX-FH20 (*Casio*) camera was used to record the images of the entire surface of the specimen. Lighting was provided by a halogen lamp with light pipes (front light and back light).

* Corresponding author: kaniowski@tu.kielce.pl

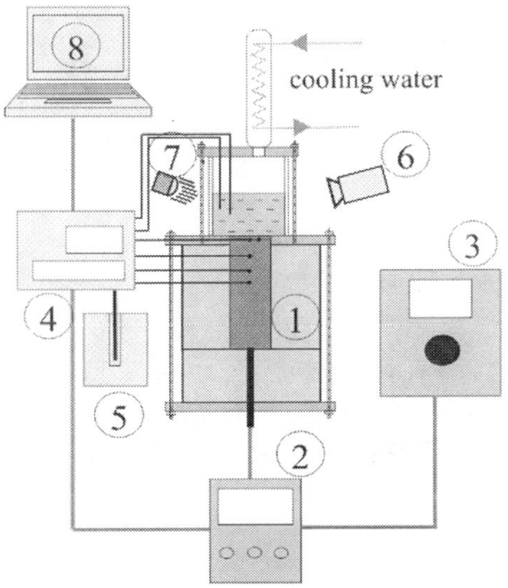

Fig. 1. Pool boiling measurement stand: 1 – main module, 2 – wattmeter, 3 – autotransformer, 4 – data logger, 5 – dry-well calibrator, 6 – high speed camera, 7 – light, 8 – PC.

Table 1. MC surface codes and specifications.

Sample code	w mm	h mm	p mm
Smooth surface	-	-	-
MC-0.2-0.5-0.4	0.2	0.5	0.4
MC-0.2-0.4-0.4	0.2	0.4	0.4
MC-0.2-0.3-0.4	0.2	0.3	0.4
MC-0.2-0.2-0.4	0.2	0.2	0.4
MC-0.3-0.5-0.6	0.3	0.5	0.6
MC-0.3-0.4-0.6	0.3	0.4	0.6
MC-0.3-0.3-0.6	0.3	0.3	0.6
MC-0.3-0.2-0.6	0.3	0.2	0.6
MC-0.4-0.5-0.8	0.4	0.5	0.8
MC-0.4-0.4-0.8	0.4	0.4	0.8
MC-0.4-0.3-0.8	0.4	0.3	0.8
MC-0.4-0.2-0.8	0.4	0.2	0.8

The specimens with test surfaces were made of copper and had parallel grooves with a constant pitch, made with an end mill of 0.2 – 0.4 mm in diameter (CNC machining process) and 0.2 to 0.5 mm deep. The test section consisted of a 32 x 32 mm^2 square copper specimen with a 27 x 27 mm^2 boiling region. Table 1 compiles the surface codes, specifications according to Fig. 2.

By Newton's law of cooling, HTC is equal:

$$\alpha = \frac{q}{\Delta T}, \tag{1}$$

According to Fourier's law:

$$q = \lambda_{Cu} \frac{dT}{dX} \cdot \frac{\pi d_c^2}{4a^2}, \tag{2}$$

The temperature gradient was calculated using three point's backward finite difference method as given below, Fig. 3.

$$\frac{dT}{dX} = \frac{3\frac{T_{T3}+T_{T4}}{2} - 4T_{T6} + T_{T7}}{2l_{T7-T3}}, \tag{3}$$

T_w was extrapolated by following equation as:

$$T_w = \frac{T_{T3}+T_{T4}}{2} - q\left(\frac{\delta_s}{\lambda_{Cu}} + \frac{\delta_{Sn}}{\lambda_{Sn}}\right), \tag{4}$$

where δ_s is distance between microchannel bottom (base) and thermocouples *T3* and *T4*.

The difference between temperatures of the heated surface and liquid ΔT (superheat) is shown by the following equation:

$$\Delta T = T_w - \frac{T_{T1}+T_{T2}}{2}, \tag{5}$$

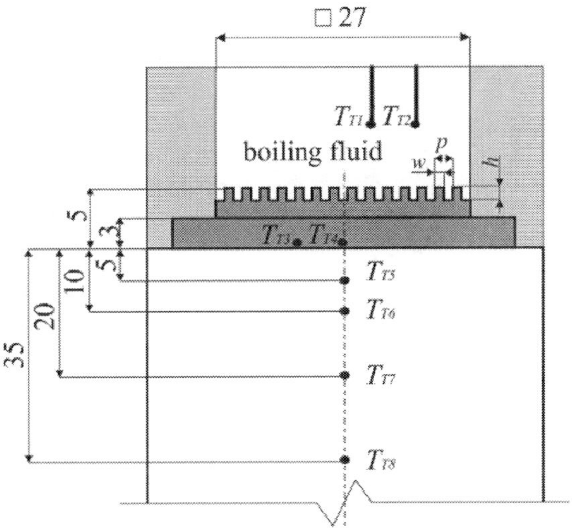

Fig. 2. Arrangemet of thermocouples.

3 Results

The investigated surfaces with open microchannels provided higher heat transfer coefficients compared to smooth surfaces. For the tested samples with microchannels, Table 1, a significant intensification of heat flux was observed, with the heat transfer coefficient depending on the depth and width of the microchannels. The best results were obtained for sample MC-0.2-0.5-0.4, where the maximum heat transfer coefficient was 56 kW/m^{2}K at superheat ΔT ~ 18.5 K, and heat flux about 1035 kW/m^2, Fig. 3a.

Analyzed surfaces with microchannels allow to obtain heat transfer coefficients within the range of 3.4 – 56 kW/m^2, which in relation to the flat, plain surface gives a 2-fold increase in HTC. The intensification of heat transfer for samples with microchannels is demonstrated by the ratio of the heat transfer coefficient for the extended surface to the coefficient for the smooth surface, Fig. 3b.

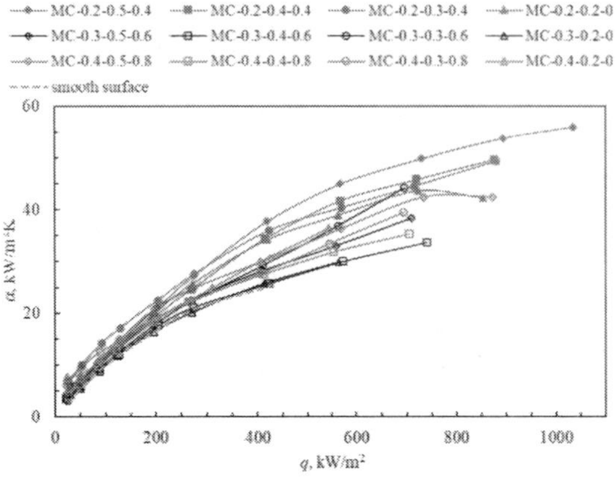

a)

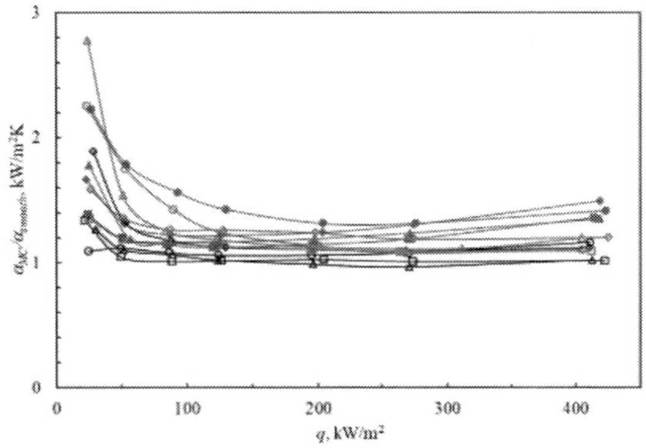

b)
Fig. 3. Boiling curves for ethanol, a) heat transfer coefficient vs. heat flux, b) data in the non-dimensional form for MC surfaces.

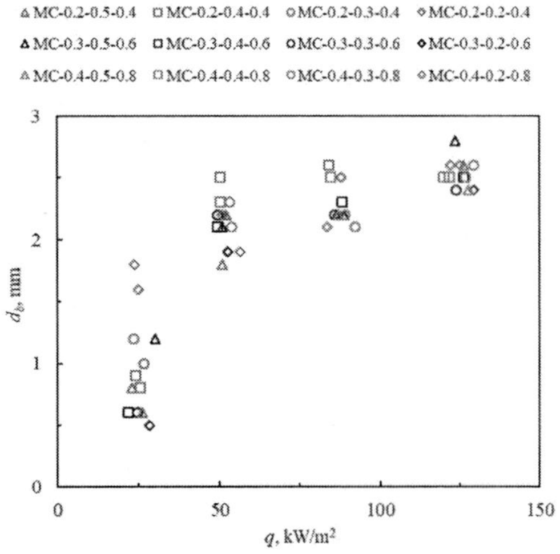

Fig. 4. Averaged departing bubble diameters for ethanol.

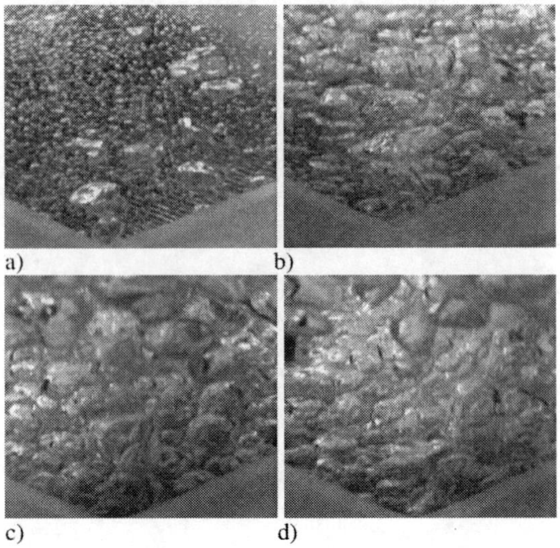

Fig. 5. Visualization observations of pool boiling of ethanol on microchannel surface MC-0.3-0.4-0.6. a) q = 21.9 kW/m², b) q = 49.3 kW/m², c) q = 88.2 kW/m², d) q = 126.2 kW/m².

Figure 4 presents the heat-flux-dependent averaged diameters of departing bubbles at boiling ethanol. Figure 5 shows bubble images on MFP surfaces for four different superheats. Vapor bubbles in microchannels generate in spaces between neighboring micro-fins, from where they move towards the fin tips, then grow and depart.

At small heat fluxes, relatively large distances between nucleation sites can be observed. At the same time, the detachable bubbles have a diameter slightly exceeding the width of the microchannel. As heat flux increases, the distances between the growing blisters decrease, and their diameters increase to 2.5 mm. It can be assumed that the initial intense increase in the diameter of the detachable blisters when increasing the heat flux from 20 to 50 kW/m² is associated with a simultaneous increase in the density of nucleation sites. On the other hand, setting the diameter of detaching bubbles at heat fluxes greater than 50 kW/m² will cause an intensive increase in the frequency of detaching bubbles.

Conclusion

The analysis of the graphs shows the following conclusions:

- The greatest HTC in the range of 3 – 56 kW/m² can be obtained using microchannels of width 0.2 mm and depth 0.5 mm.
- The small width is associated with a small pitch of microchannels, which in turn makes it possible to obtain the largest number of microchannels on the width of the sample.
- Above 800 kW/m², the analyzed surfaces with microchannels allow to obtain heat transfer coefficients in the range of 40 – 56 kW/m²K.

- The critical heat flux at boiling ethanol for surfaces with best performance exceeds 1000 kW/m^2.
- An intense increase in the diameter of the departing bubbles was observed with an increase in heat flux in the range of 20 – 50 kW/m^2.

Nomenclature

a	– specimen width, m,
CHF	– critical heat flux,
d	– diameter, m,
h	– depth, m,
HTC	– heat transfer coefficient,
l	– distance between thermocouples, m,
MC	– microchannel,
p	– pitch, m,
q	– heat flux, kW/m^2,
T	– temperature, K,
w	– width, m,

Greek symbols

α	– heat transfer coefficient, W/(m^{2}K),
δ	– thickness, mm,
λ	– thermal conductivity, W/(mK),
ΔT	– difference of temperature, K,

Subscripts

b	– bubble,
c	– cylinder,
Cu	– copper,
s	– sample,
Sn	– tin,
w	– wall.

References

1. D. Deng, J. Feng, Q. Huang, Y. Tang, Y. Lian, Int. J. Heat and Mass Transfer, **99** (2016)
2. D. Deng, , W. Wan, J. Feng, Q. Huang, Y. Qin, Y. Xie, App. Therm. Ing. **107** (2016)
3. A. Kalani, S. G. Kandlikar, proc. ASME 10th Int. Conf. Nanochannels, Microchannels and Minichannels, Rio Grande, Puerto Rico (2004)
4. R. Kaniowski, R. Pastuszko, Ł. Nowakowski, EPJ Web of Conferences **14** (2017)
5. R. Kaniowski, R. Pastuszko, EPJ Web of Conferences **180** (2018)
6. R. Kaniowski, R. Pastuszko, EPJ Web of Conferences **18** (2018)
7. R. Pastuszko, Int. J. Therm. Sci. **125** (2018)
8. M. Piasecka, K. Strąk, Heat Transf. Eng. **40** (13-14) (2019)
9. K. Strąk, M. Piasecka, B. Maciejewska, Int. J. Heat Mass Transf. **117** (2018)
10. B. Maciejewska, K. Strąk, M. Piasecka, Int. J. Numerical Method for Heat & Fluid Flow **28** (1) (2018)

EPJ Web of Conferences 269, 01025 (2022) https://doi.org/10.1051/epjconf/202226901025
EFM 2019

Comparison of aerodynamic characteristics for various slot diffusers

Michał Karch[1], *Marek* Borowski[1], *Wojciech* Rzeszut[1], *Marek* Jaszczur[2,*]

[1]AGH University of Science and Technology, Faculty of Mining and Geoengineering, 30-059 Krakow, Poland
[2]AGH University of Science and Technology, Faculty of Energy and Fuels, 30-059 Krakow, Poland

Abstract. The airflow in the room or any space has a significant impact on the proper operation of the entire ventilation system. An appropriate field of air velocity flowing through the entire room allows for climate comfort. That is why it is so important proper selection as well as proper arrangement for the air-supply and air-exhaust elements in the ventilation system. This research study presents an isothermal air stream flowing from two different slot diffusers. The supply airstream was analysed at various volumetric flow rates and the velocity of the supply air stream was measured in accordance with the PN-EN 12238: 2002 standard. In order to evaluate the optimal measuring points and to create an appropriate grid, before starting the experimental measurement smoke tests were carried out which determine the scope and shape of the supply stream. All experimental measurements were performed up-to the location where the supply air stream reaches the limiting speed of 0.2 m/s. As a result of the analysis, the obtained characteristics of two types of slot diffusers were discussed and compared.

1 Introduction

The main task of mechanical ventilation and air conditioning systems is to provide the proper amount of good quality air to the ventilated space. This is done in order to guarantee thermal comfort to users of these rooms [1-3]. To accomplish this task, the installation designer should prepare a thermal energy balance and determine the number of harmful substances and pollutants emitted in the space. However, the correctly determined amounts of air and the selection of high-quality air treatment devices do not guarantee yet the correct operation of the installation. The comfort of users in ventilated building is also determined by the correct arrangement in the air distribution. This means that the appropriate velocity field of air flowing through the entire space has to be ensure [4-6].

Depends on the geographical region ventilation plays also a significant role from the total thermal energy point of view. Heat loss due to ventilation is responsible for about 25-55% total heat loss. Taking into account this fact it has been found [7] that it is hardly possible to obtain primary class energy-efficient building without adequately designed mechanical ventilation system [8].

People spend about 80% of their time indoors. For this reason, air quality and thermal comfort are of primary issue and important aspects of the indoor environment that require particular attention. The forecasts and knowledge regarding indoor climate conditions are crucial for indoor climate optimization, thermal comfort, and energy efficiency [1]. It is known that relative humidity and temperature should be maintained within a range of 40-60% and 20-24°C respectively. Special attention should also be taken to keep the air clean and control its biological as well as chemical composition. Indoor air pressure should also be taken into consideration and airflow without recirculation is preferred.

In the last decade, a large number of studies have been conducted experimental and numerical for this topic. Nielsen [9] analysed case with wall-mounted air supply equipment. In this work, author shows air velocity distributions in the vicinity of a wall. In [10] authors used a simplified method to describe the thermal and flow conditions for 8 various diffuser types: slot diffuser, nozzle diffuser, grille diffuser, round ceiling diffuser, vortex diffuser, square ceiling diffuser, valve diffuser, displacement diffuser. Al-Hamed [11] performed CFD simulation to calculate room ventilation to predict air velocity and temperature for the cases with multiple inlets and outlets located in the ventilated space. Author use with the succeeds the k-ε turbulence model. The flow evaluated in that work was verified against experimental data. In [12] authors proposed a method for evaluating boundary conditions at the inlet of a vortex diffuser. They compared experimental visualization with numerical simulations and showed the effectiveness of the proposed method in the case of vortex diffuser. Einberga and Hagstrom [13] shown the results of

* Corresponding author: jaszczur@agh.edu.pl

 https://doi.org/10.1051/epjconf/202226901025

modelling for the case of multi-cone air diffusers. This type of diffuser is often used in industrial application. Authors compared CFD results with experimental measurements. The obtained results showed that the numerical analysis performed using the standard k-ε model can predict the non-isothermal airflow around the diffuser with good accuracy. Most of the research studies presented in the literature (experimental and numerical) have been carried out in small size spaces. It can be also seen that the studies were carried out randomly on pre-selected diffuser models and there is no comprehensive method for evaluating diffusers, for example, from the point of view of energy consumption. As a result, many problems still persist in relation to making projections on existing numerical and mathematical models, especially for low speed flows characterised by the low Reynolds numbers.

The appropriate evaluation of the air distribution requires a good knowledge of the phenomena related to the air supply flow as well as exhaust air from the ventilated space. The selection of diffuser geometry and arrangement in the space should be based on theoretical foundations and the results of tests and analyses carried out on objects similar to where the system will be implemented. In order to ensure the comfort of people staying in a ventilated space, air velocity and temperature have to be set-up so that the air in the room does not draft or stagnates. It is important that the blown air cover the entire occupied space. Therefore, the velocity and airflow should be analysed in detail already at the design stage, taking into account the impact of all possible factors affecting the distribution of air in the room [14,15].

This work presents an isothermal air stream flowing from two different slot diffusers. The supply airstream is analysed at various volumetric flow rates and the velocity of the supply air stream is measured in accordance with the PN-EN 12238: 2002 standard. In order to evaluate the optimal measuring points and to create an appropriate grid, smoke tests is also carried out.

2 Experimental setup

The measurements were carried out in the thermal technology laboratory located at the AGH University of Science and Technology. The measuring stand was constructed based on the PN-EN 12238: 2002 standard. This standard specifies the test stand and requirements for the room dimensions.

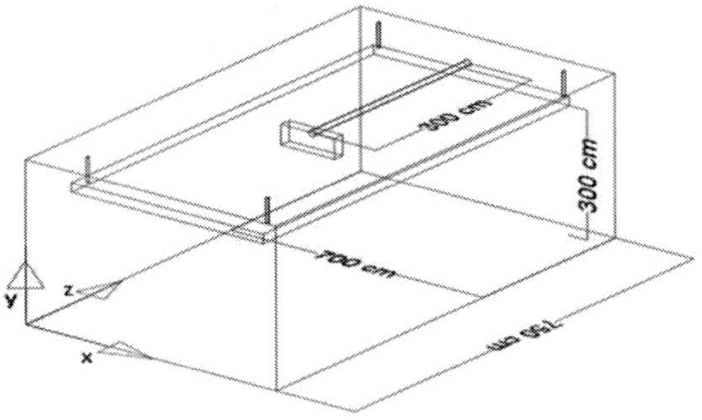

Fig. 1. Dimensions of the laboratory test section.

The room where the measurements were carried out is 7.0 m wide, 7.5 m long and 4.2 m high. A movable suspended ceiling was installed in the room, with diffusers mounted at a height of 3.0 m from the floor. The air was supplied to the expansion box with by duct 160 mm diameter and 3.0 m long (see Fig. 1).

The air was supplied by a radial fan equipped with a inverter enabling change of rotational speed. A measuring orifice was used to determine the value of supply airflow.

In order to determine the range of limiting speed L0.2 (with a value of 0.2 m/s) of the blown stream, the measuring grid presented in Figure 2 was determined. The intersection of the two lines indicates the point where the measurement was taken. The distance between the measuring points was 30 cm horizontally and 15 cm vertically. The first horizontal measuring line was 5 cm below the diffuser. At low values of the supply airflow, the measuring grid was limited most important points.

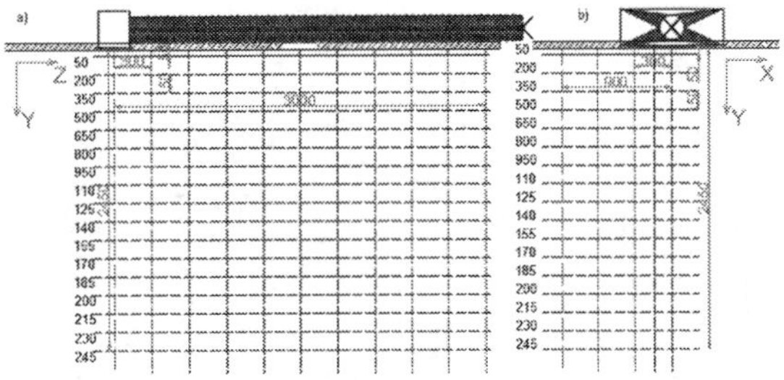

Fig. 2. Measuring grid used (a) side view, (b) front view of the diffuser.

The air velocity was measured using the AirDistSys 5000 measuring system consisting of four SensoAnemo 5100LSF series transducers with measuring probes. The measuring device has a measuring range from 0.05 m/s to 5.0 m/s and accuracy in the range of ± 0.02 m/s ± 1.5% of indication. The measuring system also includes a SensoBar 5301 transducer for measuring and automatic compensation of the influence barometric pressure on the indications of thermoanemometric sensors.

One of the two tested air diffusers was the NSAL series diffuser (Fig. 3). This diffuser is intended for installations with constant and variable airflow. It has 4 rows with 8 individually adjustable rollers. This allows for fine adjustment of the direction of the supply airflow.

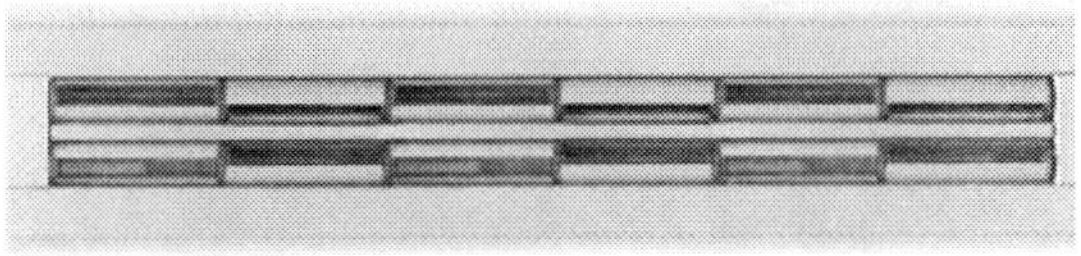

Fig. 3. NSAL diffuser photo.

The second diffuser tested was the SLN series diffuser (Fig. 4). This diffuser has four linear slots with adjustable blades allowing for the two-stage setting. Both tested diffusers had the same external dimension of 83 cm x 14 cm and were installed in identical expansion boxes.

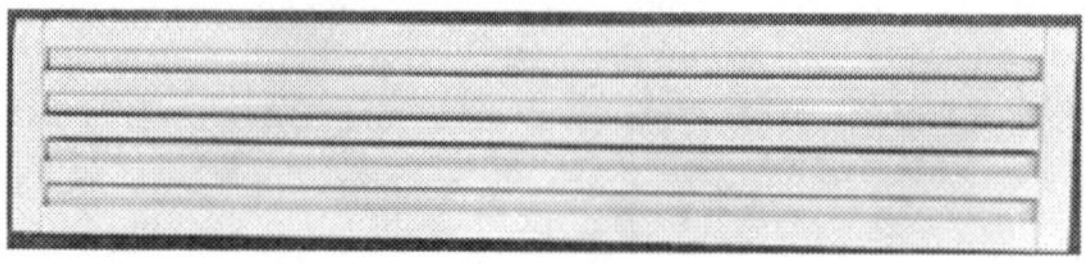

Fig. 4. SLN diffuser photo.

For both diffusers, measurements were made in two variants of air deflector settings: downward supply perpendicular to the floor and lateral supply parallel to the ceiling. The diffusers were tested at two different values of the supplied airflow 120 and 480 m^3/h.

3 Results

Figure 5 shows the NSAL diffuser during the smoke test with the outlet set vertically down for two values of the supply air stream. Figure 6 shows the SLN diffuser during the same test and at the same airflow rate. For both diffusers, the blown air stream initially flat as it moves away from the diffuser expands, falling downwards.

Fig. 5. Smoke test of the NSAL diffuser, with outflow directed downwards, for two air streams: (a) - 120 m^3/h, (b) - 480 m^3/h.

Fig. 6. Smoke test of the SLN diffuser, with outflow directed downwards, for two air streams: (a) - 120 m^3/h, (b) - 480 m^3/h.

Similarly, Figures 7 and 8 show the NSAL and SLN diffusers during the smoke test with the blown air stream parallel to the ceiling. Pictures were taken as before for two different airflow values 120 and 480 m^3/h. Smoke tests were intended to determine the shape and range of the blown streams before the final experimental anemometric measurements.

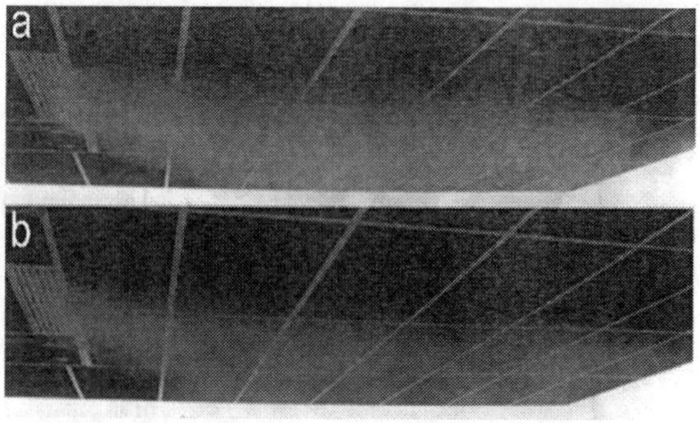

Fig. 7. Smoke test of the NSAL diffuser with outflow along the ceiling, for two air streams: (a) - 120 m^3/h, (b) - 480 m^3/h.

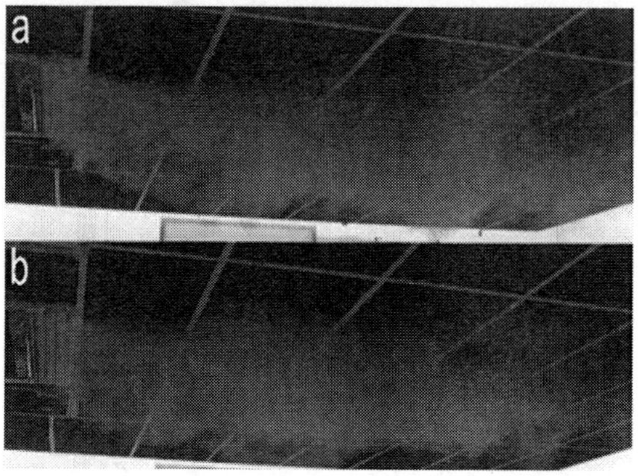

Fig. 8. Smoke test of the SLN diffuser with outflow along with the ceiling, for two air streams: (a) - 120 m^3/h, (b) - 480 m^3/h.

Figure 9 shows the velocity distributions in the NSAL diffuser axis with the outflow vertically down for different airflow values. As the distance from the supply air outlet decreases, the velocity decreases reaching 0.64 m/s (for flow 480 m^3/h) at the lowest measuring point (2.45m from the diffuser). For the lowest flow rate, the velocity of 0.2 m/s was achieved at a distance of 1.7 m from the diffuser.

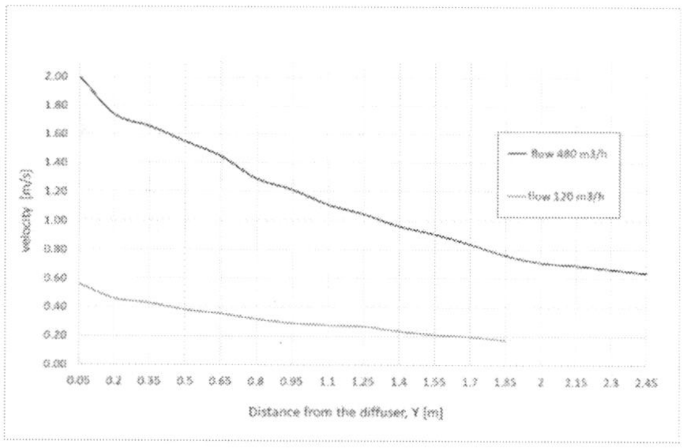

Fig. 9. Distribution of velocity along the NSAL diffuser axis for downflow.

For comparison, Figure 10 was prepared. It shows analogous velocity distributions but for the SLN diffuser. At the lowest measuring point, 0.55 m above the floor a velocity of 0.28 m/s was recorded (for a stream of 480 m^3/h). While for the 120 m^3/h streams, the airflow velocity at a distance of 1.70 m from the diffuser fell below the limiting velocity and amounted to 0.16 m/s.

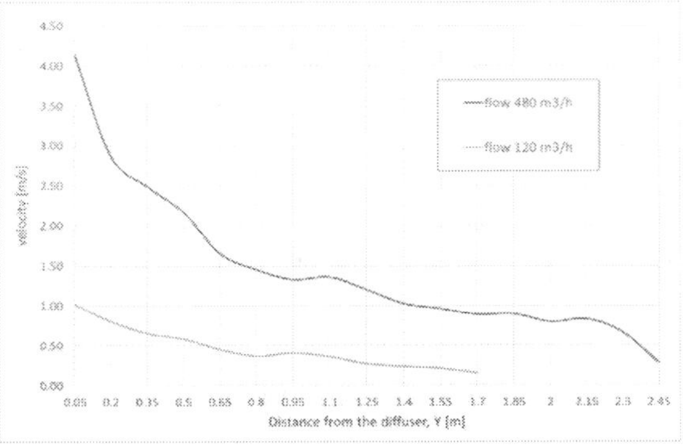

Fig. 10. Distribution of velocity along the SLN diffuser axis for downflow.

Figure 11 shows the limit velocity range 0.2 m/s (L0.2) for 480 m^3/h and 120 m^3/h streams obtained at the NSAL diffuser. The range of the stream relative to the X-axis is characterised by a decrease in width along with the distance from the diffuser. Figure 12 shows the same range statement but obtained for the SLN diffuser. The L0.2 range with a stream of 480 m^3/h in the X-axis was more evenly shaped. However, the ranges for the SLN diffuser reached much larger widths in the X-axis than for the NSAL diffuser.

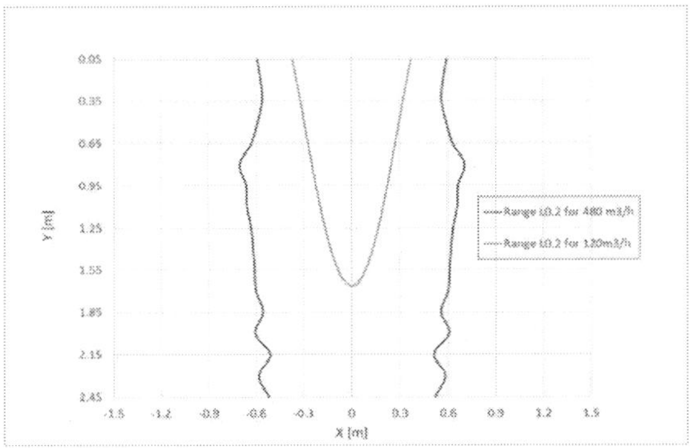

Fig. 11. Range L0.2 of NSAL diffuser flow, with the downward flow, in the X-axis.

A comparison of the measured ranges L0.2 of the tested air diffusers is shown in Figure 13. The graph shows the ranges of the air stream of the tested diffusers with lateral outflow (along with the ceiling) for the 120 m^3/h flow rate. On can infer from this figure that the NSAL diffuser has reached a larger range. The lowest point of the limiting velocity was 2.44 m from the floor at a distance of 3.0 m from the diffuser.

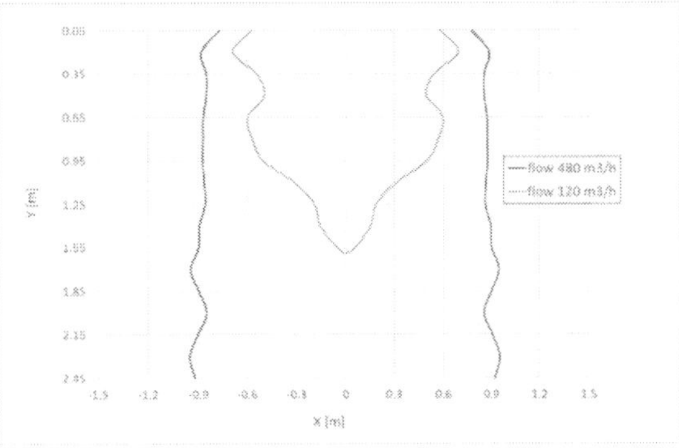

Fig. 12. Range L0.2 of SLN diffuser flow, with downward flow, in the X-axis.

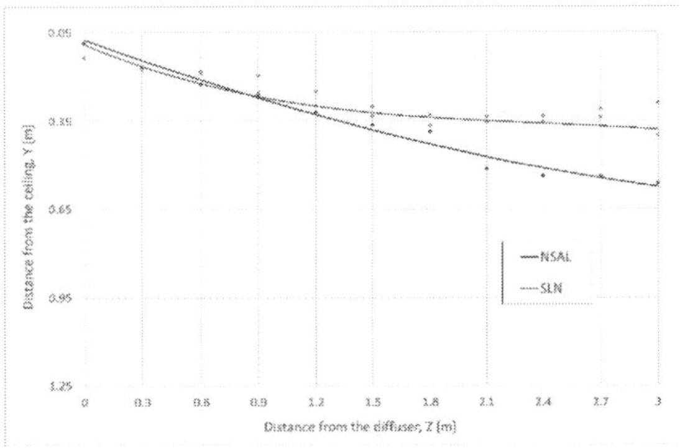

Fig. 13. Comparison of ranges of tested diffusers at 120 m^3/h flowrate, outflow lateral (along with the ceiling).

A similar comparison is shown in Figure 14. It presents the range of air streams supplied by the tested diffusers at a flow value of 480 m^3/h. This time the L0.2 range waveforms are different. NSAL diffuser has reached a much larger range. The lowest range point was 2.12 m from the floor.

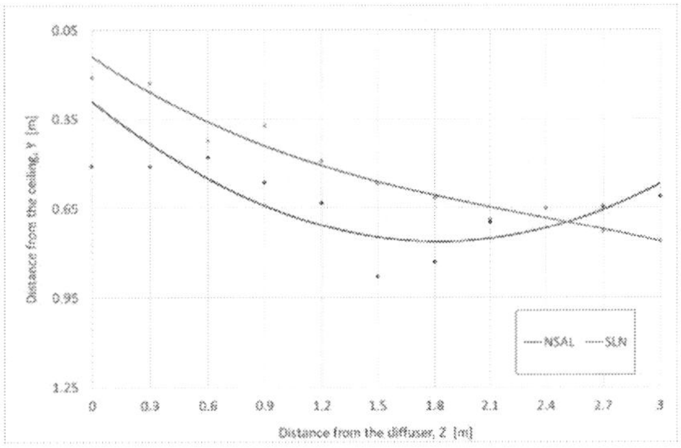

Fig. 14. Comparison of ranges of tested diffusers at 480 m^3/h flowrate, outflow lateral (along with the ceiling).

4 Conclusions

A comparative analysis of the tested slot diffusers for different construction on terms of differences in the air stream range has been done. Referring the results obtained to the conditions that should be ensured in the zone of staying of people, it can be seen that when supplying air to the ceiling direction, both diffusers meet the requirements for the limit of air velocity range 0.2 m/s. However, when the air stream is directed downwards, the limiting speed value is exceeded in the occupied zone ranged from 1.3 to 2.0 m above the floor. Therefore, it can be seen that the knowledge on the stream blown shape through the diffuser is important in the process of designing a ventilation systems.

References

1. H. Awbi, Ventilation of buidings. Taylor and Francis (2003)
2. A. Alfahaid, Effects of ventilation on human thermal comfort in rooms. Ph.D.Thesis. Norfolk: Old Dominion University (2000)
3. M. Jaszczur, M. Borowski, D. Satoła, S. Kleszcz, M. Karch,Analysis of the temperature, humidity, and total efficiency of the air handling unit with a periodic counterflow heat exchanger. Thermal Science, **23,** 1175 (2019)
4. M.A. Aziz et al, Experimental and numerical study of influence of air ceiling diffusers on room air, Energy Buildings **55** (2012)
5. E. Tavakoli, R. Hosseini, Large eddy simulation of turbulent flow and mass transfer in far-field of swirl diffusers, Energy Buildings **59** (2013)
6. M. Jaszczur, M. Branny, M. Karch, M. Borowski, Experimental analysis of the velocity field of the air flowing through the swirl diffusers, Journal of Physics: Conference Series **745**, 032049 (2016)
7. M. Jaszczur, S. Kleszcz, M. Borowski, Analysis of the anti-icing system used in air handling units with a counterflow heat exchanger. E3S Web of Conferences **108**, 01022 (2019)
8. M. Borowski, M. Jaszczur, D. Satoła, S. Kleszcz, M. Karch. An analysis of the innovative exhaust air energy recovery heat exchanger. MATEC Web of Conferences **240**, 02003 (2018)
9. P.V. Nielsen. Velocity distribution in a room ventilated by displacement ventilation and wall-mounted air terminal devices, Energy and Buildings **31**, 179 (2002)
10. J. Srebric, Q. Chen, Simplified numerical models for complex air supply diffusers, HVAC&R Research, (2002)
11. A.H. Al-Hamed. Determining Velocity, Temperature and Occupancy Comfort within a 3-D. ROOM, M.Sc. Thesis, King Fahd University (1990)
12. L. Zhou, F. Haghighat, Simplified Method for Modeling Swirl Diffusers, Department of Building, Civil and Environmental Engineering Concordia University (2007)
13. G. Einberga, K. Hagstrom, CFD modeling of an industrial air diffuser predicting velocity and temperature in the near zone, Building and Environment **40**, 601 (2005)
14. M. Borowski et al, Air flow characteristics of a room with air vortex diffuser, MATEC Web of Conferences, **240**, 02002 (2018)
15. M. Jaszczur, M. Borowski, M. Karch, M. Branny. The study of the velocity field of the air flowing the swirl diffusers using PIV method. EPJ Web of Conferences, **143**, 02047 (2017)

PIV methodology for small scale measurement

Miloš Kašpárek[1,*], *Ludmila* Nováková[2], and *Josef* Adamec[1]

[1]CTU in Prague, Department of fluid dynamics and thermodynamics, 16607 Prague 6, Czech Republic
[2]J.E. Purkyně University in Usti na Labem, Department of Energy and Electrical Engineering, 40096 Usti nad Labem, Czech Republic

Abstract. Measurement of velocity fields by PIV method is often limited by the scale of the measured model. This is doubly true for the stereo PIV method. Problems of measurement in small (but still macroscopic) scales are associated with optical distortions in the optical path of the camera. Additionally, there is a calibration problem for stereo PIV. Therefore, this paper is devoted to the description of PIV methodology for measurement in small geometrical scales. The idealized transparent model of hemodialysis vascular access was used in this paper. This model had realistic vascular access dimensions (6 mm in diameter). A special working blood-mimicking fluid (BMF) was mixed with a refractive index matching that of the model. The refraction and distortion of light passing through the interface between the model and the fluid was so small, that it was possible to calibrate the stereo PIV outside the measured model. The accuracy of this calibration is discussed and verified by measuring the velocity field on a straight section of the vascular access model for laminar flow.

1 Introduction

Method Particle Image Velocimetry (PIV) and its modifications are often used in fluid mechanics to measure velocity fields. The successful use of this measuring method is conditioned by a large number of assumptions. The sources of PIV measurement uncertainty are generally connected either with the method of data analysis or with the experimental setup. The former uncertainty sources are discussed in [12],[2],[9] the latter are, for example, the correct selection of marking particles to be added to the fluid, sufficient and uniform illumination of the measurement surface, or quality optical access to the entire measurement area. The necessity to ensure good optical access to the measurement area limits the model dimensions in which the PIV method can be used successfully. With the small dimensions of the model, it is difficult to ensure good optical access because the small dimensions of the model are associated with areas of large optical deformation. These optical deformations result in significant errors in the measurement. In the case of a modification of the PIV method, where all three components of the velocity (stereo PIV) are monitored in the measured area, it is necessary to perform calibration before the measurement itself. Calibration in the small size model is very difficult, even impossible.

Optical distortions are due to the curvature of the measured model in which the measurement is performed. Of course, the greater curvature of the model in the measured area also leads to greater optical deformations that occur at the interface of the two materials. Reflections are another optical defect associated with light reflection at the interface of two materials. These reflections introduce the signal deterioration into the measurement.

When using the stereo PIV method for the measurement, the measured area is monitored by the two cameras. These cameras track the measurement area at an angle, which allows measuring, in addition to the two velocity components, the third velocity component that is perpendicular to the measured area. Therefore, in order to evaluate all velocity components from the captured images of the measured area, the calibration is required [11], [10]. The transformation matrix can be calculated from the recording of the target image in several predefined positions in the measured model. There are several ways of calibration. The most commonly used calibration is the use of a calibration target, which is located in the measured area. For example, the calibration target consists of an array of circular dots with the precisely defined dimensions. By recording the image of the calibration target at several predefined positions in the measured model, it is then possible to calculate the transformation matrix. For small-scale stereo PIV measurements, calibration has several problems: an optical distortion problem distorting the calibration target, realizing the movement of the target in the model, and creating the calibration target itself.

Despite these difficulties, it is possible to measure velocity fields by the stereo PIV in the small size models. First of all, it is necessary to remove optical deformations from the model / fluid interface. The removal of the optical defects is dealt with in several works [15], [4], [7], [1], namely by the change of the refractive index of used fluid. Specifically, this means that the refractive index value of the fluid is changed so that the refractive index of the light of the fluid is identical (very similar) to that of

* Corresponding author: milos.kasparek@fs.cvut.cz

the refractive index of light of the model. The smaller the difference between the refractive indices of the light, the smaller the optical deformation. The refractive index of light of the fluid can be changed by adding an additive to the fluid that is soluble in the fluid. The choice of additive that changes the refractive index of light is also associated with the requirements for the properties of the resulting working fluid. The elements used for changing the refractive index of light are, for example, sodium iodide [1], [8], zinc iodide [4], potassium thiocyanate [3] or glycerine [15]. The refractive index of the light of the material from which the measured model is made must also be taken into account. The material of the model should be chosen so that it has a refractive index similar to that of untreated fluid [5]. Generally speaking, it should have the lowest possible refractive index of light. Such material is, for example, Sylgard 184. If accordance of the refractive indexes of light are very good, it is possible to make the calibration outside the measured model and then use the obtained transformation matrices for measurement in the model [14].

This thesis deals with the accurate description of the calibration for the stereo PIV measurement in the small size model and its evaluation. The vascular access model is used for the measurement, which is used as well in hemodialysis, maintaining the realistic dimensions of the vein and the dialysis needles. The diameter representing the diameter of the vein is 6 mm. Measurement of the velocity field for the comparison with the theory and thus confirmation of the success of the calibration is performed on the straight part of the model, where the flow is not influenced by the inserted needles.

The vascular access model was made using the casting technology, where the template representing the internal geometry was made with the water-soluble material. The template was made by using 3D printing and its surface was modified to be hydraulically smooth.

2 Removal of the optical defects

Changing the refractive index of the light of a fluid is difficult for several reasons. Substances added to the fluid can fundamentally alter its rheological properties, often being chemically hazardous to the environment or they are expensive. Therefore, it is advisable to select the material of the model so that the refractive index of the light is as close as possible to the base component of the working fluid solution. The distilled water is the most often used liquid as the base component of the working fluid. The refractive index of the light of the distilled water is 1.33 (1). In this work, the measured model was made by silicone elastomer with the trade name Sylgard 184. This material was chosen due to its low refractive index of the light compared to the other commonly used materials for the model making. The refractive index of the light of this silicone elastomer is 1,414 (1) [15]. The working fluid was mixed from the distilled water, glycerine and sodium iodide. The weight ratio of these components was 47.38% distilled water, 36.94% glycerine and 15.68% sodium iodide [15]. The resulting refractive index of the light of the working fluid was 1.413 (1). In Figure 1 there is a comparison of the optical deformation in the model with the water-glycerine solution (refractive index of light 1.387 (1)) and the optical deformation in the model with the working fluid. The model with the given solutions was placed on a black and white checkerboard. This makes it possible to assess the optical deformation. Although the refractive index of the water-glycerine solution is close to the refractive index of the material of the model, optical distortion is still visible in the image. The second solution (working fluid) shows that there is no visible optical deformation of the measured area. The refractive index of the working fluid and the model material was practically identical.

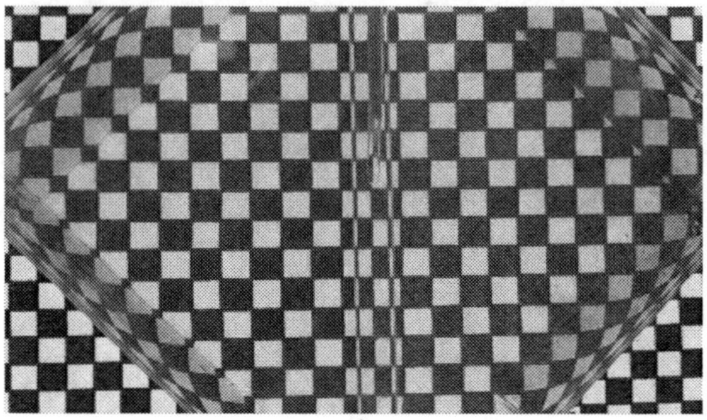

a.)

b.)

Fig. 1. Optical deformation of the measured area. a.) the model with the solution of distilled water and glycerine. b.) the model with the solution of distilled water, glycerine and sodium iodide.

3 Calibration

When the refractive index of the measured model and the working fluid is identical (very similar), the optical distortions are removed. This makes it possible to calibrate the stereo PIV as if it was an external flow. This means that the model is replaced by the calibration target.

In this work the calibration target consisting of the circular dots was used (Fig. 2). The central dot had a diameter of 0.6 mm, the four dots determining the direction of the „x" axis and the „y" axis had a diameter of 0.3 mm, and the remaining dots had a diameter of 0.45 mm. The spacing between the dots is 0.75 mm. The total number of dots on the calibration target was 225 dots. The calibration target was placed in a equipment that delimited the location of the target in the „y" direction. The equipment was placed on the two feets. The first feet allowed movement of the calibration target in the „x" direction. The second feet allowed movement of the calibration target in the „z" direction. The calibration target placed in the equipment was positioned by the displacements to the starting position. The starting position was in the measured plane and the calibration target was positioned so that the central dot was in the tube axis of the measured model. The pair of calibration target images were created in this position. The target was then shifted twice 0.5 mm in the direction of „z" axis in the positive and negative directions from the starting position. The pair of calibration target images were obtained at each new position. The transformation matrices for both cameras were evaluated from the obtained images of the calibration target at all positions by using the Dantec Dynamic studio software. After calculating the transformation matrices, a self-calibration was performed to improve the transformation matrices. The self-calibration has partially suppressed errors that could have been caused by improper placement of the calibration target in the equipment (eg small rotation of the calibration target).

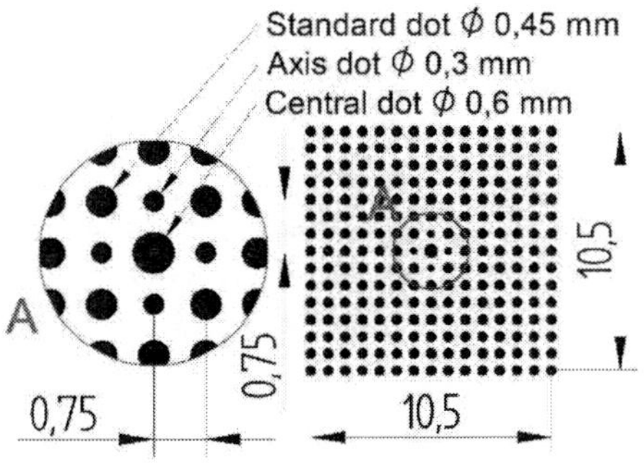

Fig. 2. Calibration target.

4 Setup of the experiment

Scheme of the experiment for the verification measurements is shown in Figure 3. The measured area of the model is illuminated by a laser sheet. The cameras

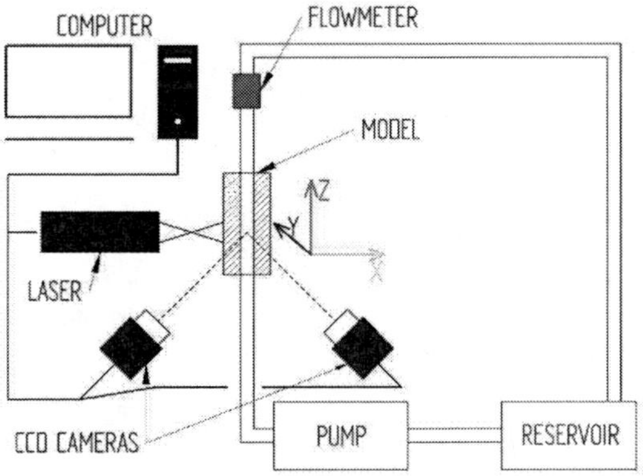

Fig. 3. Scheme of experiment.

capture the measured area at an angle. The angle between the cameras was 90 °. The Scheimpflug condition was met. The volumetric flow rate was adjusted using a flowmeter and a valve downstream of the flowmeter. The measured model was made of Sylgard 184, which has a refractive index of 1,414 (1). The working fluid was a mixture of the distilled water, glycerine and sodium iodide [15]. The refractive index of the light of the working fluid was 1.413 (1) and the dynamic viscosity was 4.31 mPa.s [15]. The measurement confirming the feasibility of the calibration outside the measured model was performed on a straight section of the vascular access model. The measured mode was characterized by Reynolds number Re = 1000.

5 Results

Figure 4 shows images of the calibration target from both cameras and their transformed images. The transformation matrices obtained by the calibration were used to transform the image. These images can be used as a first indicator of whether the calibration itself has been performed correctly. It is clear from the recorded images of the target that the spatial deformation occurs due to the angle at which the cameras capture the area with the calibration target. Transforming these images leads to the elimination of the spatial deformation and the calibration target takes the shape and the dimensions, like when viewed perpendicularly to the calibration target. The transformed image of the calibration target indicates that the computed transformation matrices work. The success of the image transformation (or transformation matrices obtained from calibration) was determined by the difference of distances („Δx") between individual dots in the x-axis direction and the difference of distances („Δy") between individual dots in the x-axis direction „ y"

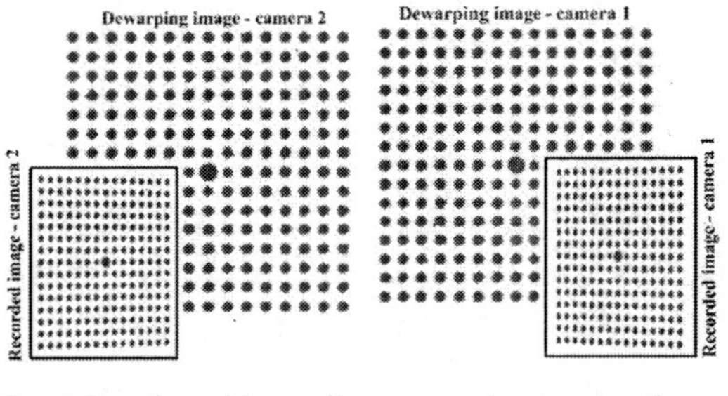

Fig. 4. Transformed images from camera 1 and camera 2.

Figure 5 shows the deviation of the distance „Δx" from the transformed image of the calibration target from the actual distance „Δx". The deviation is on the vertical axis of the graph and it is given in percentage. The horizontal axis shows the columns between which the deviation is calculated. Individual deviations then correspond to the rows of the calibration target. The graph shows that the shape of deviation that oscillates around zero is the same for all rows of the calibration target. The dots in the outer rows of the calibration target show that the deviation is increasing (green dashed line), reaching up to 6%. This is due both, the print quality of the calibration target and the fact that the endpoints of the calibration target have not been sufficiently focused. For other dots of the calibration target, the deviation is within ± 2%. The red dashed lines then show the progress and the deviation values at the point where the calibration target dots are located directly or in close proximity to the measured area. Here you can see that the deviation is below 2%. Which is a very good conformity. Figure 6 shows the deviation of the distance „Δy" from the transformed image of the calibration target from the actual

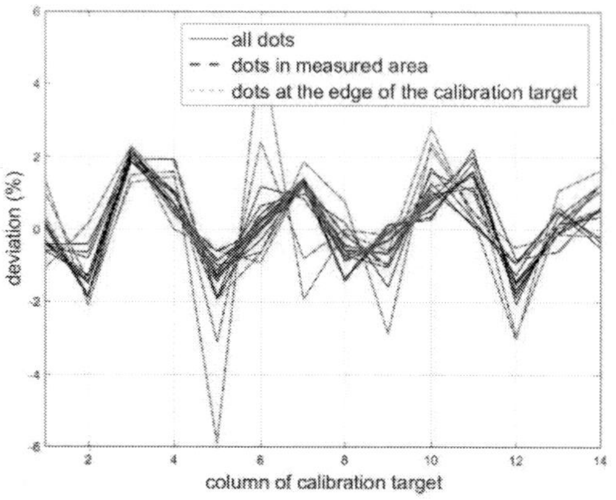

Fig. 5. Deviation of the distance between the dots of the transformed calibration target („Δx") in the „x" axis direction.

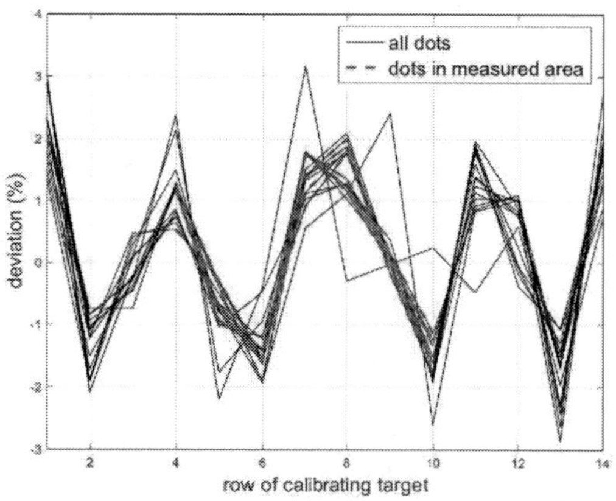

Fig. 6. Deviation of the distance between the dots of the transformed calibration target („Δy") in the „y" axis direction.

distance „y". The deviation is on the vertical axis of the graph and it is given in percentage. The horizontal axis shows the columns between which the deviation is calculated. The individual deviations then correspond to the columns where the deviation was evaluated. The graph shows that the deviation was within ± 3%. The red dashed lines then show the progress and the deviation values at the point where the calibration target dots are located directly or in the close proximity to the measured area. For these dots, the deviation is within ± 2%. The accuracy of the image transformation achieved by the calibration matrices obtained from calibration is very good due to the dimensions of the calibration target and dimensions of the measured area.

The velocity field was evaluated from the data obtained from the experiment by the transformation matrices. Two velocity profiles „w" were obtained from the velocity field. „w" is a velocity in z-direction. The first velocity profile was obtained by horizontal sectioning of the measured area. The second velocity profile was obtained by the vertical sectioning of the measured area. Figure 7a shows a comparison of the vertical section velocity profile with the theoretical laminar flow velocity profile. Figure 7b shows a comparison of the horizontal velocity profile with the theoretical laminar flow velocity profile. For both velocity profile comparisons, it can be seen that it copies the theoretical speed profile with a very

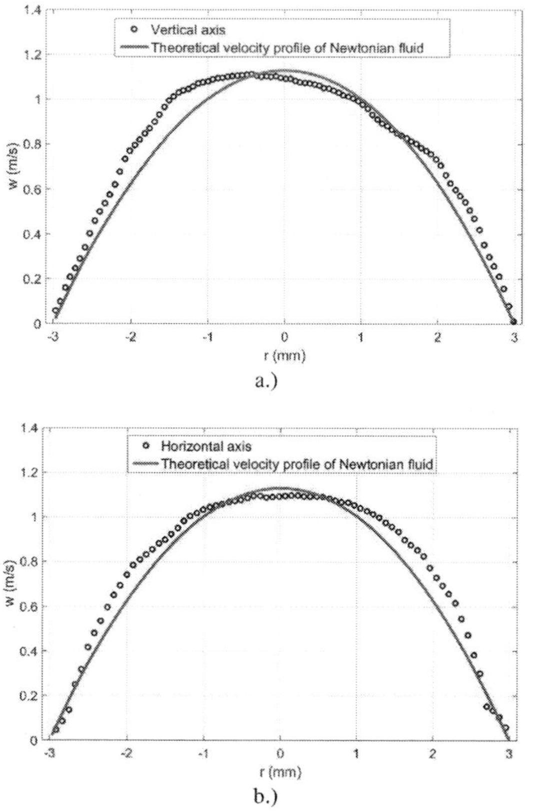

Fig. 7. Comparison of the experimentally measured velocity profile „w" with the laminar theoretical velocity profile. a.) Comparison in the vertical section through the measured area. b.) Comparison in the horizontal section through the measured area.

good accuracy. However, there is a slight asymmetry in both velocity profiles. The laminar velocity profile is extremely sensitive to small disturbances due to the heat convection or the minor misalignments of the tube segments or the minor irregularities of the tube surface. These disturbances lead to speed profile asymmetry [13].

Further the normal velocities of the profiles „u" and „v" were compared. „u" is velocity in „y" axis direction and „v" velocity in „x" axis direction. These profiles were created in the vertical section through the measured area. The comparison is made by using the percentage ratio between the „u" and „v" velocity profile and the profile of „w" velocity. As a result, there is obtained the percentage deviation that results from the measurement. This ratio is created for both „u" and „v" velocities. From the graph in Figure 8, it can be seen that for both velocities the measurement deviation was constant and the deviation increased in the walls vicinity. For the normal velocity, „u", the average deviation was 1.8%. For the normal velocity, „v", the average deviation was then 4.3%. These deviations in the measurement are associated with the calibration target, where even a slight deviation of the calibration target from the measured plane results in a deviation in the measurement [13], [6].

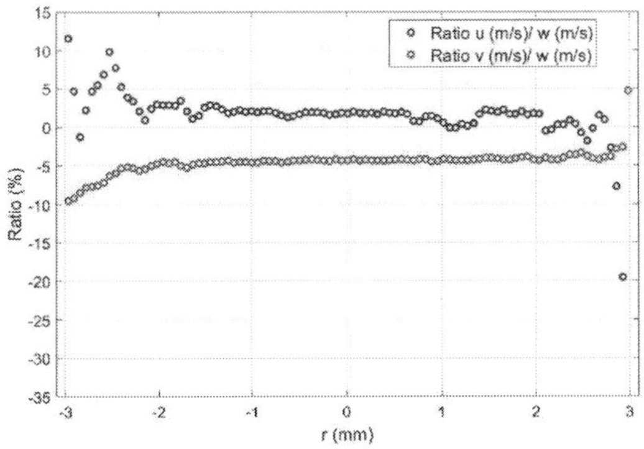

Fig. 8. Comparison of the normal velocities „U" and „V".

6 Discusion

Using the stereo PIV calibration outside the measured model, it is possible to obtain the transformation matrices that can be applied to the measured data and obtain velocity fields in the models with very small dimensions in order of millimetres. The model used in this work had a diameter of 6 mm in the measured area. This fact was confirmed by the comparison of velocity profiles obtained using these transformation matrices with the theoretical laminar velocity profile. The main prerequisite for the success of this method is in a good comparison of the refractive index of the light of the model material and the working fluid. Calibration accuracy is largely linked to the quality of the calibration target. The quality of the calibration target is more difficult to ensure due to the size of the calibration target dots. The dots must be of the size of the sufficient numbers of dots in the measured area. This method of the calibration also enables the measurement of velocity fields using the stereo PIV method in the models with very complex geometry, where there is not possible to place the calibration target in the measured area.

References

1. Bai, K., Katz, J.: On the refractive index of sodium iodide solutions for index matching in PIV, Exp Fluids, 55:1704, 2014
2. Fei, R. & Merzkirch, W.: Exp Fluids (2004) 37: 559. https://doi.org/10.1007/s00348-004-0843-x
3. Gijsen, F.J., van de Vosse, F.N., Janssen, J.D.: The influence of the non-Newtonian properties of blood on the flow in large arteries: steady flow in a carotid bifurcation model. J Biomech 32:601, 1999
4. Hassa, Y.A, Dominquez-Ontiveros, E.E.: Flow visualization in a pebble bed reactor experiment using PIV and refractive index matching techniques, Nuclear Engineering and Design 238, 3080-3085, 2008
5. Jermy M.: Making it clear: Flexible, transparent laboratory flow models for soft and hard problems, Conference on Experimental Heat Transfer, Fluid Dynamics and Thermodynamics, 2013 Lisabon, Portugal, 2013, Conference Contribution
6. Lawson, N.J., Wu, J.: Three-dimensional particle image velocimetry: error analysis of stereoscopic techniques, Meas. Sci. Technol., 8, 894-900, 1997
7. McEvoy, J., Persoons, T.: Aqueous Ammonium Thiosulfate as a working fluid for refractive index matched PIV applications, 9th World Conference on Experimental Heat Transfer, Fluid Mechanics and Thermodynamics ; 2017
8. Narrow, T.L., Yoda, M., Abdel-Khalik, S.I.: A simple model for the refractive index of sodium iodide aqueous solutions, Experiments in Fluids 28, 282-283, 2000
9. Novotny J., Novakova L. and Machovska I.: Advanced metric for particle image velocimetry accuracy estimation 19th Int. Symp. on Applications of Laser and Imaging Techniques to Fluid Mechanics, Lisbon, Portugal, 2018
10. Prasad, A.K.: Stereoscopic particle image velocimetry, Experiments in fluids, Experiments in Fluids, volume 29, 103-116, 2000
11. Raffel M., Willert C. E., Wereley S. T., Kompenhans J. : Particle Image Velocimetry A Practical Guide Second Edition, ISBN 978-3-540-72307-3 Second Edition Springer Berlin Heidelberg New York, 2007
12. Raffel M., Willert C.E., Scarano F., Kähler C.J., Wereley S.T., Kompenhans J.: PIV Uncertainty and Measurement Accuracy. In: Particle Image Velocimetry. Springer, Cham, 2018
13. van Doorne, C.W.H., Westerweel, J.: Measurement of laminar, transitional and turbulent pipe flow using Stereoscopic-PIV, Experimental Fluids, 42,259-279,2007
14. Yagi, T., Sato A., et al.: Experimental insights into flow impingement in cerebral aneurysm by stereoscopic particle image velocimetry: transition from a laminar régime, J. of the Royal Society Interface, 10, 2013
15. Yousif, M.Y., Holdsworth, D.W., Poepping T.L. : A blood-mimicking fluid for particle image velocimetry with silicone vascular models, Exp Fluids 50:769–774, 2011

An experimental investigation of the thermal efficiency and pressure drop for counterflow heat exchangers intended for recuperator

Sławosz Kleszcz[1,2], *Piotr* Mazur[2], *Marcin* Zych[1] and *Marek* Jaszczur[1,*]

[1]AGH University of Science and Technology, Kraków, Poland
[2]Frapol Sp. z o.o., Kraków, Poland

Abstract. Ventilation systems are responsible for a large part of total energy consumption in operated buildings. The possibility of significant energy savings has been noticed when using mechanical ventilation. In order to decrease energy consumption, devices called recuperators are used and their main element responsible for heat recovery is the heat exchanger. In this work, thermal efficiency and pressure drop at individual heat exchangers designed for recuperators were examined. The results of the analysis and the influence of parameters on the device operation are presented. Finally, the thermal efficiency of the entire unit was evaluated. The results were compared and the phenomena occurring in the device itself affecting the differences in calculated efficiency were described.

1 Introduction

Ventilation is a process designed to bring fresh air into rooms while extracting used air. Due to deterioration of the microclimate, including air quality inside the residential and commercial buildings, the application of mechanical ventilation becomes necessary. However, typical ventilation and air conditioning systems can consume as much as 50% of the total electricity demand [1]. By increasing the number of air changes in a ventilated space, the quality of air inside it is significantly improved but at the expense of increased energy requirement for cooling and heating. The two main strategies to reconcile these contradictions are the use of heat recovery and the implementation of controlled ventilation [2-4]. By increasing global energy consumption, legal regulations and environmental protection, manufacturers of ventilation devices are forced to use energy-savings solutions [5].

One of the solutions to fulfil current energy efficiency requirements is to use heat exchangers in air distribution devises. Thanks to this approach, along with energy-saving, we provide sufficient fresh air, better microclimate control and high energy efficiency [6]. Studies show that using heat recovery equipment, total annual energy consumption can be reduced in cold climates by up to 20% [7]. A typical mechanical ventilation system consists of air ducts, fans and a heat exchanger in which thermal energy is transferred between streams. Fans ensure air movement through the exchanger ensuring proper flow. They also give the possibility to regulate the mass flow and thus the number of air changes in the building.

Current heat recovery systems are able to recover around 60-95% of the energy contained in waste air, which is a significant value. This shows huge potential for reducing the heat needs of buildings incurring relatively low investment costs [8]. At present, plate heat exchangers are very popular due to their compactness and high heat transfer coefficient [9]. They form the basis for very efficient heat recovery because their high efficiency in combination with counter-flow allows obtaining temperatures close to the final values which is the temperature of the air in ventilated space [10-13].

Heat recovery efficiency is defined in standard rated conditions for a given device. Usually specified in relevant standards. E. Juodis [14] noticed that the actual efficiency of heat recovery depends not only on the heat recovery unit, but also on the heat loss and gain ratio in the building.

In [15] as much as 13 ventilation units were tested and the 3 best cases the actual heat efficiency was in the range of 60-70% and concerned units with a nominal efficiency of 80%. For the 3 worst results, the actual efficiency was less than 10%, so the system consumed more energy than recovered. The differences were mainly caused by leaks both inside the building and in the ventilation unit itself. M.K. Drost [16] conducted an analysis of the thermal efficiency of heat recovery systems in 38 houses. During studies, it was estimated that the devices worked on average about seven hours a day with an average thermal

* Corresponding author: jaszczur@agh.edu.pl

efficiency of 52%. Such a low value suggests that design ventilation indicators were not provided.

Research carried out by Shah and Skiepko [17] clearly showed that even with relatively small leaks (around 5%), a significant reduction in heat exchange between cold and warm air could be observed. In [18] experimental and numerical studies were carried out to determinate the relationship between various heat exchanger parameters. It has been shown that the flow conditions and heat transfer coefficient, which depend on the geometry of the exchangers, affect the efficiency of devices. The thermal efficiency of plate heat exchanger decreases as the flow rate increases. In similar research studies [19], authors noted that when the flowrate increase from 50 m^3/h to 175 m^3/h, the efficiency decreased from 94% to 78%. When the air velocity through the exchanger increases, the air resistance increases as well. Pressure loss as a function of flow directly depends on the geometry of the exchanger.

Based on the literature review it can be concluded that the heat exchange efficiency and airflow resistance are affected not only by the unit design (geometrical parameters) but also depends on unit operating parameters. In addition, counterflow heat exchangers have the highest heat recovery potential. For this reason, presented research in this work focuses on this type of heat exchangers. However, counterflow heat exchangers varies considerably depending on the design, material used and method of manufacture. In this work, the main attention is paid to thermal efficiency, which directly affect the thermal comfort of people in the ventilated space. In addition to that, the focus is on the pressure drop, which causes a significant increase in energy need to be supplied to the system. Industrial companies are interested in solutions improving the efficiency of energy systems without significant investments. They are also aimed at minimizing energy losses and maximizing energy efficiency [20,21]. The paper presents the results for two different counterflow heat exchangers produced by the same manufacturer. One of the analysed exchangers was installed and tested in a real system device unit. The presented measurements were carried out according to 13141-7 and PNEN 308 standard.

2 Experimental methodology

The experimental measurement was performed to determinate the temperature and sensible efficiency of the heat exchanger. The measurement procedure has been done under conditions presented in PNEN 308 standard (Table 1). The test was carried out on the test stand presented in Figure 1 according to Figure 2.

Table 1. Measurement conditions.

Extract air		Outdoor air	
T_{11} [°C]	T_{w11} [°C]	T_{21} [°C]	T_{w21} [°C]
25	< 14	5	-

In Tab.1. the T_w is the wet-bulb temperature of the air while indexes are as follows:

22 - the supply air	11 - the extract air (room)
21 - the fresh air	12 - the exhaust air

Parameters such as temperature (T_{21}, T_{22}, T_{11}, T_{12}) were measured using PT1000 sensors and pressure (p_{21}, p_{22}, p_{11}, p_{12}) by using PT2500 – R8 sensor. The air flowrates (V_{21}, V_{22}, V_{11}, V_{12}) were measured by using the Venturi tube. During experiments, the flow was set to 600 kg/h. The measurement was recorded by Jumo Logoscreen and the Mitsubishi Electric FX5U acquisition system with the MAPS HMI 750 panel. The measurements were recorded with time step 1 second and averaged every 60 seconds.

Fig. 1. Photo of test section.

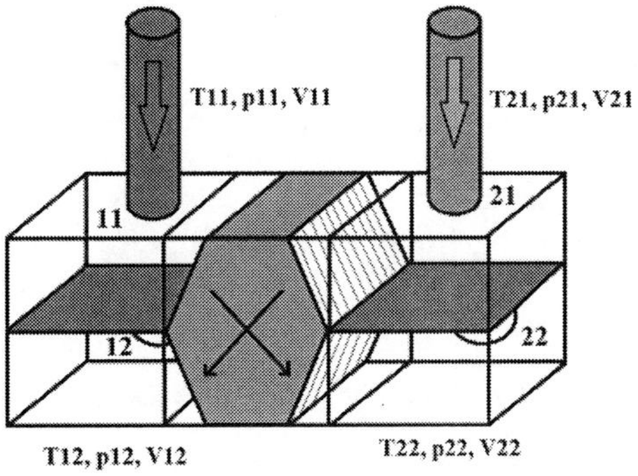

Fig. 2. Diagram of the test section.

The outdoor air was drawn to the section presented in Figures 1-2 by channel 21, and the warm air which simulated the conditions inside the ventilated space was drawn by channel 11. Respectively the channel 22 contains the fresh air while channel 12, waste air. The temperature efficiency of a heat exchanger was calculated as the ratio of the heat flux recovered by the heat exchanger transferred to or extracted from the supply air in reference to the total heating power required to heat the outside air to indoor air temperatures.

Efficiency ratio on the supply air side can be calculated using the following equation:

$$\eta_{t,su} = \frac{T_{22}-T_{21}}{T_{11}-T_{21}} \cdot \frac{q_{m22}}{q_{m11}} \quad (1)$$

where T is the temperature of air and q_m is the mass flow of air (for well-balanced units, the flow ratio of q_{m22} and q_{m11} should be up not higher than 3%).

3 Results and discussion

The experimental measurements were carried out in order to evaluate key thermal and fluid flow parameters and select the appropriate heat exchanger device for the newly designed system unit. Two different exchangers from the same manufacturer were tested. The primary to evaluate parameters were temperature efficiency and pressure drop.

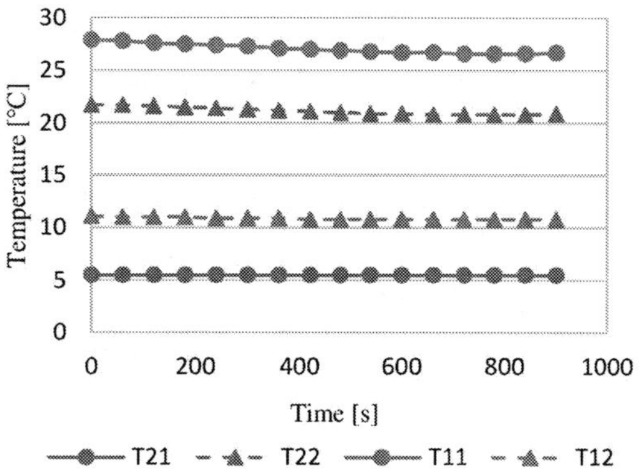

Fig. 3. The air temperature through the heat exchanger HEX1.

Based on the measured pressure, Figure 4 was plotted showing the pressure drop in the exchanger. It can be noticed that the airflow is asymmetrical, higher resistance is generated on the fresh-supply channel than on the exhaust-extract section.

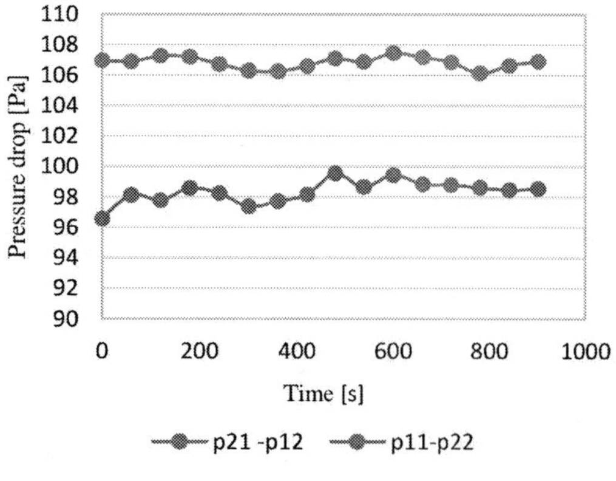

Fig. 4. Pressure loss through the heat exchanger HEX1.

Figure 3 shows all 4 streams (22 - the supply air; 11 - the extract air; 21 - the fresh air; 12 - the exhaust air) temperature during the experimental measurement. The first analysed heat exchanger HEX1 obtain an average thermal efficiency about 72.5% with a mass flow equal to 600 kg/h.

For the second heat exchanger HEX2, the same analysis was carried out. The evaluated average temperature efficiency in this case about 83.3% was achieved for the same mass flow rate. Figure 5 presents the air temperature for the heat exchanger HEX2.

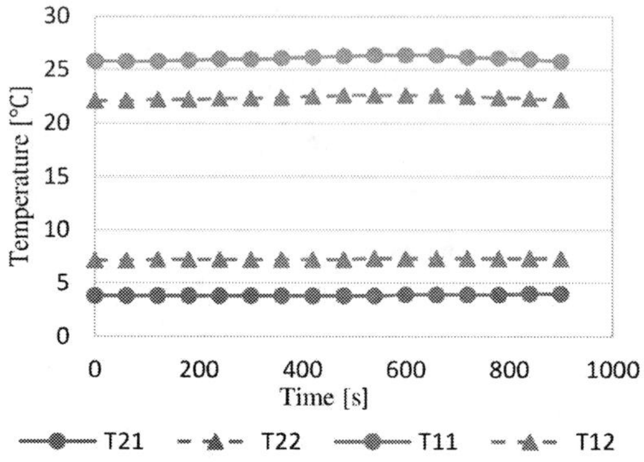

Fig. 5. The air temperature through the heat exchanger HEX2.

Also, in the case of this exchanger, the flow turned out to be asymmetrical, and the pressure difference between the inlet of supply air and exhaust air was about 30 Pa. The measurement results are presented in Figure 6.

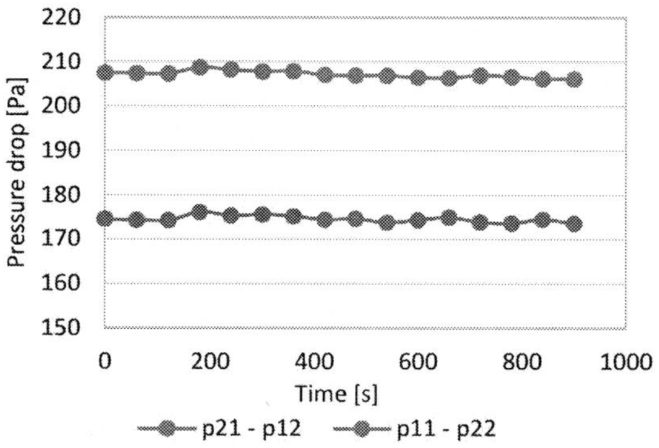

Fig. 6. Pressure loss through the heat exchanger HEX2.

Due to the two times lower pressure loss in the exchanger, the HEX1 heat exchanger was selected for assembly in the prototype system unit. This choice was caused by a significant energy saving during the operation of fans installed in the device. After assembly of the selected heat exchanger, the entire device was tested according to 13141-7 standard.

During the experimental measurement, the device was not completely sealed. Figure 7 shows the mass flow of the air over time. The ratio of q_{m22} and q_{m11} is equal to 0.995. However, the $q_{m21} \neq q_{m22}$ and the $q_{m11} \neq q_{m12}$ what suggests the internal leakage in the manufactured unit equal to $q_{ml} = 40.8(m^3/h)$.

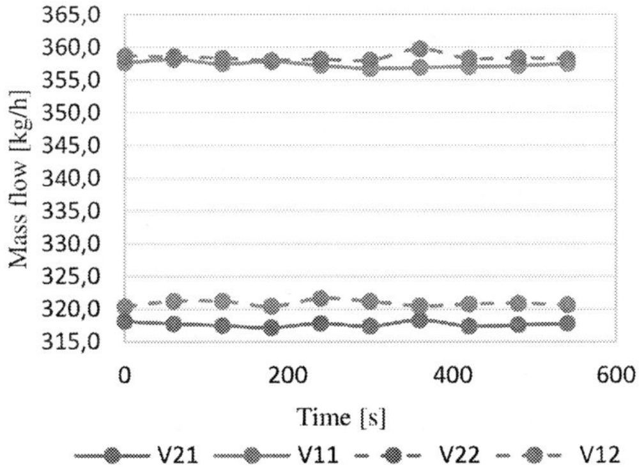

Fig. 7. The airflow through the recuperator.

There are two possible reasons for internal leakages. The first one is from room air to the fresh air and second from waste air to ambient air. According to Tab. 2, the leakage from waste to ambient air is more likely due to occurring pressure on each side of the heat exchanger unit.

Table 2. Measure pressure on each side of the heat exchanger.

P_{11} [Pa]	P_{12} [Pa]	P_{21} [Pa]	P_{22} [Pa]
-28.3	26	-31.2	19.1

Despite the internal leakage, it was decided to continue the research. Figure 8 presents the temperature course overtime during the test.

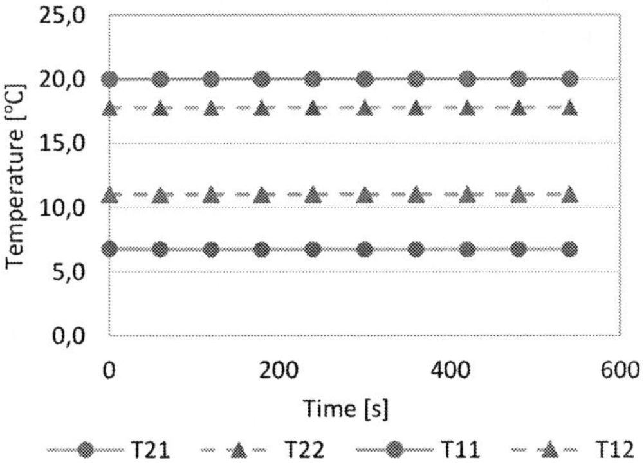

Fig. 8. The air temperature through the recuperator.

The temperature efficiency of heat exchanger and the unit was calculated on the basis of equation (1). The results are shown in Figure 9. It is easy to see that despite the use of lower efficiency, the device has achieved higher efficiency than anticipated. The average efficiency of the unit oscillates around 83.5% and this value is about 10 p.p. (percentage points) higher than the values obtained from the study of exchanger HEX1. The occurrence of the internal leakage presented above had the greatest impact on this phenomenon.

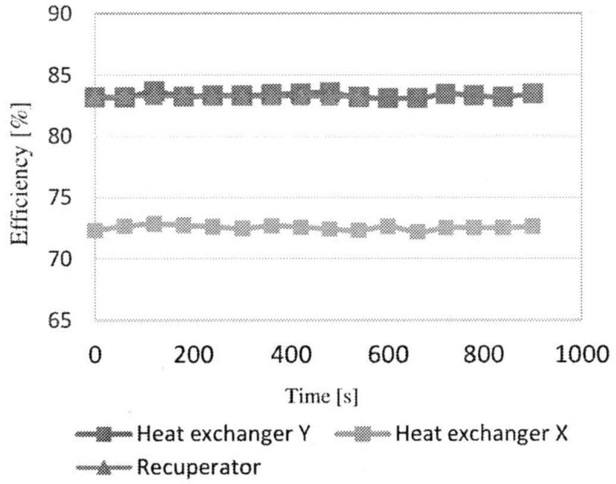

Fig.9. The efficiency over time.

4 Conclusion

One of the assumptions for designing the device was the mass flow equal to 600 kg/h. To fulfil this assumption two exchangers were preselected and tested. The most important from the efficiency point of view parameters were evaluated i.e., temperature efficiency and pressure drop across the exchanger. Despite the fact that one of the heat exchangers had a thermal efficient higher by about 10 p.p. the pressure losses generated by it almost doubled caused the decision to abandon this solution. The complete system unit thermal efficiency obtained from experimental measurement is 83.5%. From the analysis of (internal) heat exchanger, it has been calculated that the thermal efficiency is equal to 72.5% (Fig. 9).

The results can be explained by the following phenomena:

- based on the experimental measurement some internal leakage was observed which artificially raised this parameter,
- it can be assumed that other physical phenomena such as heat convection through the steel layer increase the overall efficiency of the unit,
- due to lack of space, smaller fans were used, therefore the mass flow of the tested device was lower,
- the results obtained may be affected by the fact that PNEN 308 and 13171-7 standard indicate different temperatures for testing heat exchangers and equipment.

To sum up, when specifying the parameters of the designed device, detailed tests confirming its parameters should be performed. It is not possible to rely only on the data presented by the manufacturer. A separate examination of the exchanger itself is recommended. During the operation of the device, various physical phenomena occur that may affect energy and temperature efficiency. The tested prototype must be redesigned due to the fact that there is not a possibility to evaluate the precise efficiency of the unit where the (internal or external) leakage is observed. In the first step for the accurate determination of the unit efficiency the devices should be tested on tightness and then if all streams are equal the unit is ready for final efficiency test.

During system design, one should also pay attention to the asymmetrical airflow through the exchanger. At the same power delivery to the fans, quite different airflow rates are possible. As a result, this situation may significantly disturb the air exchange in the building and increase or decrease (depends on case) the efficiency of the device.

Acknowledgement: This research is supported in the Smart Growth Operational Programme 2014-2020 by the Polish National Centre for Research and Development and by the Polish Ministry of Science 11.11.100.005.

References

1. H. Caliskan, I. Dincer, A. Hepbasli, *Exergetic and sustainability performance comparison of novel and conventional air cooling systems for building applications*, Energy Build. **43**, 1461 (2011)
2. J. Laverge, N.V.D. Bossche, N. Heijmans, A. Janssens, *Energy saving potential and repercussions on indoor air quality of demand controlled residential ventilation strategies*, Build. Environ. **46,** 1503 (2011)
3. W. Cho, D. Song, S. Hwang, S. Yun, *Energy-efficient ventilation with air-cleaning mode and demand control in a multi-residential building*, Energy Build. **90,** 14 (2015)
4. S. Guillen-Lambea, B. Rodriguez-Soria, JM. Marin, *Control strategies for energy recovery ventilators in the south of Europe for residential nZEB-quantitative analysis of the air conditioning demand*, Energy Build. **146**, 271 (2017)
5. Commission Regulation (EU) No 1253/2014 of 7 July 2014 implementing Directive 2009/125/EC of the European Parliament and of the Council with regard to ecodesign requirements for ventilation units Text with EEA relevance. (2014)
6. L. Shao, SB. Riffat, G. Gan, *Heat recovery with low pressure loss for natural ventilation*, Energy and Buildings **28**, 179 (1998)
7. M. Fehrm, W. Reiners, M. Ungemach, *Exhaust air heat recovery in buildings*, International Journal of Refrigeration 25, 439 (2002)
8. A. Mardiana-Idayu, S.B. Riffat, *Review on heat recovery technologies for building applications*, Renew Sustain Energy Rev **16,** 1241 (2012)
9. J.J. Klemes, O. Arsenyeva, P. Kapustenko, L. Tovazhnyanskyy, *Compact heat exchangers for energy transfer intensification: low grade heat and fouling mitigation*, CRC Press, (2015)
10. B.R. Lamb, *Plate heat exchangers – a low cost route to heat recovery*, Journal of Heat Recovery Systems **2**, 247 (1982)
11. M. Jaszczur, M. Borowski, D. Satoła, S. Kleszcz, M. Karch, *Analysis of the temperature, humidity, and total efficiency of the air handling unit with a periodic counterflow heat exchanger*, Thermal Science **23**, 1175 (2019)
12. M. Jaszczur, S. Kleszcz, M. Borowski, *Analysis of the anti-icing system used in air handling units with a counterflow heat exchanger*, E3S Web of Conferences **108,** 01022 (2019)
13. M. Borowski, M. Jaszczur, D. Satoła, S. Kleszcz, M. Karch, *An analysis of the innovative exhaust air energy recovery heat exchanger*, MATEC Web of Conferences **240**, 02003 (2018)
14. E Juodis , *Extracted ventilation air heat recovery efficiency as a function of a building's thermal properties,* Energy Build. **38**, 568 (2006)
15. C.-A Roulet, F.D Heidt, F Foradini, M.-C Pibiri, *Real heat recovery with air handling units,* Energy and Buildings Volume **33**, 495 (2001)
16. M.K. Drost, *Air-to-air heat exchanger performance*, Energy and Buildings Volume **19**, 215 (1993)
17. R.K. Shah, T. Skiepko, *Influence of leakage distribution on the thermal performance of a rotary regenerator*, Applied Thermal Engineering Volume **19**, 685 (1999)
18. Hu. Zhuohuan, He Xie, Ye. Li, Mo Yang, Qin. Guihua, *Full-scale research on heat transfer and pressure drop of high flux plate heat exchanger*, Applied Thermal Engineering **118,** 585–592, (2017)
19. José Fernández-Seara, Rubén Diz, Francisco J. Uhía, Alberto Dopazo, José M. Ferro, *Experimental analysis of an air-to-air heat recovery unit for balanced ventilation systems in residential buildings*, Energy Conversion and Management Volume **52**, 635 (2011)
20. D.M. Van de Bor, C.A. *Infante Ferreira Quick selection of industrial heat pump types including the impact of thermodynamic losses*, Energy **53**, 312 (2013)
21. A.A. Kiss, S.J. Flores Landaeta, C.A. Infante Ferreira, *Towards energy efficient distillation technologies – making the right choice,* Energy **47**, 531 (2012)

Experimental investigation on effect of partial flexibility at low aspect ratio airfoil - Part II: Installation both on suction and pressure surface

Kemal Koca[1,*], *Mustafa Serdar* Genç[1], *Halil Hakan* Açıkel[1]

[1]Wind Engineering and Aerodynamic Research Center, Department of Energy Systems Engineering, Erciyes University, KAYSERİ, TURKEY

Abstract. In this experimental study, flow over NACA 4412 airfoil with flexible membrane mounted both on suction surface and pressure surface was investigated at Reynolds number of $5x10^4$ and different angles of attack. The smoke-wire visualization method was used for flow visualization to demonstrate flow phenomena as laminar separation bubble (LSB), leading edge separation and tip vortices. A constant temperature anemometer (CTA) was used for measuring flow over the partially flexible airfoil. Concerning the flow visualization, smoke-wire experiment was been conducted at z/c=0.1 and z/c=0.4. Besides, hot-wire experiment by means of CTA was employed to measure flow properties including values of velocity, turbulence statistics and Reynold stress over both uncontrolled and controlled airfoil. The results showed that partially flexible airfoil had several benefits compared with the uncontrolled airfoil. The results for this partially flexible airfoil were mainly lower size of LSB, higher stall angle of attack, ensuring more stable flow characteristics and more aerodynamic performance.

1 Introduction

Air vehicles such as Unmanned Air Vehicles (UAVs) and wind turbines operate at lower Reynolds (Re) number [1-5]. At low Re number flows, the prediction, determination and controlling of LSB and transition phenomenon is important, because these low Re number flow phenomenon cause aerodynamic performance to decrease, blade/wing vibration to increase [6-24]. Therefore, nowadays researchers have focused experimental [18-21] and numerical [25-28] studies on the prediction, determination and controlling of LSB and transition phenomenon. Flexibility recently developed flow control technique [23-24], has been begun attention recently. Stall delaying and performance increasing as well as blade/wing stability can be provided with the utilizing of flexibility over air vehicles [23].

This experimental study is on a NACA 4412 wing with partial flexibility on both suction and pressure surfaces at $Re=5x10^5$ and different angles of attack. The purpose of this study is to determine controlling of the flow using partially flexibility.

2 Experimental Setup

It was conducted at the wind tunnel with test section of 500mm by 500mm in Erciyes University. Free-stream turbulence intensity is under 0.7% [13]. The wing is with an 18 cm chord and 18 cm span length. The partial flexible wing was manufactured using a 3D printer. Figure 1 shows the wing with flexibility mounted between x/c=0.2 and x/c=0.7 because of the control of LSB and separation affects. The flexible membrane is latex rubber sheet [23] with the thickness of 0.2 mm. It was glued to the wing with double side tape.

The experimental study was carried out with the smoke wire flow visualization and flow measurements using a hot-wire system. Flow visualization was captured along z/c=0.1 (to see tip vortices) and z/c=0.4 (to see LSB formation) using a camera and illumination system. X wire probe was used in the hot-wire system at the over and wake of the wing. 20480 data were collected with a sample rate of 2 kHz at selected positions. Experimental details have been explained in Part-I of the study.

* Corresponding author: kemalkoca@erciyes.edu.tr

https://doi.org/10.1051/epjconf/202226901028

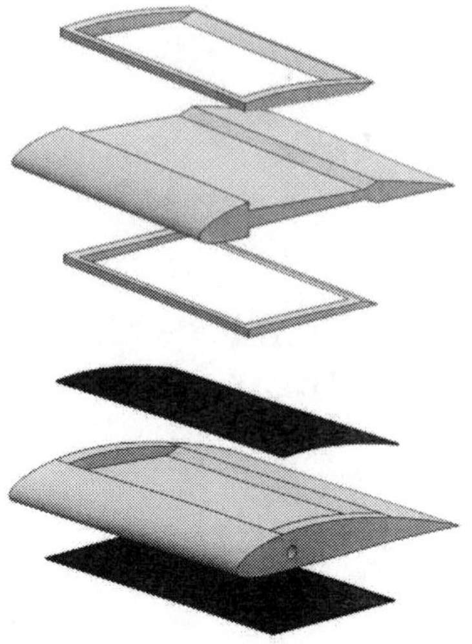

Fig. 1. The wing with flexibility on both suction and pressure surfaces.

3 Results

3.1 Velocity measurement results

Figure 2 and Figure 3 show dimensionless velocity distribution around the partial flexible wing at the about middle and tip of the wing. As seen from these figures, there is a short LSB over the wing, and the trailing edge separation formed at the rear of the wing. It was shown in Part-I that the flow separated at x/c=0.2 and reattached at x/c=0.8 in the uncontrolled case. The LSB was suppressed with the partially flexibility, and it is clearly seen that size of LSB was reduced. Flexible membrane vibration caused the flow to control with momentum transfer.

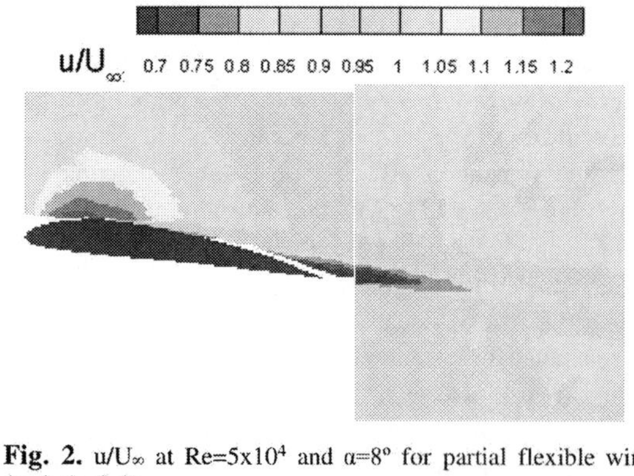

Fig. 2. u/U_∞ at Re=$5x10^4$ and α=8° for partial flexible wing (and z/s=0.4).

The long LSB formed due to adverse pressure gradient in uncontrolled case, controlled via flexible membranes on both suction and pressure surfaces. This causes the increase of lift and aerodynamic performance.

Distributions of turbulent intensity (Tu) and Reynolds stresses (Re stress) for partially flexible wind are given in Figure 4. Different contour colours at both Reynolds stress and turbulent intensity indicate changing in the flow-field. The transition to turbulence phenomenon formed at the middle of the wing during the streamwise location. In addition, the trailing edge separation was shown at the rear via the increasing in distribution of both Reynolds stress and turbulent intensity. Unlike the uncontrolled case, fluctuations because of vortex sheds at near wake are less, meaning less drag force and steadier flow in terms of aerodynamic performance.

Fig. 3. u/U_∞ at Re=$5x10^4$ and α=8° for partial flexible wing (and z/s=0.1).

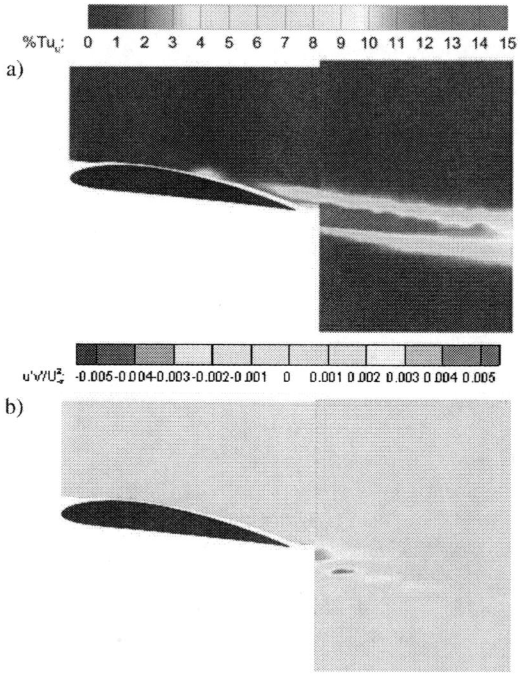

Fig. 4. a) Tu and b) Re stress at Re=$5x10^4$ and α=8° for partial flexible wing (and z/s=0.4).

3.2 Flow visualization results

In the smoke-wire flow visualization experiment, the flow pictures around the wing were taken by the camera.

In Figure 5, the photos for partial flexible wing at α=8° in the middle of the wing (z/s=0.4) and the tip of the wing (z/s=0.1) were given. The long bubble seen in the uncontrolled case was suppressed with the partially flexibility, and the trailing edge separation formed at the rear of the wing. As the angle of attack increased, the flow over the wing started to separate and the tip vortex moved towards the leading edge of the wing as seen in Figure 6 (α=12°). Then, the flow separation enlarged at the higher angles of attack (α=20°) in Figure 7. Besides, the tip vortex grew up and the flow separation at the tip of the wing suppressed. Consequently, using the partial membrane over both two surfaces of the wing caused aerodynamic performance to increase.

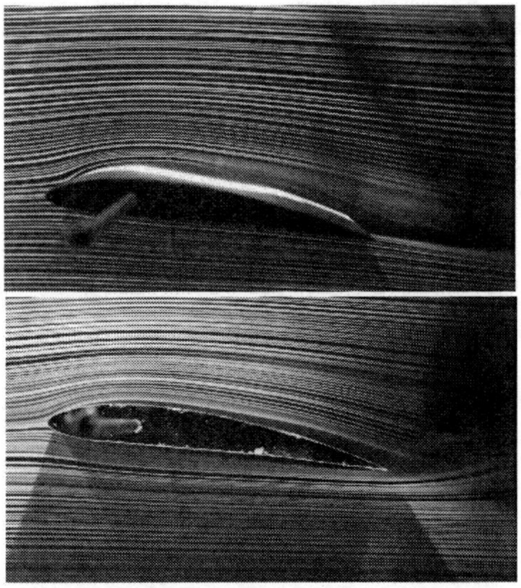

a) b)

Fig. 5. At α=8°, smoke-wire photos for partial flexible wing at the plane of a) z/s=0.4 b) z/s=0.1.

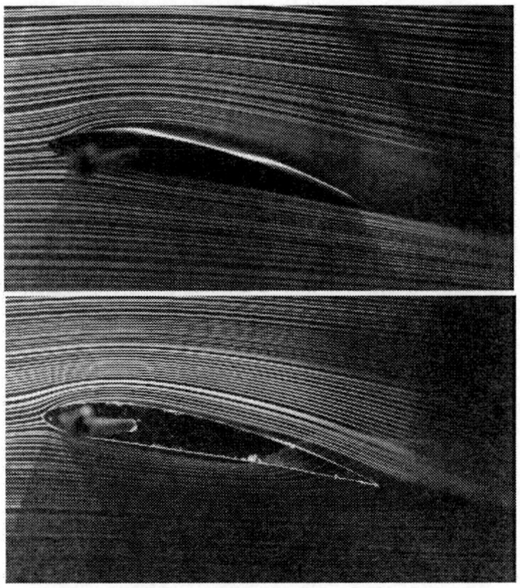

a) b)

Fig. 6. At α=12°, smoke-wire photos for partial flexible wing at the plane of a) z/s=0.4 b) z/s=0.1.

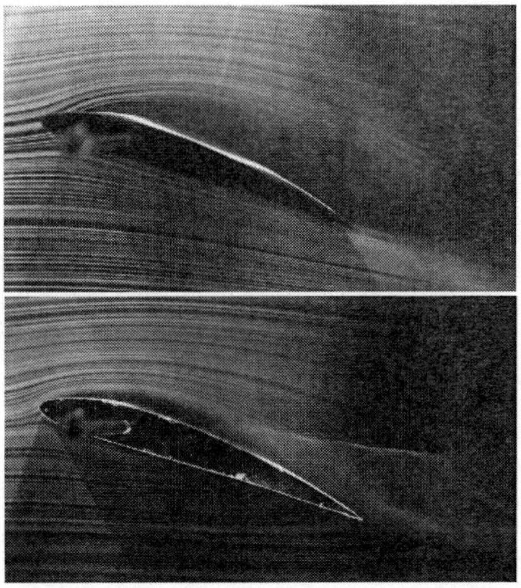

a) b)

Fig. 7. At α=20°, smoke-wire photos for partial flexible wing at the plane of a) z/s=0.4 b) z/s=0.1.

4 Conclusion

In this study, the flow on AR=1 NACA4412 wing with the partial flexibility over both the suction and pressure surfaces were investigated at Re= $5x10^4$ and different angles of attack. The experiments were done using the smoke wire flow visualization technique and a hot-wire system for velocity measurement. Flow visualization was captured along z/c=0.1 for tip vortices effects and z/c=0.4 LSB formation and the leading-edge separation. In the hot-wire system, X wire probe was used at the over and wake of the wing. Although the laminar separation bubble and flow separation formed due to adverse pressure gradient were observed in the uncontrolled case at α=8°, α=12° α=20°, the controlled cases showed that the flow was controlled and suppressed by using the partial flexible membrane material between x/c=0.2 and x/c=0.7 over both upper and lower surfaces of the wing. Unlike the partial flexibility over the suction surfaces of the wing, the flexibility on the pressure surfaces caused extra eddies to form at the lower region of the wing as seen in Figure 4b. These extra eddies provided to occur a vacuum effect at the wake of the wing, which supplied to control the flow.

References

1. Tani I., Prog Aerosp Sci, 5: 70–103, (1964).
2. Gaster, M. Aeronautical Research Council Reports and Memoranda, No: 3595, London, (1967).
3. T.J. Mueller, J.D. DeLaurier. Annual review of fluid mechanics, **35(1)**, 89-111 (2003)
4. Greenhalgh, S., Curtiss, H. C., Smith., AIAA journal, **22(7)**, 865-870 (1984).

5. M.S., Genç, I. Karasu, H.H. Açıkel, M.T., Akpolat. In Low Reynolds number aerodynamics and transition. InTech. (2012)
6. Ke, S., Zhigang, W., Chao, Y., Chin. J. Aeronaut. **21(5)**, 402-410 (2008).
7. Zhang, Wei, Rainer Hain, and Christian J. Kähler. Experiments in fluids, **45, 4** 725-743 (2008).
8. Savaliya, Samir B., S. Praveen Kumar, and Sanjay Mittal. International Journal for Numerical Methods in Fluids **64(6)**, 627-652, (2010).
9. Hain R., Kahler C., Radespiel J., Journal of Fluid Mechanics, **630**, 129–153 (2009).
10. Mizoguchi, M., and Itoh H. AIAA journal **51(7)**, 1631-1639 (2013).
11. Shields, M. and Mohseni K. AIAA journal **50(1)** 85-99 (2012).
12. Y. Lian, W. Shvy, D. Viieru, B. Zhang. Progress in Aerospace Sciences, **39(6-7)**, 425-465 (2003)
13. M.S. Genç, I. Karasu, H.H. Açıkel, Experimental Thermal and Fluid Science, **39**, 252-264, (2012).
14. K. Koca, M.S. Genç, H.H. Açıkel, M. Çağdaş, T.M. Bodur. Energy, **144**, 750-764 (2018)
15. L. E. Jones, R. D. Sandberg, N. D. Sandham. J. Fluid Mech., **602**, 175-207, (2008).
16. M. S. Genç, K. Koca, H. H. Açıkel, G. Özkan, M. S. Kırış, R. Yıldız. In EPJ Web Conf., **114**, 2029, (2016).
17. A. Ducoin, J. C. Loiseau, J. C. Robinet. Eur. J. Mech: B/Fluids, **57**, 231-248, (2016).
18. H.H. Açıkel, M.S. Genç. Proc IMechE, Part G: Journal of Aerospace Engineering, **230(13)**, 2447-2462 (2016).
19. M.S. Genç. H.H. Açıkel, M.T. Akpolat, G. Özkan, İ. Karasu, Journal of Aerospace Engineering-ASCE, 29 (6), 04016045, 2016.
20. H. Wang, B. Zhang, Q. Qiu, X. Xu. Energy, **118**, 1210-1221 (2017)
21. M.S. Genç, K. Koca, H.H. Açıkel. Energy, **176**, 320-334 (2019)
22. M.S. Genç, G. Özkan, M. Özden, M.S. Kırış, R. Yıldız. Proc IMechE, Part C: Journal of Mechanical Engineering Science, **232(22)**, 4019-4037 (2018)
23. H.H. Açıkel, M.S. Genç. Energy, **165**, 176-190 (2018)
24. H. Demir, M.S. Genç, European Journal of Mechanics B/Fluids, **65**, 326-338 (2017).
25. M.S. Genç, U. Kaynak, H. Yapıcı, European Journal of Mechanics B/Fluids, **30(2)** 218-235 (2011).
26. M.S. Genç, Proc IMechE, Part C- Journal of Mechanical Engineering Science, **Vol 224 (10)**, pp. 2155 – 2164, 2010.
27. I. Karasu, M. Özden, M.S. Genç, Journal of Fluids Engineering-Transactions of The ASME, **140(12)** 121102-1-121102-15 (2018).
28. K. Koca, M.S. Genç, H.H. Açıkel. Çukurova Üniversitesi Mühendislik-Mimarlık Fakültesi Dergisi, **31(ÖS2)**, 127-134 (2016)

 https://doi.org/10.1051/epjconf/202226901029

Shape modification of the optical probe for wetness measurement using CFD calculations

Michal Kolovratník[1,*], *Guk chol* Jun[1], *Václav* Novotný[1]

[1]CTU in Prague, Department of Energy Engineering, Technická 4, Prague 6, 166 07, Czech Republic.

Abstract. An optical extinction probe is usually used for measurements of liquid phase parameters of wet steam in steam turbines. Since the probe is relatively large, its interaction with the flow field negatively affects the accuracy of the measurements. A shape modification of the optical probe has been proposed to reduce this negative impact of the probe-flow field interaction. The benefit of the proposed shape modification of the probe has been verified using numerical (CFD) calculations. In these CFD calculations, a non-equilibrium steam condensation which enables to simulate a formation and development (growth) of fine droplets has been considered.

1 Introduction

Research of wet steam expanding in turbines is using for nearly 40 years optical extinction probes [1]. Czech Technical University has a long tradition of development and applications of these extinction probes [2]. During this development there was and still is a large effort focused improvement of measurement HW and SW, i.e.:

- instrumentation itself,
- measurement procedures,
- methodology of analysing the wet steam structure from the measured data.

All of those have a goal to increase the precision of the evaluated parameters of the wet steam structure.
Optical measurement of the liquid phase structure of the wet steam is indirect. Extinction probes obtain transmittance I/I_0 of the light passing through a measuring space of the probe. Using the Mie scattering theory [3] the required information about the liquid phase structure (about droplets causing the light attenuation) can be evaluated. Typically, the liquid phase has a polydisperse structure characterised by a spectre of droplet dimensions $\varphi(D)$ and their volume concentration N_v. Transmittance of the light is further dependent on other parameters such as light wavelength λ and thickness of the wet steam illuminated layer ℓ, i.e. on the length of the measuring beam and last but not least thermodynamic and optical properties of the wet steam. In general, this relationship can be described as:

$$I/I_0 = f(\varphi(D), N_v, \lambda, \ell, \ldots)$$

In recent years has been the attention in the probe development focused to another phenomenon with a potential of improving the quality of the extinction measurements. This is an interaction of the probe with the flow field in the turbine and its impact on the measured parameters of the wet steam. Our previous results on character and intensity of the flow field parameters alteration partly published in [4] were used in this work for modification of the extinction probes' shapes. The shape modifications are altogether focused on reduction of the difference between the transmittance measured with the probe $(I/I_0)_{EXP}$ and transmittance of the flow field unaffected by the probe$(I/I_0)_{TUR}$.

$$(I/I_0)_{EXP} \,/\, (I/I_0)_{TUR} \rightarrow 1$$

2 Simulation of the interaction between the probe and the flow field

The probe is during the measurements inserted between turbine stages. Simple put it is a cylindrical carrier with diameter around 35 mm and a considerable length which depends on the radial position of the measurements (200 – 1200 mm). On the tip of the carrier is placed a probe head with a measuring space, which shape and dimensions are shown in Fig. 1.

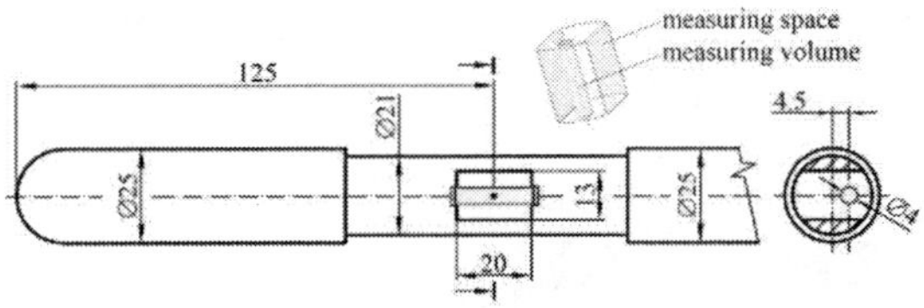

Fig. 1. The shape design and dimensions of the extinction probe head.

Given the probe dimensions, its impact on the flow field can't be neglected. Furthermore there isn't a possibility of

[a] Corresponding author: michal.kolovratnik@fs.cvut.cz

dynamic calibration of the probes. Therefore other methods are necessary to evaluate the probe's impact on the flow field. CFD calculations showed to be a suitable tool and our analyses were focused on two outcomes:

- formulation of a methodology for measured data correction based on the interaction,
- search for a new probe / probe head geometry with respect to its dimensional and strength limitations.

In our first approach of CFD simulations a simplified procedure was applied, which substituted the real flow section of the turbine by a constant cross-section channel, with idealized working fluid and uniform inlet and outlet boundary conditions. This approach has proven to be unsuitable. Thus a more complex approach was adopted. It reflects real geometry of the LP section of the 1000 MW steam turbine, the wet steam properties are defined by the IAPWS standard and phase change in the expanding steam is modelled by considering a non-equilibrium steam condensation in the turbine [5].

Character and intensity of the probe impact on the flow field is evaluated based on a comparison between the results of the CFD simulations of the steam flow without the probe presence *Var0* and with the probe *VarP*. The differences between the simulations are then analysed and compared for the region of the measuring space within the probe head, specifically for the measuring volume defined by a cylinder of a measuring light beam (yellow cylinder in Fig. 1). Placement of the probe within the turbine and the numerically modelled region is shown in Fig. 2.

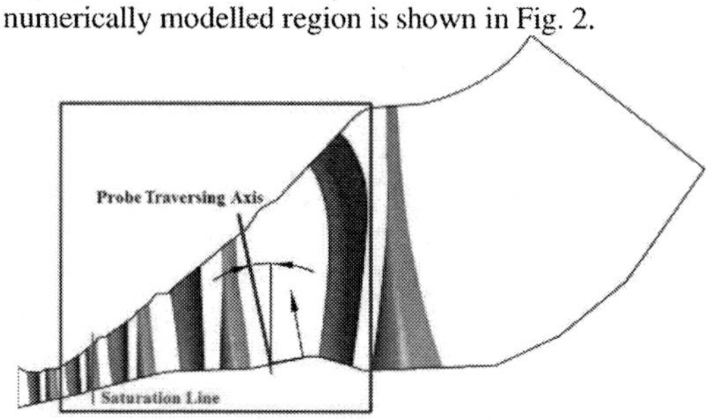

Fig. 2. The placement of the probe in the LP turbine.

The computational model is defined on the basis of a compromise between the accuracy and computation time of the simulations. The computaion model consists of 8 domains. Detailed description of the domains of this model is provided in [4]. The same model is used also for the analyses in this work. All domains for the *VarP* are introduced in Fig. 3.

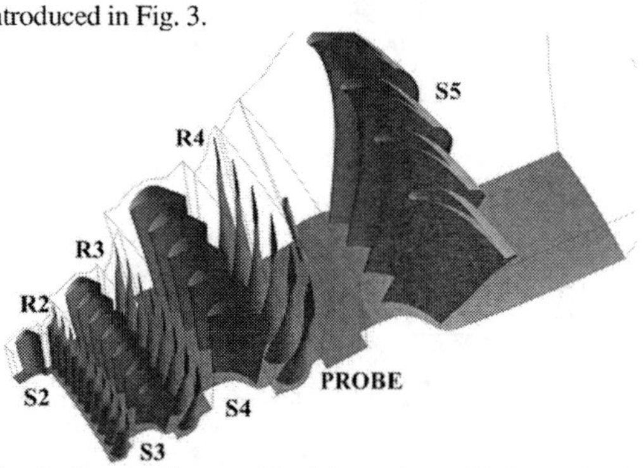

Fig. 3. Computation model of the variant with the extinction probe.

The results of the complex CFD simulation presented in detail in [4] unambiguously confirmed the expected significant impact of the current geometry of the probe on the flow field. However, at the same time there was found a non-uniformity of this impact in various positions of the probe along the blade length. This practically ruled out a possibility of preparing a universally applicable data correction methodology, which could be applied for various turbines, operation regimes and probe positions in the turbine. Currently it is only possible to correct the impact of the probe – flow field interaction for a specifically simulated cases without means of generalization. It is caused by a range of phenomena. The probe presence impacts practically all factors affecting the measured transmittance $(I/I_0)_{EXP}$, i.e. droplets size and concentration, wet steam density and thus steam wetness.

As a result, the research activities were further focused on probe head geometry modification. An example provided in Fig. 4 presents one of the tested modifications. Upper part of the figure indicates an "ideal" case, when the length of the measuring light beam is determined by cylindrical optical collimators with an outer diameter of 5 mm, close to the beam diameter. Such solution could be expected to provide very low impact on the flow field, however it can't be created in the real case. Proposed shape of the entire probe head, further marked *VarP2*, is then shown in the bottom part. Here, there were added cylindrical endings of the optical collimators into a probe shape that has been verified from a strength point of view. Length of the measuring beam of 20 mm remains constant with the previous case.

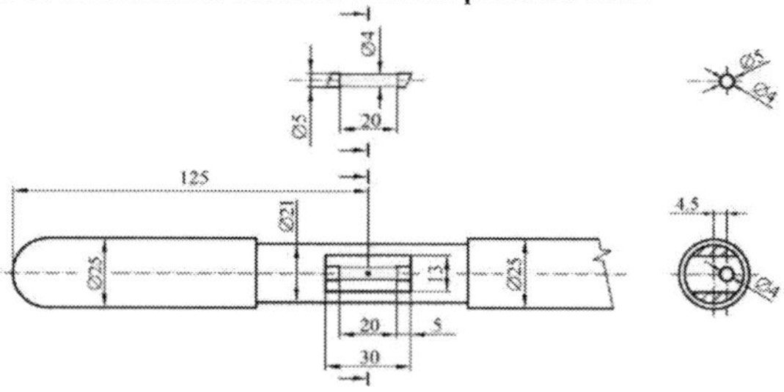

Fig. 4. The modified shape design and dimensions of the extinction probe head – *VarP2*.

This modified head has been tested by new simulations of the probe – flow field interaction. Same geometry of the turbine flow sections as the one presented in Fig. 3 has been used. Comparison of the mesh within the measurement area between the original *VarP* and modified *VarP2* head is provided in Fig. 5.

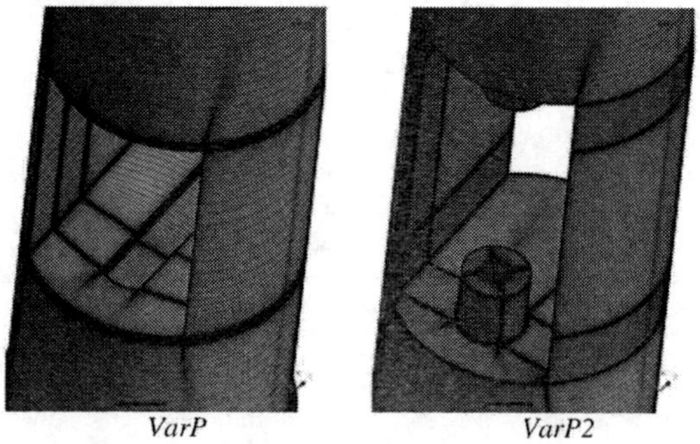

Fig. 5. Surface mesh detail of the extinction probes.

https://doi.org/10.1051/epjconf/202226901029

3 Results of the CFD simulations

Following figures 6 and 7 provide a comparison of the results of the probe – flow field interaction between the original and the new probe head geometries.

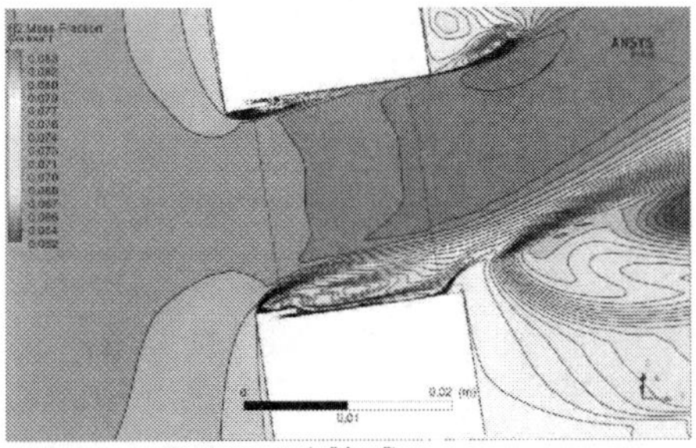

a) *VarP*

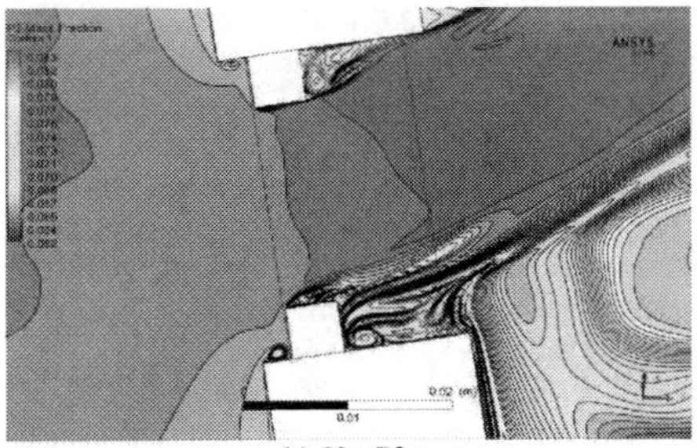

b) *VarP2*

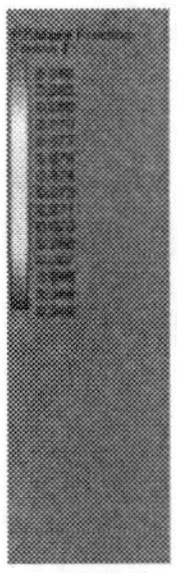

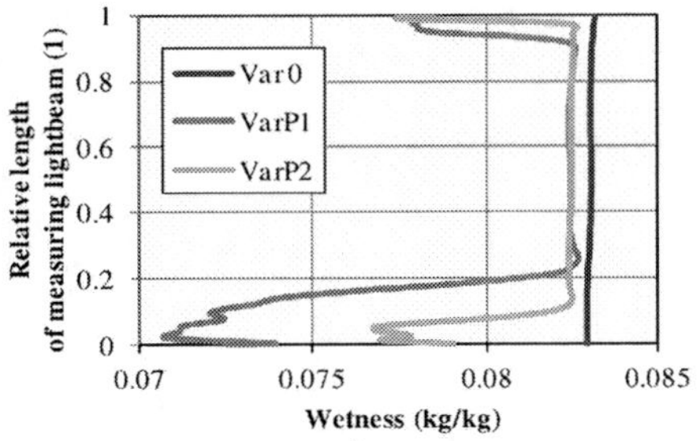

c) *Var0* d)

Fig. 6. Results of the CFD simulations – steam wetness, a), b), c) Contours of steam wetness on the cross-section of the domain PROBE, d) Distribution of average steam wetness along the measuring beam.

For clarity, the contours of the monitored parameters, i.e. steam wetness and the fine droplet diameter for the calculation variants *VarP* and *VarP2* in the probe measuring space and its surroundings are compared also with nearly homogeneous parameters of calculation variant *Var0* in the same region (see figures 6 c) and 7 c)). These contours are supplemented by distributions of the averaged corresponding parameters along the light measuring beam for the calculation variants *VarP* (red) and *VarP2* (green) in comparison with that for the calculation variant *Var0* (blue).

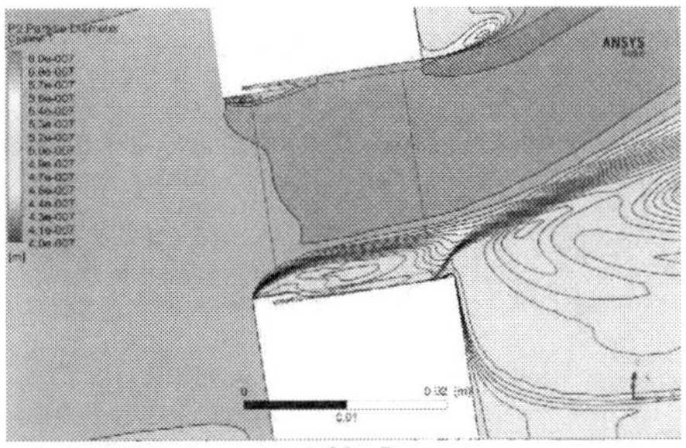

a) *VarP*

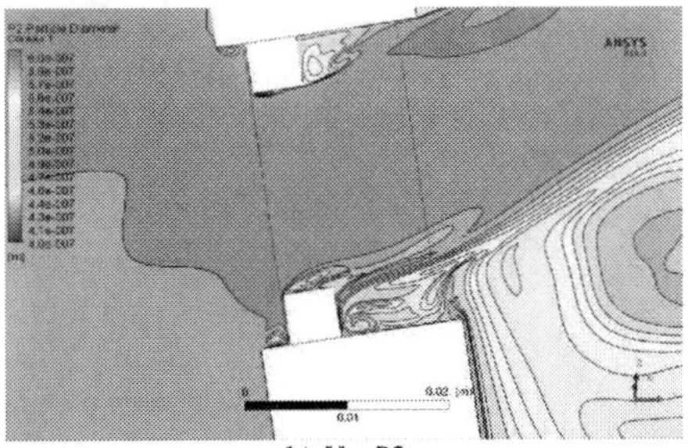

b) *VarP2*

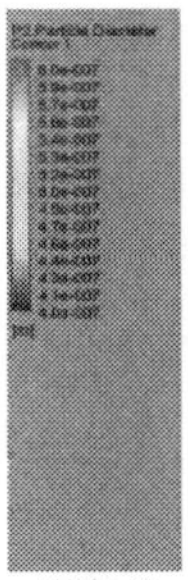

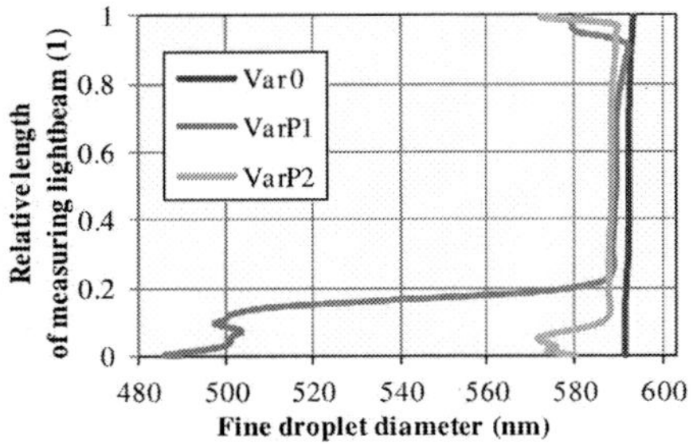

c) *Var0* d)

Fig. 7. Results of the CFD simulations – fine droplet diameter, a), b), c) Contours of fine droplet diameter on the cross-section of the domain PROBE, d) Distribution of average droplet diameter along the measuring light beam.

Fig. 8 shows a cumulative number of the fine droplets along the measuring beam for all investigated variants.

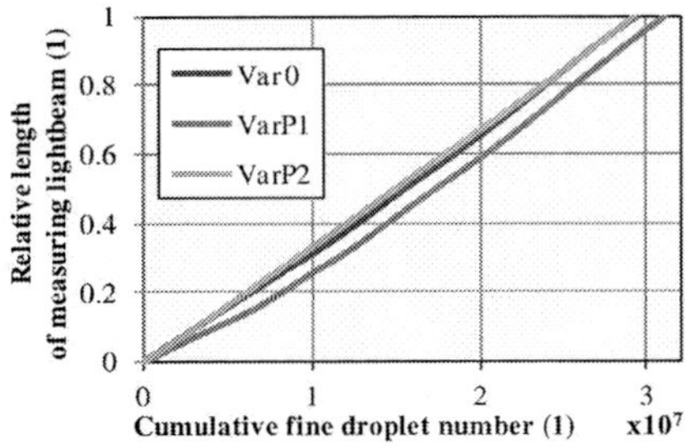

Fig. 8. Cumulative fine droplet number distributions inside the active cylindrical measuring volume.

4 Discussion

Fig. 6 compares the steam wetness contours for *VarP*, and *VarP2* shown on the PROBE domain cross section. It

can be seen that even for *VarP2*, the wetness distributions near the upper and bottom walls are different from that in the centre of the measuring space due to the interaction of the modified probe with its surroundings. However, the affected areas near the both surfaces limiting the measuring beam are in this case smaller than those for *VarP*. This fact is confirmed by the distributions of average steam wetness along the measuring beam for all variants shown in Fig. 6 d).
Similar trend can be observed in Fig. 7 for the primary droplets distribution along the measuring beam.

Significant benefit of the probe shape modification on reducing the probe – flow field interaction impact is confirmed by the distribution of the cumulative number of droplets along the measuring beam, as seen in Fig. 8. In case of the current probe head *VarP* there is a part of the droplets taken by a backflow generated downstream of the probe back into the measuring space of the probe, especially in the region near the bottom end-wall (see Fig. 9a).

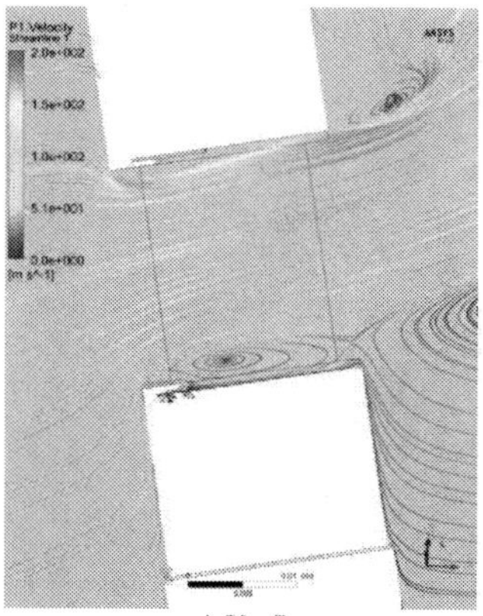

a) *VarP*

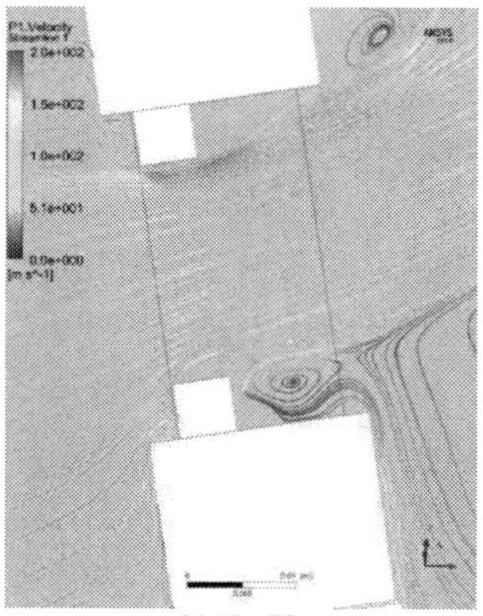

b) *VarP2*

Fig. 9. Surface streamline on a cross-section of the domain PROBE.

This is at the same time the reason of the significant difference in cumulative droplet distribution along the measuring beam between *Var0* and *VarP* in the lower part of the measuring space. Strong backflow originates also behind the new probe head design. However in this case the backflow impacts the measuring volume in lower extent than in case of the current probe head (see Fig. 9b). Therefore the distribution of the cumulative droplet count along the measuring beam in case *VarP2* has a lower deviation from the unaffected flow field *Var0*, than the *VarP* has.
In summary the partial benefits of the *VarP2* illustrated in figures 6 – 9 can be expressed by the following comparison with *Var0* on values of relative transmittance of light at 1000 nm wavelength in the modelled structure of the liquid phase.

$(I/I_0)_{EXP\ VarP} / (I/I_0)_{Var0}$	$(I/I_0)_{EXP\ VarP2} / (I/I_0)_{Var0}$
0.948	0.974

5 Conclusions

Using CFD calculations, the interaction of the optical probe head with the flow field in its surroundings has been analyzed. Using these analyses, it has been proven that the liquid phase parameters of wet steam in the probe measuring space are different from that which would have been in the case without the probe. The possibility of the shape modification of the probe which led to reduction in the probe-flow field interaction has also been found out. The benefit of the proposed shape modification of the probe has been verified using CFD simulation.
The shape modification of the probe described in this paper is only a conceptual design. There are many technical details that need to be addressed to implement this conceptual design, for example, how to prevent the formation of water film on the glass windows of the probe.

Acknowledgements

We gratefully acknowledge the support by project no. TK01020029 of the Technology Agency of the Czech Republic.

References

1. Walters, P.T.: *Practical Applications of Inverting Spectral Turbidity Data to Provide Aerosol Size Distributions.* Applied Optics, 1980,Vol. 19, No. 14, pp. 2353-2365.
2. Petr, V.: *Návrh fotometrické sondy pro měření primárních a sekundárních kapek ve stupni parní turbiny [Design of a photometric probe for fine and coarse droplets measurement in a steam turbine stage]*, ČVUT-FS-K215, Prague, Czech Republic, 1982
3. Bohren, C.F., Huffman, D.R.: *Absorption and Scattering of Light by Small Particles*, John Wiley & Sons, Inc., USA, 1998
4. Kolovratník, M., Jun, G., Bartoš, O.: *CFD analysis of an optical probe interaction with wet steam flow field.* EPJ Web of Conferences 180, 02045, 2018
5. Starzmann, J. et all: *Results of the International Wet Steam Modeling Project.* Proc Inst *Mech* Eng, Part A: Journal of Power and Energy, V232, 5, 2018, pp. 550-570

https://doi.org/10.1051/epjconf/202226901030

Quarter wavelength resonator in experiment and simulation

Jan Kůrečka[1,*], *Ondřej*Čepl[1], and *Vladimír* Habán[1]

[1]Viktor Kaplan Department of Fluid Engineering, Brno University of Technology, Technická 2896/2, Brno, 616 69, Czech Republic

Abstract. Quarter wavelength resonator provides a mean to create a pressure drop at a specified frequency. Design of hollow cylinder connected radially to a pipe was used and different lengths of this tuned resonator were measured. Correlation with computational approach using finite element method analysis (FEM) with ANSYS acoustics capabilities was achieved. Analytical formula for a length of a quarter wavelength resonator is compared to an experimental measurement data and simulation.

1 Introduction

Pressure oscillations, noise propagation and vibrations in piping systems are problems that will require for a foreseeable future effort of engineers so the machinery will not suffer from unneeded strain. These problems might be more or less complicated to solve, depending on a source of these pulsations. Self-excited vibrations are difficult kind of these problems since the change in a system might only lead to a change of frequency. [1]

Some of the problematic vibrations can be silenced using mufflers and resonators.[2] Resonance of acoustics cavities was studied for example in [3] in relation to various length and cross-section ratios.

This contribution aim is description of dynamic properties of short quarter wave resonators specifically resonators of geometry of a hollow tube installed on another tube so the findings could be further used in practice for example on pipeline systems, pneumatic, hydraulics, gas lines and so on. In process there is also motivation to validate simulation abilities for such purpose.

2 Experiment

Experiment was set up to measure preselected lengths of resonator tube. Basic description of a measurement stand was following: Two meter long tube was split in half and the quarter wave resonator was connected in this opening. The opening was constant for all measured lengths of resonator. Speaker was chosen to excite the air in a tube and to measure the response the set of four microphones was positioned along the length of tube. Main dimensions of this experimental quarter wave resonator system are listed in Tab. 1 and depicted in Fig. 3.

Motivation for a choice of tube length was to have size for at least three wavelengths along the length of tube on a studied frequency. Clearly this self imposed rule limits the lower boundary of studied frequencies. Reasoning behind dimension of resonator was in a way similar since the length of resonator is inversely proportional to its natural frequency. The shortest length of resonator is double of a opening into a tube. Other lengths were arbitrary chosen to cover spectrum of these short quarter wave resonators. Opening size makes it wide enough to not to create a resistance for a acoustic wave. Chosen diameter of a resonance cavity provides resonator cross-section close to a half of a tube cross-section.

Table 1. Summary of dimensions.

Parameter	Symbol [Unit]	Value
overall length	L_{tube} [m]	2.012
tube inner diameter	D_i [mm]	53.6
cavity outer diameter	D_{ro} [mm]	76.6
cavity inner diameter	D_{ri} [mm]	63
resonator opening	w [mm]	12
wall thickness	s [mm]	6.7
resonator lengths	L [mm]	24; 29; 35; 44; 54; 74; 94

Three main variations were measured. One was tube without resonator, for the purpose of reference and result processing. Two other consisted of change of opposite end of tube it was either closed or opened. Specified quarter wave resonator length were tested in a spectrum of frequencies. Process was to excite a specific frequency, mea-

*e-mail: 133644@vutbr.cz

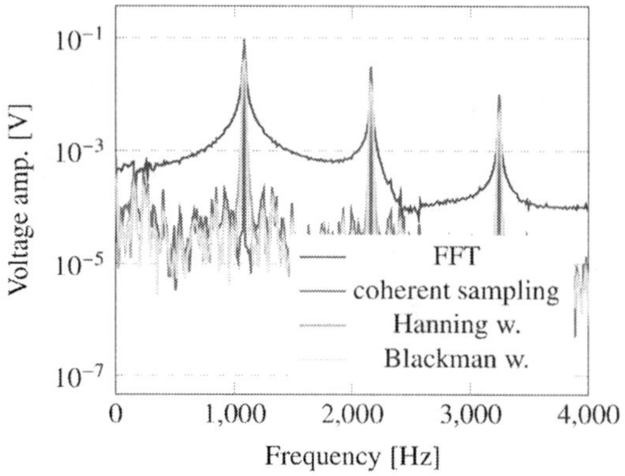

Fig. 1. Processed data.

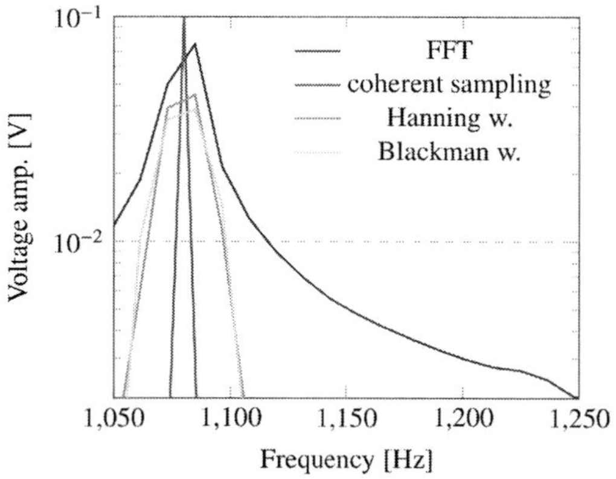

Fig. 2. Processed data, peak value detail.

sure and move to next one. This was repeated for all respective lengths of resonator and end of tube variations.

Excitation was equal through whole measurement and the loudspeaker was generating sinusoidal wave on an appropriate frequency. Frequency was upped by 20 Hz between steps and measured interval was from 100 to 8000 Hz. Sampling frequency of 108 kHz was used for a two second measuring time. Unspecified pause to system stabilization proceeded this data acquisition.

2.1 Data processing

Format of gathered data was time sampled amplitude of sensor voltage. Fast Fourier transform is needed to get amplitude-frequency characteristics from sensor recording. There is number of ways to further process data to get spectral space data readings with high level of accuracy and significance.[4] On data sets captured from aforementioned measurement there were tested four approaches to evaluate their suitability. Raw data from FFT, Hanning window, Blackman window and coherent sampling.

IEEE norm 1241 states formula (1) to be used for coherent sampling, where f_i is frequency of interest, f_s is sampling frequency, M is record length and J is integer relatively prime to M. Relative prime is defined as relation between two numbers with greatest common division of one. [4]

$$f_i = \frac{J}{M} \cdot f_s \tag{1}$$

In Fig. 1 comparison between four methods of data processing is presented. Two parameters of adequacy of method is considered. Spectral leakage and signal loss occurs and depending on further data use the appropriate method choice differs. Both mentioned effects on data are visible in Fig. 2. Clearly the coherent sampling method provides the least loss of signal and its frequency bin division provides lowest spectral leakage in comparison to the rest of tested methods. This reasoning justifies the choice of coherent sampling as suitable tool to process measured data.

Experimentally obtained data were compared to analytical formula (2) for selected resonator lengths. This is depicted in Fig. 4. Second and third shortest resonator lengths show bit of a difference to analytical value, but overall the measurement data fit very nicely with theory.

$$f = \frac{a}{4 \cdot L} \tag{2}$$

3 Simulation

To simulate quarter wave resonator FEM ANSYS package with acoustic capabilities was chosen. First difference to experiment was use of Harmonic analysis, so the simulation was performed in a frequency domain. This approach is valid since the simulated effect of resonator is response to a frequency at which it is excited by a loudspeaker.

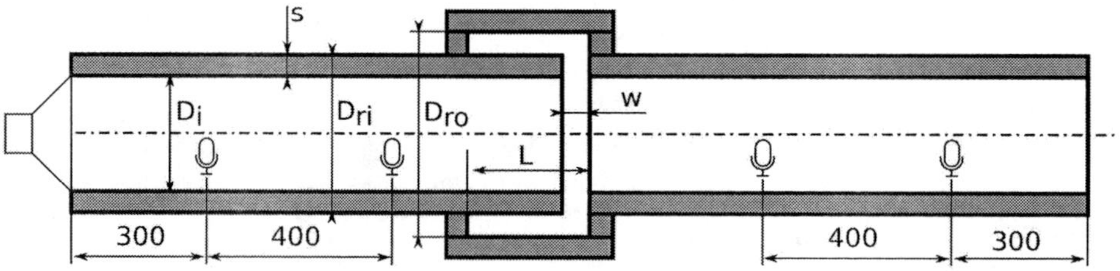

Fig. 3. Measurement apparatus schematics with positions of microphones installed in tube (dimensions are not up to scale to better illustrate geometry.

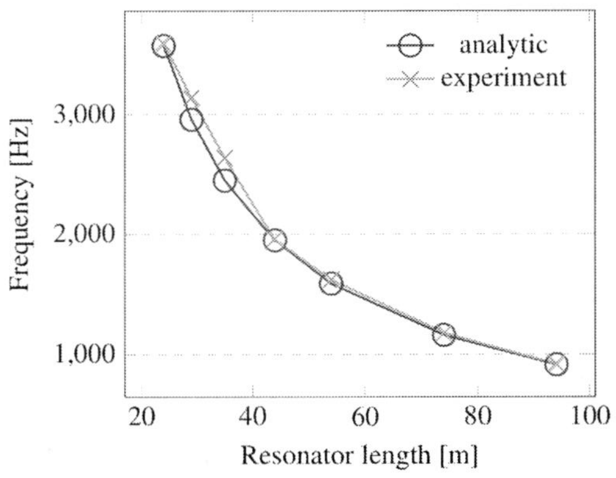

Fig. 4. Measured frequencies and analytically obtained values.

Geometry of simulated domain matches the experimental setup (as depicted in Fig. 1) but it is limited to internal volume. Tube walls are thick compared to diameter and the effect of resonator is dominant process order of magnitude stronger than effects of wall deformations and vibrations. So the structural part was not simulated and the computational domain consisted of acoustic body only.

Tube ends get replaced by a boundary conditions, harmonic pressure excitation substitutes loudspeaker and on other end either rigid wall or non reflective radiation boundary condition (open ended case). Rest of domain surfaces have default boundary condition of rigid wall. Parameters of acoustic medium are chosen to match experimental environment. Speed of sound is defined using formula (3) with temperature being 20°C, this was also used to calculate natural frequencies for resonators in Fig. 4 using formula (2). Values of media properties are stated in a Tab. 2.

$$a = 331.6 \cdot \sqrt{1 + \frac{t}{273.15}} \tag{3}$$

Mesh size of a ten thousand quadratic HEX20 elements was used to meet recommended number of elements [5] (6 quadratic elements in acoustic domain per wavelength) and further mesh density increase did not change evaluated parameters.

In Fig. 5 correlation between experimental and simulated data is visible. Experimental values are represented only by a natural frequency for respective resonator length and simulation of quarter wave resonator is presented by a acoustic pressure damping interval across frequency range. There is also a visible downward trend of acoustic pressure response with ascending frequency. This is caused by a frequency independent bulk viscosity and the reference of simulated damping by a bulk viscosity in a tube without resonator can be seen as a red dotted curve. In this case loudspeaker was substituted by a acoustic pressure excitation with amplitude of 1 kPa.

Table 2. Simulation parameter settings.

Parameter	Symbol [Unit]	Value
density	ρ [kg$\cdot$ m^{-3}]	1.225
sound speed	a [m $\cdot$s^{-1}]	343.5
viscosity	ν [Pa $\cdot$s]	$2 \cdot 10^{-4}$
bulk viscosity	ν_b [Pa $\cdot$s]	0.3

4 Transfer matrix

Simulated case provides velocity and pressure field needed for determination of another dynamic property of quarter wave resonator. Transfer function, which can be used to model such system using Transfer Matrix Method (TMM). TMM in its simplest form consist of two state vectors and transfer matrix, that enables to a variable from one state vector to change its value to the corresponding value of respective variable in second state vector - formula (4).

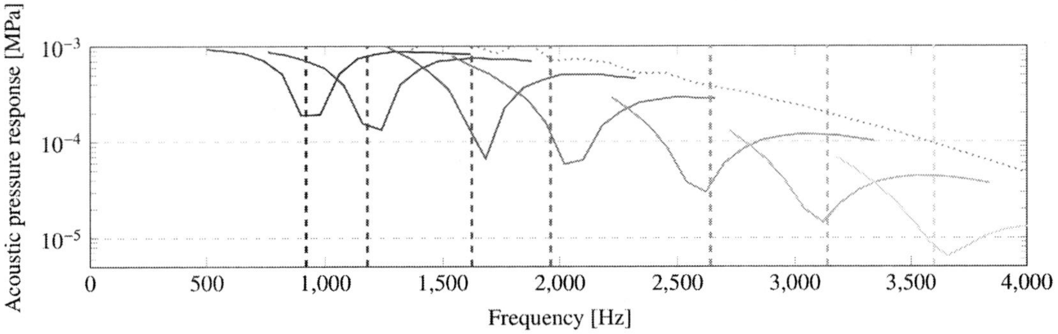

Fig. 5. Natural frequencies of resonators evaluated from experiment (dashed line) compared to computed acoustic pressure response to excitation at opposite end of tube (solid lines). Color pairs are to match same resonator lengths.

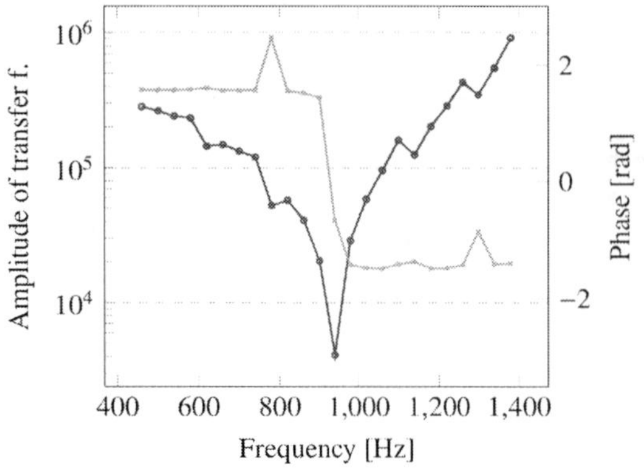

Fig. 6. Transfer function.

$$\begin{pmatrix} a \\ b \end{pmatrix}_{i+1} = \mathbf{P}_T \begin{pmatrix} a \\ b \end{pmatrix}_i \tag{4}$$

Presented case can be written as state vector of flow and pressure and transfer matrix consisting of one variable - formula (5). This assumes pressure change of state vectors to be zero. For purpose of finding the transfer function, the state vectors are set at positions at cross-section of before and after the opening of the resonator into the tube. Assumption of zero pressure change is not completely true, but compared to velocity difference the pressure change is much smaller. This is in agreement with knowledge that in resonator there is node of velocity oscillation and anti-node of pressure at closed end and vice versa for open end - the opening into the tube. All variables in this equation are complex numbers.

$$\begin{pmatrix} Q \\ p \end{pmatrix}_{i+1} = \begin{pmatrix} 1 & X \\ 0 & 1 \end{pmatrix} \begin{pmatrix} Q \\ p \end{pmatrix}_i \tag{5}$$

By using eq. 5 transfer function X can be solved. Inputs are flow and pressure field on specified surfaces at opening of resonator and then the amplitude and phase can visualised in chart with x axis being frequency. In Fig. 6 There is visible transfer function amplitude extreme at frequency that is unsurprisingly the natural frequency of simulated resonator. Phase change can be found at the same position as well.

5 Conclusion

In this contribution measurement of quarter wave resonator was studied and there was a focus on specific geometry of resonators. Dimension of length was also defining factor for this contribution, namely that respective resonators were chosen to be short. Chosen lengths of were measured and acquired data were processed using various techniques to obtain frequency and amplitude of acoustic waves. This testing concluded that coherent sampling provides best result in terms of spectral leakage and signal loss.

In next step the Finite Element Method simulation had shown that it is possible to model damping system of quarter wave resonator with a acceptable level of precision. Fields of acoustic pressure and flow allows to further study dynamics of this setup and so the transfer matrix method was used to quantify transfer function for a interval of measured frequencies. Considering the equation (5) the transfer function amplitude describes the effect of change between selected positions of defined state vectors. So the lowest value of transfer function matching the natural frequency of resonator is in agreement with a measurement where at this frequency the least energy of acoustic waves is transferred between the excited and opposite tube's end.

These findings are interesting with respect to a possible use of quarter wave resonator to damp and suppress acoustic waves, vibrations or to reduce noise. Advantages of such resonator are price, since it is only passive element and also very low pressure drop since it does not create obstruction for a fluid flow.

The presented work was performed with support of TA ČR project TE02000232 Special Rotary Machines Engineering Centre and Faculty of Mechanical Engineering of Brno University of Technology project id. FSI-S-17-4615.

References

1. HIMR, D., HABÁN V., 2014. Damping of Self-Excited Pressure Pulsations in Petrodiesel Pipeline. *Applied Mechanics and Materials*. 630, 375-382. DOI: 10.4028/www.scientific.net/AMM.630.375. ISSN 1662-7482.
2. KHAIRUDDIN, M., SAID, M., DAHLAN, A., KADIR, K. (2018). Review on Resonator and Muffler Configuration Acoustics.*Archives Of Acoustics*, 43(3), 369–384. doi:10.24425/123909
3. RONA, A, 2016. The Acoustic Resonance of Rectangular and Cylindrical Cavities.*Journal of Algorithms & Computational Technology* 1(3), 329-356. DOI: 10.1260/174830107782424110. ISSN 1748-3018.
4. *IEEE Standard for Terminology and Test Methods for Analog-to-Digital Converters*: IEEE Std 1241™-2010, 2011. Revision of IEEE Std 1241-2000). New York, USA: The Institute of Electrical and Electronics Engineers.
5. ANSYS Inc., Introduction to acoustics (product manual), ANSYS Acoustics Lectures, 2016

Numerical calculations of heat transfer during FC-72 flow in minichannels performed with ANSYS software

Kacper Kuta[1], *Magdalena* Piasecka[2,*], and *Slawomir* Blasiak[3]

[1,2,3] Kielce University of Technology, Faculty of Mechatronics and Mechanical Engineering
Al. Tysiaclecia Panstwa Polskiego 7, 25-314 Kielce, Poland

Abstract. The main aim of the paper was to apply numerical calculations in ANSYS software to solve problems related to steady-state convective heat transfer during FC-72 flow in a minichannel heat sink. A mathematical model of fluid flow in a minichannel heat sink was proposed. Data from research conducted on the experimental stand with asymmetrically heated test section were selected. The element cooled by fluid flowing in minichannels was a thin foil. The distribution of temperature on the outer heated foil surface was measured using infrared thermography. Thermocouples and pressure meters were installed at the inlet and outlet of the minichannels. A CAD model was prepared with respect to the geometry of the test section. The mesh was created in ANSYS Mesher. A suitable model of turbulences was selected and materials temperature dependencies were considered. The iterative process of creating the model and comparing the results obtained by the CFD calculation was used to capture the fluid behavior. The velocity magnitudes and temperature distributions were presented as the main results of numerical calculations.

1 Introduction

Extensive experimental and theoretical studies are required to understand the boiling phenomena in the group of small hydraulic diameter channels called micro- or minichannel heat sinks. The phase change which accompanying boiling processes allows the highest possible heat fluxes to be obtained at low temperature differences between the heated surface and the working fluid over a small heat transfer area. Scientific studies on heat transfer in cooling liquids flowing through a group of channels of small hydraulic diameters are being conducted intensively [1-9]. This topic is not popular in research mainly due to practical difficulties. The results concerning heat transfer in flows boiling in a group of mini- or microchannels are inconsistent, and even contradictory. The calculation methods of heat transfer determination in the flow boiling in a group of mini- or microchannels often represent only numerical simulations, which are not based on the results of own experimental studies, calculations are often based on the results of experimental results obtained by other researchers. It should be emphasized that the experiment is the most important aspect which contributed most to the direct describing of boiling heat transfer. It is especially significant for complicated heat transfer systems, in which the size of bubbles can be larger than the size of the channel because any forecasts on heat transfer are difficult under such circumstances.

Creating a coherent, predictable and accurate computational model of heat transfer during fluid flow through a minichannel is both a designer and a scientific challenge. Such a model, computationally simple and easy to use, would allow greater control over design of small-scale, effective heat exchangers. Sometimes it is necessary to build an experimental set-up for every design idea. A standardized model for a compact heat exchanger with minichannels of various geometries, created using numerical computing, would lead to reduction of design time in many practical applications.

Based on experimental data and analytical and analytical numerical methods with using Trefftz functions, the values of the heat transfer coefficient on the heating surface that contacts the fluid flowing in a minichannel were determined in earlier works [10-13]. Some attempts have been made to predict the behavior of cooling fluids [14, 15] flowing in minichannels due to numerical calculations. To verify CFD modeling, selected data from experiments were applied in numerical calculations using ANSYS CFX [14] and ADINA software [15]. It is worth mentioning that ANSYS CFX software, a computational fluid dynamics program, is often used to simulate and optimize fluid flow as in [16].

2 Experimental stand and methodology

In investigations conducted at the Kielce University of Technology, several experimental stands were built and furnished with equipment for investigation of flow boiling heat transfer in minigaps. The essential part of each stand is a test section of different geometries and spatial orientations. A view of the experimental stand with a test section with rectangular minichannels or an annular minigap is shown in Fig. 1.

*Corresponding author: tmpmj@tu.kielce.pl

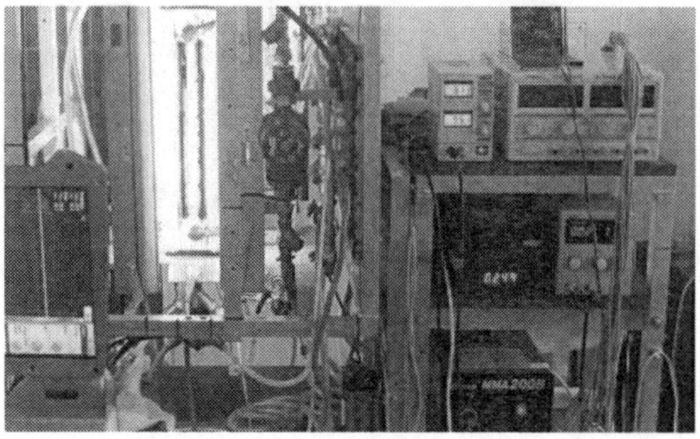

Fig. 1. A view of the experimental stand with a test section with rectangular minichannels or an annular minigap.

A new experimental stand with a test section with a group of rectangular minichannels or microchannels is shown in Fig. 2

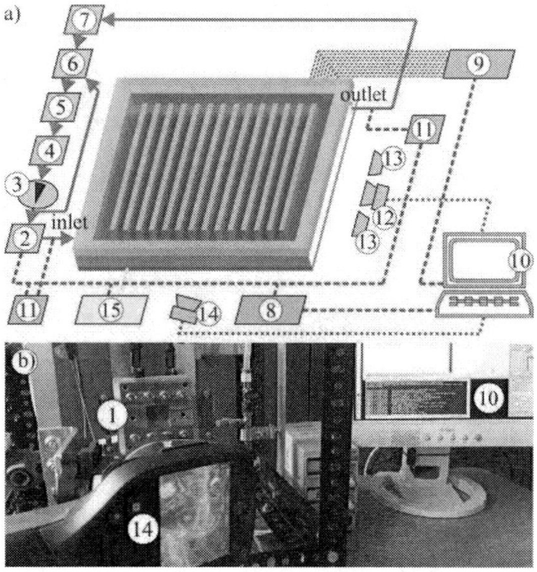

Fig. 2. The new experimental stand with a test section with a group of rectangular mini- or microchannels (a heat sink): a) a view, b) a schematic diagram; 1 – a test section, 2 – a mass flow meter, 3 – a circulating gear pump, 4 – a filter, 5 – a deaerator, 6 – pressure meter, 7 – plate-type heat exchanger, 8,9 – data acquisition station, 10 – a PC computer, 11 – a pressure meter, 12 – a high-speed camera, 13 – LEDs, 14 – an infrared camera, 15 – a power supply, 16 – a front cover, 17 – a group of mini- or microchannels, 18 – a channel body, 19 – an inlet/an outlet chamber, 20 – a glass, 21 – a heated foil.

In the flow loop, Fig. 2a, a cooling fluid (such as Novec 7200, 7100, 649, Fluorinert FC-72) is recirculated. The flow loop consists: a test section with a group of mini- or microchannels (a heat sink), a mass miniflow meter with a controller for flow stabilizing at the inlet to the test section, a circulating pump, a filter, a deaerator, a compensating tank/a pressure regulator, and a tube type heat exchanger, with water as a coolant. The data and image acquisition system, designed to collect the measurement data, includes data acquisition stations, an infrared camera and a high-speed camera with a PC computer with appropriate software. The lighting system comprises high power LEDs to provide light for the two-phase flow structures. The temperature and pressure of the working fluid of the fluid at the inlet and outlet of the test section are measured by thermocouples or pressure meters, respectively. The supply and control system contains a power supply unit and a shunt.

The essential part of the experimental stand is a replaceable test section with a group of rectangular mini- or microchannels. The test section view is shown in Fig. 3a and the transverse cross-section of the test section - in Fig. 3b. The number, width and length of the channels can vary. The heated element for working fluid flowing along minichannels is a thin mretal foil made of Haynes-230 alloy.

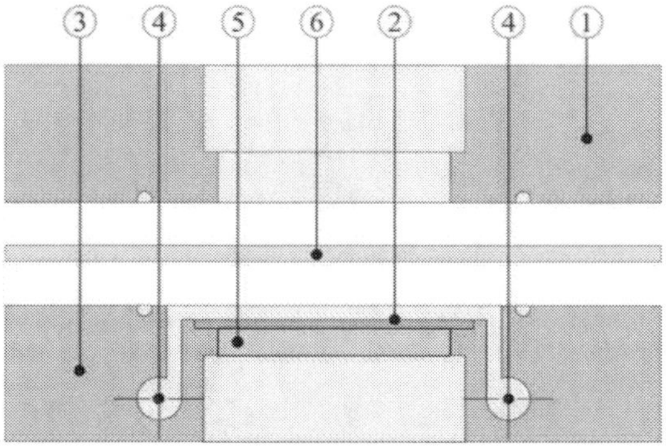

Fig. 3. A transverse cross-section of the test section with a minichannel heat sink, 1 – a front cover, 2 – a group of mini- or microchannels, 3 – a channel body, 4 – an inlet/an outlet chamber, 5 – a glass, 6 – a heated foil.

During the experiment, after deaeration, there is a flow of the working fluid in the main loop of the experimental stand. The liquid at the temperature below its boiling point flows in the test section. The heat flux supplied to the heated foil is gradually increased by adjustment of supplied heating power until developed nucleate boiling occurs. The schematic diagram of the fluid flow in the test section with a channel heat sink is shown in Fig. 4.

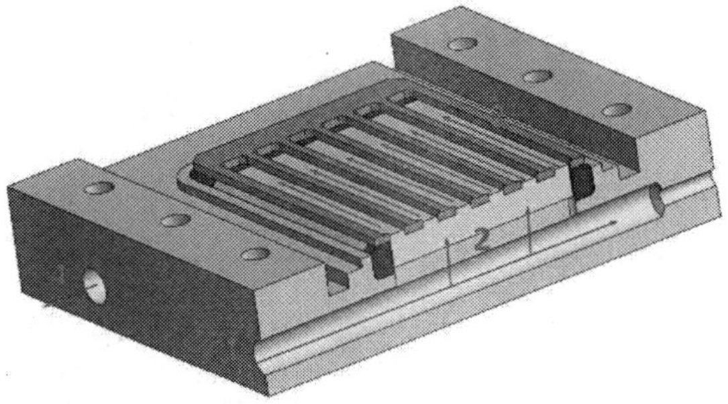

Fig. 4. The schematic diagram of the fluid flow in the test section with a minichannel heat sink, 1 – an inlet collector, 2 – inlet to the channels, 3 – a singular minichannel, 4 – an outlet collector.

For the present work, several versions of a test section construction were created in CAD software. An overall construction of the test section is depicted in Fig. 5.

a)

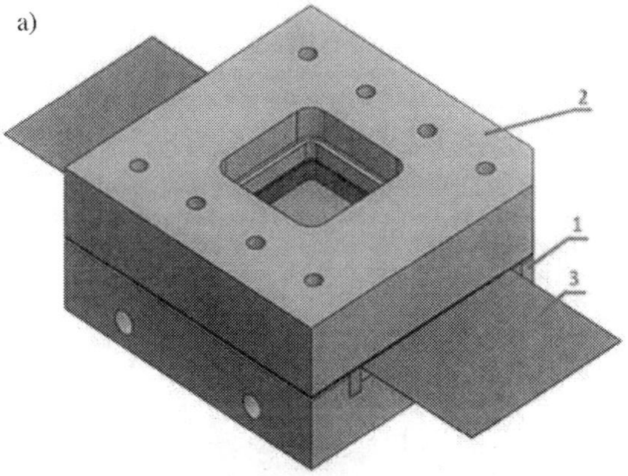

b)

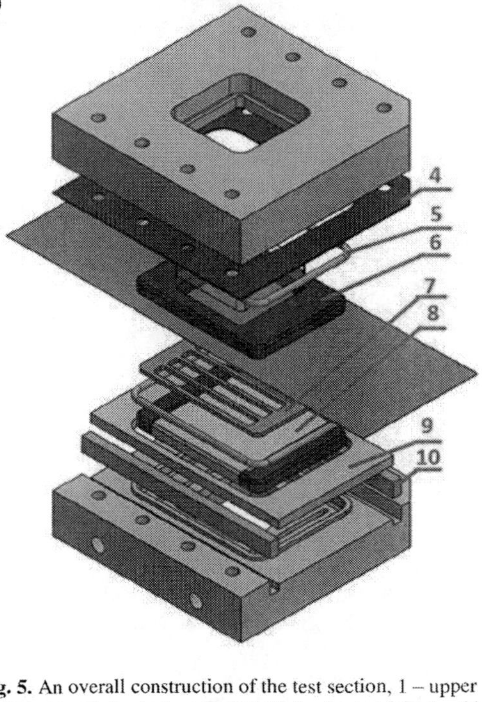

Fig. 5. An overall construction of the test section, 1 – upper body (a cover), 2 – bottom body half, 3 – heated foil, 4 – thin top insulation plate, 5 – top O-ring, 6 – thick insulation plate, 7 – minichannel plate, 8 – glass plate, 9 – bottom insulation, 10 – bottom insulation wedges.

Currently, the reduced CAD model is being built for CFD calculations. All measuring equipment and the environment outside the geometry of the test section were rejected to reduce a computational domain. Schematic diagrams of internal flow volume geometry were shown in Fig. 6. One of them was prepared for visualization purposes (Solid Works, Fig. 6a) and the other for numerical calculations (Design Modeler, an ANSYS module, Fig. 6b).

a)

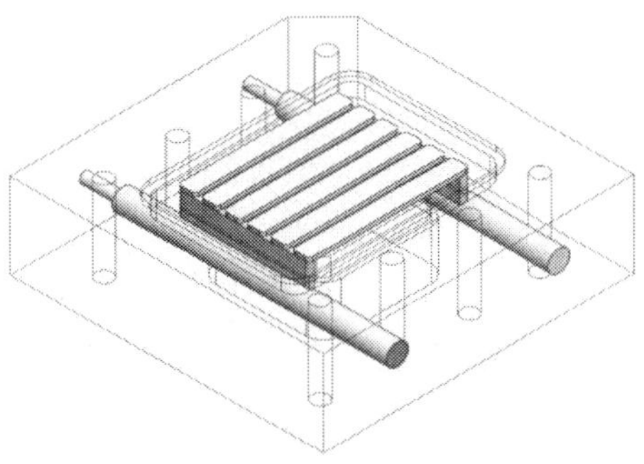

b)

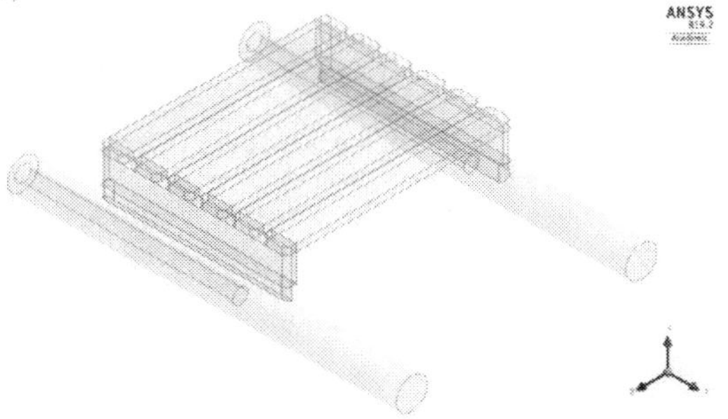

Fig. 6. Schematic diagrams of internal flow volume geometry created in programs: a) SolidWorks, b) Design Modeler, an ANSYS module.

3 Numerical calculations

3.1 Assumptions and calculation preparation

To simplify the problem, some assumptions were made. Most of them was considering incompressible flow with relatively low turbulence. Radiance was considered negligible and it was assumed that all properties of materials linearly changing with respect to temperature. No viscous heating effect was calculated. The fully developed flow at the inlet was considered.

In ANSYS, the Fluent SIMPLE algorithm was selected for solving (simplified) Navier-Stokes equations. For this algorithm, boundary conditions were applied. It was assumed that heat transfer through the walls is performed by conduction in materials according to Fourier's law.

Heat transfer in regions, exterior to computational domain, is treated as 1D (assuming the direction of heat transfer), for each wall separately. Each material has assigned its respective thermal conductivity that was taken directly from the manufacturers' data in tabular form with respect to temperature.

The fluid in a minichannel is heated through its top wall by a heated foil. It is considered uniformly heated with a heat flux value equal to 35 kW/m^2. Roughness parameters were assumed for the side walls of the minichannel and heated foil. These data, coupled with the viscosity of the working fluid, were used initially to calculate the velocity profile and then heat transfer near the walls.

The main parameters of the experimental series taken for the calculations were as follows:

- overpressure at the inlet:139 kPa,
- overpressure at the outlet:130 kPa,
- fluid temperature in the inlet chamber: 294 K.

Changes in temperature outside of the computational domain were neglected. A standard SST model of turbulence for low Reynolds flows was assumed, with energy equations and pressure-velocity coupling. After few initial iterations (~150) to initialize calculations, a second order upwind was selected to stabilize a solution.

3.2 Mesh preparation

All meshes used were hexahedral because they are relatively easy to control and allow for high tolerance of big aspect ratio cells and fit the geometry. Characteristics of meshes was specified in Table 1. Meshes were refined with the inflation method with specified thickness of the first layer. Meshing of this object, performed in Mesher, was focused on altering the following parameters: global size, inflation direction, height of the first layer and growth rate.

Table 1. Details of mesh refinement.

Mesh No.	Global size [mm]	Geometry selection	First layer height [μm]	Max. layers	Growth rate	Nodes	Elements
1	1	Top surface	42.5	10	1.1	81209	71680
2	1	Top surface	4.25	40	1.05	224519	206080
		Bottom surface	20	5	1.1		
3	0.5	Top surface	4.25	20	1.05	499851	465920
		Bottom surface	20	5	1.10		
4	0.5	Top surface (+bias)	4.25	40	1.05	463239	427896
5	0.5	Top surface (+bias)	4.25	50	1.02	513264	487296

Meshes from No. 1 to No. 3, Table 1, were selected to the test to create a sufficiently dense boundary layer. In next attempts, the roughness of the border walls was taken into consideration to ensure a physically correct velocity profile biased sizing toward the aluminum walls, Fig. 7.

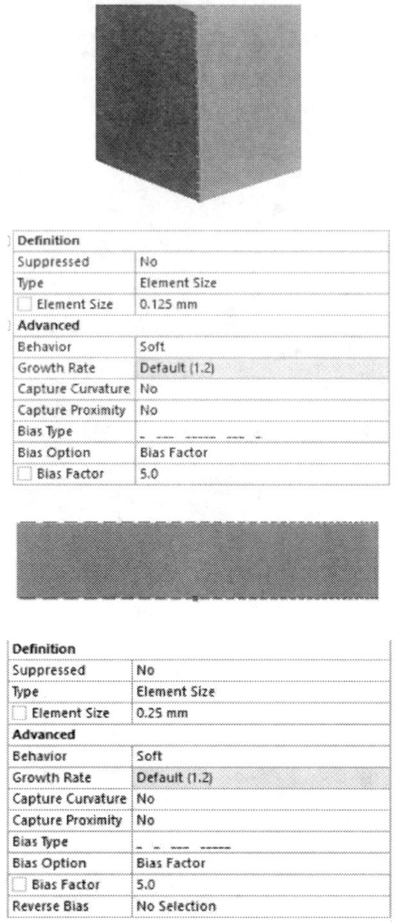

Fig. 7. Bias details.

After reaching the ANSYS Educational limit of 512k nodes + elements, another simplification was assumed for mesh No. 5 – singular minichannel symmetry plane. For the last mesh, only half of the geometry was considered. To use an additional quota of cells + elements, an increased bias strategy was added to increase the density of cells toward surfaces in contact with heated foil ('top surface'), the aluminum wall ('outer surface') and the glass plate ('bottom surface').

A cross-section and isometric view of the final mesh (No. 5, Table 1) is shown in Fig. 8. There was no significant difference between the results obtained with using meshes No. 4 and No. 5 which means reaching mesh independent solution.

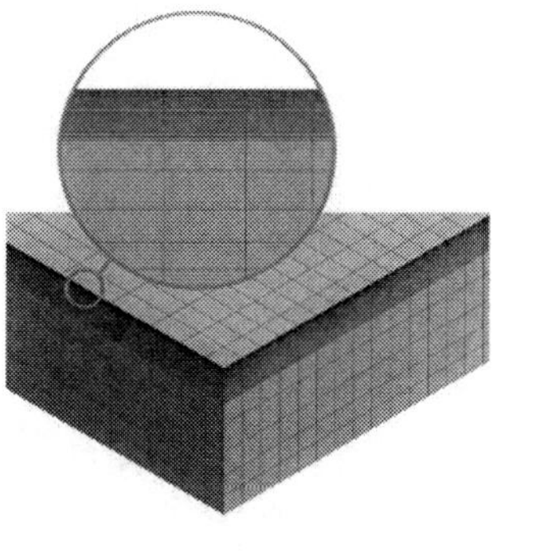

Fig. 8. Cross section of mesh No. 5.

The properties of the phase materials within the range of temperatures had to be considered during the calculation procedure. The heated foil was made of Haynes-230 alloy [17]. Fluorinert FC-72 (3M) [18] was the working fluid. Minichannel walls were made of aluminum (PA6) and typical glass material was selected for the glass panel. Following information provided by the manufacturers, the properties of all listed materials were loaded into the ANSYS Fluent program.

4 Results and discussion

The velocity magnitudes and temperature distributions are presented as the main results of numerical calculations. Planes selected for cross sections are symmetrical (for velocities) to compare numerically computed velocity profile with analytically computed average velocity obtained on the basis of the experimental data. For temperatures there is a top view showing first layer of fluid (contacting the heated foil), one solutions graph drawn on the facets of mesh, close to the end of the minichannel, and one view on cross section in the symmetry plane.

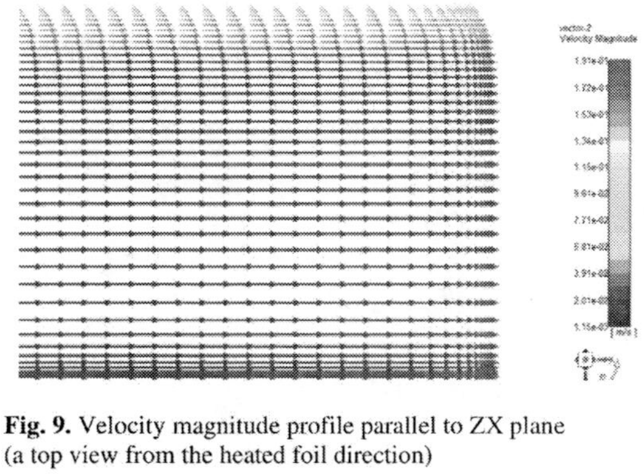

Fig. 9. Velocity magnitude profile parallel to ZX plane (a top view from the heated foil direction)

Analyzing the results obtained from calculations using the final dense mesh, a drop of velocity towards wall of the minichannel can be noticed, Fig. 9. This corresponds both with observable effects and analytical computations. Very crude cells in the middle of the image emphasize intense central flow and densely packed vectors close to the top and bottom have parabolic curve expected in such a setup. Moreover, the velocity magnitude vectors are almost constant on given depth up to last 10% width from side walls, where they parabolically decrease to reach minimal values of around 1.15e-02 m/s. From analytical calculations, an average velocity of approximately 0.12 m/s was expected (according to the experimental data). Since an average value close to that was obtained, the profile was considered physically feasible.

With changed global setting of mesh, more cells are generated in the middle of the flow, and a more pronounced parabolic profile also on top view. This barely affects maximum values of the velocity magnitude plot, but allows for the observation of the the distribution of velocity further from the top and bottom walls. The values on the top and bottom surfaces (Fig. 10) are close to 0 (0.04 m/s), which is consistent with the model. Minimal asymmetry, showing a difference of 0.15 m/s on the edges, shows small differences in roughness between the glass plate and the heated foil surfaces.

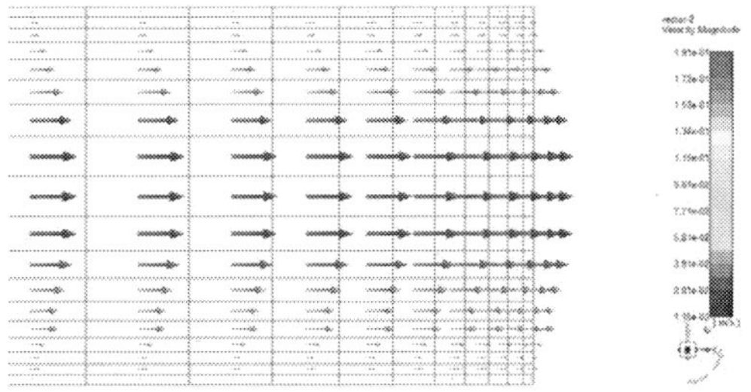

Fig. 10. Velocity magnitude profile parallel to XY plane (a side view from symmetry plane)

The static temperature distribution on the contact surface between the working fluid and the heated foil is illustrated in Fig. 11. Narrow stripes of hotter regions are observed along the whole minichannel (seen as red lines). Consistency of relative temperature of stripe and middle flow vary with different mesh densities.

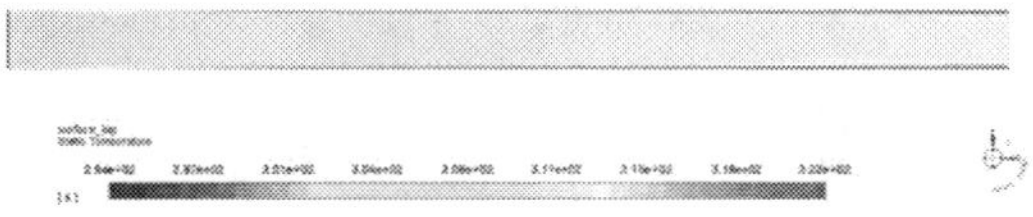

Fig. 11. Static temperature distribution on the contact surface between the working fluid and the heated foil.

An isometric view of the temperature distribution in the fluid up to the contact surface with the heated foil is presented in Fig. 12.

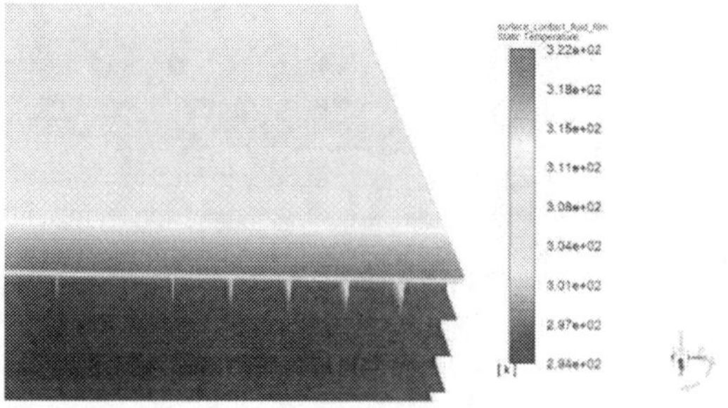

Fig. 12. Isometric view of a temperature distribution in the fluid up to contact surface with heating film, drawn on mesh facets.

Despite a denser mesh and good residual values, the temperature distribution (Fig. 12) shows remarks of not completely converged solution. It may be the result of the SIMPLE solver or not enough iterations. Since computational time was limited and residuals generally stable within range of 200 iterations, calculation was stopped, solver changed and results selected for validation after converging.

5 Conclusions

The main aim of the paper was to apply numerical calculations in ANSYS software to solve problems related to steady-state convective heat transfer during the cooling liquid flow in a minichannel heat sink. A mathematical model of fluid flow in a selected channel was proposed for the calculations. It was verified as a result of the experimental data. The essential element of the research stand was the test section with minichannels. The distribution of temperature on the surface of the heated outer wall surface of the minichannels was measured by an infrared camera. The fluid temperature and pressure at the inlet and outlet of the test section were also collected simultaneously.

A CAD model for calculations was prepared with respect to the geometry of the test section. The mesh was created in ANSYS Mesher with an optimization strategy to avoid averaging between the cells. The computational volume was limited to representation of the modeled minichannel and its surroundings to reduce the required power and time of calculations. Suitable model of turbulences was selected and materials temperature dependencies were considered to closely reproduce the behavior of fluid in experiments. The iterative process of creating the model and comparing the results obtained by the CFD calculation was used to capture the fluid behavior.

The results were presented graphically as the velocity magnitudes and temperature distributions as the main results from numerical calculations.

Acknowledgements

The research reported herein was supported by a grant from the National Science Centre, Poland, No. UMO-2018/31/B/ST8/01199.

References

1. S.G. Kandlikar, J. Heat Transf., **138**(2), 021504 (2016)
2. S.G. Kandlikar, Int. J. Therm. Sci. **49**(7), 1073-1085 (2010)
3. S.S. Bertsch, E.A. Groll, S.V. Garimella, Nanoscale Microscale Thermophys. Eng. **12**(3), 187-227 (2008)
4. D. F. Sempertegui-Tapia, G. Ribatski, Exp. Therm. Fluid Sc. **89**, 98-109 (2017)
5. R. Mastrullo, A. W. Mauro, J. R. Thome, G. P. Vanoli, L. Viscito, Int. J. Therm. Sci. **106** 1-17, (2016)
6. A. Krishnan, K. Balasubramanian, S. Suresh, Int. J. Heat Mass Transfer **120**, 1341-1357 (2018)
7. S.-M. Kim, I. Mudawar, Int. J. Heat Mass Transf. **77**, 627-652 (2014)
8. W. Qu, I. Mudawar, Int. J. Heat Mass Transf. **46**(15), 2755-2771 (2003)
9. G. Hestroni, A. Mosyak, Z. Segal, G. Ziskind, Int. J. Heat Mass Transf. **45** (6), 3275-3286 (2002)
10. B. Maciejewska, M. Piasecka, Int. J. Heat Mass Transf. **107**, 925–933 (2017)
11. B. Maciejewska, M. Piasecka, Heat Mass Transf. **53**, 1211–1224 (2017
12. K. Strąk, M. Piasecka, B. Maciejewska, Int. J. Heat Mass Transf. **117**, 375–387 (2018)
13. B. Maciejewska, K. Strak, M. Piasecka, Int. J. Numer. Methods Heat Fluid Flow **28**(1), 206–219 (2018)
14. B. Maciejewska, S. Błasiak, M. Piasecka, EPJ Web of Conf. **143**, 02071, 5 pages (2017)
15. B. Maciejewska, P. Łabędzki, A. Piasecki, M. Piasecka, EPJ Web of Conf. **143**, 02070, 5 pages (2017)
16. S. Blasiak, J. E. Takosoglu, P. A. Laski, In: *Eng. Mechanics*, Fuis V (ed), 96–99 (2014)
17. HAYNES, *HAYNES ® 230 ® alloy* (2019)
18. 3M Performance Materials, *3M Fluorinert^{TM} Electronic Liquid FC-72*, 1–4 (2002)

Simulation of the mixture flow over terrain wave

Martin Kyncl[1,*], *Jan* Česenek[1], and *Jaroslav* Pelant[1]

[1]VZLÚ - Czech Aerospace Research Centre, Beranových 130, 199 05 Praha - Letňany, Czech Republic

Abstract. We work with the system of equations describing the non-stationary compressible turbulent multi-component flow in the gravitational field. We assume the mixture of perfect inert gases. The flow over rough terrain is simulated with the use of the finite volume method. The modified Riemann problem is solved at the boundary faces. The roughness of the surface is simulated using the alteration of the specific dissipation at the wall, and with the use of the wall functions. Velocity profile is compared with the experimental results. Own-developed computational code is used.

1 Introduction

Our aim is to numerically simulate the flow of the perfect gas mixture. We consider the viscous compressible gas flow, described by the Reynolds- Averaged Navier-Stokes equations with the k-ω model of turbulence. This system is equipped with the equation of state in more general form, and with the mass conservation of the additional gas specie. The expected result is to construct a method suitable for the estimates of the pollutant concentrations caused by the given source. This contribution follows previous work [1–4].

2 Formulation of the Equations

We consider the conservation laws for viscous compressible turbulent flow of ideal gas with the zero heat sources in a domain $\Omega \in \boldsymbol{R}^N$, and time interval $(0, T)$, with $T > 0$. The system of the Reynolds-Averaged Navier-Stokes equations in 3D has the form

$$\frac{\partial \boldsymbol{w}}{\partial t}+\sum_{s=1}^{3} \frac{\partial \boldsymbol{f}_s(\boldsymbol{w})}{\partial x_s} = \sum_{s=1}^{3} \frac{\partial \boldsymbol{R}_s(\boldsymbol{w}, \nabla \boldsymbol{w})}{\partial x_s}+\boldsymbol{S} \quad \text{in } Q_T = \Omega\times(0, T). \tag{1}$$

Here x_1, x_2, x_3 are the space coordinates, t the time, $\boldsymbol{w} = \boldsymbol{w}(x, t) = (\varrho, \varrho v_1, \varrho v_2, \varrho v_3, E)^T$ is the state vector, $\boldsymbol{f}_s = (\varrho v_s, \varrho v_s v_1 + \delta_{s1} p, \varrho v_s v_2 + \delta_{s2} p, \varrho v_s v_3 + \delta_{s3} p, (E + p)\, v_s)^T$ are the inviscid fluxes, $\boldsymbol{R}_s = (0, \tau_{s1}, \tau_{s2}, \tau_{s3}, \sum_{r=1}^{3} \tau_{sr} v_r + C_k \partial\theta/\partial x_s)^T$ are the viscous fluxes, $\boldsymbol{S}$ are additional sources. $\boldsymbol{v} = (v_1, v_2, v_3)^T$ denotes the velocity vector, ϱ is the density, p the pressure, θ the absolute temperature, $E = \varrho e + \frac{1}{2}\varrho \boldsymbol{v}^2$ the total energy. Further $\tau_{ij} = (\mu+\mu_T) S_{ij} - \delta_{ij}\frac{2}{3}\varrho k$, with $S_{11} = \frac{2}{3}\left(2\frac{\partial v_1}{\partial x_1} - \frac{\partial v_2}{\partial x_2} - \frac{\partial v_3}{\partial x_3}\right), S_{12} = \frac{\partial v_1}{\partial x_2} + \frac{\partial v_2}{\partial x_1}, S_{13} = \frac{\partial v_1}{\partial x_3} + \frac{\partial v_3}{\partial x_1}, S_{21} = S_{12}, S_{22} = \frac{2}{3}\left(-\frac{\partial v_1}{\partial x_1} + 2\frac{\partial v_2}{\partial x_2} - \frac{\partial v_3}{\partial x_3}\right), S_{23} = \frac{\partial v_2}{\partial x_3} + \frac{\partial v_3}{\partial x_2}$, $S_{31} = S_{13}, S_{32} = S_{23}, S_{33} = \frac{2}{3}\left(-\frac{\partial v_1}{\partial x_1} - \frac{\partial v_2}{\partial x_2} + 2\frac{\partial v_3}{\partial x_3}\right)$, where μ is the dynamic viscosity coefficient dependent on temperature, μ_T is the eddy-viscosity coefficient. For the specific internal energy $e = c_v\theta$ we assume the caloric equation of state $e = p/\varrho(\gamma - 1)$, c_v is the specific heat at constant volume, $\gamma > 1$ is called the *Poisson adiabatic constant*. The constant C_k denotes the heat conduction coefficient $C_k = (\frac{\mu}{P_r} + \frac{\mu_T}{P_{r_T}}) c_v \gamma$, and P_r is laminar and P_{r_T} is turbulent Prandtl constant number. In our application of flow in the gravitational field we set the source terms to $\boldsymbol{S} = (0, \varrho g_1, \varrho g_2, \varrho g_3, \varrho \boldsymbol{g} \cdot \boldsymbol{v})$, where $\boldsymbol{g} = (g_1, g_2, g_3)$ is the gravity vector. For the gas mixture with two species we use the Dalton's law for the total mixture pressure

$$p = p_1 + p_2,$$

where p_1 and p_2 are the partial pressures of the first and second component gas. Let ϱ_1 and ϱ_2 be the mass density of these components. Then the total mass density of mixture is

$$\varrho = \varrho_1 + \varrho_2.$$

Temperature θ is same for all gases in the mixture, and the equation of state holds

$$p_i = \varrho_i R_i \theta, \quad R_i = \frac{R_g}{m_i},$$

where $R_g = 8.3144621$ is universal gas constant, and m_i denotes the mollar mass of the ith specie. We can introduce the species mass fractions Y_1, Y_2, with $Y_1 = \frac{\varrho_1}{\varrho}, Y_2 = \frac{\varrho_2}{\varrho}$, it is $Y_1 + Y_2 = 1$. The thermodynamic constants of the mixture satisfy (using the decomposition of the internal specific energy and enthalpy)

$$\varrho c_p = \sum \varrho_i c_{pi}, \quad \varrho c_V = \sum \varrho_i c_{Vi},$$

then the adiabatic constant γ, needed in the solution of (1), can be written as

$$\gamma = \frac{c_p}{c_V} = \frac{\sum Y_i c_{pi}}{\sum Y_i c_{Vi}}.$$

*e-mail: kyncl@vzlu.cz

The system (1) is then extended with the conservation law of the mass for one gas component (specie)

$$\frac{\partial \varrho Y_1}{\partial t}+\frac{\partial \varrho Y_1 v_1}{\partial x_1}+\frac{\partial \varrho Y_1 v_2}{\partial x_2}+\frac{\partial \varrho Y_1 v_3}{\partial x_3}=$$
$$\frac{\partial}{\partial x_1}\left(\sigma_C\mu_T\frac{\partial Y_1}{\partial x_1}\right)+\frac{\partial}{\partial x_2}\left(\sigma_C\mu_T\frac{\partial Y_1}{\partial x_2}\right)+\frac{\partial}{\partial x_3}\left(\sigma_C\mu_T\frac{\partial Y_1}{\partial x_3}\right). \tag{2}$$

Here σ_C is diffusion coefficient. The mass conservation for the second specie is automatically satisfied via the system (1).

Here we assume the system (1),(2) equipped with the two-equation turbulent model $k-\omega$ (Kok), described in [5].

3 Numerical method

For the discretization of the system we proceed as described in [6–8]. We use either explicit or implicit finite volume method (FVM) to solve the systems (1), (2) sequentionally. Other possible discretizations were shown in [9, 10]. The used FVM discretizations were presented in [4], we use the Riemann solver [11] for system (1), and Vijayasundaram numerical flux in (2).

4 Example: 3D diffusion of the passive mixture

Here we present a simple case for the test of the used numerical method. Let us assume the stationary flow with $\boldsymbol{v}=(0,0,0)$, and with constant $\sigma=\frac{\sigma_C\mu_T}{\varrho}=1$. Let us solve the problem (2) with the boundary condition

$$\nabla Y_1\cdot\boldsymbol{n}=0,$$

and initial condition given at time $t_0=0.001\ s$

$$Y_1(x_1,x_2,x_3,t_0)=\frac{1}{8\,(\pi\,\sigma\,t_0)^{3/2}}\exp\left(-\frac{x_1^2+x_2^2+x_3^2}{4\sigma t_0}\right),$$

The analytical solution is known and it can be written as

$$Y_1(x_1,x_2,x_3,t)=\frac{1}{8\,[\pi\,\sigma\,(t+t_0)]^{3/2}}\exp\left(-\frac{x_1^2+x_2^2+x_3^2}{4\sigma(t+t_0)}\right).$$

The figure 1 shows computational results for unit domain, and comparison with analytical solution.

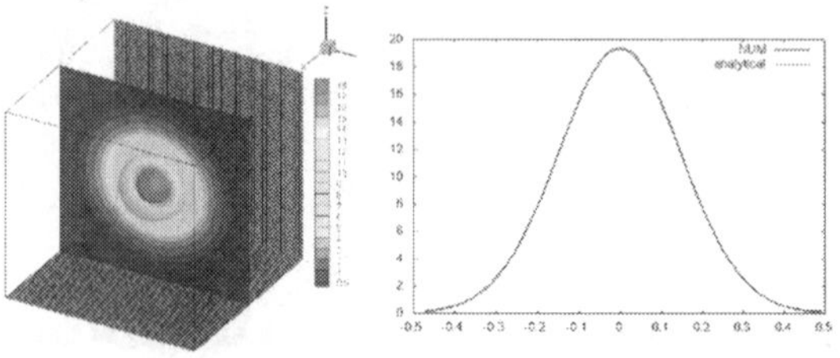

Fig. 1. 3D diffusion: simulation and comparison with the analytical solution at time $t=0.01\ s$, x_1-cut through the center.

5 Example: the flow over terrain wave

We chose to simulate the flow over the rough terrain with the sinus wave geometry, accordingly to experiments shown in [12]. Lenght of the wave was set to 0.800 m, height to 0.114 m. The experimental work was done in the boundary layer wind tunnel (BLWT) in VZLU,a.s. The test section for boundary layer development was 13.6 m long. The boundary layer developed over the entire test section with plastic sheet with high truncated cones 7mm high and a rectangular barrier 140 mm high, which served as turbulence generator. The horizontal velocity profile $U(y)$ in the start of the test section corresponds with its typical logarithmic shape estimated by the law of the wall (see [3, 5])

$$U(y)/u_\tau=\frac{1}{\kappa}ln(\frac{y}{z_S}+1), \tag{3}$$

where z_S is the wall roughness, y denotes vertical coordinate, and u_τ is the friction velocity, $\kappa\approx 0.41$ (Karman's constant). The example of measured velocity profiles is shown in figure 2. The least squares method was used for the estimation of the values for u_τ and z_S.

For the simulations of the compressible gas flow in the atmosphere we assume the change of variables with the vertical coordinate y. We use the following initial conditons for the horizontal velocity component u, temperature θ, pressure p, density ϱ

$$\theta_0=\theta_{00}-0.0065y,$$
$$p_0=p_{00}\left(1-\frac{0.0065}{\theta_{00}}y\right)^{\frac{g}{-0.0065R}},\quad g=-9.81,$$
$$u(y)=min\left(\frac{u_\tau}{\kappa}log(\frac{y}{z_S}+1),u_\infty\right),$$
$$\theta=\theta_0\left(1-\frac{\gamma-1}{2}\frac{u^2}{\gamma R\theta_0}\right),$$
$$p=p_0\left(\frac{\theta}{\theta_0}\right)^{\frac{\gamma}{\gamma-1}},$$
$$\varrho=p/(R\theta).$$

Here θ_0 denotes total temperature, p_0 total pressure, $\theta_{00}=293.15\ K$ is the chosen total temperature and $p_{00}=101325\ Pa$ total pressure at $y=0\ m$, $u_\infty=10\ m\cdot s^{-1}$ is chosen velocity regime, and $R=287.04\ J\cdot kg^{-1}\cdot K^{-1}$ is the gas constant. Here we used the law of wall and the adiabatic approximations for perfect gas.

In order to to get the state variables at each boundary face we solve the local boundary problem with the use the original analysis of exact solution of the Riemann problem. This approach was shown and described in [13, 14] and analyzed also in [8], [15–17]. Using the thorough analysis of the Riemann problem we have shown, that the missing initial condition for the local problem can be partially replaced by the suitable complementary conditions. We suggest such complementary conditions accordingly to

 https://doi.org/10.1051/epjconf/202226901032

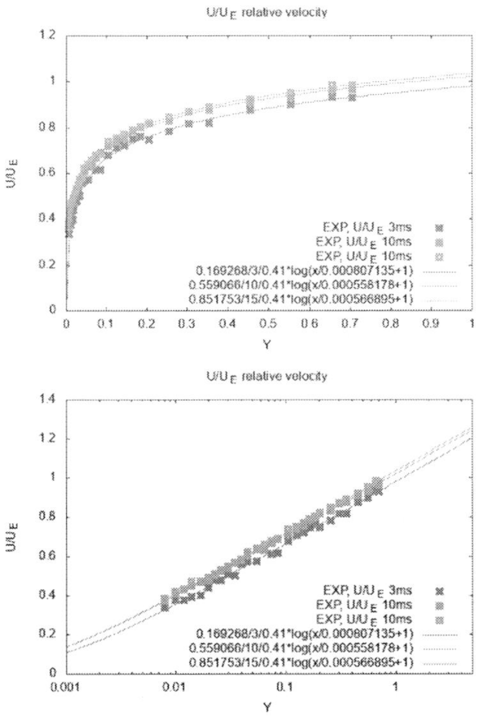

Fig. 2. The flow over rough plate: measured relative velocity profile, and estimate by the logatithmic profile $U(y)/u_\tau = \frac{1}{\kappa}log(\frac{y}{z_S} + 1)$. Constants u_τ, z_S were computed using the least-squares approximation of measured data. Velocity regimes $U_E = 3\ m \cdot s^{-1}, 10\ m \cdot s^{-1}$ and $15\ m \cdot s^{-1}$.

the desired preference. This way it is possible to construct the boundary conditions by the preference of total values, by preference of pressure, velocity, mass flow, temperature. On the contrary to the initial-value Riemann problem, the solution of such modified problems can be written in the closed form for some cases. Moreover, using such construction, the local conservation laws are not violated. Here we used the boundary conditions by preference of known velocity at the outlet, as described in [13, 15]. This boundary condition was modified at the inlet by adding the known (estimated) total temperature, shown above.

For the simulation of the rough wall we use either the modification of the specific turbulent dissipation ω or the so-called **wall functions** for the estimation of values at the first cell (near wall). We estimate the friction velocity u_τ, assuming the law of the wall (see [3, 5]) to be valid at the closest volume, i.e. solve the equation

$$U_P/u_\tau = \frac{1}{\kappa} ln(\frac{y_P}{z_S} + 1),$$

where U_P is the horizontal velocity at the considered cell, and y_P denotes the cell center distance from surface. The wall friction $\tau_w = \mu \frac{\partial U}{\partial y}|_w$ is then estimated as $\tau_w = \varrho_w u_\tau^2$.

The momentum equations are solved with the modified effective viscosity μ_e at the wall: $\mu_e \frac{U_P}{y_P} = \tau_w$. The values for the k, ω at the first cell are set using log layer equations

$$k_P = \frac{u_\tau^2}{\sqrt{\beta^*}}, \quad \omega_P = \frac{u_\tau}{\sqrt{\beta^*} \kappa y_P}.$$

We use further modifications of this wall function shown in [4] in order to avoid the problem with possible zero velocity U_P.

The regime constants for the considered flow above the rough terrain wave were chosen as

$$u_\infty = 10\ m \cdot s^{-1},$$

$$z_S = 0.000558178\ m,$$

$$p_{00} = 101325\ Pa,$$

$$\theta_{00} = 293.15\ K.$$

The friction velocity forming the initial condition was set to $u_\tau = 0.559066\ m \cdot s^{-1}$. The figure 3 shows computed velocity profiles together with the experimentally obtained results.

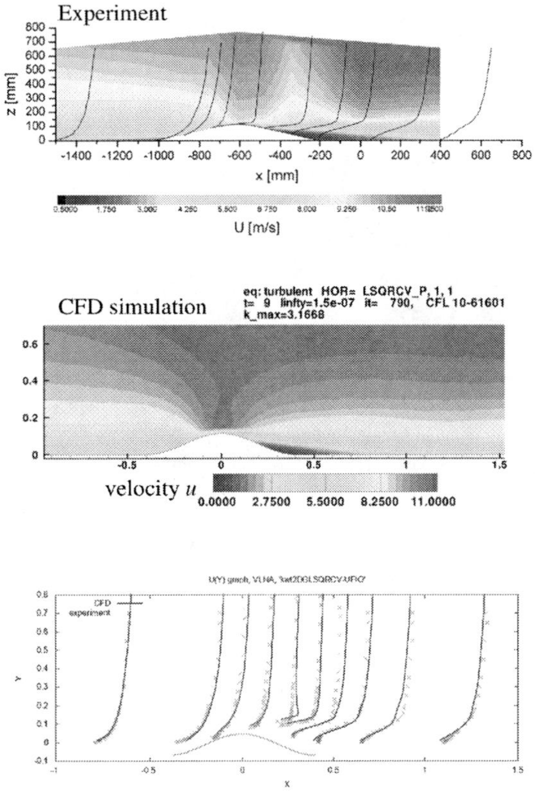

Fig. 3. The flow over the terrain wave: measured horizontal velocity profile at given vertical lines, and numerical simulation. Direct comparison is shown in the bottom picture.

Further we simulated dispersion of the gas mixture. The emmision source was located at the top of the wave, on winward and on leeward side. The source was simulated using marked air with chosen mass fraction $Y_1 = 0.1$ at the chosen elements. The computed results are shown in figure 4.

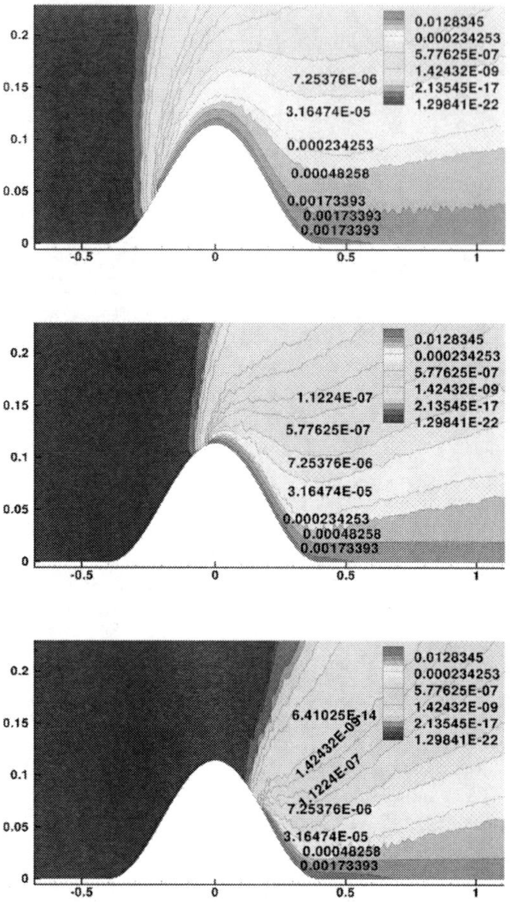

Fig. 4. The flow over the terrain wave: numerical simulation, mass fraction Y_1 of the emmision for various source placement.

Conclusion

This paper is focused on the numerical simulation of the mixture of two inert perfect gases in the gravitational field. The finite volume method is applied for the solution of the system of equations. The modification of the Riemann problem and its solution was used at the boundaries. All codes were implemented into the own-developed software. The numerical examples of the flow above the rough surface simulating terrain wave were presented.

Acknowledgements. The results originated with the support of Ministry of the Interior of the Czech Republic, project VI20172020092 - Simulation of fire and smoke dispersion in critical infrastructure area in case of accident or aircraft attack, project SCENT, Grant MSM 0001066902 of the Ministry of Education of the Czech Republic, and Ministry of Industry and Trade of the Czech Republic for the long-term strategic development of the research organization. The authors acknowledge this support.

References

1. M. Kyncl and J. Pelant, *EPJ WoC*, **114**, 02064, (2016)
2. M. Kyncl and J. Pelant, *ENGINEERING MECHANICS 2016*, **22**, p.367-370, (2016)
3. M. Kyncl and J. Pelant, *EPJ WoC*, **92**, 02044, (2015)
4. M. Kyncl and J. Pelant, *EPJ WoC*, **213**, 02050, (2019)
5. C. J. Kok, *AIAA Journal*, **38**, 7, (2000)
6. M. Feistauer, J. Felcman, and I. Straškraba, *Mathematical and Computational Methods for Compressible Flow*. Oxford University Press, Oxford, (2003)
7. M. Kyncl and J. Pelant, *Implicit method for the 3d RANS equations with the k-ω (Kok) Turbulent Model*. Technical report R5453, (2012)
8. M. Kyncl and J. Pelant, *Implicit method for the 3d Euler equations*. Technical report R5375, VZLÚ, (2012)
9. J. Česenek, *EPJ WoC* **114**, 02012 (2016)
10. J. Česenek, *EPJ WoC* **180**, 02016 (2018)
11. E. F. Toro, *Riemann Solvers and Numerical Methods for Fluid Dynamics*. Springer, Berlin, (1997)
12. P. Michálek and D. Zacho, *EPJ WoC*, **92**, 02051, (2015)
13. M. Kyncl and J. Pelant, *EPJ WoC*, **143**, 02061, (2017)
14. M. Kyncl and J. Pelant, *Proc. ECCOMAS 2016, 5.-10.6.2016*, Greece (2016)
15. M. Kyncl. *Numerical solution of the three-dimensional compressible flow*. Doctoral Thesis, Prague, (2011)
16. M. Kyncl and J. Pelant, *Proc. ECCOMAS 2014, ECFD VI, 20.-25.7.2014*, Spain (2014)
17. M. Kyncl and J. Pelant, *Appl. Mech. and Mat.*, **821**, 70-78, (2016)

EPJ Web of Conferences **269**, 01033 (2022)
EFM 2019 https://doi.org/10.1051/epjconf/202226901033

Intensification of heat dissipation from electrical elements to closed loop heat pipe

Richard Lenhard[1,*], *Katarína* Kaduchová[1], *Peter* Ďurčanský[1], *Milan* Malcho[1], and *Andrej* Kapjor[1]

[1]University of Zilina, Faculty of Mechanical Engineering, Department of Power Engineering, Univerzitna 8215/1, 010 26 Zilina, Slovak republic

Abstract. Article deals with the problem of heat transfer by condenser and evaporator for a closed loop heat pipe for cooling a hermetically sealed electrotechnical box. In an article can by find the design of the heat exchanger for condenser and evaporator for optimal cooling solution, construction and description of experimental state for measuring their performance. Air and heat exchanger temperatures, system pressure, and cooling air flow data were recorded. The measured values are then evaluated and the cooling capacity of the heat exchangers is expressed by means of a sample calculation. The performances are then compared by means of relevant graphs and verified at the end.

1 Cooling of electronic components

Electronic components produce heat as a by-product of their normal operation. When an electric current passes through a semiconductor or passive device, some of the energy is dissipated into the environment in the form of thermal energy.

It is important to keep the room temperature below the maximum allowable operating temperature specified by each electronic component manufacturer. If the maximum operating temperature is exceeded during operation, the service life, performance and reliability of the individual components decreases (Fig. 1).

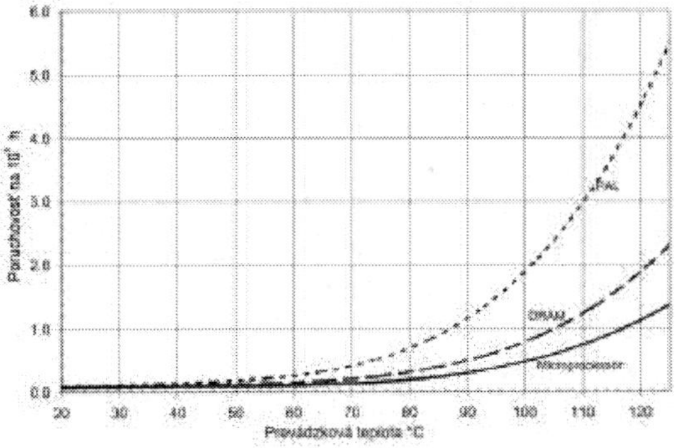

Fig. 1. Failure depending on operating temperature for PAL (Programmable Logic Array), DRAM (Dynamic RAM), and Microprocessors, [1].

Increased heat fluxes from electronic components due to increased power and reduced space requirements require more efficient and reliable cooling to ensure optimum conditions. For most electronic components it is necessary to keep the ambient temperature below 85 °C.

Due to increasing heat fluxes from electronic components, conventional types of cooling have reached their limit values. Conventional air circulation due to cooling capacity, heat transfer rate and dustiness (albeit minimal) gets into the background. The development of cooling systems focuses on cooling with liquid convection without phase change (water coolers), but also with phase change (heat pipe cooling) (Fig. 2) [2].

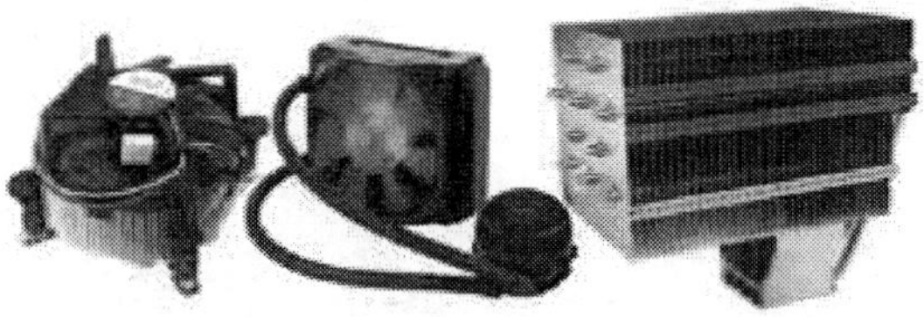

Fig. 2. Types of cooling: air circulation, water cooling, heat pipe cooling.

2 Closed loop heat pipe

The closed-loop heat pipe is used to transfer heat over relatively long distances without additional mechanical devices. The difference between the heat pipe and the closed loop heat pipe is that the fluid flow and the steam flow are conducted in separate conduits. Thus, the elimination of the interaction limit of the heat pipe was achieved.

In our experiment two-phase closed loop heat pipe is used – the working substance is brought to boiling and condensation. The construction also includes an

* Corresponding author: richard.lenhard@fstroj.uniza.sk

expansion tank to eliminate pressure build-up in the tube (Fig. 3).

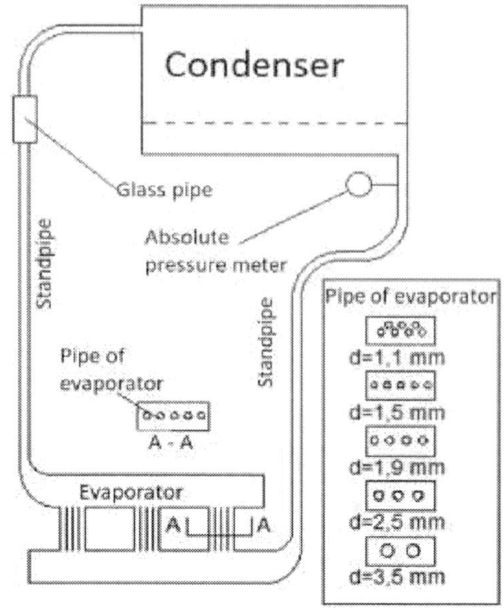

Fig. 3. Two-phase closed loop heat pipe [3].

Closed loop of heat pipe partially filled with working substance – working substance is brought to boiling and condensation. The design does not include an expansion tank to eliminate pressure build-up in the tube [4, 5].

3 Design of condenser and evaporator for closed loop heat pipe

The design of condenser and evaporator is to reduce the material and financial costs of production and achieve a cooling capacity of 1 000 W.

The condenser design consists of registers of copper coils connected to a common manifold (steam) and collector (condensate) (Fig. 4).

Fig. 4. Construction of copper coil registers for condenser.

The condenser heat exchanger consists of ninety-five 5/8" x 0.4 copper pipes on which fifty-five aluminium lamellas with dimensions 750 x 300 mm, thickness 0.25 mm and rib spacing 5.5 mm are placed (Fig. 4).

The design of the evaporator was based on the requirement for the service life of the designed equipment, the correct design and the limited space inside the electrotechnical box. The design must be able to extract the heat dissipated to maintain the components in the electrical enclosure at the optimum operating temperature. Increasing the coefficient can be achieved by changing the parameters, either on the air side or on the boiling side of the liquid. The figure (Fig. 5) show the final design for a loop thermosiphon.

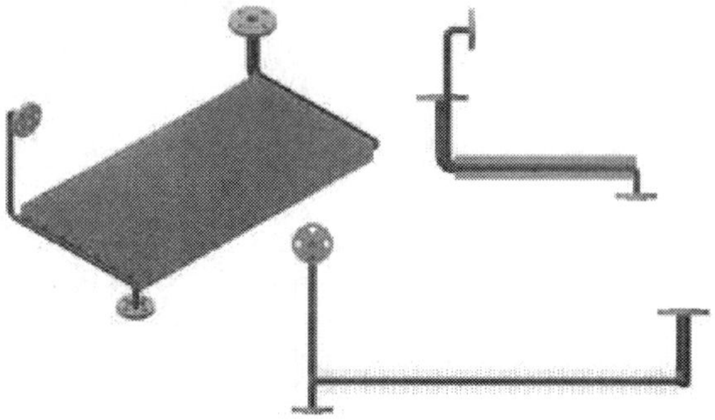

Fig. 5. The structural design of the evaporator.

4 Measurement of cooling capacity

To verify the correctness of the design, measurements were performed on an experimental state with a thermal load of 500 W, 1 000 W and 1 500 W. To achieve the maximum heat loss of 1 500 W in the interior of the electrotechnical box was used 6 resistive bodies (Fig. 6).

The following types were used as working substances: water, Novec, alcohol, acetone with a fill volume of 20 %, 40 %, 60 %, 80 %, and 100 % – the total volume after filling was 1.29 litters.

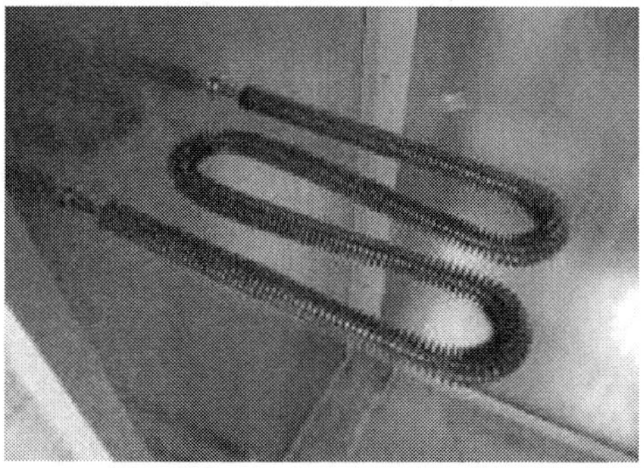

Fig. 6. Simulation heat source – resistive body.

For better heat transfer inside the electrotechnical box from the heat source to the evaporator, six computer fans 120 x 120 mm were installed with an air flow rate of 2 $m.s^{-1}$ for forced convection. The fans were connected in parallel to a separate laboratory power supply. The air flow rate was set by limiting the output current at the source. The air flow rate was measured using an anemometer. Pt100 temperature sensors were installed to record the measured temperatures in the experimental equipment. The entire experimental equipment in laboratory is shown in (Fig. 7 part with condensator a Fig. 8 part with evaporator).

Fig. 7. Experimental equipment in laboratory – part with condensator: P1 – measurement of the pressure in the heat pipe, V1 - V2 – measurement of cooling air flow rate during natural convection, T1 - T4 – measurement of the cooling air temperature after the condensator, T5 - T13 – temperature measurement on condensator, T14 - T16 – measurement of the cooling air temperature in front of the condensator, T17 – temperature measurement on steam pipe, T18 – temperature measurement on condensation piping.

Fig. 8. Experimental equipment in laboratory – part with evaporator: Tv1 – temperature measurement at the top of the box, TT2-4 –temperature measurement on evaporator tubes, Tv2 –temperature measurement in the central part of the cabinet, TT5 – vapor pipe temperature measurement, Tv3 – temperature measurement at the bottom of the cabinet, S1-6 – resistance spirals.

Cooling capacity was calculated using a calorimetric equation based on air velocity measurement and temperature measurement at inlet T11 and outlet T12.

Calorimetric equation:

$$Q = m.\ c_p.\ (T_{11} - T_{12}) \tag{1}$$

$$Q = V.\ \rho.\ c_p.\ (T_{11} - T_{12}) \tag{2}$$

$$Q = \upsilon.\ S.\ \rho.\ c_p.\ (T_{11} - T_{12}) \tag{3}$$

5 Results of cooling capacity

Measurement of the cooling capacity of the condenser in combination with the evaporator, the measured cooling capacities were shown in the graphs according to the percentage of filling and heat load (Fig. 9, Fig. 10, Fig. 11, Fig. 12).

The heat exchanger functionality is obvious from the measured results. At 80 – 100 % loading, the power was sufficient to maintain the maximum temperature requirement of 80 °C, except for acetone at a volume of 100 %. The maximum average velocity in the air flow through the condensator was 0.47 m.s^{-1}.

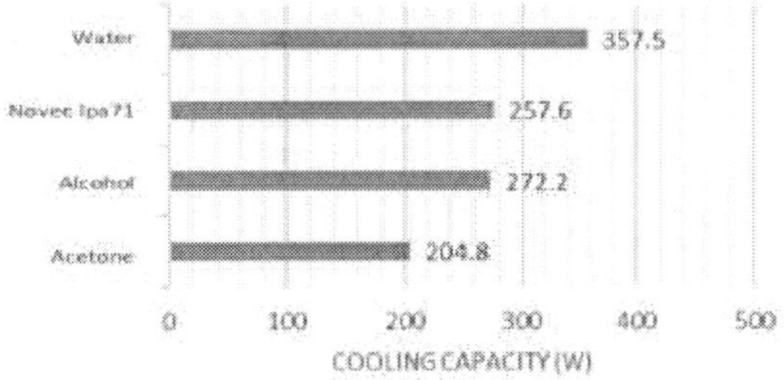

Fig. 9. Comparison of cooling capacity, load 500 W, volume 80 %.

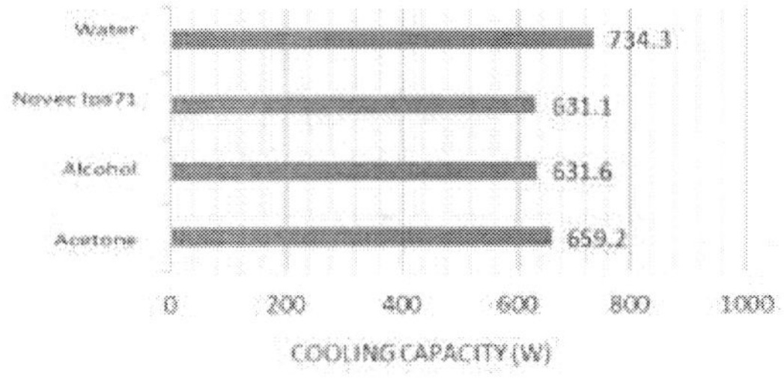

Fig. 10. Comparison of cooling capacity, load 1 000 W, volume 80 %.

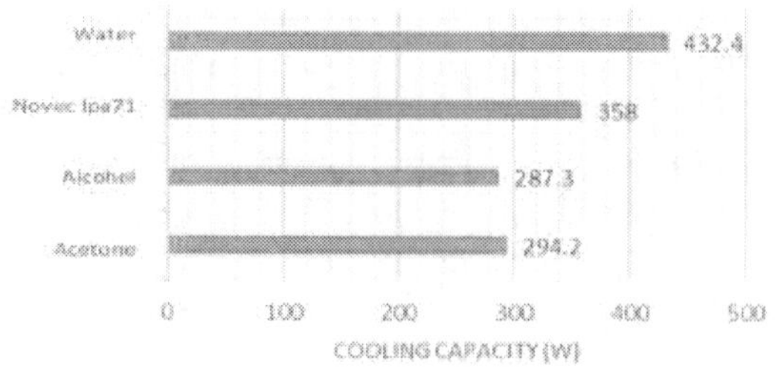

Fig. 11. Comparison of cooling capacity, load 500 W, volume 100 %.

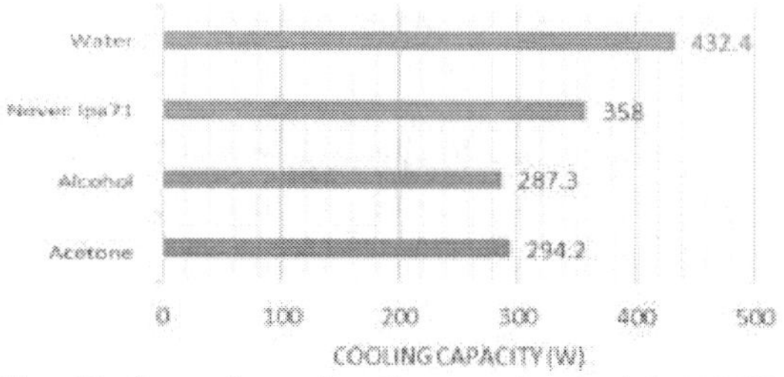

Fig. 12. Comparison of cooling capacity, load 1 000 W, volume 100 %.

5.1 Results of the temperatures in the electrotechnical box

The evaluation of the measured quantities was considered at the inlet of the heat exchanger with a homogeneous temperature field, with the same air velocity in cross-section. However, such conditions cannot be achieved in the measured device. This is also evidenced by the measured temperatures at the exchanger outlet. In Fig. 13 is an exemplary graph of temperature stabilization inside the electrical enclosure for a second evaporator using a natural convection and distilled water condenser as a working medium with a 60 % load and a 500 W load.

The total mean values of the temperatures in the box space after the measurement of individual amounts of fillings are stabilized are given in the tables (Table 1, Table 2).

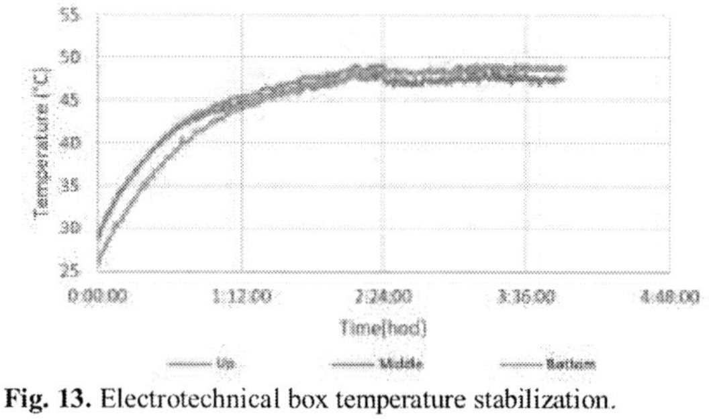

Fig. 13. Electrotechnical box temperature stabilization.

Table 1. Temperature in the electrotechnical box, load 500 W.

	20 %	40 %	60%	80 %	100 %
Acetone	-	-	59 °C	58 °C	63 °C
Alcohol	-	-	48 °C	44 °C	41 °C
Novec IPA71	80 °C	80 °C	64 °C	57 °C	-
Water	75 °C	58 °C	56 °C	45 °C	48 °C

Table 2. Temperature in the electrotechnical box, load 1000 W.

	20 %	40 %	60 %	80 %	100 %
Acetone	-	-	81 °C	76 °C	-
Alcohol	-	-	70 °C	66 °C	61 °C
Novec IPA71	-	-	-	79 °C	79 °C
Water	-	-	69 °C	61 °C	64 °C

6 Conclusion

Cooling of power converters by convection and conduction methods is affected by airborne dust. Gradual dusting of the drive components results in an increase in the internal temperature of the entire system due to insufficient heat flux, causing damage to the power converter. The heat generated by the system is an undesirable factor that shortens the life of individual components.

Experimental measurements have confirmed that the proposed evaporator solution for electrical cabinet cooling and the use of a loop thermosiphon for component cooling is a suitable solution. The Novec IPA71 working medium is recommended for direct cooling, or a different pressure in the loop thermosiphon system should be selected.

This article was created with financial support of the project: KEGA-063ŽU-4/2018 „Depositing hydrocarbon gases into hydrate structures as an alternative energy storage method“. VEGA-1/0738/18 „Optimization of energy inputs for the rapid generation of natural gas and biomethane hydrates for the accumulation of high potential primary energy“.

References

1. R. Remsburge, Thermal design of electronic equimpent. CRC Press LLC (2001)
2. P. H. Chen, S. W. Chang, K. F. Chaing, J. Li, High Power Electronic Component, Recent Patents on Engineering (2008)
3. R. Khodabandeh, B. Palm, Int. J. Therm. Sci., **41** (2002)
4. R. T. Dobson, J. C. Ruppersber, Journal of Energy Southern Africa (2007)
5. P. Nemec, R. Lenhard, *SGEM*, **19** (2019)
6. T. Puchor, Thesis (2019)
7. R. Banovčan, Thesis (2019)
8. T. Brestovic, M. Carnogurska, R. Pyszko, M. Kubik, *AEaNMiFMaE* 19-26 (2012)
9. J. Soukup, F. Klimenda, B. Skočilasová, P. Jirava, *AIP Conference Proceedings* **2000** (2018)

https://doi.org/10.1051/epjconf/202226901034

The flow of micrometre-sized glass fibres in a replica of the human trachea

Frantisek Lizal[1,*], *Milan* Maly[1], *Matous* Cabalka[1], *Elena* Lizalova Sujanska[2], *Arpad* Farkas[1], *Pavel* Starha[1], *Kristyna* Kedajova[1], *Ondrej* Misik[1], *Ondrej* Pech[1], *Jan* Jedelsky[1], and *Miroslav* Jicha[1]

[1]Brno University of Technology, Faculty of Mechanical Engineering, Energy Institute, Technicka 2896/2 Brno, 616 60, the Czech Republic

[2]Department of Pulmonology and Allergology, Hospital Kromeriz, Havlickova 660/69, 767 01, Kromeriz, the Czech Republic

Abstract. Inhaled fibres can potentially cause inflammation of the lung tissue and interstitium which, after long-term exposure, may lead to lung cancer, malignant mesothelioma or pulmonary and pleural fibrosis. For risk reduction and correct setting of occupational hygiene regulations, it is important to be able to precisely calculate the fate of inhaled fibres depending on their physical characteristics and inhalation conditions. As there is a lack of experimental data on the orientation of fibres, a new test rig has been assembled for visualization and recording of flowing fibres in a replica of the human trachea. Fibres prepared from regular glass fibres produced commercially for blown thermal insulation have been processed, dispersed and introduced into the glass tube with dimensions of the trachea. Visualization was performed using a powerful LED light and a high-speed camera. Angles of the fibres have been evaluated for six different flowrates and the dependence of the angles on the flow Reynolds number was searched for. The angles of fibres agreed with expected values, i.e. only vertically and horizontally oriented fibres were recorded. However, the number of vertically and horizontally oriented fibres did not seem to be correlated with the flow Reynolds number.

1 Introduction

Inhaled fibres are recognized as pathogens that can potentially cause inflammation of the lung tissue and interstitium[a]. Pathogenicity of different fibres is not equal and there are many questions about the effects of man-made fibres and nanofibres. Generally speaking, the inhalation of fibres can lead to numerous adverse responses of the body, and the specific reaction depends on duration and intensity of the exposition, and character of fibres (their geometry and biopersistence[b]). The evidence reported e.g. in [1] shows that toxicity depends strongly on dimensions of fibres and less on the chemical composition of fibres.

Inhaled fibres tend to avoid the most efficient mechanism of lung clearance, i.e. slow upward movement of deposited particles together with mucus driven by the synchronised motion of cilia, which transports the fibres to the throat and the digestive system [2]. Some of the inhaled fibres penetrate and deposit beyond the ciliated airways, which, after the long-term exposure may lead to pulmonary and pleural fibrosis[c], i.e. asbestosis, pleural plaques[d], bronchogenic carcinoma[e] or malignant mesothelioma[f].

[a] An interstitial space within a tissue or organ. Here specifically meant the tissue between the pulmonary alveoli and the bloodstream.

[b] The ability to remain inside an organism, resistance to being forced out or digested.

[c] Pulmonary or pleural fibrosis is a diffuse accumulation of fibrous/scar tissue in the interstitium of the lung or in the pleura. Pulmonary fibrosis aggravates the diffusion of gases from the blood to the airspaces and restricts lung expansion. Pleural fibrosis may lead to pleural effusion - an accumulation of fluid in the pleural cavity.

[d] Pleural plaques are raised hard fibrous lesions (any damage or abnormal change in the tissue of an organism, usually caused by disease or trauma) that arise on the parietal pleura that are smooth and composed of a basketwork of pure collagen; they are almost acellular and do not seem to cause any adverse effects on the lungs.

[e] Bronchogenic carcinoma is a malignant cancer of the cells lining the bronchi or airways and is the same as the lung cancer that is common in smokers.

* Corresponding author: lizal@fme.vutbr.cz

Fibres can provoke an inflammatory reaction, but they can also directly start certain chemical processes. Acute or chronic inflammation is associated with oxidative stress and production and release of mediators[g]. [3] Fibres stimulate proliferation (rapid reproduction) of epithelial or mesothelial cells, and simultaneously during the inflammation, some chemical mediators attract fibroblasts[h], and it is deposited in the lung parenchyma in the area of the offending fibre. Fibres and mediators of inflammatory reaction can also lead to deletion of the gene coding or damage DNA (genotoxic effect) of these cells. These pathobiological processes can culminate in fibroplasia[i] and neoplasia[j].

The reason why the fibres manage to pass to the lower parts of the airways grounds in the fact that their aerodynamic diameter d_{ae} can be very low although their length may reach 30 or 50 μm. The d_{ae} is defined as a diameter of an equivalent sphere with the density of water (1000 kg/m^3) which has the same settling velocity as the particle in question. The equation that can be used for calculation of the d_{ae} (for cylindrical fibre) was derived by Griffiths and Vaughan [4]:

$$d_{ae} = d_f \sqrt{\frac{\rho_f \beta_a}{\rho_0 \chi}} \left(\frac{3}{2}\right)^{\frac{1}{3}}, \tag{1}$$

where d_f is the physical diameter of the fibre, ρ_f is the fibre density, ρ_0 is the density of water (standard density), β_a is the aspect ratio, i.e. ratio of length to the diameter of the fibre, and χ is the dynamic shape factor, which has been defined as a ratio of the drag force on a non-spherical particle (fibre in our case) and the drag force of the volume equivalent sphere. Obviously, the fibre diameter influences the aerodynamic diameter more significantly than its length.

A troublesome situation is induced by the dynamic shape factor. Its value depends on the direction of the fibre motion with respect to the major axis of the fibre. Fuchs [5] has presented two equations, for parallel and perpendicular motion, respectively:

$$\chi_{\parallel} = \frac{\frac{4\left(\beta_a^2 - 1\right)}{3}}{\left\{\beta_a^{1/3} \frac{2\beta_a^2 - 1}{\sqrt{\beta_a^2 - 1}} \ln\left(\beta_a + \sqrt{\beta_a^2 - 1}\right) - \beta_a^{4/3}\right\}}, \tag{2}$$

$$\chi_{\perp} = \frac{\frac{8\left(\beta_a^2 - 1\right)}{3}}{\left\{\beta_a^{1/3} \frac{2\beta_a^2 - 3}{\sqrt{\beta_a^2 - 1}} \ln\left(\beta_a + \sqrt{\beta_a^2 - 1}\right) + \beta_a^{4/3}\right\}}. \tag{3}$$

In the cases, where the random orientation of fibres can be expected, following formula for the resulting dynamic shape factor can be used:

$$\frac{1}{\chi_{ran}} = \frac{1}{3}\left(\frac{1}{\chi_{\parallel}} - \frac{1}{\chi_{\perp}}\right). \tag{4}$$

However, several papers (e.g. [6]) suggest that fibres tend to align with the flow in human lungs. Hence, parallel orientation should be preferential in human airways. On the other hand, Kulkarni [7] suggests, that the perpendicular orientation is preferential during gravitational settling and acceleration for particle Reynolds numbers $0.01 < Re_p < 100$ and that fibre motion is often treated as a combination of these two cases. In short, there is no general rule that would predict the number of fibres with parallel and perpendicular orientation as a simple function of flow and fibre characteristics. It should be noted, that this information is missing in models for prediction of deposition sites of inhaled fibres [8]. To improve the accuracy of such models, it is necessary to supplement them with precise data on the orientation of fibres.

The main aim of this work is to provide the missing experimental data on the preferential orientation of fibres in human airways. A new experimental rig has been assembled to allow visualization of the flow of micrometre-sized glass fibres originating from regular commercially produced glass wool which is applied routinely for blown thermal insulation of buildings. The test rig and the first results of the visualization performed on a replica of the human trachea are described in this contribution.

2 Materials and Methods

The glass wool Supafil® Loft (Knauf Insulation GmbH, Simbach am Inn, Germany) was used as a source of fibres. The wool has been disintegrated in a mechanical press in order to produce separate fibres. The diameter of

[f] Mesothelioma is an unusual tumour arising from the mesothelium lining the pleural space that is slow-growing and that does not metastasise and so is well-advanced before diagnosis leading subsequently to a very poor prognosis [2].

[g] Such as lysosomal enzymes, intermediates of arachidonic acid production, glutathione S-transferases, proteases, cytokines, growth factors, and reactive oxygen species (ROS - hydrogen peroxide, superoxide anion, hydroxyl radical) from pulmonary macrophages, neutrophils, and other inflammatory cells. ROS can be produced also in cell-free systems - as a direct chemical reaction between fibre surface and extracellular fluids

[h] A cell in connective tissue which produces collagen and other fibres.

[i] The formation of fibrous tissue, as normally occurs in the healing of wounds.

[j] An abnormal growth of tissue. Neoplasms may be benign (not cancer), or malignant (cancer).

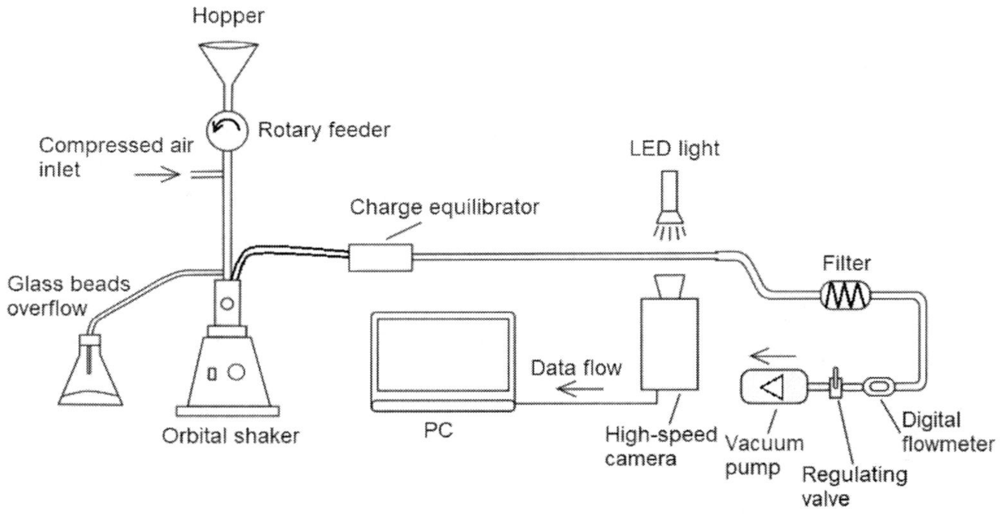

Fig. 1. A scheme of the experimental rig for visualization of flow of glass fibres in a replica of the human trachea.

fibres was 3.8±1.4 μm and the length was 34.1±19.0 μm. Then they were mixed with glass beads (Ballotini impact glass beads, Potters Industries Inc.) and sieved in order to deagglomerate the fibres and produce a homogenous mixture suitable for dispersing the fibres. The room air temperature during experiments was 20 to 21°C, hence the kinematic viscosity used for calculation of Reynolds number was 1.5×10^{-5} m^2/s, and density was 1.21 kg/m^3. The density of glass is approximately 2500 kg/m^3.

2.1 The experimental rig

The experimental rig (see Fig. 1) consisted of a reservoir of the mixture of glass beads and fibres (a hopper), a rotary feeder that provided a continuous supply of the mixture to a fluidized-bed-type disperser created from an orbital shaker Wisd NM-10 (Witeg Labortechnik GmbH, Wertheim, Germany) with an in-house extension. Fibres were charge equilibrated by ^{85}Kr neutralizer NEKR-10 (Eckert & Ziegler CESIO, Prague, the Czech Republic). The fibres than flowed through 1.5 m long glass tube to ensure the developed flow and were visualized by a high-speed camera. The flow was induced by a vacuum pump (Busch PA0008C, Busch Vakuum s.r.o., Brno, the Czech Republic) and measured by a digital flowmeter TSI 4040 (TSI, Shoreview, Minnesota, USA).

Table 1. An overview of the camera and light settings for all measured flowrates

Flow rate [l/min]	Image resolution (w×h) [px]	Frames per second [1/s]	Light pulse duration [ns]
2 l/min	1024×1024	5000	800
6 l/min	1024×1024	20000	800
12 l/min	1024×1024	20000	800
24 l/min	512×1024	35000	600
36 l/min	512×712	50000	500
48 l/min	512×512	60000	400

A high-speed camera Photron SA-Z was used to inspect the fibre flow inside a straight glass tube with an internal diameter of 15 mm. The long-range microscope lens 12X Zoom (NAVITAR, New York, USA) provided images with spatial a resolution of 1.5 μm per pixel. The flow was illuminated by a pulse LED light model HPLS-36DD18B (Lightspeed Technologies, USA). Since a wide range of the flow rates was used, the camera resolution and frame rate were changed in accordance with Table 1. Similarly, the light pulse duration varied in range from 400-800 ns.

Two cases were measured – at first a simple straight tube, then, the same tube was extended by a bifurcation (in the dimensions of the semi-realistic replica of the first bifurcation reported in [9]) and measured immediately upstream of the bifurcation. All measurements were performed in the axis of the tube, 20 mm upstream of the end of the straight tube.

The fibres appearing in a horizontal position in the images are parallel to the flow, the fibres recorded as vertical are perpendicular to the flow streamlines. Only fibres with either horizontal or vertical orientation were observed (in other words, there were no fibres with random orientation different from simply horizontal or vertical).

2.2 Statistical tests

The random variable X describing the quantity of horizontally oriented fibres is expected to have alternative (i.e. Bernoulli) distribution, or binomial distribution $Bi(1,p)$, respectively, where parameter p is the probability of horizontal fibre orientation. Hypothesis H_0: $p \geq 0.5$, therefore horizontally to vertically oriented ratio is higher than 1:1, is tested on significance level $\alpha = 0.05$. The following testing statistic is used.

$$T = \frac{\frac{X}{n} - p}{\sqrt{p(1-p)}} \sqrt{n}, \quad (5)$$

T is the value of the testing statistic, *X* is the measured quantity of horizontally oriented fibres, *p* is parameter from the given hypothesis and *n* is total amount of measured fibres. For the given significance level α, a testing interval is $\overline{W_\alpha} = < -1.645, \infty)$ and if $T \in \overline{W_\alpha}$, the null hypothesis is not rejected.

3 Results and discussion

The following tables present data and results of the statistical analysis for measurements in a simple straight tube (Table 2) and in the tube upstream of a bifurcation (Table 3). The columns of the table contain flow rates, flow and particle Reynolds number, numbers of recorded horizontally and vertically oriented fibres, the total number of fibres, and the value of the testing statistics *T*, respectively.

Table 2. Results of the measurement and statistical analysis in the simple straight tube.

Flow rate	Re_f	Re_p	Horizontal	Vertical	Total	T
2 l/min	189	0.08	95	74	169	1.615
6 l/min	566	0.23	138	116	254	1.380
12 l/min	1132	0.45	226	194	420	1.561
24 l/min	2264	0.91	392	352	743	1.504
36 l/min	3395	1.36	146	116	262	1.853
48 l/min	4527	1.81	216	179	395	1.861

The column with values of testing statistics shows, that the null hypothesis (that there is more than 50% of horizontally oriented fibres) for any of the given flow rates in the simple straight tube can't be rejected on the significance level α.

Table 3. Results of the measurement and statistical analysis in the straight tube equipped with the bifurcation (the measurement was performed upstream of the bifurcation).

Flow rate	Re_f	Re_p	Horizontal	Vertical	Total	T
6 l/min	566	0.23	153	119	272	2.062
24 l/min	2264	0.91	182	142	324	2.222
36 l/min	3395	1.36	155	115	270	2.434

Consequently, the ratio of the number of horizontally and vertically oriented fibres (H/V ratio) was calculated for each flowrate and plotted against the flow Reynolds number (see Fig. 2). Obviously, there was no significant influence of the flow Reynolds number on the orientation of the fibres. The plot of H/V ratio as a function of particle Reynolds number gives almost identical results, as only the characteristic dimension differs (diameter of the tube switches to df).

It must be noted, that in the current setup only planar information can be acquired, i.e. seeing the fibres oriented along the flow or perpendicular to it means that the fibres do not rotate around the axis perpendicular to the visualized plane, however they may still rotate around the two in-plane axes (around an axis parallel to the flow when the fibre is seen perpendicular to the flow and around an axis perpendicular to the flow when seen aligned with the flow). This potential rotation cannot be detected.

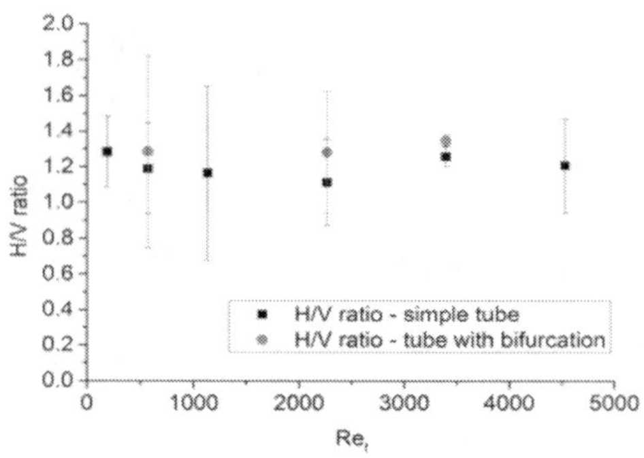

Figure 2. The ratio of horizontally and vertically oriented fibres as a function of the flow Reynolds number. The error bars show standard deviations.

4 Conclusions

The performed visualization proved that only fibres with orientation parallel or perpendicular to the streamlines appear in the axis of the simulated simplified trachea with the developed flow. Although the flow regimes ranged from laminar to turbulent, no significant dependence of the ratio of these two orientations on the Reynolds number was observed.

The null hypotheses that there is more than 50% of horizontally oriented fibres was not rejected in any of the measured cases, which suggests that the fibres tend to align with the flow which facilitates their penetration into the deep regions of the lungs. Further experiments will be performed to study the effects of harmonic breathing patterns on the orientation of the fibres.

Acknowledgement: This work was supported by the Czech Science Foundation under the grant GA18-25618S, and by the project FSI-S-17-4444.

References

1 K. Donaldson, F. A. Murphy, R. Duffin, and C. A. Poland, Particle and fibre toxicology 7 (2010).
2 K. Donaldson, C. A. Poland, F. A. Murphy, M. MacFarlane, T. Chernova, and A. Schinwald, Advanced Drug Delivery Reviews 65 (15), 2078 (2013).
3 O. Y. Osinubi, M. Gochfeld, and H. M. Kipen, Environ Health Perspect 108 Suppl 4, 665 (2000).
4 W. D. Griffiths and N. P. Vaughan, Journal of Aerosol Science 17 (1), 53 (1986).
5 N. A. Fuks, The mechanics of aerosols, Rev. and enl. ed. (Macmillan, New York,, 1964).
6 B. T. Mossman, M. Lippmann, T. W. Hesterberg, K. T. Kelsey, A. Barchowsky, and J. C. Bonner, J Toxicol Env Heal B 14 (1-4), 76 (2011).
7 P. Kulkarni, P. A. Baron, and K. Willeke, Aerosol measurement : principles, techniques, and applications, 3rd ed. (Wiley, Hoboken, N.J., 2011).
8 Á. Farkas, F. Lizal, J. Elcner, J. Jedelsky, and M. Jicha, Journal of Aerosol Science 136, 1 (2019).
9 F. Lizal, J. Elcner, P. K. Hopke, J. Jedelsky, and M. Jicha, Proceedings of the Institution of Mechanical Engineers Part H-Journal of Engineering in Medicine 226 (H3), 197 (2012).

https://doi.org/10.1051/epjconf/202226901035

Inducer with Variable Pitch

Patrik MARCALÍK[1*], *Lukáš ZAVADIL*[2], *Milada KOZUBKOVÁ*[1] *and Jana JABLONSKÁ*[1]

[1]Vysoká škola báňská – Technická univerzita Ostrava, 17. listopadu 2172/15, 708 00 Ostrava-Poruba, Czech Republic
[2]SIGMA Výzkumný a vývojový ústav, s.r.o., ul. Jana Sigmunda č. 313, Lutín, Czech Republic

Abstract. This paper deals with CFD incorporation in a process of designing inducer with variable pitch. The introductory part is devoted to the theory of centrifugal pumps, cavitation and inducer itself followed by a description of the mathematical models of turbulence, multiphase fluid and cavitation. The main part of this paper consists of CFD evaluation, first, of inducer itself, later in application with centrifugal pump with regards to cavitation properties with and without the aplication of inducer, for interest, the comparison with different type of inducer, with constatnt pitch, is also included. At the end of this paper the advantages and some disadvantages are given for CFD based design of the hydrodynamic machines.

1 Introduction

At present, CFD software are inseparable tool for most fluid engineers, especially for those that deal with design of various hydrodynamic machines, such as pumps, for example [1], as will become quite evident by the end of this paper.
The importance of correct estimation of operation parameters of the resulting proposal cannot be stressed enough, which is especially true at instances where cavitation occurs. As for cavitation occurring during flow of water, which is what this paper deals with, it is phenomenon well documented in literature, both theoretically [4 to 6] and in connection to technical practice [7 to 10].

CFD methods can be used for the optimization of the machine geometry itself [11], whenever possible, or, as if shown in this paper, it can be used to design other devices which can solve the cavitation problem.

2 Centrifugal Pumps

Centrifugal pumps are a very important tool of the technical world, be it specifically in technical practice, or in everyday life, which illustrates the worldwide centrifugal market with a volume of tens of billions of dollars a year [2] with applications across all technical sectors.

They are designed with flow rates ranging from values of thousandths to several tens of $m^3 \cdot s^{-1}$ for conveying heights up to approximately 5000 m at speeds ranging from several hundreds to several tens of thousands rpm [2].

Fig. 1. Radial Impeller [3]

Centrifugal pumps are hydrodynamic machines used for transportation of liquids by raising a volume flow to a specified pressure level. The impeller transfers the mechanical energy into kinetic energy necessary to transport the fluid and accelerates it. This causes the static pressure to increase. The fluid exiting the impeller is decelerated in the volute and the following diffuser in order to utilize the greatest possible part of the kinetic, which is transformed into hydraulic energy, for increasing the static pressure.

They come at variety of types according to the shape of the impeller. As such they can be divided into 3 general types: radial, which would be used later for CFD purposes, see Fig. 1, semi-axial and axial. By the number of impellers themselves we can further divide them into single/multi-stage pumps, see Fig. 2.

* Corresponding author: patrik.marcalik@vsb.cz

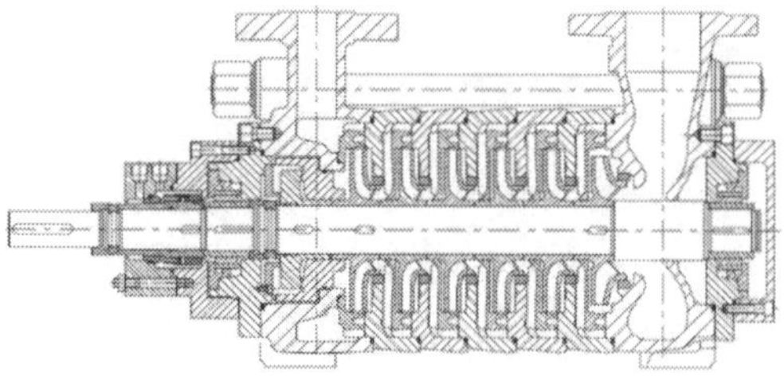

Fig. 2. Multistage segmental pump [2]

Another possible criterion is number of entries to the impeller: single/ double entry, see Fig. 3.

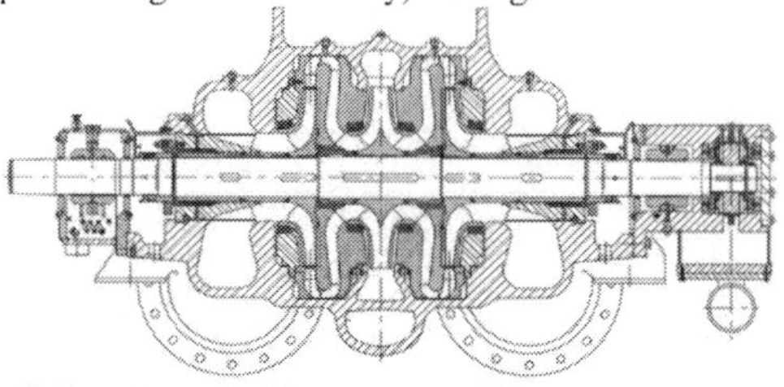

Fig. 3. Two-stage, double-entry pump [2]

For further information on centrifugal pumps and other possible divisions see [2].

3 Cavitation

In this section the problem of cavitation will be outlined. First, its physical nature will be discussed, and then solution via inducer will be presented.

3.1 Cavitation

By cavitation local multiphase flow is understood. Locally produced cavity filled with water vapor (gas phase) is transported from areas, where the static pressure drops below the saturation pressure due to the excessive velocity of the flowing liquid, downstream to regions, where the static pressure is again greater than the saturation pressure at which point cavity implodes, i.e. the vapour, that fills cavity, suddenly condenses back from the gas phase into the liquid phase, the surrounding liquid suddenly fills the newly formed space in the cavity. This phenomenon is accompanied by the existence of a significant pressure wave, possibly 100 MPa [2].

For hydrodynamic pumps typical cavitation forming area is found at the suction of the pump impeller where the liquid takes up high velocities around the wall of the suction side of the blade, cavitation cloud is then transported through the impeller canal until implosion occurs near the blade wall resulting in cavitation erosion, if left unnoticed resulting damage to the impeller might become fatal, such as in fig. 4.

Fig. 4. Cavitation erosion [3]

Other detectable outcomes of cavitation are increased noise and vibration levels. Noise levels are particully worth mentioning as they may raise up to 90 dB and more depending on the severity of cavitation [11].

3.2 NPSH

As was mentioned earlier, the criterion for the occurrence and extend of cavitation is existence of pressure within the flow that drops below the saturated pressure. In technical applications however such pressure cannot be measured by any simple means. To that end simple quantifying magnitude was introduced – *NPSH* [m].

$$NPSH = \frac{\Delta y}{g} = \frac{p_{sHG} - p_s}{\rho g} + \frac{c_{sHG}^2}{2g} \qquad (1)$$

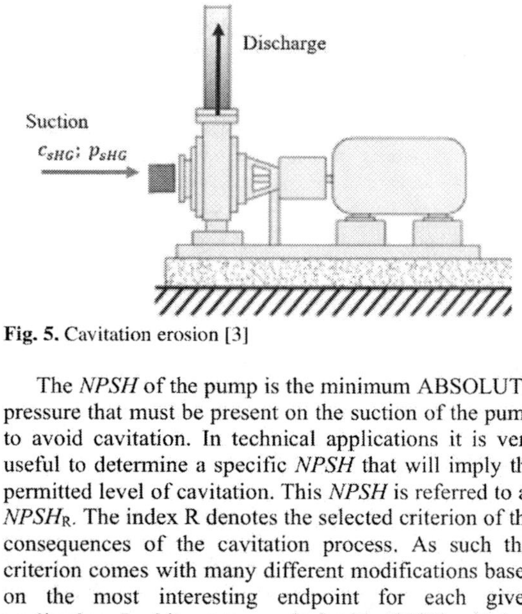

Fig. 5. Cavitation erosion [3]

The *NPSH* of the pump is the minimum ABSOLUTE pressure that must be present on the suction of the pump to avoid cavitation. In technical applications it is very useful to determine a specific *NPSH* that will imply the permitted level of cavitation. This *NPSH* is referred to as $NPSH_R$. The index R denotes the selected criterion of the consequences of the cavitation process. As such this criterion comes with many different modifications based on the most interesting endpoint for each given application. In this paper we deal with $NPSH_3$ which is criterion of maximum allowable reduction of the pump head by 3%, which in industrial application settings is usually measurable.

4 Mathematical models of multiphase flow with cavitation

In this section we will deal with basic description of the mathematical models used in later CFD calculation, first basic flow models modified by turbulent flow will be discussed, then cavitation model will be introduced.

4.1 Turbulent flow model

The basic mathematical model consists of a spatial multiphase flow model, which uses equations expressing the basic physical laws of mass and momentum conservation defined for the mixture and equations of volume fractions of individual phases [3].

In centrifugal pump applications, turbulent flow always occurs, that is flow characterized by high Reynolds numbers. Because of that all the equations must be modified by statistical method of time averaging [3].

For the CFD calculation based on the experience when dealing with CFD evaluation of centrifugal pumps the k-ω SST transient turbulent model was selected from the many available models in today's commercial software. Because of the turbulent model all evaluations must be done with time averaged values.

4.2 Multiphase model

As was said earlier cavitation is treated as a multi-phase flow, assuming in our case a local two-phase flow of water (H_2O) as liquid phase and water vapor as compressible gas phase. Because of this the cavitation model must therefore contain equations for the transport of the gas phase and phase change between vapour and water [3].

4.3 Cavitation Model

The Zwart - Gerber - Belamri cavitation model was selected again based on previous experience. This model assumes a constant bubble radius. With this assumption the relationship between the total rate of phase change in the unit of volume and the bubble density number was obtained. Through this relationship and subsequent modification, the equations for growth and implosion of the vapor bubble were obtained [3].

5 Inducer

By the term inducer we understand axially positioned impeller upstream of the main impeller, see Fig. 6.

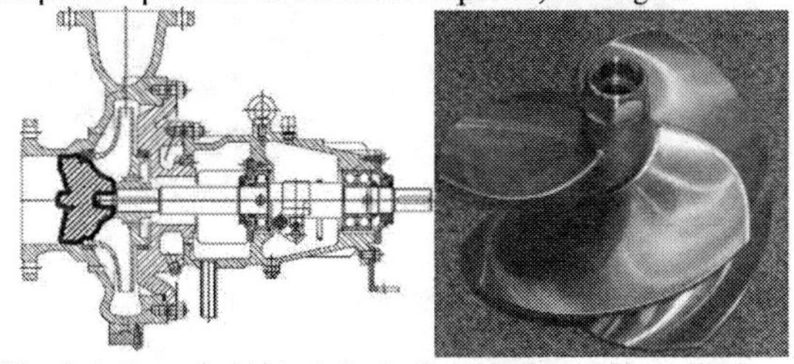

Fig. 6. Inducer (bold black line) placed in front of impeller in centrifugal pump [2] and photo of inducer

Its main function is to lower the required $NPSH_R$ up to half the value without its application by way of increasement of the static pressure in front of the pump impeller, thereby reducing the formation of vapor bubbles on the impeller blades, proportionally up to complete suppression. As a result, the pump can be operated at higher speeds, possibly with a lower suction pressure.

The profile of the blades can be designed with a constant pitch as a helical surface in an easier variant.

+ the main advantage lies in its advantageous simple manufacturing
- this approach has its limitations in achieving the necessary static pressure.

On the other hand, the design using variable pitch consist of blade angle increasement in its cylindrical cross sections in the flow direction from the inlet to the outlet in contrast to the first variant, where the blade angle over the entire length of the cross sections. This change essentially reverses the pros and cons with comparison to the first variant but is the only way to sufficiently increase static pressure before main impeller provided that reasonable dimensions are maintained.

6 Design methodology of inducer with variable pitch

In this section the methodology of inducer design process will be introduced.

6.1 Theoretical design methodology of inducer blade geometry

According to the dimensions of the pump, in the first step the basic geometrical dimensions and parameters such as inlet and outlet diameters, number of blades etc. were chosen, after which other parameters were calculated, based on [2], such as blade angle progress etc., for more information see [3], this comprises a conventional approach to the design of the inducer.

6.2 CFD evaluation and design refinement of inducer blade geometry

With thus obtained parameters [3], CFD software was approached, and 3D geometry was generated based on those parameters, after which fine enough computational mesh was generated considering the need for fine mesh for cavitation calculations (the need to capture the movement of individual bubbles).

At this point CFD analysis and subsequent evaluation itself could be realized at created computational domain, see Fig. 7, with boundary conditions (BC onward) according to the Table 1.

Table 1. BC for inducer alone CFD calculation.

Number of Phases	1 - water (H_2O)
Inlet BC	total pressure p_{total}
Outlet BC	mass flow Q_m

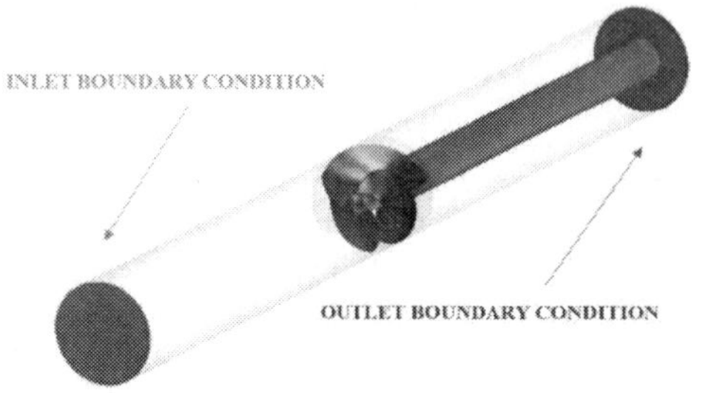

Fig. 7. Computational domain of inducer alone CFD calculation [3]

At this point the calculation was set without cavitation, as this was not yet important. Only simple single-phase, turbulent model was set with the required parameter of static head. The calculation was set as time-dependent with a time step dt = $1.11857 \cdot 10^{-4}$ s. To determine the time step, the multiplier of the inducer blades was used. After several changes of design, the optimal variant was found. The end result of a static height was over 10% better in comparison to the conventional approach [3]

7 CFD cavitation evaluations

In this section methodology of CFD cavitation evaluation will be introduced. First the computational domain will be shown with its settings after which evaluation of computed data will be presented.

7.1 Computational domain

Based on the previously obtained computational mesh of inducer and the pump itself [3] the whole computation domain could be created, see Fig. 8.

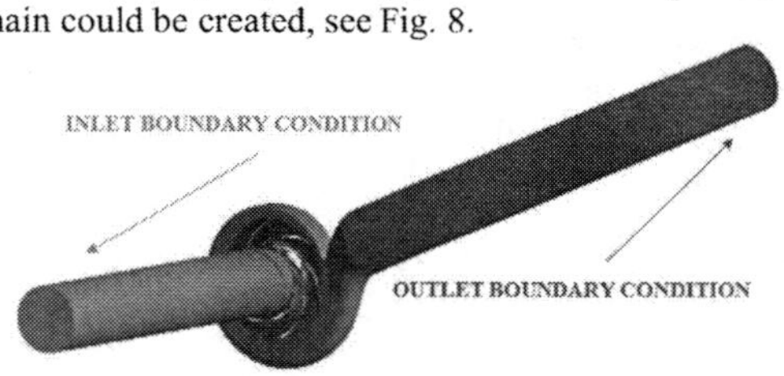

Fig. 8. Computational domain of inducer with pump [3]

Calculation was set as multiphase (two phases – liquid phase as water (H_2O) and gas phase as water vapor) with cavitation model enabled, see 3, with BC according to the Table.2.

Table 2. BC for inducer with pump CFD calculation.

Number of Phases	1 - water (H_2O), 2 - water vapor
Inlet BC	total pressure p_{total}
Outlet BC	mass flow Q_m

The calculation was made for the pump optimal mass flow and for 2 subsequent suboptimal and above optimal mass flows with relatively high value of static pressure on inlet BC to ensure non-occurrence of cavitation with gradual decreasement right up to the values where $NPSH_3$ was not met. The deciding factor was static head.

7.2 Evaluation of chosen CFD results

For the purpose of evaluation of pressure distribution and vapor volume fraction only one variant (optimal mass flow at Outlet BC with Inlet BC pressure just before arriving at the critical value for the $NPSH_3$) out of the many variants of calculation was chosen.

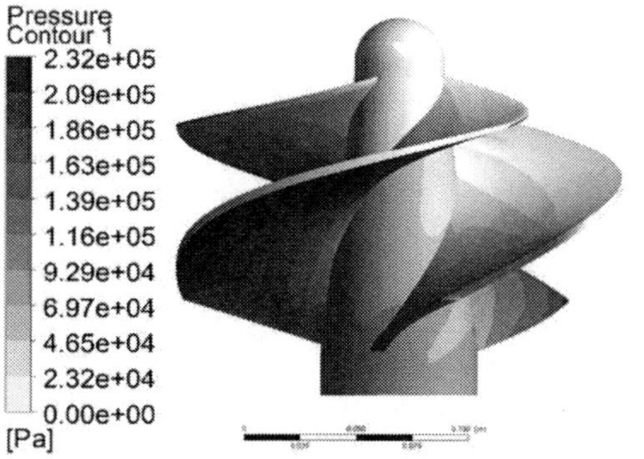

Fig. 9. Absolute static pressure at inducer domain [3]

As can be seen at Fig. 9, pressure distribution on the inducer blades clearly indicates presence of cavitation as the saturated pressure has been reached, which can be seen at Fig. 10, where the vapor volume fraction (white) of value 0.1 is shown.

Fig. 10. Vapor volume fraction of value 0.1[3]

Significant volume of vapor at the domain of inducer can be seen. In the domain of the main impeller the volume of the vapor is not as significant. This variant nears $NPSH_3$, so this is very good result which indicates that the inducer fulfils its function, which is more elaborated in next subchapter.

For interest, on the Fig. 11, comparison of the two different types of inducers is drawn:

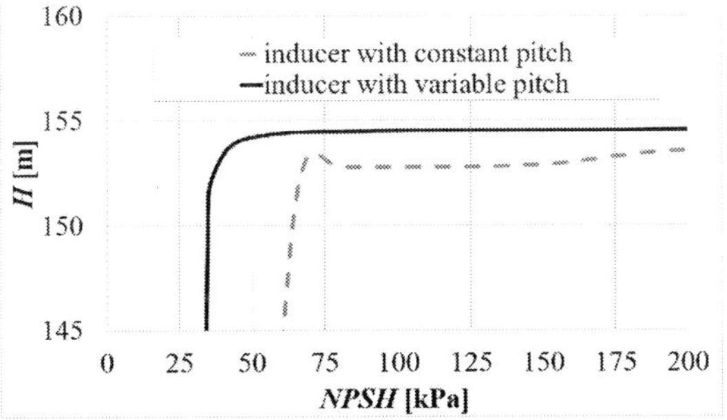

Fig. 11. Vapor volume fraction of value 0.1[3]

As can be seen from Fig. 11, inducer with variable pitch effectively doubles minimum suction pressure for this operation conditions under which the pump is still able to work properly, in contrast to inducer with constant pitch.

7.3 Evaluation of pump characteristics

On Fig. 12 a comparison of the pump static head is drawn with and without inducer, different types of inducers are also considered.

As can be seen, there are no discernible differences as the change in static pressure obtained by inducer is not as significant from this viewpoint, but inducer with variable pitch still shows best results.

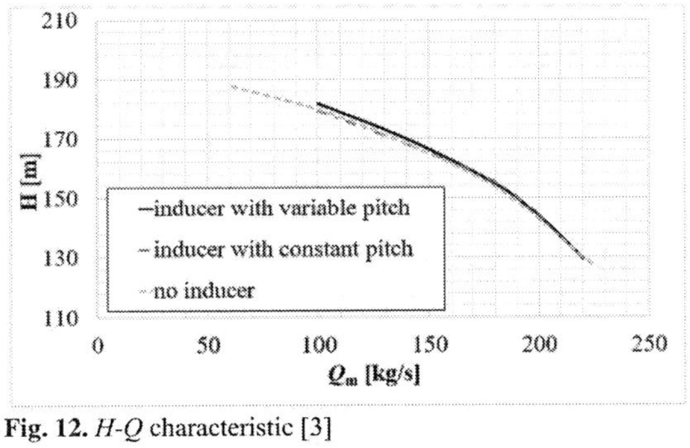

Fig. 12. *H-Q* characteristic [3]

The real difference is seen on the Fig. 13, where $NPSH_3$ is drawn. The minimal pressure at the suction is effectively doubled in comparison to pump working alone or with inducer with constant pitch across almost whole working area.

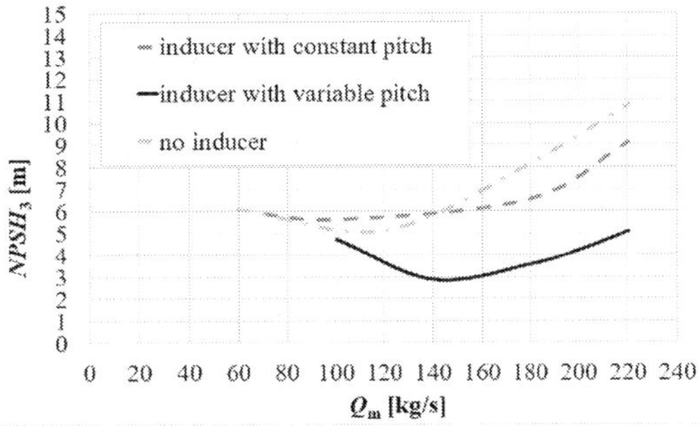

Fig. 13. $NPSH_3$ characteristic [3]

Inducer with variable pitch can raise static pressure so considerably in comparison to the inducer with constant pitch that its production costs are more than justifiable. The pump can now be operated at much lower suction pressures (higher speeds) given that good enough efficiency is still met.

8 Conclusion

This paper dealt with the incorporation of the CFD analysis into hydrodynamical design of the inducer with variable pitch.

As shown above and in more detail in [3], CFD analysis helped with the process of design significantly. If designed only by standard approaches the result would not be as effective as by incorporation of CFD methods.

From these certain conclusions can be drawn, such as:

- CFD analysis can be advantageously used in the geometric design of inducer (or any other hydrodynamic machines)
- To estimate the performance parameters with much better precision than by standard calculations alone, hence arriving at much better results by making the necessary modifications
- Integration of the common CAD formats with ANSYS
- Easily incorporable modifications to CFD geometric model based on the previously calculated results

There are also some disadvantages like:

- Need for CFD qualified worker
- Relatively long calculation time (dependent upon the geometrical complexity, mesh fineness, used mathematical models, PC hardware etc.)

CFD calculations have already proven to be inseparable part of the design of hydrodynamic machines on many levels of the process and as computer hardware progresses more and more complex calculations are possible.

Acknowledgement

This work has been supported by The Ministry of Education, Youth and Sports of the Czech Republic from the Specific Research Project SP2019/123 (Numerical Modeling of Dynamic Phenomena in Fluid Mechanics with Support of Experimental Research).

The authors thank to the project CZ.02.1.01/0.0/0.0/17_049/0008408 "Hydrodynamic Design of Pumps", provided by the Czech Ministry of Education, Youth and Sports

References

1. A. Ramadhan Al-Obaidi. Monitoring the Performance of Centrifugal Pump under Single-Phase and Cavitation Condition: A CFD Analysis of the Number of Impeller Blades. 2019. Journal of Applied Fluid Mechanics, Vol. 12, No. 2, pp. 445-459.. ISSN 1735-3572, EISSN 1735-3645. DOI: 10.29252/jafm.12.02.29303
2. Gülich, J. F. Centrifugal Pumps 3rd edition 2014. ISBN 978-3-642-40114-5
3. Marcalík, P. Design of Inducer with Variable Pitch: Diploma Thesis. Ostrava: VŠB –Technical University of Ostrava, Department of Hydrodynamics and Hydraulic Equipments, 2019, 65 p. Thesis head: Kozubková, M. (in Czech)
4. Csuka, Z., Olšiak, R., & Fuszko, Z. Research of cavitation at high shear stress. 2016. Strojnicky Casopis, 66(1), 7-16. DOI:10.1515/scjme-2016-0006
5. Brennen C.E. Cavitation and Bubble Dynamics. 270p. 2013 ISBN 13: 978-1-107-64476-2
6. Kyuichi Yasui Acoustic Cavitation and Bubble Dynamics. 2018. 131p. ISBN 13: 978-3-319-68237-2.
7. Knížat, B.; Olšiak, R.; Mlkvik, M. Observation of development of cavitation damage. 2014. In EPJ Web of Conferences (Vol. 67, p. 02053). EDP Sciences, DOI: https://doi.org/10.1051/epjconf/20146702053.
8. Luca d'Agostino, Maria Vittoria Salvetti (eds.) Cavitation Instabilities and Rotordynamic Effects in Turbopumps and Hydroturbines: Turbopump and Inducer Cavitation. 2017. Experiments and Design. 288p. ISBN 13: 978-3-319-49719-8
9. Poullikkas, Andreas. Two phase flow and cavitation in centrifugal pump: a theoretical and experimental investigation. 2019. figshare. Thesis.. https://hdl.handle.net/2134/11344
10. Noskievič, J. and collective. Cavitation in hydraulic machinery and equipment. 1990. Praha: SNTL. (in Czech)
11. Kratky, Tomas & Zavadil, Lukas & Doubrava, Vit. CFD and surrogates-based inducer optimization. 2016. International Journal of Fluid Machinery and Systems. 9. 213-221. 10.5293/IJFMS.2016.9.3.213.
12. Kozubková, M. Matematické modely kavitace a hydraulického rázu. 2009. Ostrava. 130 s. ISBN 978-80-248-2043-9. (in Czech)

Numerical modelling and experimental investigation of the influence of selected geometric elements on the working characteristics of radial compressor stages with prismatic blades

Richard Matas[1,*], *Tomáš* Syka[1], *Lukáš* Hurda[1], and *Zuzana* Linhartová[2]

[1]University of West Bohemia in Pilsen, New Technologies Research Centre, Univerzitní 2732/8, 301 00 Plzeň, Czech Republic
[2]RotMach s.r.o., Rybná 716/24, 110 00 Praha 1, Czech Republic

Abstract. The article deals with a description of basic results of research of radial compressor stages with 2D e.g. prismatic rotor blades. The experimental facility and the measurement and evaluation process are described briefly in the first part. The second part of this contribution is the numerical simulation and evaluation of the influence of the technological holes in the 2D rotor blades to compressor stage characteristics. The influence of the labyrinth seals clearance on the working characteristics can be found in the last part. The comparison of measured and computed values is presented and discussed.

1 Introduction

Centrifugal compressors are widely used in the industry. The description of processes, operations and performance of the centrifugal compressors are described in many publications, for example in [1].

The verification of the new stages properties requires detailed measurement of performance characteristics to obtain reliable data for compressor design. Measured data are also important for validation of CFD numerical models. Validated numerical models can give also the complex information about the flow in the stage and help to design the channel profile.

The article authors participated on the large research and development project of new compressor stages design. The main goal of the project was to achieve higher stage efficiency and better operational properties. In addition to the basic working characteristics of the investigated stages a certain amount of supplementary information was acquired. This work briefly presents the influence of some geometric elements on the characteristics of stages with prismatic blading.

2 Research facilities

2.1 Experimental rig

The described development of the centrifugal compressor stages was a continuation of the systematic research and long term development program; see for example the paper [2]. Experimental measurements were realized in the development test facility Howden ČKD Compressors Ltd. in Pilsen from 2014 to 2018, where the single-stage test compressor DARINA IV with the exchangeable flow part was placed. The compressor consisted of an inlet cone, an impeller, a diffuser, a return bend, a de-swirl channel and an exit part.

The compressor was driven by an electromotor with maximum input wattage 1.2 MW with frequency converter for a continuous speed control. The test rig is shown in Fig. 1.

The test rig was designed for impellers with diameter of *440 mm* and for rotation speed from *4,000* to *18,000 rpm*. A set of throttle-valves for the aerodynamic resistance adjustment was placed in the discharge piping. Mass-flow rates of the air through the suction and discharge piping were measured by orifices.

Fig. 1. Photo of the test rig for centrifugal compressor stages.

* Corresponding author: mata@ntc.zcu.cz

2.2 Experimental equipment

Parameters of the measured stage were obtained by the complex measurement and data acquisition system.

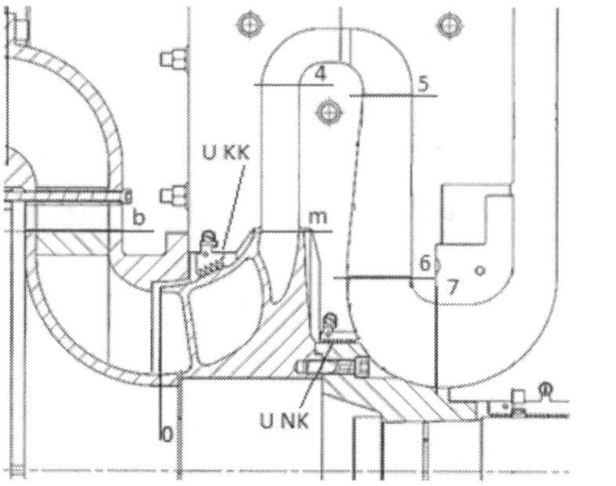

Fig. 2. The cross-section of the compressor with evaluated planes.

The flow parameters are measured in large number of points on the evaluation planes. The labels of the planes in the stage are long-term used, see [2], and are depicted in Fig. 2; the main evaluation planes or so called measuring stations for determination of the stage parameters are the stations 0 and 7. There is a number of static and total pressure and total temperature probes (multiple sensors are placed in the stations 0 and 7). The torquemeter gives the value of torque and rotating speed of the compressor rotor.

Described research was carried out with stages RTK06 and RTK04 with various blade lengths which are designed for medium to lower flow rates. The impellers and stator parts of the stages RTK06W and RTK04N are depicted in Fig. 2.

2.3 Evaluation of measurement

The measurement campaigns are established to produce curves of operational characteristics – branches of the compressor map. The whole process of the measurement ant evaluation is described in detail in the paper [3].

The aim is to measure one operational curve keeping the reference (blade) Mach number (1) constant. This number is based on the speed of sound in the gas at stage intake state and the velocity of the blades at the outer diameter of the impeller:

$$Ma_{2u} = \frac{2 \cdot \pi \cdot \Omega \cdot D_2}{a_0} = \frac{u_2}{a_0} \quad (1)$$

This means the rotational speed is being adjusted slightly based on the actual intake conditions.

Besides the whole stage evaluation, a part-by-part analysis can be carried out using the data acquired from the measuring stations visible in Fig. 2. Mean velocity vectors, performance characteristics of the diffuser or the losses of the de-swirl channel could be addressed.

Fig. 3. Models of the stages RTK06W and RTK04N in the test compressor.

2.4 Parameters of stages

Performance of the centrifugal compressor stages are analysed through two of the main properties: the pressure ratio and the efficiency. These two parameters vary with the mass flow-rate and with the Mach number.

The evaluation of various compressor stage parameters was done according to the formulas (2) to (6). Values of the evaluated parameters were averaged by the mass flow rate in postprocessing. The parameters are: specific total work (2), specific isentropic work (3), pressure ratio (4), isentropic efficiency (5) and flow coefficient (6).

$$\Delta h_t = c_p \cdot (T_{t7} - T_{t0}) \quad (2)$$

$$\Delta h_{is} = r \cdot T_{t0} \cdot \frac{\kappa}{\kappa - 1} \cdot \left[\left(\frac{p_{t7}}{p_{t0}} \right)^{\frac{\kappa - 1}{\kappa}} - 1 \right] \quad (3)$$

$$\Pi = \frac{p_{t7}}{p_{t0}} \quad (4)$$

$$\eta = \frac{\Delta h_{is}}{\Delta h_t} \tag{5}$$

$$\varphi_0 = \frac{4 \cdot Q_0}{\pi \cdot D_2^2 \cdot u_2} \tag{6}$$

The results from stages CFD simulations were compared to the measured data converted to the standard inlet conditions. Obtained characteristics include the confidential technical data and only relative values are presented in the following sections. The values PI_N, ETA_N and m_N are normalized by the design point values in the measured characteristic.

3. CFD software and models

CFD simulations are nowadays an essential tool for the design of all types of devices in turbomachinery. They enable engineers to gain a lot of time between the concept and the test. The verification of simulations is carried out in all turbomachinery development groups. For the purpose of this work, many CFD simulations of various compressor stages were run in the NUMECA FINE/Turbo (NFT) and ANSYS CFX (ACFX) software environments.

The methodology of CFD simulations for the RTK stages and properties of the model has been evaluated and improved; see e.g. [3, 4, 5, 6]. The air, modelled as perfect gas, was used as the flow medium, k-ω SST turbulence model was used for the described simulations.

4 Simulations of the technological holes influence for a 2D compressor stage

4.1 Technological holes description

The problem of the technological holes for the stages with 3D rotor blades was described in the contribution [3]. The manufacturing technology of the stages with prismatic blading is different. Rotors are split into two parts and it is possible to machine the parts on a simple milling machine. Both parts are then welded or soldered together. The soldering process gives better quality thanks to low thermal stress introduction and better surface quality of the joint, but the welding is often cheaper and preferable.

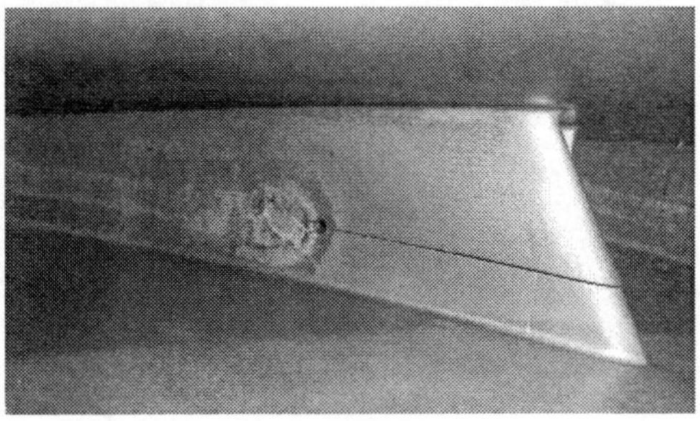

Fig. 4. An example of a technological hole.

The welding of impellers requires the technological holes in the blade to reduce mechanical stress concentration at the end of the weld, see Fig. 4. The splitting plane is located near the middle of the blade height; the weld takes place in the aft part – from the hole to the trailing edge. The goal is to find the suitable location of the hole depending on the blade type and required stage efficiency. The stage RTK04W was selected for the case study.

4.2 Model of the holes

The model of the stage with holes was based on the standard stage model described in the paper [6]. The stage dimensions correspond to the stage simulated and experimentally tested.

The model hole position was varied to five positions along the blade length marked from 1 to 5 and to five positions along the blade height marked from A to E. The positions of the holes in the model are shown in Fig. 5. Such setup yields 25 simulation sets to produce performance curves at the constant rotational speed.

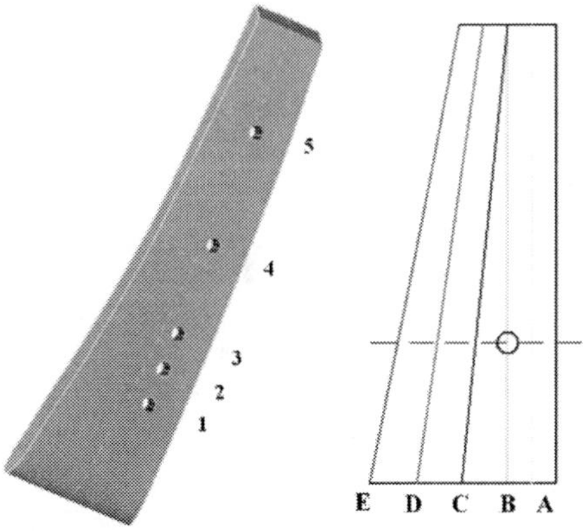

Fig. 5. Positions of holes in investigated variants.

The diameter of the circular holes was set to 4 mm except the case E, where the shape was semi-circular and the diameter was set to 5.6 mm - the cross-sectional area is then similar.

4.3 Results of the simulations

The results of investigated variants were evaluated in the same manner as for the stage without holes. Fig. 6 illustrates the pressure field on the blade surface and the contours of the relative Mach number in the slice through the hole in the blade.

The Fig. 7 shows the influence of differently positioned hole to the pressure. The results for positions along lines not shown (A, C, D) have been very similar to the positions along line B. The effect of the holes on the pressure and velocity field in the stage was assessed to have just a "local influence". The most significant influence of the holes was encountered for positions along line B and E.

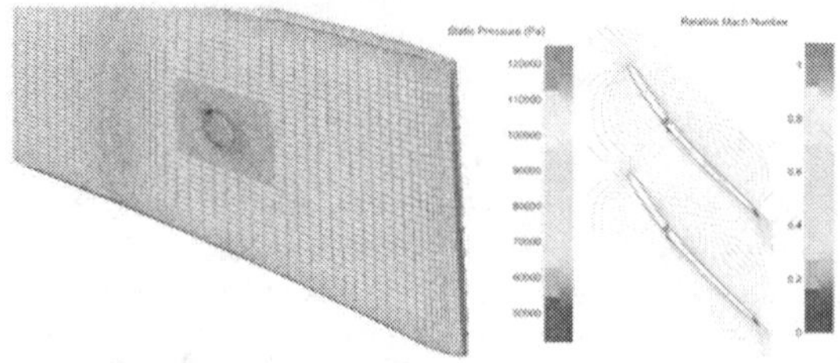

Fig. 6. Surface mesh and static pressure (left) and the relative Mach number (right).

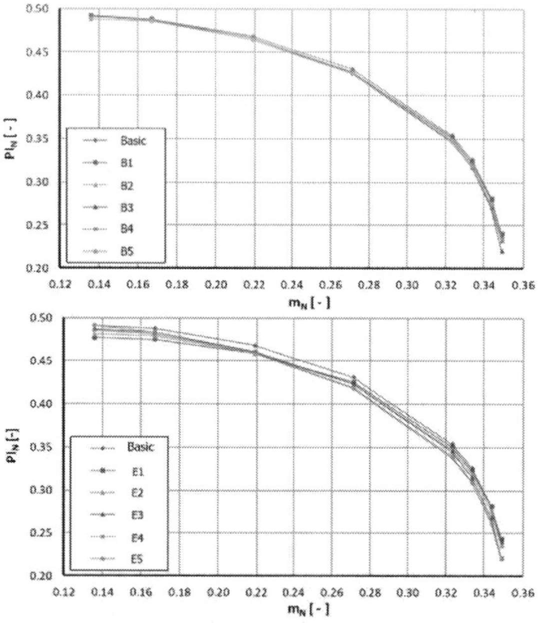

Fig. 7. The influence of the hole to the pressure ratio.

The influence on the efficiency is more important for the stage design - a selection of results for rotational speed of *12,666 rpm* (Ma_{2u} = *0.8*) is shown in Fig. 8. It was concluded that the influence of the hole position is observable but not significant. The highest negative influence on the working characteristics is seen in the case of technological holes located on the shroud (line E) and the efficiency decreases as the hole is moved towards the shroud.

The holes are usually positioned on lines B or C and in positions 1 or 2, where significant drop of efficiency is not predicted by the presented results. The influence on the efficiency for the design conditions flow-rates (noted by black dots in Fig. 8) is from *0.2* to *0.8 %* compared to the design with no hole.

5 Influence of the reduced clearance of labyrinth seals to the stage working characteristics

5.1 Description of the problem

The seals inside the compressor create the resistance to flow between volumes at different pressures. Since one of the walls bounding this volume of fluid rotates at a high peripheral speed, the use of contact seals is excluded. Contactless seals always allow a certain adverse flow that negatively affects the machine efficiency. The labyrinth seals are typically used. The Fig. 9 shows examples of seals used in radial compressors – a stepped labyrinth for impeller shroud disc and a straight-through labyrinth for sealing the shaft-diaphragm gap.

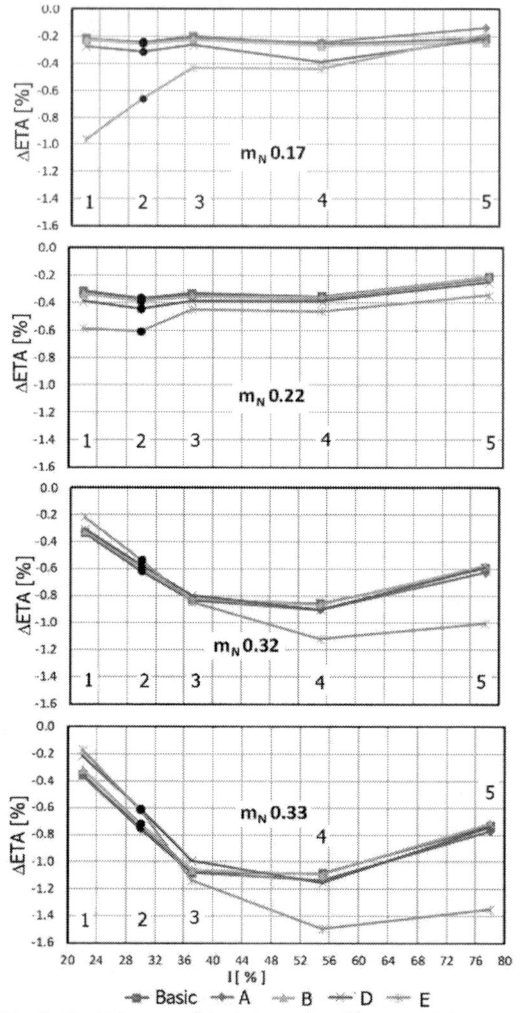

Fig. 8. The influence of the hole on the efficiency of the stage.

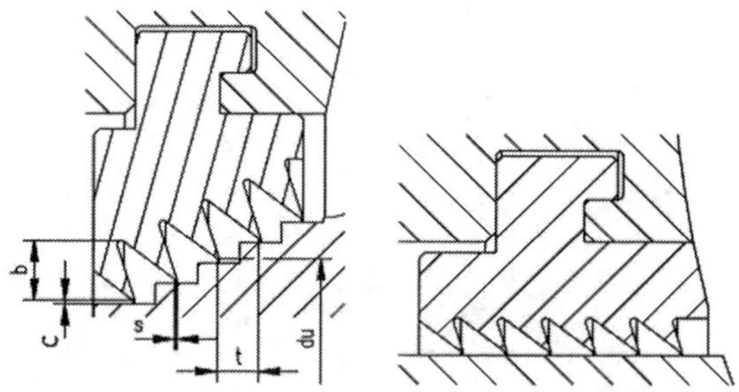

Fig. 9. The labyrinth seals for a radial compressor stage.

The basic parameter of the labyrinth is the width of clearance *c*, see left part of Fig. 9, which influences the leakage flow rate in the seal and consequently raises the losses in the stage.

A research study was conducted to determine the influence of reduced clearance to the working characteristics of the standard stage RTK06W (without holes in blades). It is published in [6]. Both experimental measurements and numerical simulations were used for the analysis. In this chapter, the analysis is focused on both seals in the stage and their equally large clearance adjusted together. The location of the seals is denoted by the flags *'U KK'* and *'U NK'* in Fig. 2.

The study was conducted in the environment of an industrial research process, which is primarily concentrated on producing compressor maps for various stages and a wide range of reference Mach numbers. Due to the stiffness and tight schedule of the process, the analysis of the seal leakage were secondary and certain trade-offs were necessary. That is why the results of measurements were only interpolated to the design speed of the stage RTK06W. On the other hand, CFD is not so expensive to prepare, but computing many operating regimes may become prohibitive when including complexities, like seals, in the cases.

5.2 Experimental measurement

The experiments were focused on three typical reference Mach numbers of the stage - Ma_{2u} = *0.7, 0.9* and *1.1* (*10,550, 13,568* and *16,377 rpm*). Two repeats of measurements with classical aluminium alloy seals (AL) and one or two runs with plastic seals (PL) with significantly smaller clearance *c* were administered. The resulting curves of both aluminium seals measurements were practically identical.

The standard clearance of the alloy seals is from *0.4* to *0.5 mm*. Both cases are analysed, while *0.4 mm* clearance is taken as a norm. The plastic seals allow the clearance to be lowered down to *0.15 mm*.

Fig. 10 shows the whole measured pressure characteristic of the stage and the limits of practical operating range. These limits are imposed upon Fig. 11, which shows the efficiency characteristic at $Ma_{2u} = 0.9$.

The Table 1 shows the averaged efficiency rise in the operating range for the stage with plastic seals based on polynomial regressions from Fig. 11. The average increase of efficiency ΔETA was *1.28 %*. For $Ma_{2u} = 0.7$ and *1.1* the efficiency increased on average by *0.81* and *1.21 %* respectively. The influence was larger at lower compressor mass-flow rates, as also seen in Table 1.

Table 1. Increase of the efficiency for $Ma_{2u} = 0.9$.

m_N [-]	ΔETA [%]
0.17	1.20
0.16	1.22
0.14	1.26
0.13	1.33
0.11	1.42

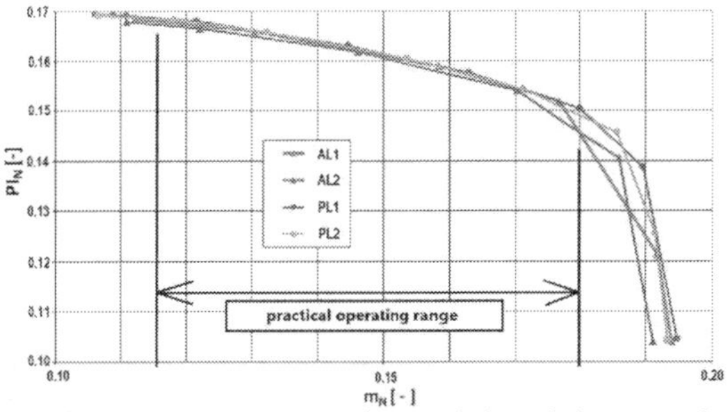

Fig. 10. Measured pressure characteristics of the stage for *13 568 rpm*.

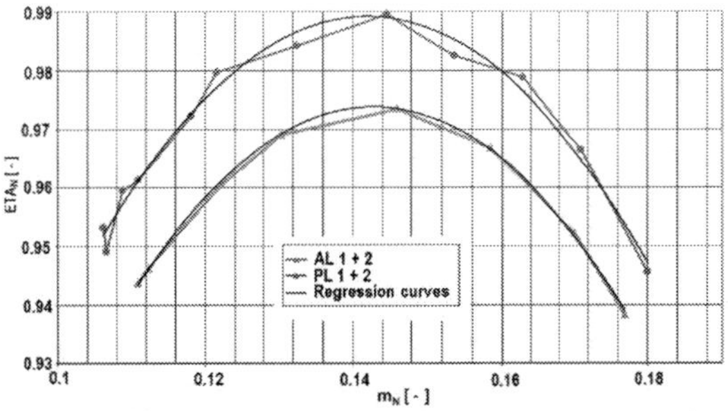

Fig. 11. Efficiency characteristics in the practical operating range of the stage at $Ma_{2u} = 0.9$. *Polynomial regression curves added.*

5.3 Computational model of the stage

The model of the stage with holes was based again on the standard stage model described in the paper [6]. The conditions of the CFD study were slightly different to the measurement because only cases for the design speed of *11,921 rpm* ($Ma_{2u} = 0.8$) were run.

A set of simulations was prepared considering various seal clearance. The value of clearance was set to *0.1, 0.25, 0.4* and *0.5 mm*, always equal for all the labyrinth teeth.

In Fig. 12 and 13 the working characteristics can be observed. An obvious but limited impact to the pressure and also efficiency characteristics is deduced.

The violet dashed line in Fig. 12 stands for the standard seal design (*0.4 mm*). Here, the complete range of the pressure characteristic curve is drawn, including also the right-most point where the stage is choked. This point is excluded from the efficiency curve in Fig. 13.

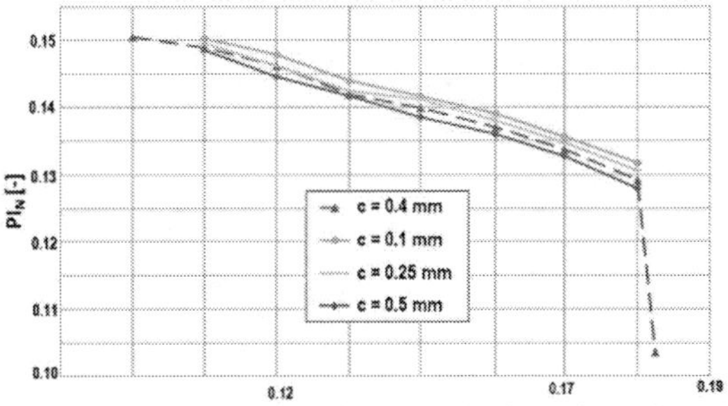

Fig. 12. Computed pressure characteristics for various width of seal clearance.

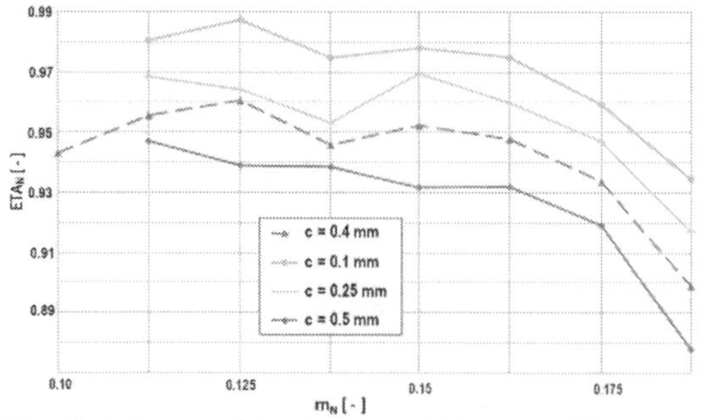

Fig. 13. Influence of the clearance width to the efficiency of the stage.

5.4 Comparison of results

Table 2 shows the averaged departure of efficiency caused by varying clearance width. The efficiency with a clearance of *0.4 mm* is taken as a nominal value.

Table 2. Deviations of the efficiency.

clearance [mm]	ΔETA [%]
0.4	0
0.1	2.83
0.25	1.25
0.5	-1.54

The values of the efficiency from CFD simulations show very strong dependence on the clearance width. A linear interpolation of presented CFD results for a clearance of *0.15 mm* yields an increase of the efficiency by *2.3 %*.

When revisiting this guess using experimental data, one needs to interpolate between the data for $Ma_{2u} = 0.7$ and 0.9 to obtain the guess for the desired $Ma_{2u} = 0.8$. Interpolating the data presented in paragraph 5.2, the increase of the efficiency for the case of *0.4 mm* clearance width and $Ma_{2u} = 0.8$, the efficiency departure reaches slightly above *1 %*.

There could me many reasons for such a large difference (2.3 % vs. 1 %). One of them can be the contact of PL seals with the rotor, which could and did occur due to the large compressor shaft bearing clearance (around *0.2 mm*). This contact could have caused a growth of the seal clearance to values of approximately *0.2 – 0.3 mm*. When the interpolation of CFD results is repeated for such clearance width, a value of *1.25 %* efficiency increase could be expected. Unfortunately, it was not possible to measure the exact value of the seal clearance after the experiment due to technical problems.

6 Influence of an increased clearance of labyrinth seals to the stage parameters and the flow-rate through the seal

6.1 Description of the problem

One of the last research projects based on the experimental compressor DARINA IV was carried out with the RTK04N stage. The goal was to assess the influence of an increased clearance of the stepped labyrinth seal on the outer side of the impeller shroud disc. The project was primarily focused on the leakage flow itself and the measured characteristics were just a supplement to it. Detailed study supported by CFD can be found in [7].

Unlike the study described in Chapter 5, only the shroud disc seal was modified. The clearance width was gradually increased from the basic value $c = 0.4$ *mm* to the values *0.6 mm* and *0.8 mm*. Again, an equal clearance was set for all the teeth of the seal.

6.2 Experimental results

In Fig. 14 you can see the measured pressure characteristics for the mentioned values of clearance at $Ma_{2u} = 0.9$ *(13,568 rpm)*. There is clear evidence that the influence of the clearance width to the pressure characteristic is non-linear. A quadratic dependence is expected as the leakage flow cross-section grows with the clearance width to the power of 2.

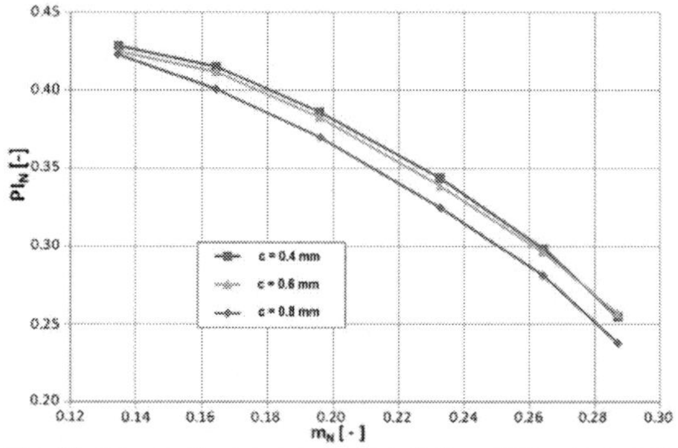

Fig. 14. Measured pressure characteristics of the stage with various clearance width of the impeller shroud disc seal at $Ma_{2u} = 0.9$.

The decrease of efficiency for the same set of cases is shown in Table 3, where m_N stands for an arbitrarily normalized compressor flow rate. The differences are higher for higher flow-rates. This phenomenon was already observed and described in chapter 5.2 for the stage RTK06W.

The study also took care to measure the leakage influences at various rotational speeds. Besides already mentioned $Ma_{2u} = 0.9$, values of 0.7 and 1.1 were set to be measured too. The decrements of the efficiency, averaged over all of the compressor flow rates for given speed, are presented in Table 4. The efficiency with standard clearance of $c = 0.4\ mm$ is taken as a norm.

Table 3. Decrease of the efficiency at $Ma_{2u} = 0.9$ for higher values of clearance width.

m_N [-]	ΔETA [%]	
	c = 0.6 [mm]	**c = 0.8 [mm]**
0.6	-0.4	-0.31
0.67	-0.56	-0.92
0.75	-0.63	-1.21
0.88	-	-1.4

The values for $Ma_{2u} = 0.7$ and 0.9 are very close to each other for a given clearance width. The highest speed, however, presents significantly larger decrease of efficiency, which seems not to vary a lot with an increase of the clearance width.

Table 4. Efficiency decrements averaged over compressor flow rates at different rotational speeds.

Ma_{2u} [1]	**average ΔETA [%]**	
	c = 0.6 [mm]	**c = 0.8 [mm]**
0.7	-0.6	-1.05
0.9	-0.53	-0.96
1.1	-1.26	-1.20

7 Conclusions

The paper describes three effects of the selected geometric elements on the working characteristics of the radial compressor stages. The effects were solved in various research projects related to the development of the new family of radial compressor stages.

At first, a CFD study shows the influence of the technological holes in blades of welded impellers. The predicted influence to the efficiency of the stage is about *0.5 %* of the overall efficiency. The results also show that the locations of the holes typically used by the structural designers correspond to the locations with low influence to the stage working characteristics.

The second complex study shows the influence of a concurrent reduction of the clearance width in both of the labyrinth seals present in the stage. The practical issues obstruct direct comparison between experimental measurement and numerical simulation. Despite that, the methods used for comparison do not pose a significant uncertainty to the results. The trends can still be deduced clearly. The uncertainties connected with both experimental and computational methods themselves would still be high above negligible.

The third study is just a complement of the previous. It supplies experimental results for the higher values of the clearance in the impeller shroud disc labyrinth. The original study is more complex than presented here. A detailed analysis of the leakage flow in the stepped labyrinth is to be published in a subsequent contribution.

Acknowledgement

The results were developed within the CENTEM PLUS (LO1402) by financial means from the Ministry of Education, Youth and Sports under the "National Sustainability Programme I." and specific research (UWB SGS-2019-021). The authors would like to thank the Howden ČKD Compressors Ltd. for permission to publish this paper and all colleagues who participated on this work.

Nomenclature

a	Speed of sound
c	Clearance of the leakage
c_p	Isobaric specific heat capacity
D	Impeller outer diameter
ETA	Isentropic efficiency
FI0	Flow coefficient for intake volumetric flowrate
l	Hole position measured from blade tip
m	Mass flow rate
M2u	Blade Mach number
PI	Total pressure ratio
p	Total pressure
r	Specific gas constant
T	Total temperature
u_2	Blade speed at impeller exit
Δh	Specific enthalpy difference
κ	Ratio of specific heats
Ω	Rotational speed of the impeller

Common indices:

0	Measuring station 0 (stage intake)
2	Impeller exit
7	Measuring station 7 (stage discharge)
is	Isentropic
N	Normalized
t	Total

 https://doi.org/10.1051/epjconf/202226901036

References

1. M. P. Boyce, Principles of Operation and Performance Estimation of Centrifugal Compressors, Proceedings of the twenty-second turbomachinery symposium, Dallas, USA. (1993)
2. J. Paroubek, V. Cyrus, J. Kynčl, The Influence of Outlet Stator Part Surface Roughness on the Performance of a Very Low Flow Coefficient Centrifugal Compressor Stage, ASME Paper **96-OT-179**, (1996)
3. T. Syka , R. Matas, L. Hurda, Experimental investigation and numerical modelling of 3D radial compressor stage and influence of the technological holes on the working characteristics, EPJ Web of Conferences, **180**, article number 02060 (2018)
4. T. Syka, R. Matas, O. Luňáček, Numerical and Experimental Modelling of the Radial Compressor Stage, AIP Conference Proceedings, **1745**, article number 020059 (2016).
5. T. Syka, Geometry structure influence on the radial compressor stage efficiency, Ph.D. thesis, in Czech, University of West Bohemia in Pilsen (2015)
6. R. Matas, T. Syka, O. Luňáček, Numerical and experimental modelling of the centrifugal compressor stage – setting the model of impellers with 2D blades, EPJ Web of Conferences, **143**, article number 02073 (2017)
7. Z. Linhartová, Numerical simulations of flow in a compressor seal, Master thesis, in Czech, Czech Technical University in Prague, 2019

A Comparison of Various Images Processing Methods Used in Evaluation of Liquid Film Disruption

Lucie Měšťanová[1,*] and *Ondřej* Bartoš[1]

[1]CTU in Prague, Energy Engineering Department, Technická 4, Prague 6, 166 07, Czechia

Abstract. The aim of the paper is to introduce and compare various methods which can be used for an image processing. Studied methods include the image pre-processing and the image segmentation based on the Gaussian mixture model and the active contour. The images are acquired within the measurement of the liquid film disruption on the aerofoil's trailing placed edge in the wind tunnel. As the result is suggested sequence of the particular image processing techniques for an optimal result. This work is the part of the wider project devoted to the coarse water droplets which have a negative effect on the steam turbines efficiency and reliability.

1 Introduction

In the long term the Energy engineering department at CTU is concerned with experimental research in the liquid phase structure in steam turbines [1, 2]. In steam turbines, there are presented two groups of water droplets.

The nucleation of droplets and their subsequent growth due to the surface condensation is responsible for the presence of fine droplets. Their expected diameters range is between 200 nm and 800 nm. Coarse droplets are formed by the disruption of water films on the blades and casing surfaces in the low-pressure turbine stages. Their diameters range is very wide, from a few micrometres to one millimetre. They have the negative effect on the reliability and the efficiency of the steam turbines and they are also responsible for unfavourable effects like the additional energy losses or reduction of the blades lifetime. The present knowledge about the coarse water droplets formation, their flow field and the size droplets distribution in turbines is still insufficient. A better understanding of the coarse droplets formation and their properties may provide future enhancements for the steam turbine and other similar applications. Previous measurement of coarse droplets directly in the steam turbine made it necessary to build an experimental tunnel for the study of the droplets formation and their properties [3, 4] in the laboratory.

2 Testing facility

Due to difficulties connected with making measurements in an operating steam turbine, a wind tunnel was designed and manufactured for the analysis of the coarse droplets formation from the liquid films. The goal of the design was to achieve the conditions in the steam turbine as closely as possible together with the advantages of working in the laboratory.

Fig. 1. Wind tunnel test section with camera.

The tunnel is designed as a classical CD nozzle, but the typical operational regime is subsonic or transonic (Fig. 1). The aerofoil, placed in the nozzle, simulates the blade in the turbine and it is possible to remove or replace it. On the aerofoil is a groove for supplying liquid on its surface. The liquid is pumped to the aerofoil surface through the dosing pump with a flow from 1 ml/min to 500 ml/min. The tunnel is equipped with four large optical windows to have a good optical access for the measurements. It is possible to operate the tunnel with the steam in continuous mode or with the compressed air in periodic mode.

Measurement methods

The time of the measurement with the air in the wind tunnel depends on the pressure in the compressor storage vessel. The pressure and temperature of entire

* Corresponding author: lucie.mestanova@fs.cvut.cz

https://doi.org/10.1051/epjconf/202226901037

measurement sequence is acquired by the NI SCXI data acquisition system for the flow conditions description. Two methods of droplets size measurement are available. The light scattering method is used for small particles and the photogrammetry is used for bigger droplets and for droplets formation on the aerofoil's trailing edge. The measurement schema is shown in Fig. 2.

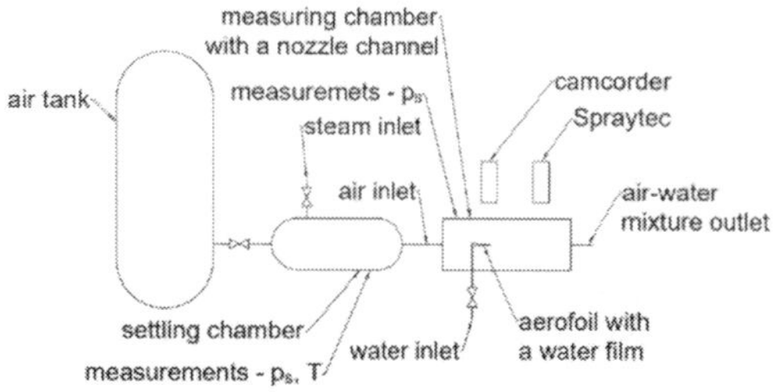

Fig. 2. Measurement schema of the wind tunnel.

The used camera XIMEA MC050CG-SY contains color CMOS digital image sensor Sony IMX250 LQR-C. The sensor has 2464x2056 active pixels and 3,45 μm pixel pitch.

Calibration is required for correct pixel/dimension ratio. The ratio was determined as 1,7 μm/px with help of the calibration target. The light source with variable length of the flash at the order of 0,5-17μs was used.

On the figures below one can see efficiency of the camera chipset and profile of the emitted light from LED.

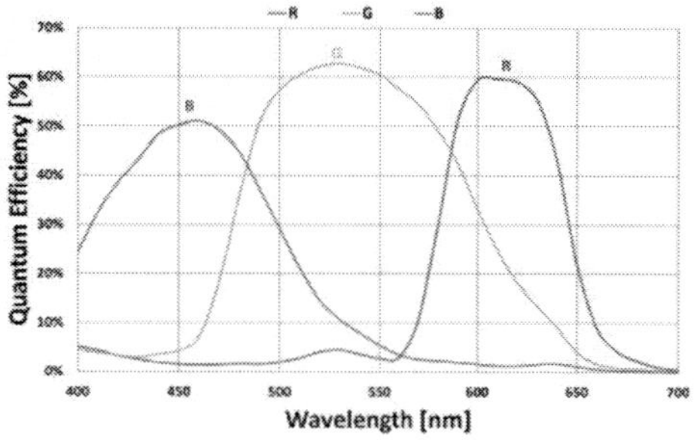

,

Fig. 3. Quantum efficiency of the camera sensor.

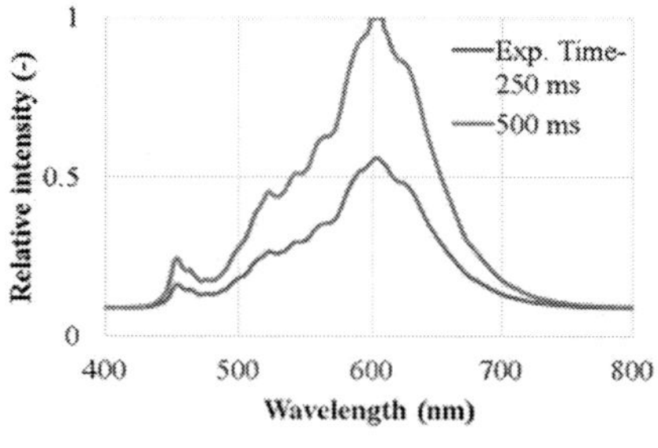

Fig. 4. Characteristic of the used diode with different time of the spectrometer.

The characteristic of the LED light source together with the quantum efficiency of the camera chipset and with involving the light scattering [5] of the light on the liquid leads to the expectation that the best visibility of the liquid film is for green or red channel of the captured image.

3 Image processing methods

The camera captures the images with the size of 3,5x4 mm in the vicinity of the aerofoil. These images are further processed in Matlab. The aim is to evaluate the amount of the water on the image for the description of the disruption process and evaluation of the droplets size distribution. Three segmentation methods were used for this purpose. Enhancement of the histogram was used for the image pre-processing.

3.1. Pre-processing (+entropy)

Enhancement of histogram by linear intensity stretching was used for image pre-processing. Entropy of the image and its parts was determined for the evaluation of the image processing methods. Entropy is a measure of randomness of the image and it is defined as:

$$H_e = -\sum_i p_i \cdot \log_2 p_i \text{ (bit)} \tag{1}$$

where p_i is a probability of intensities.

Entropy characterizes the texture of the image. As lower the entropy value is as less noisy is the image. This property is suggested as a value for the compares of the methods

3.2. Thresholding

Thresholding is the simplest segmentation method based on the classification of pixel intensity values. It is necessary to select the most accurate threshold value from the image histogram. Then, pixels with a value below the threshold are classified as the first segment and vice versa.

$$g(x,y) = \begin{Bmatrix} 0 \text{ for } f(x,y) \leq T \\ 1 \text{ for } f(x,y) > T \end{Bmatrix} \tag{2}$$

where $g(x,y)$ denotes segmented image and $f(x,y)$ its histogram.

3.3. Active contour

Active contour is sophisticated segmentation method also known as the snake method. The input to this method is a closed curve that surrounds the object to be segmented. The curve is evolving, it moves toward its interior and it stops at the object boundary.

The snake model [7] is: $\inf_C J_1(C)$, where

$$J_1(C) = \alpha \int_0^1 |C'(s)|^2 ds + \beta \int_0^1 |C''(s)|^2 ds - \lambda \int_0^1 \left|\nabla u_0(C(s))\right|^2 ds \tag{3}$$

α, β and λ are positive parameters. The first two conditions control the smoothness of the contour and the third condition push the contour toward the object in the image. The stopping term is based on Mumford-Shah segmentation techniques.

Level set formulation: $C \subset \Omega$ is represented by the zero level set of a Lipschitz function $\phi: \Omega \to \mathbb{R}$, such that

$$\begin{cases} C = \partial\omega = \{(x,y) \in \Omega: \phi(x,y) = 0\} \\ inside(C) = \omega = \{(x,y) \in \Omega: \phi(x,y) > 0\} \\ outside(C) = \Omega \setminus \varpi = \{(x,y) \in \Omega: \phi(x,y) < 0\} \end{cases}$$

3.4. Gaussian mixture model (GMM)

Another segmentation method is connected with Gaussian mixture model. In this method, it is necessary to create a mask which specify foreground and background of the picture. This mask covers the image and then based on this input. Further the probability is calculated to get a distribution based on GMM from the pixels under the mask. K-means [7] algorithm is used for clustering.

$$\sum_{k=1}^{K} \phi_k \mathcal{N}(x; \mu_k, \Sigma_k) = \sum_{k=1}^{K} \frac{\phi_k}{\sqrt{(2\pi)^D |\Sigma_k|}} e^{-\frac{1}{2}(x-\mu_k)^T \Sigma_k^{-1}(x-\mu_k)} \quad (4)$$

$$p_y(x_t) = p(x_t, \theta_y) = \sum_{k=1}^{K} \phi_{y,k} \mathcal{N}(x_t; \mu_{y,k}, \Sigma_{y,k}) \quad (5)$$

where ϕ_k is prior probability, μ_k is mean vector, Σ_k is covariance matrix, $\mathcal{N}(x; \mu_k, \Sigma_k)$ is multivariate normal distribution.

4 Results

4.1. Pre-processing (+entropy)

The graph below (Fig. 5.) shows the entropy value of original image and its variations. The green channels have the lowest value so it means that segmentation in this channel should give the best results, in case of the low level of the natural noise.

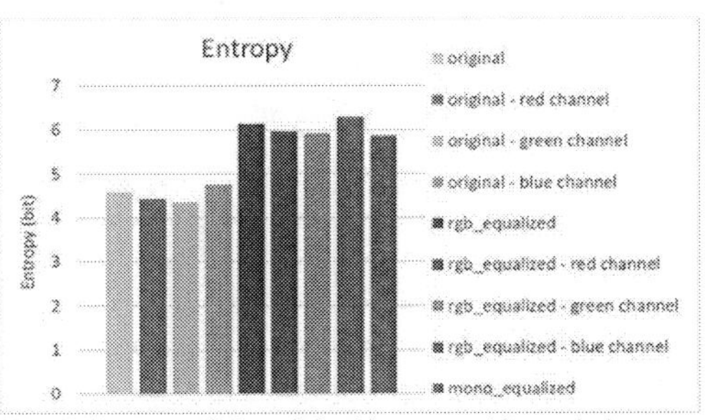

Fig. 5. Entropy of the original image and its variations.

On Fig. 7 is shown the histogram of the pixels under the mask from the original image (Fig. 6.) and on Fig. 9. is histogram of the equalized image (Fig. 8.).

Fig. 6. Original image with mask.

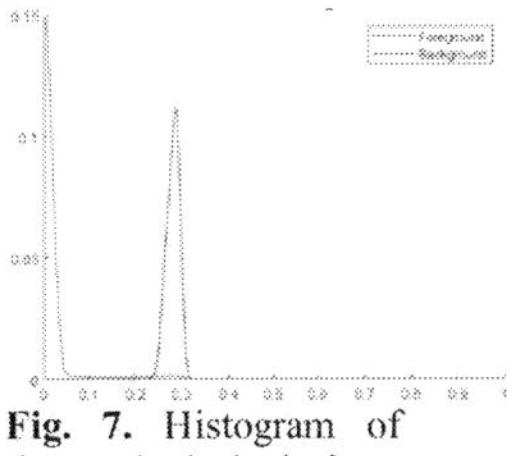
Fig. 7. Histogram of the masked pixels from the original image.

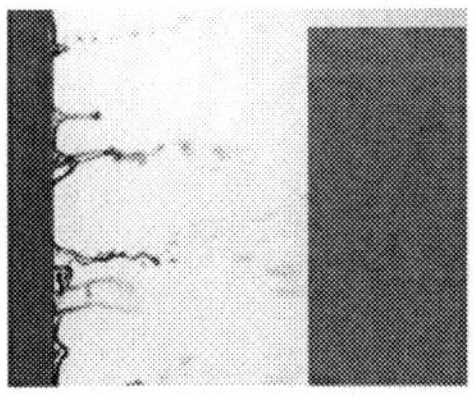
Fig. 8. Equalized image.

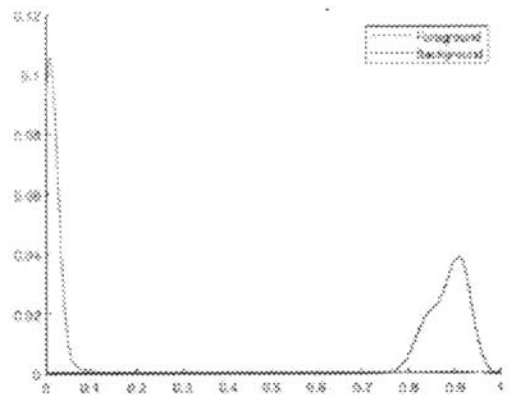
Fig. 9. Enhanced histogram.

4.2. Thresholding (TH)

In segmentation using thresholding is well visible difference between the RGB channels of the image. On Fig. 10. One can see noise on the right side of the image in compare with Fig. 11. And result from red channel gives the lowest number of segmented pixels.

Fig. 10. Segmented blue channel of original image using thresholding (threshold = 0,74).

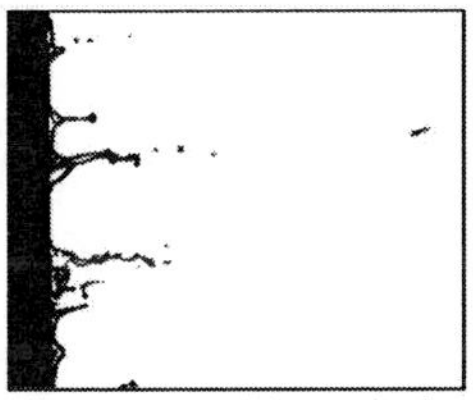
Fig. 11. Segmented green channel of original image using thresholding (threshold = 0,74).

On the graph below (Fig. 12.) one can see liquid content ratio (LCR) for images segmented by thresholding and GMM.

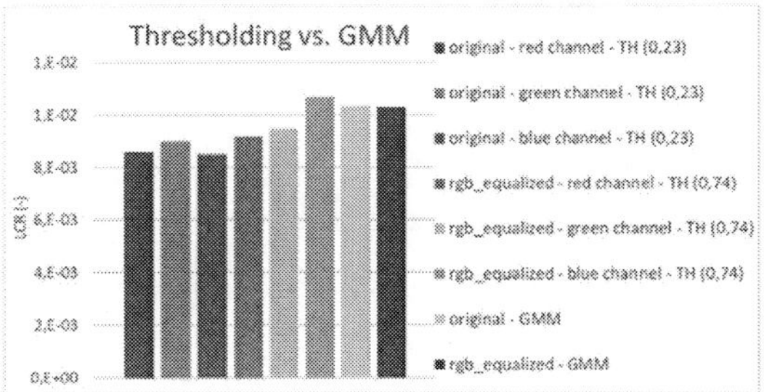

Fig. 12. Liquid content ratio of the images.

4.3. Active contour (AC)

The Fig. 13. and Fig. 14. show segmented images using active contour. One can see that it differs from segmented image using thresholding and using GMM.

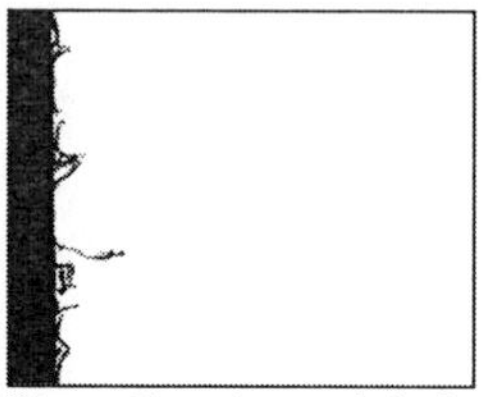

Fig. 13. An original image segmented using active contour.

Fig. 14. An equalized image segmented using active contour.

On the graph below (Fig. 15.) one can see liquid content ratio (LCR) for images segmented by active contour and GMM. The LCR value does not differ within AC segmentation, so it means that enhancement of the image does not impact the final segmentation, nor within GMM segmentation.

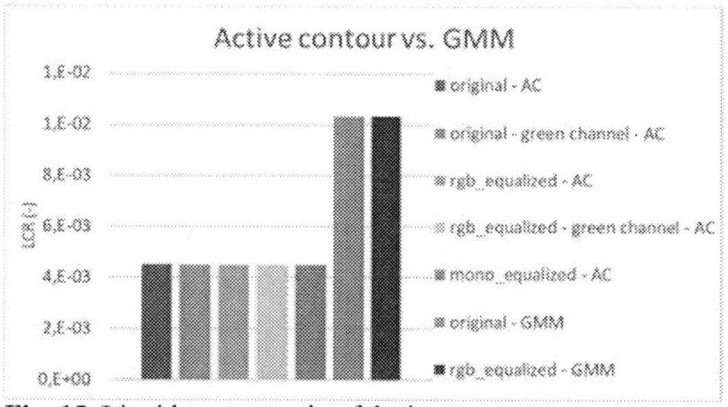

Fig. 15. Liquid content ratio of the images.

Different number of iteration was tested, started at 100 ended at 1500. After 500 iteration the results stopped varying.

4.4. Gaussian mixture model (GMM)

Segmentation using GMM gave the same results for original image and for enhanced image, even for different number of iterations. This method seems to be the most stable.

Sample RGB values and estimated GMMs. are shown on Fig. 16. The Fig. 17. and Fig. 18. show the segmented images.

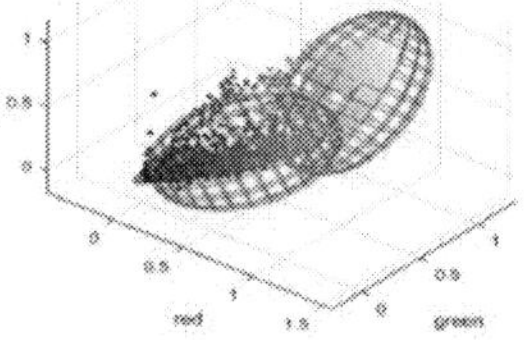

Fig. 16. Sample RGB values and estimated GMMs.

Fig. 17. An original image segmented using GMM.

Fig. 18. An equalized image segmented using GMM.

5 Conclusion

In this contribution are presented several image processing methods used for the description of the liquid films disruption. The knowledge of the liquid disruption in the steam turbines is described mostly phenomenologically [8] and for the future development would be useful to have some tool for quantitative analysis. The entire image capturing process is difficult due to the high velocity of the flowing air over the airfoil. The exposure time is very short and this reasons limiting the quality of the mages. The suggested image processing methods are suitable for the analysis of the images of the liquid atomization but seem to be necessary to find another combination of the methods parameters. One possibility is application of the nonlinear equalization.

We gratefully acknowledge the support by the Grant Agency of the Czech Technical University in Prague, Grant No. SGS 13/18 and National Centre for Energy TN01000007.

References

1. O. Bartoš, X. Cai, M. Kolovratník, EPJ Web of Conferences. 2014, vol. 67, ISSN 2100-014X
2. O. Bartoš, X. Cai, M. Kolovratník, *Experimental investigation of coarse water droplets in steam turbines by the adapted photogrammetry method.* In The Baumann Centenary Conference … Cambridge: ISBN 978-0-903428-35-4
3. M.Hoznedl, at al.. Experimental research on the flow at the last stage of a 1090 MW steam turbine. *Proceedings of the Institution of Mechanical Engineers, Part A: Journal of Power and Energy.* 2018, DOI: 10.1177/0957650917749692.
4. M. Kolovratník, at al: *Nanoparticles found in superheated steam – A quantitative analysis of possible heterogeneous condensation nuclei.* Proceedings of the Institution of Mechanical Engineers Part A - Journal of Power and Energy. 2014, vol. 228, no. 2, p. 186-193. ISSN 0957-6509
5. C.F. Bohren, D.R. Huffman *Absorption and Scattering of Light by Small Particles.* John Wiley & Sons, Inc., USA (1998)
6. Z. Fu, L. Wang, *Color Image Segmentation Using Gaussian Mixture Model and EM Algorithm.* In: Wang F.L., Lei J., Lau R.W.H., Zhang J. (eds) Multimedia and Signal Processing. CMSP 2012. Communications in Computer and Information

Science, vol 346. Springer, Berlin, Heidelberg, ISBN 978-3-642-35285-0
7. T. F. Chan, L. A. Vese, *Active contours without edges*. IEEE Transactions on Image Processing, Volume 10, Issue 2, pp. 266-277, 2001
8. H. Liu, *Science and engineering of droplets: fundamentals and applications*. Norwich, NY: Noyes Publications, c2000. ISBN 0-8155-1436-0

https://doi.org/10.1051/epjconf/202226901038

In-vitro measurement of aerosol generated by Respimat SMI penetrating through upper airways

Ondrej Misik [1*], *Frantisek* Lizal[1], *Milan* Malý[1], *Jakub* Karas[2], *Jan* Jedelsky[1], and *Miroslav* Jicha[1]

[1]Faculty of Mechanical Engineering, Brno University of Technology, Technická 2896/2, 616 69 Brno
[2]Faculty of Pharmacy, University of Veterinary and Pharmaceutical Sciences, Palackého 1946/1, 612 42 Brno

Abstract. Respimat soft mist inhaler (SMI) is an inhaler producing a soft mist with low velocity. The liquid drug solution is emitted by the mechanical energy of compressed spring, so the device is propellant-free. This paper describes the measurement of particle fraction penetrating through the model of the upper respiratory tract (mouth cavity, pharynx, larynx trachea). The solution inside the cartridge was marked by fluorescein in order to evaluate the regional particle deposition in the model. After the exposition, segments of the model, as well as the filter, were washed out and the deposited fluorescein was extracted into water. The analysis showed that the fraction of particles penetrating through the upper respiratory tract was 64.7±6.7 %. However, the real inhalation pattern was not simulated and measurement was provided with continual "inhalation" without "exhalation".

1 Introduction

An inhalation treatment grew in popularity in the last decades. It is a non-invasive and effective way of drug delivery and it is attractive for patients as well as caregivers. Today the inhalation therapy is commonly used for the treatment of pulmonary diseases but there are already developed inhalation systems delivering the therapeutic aerosol into the circulatory system via alveoli as well [1]. Chronic Obstructive Pulmonary Disease (COPD) was the fifth most frequent cause of death in 2002 and according to the World Health Organisation (WHO), it will be the fourth one in 2030. COPD nowadays affects approximately 210 million people [2].

The efficacy of the inhalation treatment obviously depends on the amount of drug deposited in the patient lungs. This can be influenced by parameters of pharmaceutical aerosols, such as its velocity and particle size [3]. The commonly used inhalation systems are divided into 3 main categories: Dry Powder Inhalers (DPI), Metered Dose Inhalers (MDI) and nebulizers [4]. The DPIs are portable inhalers used for inhalation of drug formulation in a powder form. They are passive devices, it means, it is necessary to create a forced deep inspiration to suck the powder from the inhaler. A patient needs to create a sufficient inspiration flowrate to ensure the proper deagglomeration of particles. It can be complicated for children, old people or patients in the late stage of the disease. On the other hand, the passivity of DPIs makes the coordination of inspiration and particle dispersion easier in comparison with MDIs [5].

The nebulizers are usually a bit larger devices nebulizing the liquid drug solution. Commonly, they are not portable because they need to be plugged into electricity. The most common technology used for nebulization of liquid drug solution is the compressed air. However, there are other nebulization technologies, such as vibrating mesh or ultrasonic vibrations [6]. However, the using of nebulizers is more time-consuming in comparison with MDI and DPI; one dosing takes approximately 7 minutes [5].

The MDIs are also portable devices emitting a precisely metered dose of the drug formulation. MDIs use a propellant (nowadays Hydrofluoroalkanes – HFA) to disperse the formulation into the aerosol form. MDIs contain a reservoir with a mixture of drugs, liquidized propellant, and other components. After actuation, this pressurized formulation (metered in the metering valve system) is emitted through the nozzle [7]. It creates an aerosol with high velocity, where many compounds of formulation evaporate immediately after dispersion [8]. A major fraction of particles generated by these devices deposits in upper airways (mouth cavity, pharynx larynx) due to the significant effect of inertial impaction. It is gaining in importance with the increase of particle size and particle velocity. This aerosol is often cold and it can decrease the effectivity of treatment in patients with cold air induced bronchospasm [9]. The small deposition fraction within the lungs in case of MDI was shown many times. *Newman et al.* [10] in a wide-frame study of fenoterol particles deposition found that the deposition fraction (a mass of particles deposited in selected area expressed as a percentage of total deposited mass) in the lungs using the pMDI is lower than 20 %. Similar results were acquired also in his previous study [11] comparing lung deposition with and without using the auxiliary devices (spacers) [11]. According to the *Egan's Fundamentals of Respiratory Care* [9], in general,

* Corresponding author: Ondrej.Misik@vutbr.cz

approximately 80 % of aerosol emitted by MDIs deposit in the oropharynx.

This was the reason for the development of the Respimat SMI. Low aerosol velocity and longer time of aerosol emission should cause lower oropharyngeal deposition. The Respimat SMI (Boehringer Ingelheim, Germany) is a propellant-free inhaler applied for asthma and COPD therapy. This system is very popular nowadays and it is becoming the preferable inhaler over the pressurized Metered Dose Inhalers (pMDI) or Dry Powder Inhalers (DPI) [12]. The Respimat uses a special system for nebulizing the drug solution, which enables to produce the aerosol stream with low velocity. The drug solution is stored in an aluminium cartridge. By rotation of the clear base of the Respimat by 180°, the spring compresses, moves the cartridge against the capillary tube and transfers the precisely metered amount of drug solution from the cartridge into the dosing chamber through the capillary tube [12]. A most important part of this device is the uniblock. It consists of two nozzle outlets placed in extremely fine filter channels. After pressing the dose release button, the liquid is forced through these nozzles, it means the solution is split into two streams within two nozzles converging in the precisely controlled angle. The collision of these two streams causes the atomization of the liquid into slow-moving aerosol – it means 0.8 m/s [13,14]. This device produces a high fine particle fraction 65 — 80 % (particles with diameter < 5.8 μm) [3]. The time period of the aerosol emission is much longer than in the case of pMDIs. It facilitates the coordination of actuation of the device and inspiration.

Pitcairn et al. [15] performed measurements with radiolabelled aerosol produced by Respimat SMI and selected DPI and pMDI. The aerosol was inhaled by fourteen patients with mild to moderate asthma and ages between 19 and 65 years. The flow rate of 30 L/min was used for all devices except for DPI, for which the study was provided in both 30 and 60 L/min flowrates [15]. In the case of Respimat SMI was the dose deposited in lung determined to be 51.6 (46—57) %, then 19.3 % of the dose deposited in the oropharynx, and 17.7 % in the device. The fraction exhaled from the respiratory system

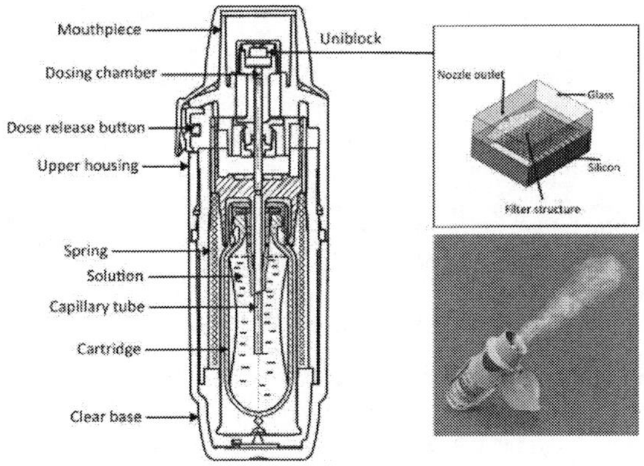

Fig. 1: Scheme of Respimat SMI [12].

was 10 % [15]. For cases of DPI (60 L/min), was the percentage of the dose deposited in lungs 28.5 %, for DPI (30 L/min) it was 17.8 % and for pMDI only 8.9 % [15]. These data show very good effectivity of the Respimat SMI device in comparison with commonly used inhalers.

This contribution reports similar measurements of Respimat SMI, however, our experimental setup included a mechanical model (replica) of human upper airways created by *Lizal et al.* [16]. This approach is supposed to provide better repeatability of the results and hence can be suitable for validation of potential computational simulations.

2 Methods

Firstly, the liquid was removed from the canister of the Respimat SMI placebo and replaced by fluorescein solution with a concentration of 3 mg/ml. Because number of doses was limited the concentration had to eb so high. [17]

For the measurements of aerosol deposition the Respimat SMI Placebo with refilled cartridge was used. The parameters of aerosol produced by this device with the labelled formulation were validated by Laser Doppler Anemometry/Phase Doppler Anemometry (LDA/PDA) and compared with parameters of Respimat SMI Spiolto (the real drug formulation) measured by the same methods. The differences between parameters (velocity and particle size distribution) of aerosols generated from these two formulations were negiligible. Since the differences between parameters of these two solutions were negligible the measurements of aerosol deposition could be provided [17].

Subsequently, the experimental rig for the measurement of the particle deposition was set. It consisted of the oropharyngeal region segment and trachea segment connected together (illustrated in Figure 2 and in [16]), the hose with a nitrocellulose membrane filter connected to a vacuum pump and all the contacts were sealed well. [17]

The flow rate was set to 30 L/min. This flow rate was steady, it means the inhalation and exhalation were not simulated. [17]

The Respimat SMI was placed coaxially with the cylinder on the front side of the mouth cavity. Two series consisting of three consecutive repeated measurements were performed. In the first series, 10 doses were emitted per one measurement. In the second series, 20 doses were emitted per one measurement [17].

After the exposure described above, it was necessary to extract the deposited aerosol into distilled water. The membrane filter was placed into the ultrasonic cleaner for 5 minutes. The concentration of fluorescein was determined by fluorimeter (Quantus^{TM} Fluorometer, Promega) and deposited mass was calculated according to the fluorescein concentration and sample volume. [17]

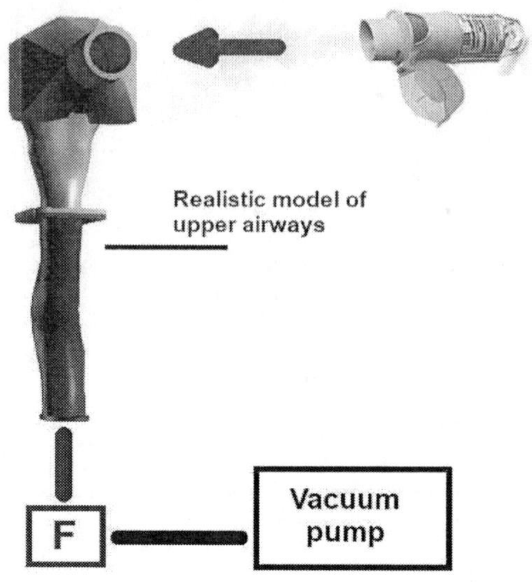

Fig. 2 : Scheme of the assembly.

3 Results

The deposited particle mass in each segment was determined. This mass was transformed into the deposition fraction defined as the mass of particles deposited in one segment divided by the total deposited mass. From the acquired results it is visible that the marginal part of aerosol was able to penetrate into the tracheobronchial tree. The fraction measured on hoses and filter (it means in the parts after oropharyngeal region and trachea) in all of the measurements exceeded 50 % and it was higher than 60 % in most of the measurements. The fraction of particles deposited within the model of the trachea was very low in comparison with the oropharyngeal region and hoses with filters, as expected.

Results of the deposition fraction in the oropharyngeal region and in the last segment (hoses and filter) were quite variable as can be seen from standard deviations shown in Table 1. However, although the average results of deposition fraction in the last segment differ between measurements with 10 doses and 20 doses, it seems the number of actuations did not have a

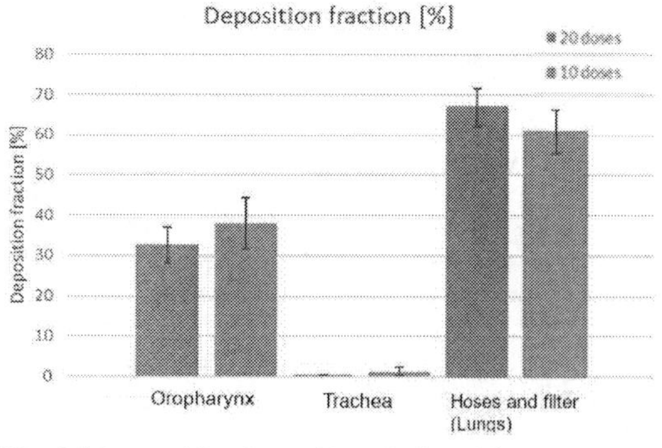

Fig. 3: Measured fractions of deposited particles.

significant impact on the results. In order to test the effect of the number of used doses, it would be necessary to repeat the measurement much more times. If we assume that the experiments with 10 and 20 doses are equivalent, the average value of deposition fraction in segments below oropharyngeal region is 64.7±6.7 %.

4 Discussion

Data gained by measurements presented in this paper show a significant difference in comparison with the oropharyngeal deposition of common MDI. The oropharyngeal deposition in these measurements was approximately 35% and fraction penetrating into lower airways approximately 63 %. Even better deposition data were gained during *in vitro* measurements presented in the source [18], where the oropharyngeal deposition fraction was determined as 27 — 29 %.

During the *in vivo* measurement performed using radiolabeling provided by *Pitcairn et al.* [15], the lungs deposition fraction was a bit lower - 51,6 % (46-57 %) than in the measurements described in this paper. However, during our measurements, the aerosol was delivered into the model by steady airflow of 30 l/min, whereas during the above mentioned *in vivo* measurements, the whole inhalation and exhalation were provided.

5 Conclusion

Since the Respimat SMI is an important novel device for an inhalation treatment, measurements of oropharyngeal aerosol deposition of particles produced by this device were performed on the realistic model of upper airways. The liquid solution of fluorescein was atomized by Respimat SMI and inhaled into the model with constant flow rate of 30 l/min. In general, current results are in good agreement with previous measurements. The deposition fraction published in *Delvadia* thesis of dissertation [19] measured with the same steady flowrate (30 l/min) was even a bit higher. However, the *in vitro* measurements performed on Alberta throat model [20] compared the lung deposition of Respimat aerosol using different breathing cycles (moderate and very severe COPD) and significant variability between breathing patterns were shown. Certain differences were found after comparison with previously published measurements of *Pitcairn et al.* [15], who measured *(in vivo)* the whole unsteady inhalation - exhalation cycle. Nonetheless, in their case, the exhaled fraction was approximately 10 %, which accounts for the difference between the our measurements. For more accurate determination of the lung deposition within the human lungs, the model with breathing simulator created at the Brno University of Technology (BUT) [16] will be used. The measurements described in this paper form the basis for the future measurement of regional lung deposition of aerosol generated by Respimat SMI.

6 Acknowledgement

This work was supported by the Czech Science Foundation under the grant GA18-25618S, and by the project FSI-S-17-4444.

7 References

1. J. S. Patton and P. R. Byron, Nat. Rev. Drug Discov. **6**, 67 (2007).

2. World Health Organisation, *Global Surveillance, Prevention and Control of Chronic Respiratory Diseases: A Comprehensive Approach* (World Health Organization, 2007).

3. R. Dalby, M. Spallek, and T. Voshaar, Int. J. Pharm. **283**, 1 (2004).

4. M. Ibrahim, R. Verma, and L. Garcia-Contreras, Med. Devices (Auckl). **8**, 131 (2015).

5. N. Yadav and A. Lohani, **3**, 142 (2013).

6. O. NERBRINK, Inhalation **10**, (2016).

7. D. J. W. Grant and F. Group, *Metered Dose Inhaler Technology* (1997).

8. A. R. Martin, D. Y. Kwok, and W. H. Finlay, J. Aerosol Med. **18**, 218 (2005).

9. R. M. Kacmarek, J. K. Stoller, A. J. Heuer, and D. F. Egan, *Egan's Fundamentals of Respiratory Care*, 10th ed. (Elsevier/Mosby, St. Louis, Mo, 2013).

10. S. P. Newman, G. Pitcaim, K. Steed, A. Harrison, and J. Nagel, J. Allergy Clin. Immunol. **104**, 253 (1999).

11. S. P. Newman, A. R. Clark, N. Talaee, and S. W. Clarke, Thorax **44**, 706 (1989).

12. A. J. Hickey, *Pharmaceutical Inhalation Aerosol Technology* (2019).

13. A. J. Hickey, *Pharmaceutical Inhalation Aerosol Technology, 3rd Edition* (1966).

14. R. N. Dalby, J. Eicher, and B. Zierenberg, Med. Devices (Auckl). **4**, 145 (2011).

15. G. Pitcairn, S. Reader, D. Pavia, and S. Newman, J. Aerosol Med. **18**, 264 (2005).

16. F. Lizal, J. Elcner, P. K. Hopke, J. Jedelsky, and M. Jicha, *Development of a Realistic Human Airway Model* (2012), pp. 197–207.

17. O. Mišík, Master Thesis, Brno University of Technology, 2019.

18. P. W. Longest and M. Hindle, J. Aerosol Med. Pulm. Drug Deliv. **22**, 99 (2009).

19. R. Delvadia, Thesis of Dissertation, Virginia Commonwealth University, (2012).

20. A. M. Ciciliani, P. Langguth, and H. Wachtel, Int. J. COPD **12**, 1565 (2017).

Impact of baffle geometry on the fluid motion in the stirred vessel

Anna Młynarczykowska[1,*], *Simone* Ferrari[2], *Luana* Demurtas[2], and *Marek* Jaszczur[3]

[1]AGH University of Science and Technology, Faculty of Mining and Geoengineering,, A.Mickiewicza 30, 30-059 Kraków, Poland
[2]University of Cagliari, Department of Ingegneria Civile, Ambientale e Architettura, Via Marengo 2, 09123 Cagliari, Italy
[3]AGH University of Science and Technology, Faculty of Energy and Fuels, A.Mickiewicza 30, 30-059 Kraków, Poland

Abstract. The fluids mixing is a crucial operation in a large number of engineering systems. It has major significance in chemical engineering, food, cosmetics and pharmaceutics production, biotechnology, wastewater treatment engineering, and countless other applications. Among many available systems online mixing with static mixers and stirred tanks plays a primary role and has been developed to meet several processing objectives. The effectiveness of the mixing process depends on a number of parameters, i.e. impeller shape, mixing phases properties, process conditions as well as stirred vessel design - in particular, number and baffles' design. The optimal baffles geometry is still an open issue and is usually design with trial and error methods. In this study, the focus is on the experimental investigations of the baffle geometry on the fluid flow and mixing phenomenon as well as on the required by the mixer power. In order to evaluate velocity field and mixing parameters, particle image velocimetry measurement is used whereas to obtain the power consumption by the stirred vessel precise torque meters were used. Measurements are carried out for different Reynolds numbers, to determine the most efficient process parameters. It has been shown that for the analyzed range of Reynolds numbers, the baffles design significantly influences fluid flow motion, mixing phenomena and the pumping number power number, but not affect the power number.

1 Introduction

Agitated tanks are widely used in chemical processes, food industry, pharmaceuticals, wastewater treatment, mineral and oil processing. The mechanical mixing is essential in selected processes as gas dispersion in liquids, the formation of suspensions and slow-sedimenting mixtures, prevention of sediment aggregation [1,2]. Depending on the type of mixed substances and the achievement of the assumed technological effect, different types of stirred vessel and impeller are used. Mixing, in general, fulfils two main objectives: technological and research. The technological aspect is the preparation of mixtures and the increasing efficiency of simultaneously occurring phenomena, chemical reactions, bioprocesses, dissolution, crystallization, cooling or heating. The research aspect is the creation of mathematical and physical models and optimization of the mixing process. The practical effect of the research is the development of new constructions of agitated vessels and rotors. Depending on the geometry and impeller speed, mixer produces a unique velocity profile: radial, axial or mixed. The most commonly used mixers are: turbine, propeller, foot, frame, etc. Their use depends primarily on the type of the mixing liquid and in particular on the liquid viscosity. Turbine and propeller impellers have the widest range of applicability. They can mix liquids or gas-liquids mixtures with an average viscosity range from 1 to about 10^4 mPa·s.

For this reason, the most-high-speed turbine agitators are commonly used in industry. Thus, they produce radial or radial-axial fluid flow. A special type is a disc agitator with blades - a Rushton turbine. Because of the high risk of the hopper formation and turbulence, in mixers quite often the baffles are used (usually two to four), which separate stirring areas. Researchers obtain a very good agreement between experimental and predicted mixing time over a wide range of impellers geometries or diameter and in the area of number blades, blade width and blade angle[3,4]. Thus, in addition to the stirred vessel geometry, the number, shape and position of the blades also have a significant impact on the power, efficiency and quality of mixing. It has been found that the optimal

* Corresponding author: mindziu@agh.du.pl

number of blades for turbine impeller is six. A larger number of blades in the turbine only slightly increasing the power mixing and has negligible impact on mixing phenomena.

Many researchers focus on the mechanical mixing. In the past, different solutions about the shape of the tank and rotor were tested only experimentally. Currently, thanks to advanced computational technology and based on mathematical models, the mixing process hydrodynamics can be predicted for more and more complex systems, replacing the time-consuming and expensive real experiments. However, due to the complexity of mixing processes, experimental verification of numerical modelling is still highly required as a final step of analysis or optimisation process [5].

In the literature large number, experimental and numerical investigation can be found. Scargiali et al. [6,7] studied the consumption of energy in the mixing process. Authors showed that the homogenisation process for mixture in a stirred vessel with rotation impeller without baffles does not require much electrical power. Jaszczur et al. [4] analysed the process of the fluid flow in the mechanically agitated vessel without baffles with new impeller type versus the Rushton turbine. The authors, as the basis for the assessment of the intensity degree and efficiency of mixing, used the analysis of velocity vectors distribution and power number. Based on an experimental and numerical study carried out for various stirred process parameters, it was demonstrated that the power number is at least 6-10 times higher for the Rushton turbine then for the novel impeller. This results in larger power consumption at the same rotational speed. On the other hand, the pumping number was calculated which depends significantly on the Reynolds number. Both these values show that efficiency defined as a ratio of pumping capacity Q_c to the power number N_p can be substantially higher for the tested novel impeller than for the Ruston turbine. In the literature, not much information can be found related to the universal applications of the stirred vessels with or without baffles, particularly for the very viscous fluids. In the case of very viscous fluids and when very important is high product quality not exist a lot of studies about using the vessel with baffles [8].

Godlewska and Karcz [9] for low viscous fluids showed that the value of the power number depends on the length L of baffles in the stirred vessel and that the Newton number is increasing as the length of the baffles increases. The influence of non-standard the baffles located inside the vessel was described in research studies Major i Karcz [15]

In the work of Nishikawa et al. [3], the correlation of mixing time with the number of baffles was examined by measuring the energy consumption by the rotor. This relationship was defined as a constant and characteristic for mechanical mixing. Based on the results, the authors concluded that if the width of the baffles is less than one-tenth of the length of the tank diameter, the optimal number of baffles are not less than three. The influence of the size of the baffle, the type of mixed fluid or the degree of aeration was studied by Lu et al. [10]. The presented numerical and experimental analyses showed that the uniform distribution of baffles and lack of air increases the quality of mixing. The authors also showed that there is not a universal number of baffles used that will be good for any mixing process.

Researchers agree, however, that the main task of baffles is to reduce angular velocity (fluid rotation in the vessel) and to cause additional turbulence which causes that the mixing process will be more intensively. The key, therefore, is to combine the high mixing enhancement with the mixer operating parameters, such as power consumption; type, amount and positions of baffles as well as fluid parameters like liquid viscosity [8]. Many studies describe laminar mixing, which can be a better solution in reference to high-speed turbulent more destructive mixing with typically required higher consumption of the power. In interesting studies, Ferrari and Rossi [11] measured the mixing efficiency in a quasi-turbulent flow via the experimental measurements of power input and output in the flow.

The mixing process can also be described utilizing numerical modelling by determining the distribution of mean and turbulent kinetic energy[13,1].

In literature, many research papers can be found in which researchers presented assessments of various turbulent models. It was found that the standard k–ε model under-predict the turbulent kinetic energy in the region of the rotor and failed to predict the mean flow associated with the strong vortex. The reason is that the models do not correctly describe the lack of fluid movement in the rotor area.

This paper presented selected results of the extensive analysis of different baffles configurations for variable rotor speeds and its optimal position. The goal was to obtain the best mixing process effectiveness. The presented analysis is based on the experimental measurements and include power number N_p, pumping capacity Q_r (two key parameters which describe mechanical mixing) as well as velocity fields for a selected cross-section from Particle Image Velocimetry (PIV) measurement.

2 Experimental set-up and methodology

Figure 1 presents the mechanically agitated vessel configuration. The dimensions of the chamber with the baffles used in experiments and also their location and the distance from the vessel bottom, are shown in Table 1.
As a working fluid 100% glycerine (ρ=1261.1 kg/m^3, μ=1.4101Pa·s at 20°C) as used.

Abbreviations used in Figure 1 are as follows: H-fluid level, D-inner diameter of stirred vessels, d-diameter of the agitator, h-distance between the agitator and bottom of the vessel, L-length of the baffle, s-distance between lower edge of the baffle and bottom of the vessel, w-width of the baffle, b-width of the agitator blade, a-height of the agitator blade. The thickness of the baffles and impeller was around 2 mm. The distance between the impeller and the bottom of the tank h was fixed and set at $\frac{1}{3}$H.

Table 1. Dimensions of the stirred vessels.

H	h	D	d	w	a	b	s	L
mm	mm	mm	mm	mm	mm	mm	mm	mm
170	57	170	57	20	12	15	0-170	170-0

The agitated vessel investigated in this work has smooth cylindrical shape walls with the possibility of installing up to four baffles of various lengths and distances from the vessel bottom. As an impeller six-bladed Rushton turbine was used.

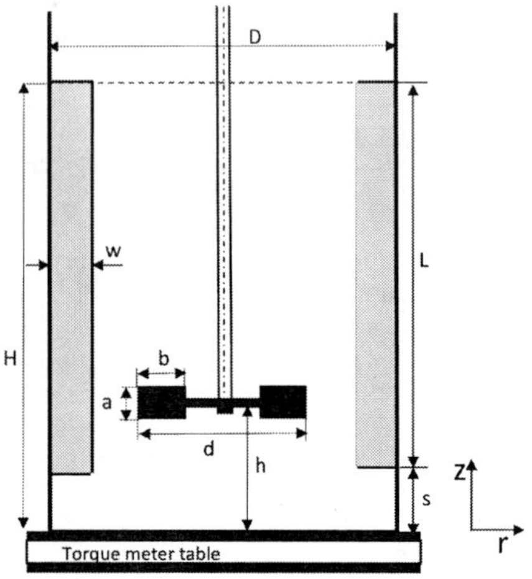

Fig. 1. Schematic diagram of the turbine and stirred tank.

In a first step the torque was measured by placing the impeller, by installed the stirred vessel on the measuring device with torque meter FSA-2 (AXIS, max 2Nm) enabling the measurement of the analysed impeller torque with the precision of 0.001Nm and sampling frequency of 1000 Hz. For verification of measurements on the rotor shaft was installed an additional calibrated torque transducer DATAFLEX 16 (KTR, max 10Nm, inaccuracies 0.001) connected to the data acquisition system National Instruments DAQ.

In the second step, velocity fields were measured using the Particle Image Velocimetry (PIV) method.

For evaluating, mixing efficiency the following parameters were determined from the experimental measurements Newton number (non-dimensional power number), flow field and pumping capacity Q_r. The instantaneous velocities in the vertical direction were measured in the cylindrical vessel, which was placed inside of the square tanks filled with water. PIV measurements were carried out for the constant position of the impeller and variable rotor revolutions and the presence or absence of baffles inside the vessel. All the measurement were performed for the fluid high H=D.

Because the fluid level increases with the agitator speed, therefore the only sub-critical range of impeller speeds was analysed to prevent air bubbles in the fluid.

The stirred vessel was illuminated with a double-pulse Nd:YAG laser of the energy of about 30 mJ per pulse. A vertical laser sheet of height 200 mm and the thickness about 1 mm was located through the centre of the tank. At current configuration, single CCD camera - La Vision with a resolution of 2048x2048 pixels was used. The PIV set-up after calibration procedures under fluid flow condition was testes in order to evaluate the minimum number of images required to ensure accurate mean velocity field. After analysis for laminar flow, the number of frames acquired with the CCD camera was set to 100 as a sufficient for low Reynolds number analysis (Re=10-160) presented in this paper. The methodology of PIV measurement described in detail in the paper Młynarczykowska et all. [12]

In the final step of experimental analysis based on the results two keys criteria numbers were evaluated: Power number N_p and Reynolds number Re. Those parameters designated according to the equations:

$$N_P = \frac{P}{\rho N^3 d^5} \quad ; \quad Re = \frac{N d^2 \rho}{\mu} \qquad (1)$$

where: ρ - density (kg/m^3), μ- viscosity (kg/m·s), N - agitator speed (1/s) , d - impeller rotor maximum diameter (m), P – power (W) required to rotate the impeller with the specified rotational speed (depends on the torque). Power was calculated from torque measurement as follows :

$$P = 2\pi \cdot N \cdot \tau \qquad (2)$$

where τ is the torque (N·m).
The last relevant parameter is the pumping capacity Q which dependents on the Reynolds number. It can be calculated for various Reynolds numbers through a cylindrical surface with a specified radius r or through the horizontal surface at specified heights z.
In this work due to impeller height variance more relevant is radial pumping capacity calculated through the cylindrical surface with a set point of radius r= 35 mm
Pumping capacity Q_r was evaluated through a vertical surface (radial pumping capacity and for the height z_1=0 and z_2=0,17 m using the relationship:

$$Q_r = \int_{z_1}^{z_2} |u_r| dS \qquad (3)$$

In the literature of the mixing, this parameter is also quite often normalise by the component $N{\cdot}d^3$. This quotient is the flow number, which represents the global flow characteristic.

3 Results and discussion

In order to correctly assess the efficiency and economy of fluid mixing the required power evaluation for the mixing process as a major parameter. Figure 2 shows the power number N_p evaluated for the Rushton turbine and for fixed impeller distances from the tank bottom equal to 1/3H i.e. about 57mm.

For this analysis, the impeller distance was fixed but the position of the four installed baffles above the bottom of vessel *s* was set at: ¼H (42mm),1/3H (57mm), ½H (85mm). Thus, the (active) length *L* of the baffles located inside of the cylindrical vessel (on the walls) was respectively, 3/4H, 2/3H, 1/2H.

Based on the results obtained, it can be indicated that the power number decreases with the position ratio s/H and increasing with the length (L) of the baffles for fixed rotor revolutions. However, the changes are very small and clearly visible only for the lowest Reynolds number analysed here. For the Reynolds numbers between 10 and 30 power number is almost independent on the length nor height of the baffle.

One can infer from this figure that the presence of baffles in the tank, regardless of their position during the measurement, did not cause significant changes in the evaluated power number for any of of the speed rotor analysed here. Similar results were observed for the case without baffles (L=0). Increasing Reynolds number, Power number N_p decreases significantly for all investigated cases including no baffle case.

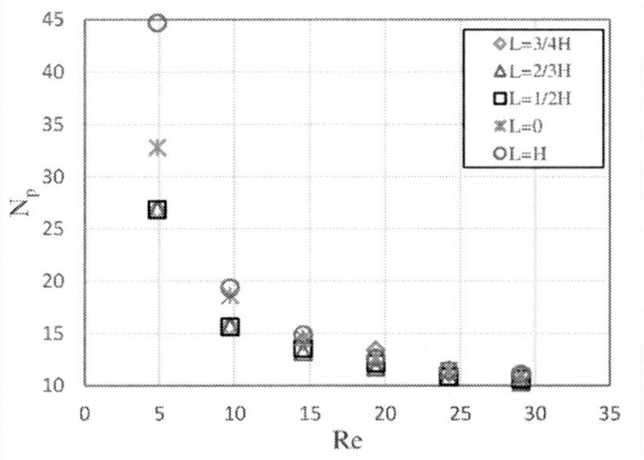

Fig. 2. Power number N_p for Rushton turbine and fixed impeller distances from the tank bottom h=1/3H and various baffles lengths or no baffles (L=0).

Figure 3 presents a radial of normalized pumping capacity Q_r/Nd^3. As can be seen, the pumping number values are small in the range from 0.099 to 0.20. The exception is the calculated and significantly different by remaining, Qr <1 value for Re> 14 for N = 300rpm

Analogy conclusion was presented Godlewska i Karcz [14] own research carried out for liquids of low viscosity. They tested the effects of geometrical parameters of the baffles in different positions from the flat bottom of the vessel. It showed that for the Rushton turbine function Np=f(L/H) decreases with shorter length L of the baffles. Worth noticing, that the power number of the standard Rushton turbine (the angle of blades is 90°) for an agitated vessel without of baffles it can be even twice as higher rather than for different angles of blades. However, overall they are numerically low values, especially for low viscosity liquids.

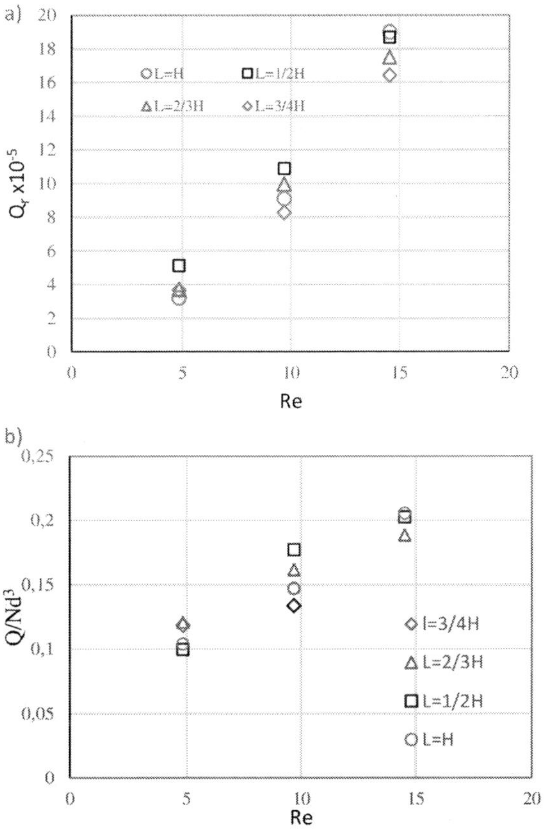

Fig. 3. (a) Radial pumping capacity Q_r for Rushton turbine in function of Reynolds number and for different position of baffles from tank bottom, (b) normalized pumping capacity $Q_r/N{\cdot}d^3$.

Figures 4-6 presentation of the velocity vectors evaluated with Particle Image Velocimetry in *z-r* cross-section plane and for tree rotational speeds *N*=100, 200, 300 rpm, for the 4 positions of baffles from bottom of the vessel s =0, s=¼H, s=⅓H, s=½H, the length of the baffle respectively is: L=H, L=3/4H, L=2/3H, L= 1/2H. The fluid intensive recirculation zone is seen in the area of rotor activity regardless of the position of the baffle. Clearly intensive radial and mixed flow can be observed as the rotor speed increases.

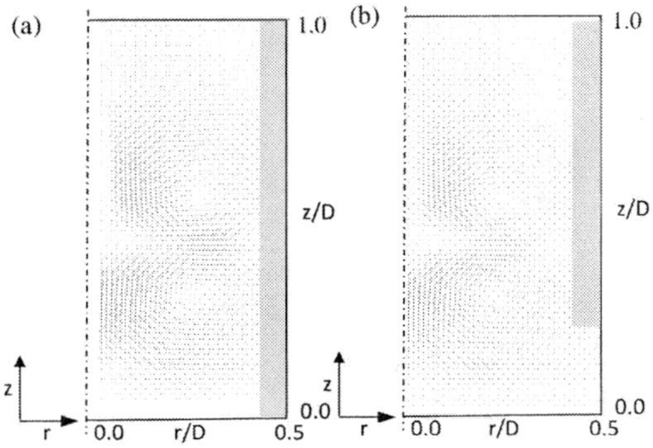

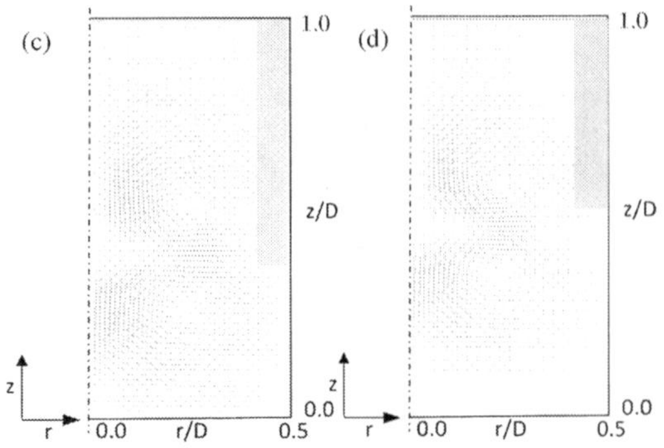

Fig. 4. Velocity field for h/H=1/3 Rushton turbine for rotational speed N=100 rpm and lenght of baffles: (a) L=H, (b) L=3/4H, (c) L=2/3H, (d) L=1/2H.

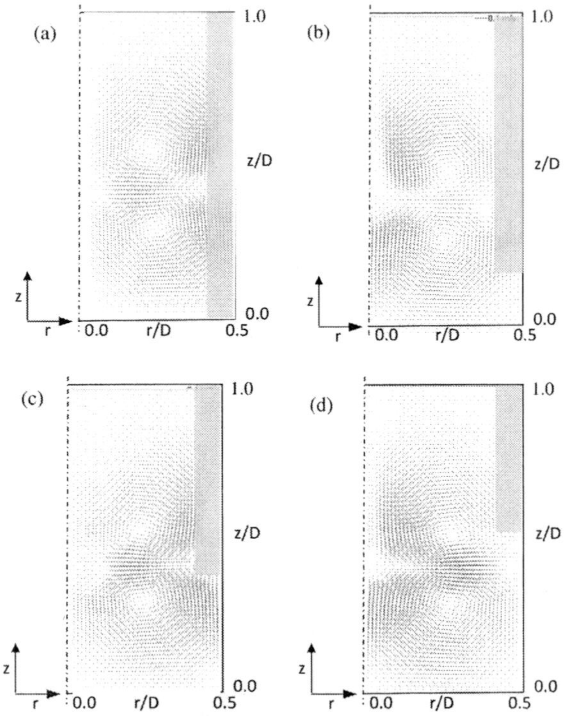

Fig. 5. Velocity field for h/H=1/3 Rushton turbine for rotational speed N=200 rpm and length of baffles : (a) L=H, (b) L=3/4H, c) L=2/3H, d) L=1/2H.

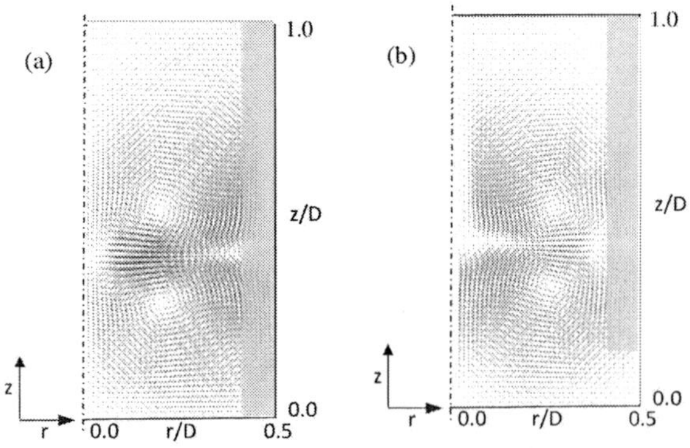

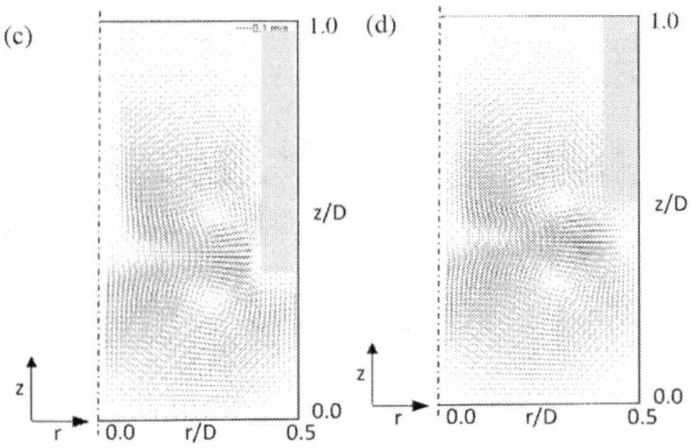

Fig. 6. Velocity field for h/H=1/3 Rushton turbine for rotational speed N=200 rpm and lenght of baffles (a) L=H, (b) L=3/4H, (c) L=2/3H, (d) L=1/2H.

4 Conclusions

This paper shows an experimental investigation of the mixing process for the high-viscosity liquid - glycerine. It was presented that the power number N_p depends on the Reynolds number (as the result of rotor speed) and only weakly on the length of the baffle inside the vessel.

The most substantial effect of tank geometry was found for the agitated vessel equipped with the baffles (L=H), where the power number decreases as an exponential function

Analysing the calculated values for pumping power and power number, it might seem that the presence of the baffles in the vessel (regardless of the location) does not have the important effect of mixing efficiency nor required power. However, PIV measurements clearly showed that the despite high viscosity of liquid, the fluid was mixed. Therefore, it is worth considering the use of a custom description for the fluid flow of different viscosity take into account the dynamic and mechanical parameters of the stirrer. The authors are going to investigate this issue in the subsequent studies.

Acknowledgements The present work was partially supported by the Polish Ministry of Science (Grant MNiSW/2019/162/DIR).

References

1. M. Jaszczur, A. Młynarczykowska, R. Hanus, *An analysis of the velocity field distribution inside the flotation chamber*, J. of Physics. Conf. Series, **745**, 1 (2016)
2. M. Jaszczur, A. Młynarczykowska, L. Demurtas, *Effect of impeller design on power characteristics and Newtonian fluids mixing efficiency in a mechanically agitated vessel at low Reynolds number*, Energies, **13**, 640 (2020), doi:10.3390/en13030640
3. M. Nishikawa, K. Ashiwake, N. Hashimoto, S. Nagata, *Agitation power and mixing time in off-centering mixing*, Int. Chem. Eng. **19**, 153 (1979)

4. M. Jaszczur, A. Młynarczykowska, L. Demurtas, *An experimental and numerical analysis of the fluid flow in mechanical agitated vessel,* ICCHMT, E3S Web of Conferences 128, 08002 (2019)
5. A. Delafosse, A. Line, J. Morchain, P. Guiraud, *LES and URANS simulations of hydrodynamics in mixing tank: comparison to PIV experiments*, Chem. Eng. Res. and Design, **86**, 1322 (2008)
6. Scargiali F, Busciglio A,Grisafi F,Tamburini A, Micale G, *Power consumption in uncovered unbaffled stirred tanks: influence of viscosity and flow regime,* Ind. & Eng. Chem. Res., **52**, 14998 (2013)
7. Scargiali F, Tamburini A, Caputo G, Micale G, *On the assessment of power consumption and critical impeller speed in vortexing unbaffled stirred tanks*, Chem. Eng. Res. and Design, **132**, 99 (2017)
8. Basavarajappa M, Draper T, Toth P, Ring T. A, Miskovic S, 2015, *Numerical and experimental investigation of single phase flow characteristics in stirred tanks using Rushton turbine and flotation impeller*, Minerals Engineering, **83**, 156
9. M. Major- Godlewska, J. Karcz, *Power consumption for an agitated vessel equipped with pitched blade turbine and short baffles*, Chem Zvesti., **72**, 1081 (2018)
10. W. Lu, H.Wu, M. Ju, *Effect of baffle design on the liquid mixing in aerated stirred tank with standard Rushton turbine impellers*, Chemical Engineering Science , **52** (1997)
11. S. Ferrari, L. Rossi, *Particle tracking velocimetry and accelerometry (PTVA) measurements applied to quasi-two-dimensional multi-scale flows,* Exp Fluids, **44**, 6 (2008)
12. A.Młynarczykowska, S.Ferrari, L. Demurtas, M. Jaszczur, *An experimental investigation on the fluid flow mixing process in agitated vessel*, EFM 2019 (conference material)
13. J.H. Rushton, E. W. Costich, H.J. Everett, *Power characteristics of mixing impellers Part II*, Chem. Eng. Progress, **46**, 467 (1950)
14. J. Karcz, J. M. Major, *An effect of the baffle length on the power consumption in an agitated vessel*, Chem. Eng. Proc., **37**, 249 (1998)
15. J. M. Major, J. Karcz, *An effect of the tubular baffles configuration in an agitated vessel with a high-speed*, Chemical Papers, **72**, 2933(2018)

An experimental investigation on the fluid flow mixing process in agitated vessel

Anna Młynarczykowska[1], *Simone* Ferrari[2], *Luana* Demurtas[2], and *Marek* Jaszczur[1,*]

[1]AGH University of Science and Technology, Faculty of Energy and Fuels A.Mickiewicza 30, 30-059 Kraków, Poland
[2]University of Cagliari, Department of Ingegneria Civile, Ambientale e Architettura, Via Marengo 2, 09123 Cagliari, Italy

Abstract. The fluid mixing process is a common supportive phenomenon that often occurs in a large number of industrial systems. This phenomenon is the subject of many numerical and experimental analysis. The mixing process effectiveness depends on: mixing tank construction, mixing phases viscosity, temperature, density of liquids and, what is crucial, the impeller shape. The optimal design of impeller geometry is still an open issue. In this research work, the main objective is experimental investigations of the influence of the newly constructed impeller type on the fluid flow motion phenomena and energy consumption. Flow field values were evaluated using PIV measurement and the power consumption using precise torquemeter. The comparison between the Rushton turbine and a novel impeller is presented and discussed. The basis for the assessment of the intensity degree and efficiency of mixing was the analysis of velocity vectors distribution and power number. Results show that the power number for both impellers are similar but the fluid motion is quite different. The pumping capacity Q_z for the novel impeller in reference to the Rushton turbine is for many cases at least one order of magnitude higher. This shows that the proposed impeller can be a very promising alternative to the classic blades and non-blades based impeller types.

1 Introduction

Agitated vessels are widely used in a large number of the process industries for example chemical, food, oil, pharmaceuticals as well as wastewater treatments. The flow and mixing phenomena of fluids in a vessel were always of particular attention among scientist and engineers. To obtain good quality products with economical way it is crucial to evaluate mixing performance, quality and fluid motion in the vessel. It is known that a large amount of electrical energy is required for a mixing process and that generates significant expenses. For this reason, many studies have been done to obtain various impellers shapes that could guarantee a high mixture quality with low axial velocity, low internal heat generation and with low power consumption.

The flow phenomena inside of agitated vessels have been analysed using various types of measurements method such as LDV, LIF, PIV. Impellers may create highly turbulent and very complex 3D flow as a result of interaction between blades and tank walls. Up to present, many analyses performed for single-phase flow allowed researchers to achieve a significant degree of awareness on the elementary mechanism of mixing. In analysis particular effort has been paid to the impeller and vessel shape. That results in several new types for the design of the impeller. Unbaffled vessel in the past was regarded as less effective to that in which baffles were installed. However recent analyses show that unbaffled vessel with properly designed impeller can be a good alternative to the baffled one in a large number of industrial processes. [1-3]. The impeller rotation for a complete mixture homogenisation can be much lower and as a consequence, electrical power required for the process is also much lower in unbaffled vessels [4,5]. Baffles are not used in the case of very viscous fluids and in applications where large shear can influence product quality (enzymes, vitamins) [6]. In the literature, there is still insufficient information on the optimal impeller design nor about fluid behaviour in unbaffled vessels. For every process industry, the target is to find an optimal way of mixing using as low power as possible and to keep high product quality.

In the past several experimental and numerical studies concentrate on laminar mixing [7] because this type of mixing can be a promising alternative to the more destructive and require higher power mixing. It has been found that the required power can be significantly decreased improving the shape of the impeller. Ferrari and Rossi [8] and Rossi et al. [9] measured the mixing efficiency in a quasi-turbulent flow via the experimental measurements of power input and output in the flow. The study of T. Su et al. [10] compared the hydrodynamics of the fluid and mixing process using the classical Rushton

* Corresponding author: jaszczur@agh.edu.pl

turbine and modified one. The results showed that 18% power decrease in the case of modified Rushton turbine. New design also managed to enhance the local flow motion and obtain flow structure which promotes mixing. As a result, shorter mixing time was required.

The flow pattern is typically impacted by the impeller shape and its configuration in agitated vessels. In [11] authors analyse the impact of Reynolds and Froude numbers on power consumption in the case of unbaffled stirred tanks. It has been found that the essential parameter in power consumption decrease and mixing quality is the impeller shape. T. Kumaresan and J. B. Jyeshtharaj [12] performed experimental analysis using LDA measurements and Computational Fluid Dynamics (CFD) in order to evaluate the effect of impeller shape on the mixing time and flow pattern for a preselected set of impellers. Authors obtain a good prediction for the mixing time over a wide range of impellers geometries. In other numerical studies [13] authors analyse turbulent flow characteristics in laboratory-scale stirred vessel with different shape. CFD simulations were performed for the Rushton turbine and compare with the flotation impeller. The impact of the baffles inside the stirred tank was analysed in [14] and as a results authors conclude that the power number depends on the baffles length and decrease in power consumption was observed for shorter baffles. A. Delafosse et al. [15] performed a research study on the flow rates between two adjacent compartments using numerical analysis. The analysis showed that the model accurately reproduces the spatial distribution of concentrations during the mixing process. H. Ameur et al. [16] performed a three-dimensional simulation to investigate the effects of blade diameter and curvature as well as blade number and Reynolds number on the power consumption reduction. Authors conclude that the curved blades provide the most efficient solution.

Recently Large Eddy Simulation (LES) become a very attractive way of problem-solving and promising method to get high resolution at spatial and temporal scales. In the case of flows with rotating impellers in a stirred tank, LES method is able to resolve all essential the flow phenomena related to the problems unsteadiness in referent to the others methods [17]. In [18] authors use LES method to and show a low computational cost model able to resolve all key flow features in agitated vessels, without losing accuracy. Presented results have shown very good agreement with experimental PIV measurements. It has been shown that to increase quality and reliability of the CFD results it is very important to perform the numerical grid dependence study as well as test and evaluate the proper size of the rotating zone when using multiple reference frame (MRF). H. Patil et al. [19] developed a numerical model for stirred vessels. The objective was to optimise the size of the rotating fluid zone MRF and evaluate if the optimal MRF size affects significant velocity results. Authors conclude that MRF dimension is an important issue which impact solution and finally predictions were found to be with good agreement with literature results. It this work authors also analysed the optimal MRF zone size on the mixing efficiency by comparing numerical and experimental results for the power number at several Reynolds numbers.

As the mixing process depends on the distribution of mean and turbulent kinetic energy, in many research studies authors have made assessments of various turbulent models. The k–ε model was compared to the Reynolds Average Stress Model and LES by the Murthy et al. [20]. Authors have done experimental measurement and numerical CFD simulation for the flow created by the different impellers. It has been found that a standard k–ε model under-predict the turbulent kinetic energy in the region of the impeller and failed to predict the mean flow related to the strong swirl. Similar conclusion has been found by the in the study performed by the H. Singh et al. [21]. They found that the accuracy of evaluation for the mean axial and angular velocities is good for all analysed models. However, most of the models do not predict properly the decay of mean radial velocity away from the impeller region.

Presented paper focuses on the analyses of different impeller configurations and position in order to get a high efficiency mixing process. This work provides velocity field, power number N_p and pumping capacity Q_z, two key parameters which describe the mixing process.

2 Experimental set-up and methodology

The mechanically agitated vessel configuration is presented in Figure 1, and the dimensions used are shown in Table 1. For presented analysis, 100% glycerol (ρ=1261.1 kg/m^3, μ=1410.1Pa·s at 20°C) was used as a working fluid with a variable viscosity for different temperatures. The distance between the impeller and the bottom of the tank h was the case parameter and was equal to ¼H, ⅓H, ½H of the fluid initial high H.

Table 1. Dimensions of the stirred tank.

Impeller type	H, H_c mm	h mm	D/d mm	a/d_1 mm	b/d_2 mm	l mm
Rushton	170,	¼H, ⅓H, ½H	170	12	15	2
Novel	300	42, 57, 85	85	15	18	40

The H denotes fluid level while H_c is the tank height and D is tank diameter. Impeller maximum diameter independent of the impeller type is denoted as d. Additional geometrical dimensions are a, b, l height, width and thickness for Rushton turbine blades and d_1, d_2, diameter at entrance and exit to the "blade" of the novel impeller and l is the length.

The stirred tank investigated has smooth walls and has no baffled pilots. Propeller is equipped with a six-bladed Rushton turbine (for case A) or three-bladed novel impeller (for case B). The Novel impeller is a rescaled copy of proposed real industrial design. The analysed impellers are created using 3D printer and have smooth plastic walls of the thickness 2 mm.

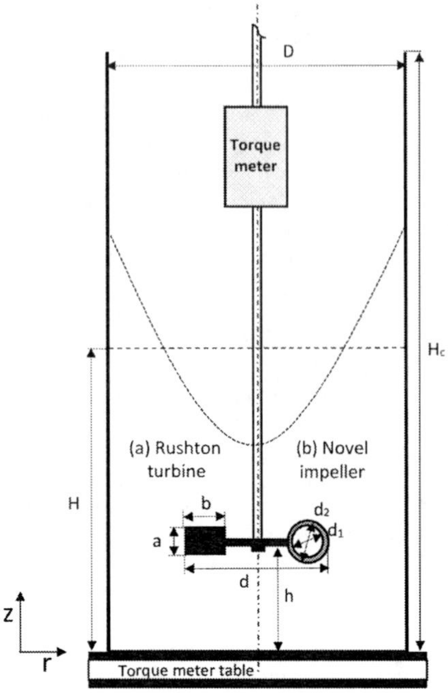

Fig. 1. Schematic diagram of the turbine and stirred tank.

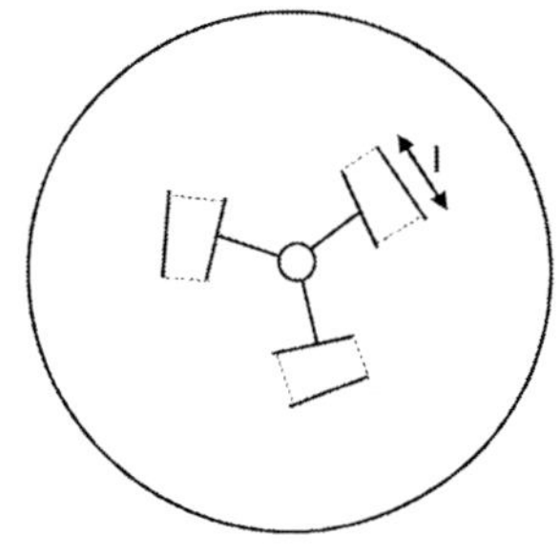

Fig. 2. Schematic diagram of the Novel impeller (top view).

The stirred tank is installed on the specially designed table equipped with torque meter FSA-2 (AXIS, max 2Nm) enabling the measurement of the analysed impeller torque with the precision of 0.001Nm and sampling frequency of 1000Hz. An additional calibrated torque transducer DATAFLEX 16 (KTR, max 10Nm, inaccuracies 0.001) was installed on the impeller shaft and connected to the data acquisition system National Instruments DAQ.

In order to evaluate mixing properties, flow field and pumping capacity, Particle Image Velocimetry (PIV) method were used. For this measurement, the cylindrical tank was placed inside of the square vessels filled with water to minimise optical distortions.

The instantaneous velocity in vertical direction were measured in the tank filed under no agitation conditions with fluid high $H=D$ (in total about 5 litres). However, it should be noticed that due to rotation fluid level close to the cylinder wall increases.

The seed particles were illuminated with a double-pulse Nd:YAG laser of energy of about 30 mJ per pulse. A vertical laser sheet of height 200 mm and the thickness about 1 mm was located through the centre of the tank directly illuminating rotor shaft and shadowing left part of the tank. At present configuration, one CCD camera (La Vision) with a resolution of 2048x2048 pixels equipped with a lens 50 mm/1.2 was used.

The flow was seeded with hollow sphere d<1μm. Reflected laser light was acquired by the CCD camera, installed perpendicular to the plane of the light-sheet. To determine the distribution of the velocity vectors, seed particles are registered by the CCD camera on two different and consecutive frames.

The measuring PIV system was subjected to testing and calibration procedures under fluid flow condition in order to evaluate the minimum number of acquired images required to ensure accurate statistics, as well as the optimal size of the interrogation window for the cross-correlation application. In each measurement, 100 frames were acquired with the CCD camera at a frequency of 5 Hz, which gives the time of single measurement equal to 20 seconds. This time and number of frames was sufficient for low Reynolds number analysis (Re=10-160) presented here. The time step Δt between 2 double-frames varied depends on Reynolds number 800-20000 μs.

During the calculations, the interrogation windows size that exhibit satisfying results varied from 32x32 pixels to 64x64 pixels. To analyse the series of double-frames and to calculate the velocity vector components the Davis software ver. 8.0 with PIV/LIF module was used.

In the mixing system analysis, there are two important criteria numbers: Power number N_p and Reynolds number *Re* defined and calculated according to the following equations:

$$N_P = \frac{P}{\rho N^3 d^5} \quad ; \quad Re = \frac{N d^2 \rho}{\mu} \tag{1}$$

where ρ is the density (kg/m^3), μ is the viscosity (kg/m·s), N is the number of rotation per second (rot/s) , d is the impeller rotor maximum diameter (m) and P is the power (W) required to rotate the impeller with certain rotational speed. It depends on the torque τ, and can be evaluated as follows:

$$P = 2\pi \cdot N \cdot \tau \tag{2}$$

where τ is the torque (N·m).
The third important parameter is the pumping capacity Q. It can be calculated for different Reynolds numbers through a cylindrical surface with a specified radius r or through horizontal surface at specified heights z. In the present configuration, pumping capacity was evaluated through a horizontal surface (axial pumping capacity Q_z) and for the height z=0.01 m. Axial pumping capacity Q_z can be obtained from PIV measurement by integrating the mean axial velocity over the whole horizontal positions, i.e. from $r_1 = 0$ m to $r_2 = 0.085$ m.

$$Q_z = \int_{r_1}^{r_2} |u_z| dS \quad (3)$$

It is also common to normalise this parameter with $N \cdot d^3$ giving one of the global flow characteristic.

3 Results and Discussion

In order to account of mixing process economics parameters related to the energy consumption and mixing phenomena need to be evaluated. The power required by the impeller during the mixing process is an essential parameter. Figure 3 shows, one of the parameter which is directly related to the power required in the mixing process - Power number N_p - calculated from eq.(1). Results are evaluated for different Reynolds numbers and for Rushton turbine. Impeller was installed at various heights above the tank bottom - equal to ¼H (42mm), ⅓H (57mm), ½H (85mm) and ⅔H (113mm). One can infer from this figure that the Power number N_p depends on the Reynolds number and this parameter is almost independent on the distance from the tank bottom h. Only for Re=25 non-negligible differences in N_p for different h were observed. For this Reynolds number, significant Power number increase was also detected. When rotational speed exceeds certain value - in this case, Reynolds number about 20 - flow motion with large recirculation zone was observed. Flow redirection required significant power and results in large power number increase for this Re. It is worth to notice that this effect has been observed for both impellers and for all considered impellers height.

In Figure 5 power number N_p for novel impeller is presented. Results are similar and for both impellers, very similar value is observed. Although the required power is very similar flow motion generated by the impellers is quite different. Figures 5-6 present mean axial pumping capacity Q_z through a horizontal surface located 1 cm above tank bottom, and dimensionalised by the $N \cdot d^3$.

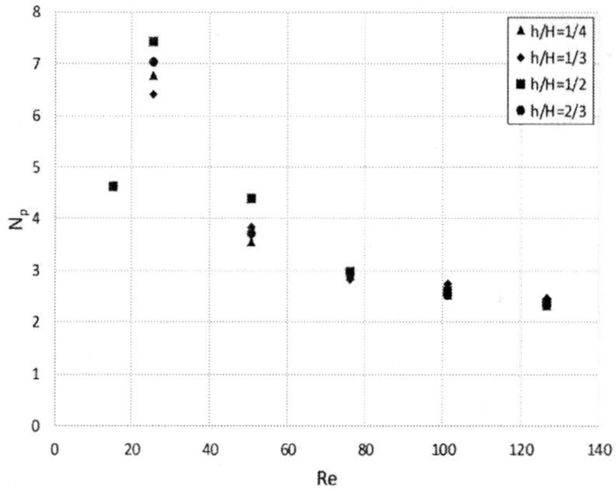

Fig. 3. Power number N_P for Rushton turbine and for different impeller distances from tank bottom.

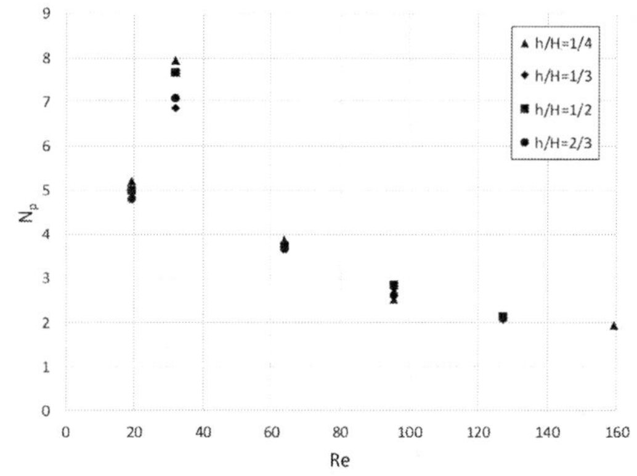

Fig. 4. Power number N_p for novel impeller and for different impeller distances from tank bottom.

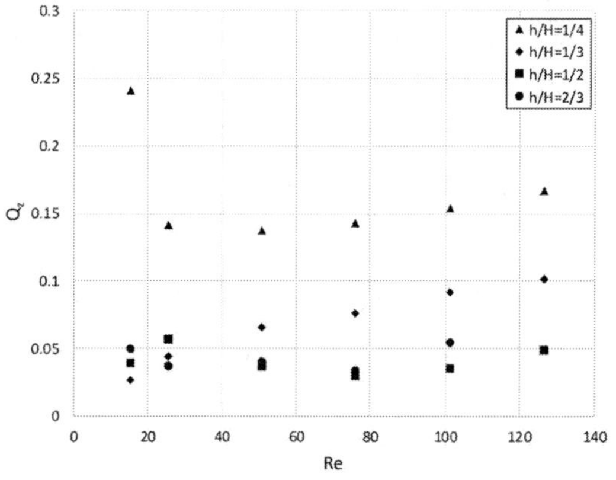

Fig. 5. Mean axial pumping capacity $Q_z / N \cdot d^3$ for Rushton turbine in function of Reynolds number and for different impeller distances from tank bottom.

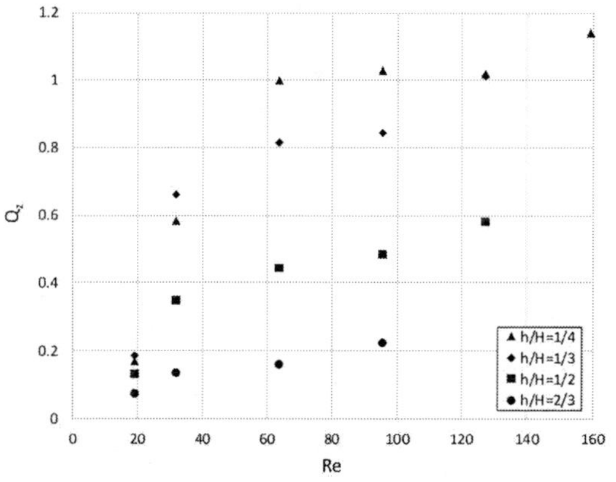

Fig. 6. Mean axial pumping capacity $Q_z / N \cdot d^3$ for novel impeller in function of Reynolds number and for different impeller distances from tank bottom.

Particularly for Reynolds numbers above 20 significant differences are observed in $Q_z/N{\cdot}d^3$ for both turbines and for all impellers heights above the tank bottom. For analysed cases, novel impeller performance is higher than the Rushton turbine and this difference can be as large as one order of magnitude. This means that proposed impeller for similar electrical power results in significantly higher flow motion which enhance the mixing process. This is particularly important for low-speed mixing which is less destructive for the mixing components and generate less heat in the fluid (a critical issue for enzymes or vitamins) and for this reason is highly promoted.

The ratio of the $Q_r/N{\cdot}D^3$ to N_p, i.e. non-dimensional pumping capacity to the non-dimensional power, can be considered as a mixing efficiency parameter. This ratio for some cases is 50 times higher for novel impeller than for Rushton turbine

In Figures 7-9 mean velocity vectors evaluated with Particle Image Velocimetry in *z-r* cross-section plane, for two rotational speeds N=100 and 300 rpm, for three impeller hights ¼H, ⅓H, ½H and for two different impellers analysed here are presented. It can be seen very high (radial and axial) fluid motion in the whole domain cause by the novel impeller. As can be seen, different impellers generate very different flow motion. The zone of impeller activity is much smaller for Rushton turbine while for novel impeller flow motion is significantly enhanced in whole mixing tank - from top surface until tank bottom (no dead zones).

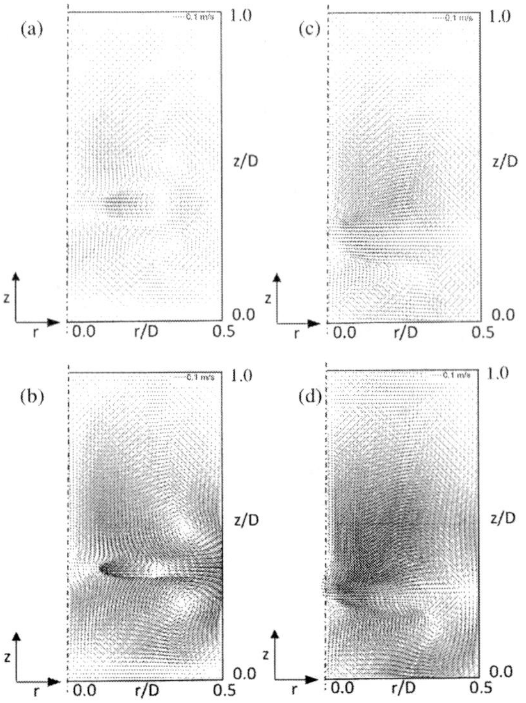

Fig. 8. Mean velocity field for h/H=1/3 and Rushton turbine (a,b), novel impeller (c,d) and rotational speed N=100 rpm (top), N=300 rpm (bottom).

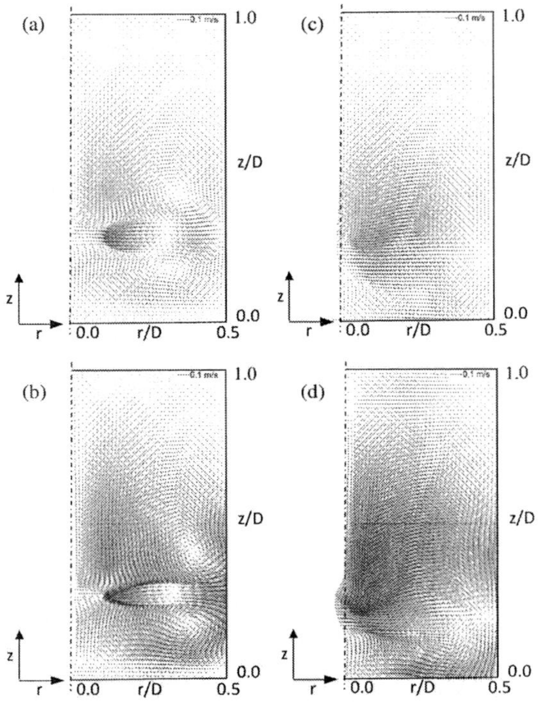

Fig. 7. Mean velocity field for h/H=1/4 and Rushton turbine (a,b), novel impeller (c,d) and rotational speed N=100 rpm (top), N=300 rpm (bottom).

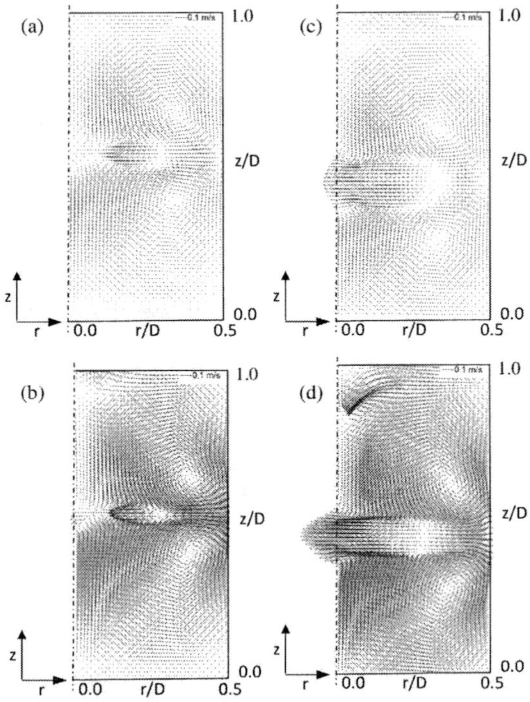

Fig. 9. Mean velocity field for h/H=1/2 and Rushton turbine (a,b), novel impeller (c,d) and rotational speed N=100 rpm (top), N=300 rpm (bottom).

4 Conclusions

In this work, an experimental investigation on the fluid flow mixing process in agitated vessel has been done. It has been shown that using the optimised impeller shape, it is possible to significantly increase mixing performance maintaining the same electrical power required. After exceeding the critical for the phenomena value for the Reynolds number (about Re=25), a significant initial increase in power number N_p is observed preceding a steady decline in this parameter. The power number values for the Rushton turbine and for novel impeller for highly viscous fluid (μ=1410.1Pa·s) are quite similar. Increasing Reynolds number the stabilization of N_p values is observed. It is worth to notice that even power number is similar the pumping capacity can be up to one order of magnitude larger for the novel impeller than for the Rushton turbine. This results in more intensive fluid motion at the same rotational speed. The analysed new impeller type is a promising alternative to the classic rotating disc or other blades and non-blades based impeller types.

Acknowledgements. The present work was supported by the Polish Ministry of Science (Grant MNiSW/2019/162/DIR).

References

1. A. Busciglio, F. Scargiali, F. Grisafi, A. Brucato, *Oscillation dynamics of free vortex surface in uncovered unbaffled stirred vessels*, Chem. Eng. Journal **285**, 477 (2016)
2. F. Scargiali, A. Busciglio, F. Grisafi, A. Brucato, *Mass transfer and hydrodynamic characteristics of unbaffled stirred bio-reactors: Influence of impeller design*. Biochemical Engineering Journal **82**, 41 (2014)
3. F. Scargiali, A. Busciglio, F. Grisafi, A. Brucato, *Free surface oxygen transfer in large aspect ratio unbaffled bio-reactors, with or without draft tube*. Biochemical Engineering Journal, **100**, 16 (2015)
4. M. Jaszczur, A. Młynarczykowska, R. Hanus, *An analysis of the velocity field distribution inside the flotation chamber*, J. of Physics. Conf. Series, **745**, 1 (2016)
5. J. Wu, S. Wang, B. Nguyen, D. Marjavaara, O. Eriksson, *Improved mixing in a magnetite iron ore tank via swirl flow: Lab-scale and full-scale studies*, Chem. Eng. Technol. **39**, 505 (2016)
6. M. M. Alvarez, J.M. Zalc, T. Shinbrot, P. Arratia, J. Muzzio, b. *Mechanisms of Mixing and Creation of Structure in Laminar Stirred Tanks*, AIChE Journal **48**, 2135 (2002)
7. H. Furukawa, Y. Kato, Y. Inoue, T. Kato, S. Hashimoto, *Correlation of Power Consumption for Several Kinds of Mixing Impellers, International Journal of Chemical Engineering*, Int. J. of Chem. Eng. **2012**, 1 (2012)
8. S. Ferrari, L. Rossi, *Particle tracking velocimetry and accelerometry (PTVA) measurements applied to quasi-two-dimensional multi-scale flows*, Exp Fluids, **44**, 6 (2008)
9. L. Rossi, S. Bocquet, S. Ferrari, J.M. Garcia de la Cruz, J.M., S. Lardeau, S., *Control of flow geometry using electromagnetic body forcing*, Int J Heat and Fluid Flow, **30**, 3 (2009)
10. T. Su, F. Yang, M. Li, K. Wu, *Characterization of the hydrodynamics of a covering plate Rushton impeller*, Chinese J. of Chem. Eng. **26**, 1392 (2017)
11. A. Delafosse, M.L. Collignon, S. Calvo, F. Delvign, M. Crine, P. Thonart, D. Toye, *CFD-based compartment model for description of mixing in bioreactors*, Chem. Eng. **106**, 76 (2013)
12. T. Kumaresan, J.B. Jyeshtharaj, *Effect of impeller design on the flow pattern and mixing in stirred tanks*, Chem. Eng. J., **115**, 173 (2005)
13. S. Malika, E. Lévêquea, L. Bouaifi, L. Gamet, E. Flottes, S. Simoëns, M. El-Hajem, *Shear improved Smagorinsky model for large eddy simulation of flow in a stirred tank with a Rushton disk turbine*, Chem. Eng. Res and Design, **108**, 69 (2016)
14. M. Major- Godlewska, J. Karcz, *Power consumption for an agitated vessel equipped with pitched blade turbine and short baffles*, Chem Zvesti., **72**, 1081 (2018)
15. A. Delafosse, A. Line, J. Morchain, P. Guiraud, *LES and URANS simulations of hydrodynamics in mixing tank: comparison to PIV experiments*, Chem. Eng. Res. and Design, **86**, 1322(2008)
16. H. Ameur, Y. Kamla, D. Sahel, *Optimization of the Operating and Design Conditions to Reduce the Power Consumption in a Vessel Stirred by a Paddle Impeller*, Per. Pol. Mech. Eng., **62**, 312 (2018)
17. J. Derksen H.E.A. Van den Akker, *Large eddy simulations on the flow driven by a Rushton turbine*, AIChE J., **45**, 209 (1999)
18. J. Malika, S, Lévêquea, E, Bouaifi, L, Gamet, L, Flottes, E, Simoëns, S, and El-Hajem, M, *Shear improved Smagorinsky model for large eddy simulation of flow in a stirred tank with a Rushton disk turbine*, Chem. Eng. Res and Design, **108**, 69 (2016)
19. H. Patil, A.K. Patel, H.J. Pant, A.V. Vinod, *CFD simulation model for mixing tank using multiple reference frame (MRF) impeller rotation*, ISH J of Hydr. Eng., **2**, 971 (2018)
20. B. N. Murthy, J. B. Joshi, *Assessent k–ε, RSM and LES turbulence models in a baffled stirred vessel agitated by various impellers*, Chem. Eng. Sci. **63**, 5462 (2008)
21. H. Singh, D. F. Fletcher, J.J. Nijdam, *An assessment of different turbulence models for predicting flow in a baffled tank stirred with a Rushton turbine*, Chem. Eng. Sci, **66**, 5976 (2011)

Multiphase numerical analysis of the vortex formation near the suction of the vertically mounted axial-flow pump with influence of the free water level

Prokop Moravec[1,*], *Lukáš* Zavadil[1], *Jakub* Stareček[1], *Tomáš* Krátký[2], and *Milan* Sedlář[1]

[1]SIGMA Research and Development Institute, Department of the hydraulic research, Jana Sigmunda 313, Lutín, Czech Republic

[2]CENTRE OF HYDRAULIC RESEARCH, Jana Sigmunda 313, Lutín, Czech Republic

Abstract. Presented paper deals with the numerical investigation of a flow in a complete hydraulic system of a pump station. The station uses a recently developed high-efficiency axial-flow pump with the specific speed n_s=680 min^{-1}. The pump is vertically mounted in the suction object with the free water level and equipped with the suction bell and the suction cone with one rib. For the numerical simulations, the lowest possible water level is considered. Such scenario is the most limiting for the pump station design and need to be analysed properly. The unsteady multiphase numerical simulations were focused on the creation of the complex vortical structures at the suction part of the mentioned axial-flow pump. This phenomenon strongly influences the pump performance (hydraulic efficiency, delivery and suction head, etc.) and operation (especially the vibrations and noise) characteristics. The unsteady flows inside the investigated domains are simulated using Ansys Fluent with appropriate solver settings. The Scale Resolving Simulations (SRS) are used to model large vortices and the interaction between stationary and rotating parts.

1 Introduction

This paper deals with the numerical analysis of the fluid flow in a complete hydraulic system of a pump station, including the effects of the free water level. Such hydraulic system usually consists of an intake object, one or more pumps, a discharge object and a piping connecting the pump runner to the outlet domain. The presented research is focused on the turbulent flow in the intake object with one pump (fig. 1). There are several references pursuing this topic, using CFD simulations or laboratory measurements [1, 2, 3 and 4].

To investigate possible vortex creation in the vicinity of the pump runner, the lowest possible water level in the intake object is considered. Such regime is the most critical, as with the low water level there is the highest chance of air entering the pump – resulting in serious problems for the pump operation.

Fig. 1. 3D model of the hydraulic pump system.

* Corresponding author: p.moravec@sigma.cz

2 Vortex creation

A creation of the turbulence (small vortices) is typical for any turbulent flow. The large-scale vortex develops in shear flows, where significant velocity gradients strongly dominate. In the vortices, the pressure decreases its value from the border region of a vortex shape to a vortex core. On the free water level, the fluid surface sinks in the vortex core due to a constant gas pressure, which affects the liquid surface [5].

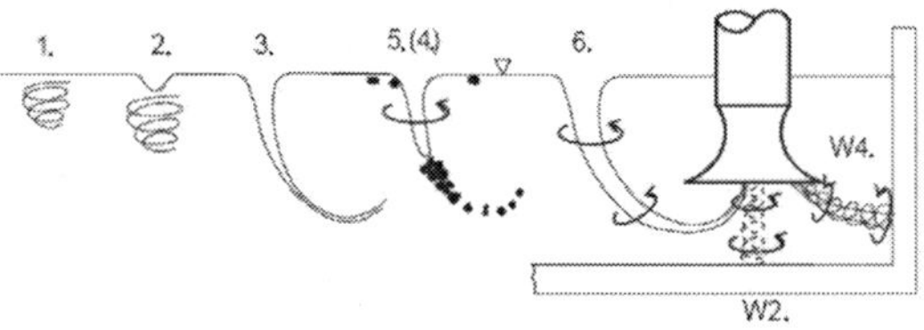

Fig. 2. Types of vortices [5].

When focusing on a vortex strength, several vortex classifications emerge (see fig. 2) [5]:

Free surface vortices [5]:

1. Swirl visible on liquid level.
2. Surface dimple.
3. Dye core to intake.
4. Vortex pulling trash (not air) to intake.
5. Vortex pulling air bubbles to intake.

6. Vortex pulling air continuously to intake.

Submerged vortices [5]:

W1. Weak swirl (no coherent core).
W2. Coherent dye core.
W3. Organized air core (air separation).
W4. Vapour core (strong cavitation noise).

The generation of the vortices has a significant impact on the pump operation and can cause following unfavourable phenomena [5]:

- Power, efficiency, head and flow rate change.
- Vibrations and noise.
- Hydraulic excitation forces increase.

The related energy dissipation means lower efficiency and has adverse effect on the pump performance. Thus, it is crucial to consider the effects of vortices for the hydraulic system design.

3 CFD

All numerical simulations were performed with commercial software ANSYS Fluent (release 2019R3 [6]). The Volume-of-Fluid (VOF) method, combined with the SIMPLE pressure-velocity coupling and PRESTO! pressure interpolating scheme, was used for the numerical solution of the free-surface flow. The simulation was fully-transient, with timestep matching 360 or alternatively 360 steps during one impeller revolution (i.e. one or two degrees of impeller blades movement). SAS model of turbulence was used. To correctly capture the fluid flow inside the investigated domains, gravity effects (acceleration constant $g = 9.81\ m/s^2$) were also considered.

The computational mesh was assembled from four parts - computational domains (intake object, pump impeller, diffuser and the piping). In the computational model, the domains are connected via interfaces. The impeller was set as moving (rotating) domain, the remaining domains were stationary. The mesh size was about 14 million nodes (7 million cells) and combined together hexahedral, tetrahedral and polyhedral elements (fig. 3).

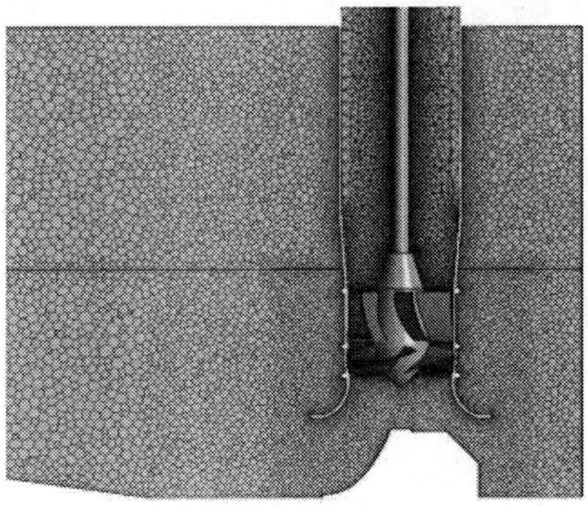

Fig. 3. Computational mesh (blades are shown in red).

3.1. Pre-processing

3.1.1 Intake object and piping

The intake object provided prescribed volume flow rate towards the pump impeller ($Q = 4\ m^3/s$). It consists of a pump chamber with suction bell, a splitter and the free water level, which is located in the distance of 2.3 m from an object intake base wall (fig. 4).

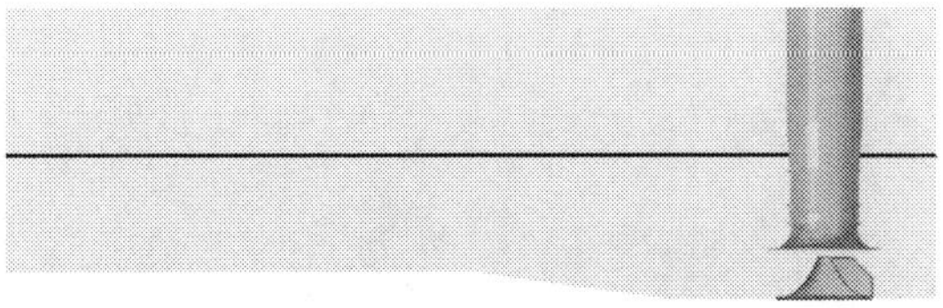

Fig. 4. Water level (the black line) in the intake object.

The connecting piping connects (fig. 5) the pump's outlet and the water tank. As the water has only negligible impact the pump intake, it is neglected in our CFD model, and replaced by an outlet boundary condition (represented by Outflow and located $15{*}D_{pump}$ from a pipe elbow). This is helping in keeping reasonable size and complexity of the computational mesh. This domain also contains moving walls of the hub and shaft. Such walls rotate with the same RPM and direction as the pump impeller.

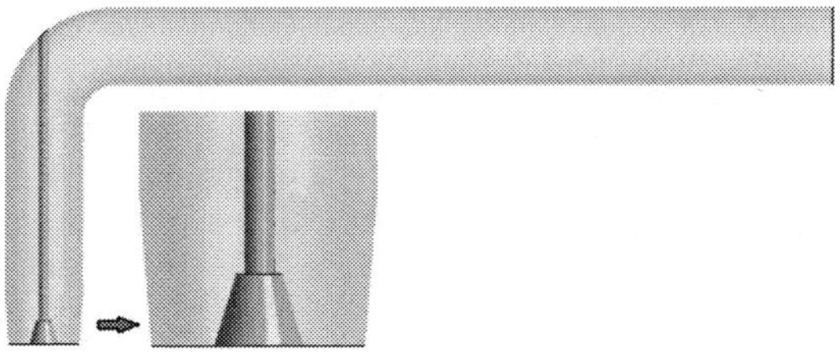

Fig. 5. Connecting piping and shaft.

3.1.2 Pump impeller and distributor

The pump station is based on a recently developed high-efficiency axial-flow pump with specific speed $n_s = 680\ min^{-1}$. It has five impeller and six distributor (diffuser) blades (fig. 6). The outer diameter of the impeller is $D_{pump} = 1072$ mm and $RPM = 380\ min^{-1}$.

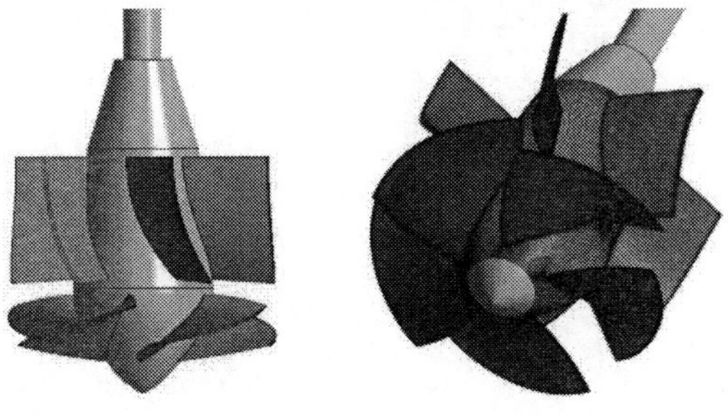

Fig. 6. Pump impeller and distributor (diffuser). Geometry (left) and surface mesh (right).

3.2 Post-processing and results evaluation

For the proper evaluation of the unsteady numerical simulations of the fluid flow inside the investigated pump hydraulic system served CFD-Post, part of the commercial ANSYS software package.

Figures 7-9 show a comparison between data from two separate timesteps of the calculation – first is from the timestep $t_1 = 0.324$ [s], the second one from $t_2 = 0.648$ [s]. A time difference equals to $\Delta t = 0.324$ [s].

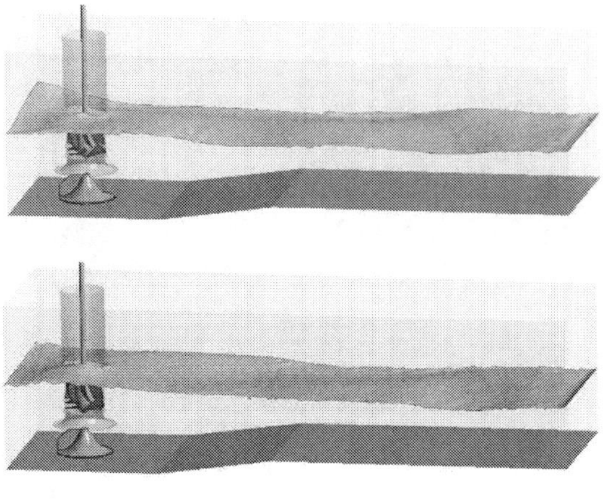

Fig. 7. Free water level – two different values of timestep.

Figure 7 illustratively shows the free water level (portrayed by the green colour) inside the intake object of the investigated hydraulic system. A visualization is done by using of an isosurface of a water fraction ($\alpha_{water} = 0.5$). On the top of the figure 7 is located the first observed timestep $t_1 = 0.324$ [s], on the bottom is $t_2 = 0.648$ [s]. Both cases show mild water level distortion in the close surroundings of the suction part of the pump. Such distortion afterwards propagates itself towards the inlet part of the intake object – a noticeable wave-like instability could be observed.

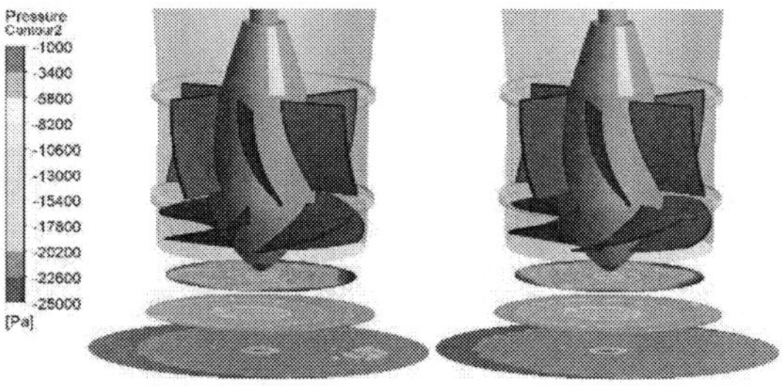

Fig. 8. Static pressure in the suction bell.

Figure 8 compares static pressure distributions inside the suction bell of the pump configuration of the two chosen observed timesteps. The static pressure was evaluated on three horizontal planes, which were equally inserted into the suction part of the pump. Both of these distributions show similar values and qualities, with relatively higher values in front of the pump hub towards the splitter (middle section of each evaluated horizontal plane).

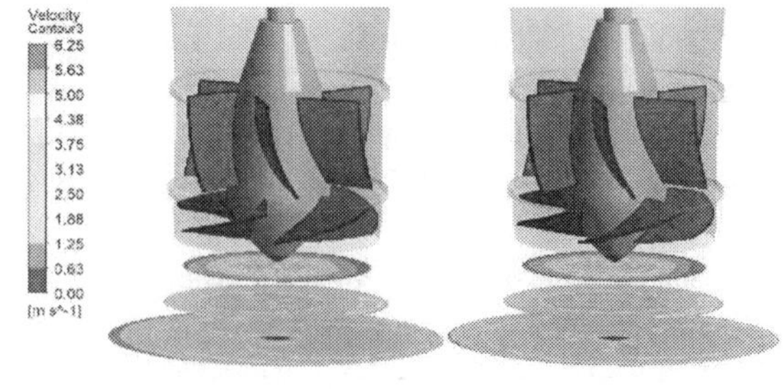

Fig. 9. Velocity in the suction bell.

On the same planes as were defined and displayed in figure 8, values of the velocity in the stationary frame were shown. The rotation of the impeller strongly influences fluid (water) behaviour in the suction part of the pump. The liquid is accelerated by this movement, which is afterwards propagated as a circular change of the velocity (fig. 9).

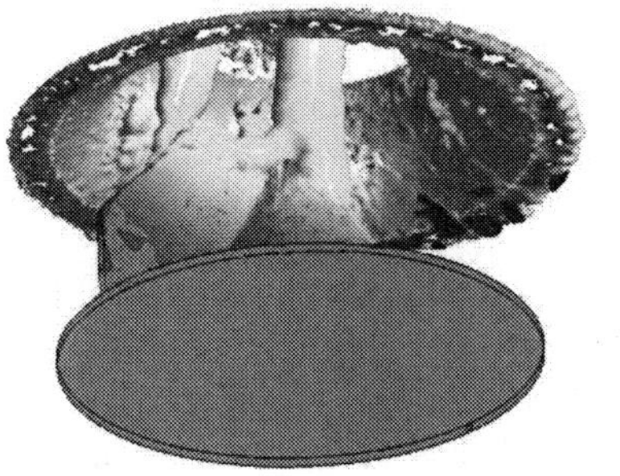

Fig. 10. Vortices in the suction bell of the pump in t_1.

As it was mentioned above, larger vortical structures usually manifest in the flow, where significant velocity gradients dominate. Figures 10-11 show isosurfaces of an absolute helicity coloured by the water velocity inside the suction bell of the pump (warmer colours represent higher velocities, colder colours the other way around). The isosurface in figure 10 uncovers the two main vortical structures: first vortical structure continuously attaches to the tip of the hub, the second structure heads toward the splitter, where it is successfully eliminated by the shape of the splitter.

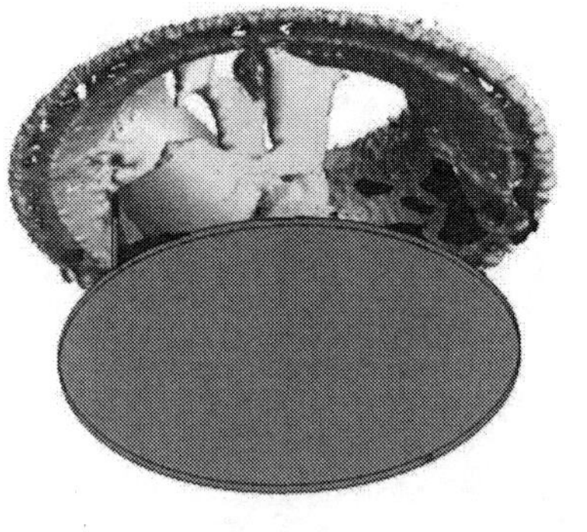

Fig. 11. Vortices in the suction bell of the pump in t_2.

https://doi.org/10.1051/epjconf/202226901041

Figure 11 captures vortical structures in the observed timestep t_2. There are three main vortices, which are situated in close surroundings of the tip of the impeller hub. One vortex is undeveloped, the rest heads toward the splitter, and where the vortex grow is limited. No vortical structures from the free water level towards the impeller were detected.

Following six figures show the pressure and velocity distributions in the horizontal plane (XY plane) that passes through a centre of the impeller of the axial-flow pump (z = 1.4 [m]).

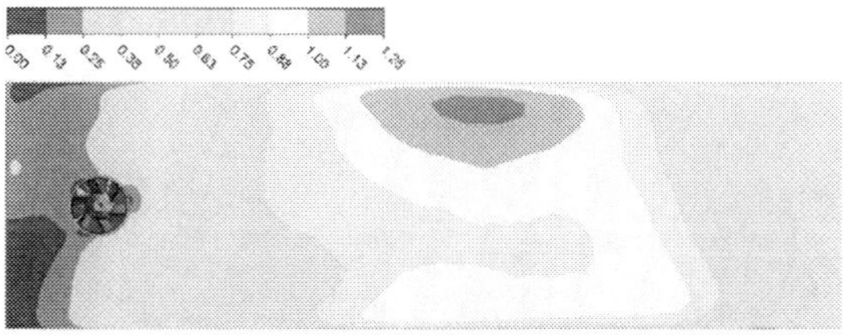

Fig. 12. Contours of the velocity in the stationary frame in t_1.

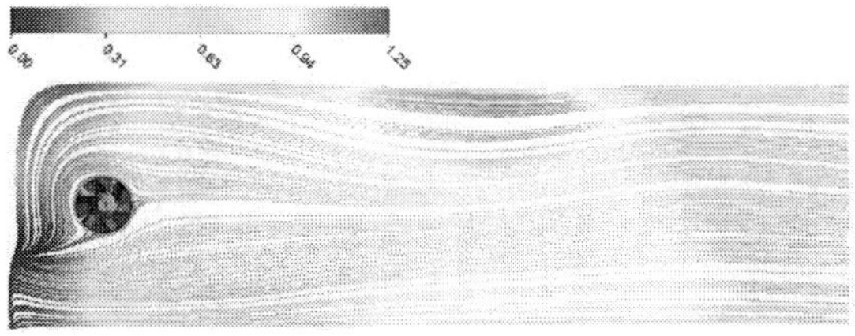

Fig. 13. 2D streamlines coloured by the velocity (timestep t_1).

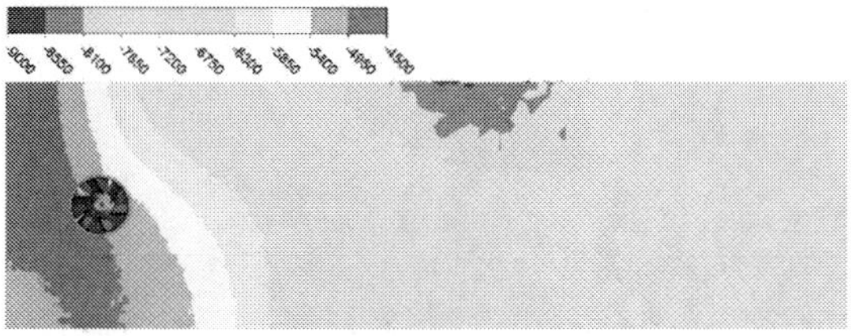

Fig. 14. Contours of the static pressure in t_1.

In the timestep t_1, higher velocity (lower static pressure) values are visible in the middle section of the intake object. This is caused by the deformation of the free water surface, which is in this location slightly lowered. Behind the impeller the water is decelerated due to an interaction with the wall of the intake object.

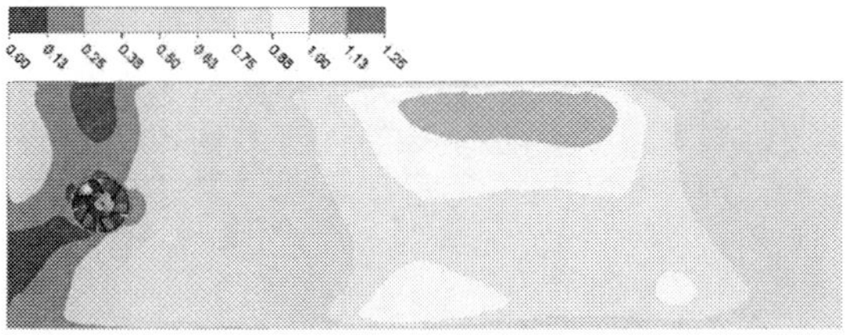

Fig. 15. Contours of the velocity in the stationary frame in t_2.

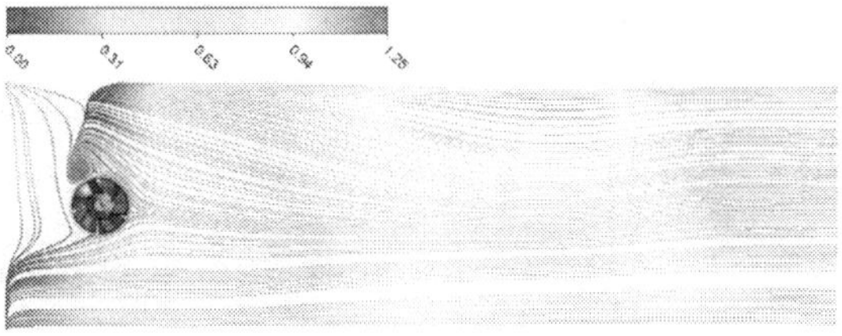

Fig. 16. 2D streamlines coloured by the velocity (timestep t_2).

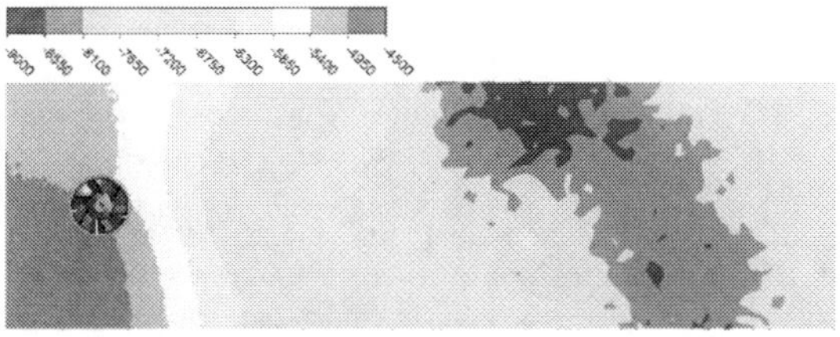

Fig. 17. Contours of the static pressure in t_2.

The timestep t_2 also indicates lowered free water level located in the middle section of the intake object, which is characterized by higher velocity (lower static pressure) values. The major difference between the two selected timesteps can be observed in a visualization of the 2D streamlines (figures 13 and 16). The plane (z = 1.4 [m]) in the timestep t_2 uncovers minor flow instability, which is formed around a part of the impeller chamber. Such instability dissipates energy and can consequently negatively influences fluid entering the suction and lower the pump performance.

4 Conclusion

Presented research was focused on the intake object of the hydraulic pump system, more precisely on the standard operation regime of the axial-flow pump in such system (the optimal volume flow rate). The examined layout was based on the recently developed high-efficiency axial-flow pump with the specific speed n_s = 680 min^{-1} (see figure 6). This pump was mounted in the intake object, where the free water level was set to its minimal possible height. The outlet domain was substituted by the piping system with the elbow. The minimal possible height of the water level is the most critical condition, in which is the highest probability of the creation of the vortical structures towards the free water level that can negatively affect the correct pump behaviour (air pulling towards the impeller, vibrations, noise, impact on the performance characteristics, cavitation).

The numerical simulations of the two-phase fluid flow using the commercial software ANSYS Fluent (release 2019R3) with chosen SAS model of the turbulence were employed. Two different timesteps were selected to demonstrate the water (and consequently air) behaviour during the examined pump run (data from 0.324 and 0.648 [s] of the pump run). The water level representation (green surface in figure 7) shows only the mild type of the deformation in the close vicinity of the impeller (only the wave-like instabilities were observed)

– no major height decrease/lowering on a local scale that could indicate the vortex creation towards the free water level.

The contours of the velocity in the stationary frame near the pump impeller (displayed on the chosen XY plane – figures 12 and 15) uncovered only the significant water deceleration caused by the interaction with the solid walls of the intake object.

The vortical structures were presented only in the suction bell area (in front of the pump impeller). Their behaviour was unsteady – the duration and count varied in time. Those structures tended to connect with the top part of the splitter. Bigger vortical structures (towards the free water level or towards the solid side of the intake object) were successfully eliminated by the splitter, which is located in the bottom of the intake object (see figures 10-11).

It must be noted that the presented numerical simulations captured only a small fraction of the real pump run that usually lasts in hours or days. The flow simulations were also free of the additional solid particles that may initiate the formation of the vortical structures.

This research was funded by the Ministry of Education, Youth and Sports of the Czech Republic, grant number CZ.02.1.01/0.0/0.0/17_049/0008408. Computational resources were supplied by the project "e-Infrastruktura CZ" (e-INFRA CZ LM2018140) supported by the Ministry of Education, Youth and Sports of the Czech Republic.

References

1. T. Okamura, K. Kamemoto, J. Matsui, CFD prediction and model experiment on suction vortices in pump sump, Proc. 9th Asian Int. Conf. on Fluid Machinery (2007) 1-10.
2. S.N. Shukla, J.T. Kshirsagar, Numerical prediction of air entrainment in pump intakes, Proc. 24th Int. Pump Users Symposium (2008) 29-33.
3. A.C. Bayeul-Lainé, G. Bois, A. Issa, Numerical simulation of flow field in water-pump sump and inlet suction pipe, Proc. 25th IAHR Symp. on Hydraulic Machinery and Systems (2010) 1-8.
4. C.G. Kim, B.H. Kim, B.H. Bang, Y.H. Lee, Experimental and CFD analysis for prediction of vortex and swirl angle in the pump sump station model, IOP Conf. Series: Materials Science and Engineering 72 (2015) 042044, doi:10.1088/1757-99X/72/4/0420.
5. J. F. Gülich, Centrifugal pumps, 3rd edition. Heidelberg: Springer, 2014, xli, p. 1116. ISBN 978-3-642-40113-8.
6. ANSYS, Inc., ANSYS CFD Documentation, Release 2019R3 (2019).

Flow structures stability analysis in a model of a cattle barn using oscillating pattern decomposition method

Štěpán Nosek[1,*], *Zuzana* Kluková[2], *Radka* Kellerová[1], *Michala* Jakubcová[1], and *Zbyněk* Jaňour[1]

[1]Institute of Thermomechanics, Czech Academy of Sciences, Dolejškova 1420/5, 182 00 Praha 8, Czech Republic
[2]Department of Atmospheric Physics, Charles University, V Holešovičkách 747/2, 180 00 Praha 8, Czech Republic

Abstract. A model of a cattle barn and a part-depth atmospheric boundary layer were scaled down to 1:50 to simulate the natural ventilation in a wind tunnel. We studied the impact of three opening configurations (entirely, half and quarterly open, which represented 0.7, 0.35 and 0.17 wall porosity, respectively), on the mean flow and the stability of the flow structures in the barn using time-resolved PIV technique and oscillating pattern decomposition method (OPD). While we observed the most-energy-containing structures (POD modes) in the case of the half opening, the most stable structures (OPD modes) correspond to the fully open case. We found that the stability of the structures increases with the opening width (wall porosity), and the highest frequency of these structures was 11 Hz.

1 Introduction

Naturally ventilated livestock buildings (NVLB), such as a cattle barn, represent one of the most significant sources of ammonia and greenhouse gases emissions. While these emissions pose a severe environmental issue worldwide, the impact of both the outdoor and indoor climatic conditions on these emissions is still not well understood [1].

One of the Environmental parameters governing the ammonia transfer is the type of flow about the manure surface. Previous studies [2–5] observed that the higher is the air velocity and turbulence above the manure surface, the higher is the emission of the ammonia. However, the prediction of such flows inside NVLB is not trivial due to relatively large openings which must ensure the ventilation. This yields in a complex interaction of two types of turbulent flows which is hard to predict and which in turn makes the estimation of ammonia emission from NVLB so difficult.

This study is aimed to shed light on the flows developed within a cattle barn. To do so, we choose the physical modelling of both the outdoor and indoor flows in a wind tunnel since the boundary conditions might be fully controlled and statistically steady. Special attention is paid to model the proper outdoor flow, so-called atmospheric boundary layer (ABL), with all essential flow characteristics such as the mean velocity and turbulence vertical profiles, and integral length scale of turbulence. While the later has never been simulated in previous wind-tunnel studies on NVLB [e.g., 6–9], these studies gave an important insight on the impact of openings configuration on the mean airflow patterns within NVLB. To bring a more in-depth insight into the problem, we captured and analysed the flow structures within a cattle barn model using time-resolved particle image velocimetry (PIV) and oscillating pattern decomposition (OPD) method, respectively. Mainly, the OPD method enables to reveal the "hidden" coherent structures of the flow, which are the most stable (last the longest time in the flow) and how they propagate in space. Compared to Proper Orthogonal Decomposition (POD), OPD gives the information about the frequency at which these structures are excited, and thus which part of turbulence spectra is essential for the flow.

2 Methods

The physical modelling of the interaction of outdoor wind (ABL) with the indoor environment of cattle barn was performed in the Environmental wind tunnel of the Institute of Thermomechanics of the Czech Academy of Sciences. The wind tunnel is an open low-speed wind tunnel with cross dimensions of 1.5 m× 1.5m, and the lengths of the development and test sections are 20.5 m and 2m, respectively.

2.1 Cattle barn model

As far as we know, the first rigorous wind-tunnel study on the interaction of ABL with the indoor environment was performed by Cermak et al. [10]. They studied the natural ventilation of building at a scale of 1:25 and found that so-called internal Reynolds number, Re_i (based on the maximum velocity in the room, which is usually equal to the air velocity at the openings, and the

* Corresponding author: nosek@it.cas.cz

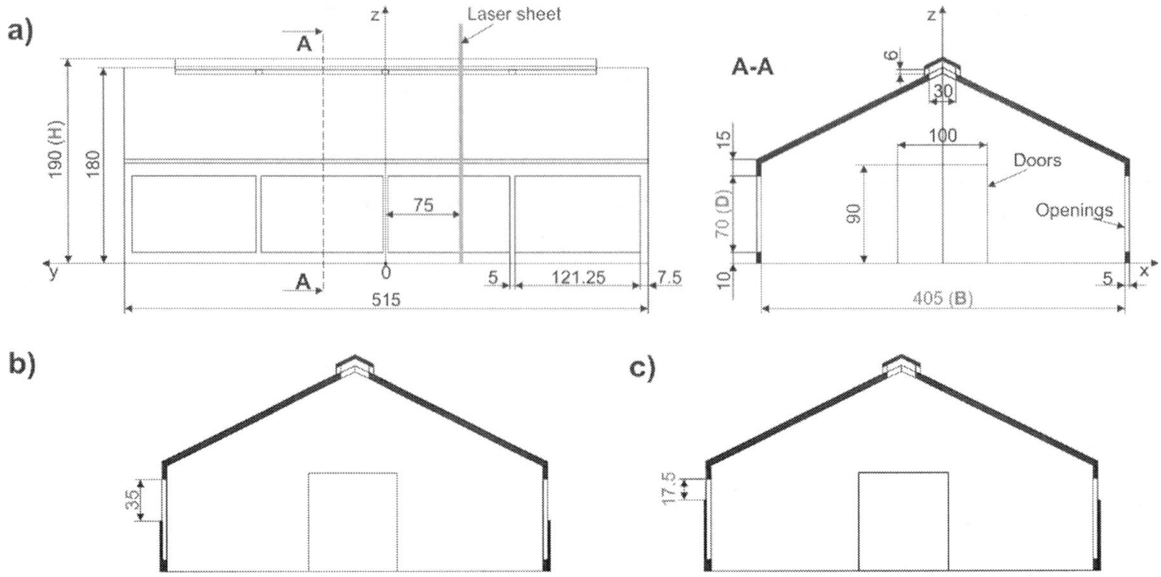

Fig. 1. Sketch of the studied cattle barn model at the scale of 1:50 with respect to the wind-tunnel coordinates: **a)** side view (left) and the cross-section of the cattle barn in the case of the fully open configuration. Cross-sections of the **b)** half and **c)** quarterly open configuration. The doors were left open during the all studied cases. All dimensions are in mm.

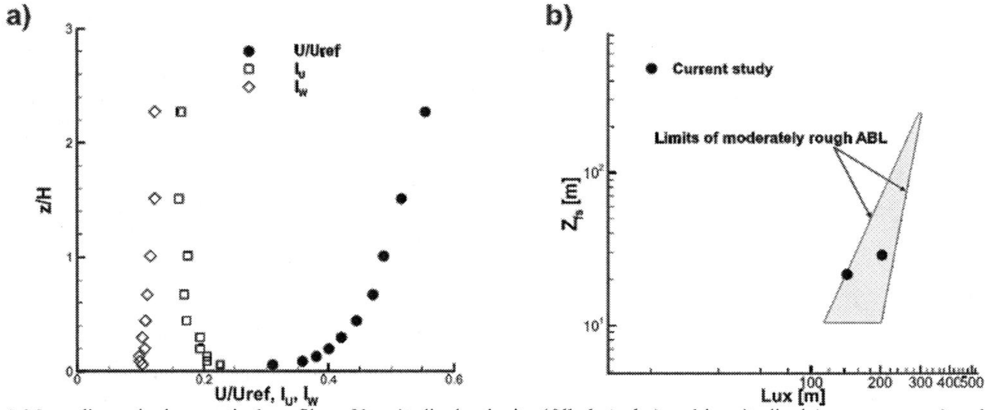

Fig. 2. a) Mean dimensionless vertical profiles of longitudinal velocity (*filled circles*) and longitudinal (*empty squares*) and vertical (*empty diamonds*) turbulence intensity of the simulated ABL; **b)** Comparison of the modelled (*filled circles*) integral length scales of turbulence (L_{ux}) in WT with those (*greyed* area) recommended for moderately rough ABLs by VDI guidelines at full scale.

smallest dimension of the room) should be at least equal to 2×10^4 to ensure the indoor flow independent on Reynolds number. The dimension of the room might be attributed to the inner height of NVLB which was used in this study. Considering Re_i criterion, criteria for ABL modelling in a wind-tunnel and the dimensions of the Environmental wind tunnel, the resulting scale of the cattle barn model yields in 1:50.

The sketch of the model with all important dimensions and all studied configurations is depicted in Fig. 1. The studied configurations reflect three regimes of natural ventilation. The first (Fig. 1a) corresponds to summer, the second (Fig. 1b) to autumn/spring and the third (Fig. 1c) to winter regime. These three regimes also reflect the porosity of the walls (0.7, 0.35 and 0.175, respectively) oriented perpendicularly to the approach wind. While the model was designed as simple as possible, all the important geometrical features (roof ridge, openings geometry, doors) of a typical cattle barn were reproduced.

2.2 ABL model

Modelling of all important ABL characteristics in a wind tunnel simultaneously is not an easy task. The problem usually begins with matching of integral length (longitudinal) scale of turbulence, L_{ux}, or turbulence spectra. The less the model of ABL is scaled down, the harder is to obtain these length scales in a wind tunnel.

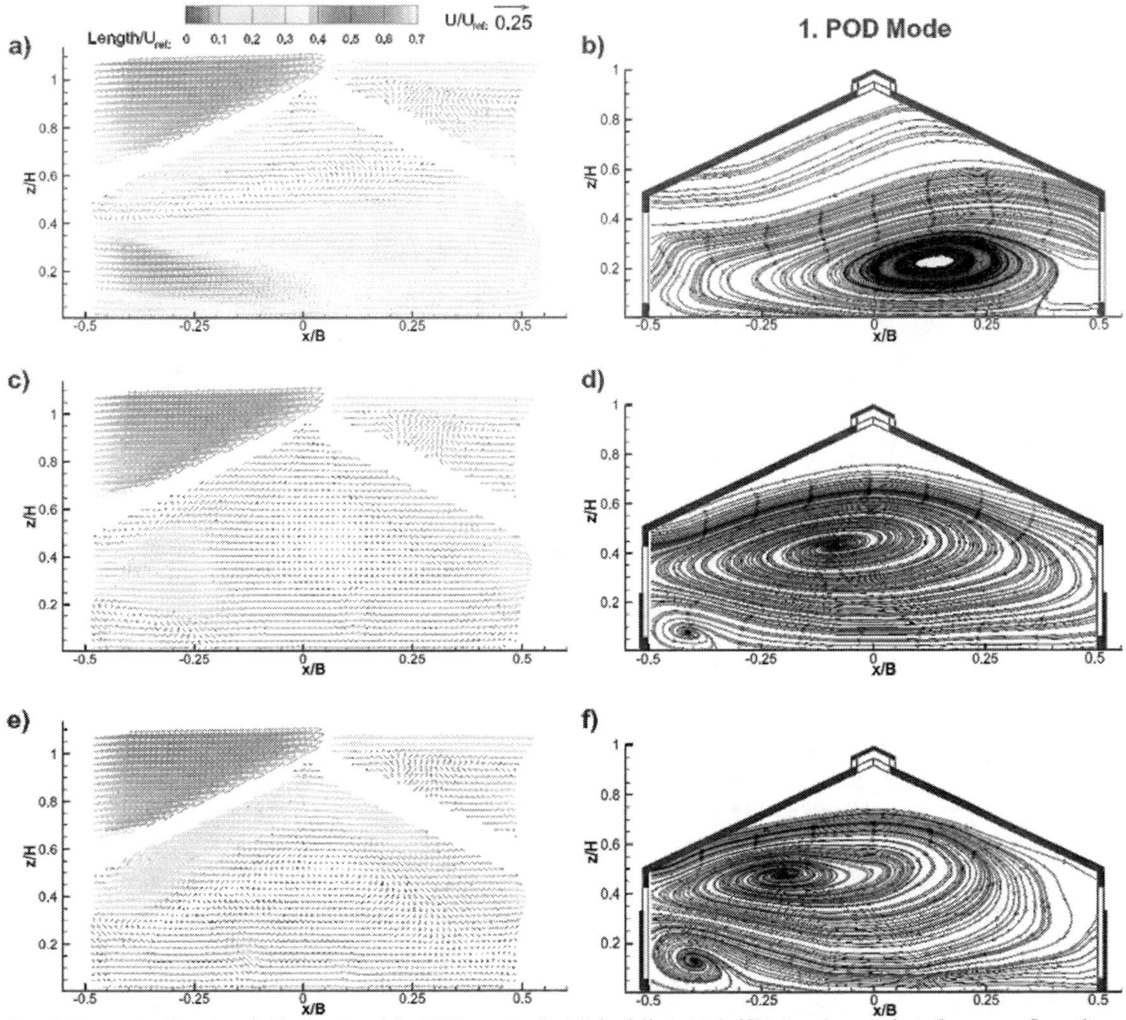

Fig. 3. Mean velocity vector fields (**a,c,e**) and first POD modes (**b,d,f**) for fully (**a,b**), half (**c,d**) and quarterly (**e,f**) open configurations. The colour of the vector corresponds to the dimensionless resultant 2D velocity (*Length/U_{ref}*). The dimensionless reference velocity vector (*U/U_{ref}*) is depicted in the top right of (a).

Cook [11] show that if L_{ux} and roughness length, z_0, is reproduced in a wind tunnel, a successful simulation of part-depth ABL might be achieved. Such par-depth simulation was also used in the present study since the chosen scale was at the low end of possible scales of the current wind tunnel.

From the velocity and turbulence profiles (Fig. 2a), and integral length scales of turbulence computed for two heights (Fig. 2a) can be clearly seen that the modelled part-depth ABL corresponded those observed for rural (moderately rough) terrains at full scale. The profiles were measured by laser Doppler anemometry 750 mm upstream of the cattle barn model, and hence 250 mm downstream of the last row of turbulent generators. The turbulent generators (thin plates of 50 mm in width and 10 mm in height) covered the bottom of the entire development section (10 m in length) of the wind tunnel in a staggered form. The integral length scale of turbulence was enhanced by vortex generators (spires of 1420 mm in height) at the beginning of the development section.

2.3 Time-resolved PIV setup

A 2D PIV system from Dantec was used at a repetition rate of 200 Hz. The wall-normal streamwise laser sheet from DualPower 30-100 laser was positioned at the middle of the openings (75 mm from the centre of the model, see Fig. 1a) to light particles carried by the flow within the barn at this plane. The particles (about 1 μm in diameter) were continuously injected by a Laskin nozzle generator upstream of the vortex generators. The CMOS camera SpeedSense VEO 410 captured 6227 consecutive images during an acquisition time of 30 s for

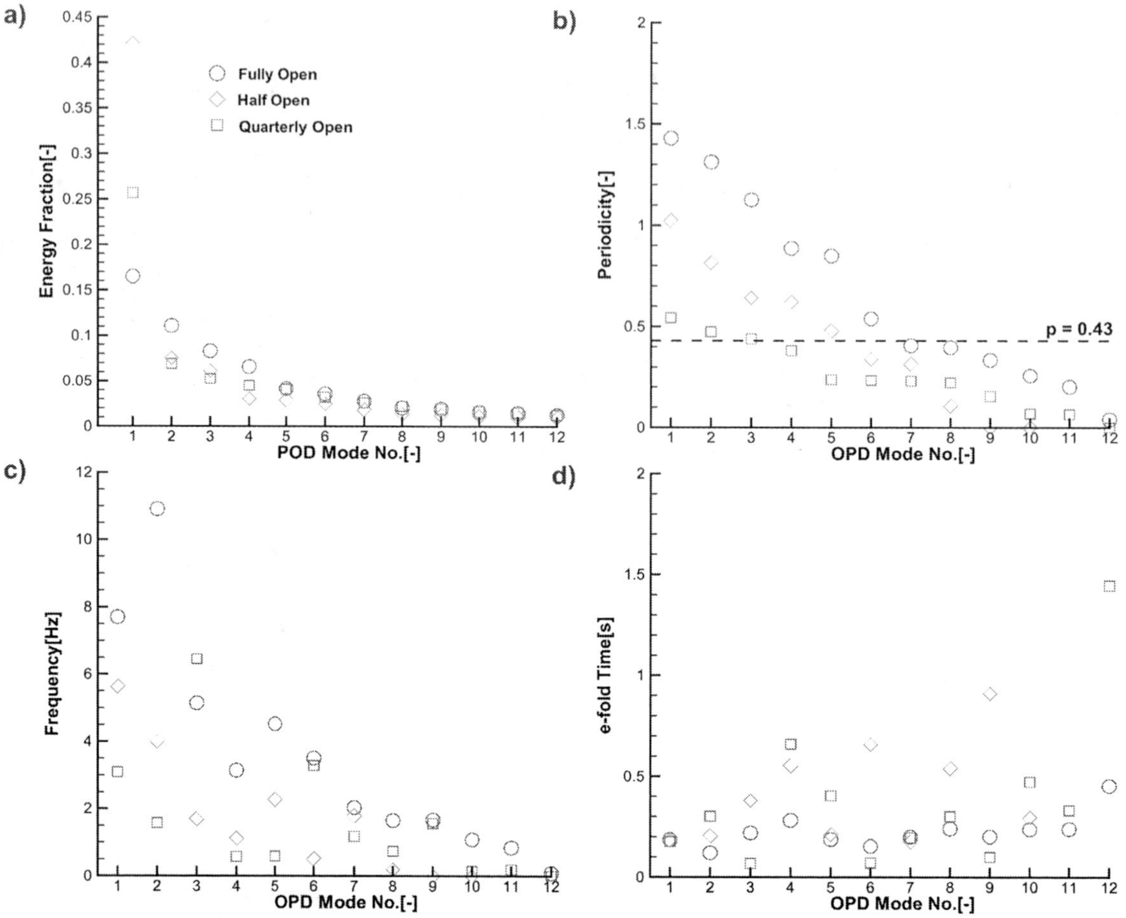

Fig. 4. Computed modes from POD (**a**) and OPD (**b,c,d**) analysis. The dashed line in (b) represents the threshold periodicity ($p = 0.43$) suggested by Uruba [12].

each studied case. The instantaneous velocity vector fields were computed using Adaptive PIV method by DynamicStudio software. The size of the interrogation area was set to 32×32 pixels with 50% overlap, which resulted in 5 mm (250 mm at full scale) of the spatial resolution of the velocity vector filed.

3 Results

3.1 Mean velocity field

Figs. 3a, 3c and 3e present the mean velocity vector fields computed from all 6227 instantaneous vectors fields for each studied case. The vectors are normalised by the mean free-stream velocity of the wind tunnel, U_{ref}, which was about 5.3 m/s and controlled by Prandtl tube during all runs. The vectors are coloured according to the normalised (by U_{ref}) resulting in 2D velocity.

From Figs. 3a, 3c and 3e one can observe that while within the cattle barn there is a substantial decrease of the velocity with the decrease of the size of the opening, there is no appreciable change in the outdoor flow. However, it should be noted that this insensitivity of the outdoor flow on the cattle barn porosity might not be actual for other areas around the barn due to the three-dimensionality of the studied cases. The lowest internal Reynolds number (Re_i = 22430) corresponds to the quarterly open configuration, but still above the threshold suggested by Cermak et al. [10].

The flow patterns within the barn differ from each case. The fully open configuration (Fig. 3a) enables the outdoor wind blows with ease through the barn to the outlet opening. This blow produces the incoming jet at the lower part of the barn and an anticlockwise-rotating vortex at the upper part. The similar pattern has been observed by Shen et al. [13], although their model of cattle barn was without the ridge openings and open doors.

In the case of the half-open configuration, the anticlockwise vortex is suppressed close to the inlet, and a new, smaller, clockwise vortex is generated just behind

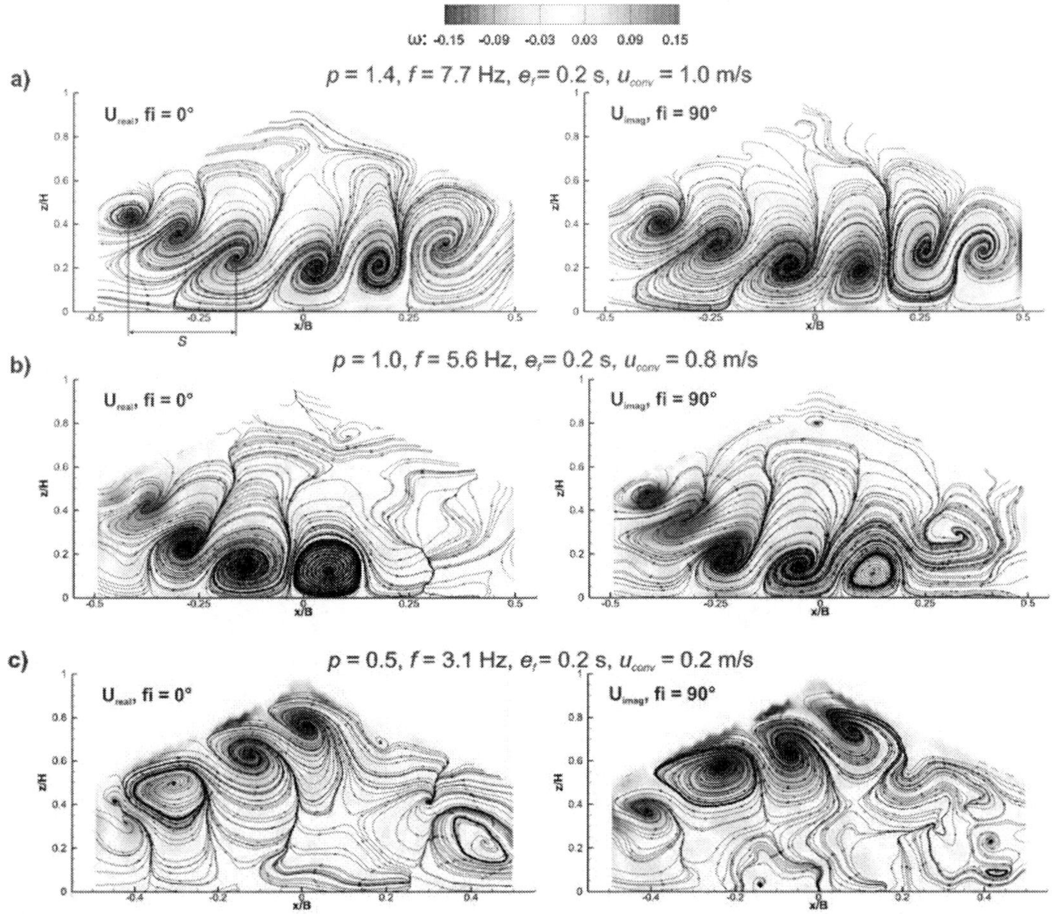

Fig. 5. Spatial OPD modes of the **a)** fully, **b)** half and **c)** quarterly open configuration having the highest periodicity. The first and second column represents real ($\varphi = 0°$) and imaginary ($\varphi = 90°$) part of these modes, respectively. The contours represent the vorticity computed from these parts. The structures' spacing is denoted by s.

the wall of the halved inlet opening (Fig. 3c). These two vortices are the result of the shear layers produced at the upper and lower edge of the incoming jet. For the quarterly open case (Fig. 3e), the incoming flow attaches to the roof and the well-known Coanda effect can be observed due to the higher static pressure in the middle of the barn. Such a flow attachment was reported by Morsing et al. [7]. Further downstream, the flow falls to the model bottom and reverses back to the inlet wall, producing a weak anticlockwise vortex in the middle of the barn.

3.2 Flow structures stability analysis

To analyse the dynamics of the flow structures, which are "hidden" due to the chaotic nature of turbulence, we applied the oscillation pattern decomposition (OPD) method. The OPD is based on the idea that complex dynamical systems of many degrees of freedom (such as turbulent flow) might be reduced to more simple dynamical models. Such reduced dynamical model, based on linear algebra, was in climatology introduced by Hasselmann [14] as Principal Oscillation Patterns (POPs). Simply speaking, the OPD method extends the POPs in the way that OPD can resolve standing oscillations [15]. Due to solving the eigenvalue problem embodied in so-called Fokker–Planck equation, the OPD modes consist of complex eigenvalues and complex eigenvectors [12]. While the complex eigenvectors reveal the OPD mode topology, the complex eigenvalue defines the frequency, f, and time (so-called e-fold time, τ_e) at which the amplitude of the oscillation decays to $1/e$ [16]. The most stable dynamic structures decomposed from the instantaneous flow filed are those which have the highest periodicity, $p = f.\ \tau_e$. Indeed, the e-fold time is the crucial parameter here, but if there is no oscillation ($f = 0$), there is no dynamics, and hence, in that case, the e-fold time characterises a pulsating structure without any propagation in space. Uruba [12] proposed the threshold value of $p = 0.43$, which means that bellow this value the OPD mode amplitude decays more than by factor 10 during the period, and hence a non-oscillating rapidly decaying mode is detected in the flow.

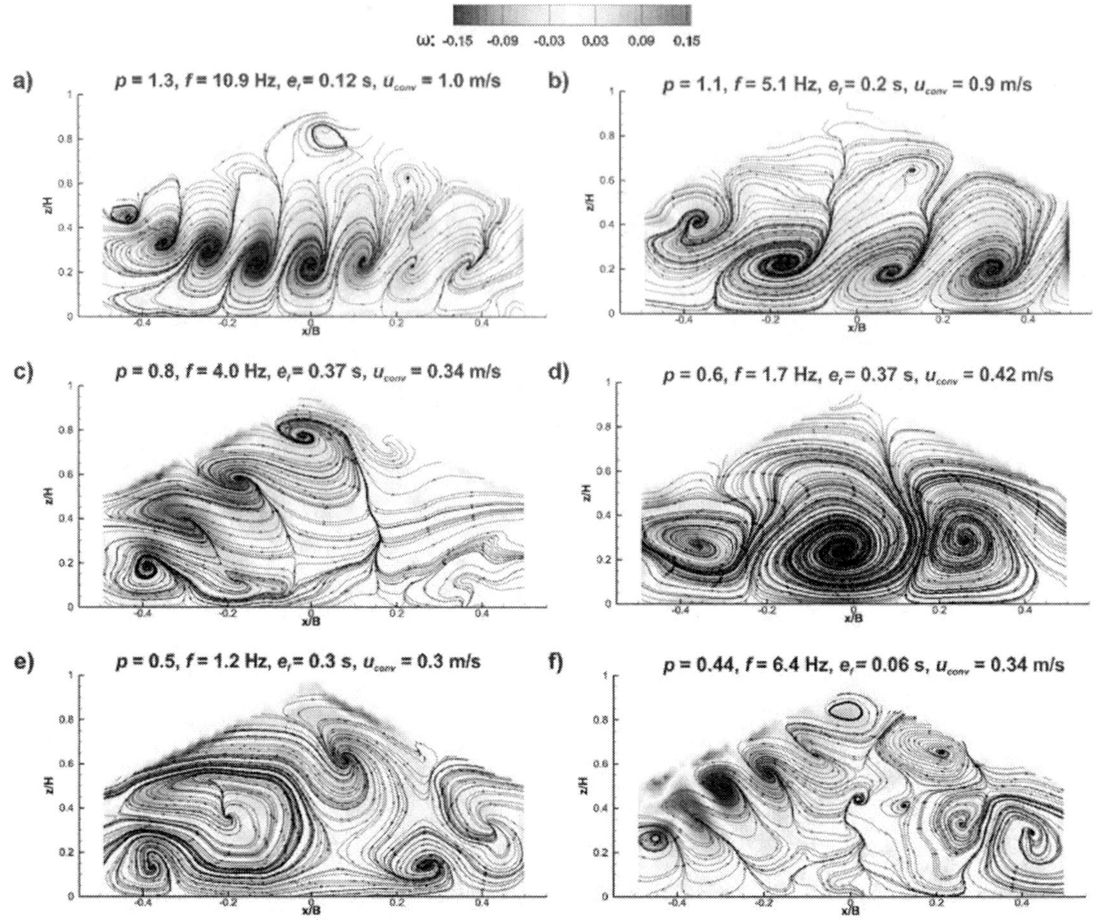

Fig. 6. Spatial OPD modes of the fully (**a,b**), half (**c,d**) and quarterly (**e,f**) opened configuration having the second (**a,c,e**) and third (**b,d,f**) highest periodicity. Only the real parts of the OPD modes at phase 0 are presented for the sake of brevity.

For the OPD analysis, we used DynamicStudio software. Due to saving the computational time, this software computes the OPD eigenvectors and eigenvalues from the already computed "topos" and "chronos" by Proper Orthogonal Decomposition (POD), or more generally by Bi-orthogonal Decomposition (BOD) [17]. Therefore, we also present these POD modes, which are ranked according to their contribution to the total turbulent kinetic energy of the flow filed and compare them with those from the OPD analysis (Fig. 4). The OPD analysis was set to compute all the POD modes which have energy fraction higher than 0.005.

The comparison of the POD (Fig. 4a) with OPD (Fig. 4b) modes shows that the rankings of the primary modes differ concerning the studied case. While the most energetic structures (the first POD mode, Fig. 4a) have been found for the halved opening, the most stables structures (the first OPD mode, Fig. 4b) correspond to the fully open configuration (the same result was obtained for the higher or lower energy fraction setting of OPD). There is no impact of the studied case on those modes (both POD and OPD) which are lower than 11th position of the ranking. However, an appreciable ($p > 0.43$) structure stability can be observed only for those modes which are higher than 7th (full opening), 6th (half opening), and 4th (quarter opening) position of the OPD ranking.

Fig. 4c shows that the highest frequency at which the structures are excited is about 11 Hz and was detected for the fully open case. With the increase of the OPD modes the frequency generally decreases. The opposite is true for the e-fold time (Fig. 4d). Interestingly, the most stable structures (1 OPD modes) have relatively very short e-fold time which is the same for each studied case, and hence the frequency is the key parameter which drives the stability/dynamics of these structures. Generally, the bigger is the opening, the higher frequency is excited.

For the sake of brevity, we compare the spatiality of the most dominant POD (Figs. 3b, 3d and 3f) and OPD (Fig. 5) modes only. From the first glance of view, one can observe the absolute difference in the spatiality between the POD and OPD modes. The most energetic structures form one (the fully open case, Fig, 3b) or maximum of two (the half, Fig. 3d, and quarterly open case, Fig. 3f) vortices. The most stable dynamic structures have at least four vortices and are more dependent on the studied case (Fig. 5). However, the sizes of these vortices remain approx. the same for each studied case. They propagate from the inlet opening to the outlet, but only in the case of the fully open configuration the vortices are also presented at the outlet opening, thus do not vanish across the entire barn width. The angle of the direction propagation is the same (approx. -30°) for both the full (Figs. 5a and 5b) and half (Figs. 5c and 5d) openings but changes about 60° (to approx. 30°) in the case of the quarter opening. The convective velocity of the vortices (computed according to [12] from the structures' spacing, s, and frequency, f, as $v = sf$) decreases with the decrease of the opening width.

The second (Figs. 6a, 6c, and 6e) and third (Figs. 6b, 6d, and 6f) OPD modes demonstrate the impact of the excitation frequency on the spatiality of the structures. The higher is the frequency, the more and smaller vortices are excited, and the higher is their convective velocity.

4 Conclusions

In this study, we demonstrate the importance of the stability analysis of turbulent flow within a model of a naturally ventilated cattle barn. The comparison between the studied cases shows that the most energy-containing structures are ranked differently than the most stable dynamic structures. Specifically, the most energetic (POD) mode has been found for the halved opening, while the most stable dynamic (OPD) mode corresponded to the fully open case. Generally, the higher is the opening width, the higher is the stability of the dynamic structures.

The OPD detected the highest excitation frequency of the dynamic structures about 11 Hz from all studied cases. Considering the Nyquist criterion, this suggests that the sampling frequency about 35 Hz should be satisfactory to provide time-resolved data for such studied case and scale ratio. Interestingly, while both the number and convective velocity of the structures increased with the increase of the opening width, the decay time and size of the most stable dynamic structures was independent of the opening width.

Acknowledgement

The authors gratefully acknowledge funding from the Ministry of Education, Youth and Sports of CR (LTC18070) and from the institutional support RVO:61388998. Authors also would like to thank prof. Vaclav Uruba for his consultations to OPD analysis, and Jan Kozohorsky for his valuable support during the experiments.

References

1. B. Bjerg et al., "Review Modelling of ammonia emissions from naturally ventilated livestock buildings. Part 1 : Ammonia release modelling," Biosyst. Eng., vol. 116, no. 3, pp. 232–245, 2013.
2. J. Arogo, R. H. Zhang, G. L. Riskowski, and D. L. Day, "Mass transfer coefficient for hydogen sulfide emission from aqueous solutions and liquid swine manure," Trans. ASABE, vol. 42, no. 3, pp. 1455–1462, 1999.
3. L. Rong, P. V. Nielsen, and G. Zhang, "Effects of airflow and liquid temperature on ammonia mass transfer above an emission surface: Experimental study on emission rate," Bioresour. Technol., vol. 100, no. 20, pp. 4654–4661, 2009.
4. J. Olesen and S. Sommer, "Modeling effects of wind-speed and surface cover on ammonia volatilization from stored pig slurry," Atmos. Environ. Part a-General Top., vol. 27, no. 16, pp. 2567–2574, 1993.
5. M. De Paepe, J. G. Pieters, L. B. Mendes, S. Van Weyenberg, B. Merci, and P. Demeyer, "Wind tunnel study of ammonia transfer from a manure pit fitted with a dairy cattle slatted floor," Environ. Technol. (United Kingdom), vol. 37, no. 2, pp. 202–215, 2016.
6. Y. Choiniere, F. Blais, and J. Munroe, "A wind tunnel study of airflow patterns in a naturally ventilated building," Can. Agri. Eng, pp. 293–297, 1988.
7. S. Morsing, A. Ikeguchi, J. C. Bennetsen, J. S. Strøm, P. Ravn, and L. Okushima, "Wind induced isothermal airflow patterns in a scale model of a naturally ventilated swine barn with cathedral ceiling," Appl. Eng. Agric., vol. 18, no. 1, pp. 97–101, 2002.
8. M. De Paepe, J. G. Pieters, W. M. Cornelis, D. Gabriels, B. Merci, and P. Demeyer, "Airflow measurements in and around scale model cattle barns in a wind tunnel: Effect of ventilation opening height," Biosyst. Eng., vol. 113, no. 1, pp. 22–32, 2012.

[9] Q. Yi et al., "Wind tunnel investigations of sidewall opening effects on indoor airflows of a cross-ventilated dairy building," Energy Build., vol. 175, pp. 163–172, 2018.

10. J. E. Cermak, M. Poreh, J. A. Peterka, and S. S. Ayad, "Wind tunnel investigations of natural ventilation," J. Transp. Eng., vol. 110, no. 1, pp. 67–79, 1984.
11. N. J. Cook, "On simulating the lower third of the urban adiabatic boundary layer in a wind tunnel," Atmos. Environ., vol. 7, no. 7, pp. 691–705, 1973.
12. V. Uruba, "Near wake dynamics around a vibrating airfoil by means of PIV and Oscillation Pattern Decomposition at Reynolds number," J. Fluids Struct., vol. 55, pp. 372–383, 2015.
13. X. Shen, R. Su, G. K. Ntinas, and G. Zhang, "Influence of sidewall openings on air change rate and airflow conditions inside and outside low-rise naturally ventilated buildings," Energy Build., vol. 130, pp. 453–464, 2016.
14. K. Hasselmann, "PIPs and POPs: The Reduction of Complex Dynamical Systems Using Principal Interaction and Oscillation Patterns," J. Geophys. Res., vol. 93, 1988.
15. V. Uruba, "Decomposition methods in turbulence research," in EPJ Web of Conferences, 2012, vol. 25.
16. V. Uruba and P. Procházka, "On interpretation of spatiotemporal data decomposition," in 15th International Conference on Fluid Control, Measurements and Visualization, 2019, no. May, pp. 1–11.
17. V. Uruba, "Energy and Entropy in Turbulence Decompositions," 2019.

CFD simulation of wettability of laser-structured surfaces

Jan Novosád[1,*], *Libor* Dvořák[1] and *Pavel* Peukert[1]

[1]VÚTS, a.s., Svárovská 619, 460 01, Liberec, Czech Republic

Abstract. This paper describes the numerical analysis of the wettability of the micro structured surfaces. The structure of the surface are the grooves created by the laser ablation. The main goal is to find the influence of the groove parameters on the wettability, specifically the contact angle. Ansys Fluent using the VOF method for multiphase simulation is used. In total, five different structures were investigated. The work started with the validation of the CFD methodology based on the experimental data for some structures. The CFD analysis was then performed. The fields of volume fraction as a results have been assessed to reach the contact angle value. The influence of structure parameters on the contact angle value were discussed. Finally, the recommendations for the optimal structure were defined.

1 Introduction

Creating functional surfaces with ultra-short pulse lasers is currently a very rapidly developing field of surface finishing. Lasers can achieve specific surface properties of various materials. It is a controlled process with high repeatability. Thus, a change in wettability, a reduction in the coefficient of friction or an improvement in biocompatibility can be achieved on the surface. Other properties that can be influenced by ultra-short pulse lasers include hardness, surface residual stress, or optical properties [1].

With sufficient intensity of the laser beam focused on the surface, material ablation occurs. Then the material evaporates and a controlled surface texture can be achieved. During the so-called "cold ablation", the material evaporates at the point of impact of the laser beam, but the surrounding material remains thermally unaffected. Hence this can be used to create very precise structures with picosecond or femtosecond lasers.

Fig. 1. Grid textured surface (grid) – fs laser, mag. 1000x.

For such surfaces, it is then necessary to analyse the achieved properties influenced by the action of the laser [1]. Also, surface texturing (see ex. on figure 1) with ultra-short laser pulses enables the creation of microstructures on a wide range of materials, including metal, ceramic and polymeric materials [2].

Titanium alloy (Ti grade V) was used for our experiment due to its mechanical properties and high biocompatibility. It is often used in biomedicine, e.g. as a joint replacement, dental implant or different types of microelectronics [3].

Implant infection and failure may occur during implementation. Infection is often caused by contamination of the surgical environment during implementation. The presence of bacterial colonies may prevent proper osseointegration. Consequently, it is important to reduce the likelihood of colonization of the implant surface through bacteria by providing these surfaces with antibacterial properties. Bacteria are sensitive to surface properties such as topography, chemical composition and wettability [2]. Our work focuses on the impact of the surface shape on wettability.

Wettability is the property of a liquid to adhere to the surface of some solids. Surface wetting is influenced by many circumstances such as surface roughness, surface finish, material, etc. Furthermore, wettability is affected by the surface tension of the liquid. Surface tension is the effect by which the surface of liquids strives to reach the lowest energy state [4]. Finally, a drop forms a spherical cap lying on the surface with a contact angle θ [4]. Then the surfaces are referred based on the contact angle value (see figure 2) to as hydrophilic ($\theta < 90°$),

* Corresponding author: jan.novosad@vuts.cz

https://doi.org/10.1051/epjconf/202226901043

hydrophobic (θ > 90°) and super hydrophobic (θ > 150°).

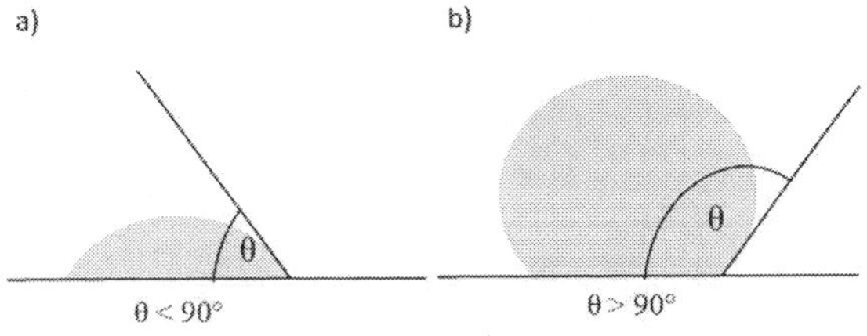

Fig. 2. Hydrophilic (a) and hydrophobic (b) surface.

The experimental investigation of the contact angle could be provided by several methods, including contact and optical systems [4], [5]. The shape of the flattened droplet is obviously measured and the contact angle is evaluated. Thus, the preparation of a sample with the surface texture is required.

Otherwise, the CFD (computational fluid dynamics) simulation could be used for investigation as the sample-free method. Wetting could be simulated as the multiphase problem when the liquid droplet falls in air onto the solid surface. The VOF (Volume of Fluid) method could be used to simulate thus type of multiphase problem [6]. The VOF method is often used successfully to model the behaviour of a drop on a solid surface. Nichita and Zun [7] have used the coupling of the VOF with dynamic contact angle. In recent years, numerical simulations have been used in research to investigate textured surfaces. The aim is to achieve hydrophobic or superhydrophobic surfaces [8], [9], [10].

2 Problem description

The main goal of our work was to prepare and validate the methodology of the numerical investigation of contact angle for the surfaces with microstructure textures. A schematic diagram of the phenomenon is shown in figure 3. Validation will be based on the experimentally investigated contact angle using the drop test. After that the validated methodology will be used to find the relationships between the parameters of the microstructure and the contact angle.

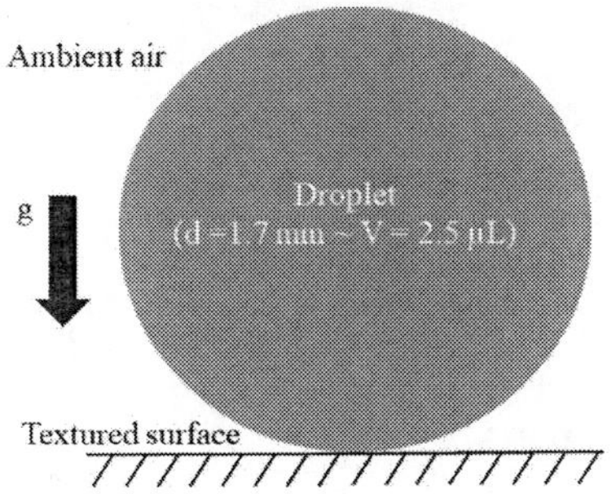

Fig. 3. Schematic diagram of falling droplet in contact with surface.

2.1 Geometry of the microstructure

This study is limited to the linear grooves structure only. The main parameters of the microstructure are shown in figure 4. Thus three parameters are defined, the pitch R, width of the groove P and height of the groove H.

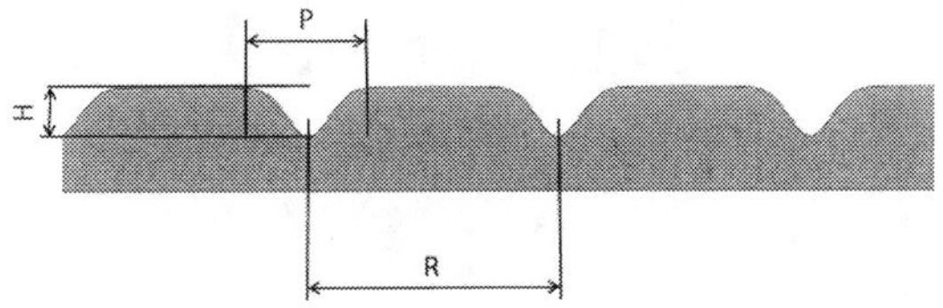

Fig. 4. Microstructure – sketch.

The shape of the groove corresponds to the Gaussian curve defined by:

$$f(x) = a\,e^{-\frac{(x-b)^2}{2c^2}} \tag{1}$$

where a = H, b = 0, c = P/6. In all solved cases the height H is 20 μm. Other parameters varies following the values in table 1. The grid of grooves was composed as two mutually perpendicular linear structures (see figure 1).

Table 1. Structures – main parameters and description.

Label	R (μm)	P (μm)	Groove type
1LR20P10	20	10	Linear
3MR50P25	50	25	Grid
1LR100P50	100	50	Linear
7LR100P20	100	20	Linear
xLR50P20	50	20	Linear

2.2 Model description

Some cases were investigated experimentally by the external laboratory in the past. The Surface Energy Evaluation System (See System [12]) was used. The volume of testing droplet was 2.5μL. As a result the contact angle value of structures 1LR20P10 (θ = 125.4° ± 3.3°) and 3MR50P25 (θ = 131.7° ± 2.2°) is known. These values can be used to validate the CFD simulation setup.

Following the situation on figure 3 the 3D domain were created respecting the parameters of the grooves. The boundaries were defined to be relevant to the reality and the spherical region was defined as the initial volume of the droplet (V = 2.5μL). Two symmetry planes were applied, therefore, the quarter model was used to speed up the computation. The dimensions of the quarter model were (2.5 x 2.5 x 2.5) mm.

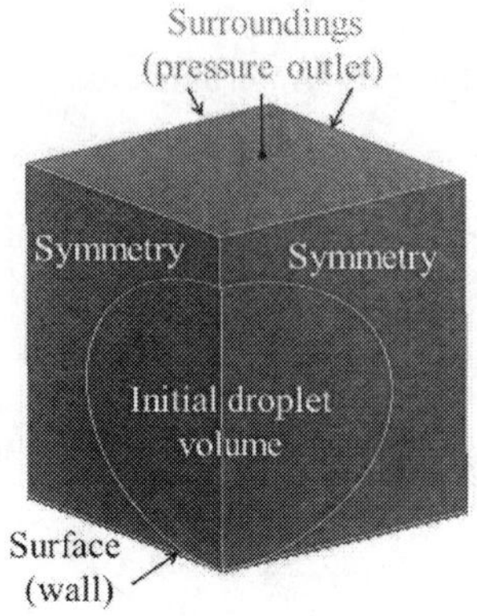

Fig. 4. Microstructure – sketch.

The tetrahedral grid was created using Ansys Meshing software. The grid refinement in the boundary layer region was used. Finally, the computational grid consists of around 2 mil. elements.

3 CFD simulation

For CFD simulations Ansys Fluent 2019 R2 were used. Multiphase flow was modelled by applying VOF model. The VOF model is a surface-tracking technique applied to a fixed Eulerian mesh. It is designed for two or more immiscible fluids where the position of the interface between the fluids is of interest. In the VOF model, a single set of momentum equations is shared by the fluids, and the volume fraction (VF) of each of the fluids in each computational cell is tracked throughout the domain [11]. In brief the volume fraction field is described by values between 0 and 1, where $VF_i = 0$ means no phase *i*, on the contrary $VF_i = 1$ means the volume is fulfilled by phase *i*.

For our case the two phase flow was modelled when the air as gas and water as liquid was used. Both with constant properties at 20°C. The laminar model is used. Gravity force ($g = 9.81$ m/s^2) was applied in the direction normal to the surface. The contact angle for flat surface (θ_0) have to been defined for the simulation. The values ranging from 72° to 84°, which are general tabular values for the titanium surface, have been applied to the textured surface.

The calculation is transient, the time-step was set to 10^{-7} s. The coupled scheme including the coupling with volume fractions was used for pressure velocity coupling. Body force weighted scheme was defined for pressure interpolation, second order scheme was used for momentum and compressive scheme for volume fraction was applied.

The criterion for convergence is defined as the area weighted average of volume fraction on the solid surface. The calculation stops, when the convergence criterion stabilizes over time. In other words the droplet falls onto the surface and forms a stable shape.

The CFD settings were unchanged for all analysed cases.

4 Results assesment

CFD simulation results are the fields of volume fraction. To be around the used count of elements the grid in outer region (far from the surface) is quite coarse, which causes relatively wide (diffuse) area of the interface between phases. For this reason the identify criterion for the interface was chosen to be $VF_{water} = 0.9$. It was supposed that the shape of the droplet could be described by two sections in planes of symmetry. In each plane the interface profile was exported as a set of points. First of all these points were fitted with a suitably selected curve that was defined by the equation:

$$f = ax^2 + bx + c, \tag{2}$$

with derivative,

$$f' = 2ax + b. \tag{3}$$

After that the derivative at the zero-point was calculated and used in the relation,

$$f'(0) = tg\,\theta, \tag{4}$$

from which the contact angle θ is evaluated. Finally the results from both symmetry planes are averaged to obtain the final value of contact angle.

The example of the set of interface points including the interpolation curve is shown in figure 5.

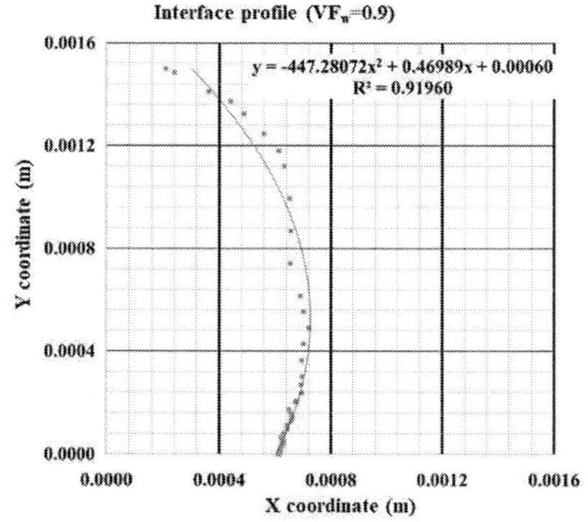

Fig. 5. Interface profile – example.

5 Results

The volume fraction $VF_{water} = 0.9$ was used to obtain the interface profile (points) from the CFD results. Based on these points the contact angle was assessed following the methodology described in chapter 4.

5.1 Validation results

3MR50P25 structure was investigated first. The values of θ_0 was set up to be {72°, 74° and 84°}. The angle θ_0 in the range between 72° to 74° gives the results close to the experimental data accordingly to the bar chart shown in figure 6. The deviation between the simulation and experimental data is lower than the measurement uncertainty, so the CFD setup could be considered as validated.

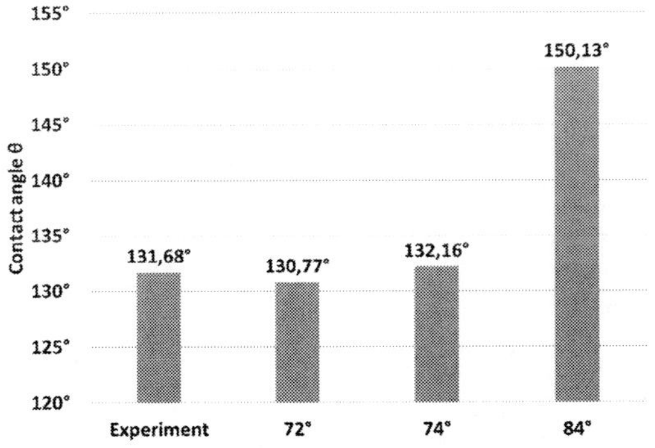

Fig. 6. Contact angle depending on θ_0 – structure 3MR50P25, θ_0 = {72°, 74° and 84°}.

To be sure with validation second calculation for 1LR20P10 structure was realized using the θ_0 = 74°. The results of validation is then shown in figure 7. To summarise, simulation results of both structures are in a good agreement with experimental data. The optimal value of the contact angle for flat surface which is set up is θ_0 = 74°.

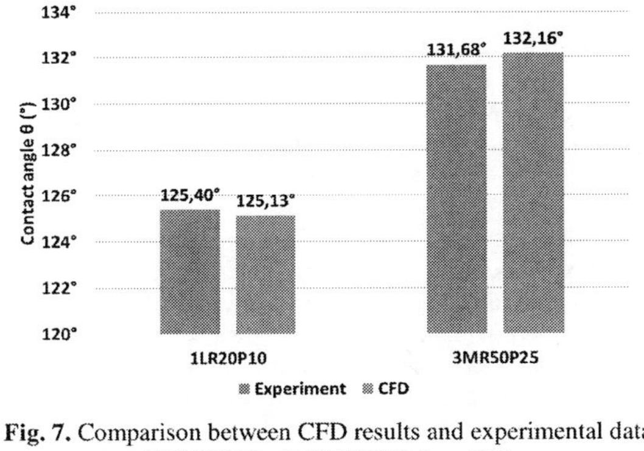

Fig. 7. Comparison between CFD results and experimental data – structures 1LR20P10, 3MR50P25, θ_0 = 74°.

5.2 Linear structures

The validated methodology (CFD setup) was used to investigate the influence of linear structure parameters onto the contact angle. Contact angle for flat surface was set up as θ_0 = 74°. Contact angle of the investigated structures 1LR100P50, 7LR100P20 and xLR50P20 was compared with the reference structure (1LR20P10). As a result the pitch and width of grooves significantly changes the value of contact angle (see figure 8).

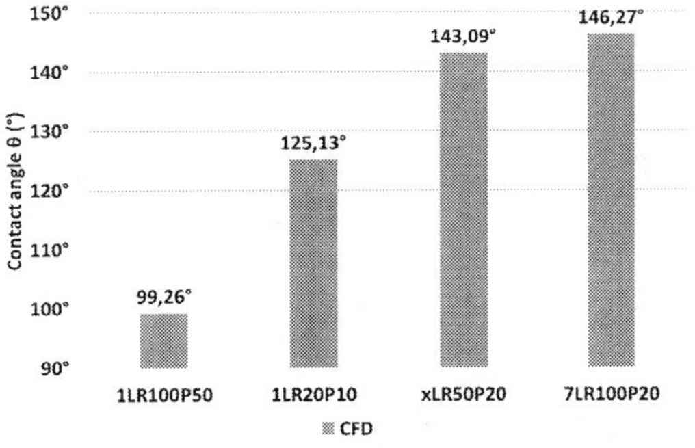

Fig. 8. Comparison of contact angle for different structures.

Unfortunately, there is not so much geometries to find the final statement about the influence of each parameter on the contact angle. However, it could show the trends and give us the ideas for choosing additional modification for testing. For this reason the influence of parameters on increase/decrease of the contact angle is our goal to design hydrophobic/hydrophilic structure.

Based on the available results (see figure 9), it can be concluded that increasing the width causes the significant decrease of contact angle. It is probably caused by decreasing the flat area in the space between grooves. Additionally, the fillet at the top of the groove where the groove merges into the straight part, is increasing. Thus, water can more easily flow into the groove, so the surface becomes less hydrophobic.

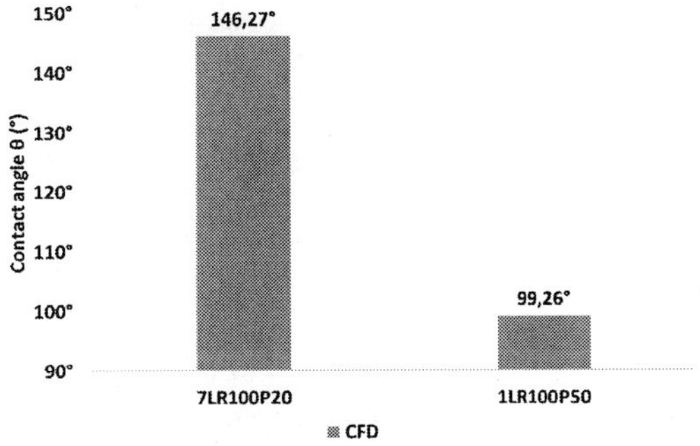

Fig. 9. Influence of the width of the grooves on contact angle (constant pitch R = 100 μm).

Otherwise the increase of the pitch (see figure 10) causes the increase of the contact angle. The contact area on the flat surface is larger with increasing the pitch.

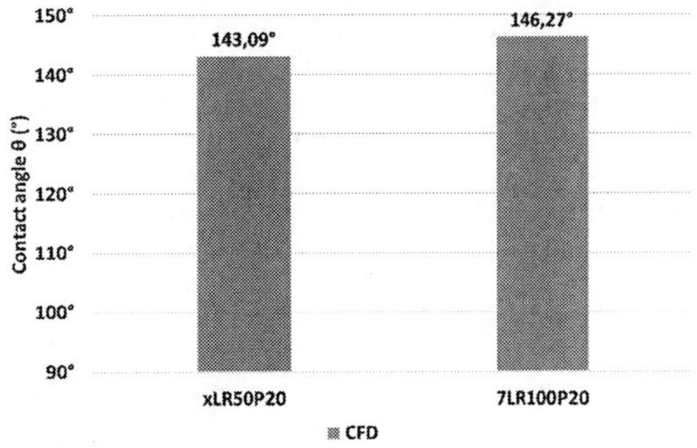

Fig. 10. Influence of the pitch between grooves on contact angle (constant width P = 20 μm).

Also the dimensionless R/P ratio could be set as a characteristic of the geometry. According to the comparison of all four cases (shown in figure 11) the increase of R/P leads to the increase of contact angle.

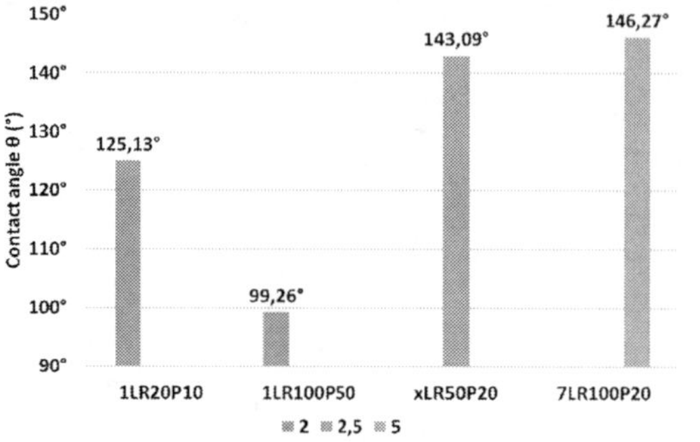

Fig. 11. Influence of the R/P parameter.

The studied cases results shows that the limit for maximum contact angle will be probably reached. Cn the contrary the infinite pitch creates the flat surface ($\theta \rightarrow \theta_0$).

6 Conclusion

For our work five different textured surface structures were selected for analysis. The main goal of the work was to analyse numerically the influence of individual structure parameters on the wettability, specifically the contact angle value. The experimental data exists for structures 1LR20P10 and 3MR50P25. Therefore the simulations of these structures had been solved firstly. CFD analysis based on the VOF method for multiphase have been created. The own simple methodology was used also for evaluating the contact angle from the field of volume fraction values. Validation of the CFD methodology was then performed by comparison with experimental data. Finally, three structures with different parameters of grooves geometry were simulated using the validated CFD setup.

The contact angle values were compared in different combinations to evaluate the effect of grooves width P, pitch R and the R/P ratio. However, the existence of results for only four variants of the line structure exist, the comparisons are always made for a very small number of variants. Therefore, it cannot be considered as meaningful for definitive conclusions.

Based on the available results, it can be concluded that increasing the width of the groove will significantly reduce the contact angle. This fact is probably related to a significantly changed geometry (larger fillet) at the top of the groove, where the groove merges into the straight part (flat surface). Thus, water can more easily flow into the groove, so the surface becomes less hydrophobic. Increasing the groove pitch, while the groove width is the same, shows a slight increase in contact angle. Also the increase of R/P ratio causes the increase of the contact angle. At the same ratio the results show a more dominant effect of the groove width. In the variants examined, the contact angle was decreased even if the pitch was increased. The influence of the groove width is probably more dominant than the pitch.

To summarise, the highest contact angle should be reached on the structure with a larger pitch R and a small slot width P should be suitable. The optimal maximum value have to be found as a limit, however very high values of pitch creates the flat surface, for which the contact angle is quite small equally to the angle for flat surface.

All the above-mentioned dependencies are analyzed on the basis of a very small number of solved structures. In order to confirm the validity of the above conclusions, it would be appropriate to continue the analysis of other structural variants. At the end the final structure sample will be prepared and investigated experimentally to confirm the results of CFD analysis.

This publication was written at the VÚTS, a.s. with the support of the project FV30265 - „Laserové technologie pro mikrostrukturování bionických a funkčních povrchů pokročilých materiálů". Our acknowledgment belongs to all donors that helped us to reach the results of the work.

References

1. Y. Cai, W. Chang, X. Luo, A. M. L. Sousa, K. H. A. Lau, Y. Qin et. al. Precision Engineering **52**, pp. 266-275 (2018).
2. A. Cunha, A. M. Elie, L. Plawinski, A. P. Serro, A. M, Botelho do Rego, A. Almeida, M. C. Urdaci, M.-Ch. Durrieu, R. Vilar et. al. Applied Surface Science **360**, pp 485-493 (2016).
3. Z. Yu, G. Yang, W. Zhang, J. Hu et. al. J. of Materials Processing Technology **255**, pp. 129-136 (2018).
4. P.-G. de Gennes, F. Brochard-Wyart, D. Quéré. *Capillarity and wetting phenomena.* ISBN 0-387-00592-7 (2004).
5. DataPhysics. In: Sessile drop method - Optical determination of the contact angle [online] (2016) [cit. 2018-08-25]. Available at: www.dataphysics-instruments.com/knowledge/understanding-interfaces/sessile-drop-method/
6. J. H. Ferziger, M. Perić. *Computational methods for fluid dynamics*, 3rd Ed. ISBN 3-540-42074-6 (2002).
7. B. A. Nichita, I. Zun, J. R. Thome. A VOF method coupled with a dynamic contact angle model for simulation of two-phase flows with partial wetting. 7th International Conference on Multiphase Flow, ICMF 2010, Tampa, FL, USA (2010).
8. S. Chun, Z. Chengchun, G. Meihong, L. Xueli, L. Yan, R. Luquan. Int. J. of Heat and Fluid Flow **74**, pp. 89-109 (2018).
9. Ch. Liu, L. Zhu, W. Bu, Y. Liang. Micron **107**, pp. 94-100 (2018).
10. X. Wang, D.-L. Sun, X.-D. Wang, W.-M. Yan. Int. J. of Heat and Mass Transfer **135**, pp. 235-246 (2019).
11. *ANSYS Fluent Theory Guide*, Release 2019 R2, Ansys Inc. (2019).
12. See System.
Available at:www.advex-instruments.cz

Uncertainty Estimation Methods in Particle Image Velocimetry

Jan Novotný[1,*], *Ludmila* Nováková[1], and *Ilona* Machovská[1]

[1]Institute of Machines and Power Engineering, the Faculty of Mechanical Engineering, J. E. Purkyně University in Ústí nad Labem, Czech Republic

Abstract. The aim of present work is focused on determining accuracy of the measurement using the Particle Image Velocimetry based on an analysis of synthetic Uniform flow test and Couette flow test. We focused on testing the Standard Cross Correlation algorithm and interdependence between the several metrics based on the ratio signal and noise peak and the measurement accuracy at each point. This work is describing the influence of the new corrected metric that is corrected by a displacement measured in the last iteration, detected mean density of the particles and the velocity gradient. Newly proposed method of calculating the metric, Lost of Particle Ratio, is based on the determination of the magnitudes of two correlation peaks. In this work there is defined and compared another metric based on the result of synthetics data test which is based on determination of main parameters as number of particles, gradient in interrogation area and displacement of particles calculated in the last step of correlation. For all presented metrics the relationship is between the metrics and the accuracy.

1 Introduction

The influence of individual recorded data on the result of the measurement error is monitored from the very beginning of the PIV measurement method development. At the turn of the 1980s and 1990s, Keane and Adrian looked at this issue, discovering the effects of the displacement and gradient of the outer investigated plane [1], followed by the other authors and detailing the effects of the displacement and magnitude from planar motion to the measurement [2], [3], [4]. Another parameter that significantly affects the resulting measurement accuracy is the average value and other parameters related to data such as the number of particles in the interrogation area and camera noise [5], [6], [7]. In addition to these two main groups, measurement accuracy affects a number of other parameters such as the size of the interrogation area and the weighting functions [8], [9] or the interpolation algorithm [10], [11]. The measurement accuracy of the PIV method has received considerable attention in recent years. Among the first who defined the correlation between the quality of the recorded data and the resulting measurement error was Timmins et al. who defined the uncertainty surface [12] to determine the resulting measurement error for each measurement. Another possibility to determine the resulting measurement error was defined by Charonko and Vlachos, who determined the influence of the correlation plane metrics and their work was subsequently developed by other authors [13], [14], [15], [16], [17]. Charonko and Vlachos followed by Xue et al. defined the influence of the metrics designated as Primary Peak Ratio PPR and Mutual Information MI and established the theoretical correlations between the metric value and the measurement error. In this work, we want to describe the work of these two authors and describe how to correct these two metrics and improve the resulting correlation between the metric value and the measurement error. Both the PPR and the MI have a direct effect on the number of particles in the interrogation area. Since the number of particles is one of the main parameters influencing the resulting measurement accuracy, the effort to use these metrics to determine the correlation between their value and the measurement error is very important. However, the effect of the number of the particles is not the only parameter that affects the shape and size of the correlation and noise peak and hence the size of the two metrics. Other parameters that significantly affect the overall measurement error include, in addition to the number of the particles, the number of lost pairs, the size of the velocity gradient and the particle diameter. The influence of these parameters on the shape of the correlation plane can be seen in Figure 1 and Figure 2. It can be seen from these images that the signal peak decreases and the PPR ratio decreases as the gradient increases. The number of the lost pairs and the amount of noise have the same effect. These parameters have a similar effect on MI as defined by [14]. Based on

* Corresponding author: novotny.jan@ujep.cz

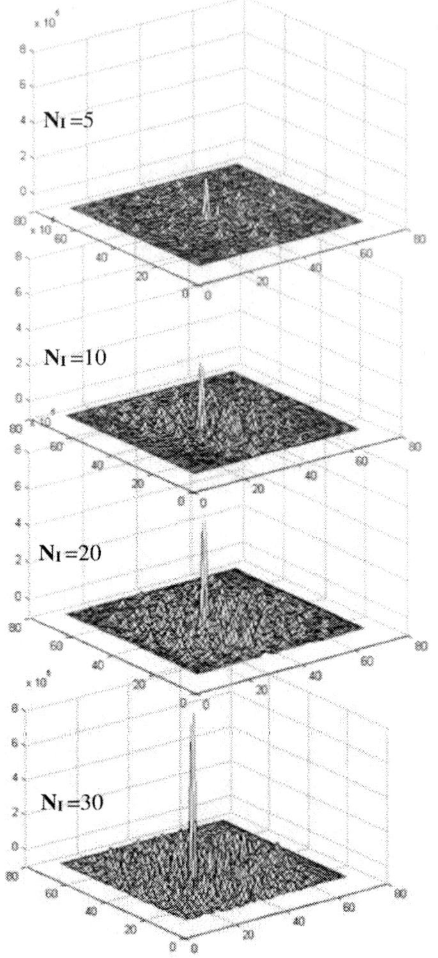

Fig.1. Correlation between seeding particle density (MI) and signal peak ratio.

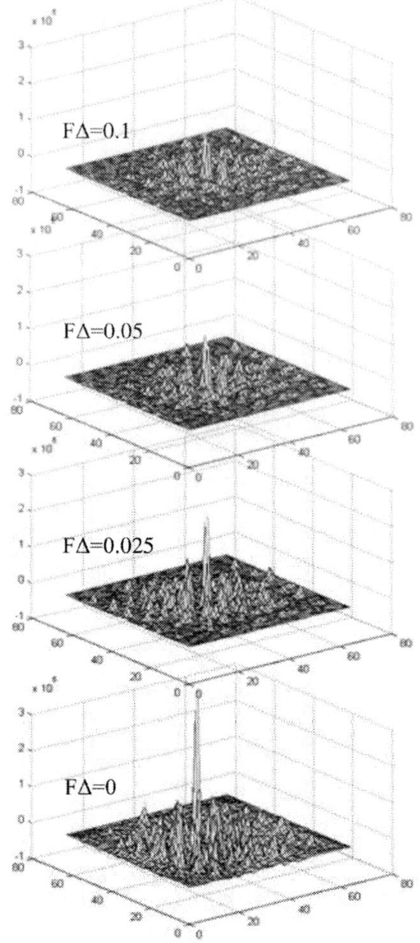

Fig. 2. Correlation between velocity gradient and signal peak ratio.

the effect of the gradient, the displacement and other parameters on the resulting size of these two metrics, we decided to find appropriate corrections that will include the effect of these parameters on the resulting measurement error and thus find a more accurate correlation between the metric value and the overall measurement error. As part of this work and based on the knowledge we have gained when searching for suitable metrics corrections, we have also defined a new metric called Lost of Pair Ratio LPR, which is based on the MI definition and corrects it by shifting the gradient and particle size.

2 Correlation plane metrics

Several metrics can be used to determine the correlation between the metric and the measurement error. In addition to the aforementioned PPR and MI, it is also possible to define the metrics defined as the Peak to Root Mean Square Ratio PRMSR and the Peak Correlation Energy PCE. In our work, we focused on finding the appropriate correction of PPR and MI. We defined the correction of these metrics based on the assumption of the knowledge of the main parameters such as particle number, velocity gradient, particle diameter size and number of lost pairs in the interrogation area. The reason why we need to look for the necessary corrections for PPR and MI is shown in Figure 3 and Figure 4. From these synthetic test results (Uniform Flow Test), a clear sensitivity of the size of the metric to both the number of particles within the interrogation area and the edge size of the interrogation area is evident.

Based on the results of synthetic tests (Uniform Flow Test UFT and Couette Flow Test -CFT), we designed the following corrected PPR and MI metrics:

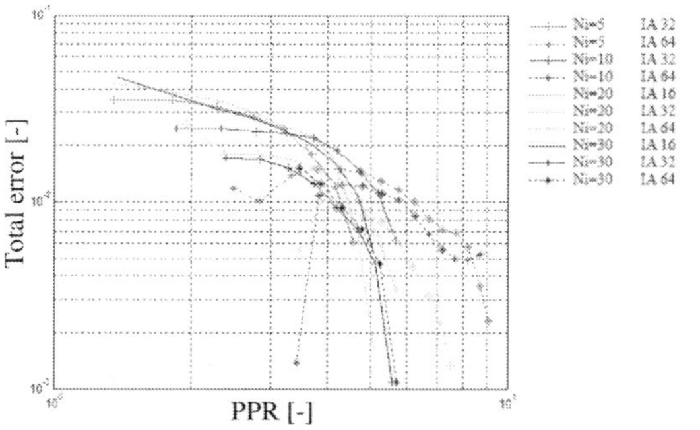

Fig. 3. Influence of seeding density and interrogation area size for PPR.

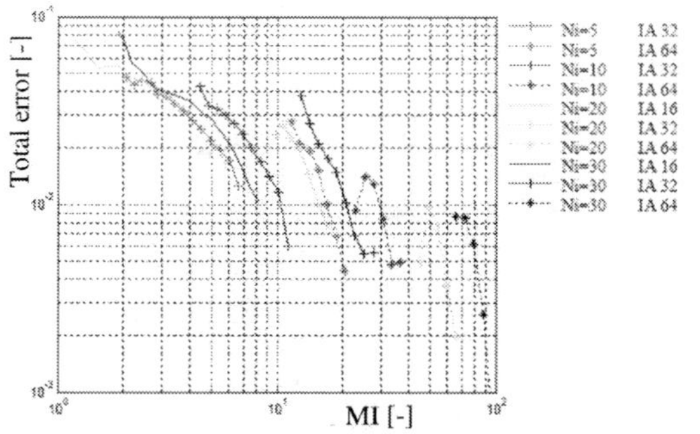

Fig. 4. Influence of seeding density and interrogation area size for MI.

PPR corrected by the size of the displacement measured in the last iteration.

$$PPR_{\Delta} = \frac{PPR - 1}{2 * (\Delta x^2 + \Delta y^2) + 0.0001} + 1 \tag{1}$$

PPR corrected by the size of displacement measured in the last iteration and the number of particles in the interrogation area.

$$PPR_{\Delta,MI} = \frac{PPR - 1}{16\sqrt{(\Delta x^2 + \Delta y^2)} * MI^{(-0.42)} + \frac{0.08}{MI}} + 1 \tag{2}$$

PPR corrected by the size of the displacement measured in the last iteration, the number of particles in the interrogation area and the gradient size in the interrogation area.

$$PPR_{\Delta,MI,GR} = \frac{PPR - 1}{\begin{pmatrix} 0.01 + \sqrt{\Delta x^2 + \Delta y^2} * MI^{-0.42} + \\ sgn(GR) * 4 * exp(\frac{MI}{20})\sqrt{\Delta x^2 + \Delta y^2} \bullet exp(GR) \end{pmatrix}} + 1 \tag{3}$$

MI corrected by the size of the displacement measured in the last iteration.

$$MI_{\Delta} = \frac{MI}{2 * (\Delta x^2 + \Delta y^2) + 0.001} \tag{4}$$

MI corrected by the size of the displacement measured in the last iteration, and the size of the gradient in the interrogation area.

$$MI_{\Delta,GR,D} = \frac{MI}{[0.5 * (0.02 + \sqrt{\Delta x^2 + \Delta y^2} + \mathrm{sgn}(GR) * MI * \sqrt{\Delta x^2 + \Delta y^2 e^{GR}})]} \tag{5}$$

Based on the corrections made and the results of the synthetic tests, we then proposed a new metric defined as the MI and NI ratio [16] as the ratio of velocity gradient to particle diameter. LPR is defined as:

$$LPR = (MI \cdot (1 - 0.4 \cdot Gr)) \cdot (0.01 + 2 \cdot \Delta_x \cdot N_i)^{-1} \tag{6}$$

As can be seen in Figure 5 and Figure 6, correction of PPR and MI by the displacement and particle number shows a significantly better correlation between the metric value and the overall measurement error compared to uncorrected metrics.

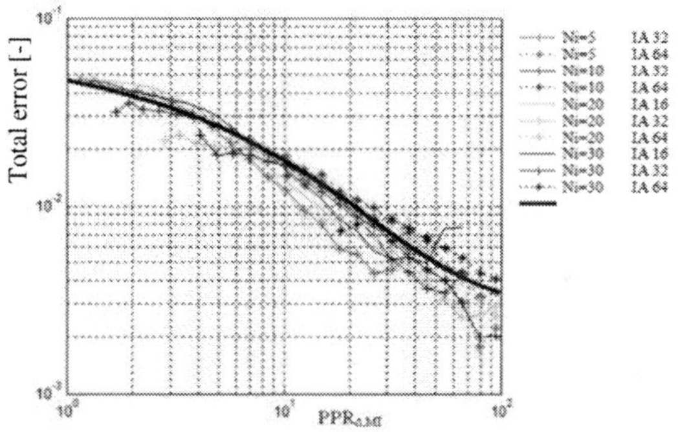

Fig. 5. Influence of seeding density and interrogation area size for PPRΔ.MI.

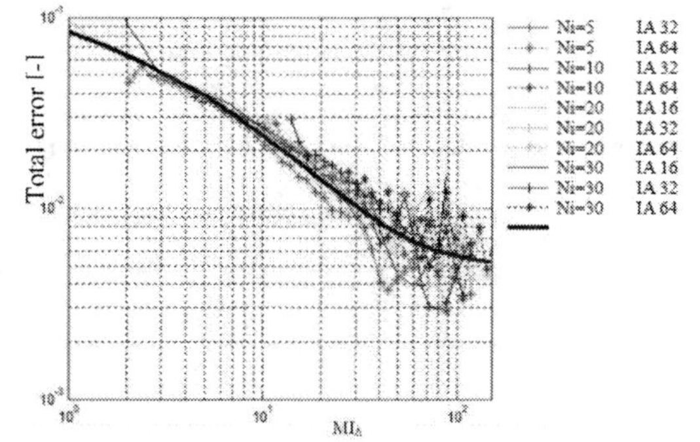

Fig. 6. Influence of seeding density and interrogation area size for MIΔ

Although the UFT results for both corrected metrics show a very good correlation between the metric value and the overall error, using the CFT test, the results for each metric already show different metric values for the metric values close to the minimum values as shown in Figure 7. For this reason, we have proposed a new metric called LPR, which does not show these shortcomings as shown in Figure 8.

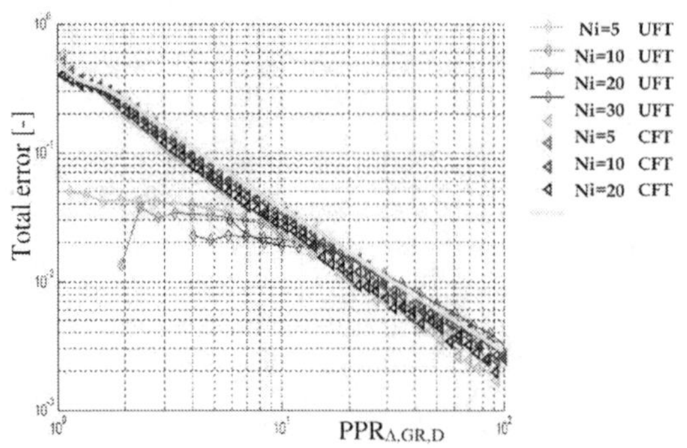

Fig. 7. Influence of seeding density and interrogation area size for PPRΔ.

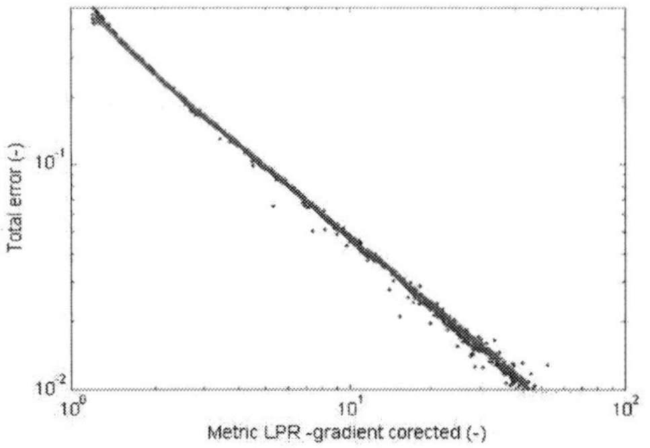

Fig. 8. Influence of seeding density and interrogation area size for fully corrected LPR metric.

3 Conclusions

The aim of this work was to compare the basic metrics used to determine the accuracy of measurements. Our goal was to describe and compare the individual corrections of metrics and on the basis of performed tests to evaluate the most advantageous metrics and then to propose further improvement of used metrics and algorithms for their determination. It is clear from the presented results that corrections of individual metrics significantly improve the resulting accuracy of the PIV measurement error determination. Using basic metrics without these corrections appears to be inaccurate and unusable for metric values near their minimum values. In contrast, corrected metrics, particularly fully corrected MIΔ, GR, D, and LPR, show a very good correlation between the metric value and the overall measurement error across the range of values. One disadvantage of the proposed procedure is the impossibility to determine the measurement error in each direction separately. This shortcoming will need to be addressed in the future work to verify the impact of the individual parameters on the measurement accuracy in each direction. Furthermore, it will be necessary to design a sufficiently robust algorithm to determine the individual parameters necessary for the calculation of corrected metrics and to verify these procedures not only on synthetic data but also on real experiments.

References

1. Keane R D and Adrian R J Optimization of particle image velocimeters. Part I: Double pulsed systems Meas. Sci. Technol. **1** 1202–15 (1990).
2. Fincham A M and Spedding G R Low cost, high resolution DPIV for measurement of turbulent fluid flow Exp. Fluids **23** 449–62 (1997).
3. Mayer S A generalized processing technique in digital particle image velocimetry with direct estimation of velocity gradients Exp. Fluids **33** (2002)
4. Novotny J Influence Of Data Quality On Piv Measurement Accuracy, Journal of Flow Visualization and Image Processing **19**, Issue 3, 215-230 (2012).
5. Willert C E The fully digital evaluation of photographic PIV recordings Appl. Sci. Res. **56** 79–102 (1996).
6. Stanislas M Okamoto K and Kähler C J Main results of the first international PIV challenge Meas. Sci. Technol. **14** 63–89 (2003).
7. Eckstein A and Vlachos P P Digital particle image velocimetry (DPIV) robust phase correlation Meas. Sci. Technol. **20** 055401 (2009).
8. Nogueira J Lecuona A and Rodríguez P A Limits on the resolution of correlation PIV iterative methods. fundamental Exp. Fluids **39** 305–13 (2005)
9. Huang H T Dabiri D and Gharib M On errors of digital particle image velocimetry Meas. Sci. Technol. **8** 1427–40 (1997).
10. Stanislas M Okamoto K Kähler C J and Westerweel J Main results of the second international PIV challenge Exp. Fluids **39** 170–91 (2005).
11. Kähler C J et al Main results of the 4th International PIV Challenge Exp. Fluids **57** 97 (2016).
12. Timmins B H Wilson B W Smith B L and Vlachos P P A method for automatic estimation of instantaneous local uncertainty in particle image velocimetry measurements Exp. Fluids 53 1133–47 (2012).
13. Charonko J J and Vlachos P P Estimation of uncertainty bounds for individual particle image velocimetry measurements from cross correlation peak ratio Meas. Sci. Technol. **24** 065301 (2013).
14. Xue Z Charonko J J and Vlachos P P Particle image velocimetry correlation signal-to-noise ratio metrics and measurement uncertainty quantification Meas. Sci. Technol. 25 115301 (2014).
15. Xue Z Charonko J J and Vlachos P P Particle image pattern mutual information and uncertainty estimation for particle image velocimetry Meas. Sci. Technol. 26 074001 (2015).
16. Novotny J Novakova L and Machovska I Advanced metric for particle image velocimetry accuracy estimation 19th Int. Symp. on Applications of Laser and Imaging Techniques to Fluid Mechanics (Lisbon, Portugal) (2018).
17. Novotny J Machovska I Primary Peak Ratio Correlation To The Measurement Accuracy Of Piv Method, Acta Polytechnica **58**(3):189–194 (2018).

On spectral/hp simulations of heat-exchange effects in low-Mach non-stationary fluid flow

Jan Pech[1,*]

[1]Institute of Thermomechanics of the CAS, Dolejškova 1402/5, 182 00 Praha 8, Czech Republic

Abstract. Non-stationary flow of fluid, whose material properties change as influenced by heating, is the subject of this study. Model of coupled Navier-Stokes-Fourier system with temperature dependent material properties is introduced together with construction of appropriate numerical scheme. Spectral/hp element method is used as the discrete approximation of the functions in the spatial coordinates. Performance of the code implemented using Nektar++ library [1] is presented on problems of forced and natural convection around the heated cylinder.

1 Introduction

Numerical simulations of fluid flows advecting a quantity, which influences the flow field itself, are demanded across various areas of research and applications. Temperature is considered as the advected quantity in presented study, however, resulting system of equations coincides also with a wide fields of turbulent models.

We introduce results from simulations using original solver of evolutionary Navier-Stokes-Fourier system with temperature dependent material properties based on spectral/hp element approximation in space together with results from simulations of heat exchange-influenced incompressible fluid flow. Convergence in sense of temporal discretisation was verified on system which includes variable viscosity, thermal conductivity and density together with buoyancy and viscous heating. Unless we constrain to situations where compressibility can be neglected, we consider non-homogeneous divergence of velocity field as a consequence to thermal expansion, which is included naturally due to temperature dependent density. Density variation has impact not only due to the buoyancy, but also in other structural changes of the flow field, what we observe in change of measurable quantities as is the Strouhal number in the case of forced convection around the heated/cooled cylinder. Due to non-constant density, model of buoyancy is not of Boussinesque type and the solution includes hydrostatic pressure, what has to be taken into account in setting the Dirichlet BC for pressure on outflow. In contrast to variable viscosity and density, effect of viscous heating do not influence studied regimes significantly, but shows the accuracy of spectral/hp element approximation. Comparison with experimental data demonstrates influence of particular terms in the model using measurable quantities, e.g. Strouhal number or Nusselt number.

*e-mail: jpech@it.cas.cz

2 Navier-Stokes-Fourier system

Dimensionless system of equations representing flow of a heated fluid, which is supposed to be Newtonian, calorically perfect and obeying Fourier law, is often called the Navier-Stokes-Fourier system. In 2D, the unknowns are velocity $\mathbf{v} = (u, v)^{\mathrm{T}}$, pressure p and temperature T

$$\rho\left(\frac{\partial \mathbf{v}}{\partial t} + \mathbf{v}\cdot\nabla\mathbf{v}\right) = -\nabla p + \frac{1}{\mathrm{Re}}\nabla\cdot(2\mu\mathbb{S}) + \frac{1}{\mathrm{Fr}^2}\rho\mathbf{g} \tag{1a}$$

$$\frac{\partial \rho}{\partial t} + \nabla\cdot(\rho\mathbf{v}) = 0 \tag{1b}$$

$$\rho\left(\frac{\partial T}{\partial t} + \mathbf{v}\cdot\nabla T\right) = \frac{1}{\mathrm{Re\,Pr}}\nabla\cdot(\kappa\nabla T) + \frac{\mathrm{Ec}}{\mathrm{Re}}2\mu\mathbb{S}:\mathbb{S}, \tag{1c}$$

where

$$\mathbb{S} = \mathbb{D} - \frac{1}{3}(\nabla\cdot\mathbf{v})\mathbb{I} \tag{2}$$

is related to the deviatoric part of the stress tensor (pure shear stress) and $\mathbb{D} = \frac{1}{2}\left[\nabla\mathbf{v} + (\nabla\mathbf{v})^{\mathrm{T}}\right]$. Term

$$\mathbb{S}:\mathbb{S} = \mathbb{D}:\mathbb{D} - \frac{1}{3}(\nabla\cdot\mathbf{v})^2 \tag{3}$$

is related to the effect of viscous heating. We do not impose any a priori restrictions to form of the temperature dependencies $\mu(T)$, $\kappa(T)$ and $\rho(T)$. Note, that this

Table 1. Definitions of nondimensional numbers used in system (1). Note, that the Eckert number in this form belongs to temperature normalised by value T_∞^+ instead of widely used characteristic temperature difference ΔT^+. L^+ is a characteristic length, c_p^+ specific heat at constant pressure.

Name	Symbol	Definition
Eckert	Ec	$\frac{\lvert \mathbf{v}_\infty^+ \rvert^2}{c_{p\infty}^+ T_\infty^+}$
Froude	Fr	$\frac{\lvert \mathbf{v}_\infty^+ \rvert}{\sqrt{g_\infty^+ L^+}}$
Prandtl	Pr	$\frac{c_{p\infty}^+ \mu_\infty^+}{\kappa_\infty^+}$
Reynolds	Re	$\frac{L^+ \lvert \mathbf{v}_\infty^+ \rvert \rho^+}{\mu_\infty^+}$
Richardson	Ri	$\frac{\Delta\rho^+ g^+ L^+}{\rho_\infty^+ \lvert \mathbf{v}_\infty^+ \rvert^2}$

formulation avoids the Boussinesque approximation of buoyancy, since the gravitational action ($\mathbf{g}$ is a unit vector oposite to gravitational force) to momentum balance comes purely from density variation. This approach is advantageous, because it avoids specification of the thermal expansion coefficient, what is the case of Boussinesque's model. The volumetric term $\rho\mathbf{g}/\mathrm{Fr}^2$ in this form brings two effects to the solution. Beside of buoyancy it includes model of hydrostatic pressure gradient. It might be advantageous to leave only the buoyancy what follows by subtraction of some, e.g. farfield[1], state from (1a). The volumetric term in (1a) then takes form

$$\frac{1}{\mathrm{Fr}^2}(\rho - 1)\mathbf{g} \tag{4}$$

and allow comparison with the results achieved with Boussinesque approximation through general form of the Richardson number since $(\rho - 1) = (\rho^+ - \rho_\infty^+)/\rho_\infty^+ \approx \beta\Delta T$ (c.f. table 1, where dimensionless numbers are summarized).

Unless we consider only flow of air in this work, energy equation (1c) has the same structure for both the gases and liquids with the minor change, that specific heat at constant pressure c_p^+ is replaced by c_V^+ (specific heat at constant volume), c.f. redefinition of Pr by substituting $c_p \to c_V$.

3 Numerical method

3.1 Spectral/hp element method

The spectral/hp element method was chosen for approximation of functions in spatial coordinates. It combines decomposition of the computational domain Ω to elements Ω_e, $\Omega = \cup_{e=1}^E \Omega_e$, with high degree approximation on every element. Consequently, it provides high-order accuracy together with geometrical flexibility. It can be viewed as a generalisation of global, $E = 1$, spectral methods, which follows C^0 continuity among elements and finally coincides with hp-finite elements. The reason for choosing this method is in efficient achievement of high-order approximation, which significantly suppress numerical diffusion and dispersion, so the influence of particular terms of the equation system may be compared in the simulations. Other advantages of the method are in lowering the number of degrees of freedom in the algebraic system or providing an instant insight to estimates of error ([2]). Mentioned estimate of discretisation error comes from insight to a single-element spectra of expansion coefficients $\hat{f}_{pq}$, representing a function f

$$f(x, y) = \sum_{p,q=0}^{N} \hat{f}_{pq}\phi_p(x)\phi_{pq}(y) \tag{5}$$

as expanded over a basis $\mathcal{B}_N = \{\phi_{pq}\}_{p,q=0}^N$, which is in our case of polynomial type and has hierarchical structure, $\mathcal{B}_N \subset \mathcal{B}_{N+1}$(notation in (5) is for tensor-product type basis over triangles, for details see e.g. [3]). Exponentially converging spectra for a single triangular element is shown in Figure 1. An open source library Nektar++ [1] of version 3.3.0 with own substantial changes was used as a basis for implementation of the solver.

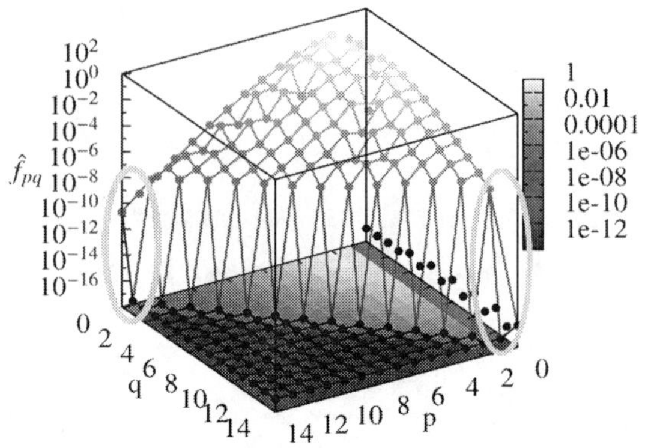

Fig. 1. Example of decaying spectra on a single triangular element for 2D polynomial expansion of degree $N = 14$. Spectral coefficients $\hat{f}_{pq}$ are only those with index $p+q <= N$ in the graph. Green ellipses show the maximal degree coefficients whose sum provides heuristical error estimate. For this case, one can conclude, that the spatial approximation *is not better than* 10^{-8}.

3.2 Discretisation in time

Computational scheme for system (1) combines multiple techniques. It is basically derived from solver for strictly incompressible, $\nabla \cdot \mathbf{v} = 0$, flow model [4]. It may be seen as a velocity-correction semi-implicit (IMEX) scheme

[1]We denote farfield/reference values by subscript ∞ and quantities with physical units are denoted by '+' in superscript.

Table 2. Coefficients of IMEX schemes up to order 3.

Q	γ	α_0	α_1	α_2	β_0	β_1	β_2
1	1	1	-	-	1	-	-
2	3/2	2	−1/2	-	2	−1	-
3	11/6	3	−3/2	1/3	3	−3	1

with a high-order approximation of Neumann type pressure boundary condition (HOPBC). However, multiple extensions of the scheme had to be performed.

The variable viscosity was implemented following approach mentioned in [5]. It is based on splitting the variable parameter, e.g. $\mu(T) = \bar{\mu}+\mu_s$, to constant-in-time part $\bar{\mu}$ and part variable both in time and space, $\mu_s = \mu_s(\mathbf{x}, t)$. This splitting impose a limit to amplitude of μ_s to ensure stability (see [5]), but it preserves structure of equations solved in the scheme. Similar approach was used for splitting the variable thermal conductivity in the energy balance, as described in [6]. Temperature-dependent density $\rho(T)$ in both the momentum and energy balance was implemented following approach proposed in [7], where this variable property was only in momentum equation. The most significant feature of proposed scheme lies in non-trivial velocity divergence, which emerges as a consequence of thermal expansion of the heated material. The detail description of discretisation to the full system, which is beyond scope of this paper, is prepared to be published. Here, we will only sketch the key points of the method.

The scheme is based on approximation of the time derivatives by *Backward Difference Formula* of order Q

$$\frac{\partial f}{\partial t} \approx \frac{\gamma f_{n+1} - \sum_{q=0}^{Q-1} \alpha_q f_{n-q}}{\Delta t}, \tag{6}$$

where f represents u, v, T or ρ. Δt denotes discrete time step, γ and $\{\alpha_q\}_{q=0}^{Q-1}$ are coefficients of the BDF formula (see table 2) and $f_n = f(t_n)$ is the function value at time $t_n = t_0 + n\Delta t$.

We denote

$$[f]^* = \sum_{q=0}^{Q-1} \beta_q f_{n-q} \tag{7}$$

the Q-th order extrapolation consistent to chosen BDF formula. The scheme follows these steps:

1. solve Pressure-Poisson equation, which is derived from application of divergence operator to (1a)

$$\begin{aligned}\nabla^2 p = {} & \frac{\gamma}{\Delta t}\left(\left[\frac{\partial \rho}{\partial t}\right]^{**}\right)\\ & + \nabla\cdot\left\{\frac{\gamma}{\Delta t}[\rho]^*\hat{\mathbf{v}}\right.\\ & + \frac{1}{\mathrm{Re}}\left[\nabla\mu\cdot\left(\nabla\mathbf{v} + (\nabla\mathbf{v})^{\mathrm{T}}\right) - \nabla\mu\cdot(\nabla\times\nabla\times\mathbf{v})\right]^*\\ & + \frac{1}{\mathrm{Re}}\left(-\frac{2}{3}[\nabla\mu]^*\langle\nabla\cdot\mathbf{v}\rangle + \frac{4}{3}[\mu]^*\nabla\langle\nabla\cdot\mathbf{v}\rangle\right)\\ & \left. + \frac{1}{\mathrm{Fr}^2}[\rho]^*\mathbf{g}\right\},\end{aligned} \tag{8}$$

where $\hat{\mathbf{v}} = \sum_{q=0}^{Q} \alpha_q \mathbf{v}_{n-q} - \Delta t\,[\mathbf{v}\cdot\nabla\mathbf{v}]^*$, $[\cdot]^{**}$ denotes extrapolation of time derivative approximated from BDF formula (6) and we used $\nabla\cdot(\rho\mathbf{v}) = -\partial\rho/\partial t$ from (1b). Forward estimate of divergence follows from extrapolation of (1b), since $\rho_n = \rho(T_n)$

$$\langle\nabla\cdot\mathbf{v}\rangle = \frac{1}{\rho^*}\left\{-[\mathbf{v}\cdot\nabla\rho]^* - \left[\frac{\partial\rho}{\partial t}\right]^{**}\right\} \tag{9}$$

Equation (8) is subject to the Dirichlet boundary condition or natural Neumann pressure boundary condition with high order extrapolation (HOPBC), which follows from projection of (1a) to boundary normal $\mathbf{n}$

$$\begin{aligned}\frac{\partial p}{\partial \mathbf{n}} = \mathbf{n}\cdot\Bigg\{\Bigg[& -\rho\frac{\partial\mathbf{v}}{\partial t} - \rho\mathbf{v}\cdot\nabla\mathbf{v}\\ & + \frac{1}{\mathrm{Re}}\left(-\mu\nabla\times\nabla\times\mathbf{v} + \nabla\mu\cdot\left[\nabla\mathbf{v} + (\nabla\mathbf{v})^{\mathrm{T}}\right]\right)\Bigg]^*\\ & + \frac{1}{\mathrm{Re}}\left(-\frac{2}{3}[\nabla\mu]^*\langle\nabla\cdot\mathbf{v}\rangle + \frac{4}{3}[\mu]^*\nabla\langle\nabla\cdot\mathbf{v}\rangle\right)\\ & + \frac{1}{\mathrm{Fr}^2}[\rho]^*\mathbf{g}\Bigg\}\end{aligned} \tag{10}$$

2. The velocity-correction step then consists of solution to an elliptic problem

$$\begin{aligned}\nabla^2\mathbf{v}_{n+1} - \frac{\gamma}{\Delta t}\mathrm{Re}\,\mathbf{v}_{n+1} = {} & \\ \Bigg\{ & -\mathrm{Re}\frac{\gamma}{\Delta t}\hat{\mathbf{v}}\\ & + \frac{1}{[\rho]^*}\Big[\mathrm{Re}(\nabla p_{n+1} - \mathbf{f}_{n+1})\\ & \quad - \left[\nabla\mu\cdot[\nabla\mathbf{v} + (\nabla\mathbf{v})^{\mathrm{T}}]\right]^*\\ & \quad + \frac{2}{3}[\nabla\mu]^*\langle\nabla\cdot\mathbf{v}\rangle\\ & \quad - \frac{1}{3}[\mu]^*\nabla\langle\nabla\cdot\mathbf{v}\rangle\Big]\Bigg\}\\ & - (\nabla\langle\nabla\cdot\mathbf{v}\rangle - [\nabla\times\nabla\times\mathbf{v}]^*)\\ & \left[\left(\frac{1}{[\rho]^*}\right)_s[\mu]^* + [\mu]_s^*\right].\end{aligned} \tag{11}$$

Above equation has many terms on right-hand side, since the technique of splitting to constant and variable-in-time part was done both to viscosity and density, c.f. [7].

3. Elliptic equation in similar form $(\nabla^2 - \lambda)T_{n+1} = \ldots$ is solved for temperature. Similarly to step 2), we split both $\kappa(T)$ and $\rho(T)$.

Unless the constant-in-time properties may be variable in space, the formulation is much more complicated and we set $\bar{\mu} = \bar{\kappa} = \bar{\rho} = 1$ for simplicity.

Convergence of the scheme was already verified on manufactured solutions, what is behind the scope of this contribution and is to be published together with more detail description of the scheme.

The buoyancy term in momentum equation (1a) is relatively easy to implement, but has substantial consequence in inclusion of complete model for natural convection. We consider $\mathbf{g}$ to be constant, and the buoyancy term is evaluated from extrapolated density in form including model of hydrostatic pressure gradient

$$\frac{1}{\mathrm{Fr}^2}[\rho]^*\mathbf{g}\,. \tag{12}$$

or without hydrostatics

$$\frac{1}{\mathrm{Fr}^2}([\rho]^* - 1)\mathbf{g} \tag{13}$$

accordingly to (4) as discussed in sec. 2.

The viscous heating, (3), in energy balance was included rather for completeness and testing. It is also evaluated as a pure extrapolation of data from previous time levels, while $\langle \nabla \cdot \mathbf{v} \rangle$ comes from (9).

4 Results

All computations presented concern air as the flowing medium. We use a power-law approximation for temperature dependencies of material properties μ, ρ and κ

$$f(T) = \frac{f^+_{ref}}{f^+_\infty}\left(\frac{T^+}{T^+_{ref}}\right)^{\omega_f} \tag{14}$$

(subscript "ref" denotes any virtual reference state following from the data-curve fit). Particular constants for $T^+_{ref} = 285\mathrm{K}$ are as follows

$$\begin{aligned} \mu^+_{ref} &= 1.78 \times 10^{-5}\,, \omega_\mu = 0.72 \\ \kappa^+_{ref} &= 0.025\,, \omega_\kappa = 0.84 \\ \rho^+_{ref} &= 1.239\,, \omega_\rho = -1.0013\,, \end{aligned} \tag{15}$$

derived from power-law regressions to data [8]. With the complete model (1) we study two cases, forced and natural convection.

4.1 Impact of variable material properties to vortex shedding frequency behind a heated cylinder

Flow around a cylinder is a widely used case for code testing and deep insight to flow physics for its simple geometry and sufficient available data. In this section, we compare our results from 2D simulations with experiments presented in [9] and also theoretically investigated in [10].

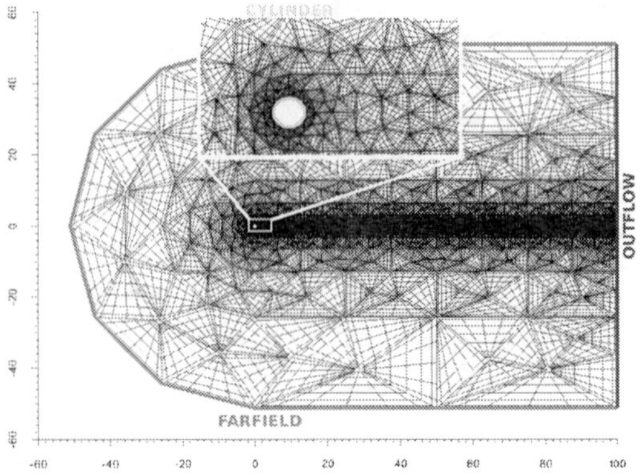

Fig. 2. Computational mesh with 2360 elements and domain boundary designation. Lines inside the domain connect quadrature points for degree 6 approximation inside triangular elements.

Data in [6] were computed with model including variable viscosity and thermal conductivity, while density was assumed constant and the velocity field was divergence free. Using the same computational mesh, boundary conditions and degree of polynomials $N = 6$ and other settings ($\Delta t = 0.001$), we recalculated the same situations. The boundary conditions are following (c.f. Figure 2):

- Farfield: $u = 1, v = 0, T = 1$, HOPBC
- Cylinder: $u = 0, v = 0, T = T^+_W/T^+_\infty$, HOPBC
- Outflow: $\partial u/\partial \mathbf{n} = 0$, $\partial v/\partial \mathbf{n} = 0$, $\partial T/\partial \mathbf{n} = 0$, $p = -g^+_\infty L^+ y/|\mathbf{v}^+_\infty|^2$

Pressure gradient set on the outflow boundary is consistent with used model of buoyancy, which includes the hydrostatic contribution. If buoyancy is modelled with (13), $p = 0$ on outflow is corresponding boundary condition. Constant fields with values $(u, v, T)^\mathrm{T} = (1, 0, 1)^\mathrm{T}$ were used as initial conditions.

High order approximation, polynomial of degree 15, was used also for representation of the deformed domain boundaries, the cylinder and upstream/inlet boundary, in the mesh. The second order scheme in time, $Q = 2$, was used as default.

Results from the new model and scheme are directly compared with experiment, [9], and previous numerical results in Figure 3. Reynolds numbers from [9] were recomputed for consistency with the relations describing material properties (15) (in [9] temperature dependence of kinematic viscosity was used in definition of Re, what is inconsistent with our model). It is worthy to emphasize, that model with constant density used in [6] strongly reflects the influence of temperature variation through variable viscosity and thermal conductivity, but is still insufficient as the results for highest temperature ratio ($\tilde{T} = T^+_W/T^+_\infty = 1.8$) coincide rather with different level of heating considered in experiment, compare coincidence of "emp. $\tilde{T} = 1.5$" and "cmp. [6] ρ = const.". Finally, the current results coincide with experimental data more

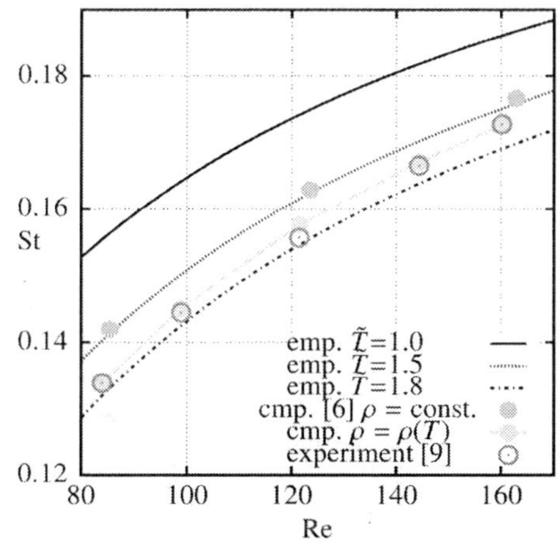

Fig. 3. Direct comparison of data from model proposed in [6] with current data (same setting of discretisation) and empirical formulas [10]. All computations "cmp." and experimental "exp." data belong to temperature ratio $\tilde{T} = T^+_W/T^+_\infty = 1.8$.

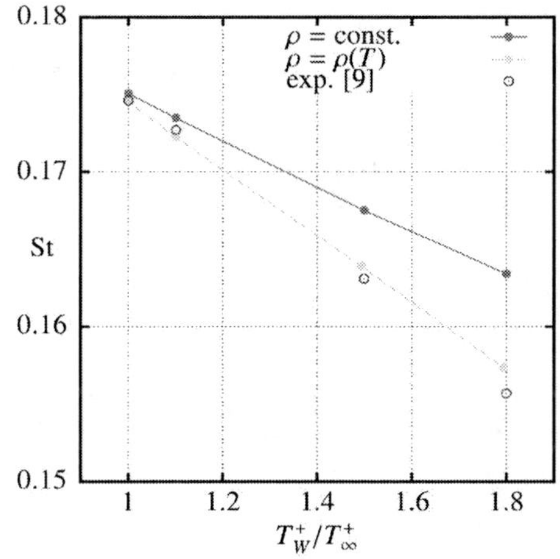

Fig. 4. Comparison of model with ρ = const. ([6]) and the current one with $\rho = \rho(T)$. The difference in dimensionless frequency of vortex shedding, St, is substantial. Computations followed Re from experiment, all being close to an average value Re ≈ 121.4.

accurately than empirically derived formula proposed in [10].

The difference between the models is even better visible in plot St(T^+_W/T^+_∞) for fixed value of Re, see figure 4. From this comparison of current data and result from [6], we can separately discern influence of variable viscosity and variable density to the flow. Note, that only the value from experiment at $T^+_W/T^+_\infty \approx 1.8$, used in Fig. 4, exhibit stronger deviation than other data also in Fig. 3.

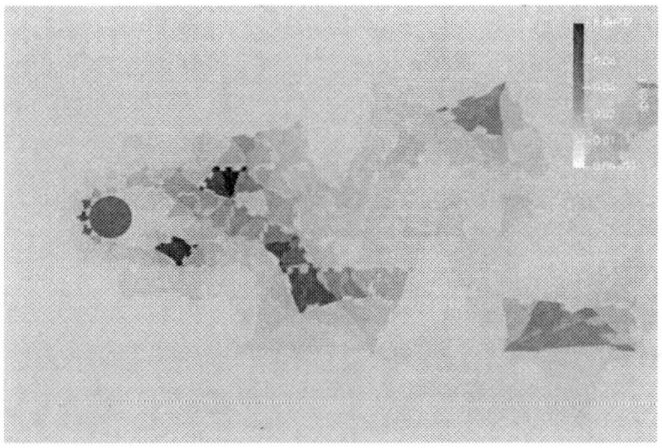

Fig. 5. Distribution of spatial error over the computational mesh. Values obtained as a sum of spectral coefficients belonging to highest degree boundary modes. Max value $\approx 5 \times 10^{-2}$.

The estimate of error in spatial discretisation as mentioned in sec. 3.1 is one of natural outputs from simulations in spectral/hp elements. Distribution of the error over the computational mesh, may be analysed at every timestep. Its evaluation is computationally cheap, since it is summation of only a few coefficients from the spectra of every element, c.f. figure 1. Figure 5 shows an example of this approach for a particular t_n on function representing temperature distribution in flow around heated cylinder. The error distribution changes as the function develops in time.

4.2 Natural convection

Flow motion induced by heating was investigated for validation of buoyancy model in presented scheme. Simple geometry of heated cylinder was chosen for comparison with previous data from literature. In [11], tabulated data are available. However, numerical method used in [11] differs substantially from the scheme proposed in sec. 3.2. Due to lack of data for this particular case in literature, we will compare results of model for steady incompressible constant-property flow with Boussinesque approximation for buoyancy with our model for evolutionary problem allowing non-homogeneous divergence, variable properties and without the Boussinesque's approximation.

Present simulation was performed with $\Delta t = 0.005$, polynomial degree $N = 24$ on mesh consisting of only 289 quadrilateral elements while the domain extent was $\Omega = [-500 : 500] \times [-500 : 500]$ for cylinder of diameter 1. Curve representing the cylinder was approximated by polynomial of degree 17. Domain dimensions were so large, that the buoyant disturbances travelling through the volume did not reach $\partial\Omega$ during whole time interval, $t \in [0 : 350]$. Therefore, the physical meaning of the solution was not strongly influenced by inaccuracy following from boundary conditions, which were as follows

- Inlet: $\partial u/\partial \mathbf{n} = 0$, $\partial u/\partial \mathbf{n} = 0$, $T = 1$, $p = 0$
- Sides: $u = 0$, $v = 0$, $T = 1$, HOPBC
- Cylinder: $u = 0$, $v = 0$, $T = T^+_W/T^+_\infty = 1.05$, HOPBC

- Outflow: $\partial u/\partial \mathbf{n} = 0$, $\partial v/\partial \mathbf{n} = 0$, $\partial T/\partial \mathbf{n} = 0$, $p = 0$

Shape of the computational domain was similar to Fig. 2 with half-circle arc on inlet. Gravity action was in direction of domains axis of symmetry. Initial conditions were $(u, v, T)^{\mathrm{T}} = (0, 0, 1)^{\mathrm{T}}$. Used model of buoyancy was without the hydrostatic pressure gradient, (13).

Simulation was performed for three values of Rayleigh number, $\mathrm{Ra} = \mathrm{Ri}\,\mathrm{Pr}\,\mathrm{Re}^2$, $\mathrm{Ra} \in \{10^2, 10^3, 10^4\}$. Material parameters for air were used, the same as in the case of forced convection over heated cylinder as discussed in sec. 4.1. Some value of Re has to be defined to the solver, however it is unknown a priori. It follows from binding of Re and Fr through L^+ and $|\mathbf{v}^+_\infty|$ (see Table 1), that lower influence of viscous term implies lower influence of buoyant term for fixed Ra. Instability of the scheme occured for values $1/\mathrm{Fr}^2$ of order 10 if $\mathrm{Re} < 50$. All simulations run with $T \in [1 : 1.05]$ and their inputs were

- $\mathrm{Ra} = 10^2$: $\mathrm{Re} = 100$, $L^+ = 0.00475$, $|\mathbf{v}^+_\infty| = 0.3 \Rightarrow \dfrac{1}{\mathrm{Fr}^2} = 0.51$
- $\mathrm{Ra} = 10^3$: $\mathrm{Re} = 100$, $L^+ = 0.01$, $|\mathbf{v}^+_\infty| = 0.14 \Rightarrow \dfrac{1}{\mathrm{Fr}^2} = 5.1$
- $\mathrm{Ra} = 10^4$: $\mathrm{Re} = 300$, $L^+ = 0.02$, $|\mathbf{v}^+_\infty| = 0.196 \Rightarrow \dfrac{1}{\mathrm{Fr}^2} = 5.6$

Note, that above values of Re are parameters chosen only for proper definition of the solved problem and stability of computation and have to be recalculated using some well defined values from solution fields for full comparison with experiment.

Comparison of local Nusselt number, $\mathrm{Nu} = \mathbf{n} \cdot \nabla T$, over the cylinder surface is in figure 6 and difference between the results is well observable. Similar difference, especially in vicinity to cylinder surface can be seen also in comparison of experimental and numerical isotherms already in [11], fig. 7. Similar difference to Kuehn's data in local Nusselt number was reported in [12], where a fourth order approximation in space was used in computation of steady model. Work [12] does not contain tabulated data and we have to conclude with only this qualitative comment to comparison with our data.

To emphasize the novelty of our scheme, we show distribution of velocity divergence, result of thermal expansion of the heated air, see figure 7. The same figure contains detail of computational mesh in vicinity of the cylinder and presents also the spatial resolution of variable u. For this field, in vicinity of the cylinder, worst element spectrum is converged to values around 10^{-6} and in elements on cylinder boundary we observe even better resolution up to 10^{-12}, what coincides with round-off tolerance in the Nektar++ library.

In contrast to cited stationary models we observe flow instability in present results as shown in Fig. 4.2 for $\mathrm{Ra} = 10^3$ and 10^4. These figures belong to the same time-step ($t_n = 250$). Wake in Fig. 4.2-A shows a pair of strong vortices, which travel from the cylinder while the wake becomes unstable, what can be seen from variation of mean

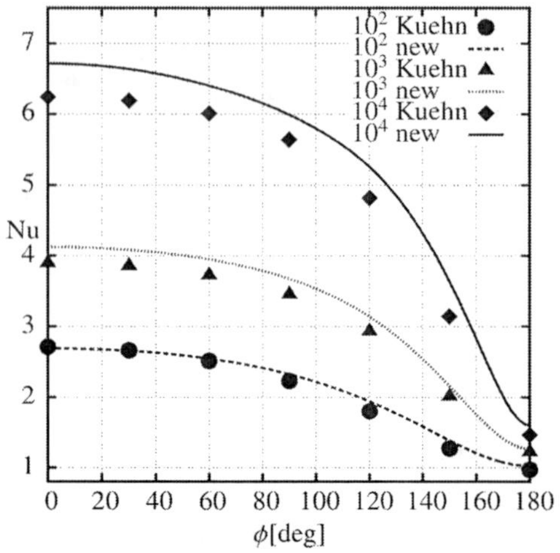

Fig. 6. Nusselt number distribution over cylinder in natural convection, comparison of data from [11] and the new scheme for $\mathrm{Ra} \in \{10^2, 10^3, 10^4\}$. Angle ϕ is measured from the lowest point of the cylinder.

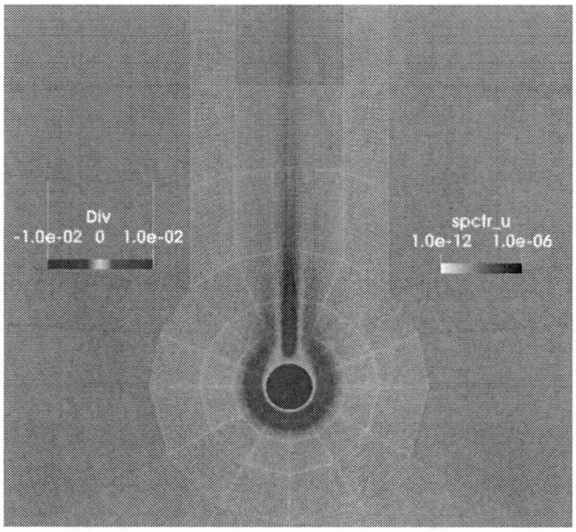

Fig. 7. Field of velocity divergence "Div" in natural convection problem with Ra = 1000. Spatial error estimate "spectr_T" for temperature field is projected onto the divergence field together with mesh grid, which belongs to approximation by polynomial of degree 24.

Nusselt number in Fig. 9. For $\mathrm{Ra} = 10^4$ we early observe development of instability and complicated vortex structures.

Obtained mean (integral) values of Nusselt numbers are

- $\mathrm{Ra} = 10^2$: $\mathrm{Nu}_{\mathrm{mean}} \approx 2.1$
- $\mathrm{Ra} = 10^3$: $\mathrm{Nu}_{\mathrm{mean}} \approx 3.3$
- $\mathrm{Ra} = 10^4$: $\mathrm{Nu}_{\mathrm{mean}} \approx 5.3$

 https://doi.org/10.1051/epjconf/202226901045

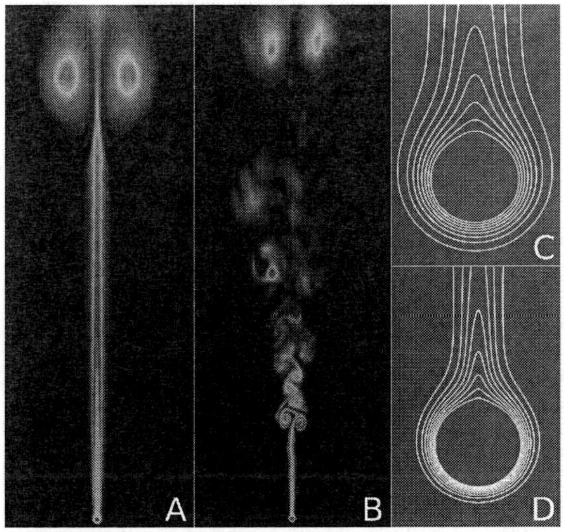

Fig. 8. Comparison of temperature fields for Ra = 10^3, (A and C), and Ra = 10^4, (B and D). 8 contours visible in the cylinder area (C, D) have regular spacing and cover range of temperature $T \in (1 : 1.05)$.

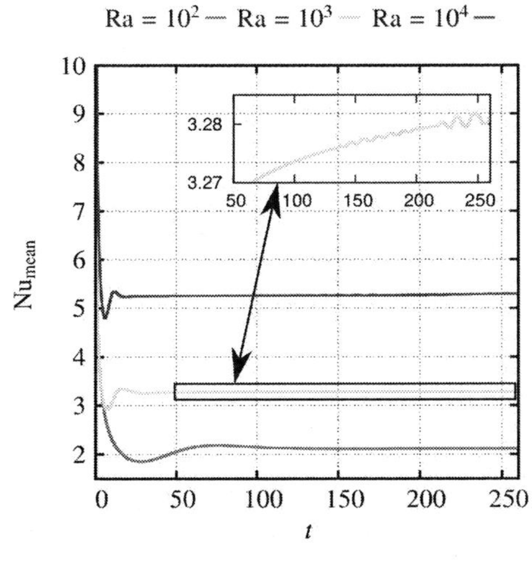

Fig. 9. Temporal evolution of mean Nusselt number Nu_{mean} for various values of Ra. Rescaled detail for Ra = 10^3.

We emphasize that above values are approximative, since the non-stationary nature of the flow is reflected in this data evaluated from the boundary layer on the cylinder. Mentioned instability is shown in a detail view inside Fig. 9.

4.3 Viscous heating

Effect of viscous heating is negligible in both studied cases, but running the simulation of forced convection case with the full model and $T_W^+ = T_\infty^+$ returns the temperature field shown in figure 10. Negligible variation being under resolution of less accurate numerical approximations is here well recognized, for more, the resulting field is smooth, unless the viscous heating term is evaluated through spatial derivatives of the velocity field and includes also the square of $\langle \nabla \cdot \mathbf{v} \rangle$, which is purely extrapolated and may accumulate numerical error.

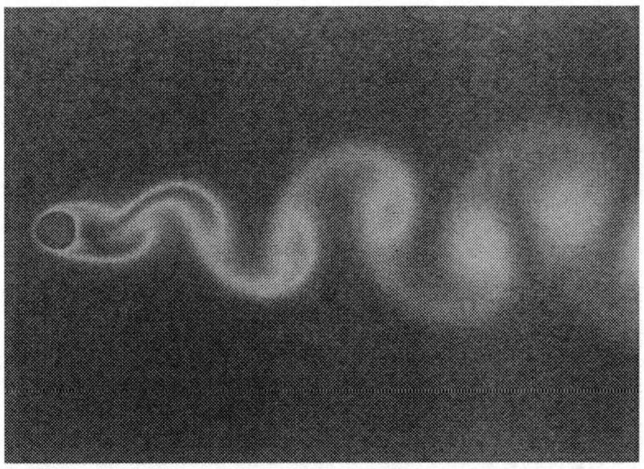

Fig. 10. Temperature field induced purely by viscous heating term (Re = 121.8, $T_W^+ = T_\infty^+$). Maximal variation of this normalized temperature is 3.8×10^{-6}.

5 Conclusion

Problem of cross-influence of heat and velocity field was studied in this work. Model for heated flow consisting of coupled system of momentum, mass and energy balance was formulated and appropriate numerical scheme was proposed. The system of equations is very general and similarities may be found also with turbulence models. Up to the authors knowledge, this is first implementation of this kind of models to spectral/hp methods. Field of numerical simulations of heated flows with variable material properties is not very wide and results found in literature seem to be mostly restrictions of the model and scheme presented in this contribution. Testing the convergence and rigorous description of the scheme is demanding and was left to specialized paper.

Performance of the scheme was tested on 2D simulations of evolutionary problems, forced convection over heated cylinder and natural convection induced by heating of the fluid on cylinder wall. Very good coincidence with experimental data [9] was achieved. Improvement of achieved Strouhal-Reynolds relationship for various temperatures of cylinder was recognized as substantial in comparison with previous model proposed in [6]. Current data very well coincide with experiments and uncover strong impact of both thermal expansion and change of viscosity to heated flow in regime of forced convection.

Numerical result achieving resolution to computer precision in cylinder area and boundary layer was achieved in problem of natural convection. It is observable from these results (e.g. figure 7), that the divergence of velocity induced by thermal expansion of the fluid reaches its maxima in certain distance from the heated surface. Divergence distribution in the boundary layer will be one of the subject for future detail investigations, since its model inclusion represents one of the novelties of introduced numerical scheme. In comparison with previous data from literature, difference in distribution of local Nusselt number over cylinder surface was observed and should be confirmed by further detail comparison with experimental data.

In studied cases, good performance of spectral/hp element method was shown, both in sense of accuracy of spatial approximation and effectivity of the algorithm for evolutionary problems. Original strategy of error estimate in spatial approximation, following from insight to coefficient spectra, was outlined.

Acknowledgements. The author wishes to acknowledge the support of the Czech Academy of Sciences, project MSM100761901.

References

1. C. Cantwell, D. Moxey, A. Comerford, A. Bolis, G. Rocco, G. Mengaldo, D.D. Grazia, S. Yakovlev, J.E. Lombard, D. Ekelschot et al., Computer Physics Communications **192**, 205 (2015)
2. J. Pech, P. Louda, *Comparison of Finite Volume and Spectral/HP Methods on Navier - Stokes Equations for Unsteady Incompressible Flow*, in *Proceedings Topical Problems of Fluid Mechanics 2018* (2018), pp. 223–230
3. G. Karniadakis, S. Sherwin, *Spectral/hp Element Methods for Computational Fluid Dynamics: Second Edition*, Numerical Mathematics and Scientific Computation (OUP Oxford, 2005), ISBN 9780198528692
4. G.E. Karniadakis, M. Israeli, S.A. Orszag, Journal of Computational Physics **97**, 414 (1991)
5. G.S. Karamanos, S. Sherwin, Applied Numerical Mathematics **33**, 455 (2000)
6. Pech, J., EPJ Web of Conferences **114**, 02089 (2016)
7. J. Pech, *Scheme for evolutionary Navier-Stokes-Fourier system with temperature dependent material properties based on spectral/hp elements*, in *Lect. Notes Comput. Sci. Eng.* (In print)
8. B. Gebhart, *Heat Conduction and Mass Diffusion*, McGraw-Hill series in mechanical engineering (McGraw-Hill, 1993), ISBN 9780071125147
9. A.B. Wang, Z. Trávníček, K.C. Chia, Physics of Fluids **12**, 1401 (2000), `https://doi.org/10.1063/1.870391`
10. F. Maršík, Z. Trávníček, R.H. Yen, A.B. Wang, in *Proceedings of CHT-08* (DOI:10.1615/ICHMT.2008.CHT.210)
11. T. Kuehn, R. Goldstein, International Journal of Heat and Mass Transfer **23**, 971 (1980)
12. T.V.S. Sekhar, B.H.S. Raju, International Journal of Physical and Mathematical Sciences **6**, 1032 (2012)

A light sculpture element under wind load – numerical FSI analysis and an experiment in a wind tunnel

Pavel Peukert[1,*], *Jaroslav* Kult[1], *Hiren* Rashiwala[1], *Pavel* Klouček[2], *Norbert* Pomp[2], Jan Hochmann[2], and *Václav* Vomáčko[1]

[1]VÚTS a. s., Department of Computation and Modeling, Svárovská 619, 460 01 Liberec, Czech Republic
[2]VÚTS a. s., Department of Measurements, Svárovská 619, 460 01 Liberec, Czech Republic

Abstract. The article introduces a numerical and experimental work to analyse the behaviour of a specific light sculpture element in a wind-load. It starts with a static analysis, which moves to a simplified unsteady analysis followed by a real experiment in a wind channel. The numerical part was not able to provide all necessary information about the light sculpture element behaviour. The experimental part was done to determine the real behaviour of the sculpture element under specified wind conditions.

1 Introduction

Fluid-structure-interaction computations get over the last years a higher level of importance. It allows predicting in a certain degree the mutual interaction between mechanical parts, parts in complex systems and fluid environments. Things becoming much more complex if unsteadiness and vibrations will appear in the analysis. For a proper analysis were often necessary parameters which hardly could be guessed and an experiment had to be done to obtain the requested data.

The investigated object here was a light sculpture concept depicted in fig. 1. The major diameter of the whole sculpture is around 8 m and it consists from more than 500 parts. The sculpture will be mounted at a ceiling of a pool bar on a beach. It will be permanently exposed to the actual wind situation. To ensure the safety of the artwork it was necessary to evaluate deflections of the light sculpture elements from the original position due to wind.

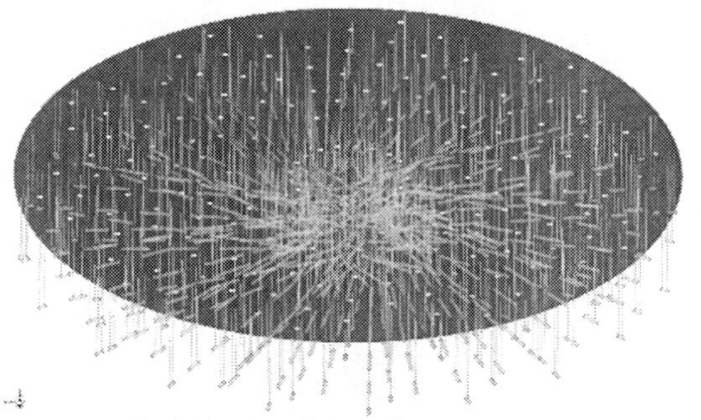

Fig. 1. Model of the whole light sculpture.

Because of the high complexity of the sculpture, it was chosen to investigate only the light sculpture elements with expectantly the highest deflections. One of the chosen elements, which will be investigated during this article in fig. 2 is the one with the largest surface area, which is exposed to the wind and has the longest support rods.

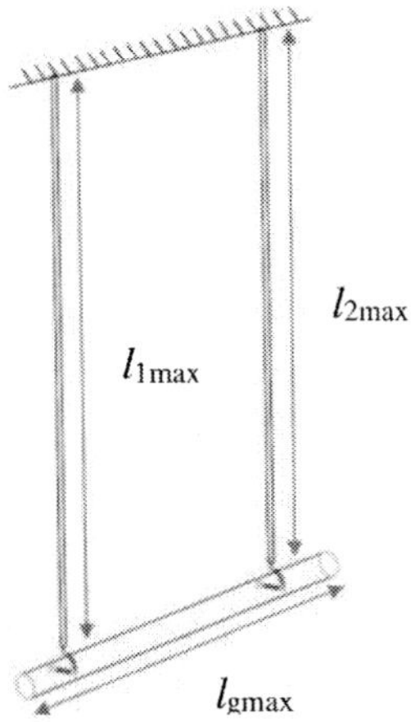

Fig. 2. Model of the whole light sculpture.

The analysis started with a one way FSI computation for the highest static deflection. The reason why the interaction that means, the flow field simulation was done, was given partially by the customer to catch quite accurately the flow, with potential disturbances etc. During this analysis was found that there could be some risk of vibrations of the sculpture elements at natural frequencies excited by vortex shedding behind the support rods or glass element. Such a proper numerical analysis would request quite high computing power and additional parameters like damping factors. Especially correctly, evaluated damping factors of the given system will influence the results significantly. For that, reason was executed an experiment in a wind channel of VZLÚ. It provided real measured data to evaluate the deflections and compare them with the numerical approach and

* Corresponding author: pavel.peukert@vuts.cz

started a future work of investigations of tuning vibrating systems.

2 Numerical analysis of the sculpture element

The problem was for the numerical simulation approach quite complex and so this part was divided in two sections. First a steady static analysis of the element was done and later a try to simulate the unsteady behavior, at least in partial steps. The analysis was done with a specified element depicted in fig. 2 and named as sample 1 (s1). A real sample with the same geometry was later used in the experimental part.

2.1 Steady static analysis

fig. 3 shows the simple configuration of the boundaries for the one-way FSI simulation. This approach was chosen because the deflections of the element were quite low in the comparison to the main sizes so that the deformation had negligible influence on the flow. During the first step the whole geometry was modeled. Later was prepared a one-way FSI coupled system in Ansys-workbench.

The CFD part consists of the outer limit brick-shaped surroundings around the sculpture element. It has one inlet, one top wall (ceiling) and four pressure outlets. The solid is represented only as walls ie. obstacles in the flow.

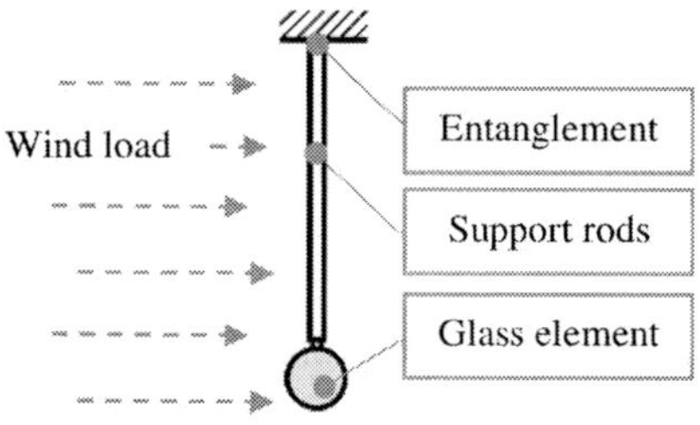

Fig. 3. Configuration for the computation.

The mechanical part was computed by a static structure analysis. It includes all analyzed solids with a defined anchoring of the light sculpture element. The pressure load is evaluated from the flow computation (fig. 4). The pressure field on the whole light sculpture surface is interpolated to the surface boundary in the mechanical computation (fig. 5). The gravity was taken in account too, because the weight of the element was relevant for the results.

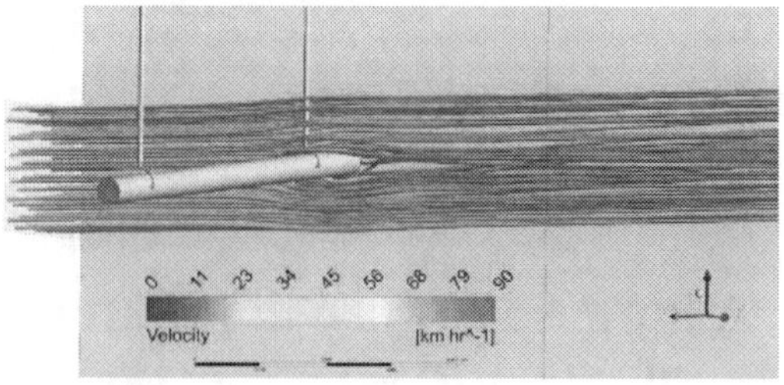

Fig. 4. Stream lines around a part of the light element.

Computations were made for a wind flow range from 20 km/h, which is a common breeze in the location of the sculpture [3], up to 80 km/h, which is aproximatelly 20 km/h higher then the observed common extreme situation at the location. The flow was handled like incompressible flow due the relative low velocity. For turbulence modeling was used the k-omega SST model.

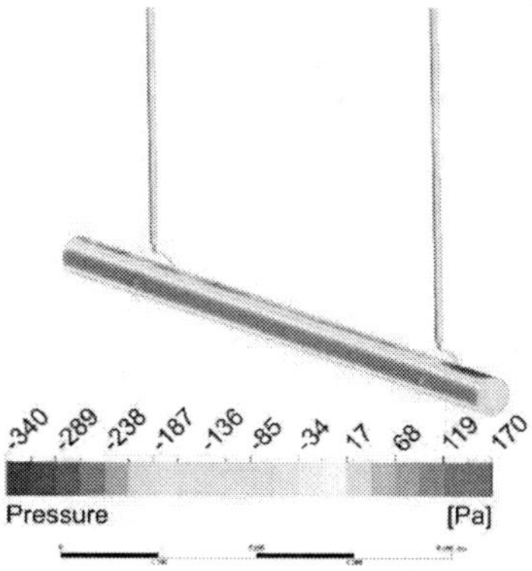

Fig. 5. Pressure load from one-way FSI simulation.

A representative result from the computation shows fig. 6. The geometrical deformation scaling is overestimated for the evaluation but the red color shows the correct maximum deformation eq. deflection of 12.4 mm. To get a wind-velocity, deflection dependence few points were computed and result in a plot in fig. 7. Like expected the dependence of deflection is aproximately quadratic on the wind velocity. It is given simply due to the fact that pressure is lineary dependent on the flow stream energy but the energy of the flow is quadraticaly dependent on the velocity.

https://doi.org/10.1051/epjconf/202226901046

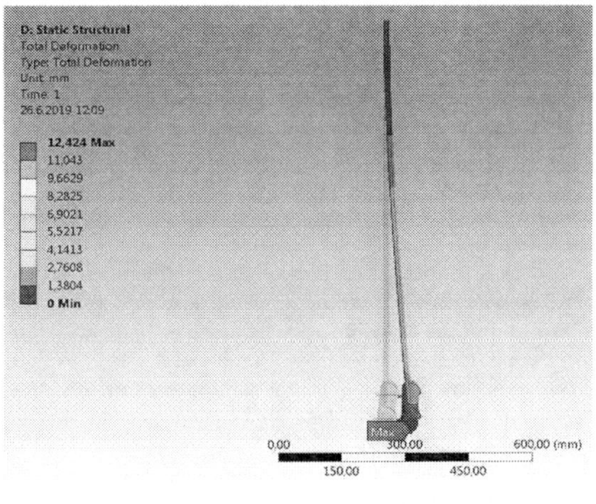

Fig. 6. Computed static deflection by one-way FSI simulation for 60 km/h wind load.

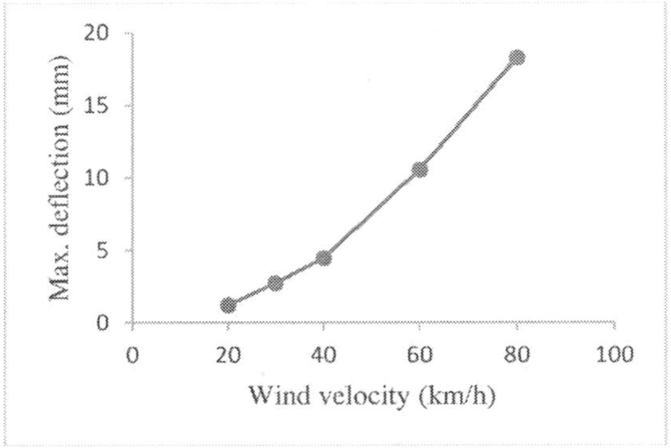

Fig. 7. Numerically evaluated static dependence of deflection ond wind load.

2.2 Unsteady behaviour analysis

The geometry of the sculpture element, which consists from two thin and long support rods, which hold the quite massive glass element, rises an uncertainty about the unsteady behaviour of it for different wind regimes. It was not clear if something like flatter could occur.

A very simplified estimation of the vortex shedding frequencies for the support rods and the glass elements was done together with a modal analysis of the light element. The comparison between the range of the vortex shedding frequencies in fig. 8 and the low mode natural frequencies of the sculpture element in fig. 9 shows some overlaping. That could cause an excitation of the system at quite high deflections.

$$Sr = \frac{fL}{U} \qquad (1)$$

To get an idea of the force, which acts on the glass element a simple 2D numerical computation of vortex shedding behind a cylinder was done. The simulation was unsteady and as expected it forms the typical von Karman vortex street, see fig. 10.

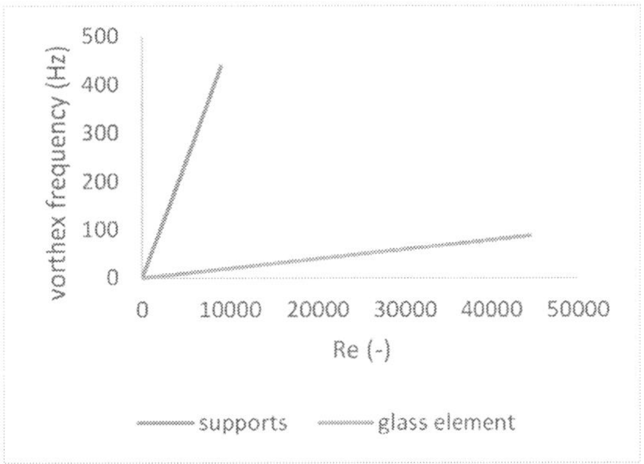

Fig. 8. Rough estimated dependence between Reynolds numbers and vortex shedding frequencies for cylinders of the diameter for the support rods and the glass element.

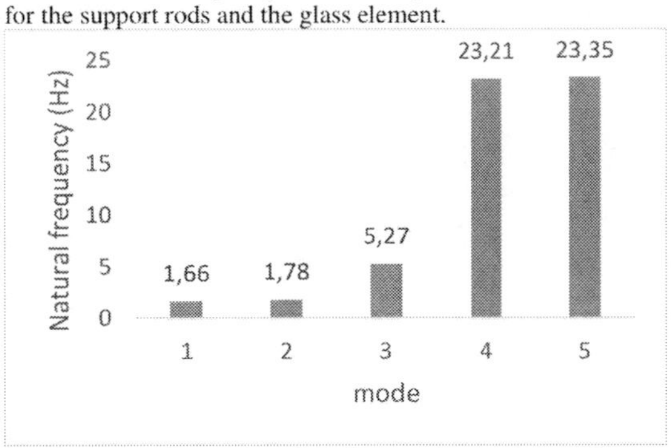

Fig. 9. Natural frequncies of the light sculpture element from a modal analysis in Ansys.

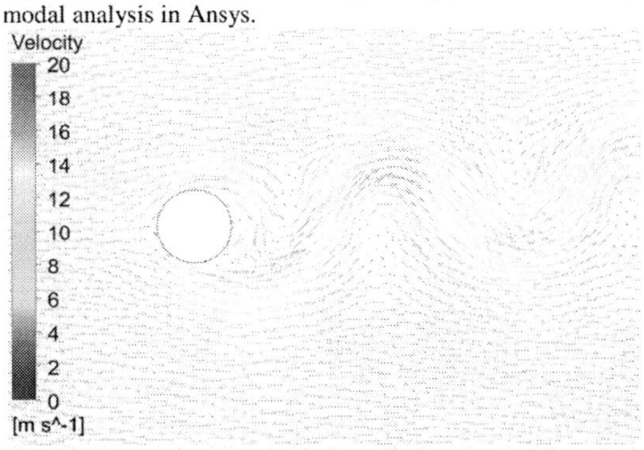

Fig. 10. Vortex shedding behind a cylinder, simple 2D analysis of the glass element cross-section.

During the 2D unsteady computation was recorded the total force acting on the cylindrical surface in the flow and perpendicular direction. The foce-time course for the glass element is in fig. 11 and for the support rod in fig. 12. The forces are directly from the 2D analysis in N for 1 m length. The lengths of the geometrical elements are around 1 m so this value is quite representative. The frequncy of the perpendicular direction is in a reasonable range compared to fig. 8 (around 50 Hz from the equation and 40 Hz from the CFD computation, most likely is the difference given by setting of an average Sr of 21). The highest chance to get in the naturaly oscilation mode of the element is given probably by the glass element, which provides vortex schedding frequencies near to the range of the natural frequencies of the sculpture element.

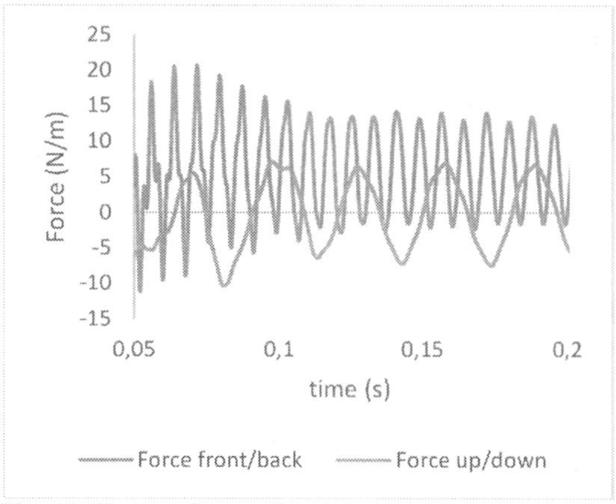

Fig. 11. Force acting on the glass element surface in flow direction and perpendiculatr to it for a flow velocity of 36 km/h.

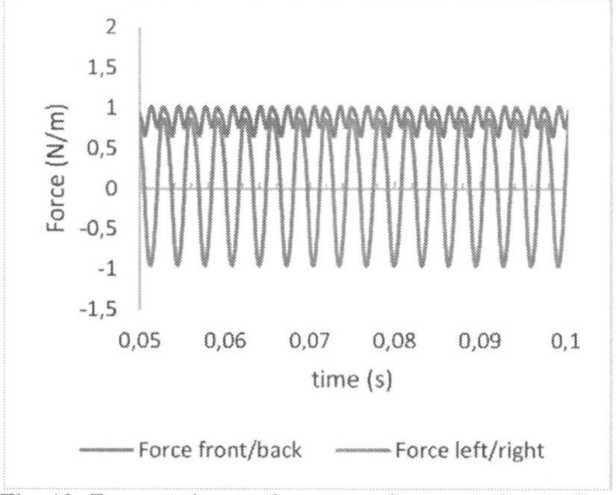

Fig. 12. Forces acting on the suppor element surface in flow direction and perpendiculatr to it for a flow velocity of 36 km/h.

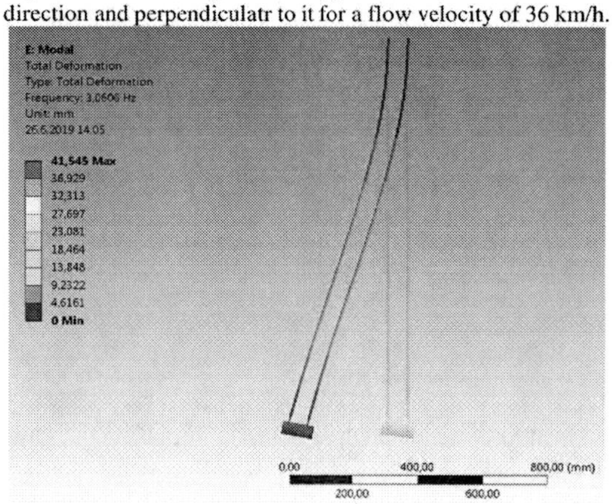

Fig. 13. Maximum deflection from a numerical test for a cyclicali waken system.

In the next step was tried to made a small test where the values of the evaluated forces from the 2D simulation were applied at their frequency on the surfaces. This approach had a lot of limitations. From fig. 11 and fig. 12 is obvious a different frequency of the forces caused by vortex shedding at different parts and also in different directions. The setting in the used software was quite limitting so that it was posible only specify the force definition and one frequency of cycling for all forces. But in the real case all forces on different part act together. Another problem was the definition of the damping factor. This is parameter is given by more factors, like material, geometrical design of the parts (welding/screw connections of parts), how the part is anchored etc. At the actual state was tested only one force applied at a given frequncy with some guesed damping factor and a results of a quite high deflection was obtained (fig. 13).

For a proper prediction of these deflections and in general how the the element would act was not really possible due the lack of input values like main different forces and the damping factor. In the end it was chosen to do an experiment how the light sculpture element would behave in near-reality states.

3 Experiment in the wind channel

The experiment was prepared and performed at the open circuit atmospheric wind channel with continuous operation at VZLÚ Letňany.

The light sculpture element was attached to a present construction beam above the channel. It was mainly tested in the arrangement with the highest surface load from the air stream eq. perpendicular to it like it is visible in fig. 15. A partially rotated configuration was also tested but the results were insignificant. The measurements were done by two approaches. One with two high-speed cameras which recorded the movement of the glass element. The second was based on the use of triaxial piezoelectric accelerometers and uniaxial capacitive accelerometers attached on the glass elements. For powering and read out the sensor was used the Dewetron analyser with the necessary modules. Just before the measurement, was the setup prepared like in fig. 15.

Before the measurement was performed a simple test for the reached airflow velocities in the wind-channel. The velocities from the steering software show the same values like the used hand flowmeter.

A strict procedure of the measurement was performed for every measurement. The airflow velocity in the wind-channel was set to a required value. After the measured velocity had stabilized, the measurement had started. First were measured the acceleration values from the sensors positioned on glass element surface for a specified time. If required, in the next step was taken a record by the cameras for few seconds. The camera record was done only for predetermined velocities or if some more vibrant state had occurred. The reason was a quite time-consuming image set download from the buffer to the storage medium and later data processing for states without "visible" behaviour.

 https://doi.org/10.1051/epjconf/202226901046

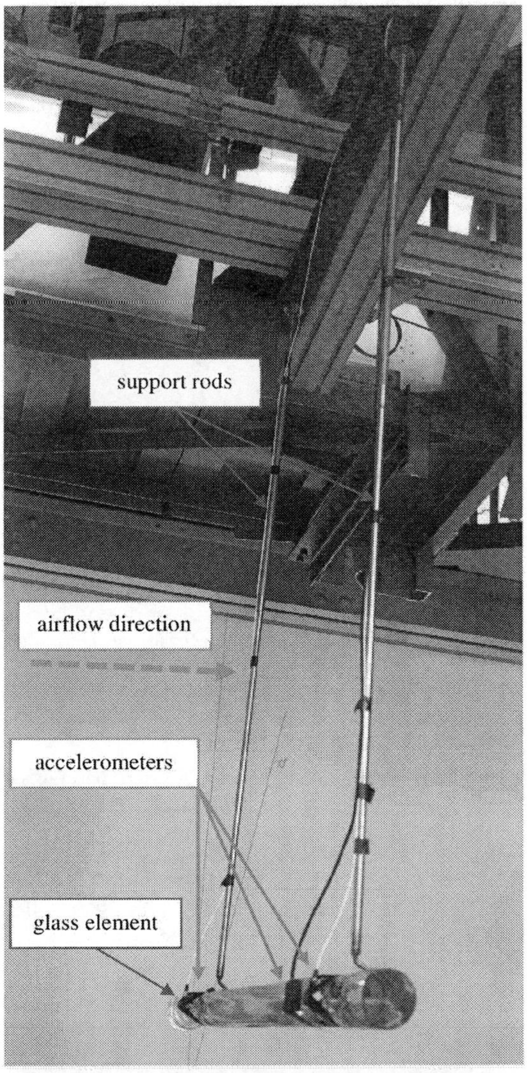

Fig. 14. Tested light sculpture element in the wind channel.

Fig. 15. Setup for measurement in the windchannel.

3.1 Results from the experiment and comparison with some relevant data from the numerical simulation

From the experiment were taken image sets of the glass element positions for various wind speeds. Fig. 16 shows the extreme case for a wind of 90 km/h.

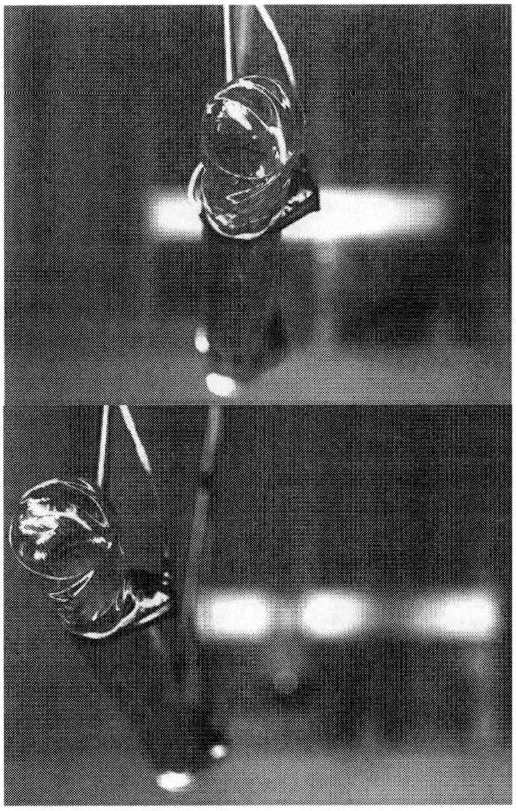

Fig. 16. Images for deflection evaluation a, calm state eq. any load b, maximum wind load.

The images from the cameras were evaluated and show a tendency commonly like in fig. 17. The central dashed line of the experiment shows a mid value of deflection while the dotted lines show the maximum and minimum values of the deflection for the given wind speed. It is evident that for slow wind speeds the deflections were too small to be captured. It was interesting that for the highest velocity the deflection would be higher about 35% than from the steady state computation.

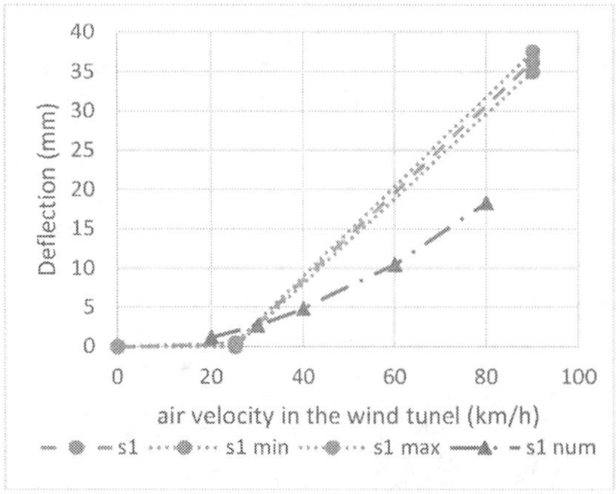

Fig. 17. Dependence between measured deflection (dot) and static numerical simulation one-way FSI simulation.

Fig. 18 shows the frequencies evaluated from the accelerometers and if they were compared with the results from the numerical modal analysis in fig. 9 a correspondence in the first two frequencies is visible and partially at the forth. That means that the sculpture element swings at the frequency 1st and 2nd natural frequency at the measured state.

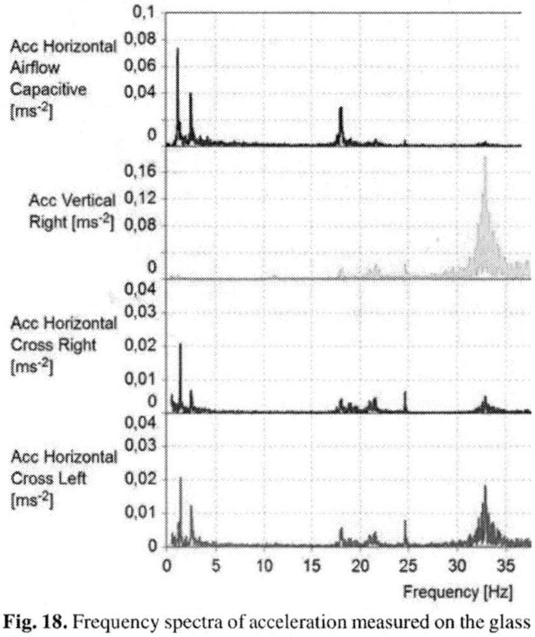

Fig. 18. Frequency spectra of acceleration measured on the glass element.

The frequencies change slightly with the measured wind speed, but mostly the swing on the 1st two natural frequencies is present everywhere.

4 Conclusions

The measurements at the wind channel had not identified any dangers regimes (flatter) for the light sculpture model during the tested air velocities. It also evaluate real deflections, which were taken in count to specify the space sizes between the light sculpture elements.

From the behaviour evaluation in the wind channel it was obvious that the factors of damping were quite high compared to the values used in the computations. An unsteady two-way FSI approach was not possible due to the lack of time/necessary computing power. The reason were geometrical ratios about more than 1:100 of the main sculpture element parts (tiny support part and long supports).

The work actually continues at VÚTS with the focus on the vibrations caused by mechanical waking. The main goal of the future research is to evaluate critical vibration modes waken by defined forces at specified frequencies numerically. That could later server for predictions of flatter for systems of diverse geometry for specified flow velocities.

Acknowledgments

The authors would like to acknowledge the material support and permission to publish this work to the company of Kolektiv Ateliers from Nový Bor, which produces apart from lighting sculptures also various other glass artwork.

This publication was further financially supported by the Ministry of Industry and Trade of the Czech Republic within the institutional funding for a long-term conceptual development for the research organization – provider Ministry of Industry and Trade of the Czech Republic, recipient – VÚTS, a.s..

References

1. Strouhal, V. Über eine Besondere Art der Tonerregung. Annalen der Physik. **241**. 216 - 251. (2006) DOI : 10.1002/andp.18782411005.
2. Brepta, R., Půst, L., Turek F. *Mechanické kmitání technický průvodce.* (1994). ISBN 80-901684-8-5
3. Meteo Blue Weather [online]. [cit 2019-07-15]. Available at: https://www.meteoblue.com
4. ANSYS Fluent Theory guide
5. ANSYS APDL Structural Analysis Guide

Simulation of temperature influence on shear stress during mineral oil flow

Tomáš Polášek[1, *], *Marian* Ledvoň[1], *Adam* Bureček[1] and *Lumír* Hružík[1]

[1]VŠB -Technical University of Ostrava, Faculty of Mechanical Engineering, Department of Hydromechanics and Hydraulic Equipment, 708 00 Ostrava, Czech Republic

Abstract. The article is focused on the mathematical simulation of temperature influence on the shear stress during the mineral oil flow. The rotary rheometer is used to measure the shear stress. It works on the principle of torque measurement when the spindle rotates in the container with the measured liquid. The rotating spindle together with the container forms the hollow cylinder which is filled by measured oil. The shear stress is evaluated for the various oil temperatures. The software ANSYS Fluent is used to evaluate the partial numerical simulations. The experimental evaluation of the oil physical properties depending on the temperature is described in the article.

1 Introduction

If the fluid flows and its two adjacent layers have different velocities, the friction and the shear stress τ occur at their interface. The cause of this phenomenon is fluid viscosity η. The single flow streamlines are parallel during the laminar flow and the fluid does not mix. The following equation applies to this flow:

$$\tau = \eta \cdot \frac{d\gamma}{dt}, \tag{1}$$

where τ is the shear stress, η is dynamic viscosity and $d\gamma/dt$ is shear rate.

The shear stress τ is linearly proportional to the shear rate $d\gamma/dt$ according to this law. The dynamic viscosity η is given by the ratio of the shear stress τ and the shear rate $d\gamma/dt$.

The ratio of shear stress τ and shear rate $\dot{\gamma} = d\gamma/dt$ is constant for the Newtonian liquid.

For the non-Newtonian liquid, the ratio is not constant and this ratio is called the apparent viscosity.

The dependence of shear stress τ on the shear rate $\dot{\gamma}$ is called the flow curve. The flow curves are determined experimentally on the rheometers. The type of liquid can be determined according to the flow curve.

2 Experimental measurement

The physical properties are determined for the mineral oil sample type HV VG46. The dependence of the dynamic viscosity η_O, the shear stress τ_O and the density ρ_O of the mineral oil on the temperature t_O is evaluated. The measurement is performed for the oil temperature range $t_O = (10 \div 80)$ °C. The mineral oil is the most widespread liquid using in the hydrostatic systems nowadays [1].

2.1 Description of experimental measurement

The rotary rheometer Brookfield LVDV-II+Pro is used to determine the dependence of the oil dynamic viscosity η_O on the temperature t_O. The spindle rotates in the container where is situated the measured oil. The oil flows in the gap between the spindle and the inner container wall due to the spindle rotation. The rotating spindle together with container forms volume which has the hollow cylinder shape.

The dependence of the oil shear stress τ_O and the spindle torque T on the oil temperature t_O is also determined using this rheometer. The generated torque T on the rotating spindle is absorbed by the calibrated spring. The spindle torque T range must be within the range $T_{\%} = (10 \div 100)$ % to ensure the measurement accuracy. The percentage value $T_{\%} = 100\%$ corresponds to the torque value $T = 0.0673$ mN·m [2]. The torque value T can be corrected using the speed s change respectively the shear rate $\dot{\gamma}$ change. The spindle speeds $s_1 = 14$ min^{-1} and $s_2 = 110$ min^{-1} are chosen for the experimental measurement. These spindle speeds correspond to the shear rates $\dot{\gamma}_1 = 13.03$ s^{-1} and $\dot{\gamma}_2 = 102.37$ s^{-1}. The Mohr-Westphal balance is used to determine the dependence of the oil density ρ_O on the temperature t_O [3].

2.2 Evaluation of experimental measurement

The experimentally determined dependence of the oil dynamic viscosity η_O on the temperature t_O is shown in Fig. 1. The experimentally determined dependence of the oil density ρ_O on the temperature t_O is shown in Fig 2.

* Corresponding author: tomas.polasek@vsb.cz

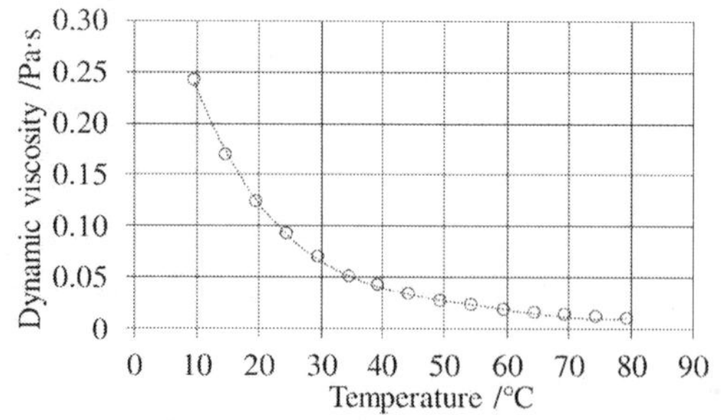

Fig. 1. Dependence of the oil dynamic viscosity η_O on the temperature t_O.

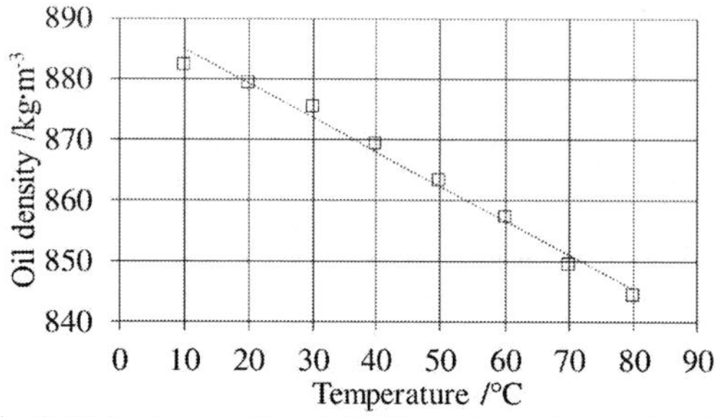

Fig. 2. Dependence of the oil density ρ_O on the temperature t_O.

Both dependencies have the decreasing character when the oil temperature t_O is increased [4]. These obtained oil parameters allow us to define the liquid in the numerical simulation.

The experimentally determined dependence of the shear stress τ_O on the temperature t_O is shown in Fig. 3. In Fig. 4, the experimentally determined dependence of spindle torque T on the oil temperature t_O is shown.

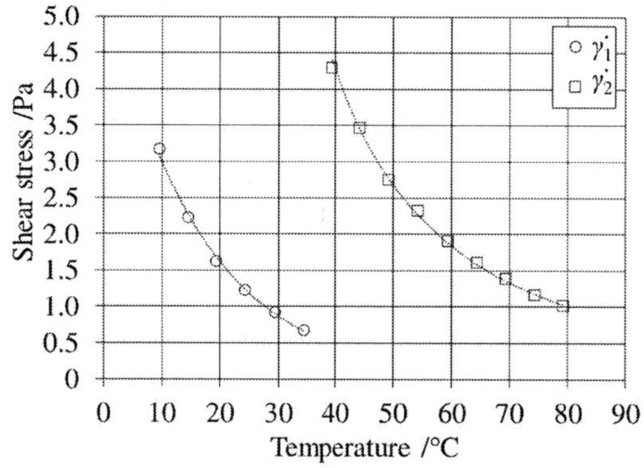

Fig. 3. Dependence of the oil shear stress τ_O on the temperature t_O, the shear rates are $\dot{\gamma}_1 = 13.03\ s^{-1}$ and $\dot{\gamma}_2 = 102.37\ s^{-1}$.

The decreasing dependence of the oil shear stress τ_O and the spindle torque T with the increasing temperature t_O can be observed in Fig. 3 and Fig. 4. The oil dynamic viscosity η_O decreases with the increasing oil temperature t_O. The minimal spindle torque $T_{\%} < 10\ \%$ is reached at the oil temperature $t_O = 40$ °C and the spindle speed $s_1 = 14\ min^{-1}$ ($\dot{\gamma}_1 = 13.03\ s^{-1}$) during the oil dynamic viscosity η_O measurement. For this reason, the spindle speed s is increased to the speed value $s_2 = 110\ min^{-1}$ ($\dot{\gamma}_2 = 102.37\ s^{-1}$).

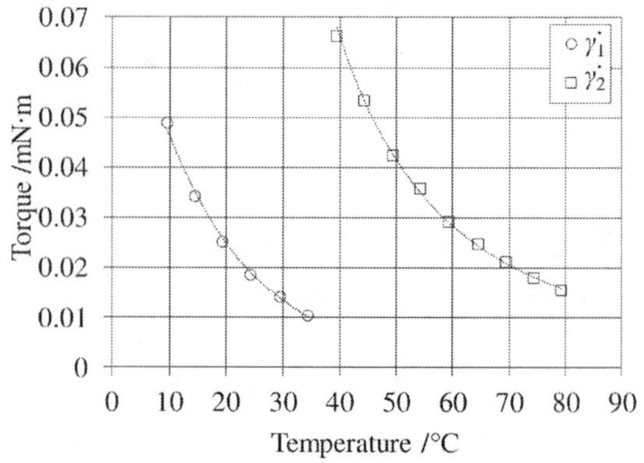

Fig. 4. Dependence of the spindle torque T on the oil temperature t_O, the shear rates are $\dot{\gamma}_1 = 13.03\ s^{-1}$ and $\dot{\gamma}_2 = 102.37\ s^{-1}$.

The dependence of the oil shear stress τ_O and the spindle torque T on the oil temperature t_O is simulated using the numerical simulation.

3 Numerical simulation

3.1 Inverse geometry and computational mesh

The DesignModeler environment is used to modify the inverse geometry. Fig. 5 shows the geometry of the solved region which is formed by the rotating spindle and the container. The structured computational mesh is also shown in Fig. 5. The mesh approximately contains 900 000 elements and it is mainly made up of hexahedron elements. In the region of the largest flow restriction, the computational mesh (dimension $a = 1.14$ mm) is designed to contain 25 elements in the row. The largest refinement of the mesh is created in the gap near the rotating wall.

3.2 Boundary conditions and mathematical model

The boundary condition stationary wall is defined on the inner container wall 1, see Fig. 5.

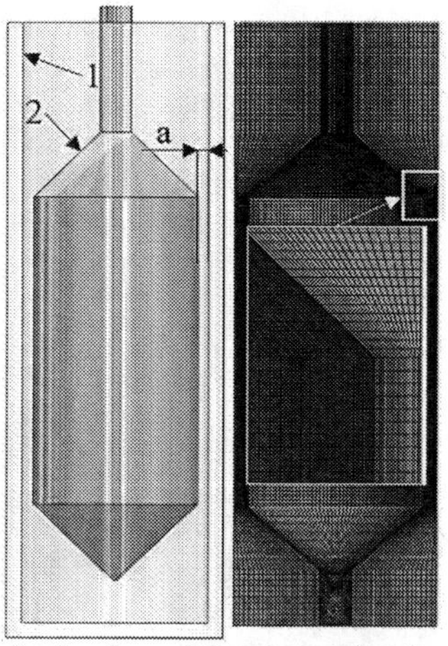

Fig. 5. The solved region and the computational mesh.

On the spindle wall 2, the boundary condition rotating wall is defined. The rotary wall speeds $s_1 = 14$ min^{-1} and $s_2 = 110$ min^{-1} are set at this boundary condition. These spindle speeds correspond to the shear rates $\dot{\gamma}_1 = 13.03$ s^{-1} and $\dot{\gamma}_2 = 102.37$ s^{-1}. The axis and the direction of the rotation are also specified.

The density ρ_O and the dynamic viscosity η_O of the flowing liquid are defined for the oil temperature range $t_O = (10 \div 80)$ °C. These parameters are determined experimentally, see Chapter 2.2. The mineral oil flows in the gap between the rotating spindle wall and the inner container wall. In this case, the mineral oil flow has the laminar character. The laminar model is used in the numerical simulation.

3.3 Evaluation of numerical simulation

Fig. 6 shows the geometry example of the rotating spindle in the container.

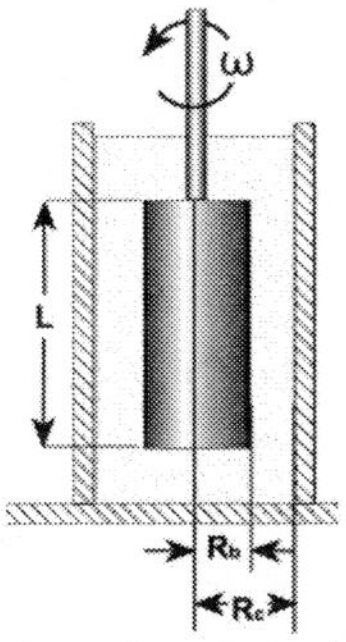

Fig. 6. The spindle and container dimensions [5].

The shear rate $\dot{\gamma}$ is formulated for the cylindrical spindle in the following equation [5]:

$$\dot{\gamma} = \left[\frac{2 \cdot R_C^{\,2} \cdot R_B^{\,2}}{x^2 \cdot \left(R_C^{\,2} - R_B^{\,2}\right)} \cdot \frac{2\pi}{60} \right] \cdot s, \qquad (2)$$

where $R_C = 9.525$ mm is the container radius, $R_B = 8.385$ mm is the spindle radius, x is the radius at which the shear rate $\dot{\gamma}$ is calculated, s is the spindle speed. The shear rate $\dot{\gamma}$ of the used spindle deals with the equation [2]:

$$\dot{\gamma} = 0.93 \cdot s. \qquad (3)$$

The resulting shear rate $\dot{\gamma}$ and shear stress τ_O are determined for the spindle wall near the gap.

Table 1. The shear rate $\dot{\gamma}$ results according to the equation (3)

	$s_1 = 14$ min^{-1}	$s_2 = 110$ min^{-1}
$\dot{\gamma}$ [s^{-1}]	13.03	102.37

The subsequent evaluation of the oil shear stress τ_O is performed for the spindle wall near the gap. The resulting oil shear stresses τ_O for the temperature range $t_O = (10 \div 80)$ °C are shown in Fig. 7. The spindle torque T results depending on the oil temperature t_O are shown in Fig. 8. Simultaneously, the simulation results and measurement results are compared in Fig. 7. and Fig 8.

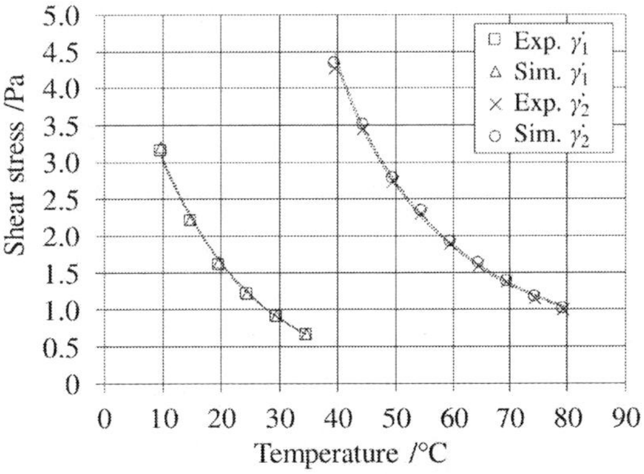

Fig. 7. Comparison of the numerically simulated oil shear stress τ_O depending on the temperature t_O with the experiment.

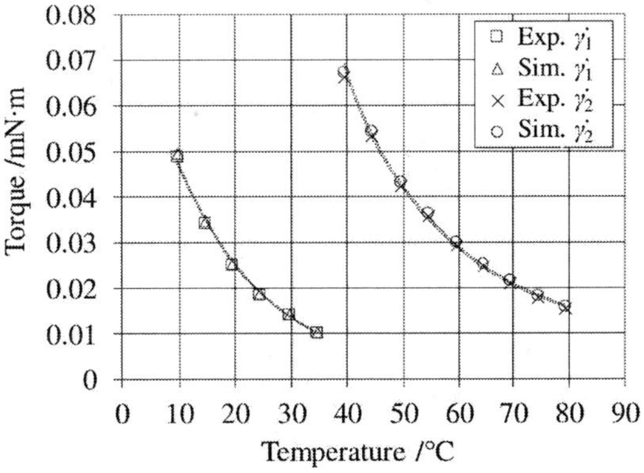

Fig. 8. Comparison of the numerically simulated spindle torque T depending on the temperature t_O with the experiment.

The oil shear stress results τ_O, determined by the numerical simulation, achieve the match with the experimentally determined results. The spindle torque results T, determined by the numerical simulation, also correspond to the experimental measurement.

The numerically simulated contours of oil shear stress τ_O on the spindle wall and the contours of the velocity v in the XY plane are shown in Fig. 9.

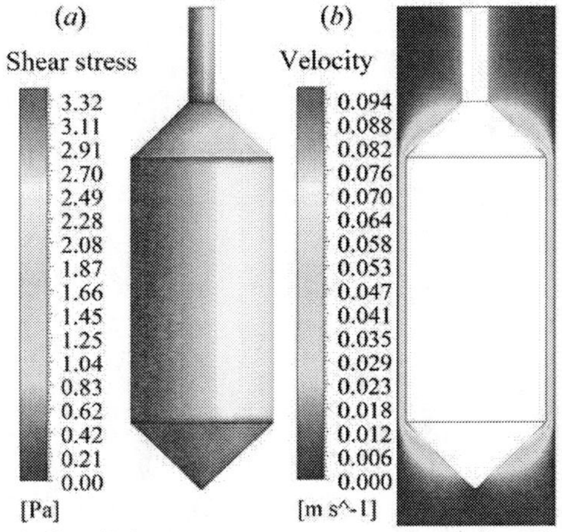

Fig. 9. The oil shear stress τ_O contours on the spindle wall (*a*), the velocity v contours in the XY plane (*b*).

The numerically simulated contours of the shear rate γ' on the spindle wall and in the XY and XZ plane are shown in Fig. 10. The graphical evaluation of the numerical model is done for the oil temperature $t_O = 50$ °C and the spindle speed $s_2 = 110$ min^{-1} in Fig. 9 and Fig. 10.

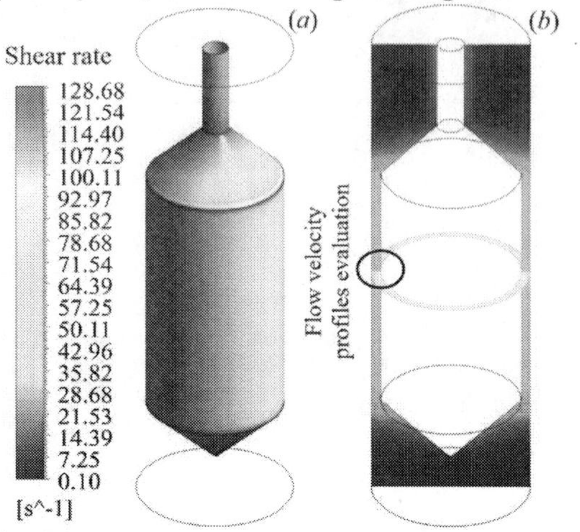

Fig. 10. The shear rate γ' contours on the spindle wall (*a*), in the XY and XZ plane (*b*).

In Fig. 11 and Fig. 12, the numerically simulated flow velocity profiles in the gap are shown. The figures show the influence of the rotating spindle on the oil flow velocity profile.

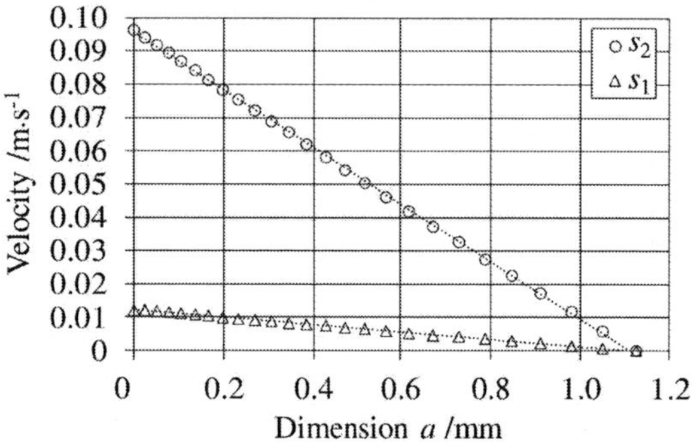

Fig. 11. The oil flow velocity profile in the gap for the spindle speeds $s_1 = 14$ min^{-1} and $s_2 = 110$ min^{-1}.

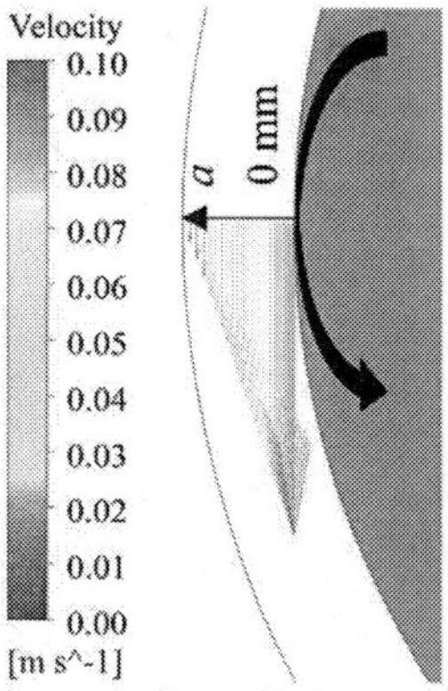

Fig. 12. The oil flow velocity profile in the gap for the spindle speed $s_2 = 110$ min^{-1}.

Conclusions

The influence of the temperature on the shear stress during mineral oil flow and the rheometer spindle torque during its rotation was numerically simulated in this article. The experimentally determined results of the oil density and the oil dynamic viscosity dependence on the temperature were summarized in the article. The dependence of the oil shear stress and the rheometer spindle torque on the temperature was also measured. Both these quantities decreased with the increasing oil temperature. The mathematical model of the flow was created in the ANSYS Fluent environment. Then the numerical simulation results were compared with the experimentally achieved results. The match of the numerical simulation and the experimental measurement was achieved.

This work was supported by the European Regional Development Fund in the Research Centre of Advanced Mechatronic Systems project, project number CZ.02.1.01/0.0/0.0/16_019/0000867 within the Operational Programme Research, Development and Education.
This work has been supported by The Ministry of Education, Youth and Sports of the Czech Republic from the Specific Research Project SP2019/123 (Numerical Modelling of Dynamic Phenomena in Fluid Mechanics with Support of Experimental Research).

References

1. M. Kozdera, and S. Drábková, Numerical modelling of the flow in the annular multi-recess hydrostatic thrust bearing using CFD methods. *The European Physical Journal*. 2013, Vol. 45. ISSN 2101-6275
2. Brookfield Engineering Laboratories, Inc., Brookfield DV-II+Pro Programmable Viscometer, Operating Instructions
3. T. Polášek, A. Bureček, and L. Hružík, EPJ Web of Conferences, EFM18, *Mathematical simulation of the temperature influence on pressure drop at pump suction line*, **213**, 02069, (2019)
4. D. Knežević, V. Savić, FACTA UNIVERSITATIS, Mechanical Engineering, *Mathematical modeling of changing of dynamic viscosity, as a function of temperature and pressure, of mineral oils for hydraulic systems*, **4** (1), 27-34 p., (2006)
5. Brookfield Engineering Laboratories, Inc., More Solutions to Sticky Problems, A Guide to Getting More from Your Brookfield Viscometer

Stereo PIV measurement of diffuser swirling flow

Pavel Procházka[1,*], *Václav* Uruba[1,2], *David* Štefan[3], and *Pavel* Rudolf[3]

[1]Institute of Thermomechanics, ASCR, v.v.i., Dolejškova 5, Praha 8, CR
[2]UWB, FME, Department of Power System Engineering, Universitní 8, Plzeň, CR
[3]BUT, FME, Viktor Kaplan Department of Fluid Engineering, Technická 2896/2, Brno, CR

Abstract. The swirling flow in a diffuser has been studied using the stereo PIV technique. The DANTEC hardware and software were used for the experiment. The water pipe flow in the diffuser inlet has been set spinning by introducing tangential jet flow. Both piping and diffuser were of circular cross-section, however diffuser casing had rectangular (square) cross-section. The plane of measurement was located perpendicular to the flow within the diffuser. The two cameras with the Scheinpflug lens mounting were used, however the angle between the lenses axes and the diffuser outer surface was far from to be perpendicular, as recommended in similar situations. This configuration has been dictated by technical conditions in the laboratory. The two-level calibration target with dots was used for the cameras calibration. The calibration covers the diffusers central region in a good quality, however regions close to the walls were not well resolved. Many combinations of axial and tangential inflow conditions have been studied. The acquisition covered about 3 seconds in real time with frequency 500 Hz. Both statistics and dynamics of the flow have been studied in details.
Results are shown for both statistical characteristics and dynamical analysis of various conditions. During experiments the air-glass-water interface was not normal to the camera viewing axis. The discussion how to solve this problem and of the bias of the calibration will be introduced.

1 Introduction

Recently, renewable energy sources are rapidly implemented into the power grid. However, their big disadvantage is highly fluctuating power output which is a big challenge to be compensated. Hydraulic turbines (able to operate in reversible mode) are adequately appropriate for this purpose. The compensation of power fluctuation in real time results in the demand that such turbine must work within a wide range of working regimes. These regimes far from designed one are accompanied by presence of pressure fluctuations and self-induced instabilities as a consequence of decelerated swirling flow developed in diverging section well-known as precessing helical vortex (vortex rope).

The considerable pressure fluctuation are propagated through the entire flow domain which could result in rotor blades damage [1] or even, when the frequency of pressure pulsation is equal to natural frequency (eigenfrequency) of the device [2], the destruction of the entire system. There are some possibilities to control this phenomenon. Passive methods (e.g. various fins, grooves etc.) have the same effect for all working regimes without the possibilities to turnig off their effect. The active method to reduce the unwanted flow instabilities can be for example jet injection along the machine axis from the runner crown [3].

Since this experimental investigation follows on previously conducted study of instabilities in the vicinity of the airfoil [4], the used experimental facility and processing methods are similar. To examine the swirl flow in a diverging pipe, the time-resolved stereoscopic Particle Image Velocimetry (TR-PIV) was used to measure all three velocity components in a given plane. Further analysis is conducted using FFT (Fast Fourier Transform) and mainly using well-known dynamical decomposition method POD (Proper Orthogonal Decomposition). This method was introduced by Lumley [5] in seventies to fluid mechanics and proved to be effective in study of vortex rope dynamics [6].

2 Experimental setup

The swirling flow in the diffuser was created by mixing of axial inflow with tangential inflow coming from the spiral casing. Various regimes were characterized by different rate of both components. Nominal flow rate is a sum of axial and tangential flow: $Q_N = Q_A + Q_T$. The Dantec hardware and software was utilized. Dynamical mode decomposition method was used.

2.1 Swirl generator apparatus

The swirling flow is generated by swirl generator apparatus depicted in figure 1a. It consists of plexiglass diffuser with 8° divergence angle. Measurements were conducted in closed-loop hydraulic circuit consisting of

* Corresponding author: prochap@it.cas.cz

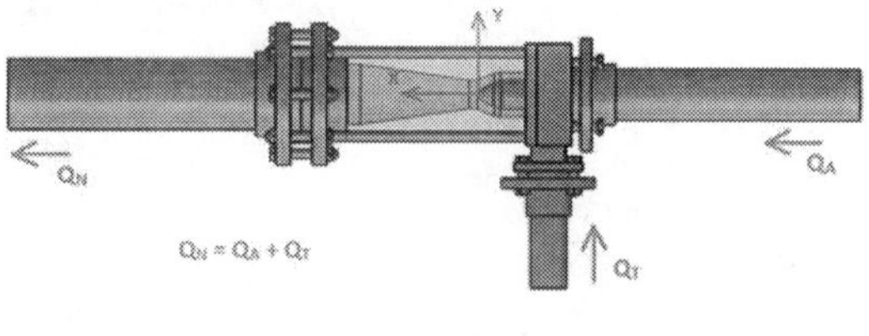

Fig. 1a. Design of the swirl flow generator.

centrifugal pump, stainless steel pipes, pressurized vessel and induction flow meters. While piping and diffuser were of circular cross-section the spiral casing of tangential inflow has rectangular cross-section. Swirl generator was designed as bladeless with tangential and axial intakes. Flow ratio was adjusted by valves. Such kind of design allowed sufficient control and the manufacturing costs were much lower than with rotating or stationary blades.

The flow parameters were measured by several sensors (Table 1). The flow rate was measured with two induction flowmeters (one in main pipe, one in tangential intake). Also pressure measurement points were set twice in inflow pipes, in outflow pipe and also in suction tank.

Tab. 1. Instrumentation.

	Brand	**Specification**
Flow meter nominal	ELA	DN 80, PN 1,6 MPa
Flow meter tangential	ELA	DN 50, Q_{max}=20 l/s
Temperature sensor	PTP 55	-5°C – 55°C
Pressure sensor	**Brand**	**Range (bar)**
Axial inflow	BD sensor	0-6
Tangential inflow	BD sensor	0-6
Outflow	BD sensor	0-2,5
Suction tank	BD sensor	0-10

Tab. 2. Operating regime of cross-sectional plane measurements.

OR	Ax. comp. (%Q_n)	Tan. comp. (%Q_n)	Q_n (l/s)	Pump rev. n (min^{-1})
1	100	0	10	500
2	90	10	10	500
3	80	20	10	500
4	70	30	10	500
5	60	40	10	500
6	50	50	10	500
6,5	45	55	10	500
7	40	60	10	701
8	30	70	10	701
8x	30	70	5	701
9x	20	80	5	701
10x	10	90	5	701
11x	0	100	5	701

Totally 13 operating regimes were examined. Each regime is characterized by different rate of axial and tangential component (Table 2, 3). The nominal flow rate through the closed-loop circuit has been set twice smaller for regimes with high portion of tangential part due to technical reasons (these regimes are marked with x). Operating regimes (OR) could be adjusted from pure axial flow (OR 1) to flow where tangential component dominates (OR 11x). Spiral vortex breakdown was pronounced for flow beginning with ratio 50:50 (OR 6).

Tab. 3. Operating regime of longitudinal plane measurements.

OR	Ax. comp. (%Q_n)	Tan. comp. (%Q_n)	Q_n (l/s)	Pump rev. n (min^{-1})
1	100	0	10	500
2	90	10	10	500
3	80	20	10	500
4	70	30	10	500
5	60	40	10	500
6	50	50	10	500
6,5	45	55	10	500
7	40	60	10	600
8	30	70	10	700
8x	30	70	5	700
9x	20	80	5	700
10x	10	90	5	700
11x	0	100	5	700

2.2 PIV setup

The main goal of this project was to detect all three velocity components in distinct planes. Classical way how to perform the stereoscopic PIV calibration is to use a calibration target which should be moved across the measured plane [4]. Such kind of experiment (circular cross-sectional pipes flooded by water) did not allow using a plane dot matrix target.

Hence, the multi-level calibration targets with white dots on the black background were used (dot spacing 10mm, level distance 2mm, zero marker diameter 4,5mm and main marker diameter 3mm). One target was used to covered as large as possible measured area for longitudinal plane (target edge touched the swirl generator hub which was the origin of coordinate system – Fig. 1b) and the second one was established to calibrate cross-sectional plane (diameter was 70mm and location in downstream direction from the origin x = 84 mm).

Stereo arrangement of PIV was used. The set consisted of a laser with cylindrical optics and two cameras. The laser New Wave Pegasus Nd:YLF has double head emitting coherent light with wavelength 527 nm by maximal frequency equal to 10 kHz. The pulse energy is optimized for 1 kHz when its value is 10 mJ (corresponding power is 10 W per head). Two CMOS cameras NanoSense MKIII were used, resolution 1280 x 1024 pixels and maximal achievable acquisition frequency about 500 Hz. The memory is 4 GB. The cameras were used with Scheinpflug lens mounting, however, the angle between lenses axes and the diffuser outer surface was far from to be perpendicular, as recommended in similar situation [7, 8]. This configuration has been dictated by technical conditions in the laboratory. As tracing particles, hollow glass spheres coated with silver were used.

The longitudinal plane oriented with diffuser axis as well as cross-sectional plane were in the centre of investigation. The first one covers the flow region from the top of the hub to x = 140 mm downstream. The calibration model (pinhole model) was applied on picture of multi-level calibration target. The reprojection error was set approximately 1 pixel which is still satisfying according recommendation. Finally, we got the velocity vector maps with 129 x 110 vectors (up to 12% of substituted vectors) corresponding to space resolution approx. 1 mm.

The second measured plane was established to cover as large cross-section as possible. The enormous optical distortion made the calibration very difficult. Although the Direct Linear Transform is unable to model any non-linear distortion, just this calibration model was able to calibrate stereoscopic arrangement. However, the reprojection error was far from recommended values, so the cross-sectional plane values should be treated as qualitative ones. The final velocity vector map has 104 x 64 vectors (up to 15% of substituted vectors).

The acquisition frequency was set to 100 Hz during 10 sec of measurement and to 500 Hz (3 sec) to evaluate the statistical quantities and for dynamical analysis, respectively.

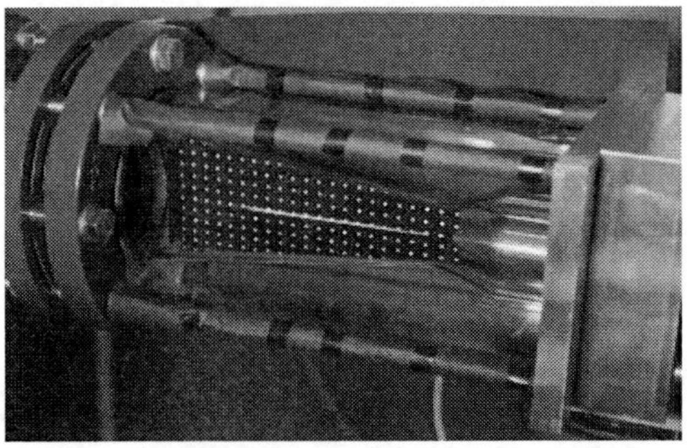

Fig. 1b. Used multi-level PIV target.

2.3 POD

The flow with vortex rope inside the diffuser is typically complex field to which it is possible to apply Proper Orthogonal Decomposition (POD). POD applied on PIV data is a powerful method of for dynamic flow features analysis. It is a good tool to interpret snapshots in order to distinguish between coherent flow patterns and random flow variations. POD analysis is very often applied on spatiotemporal 2D data; we apply it to all three components of velocity vector in our case. Most researchers apply this analysis to fluctuation part of scalar quantity (vel. components, pressure, etc.) [9]. Result is a set of orthogonal modes according to the relative kinetic energy content (mode No. zero is actually mean flow). Low-order modes have high content of energy and their topos is created by large/coherent structures while high-order modes have the energy content less than approx. 0,5% and their topos contain very often only turbulent noise or measurement bias. Further information about decomposition methods can be found in work of Uruba [10].

3 Results

The presented vector field will be plotted in dimensional form. Coordinate system origin is located at the end of the hub. Black vectors represent the velocity projected onto the plane of measurement. The color represents other scalar value (third vel. component, velocity variations etc.). The swirl motion can be detected in cross-sectional planes by black vectors orientation and also by comparison of positive and negative out-of-plane velocities in longitudinal plane. Note, that the direction of the vectors is not meaningful in the case of POD modes as the calculation comes from the vel. fluctuation. The vorticity distribution is used here to detect vortices (no shear layers are expected). Positive vorticity identifies the counter-clockwise rotation (red color).

3.1 Mean flow field

The flow inside the diffuser seems to be stable for time-averaged quantities. No secondary vortices and no flow separation close to the walls are present. The flow of regime No. 2 is characterized by flow deceleration along streamwise coordinate as the flow cross-section becomes wider. Highest velocities are located just behind the hub where the diffuser diameter is 50 mm. Mean velocity value is higher than 5 m/s in this region. Flow is spread across the diffuser almost homogeneously with boundary layer close the walls (see Fig. 3a). No significant swirl motion is detectable in the whole area (Fig. 2a). The regime No. 8, which is characterized by 70% of tangential component, is characterized by the existence of considerable helical vortex structures (Fig. 2b, 3c). The structure rotates in counterclockwise sense which is given by the connection of the tangential intake to the swirl generator. The highest velocities were detected in annular region between boundary layers and the diffuser axis. The area of backflow is getting more important with increasing of tangential component. There is a tiny backflow just in the diffuser centre for regime No. 5 (depicted in Fig. 3b). On the other hand, more significant backflow region is visible in figure 3c, where regime No. 8 is captured. Since the backflow area is increasing with tangential flow rate component, the positive streamwise vel. components are also increasing and are located more close to the walls (well visible in longitudinal planes). The integration of streamwise vel. component will lead to equal nominal flow rate for all regimes – 10 or 5 l/s according to table 2, 3.

 https://doi.org/10.1051/epjconf/202226901048

The distribution of turbulent kinetic energy (TKE) which is given by equation

$$TKE = \frac{1}{2}\left\{\overline{(u')^2} + \overline{(v')^2} + \overline{(w')^2}\right\} \qquad (1)$$

represents the fluctuation activity of the flow. The fluctuation is getting greater for higher OR and the most significant region is located in the diffuser axis at x = {20-40} mm. Just in this area, the merging of annular stream deflected by the hub is responsible for such high vales of TKE (Fig. 2c). Velocity variations are smaller further downstream as we can see in figure 3d (84 mm downstream the hub). There the fluctuation is significant mainly in annular cross-section.

Based on time-averaged PIV data, the flow topology is strictly dependent on the flow rate through the tangential branch. There is no dependency on the nominal flow rate at all (there is no difference between regime No. 8 and No. 8x).

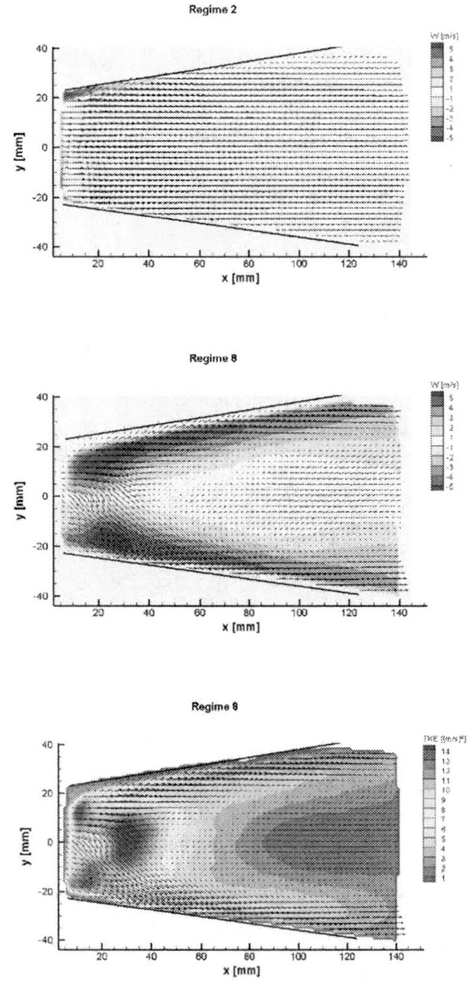

Fig. 2a, b, c. Longitudinal planes, distributions of out-of-plane velocities (a, b) and turbulent kinetic energy (c).

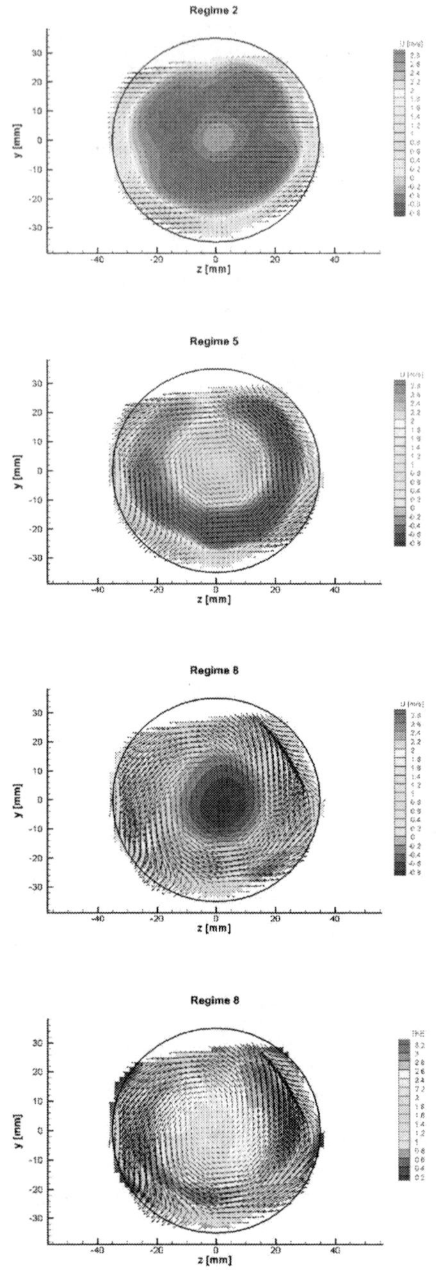

Fig. 3a, b, c, d. Cross-sectional planes, distribution of streamwise velocity component (a, b, c) and TKE (d).

3.2 Instantaneous flow field

The snapshots (Fig. 4a, b) have highly turbulent character of the flow patterns. The main helical vortex pattern is overshadowed by smaller secondary vortices located close to the diffuser walls. The flow field is completely chaotic and differs for each time step. There is an indication of rotating vortex approaching toward the upper wall and deflected by an angle approx. 30°.

The information from these snaps will be applied for POD analysis and for spectrograms. The determination of Strouhal number will be carried out only for longitudinal plane due to more precise calibration.

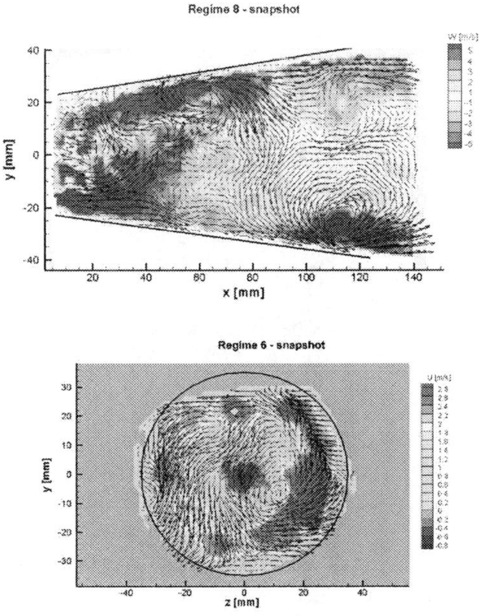

Fig. 4a, b. Instantaneous flow fields, longitudinal (a) and cross-sectional (b) planes.

3.3 Dynamical aspects

The decomposition using POD was applied on fluctuation part of all three vel. components calculated by subtraction of mean part from all snapshots. After that the temporal correlation matrix is formed, the main principle is solving the eigenvalue problem [10]. The set of POD modes is a product of this analysis. The modes are ordered according to relative energy content. Generally it can be said, that POD is not able to associate distinct frequency value to one individual POD mode (this problem solves another method). Nevertheless, an assumption for the frequency can be done very often; except the case when POD mode is associated with significant broadband.

There is a trend of kinetic energy content for longitudinal plane and for cross-sectional plane in figure 5a and 5b, respectively. The red line is plotted for energy of the very first POD mode in dependency on various OR. The blue line is actually the energy sum of first ten modes also in dependency on tangential fraction of the flow rate. The mode with highest energy fraction starts occur for regime No. 8 and higher. The first POD mode has almost 20% of relative kinetic energy. We can also see from these graphs that the sum of ten modes contains almost 50% of energy for longitudinal plane and even more than one half for cross-sectional plane. This tendency occurs from $Q_T > 50\%$. It is clear that the most energetic dominant structures stop growing for flow regime having tangential fraction of mass flow rate higher that 70%.

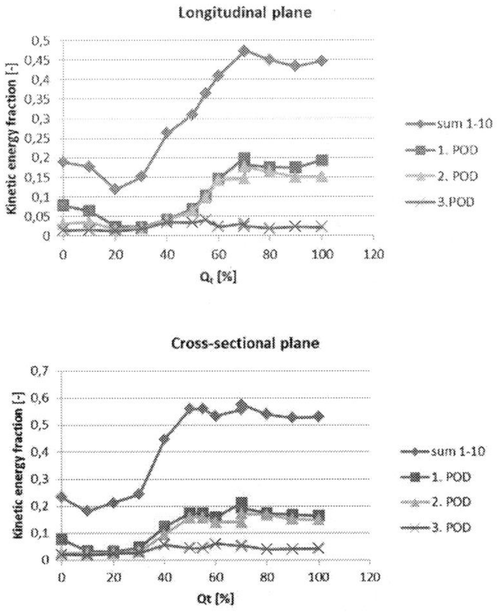

Fig. 5a,b. Kinetic energy fraction of low-order modes for both planes of measurement in dependency on tangential flow rate.

Regime No. 6 (rate 50-50) was chosen as an appropriate representative case of the mode topology (topos). Figure 6a is depicted for the very first mode. It is characterized by two pairs of counter-rotating vortices near the wall. Another two much smaller pairs can be seen near the diffuser throat where they are created. The associated frequency was estimated as approx. 29 Hz. Second mode (not plotted) is very similar, its topos is phase-shifted and has the same frequency. Mode No. 3 is plotted in figure 6b. This one is associated with higher frequency (30 Hz) and contains less energy (Fig. 5a). The topos is again composed from two pairs of counter-rotating vortices, however, now, they are oriented in perpendicular sense. The vortices are located very close to the throat. Some modes with energy less than 1% are plotted further; they are high-order modes containing smaller structures mainly responsible for chaotic behaviour of the flow.

Yet another two examples of modes are plotted for cross-sectional plane in figure 7. The topos of second mode is created from two big vortices filling the whole diameter. There is a huge diagonal flow between the vortex cores. The third mode consists of four vortices situated symmetrically with a singular point in the middle.

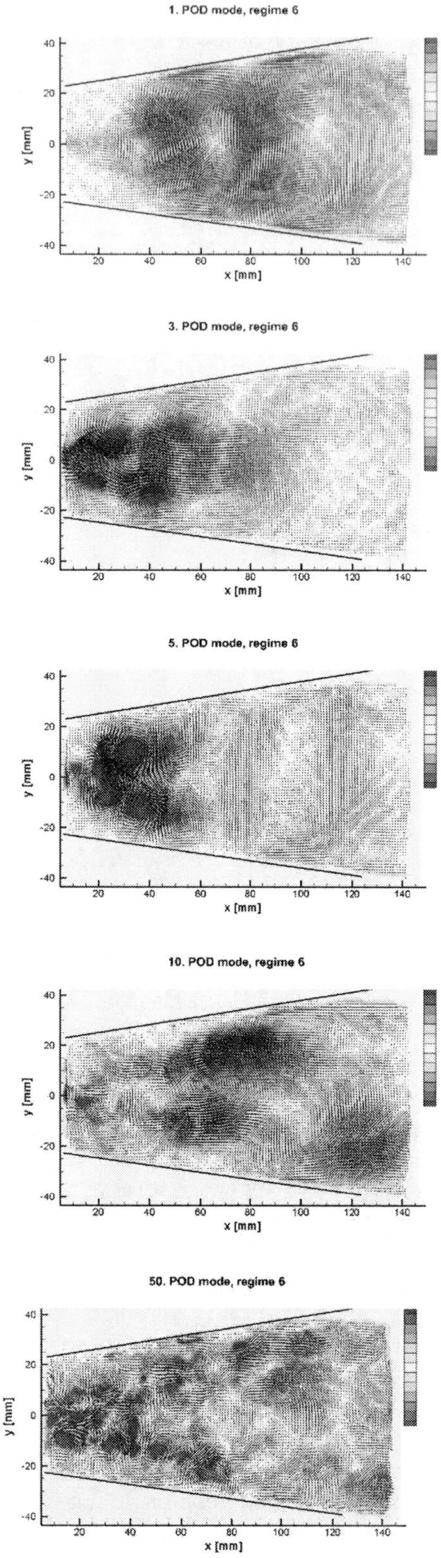

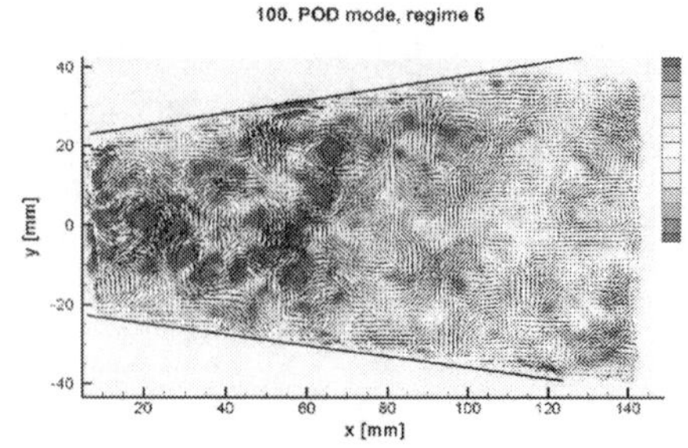

Fig. 6a, b, c, d, e, f. POD modes for regime No. 6, 1st (a), 3rd (b), 5th (c), 10th (d), 50th (e) and 100th (f), vorticity.

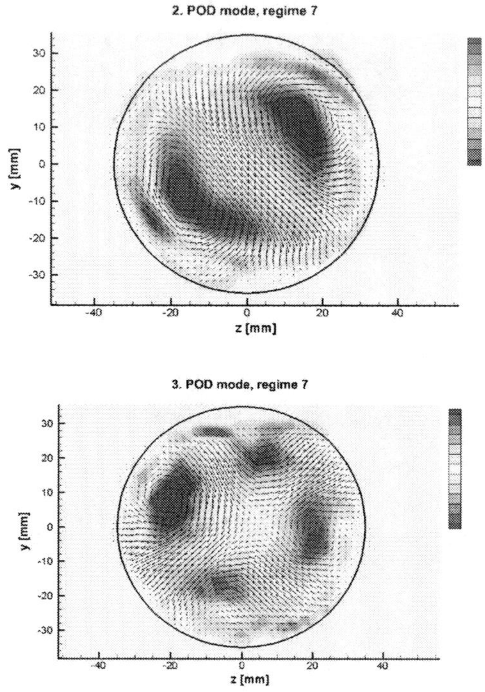

Fig. 7a, b. POD modes for regime No. 7, cross-sectional plane, 2nd (a) and 3rd (b), vorticity distribution.

3.4 Frequency analysis

The dominant frequencies occurring in arbitrary point of the flow field were investigated using FFT analysis. Only data for longitudinal plane was used. An estimation of power spectral density was conducted and the dominant frequency was found. This frequency was used to determine Strouhal number

$$Sh = \frac{f D_{ref}}{v_{ref}}, \qquad (2)$$

where D_{ref} is the diameter of throat (50 mm). Referent velocity was set according to

$$v_{ref} = \frac{4Q_N}{\pi D_{ref}^2}. \qquad (3)$$

The dominant frequency of OR No. 6 was determined as 29 Hz which is exactly the same value as the associated frequency of POD mode No. 1 and 2. The last figure 8 is plotted here to show the dependency of Strouhal number on increasing fraction of tangential flow rate. The regime No. 8 and 8x were characterized by twice as different frequency value; Sh number was identical.

It was shown that Strouhal number almost linearly depends on the flow rate, which is in an agreement with findings of other authors.

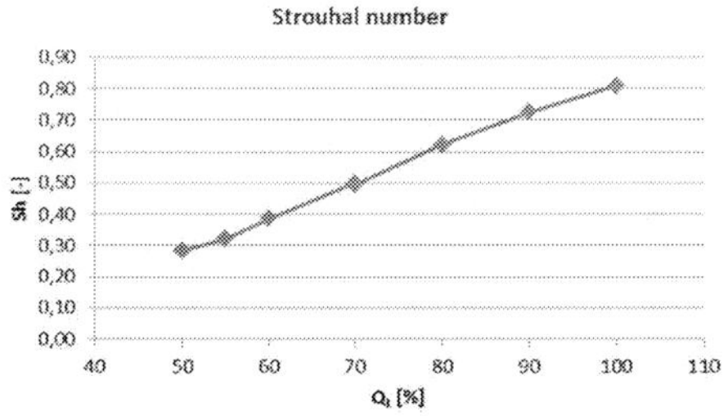

Fig. 8. Strouhal number dependency on used tangential fraction of the mass flow rate.

4 Conclusions

The stereoscopic TR-PIV measurement was conducted inside the model of swirl flow apparatus. The optical access to the measurement planes was very challenging due to complex shape of the model. The only way was to use two-level calibration target. Various calibration models were utilized to get best calibration reprojection errors. The resultant vector maps were under investigation to describe the mean flow topology and mainly dynamical flow structures and associated frequencies.

A central helical swirling vortex was detected from time-averaged flow field and snapshots. The potency of this structure is directly proportional to applied rate Q_T to Q_A. The region with highest fluctuation activity is right behind the hub. There is no flow separation near the walls.

The spatiotemporal information in the form of POD modes was presented here for different swirling flow generated under several regimes. OR No. 8 and higher ($Q_T > 60\%$) suppresses higher POD modes as the kinetic energy of the first two modes is rapidly increasing. Next studies will be devoted to more advanced method (OPD) to uncover all frequency set present in such kind of flow.

This work was supported by the Grant Agency of the Czech Republic, project No. 17-01088S.

References

1. D. Frunzaverde, S. Muntean, G. Marginean, V. Campian, L. Marsavina, R. Terzi, V. Serban, Failure Analysis of a Francis Turbine Runner, IOP Conf. Ser.: Earth Environ. Sci., **12**(1), 2010, p. 012115.
2. P. Dörfler, M. Sick, A. Coutu, Flow-Induced Pulsation and Vibration in Hydraulic Machinery, Springer-Verlag 2013, London.
3. D. Štefan, P. Rudolf, S. Muntean, R. Susan-Resiga, Proper Orthogonal Decomposition of Self-Induces Instabilities in Decelerated Swirling Flows and their Mitigation through Axial Water Injection, Journal of Fluids Engineering, Vol. **139**, 2017, 081101-1.
4. P. Procházka, V. Uruba, V. Skála, On the 3D Structure of the Flow-Field in the Vicinity of Inclined Plate, Journal of Physics: Conference Series, Vol. **1101**, 2018.
5. J. L. Lumley, Stochastics Tool in Turbulence, Academic Press, 1970, New York.
6. P. Rudolf, D. Štefan, Decomposition of the Swirling Flow Field Downstream of Francis Turbine runner, IOP Conf. Ser.: Earth Environ. Sci., **15**(6), 2012, 062008.
7. C. Tropea, A. L. Yarin, J. F. Foss, Handbook of Experimental Fluid Mechanics, Springer, 2007.
8. M. Raffel, C. E. Willert, S. T. Wereley, J. Kompenhans, Particle Image Velocimetry. A practical guide, Springer, 1998.
9. K. E. Meyer, J. M. Pedersen, O. Őzcan, A Turbulent Jet in Cross-flow Analysed with Proper Orthogonal Decomposition, J. Fluid Mech, **583**, 2007, pp. 199-227.
10. V. Uruba, Decomposition Methods in Turbulent Research, EPJ Web of Conference, **25**, 2012, 01095.

Preliminary evaluation of interferometry measurements of the flow field in the vicinity of a self-excited single profile

Pavel Procházka[1,*], *Václav* Vlček[1], and *Petr* Šidlof[1,2]

[1]Institute of Thermomechanics, Czech Academy of Sciences, Czech Republic
[2]NTI FM, Technical University of Liberec, Czech Republic

Abstract. This paper deals with the interferometry measurement of the flow field in the vicinity of the surface of the self-excited single profile NACA 0015 in subsonic air flow. The pressure and velocity on the profile surface, the total forces and total moments acting on the profile are evaluated in the range of the Mach number Ma = 0.18-0.45 and the Reynolds number Re = $(3\text{-}6)\cdot 10^5$. Since the interferogram processing was conducted with simplified assumptions of the isentropic changes of the flow field, the results have a limited accuracy especially in the case with flow separation. Results were not compared with the direct pressure measurement on the profile surface. The optical measurement method enabled the forces and moments to be evaluated separately for the suction and pressure surface of the airfoil.

1 Introduction

Flow-induced vibration (FIV) is a serious problem during operation of many technical devices such airplane wings, helicopter rotors, propellers, turbine / compressor blades, antennas or transmission wires. This phenomenon is caused due to the specific interaction between flow instabilities with elastic structures of a moveable device. Flutter is dangerous for many reasons: significant influence of lift and drag; the loss of efficiency of energy conversion; the occurrence of resonance effect if natural frequency is identical to frequency of vortex-induced vibration (VIV), etc.

A good understanding of self-excitation mechanism of single profile is necessary condition for studying much more complex arrangements – self-excited profiles located within a cascade. The oscillation of single profile with two degrees of freedom is solved quite commonly both numerically and experimentally. Classical flutter phenomenon [1] is characterized by the attached flow all the time of the airfoil movement. Unlike stall dynamic flutter [2] where airfoil has very high angle of attack (AoA) values while moving and boundary layer can be separated along all suction side resulting in rapid drop of lift force. Stall flutter has nonlinear reaction of aerodynamic forces to the airflow vibration. Thus, the basic instability and its principal features are explainable in terms of nonlinear forces and moment.

The aeroelasticity issues of profiles located in cascade are mentioned in literature rarely (e. g. [3], [4]). The information about self-induced profile inside cascade with pitch and plunge degrees of freedom was not found in the literature at all. The experimental research described in this article should help to start up a challenging investigation of self-excited profiles within turbine or compressor blade cascades [5].

This article follows previously conducted experiments with single profile NACA 0015 (e.g. [6]) which were carried out in high-speed wind tunnel of the Institute of Thermomechanics in Nový Knín. This tunnel is able to reach velocities up to Ma = 1.9 with turbulence intensity below 1%. The main result here is an analysis of interferometric measurement of non-stationary flow field in the vicinity of profile with pitch and plunge degrees of freedom. Self-induced vibration was observed during Mach number range from 0.2 to 0.45 for this particular geometry. Two distinct regimes for Ma = 0.28 and 0.45 (Reynolds number Re = $4\cdot 10^5$ and $6.5\cdot 10^5$) are presented. Both regimes are characterized by the existence of massive flow separation at suction side forthcoming by extreme pitch. Moreover, the lower velocity case is responsible for significant instability of the separation point position which has led to non-uniformity of plunge motion.

2 Experimental setup

The used profile NACA 0015 is an original design of previous experiments [6] from technical reasons. The airfoil had ideal chord length c = 67 mm and real length was 96% of ideal length. The profile shape and its support is depicted in figure 1. The profile has span dimension of 76.6 mm. The airfoil was adjusted to have a slot to install bearing and elastic element (rotary axes located at 1/3 of chord). A torsion rod was used to support elasticity of the rotation. Vertical profile guides were supported by flat linear springs enabling plunge motion. The guide was designed to have triangular cutout to not block the optical access in the vicinity of the profile surface.

* Corresponding author: prochap@it.cas.cz

 https://doi.org/10.1051/epjconf/202226901049

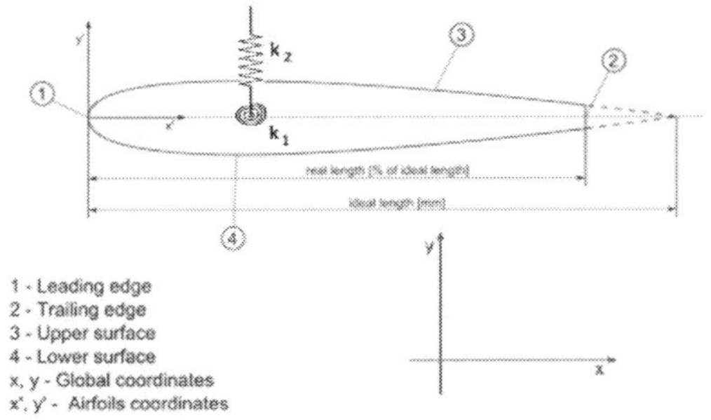

Fig. 1. Schematic view of used profile and its support using elastic elements.

The profile was fixed in a test section with dimensions 80 x 210 mm which was connected to the suction-type wind tunnel. The optical access for interferometry was provided by two glass windows with diameter of 160 mm and thickness of 20 mm. Precautions against scratching the glasses were performed. Compensating glasses are placed between the test section and the interferometer.

The interferometer of Mach-Zehnder type (Fig. 2) was utilized to measure the flow properties. The interferometer is sensitive to changing of density. The principle of this method is the phase shift variation between two collimated beams generating by splitting light from a single source (mercury lamp with wavelength 435.8 nm). One beam is guided through the compensator glasses and second one is guided through measurement section where density changes of moving air are present. An interferogram consisting of interference fringes of constant density are used then to evaluate the pressure distribution. Bigger width of measurement section is used, higher amount of fringes are present (our section had a width of 80 mm). The set of mirrors were set so that the fringe No. zero had infinity width (i. e. it corresponded to density of incoming flow). Fast CMOS camera with acquisition frequency of 1 kHz was used to capture the interferograms.

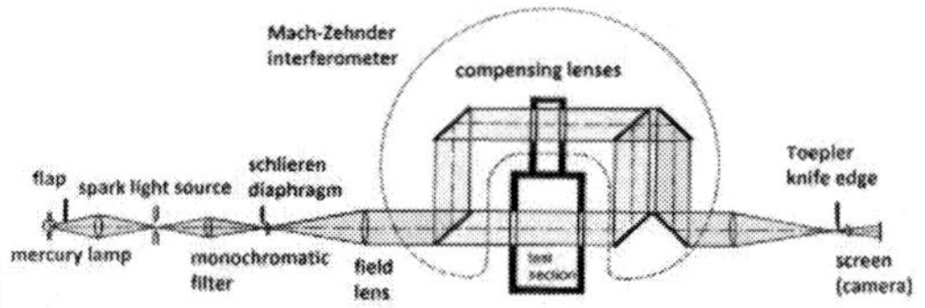

Fig. 2. The scheme of the optical measurements using Mach-Zehnder interferometer.

The evaluation of interferogram can be affected by several errors. The determination of a distinct fringe is possible with precision of a half fringe (one fringe is composed from bright and dark part). The fundamental condition is to determine correctly the fringe No. zero. The accuracy of measurement is dependent on the flow velocity (the number of fringes is decreasing with lower velocity values). The experience of an evaluator plays a crucial role as well as the contrast of interferogram itself. Sometimes several fringes can be covered by supported part of experimental device. The fringe contrast can be decreased by unsatisfactory two-dimensionality of the flow. The interferometry method does not detect the pressure loses due to dissipation of energy. Usually, this should be corrected using static pressure measurement along airfoil profile. This correction was not included because of the high technical demands. Future experiments will contain static pressure measurement along profile using small MEMS sensors.

An in-house software IFGPro was developed and used for the post-processing and evaluation of interferogram sequences. The software is semi-automatic with possibility to investigate arbitrary airfoil with two degrees of freedom. The constants used for density estimation: wavelength of light λ = 435.8 nm; heat capacity ratio κ = 1.4; specific gas constant R = 287.67 $J \cdot kg \cdot K^{-1}$; Gladston-Dale constant K = 0.0002297 $m^3 \cdot kg^{-1}$. The result is in form of pressure and velocity distribution, lift and drag force, moment, pitch and plunge coordinates etc. The example of GUI is in figure 3. One can see marked fringes (green lines) around whole profile, the atmospheric pressure value (purple line), the pressure distribution around airfoil (red line) and acting forces on the profile in the picture. This software enables to calculate forces acting on upper and lower sides separately which is crucial feature by investigation of both half-periods of movement. The asymmetry of the excitation motion can be then assessed.

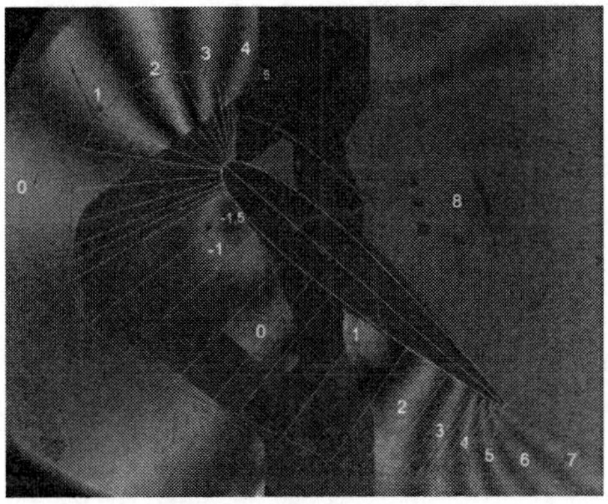

Fig. 3. The cut of GUI of post-processing.

3 Results

The self-excited profile oscillates for Mach numbers 0.45 and 0.28 with plunge frequency approximately 32 Hz and 26 Hz, respectively. This is the reason why both regimes will be compared for separately plotted graphs. Firstly the kinematics of the profiles will be solved. Consequently the acting forces and moment on the airfoil will be discussed.

https://doi.org/10.1051/epjconf/202226901049

3.1 Kinematics

The self-excited profile has performed a periodic motion in terms of pitch and plunge degrees of freedom. The pitch values during one period follow almost perfectly sinus wave without any distortion for both investigated Mach numbers. The example of the pitch-time dependency is given in figure 4. The pitch peak-to-peak value was 83° for higher Mach number and 78° for smaller Mach number. Figure 5 is plotted for plunge coordinate during one period for both examined Mach numbers. Peak-to-peak plunge values are 14 mm and 16 mm for Mach number 0.45 and 0.28, respectively. The plunge value reaches its maxima when the AoA is equal to zero. The curve of plunge motion is not perfectly smooth, especially at top dead position (for Ma = 0.45). There is one concave and one convex additional part which are superimposed to plunge curve to both sides from the main peak for lower Mach number. This may be connected with instability of the separation point and it will be discussed later.

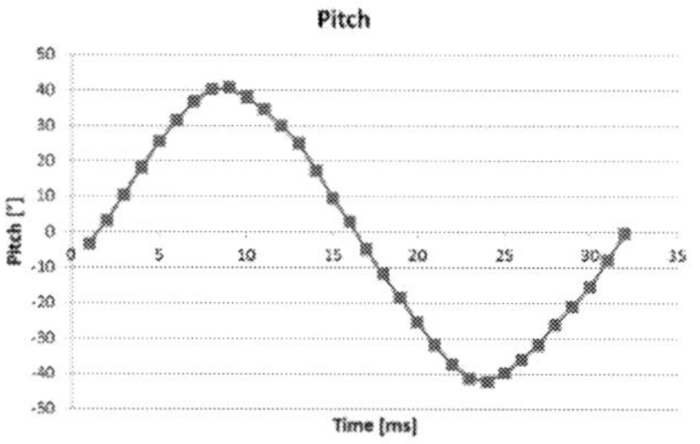

Fig. 4. Sine wave of pitch-time dependency, Ma = 0.45.

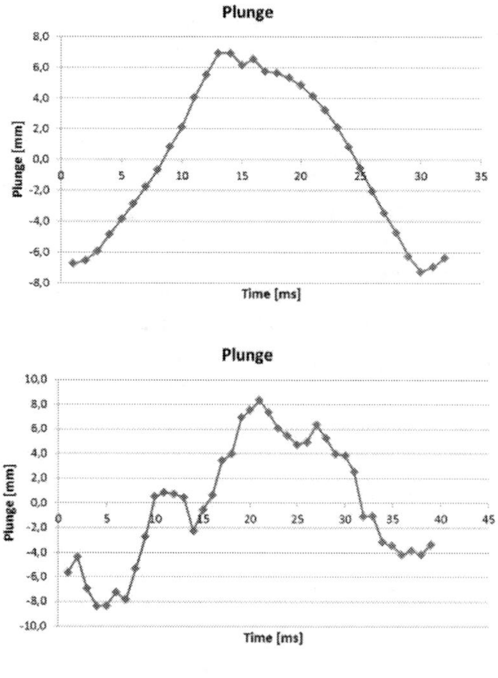

Fig. 5. Plunge-time dependency for M = 0.45 and M = 0.28.

The pitch velocity [°/s] and plunge velocity [m/s] are evaluated in figure 6 for Ma = 0.45. The biggest pitch velocity (about 7500 °/s) was observed for zero value of AoA. The maximal plunge velocity is approx. 2 m/s and the curve is wobbled for top dead position. The bottom dead position does not demonstrate this feature. This could be caused by the effect of gravity. Generally, it can be said that the profile is very sensitive to deep stall dynamic flutter under investigated flow velocities.

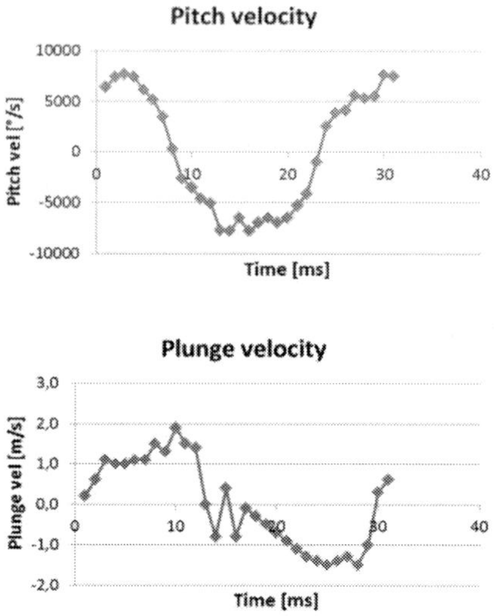

Fig. 6. Pitch velocity and plunge velocity, Ma = 0.45.

3.2 Forces acting on the surface

Henceforward, the flow of Mach number 0.45 will be described. The lift force oscillates between positive and negative values during one period of motion. The waveform is almost symmetric for both half-periods (see figure 7, 8). The peak value is about +/- 80 N. After that point, significant decrease of lift force is present due to massive flow separation. The separation is apparent for pitch values higher than approx. +/- 25° (the value differs slightly for both half-periods). Since the separation point position is more or less steady, the lift is kept at constant value of approx. +/- 60 N and then goes to zero. The drag distribution is perfectly symmetric for both half-periods. Also moment seems to be symmetric with sign of opposite sense than lift force.

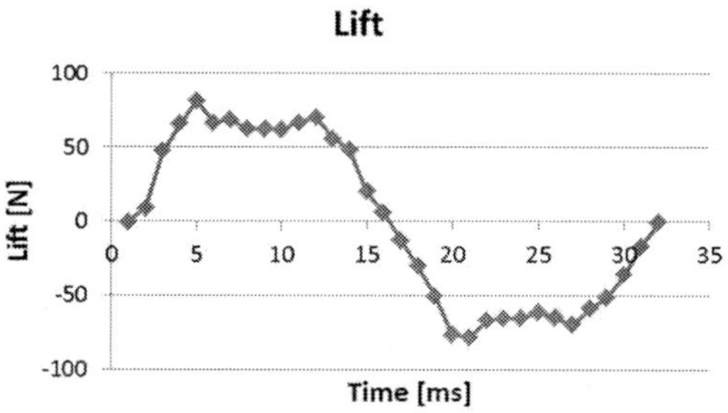

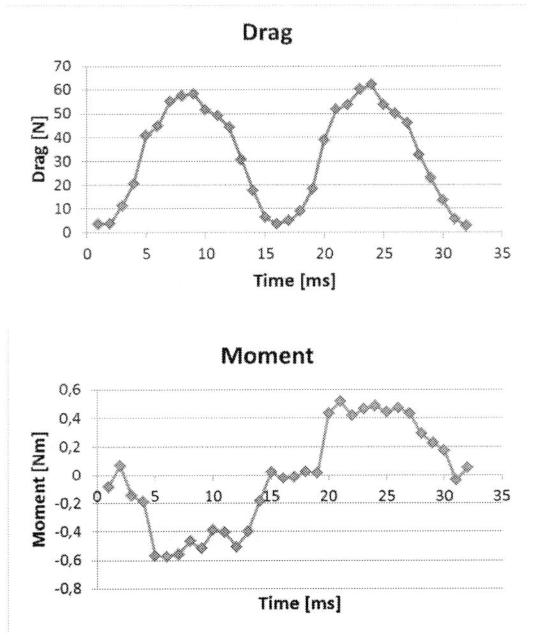

Fig. 7. Total forces: lift, drag and moment for Ma = 0.45.

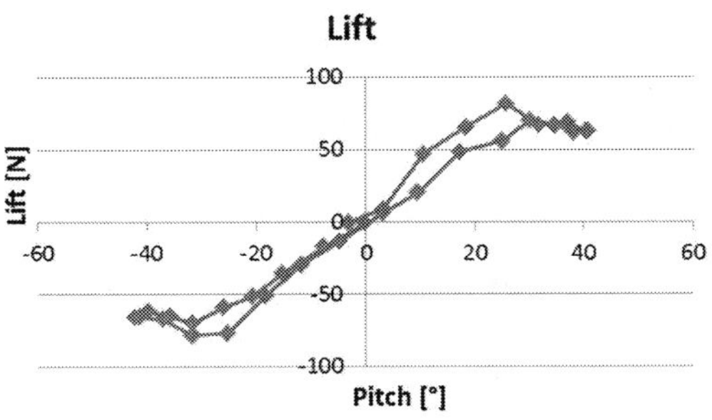

Fig. 8. Lift force in dependency on AoA, Ma = 0.45.

A useful feature of the evaluating program is a possibility to calculate acting forces separately for upper and lower part of the airfoil. Especially lift force will be presented here. Figure 9 is plotted to show the lift dependency on plunge coordinate for the entire airfoil and for both sides. It can be seen from both closed curves that the contribution from upper side does not correspond perfectly with lower side. The contribution of the lower profile part is more dominant in comparison with upper part. Lift force generated by upper and lower side in dependency on time is visible in figure 10. The asymmetry is clearly made visible here using polynomial interpolation. We can see that distance between both polynomials is not constant. Nevertheless, the contribution from upper side dominates during the first half-period while the lower side is dominating during second one.

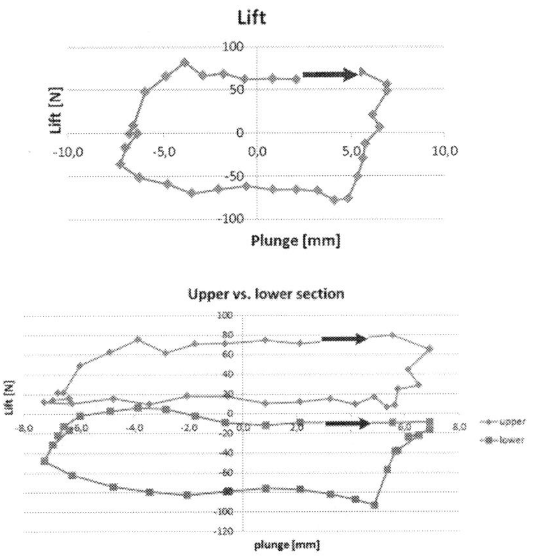

Fig. 9. Lift force in dependency on plunge coordinate, for entire shape, for upper and lower sides, Ma = 0.45.

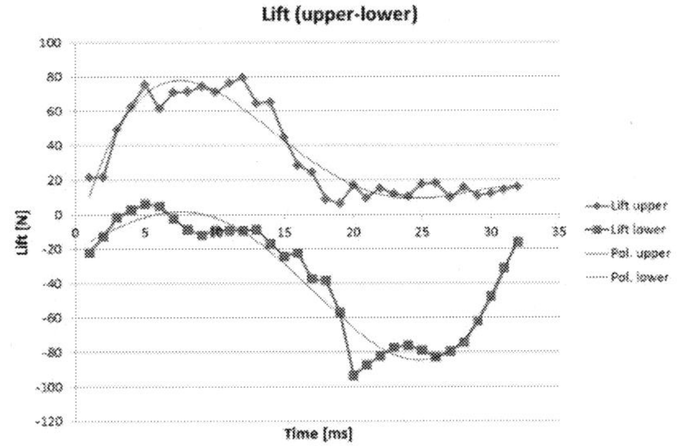

Fig. 10. Lift force divided for upper and lower sides, Ma = 0.45.

Last results will be devoted to case with Mach number equal to 0.28. The lift force, drag force and momentum are plotted in figure 11. All peak values are not so significant in comparison with previous case. More the curves are not so smooth. Especially, this is apparent for lift force divided for upper and lower side (figure 12). The curves are wobbled due to the fact that lower Mach number has led to lower number of fringes and thus worse resolution was available to calculate all quantities. Second reason, no less important, is the non-stationarity of the separation point and the whole flow field.

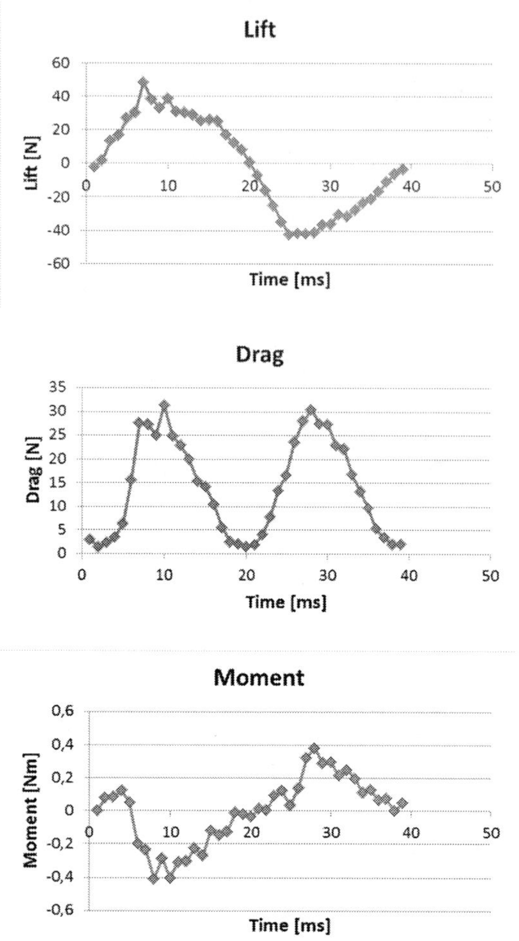

Fig. 11. Total forces: lift, drag and moment for Ma = 0.28.

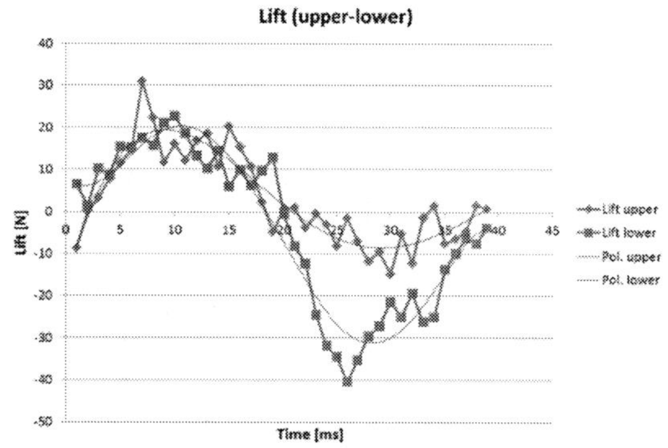

Fig. 12. Lift force divided for upper and lower sides, Ma = 0.28.

4 Conclusion

The flow around the self-excited single NACA 0015 profile was experimentally investigated using Mach-Zehnder interferometer. The acquired data were further analyzed using in-house evaluation software to detect complex information about profile movement with two degrees of freedom and about forces acting on the profile surface. It has been shown that the possibility to calculate forces separately for upper and lower airfoil side is very useful when interpreting data. Two regimes with different Mach number were compared.

The profile is very sensitive to deep stall flutter with peak-to-peak pitch value close to 80° and plunge value approximately 15 mm for both regimes. The biggest difference between both regimes is stability and stationarity of analyzed flow field. The point of separation is more stable for Mach number 0.45 since it is close to the limit of self-oscillation occurrence. On the other hand, the regime generated under Mach number 0.28 is characterized by dynamical changes of separation point location and insufficient two-dimensionality of the flow. This results in wobbled shape of the force waveforms and consequently the plunge movement is also choppy. The question is why the pitch waveform has almost perfect sinus shape (despite the fact that plunge has much higher inertia moment). The partial answer could be lower accuracy of evaluation for smaller velocities. However, the evaluation precision of pitch and plunge degrees remains the same regardless the selected regime.

There will be incorporated the effect of gravity in the future experiments. More detailed analysis is necessary to better understanding of the different contribution of upper and lower sides of the airfoil. Finally, the main future goal is to be able to examine the self-oscillation of the blades located inside a cascade.

This work was supported by the research project of the Institute of Thermomechanics "Experimental investigation of self-excited oscillations of a blade cascade".

References

1. Theodorsen T., NACA report 496, pp. 416-433 (1935)
2. Dowel, E., A Modern Course in Aeroelasticity, Springer (2015)
3. Hall K. C., Kielb R. E., Thomas J. P., Unsteady aerodynamics, aeroacoustics and aeroelasticity of turbomachines. Springer, Dordrech, (2006)
4. Stenfelt G., Ringertz U., The Aeronautical Journal, **119**, NO. 1222, (2015)
5. Vlček V., Procházka P., EPJ Web of Conferences **213**, 02095 (2019)
6. Šidlof P., Štěpán M., Vlček V., Journal of Fluids and Structures, **67**, pp. 48-59 (2016)

Validation of a simplified AQUATOX model to assess quality status of a model river in Sardinia

Luca Salvadori[1,*], *Davide* Moccia[2], *Laura* Melis[1], *Giulia* Folegnani[1], *Antonio* Pusceddu[2], *Alessandra* Carucci[1], and *Simone* Ferrari[1]

[1]DICAAR, Dipartimento di Ingegneria Civile, Ambientale e Architettura, Via Marengo 2, Cagliari, Italia
[2]DiSVA, Dipartimento di Scienze della Vita e dell'Ambiente, Via Fiorelli 1, Cagliari, Italia

Abstract. In the framework of the periodic update of the Sardinian River Basin Management and Water Protection Plans (Italian Law 152/2006, transposition of the 2000/60/EC Directive), we investigated the AQUATOX model feasibility as a tool for assessing the relationships between Driving forces, Pressures, States, Impacts and Responses (DPSIR scheme). To do this, we applied the AQUATOX model to forecast the response of a Sardinian (Italy) river quality status to a set of different environmental perturbations, including nutrient load and water flow variations. The final aim of this work is to study a feasible model, based on a DPSIR scheme, for identifying the necessary and sustainable measures to tackle the criticalities resulting from a set of known environmental perturbations on Sardinian rivers.

1 Introduction

Since Water Framework Directive (2000/60/EC) enactment, River Basin Management Plans have become one of the most important controlling tools for surface water bodies safeguard, and the scientific community started to focus greater attention to the effects of anthropogenic pressures on riverine and marine ecosystems (1). In particular, a lot of scientific interest is addressed on rivers and streams study, in terms of qualitative and quantitative responses to anthropic pressures on the system, applying the Driving forces, Pressures, States, Impacts and Responses (DPSIR) scheme. However, many uncertainties occur when trying to reach such an objective: first of all, rivers and streams, as every environmental system, have to be considered as open systems and present a remarkable sensitivity to many external characteristics, such as water flow rate, temperature, wind intensity. Among available numerical models created to reach such ambitious objective, AQUATOX was chosen in this study, because of its good evaluations in review literature (2) and its simplicity in interfaces.

AQUATOX is an ecological risk model for water bodies, developed by U.S.A. EPA (Environmental Protection Agency) for water quality study and assessment of American streams, lakes, estuaries and lagoons (most recent version available since September 2018). It implements about 450 equations, including a broad set of parameters and variables in its solving process. The model simulates a whole ecosystem, including nitrogen, phosphorus and organic matter cycles, along with the variability of the abundance of plants, invertebrates and fish at different life-cycle. AQUATOX is the last development of a model series which includes CLEAN, CLEANER and PEST (3), and it is different from these because of the additional implementation of the prediction of the effect of nutrients and the toxicants determination (4). It is able to simulate various water body typology, each of them with specific parametrisation and model setting; it usually works with a large amount of data, but not all of them are strictly necessary and can be used to improve model performance, rather than to achieve a rough representation of pressures-response scenarios. This is an important fact in relation to the main task and the central question of this study: the determination of the possibility to use AQUATOX in other contexts apart from the U.S.A. main calibration and validation case studies (5), and in situations of lack of information and data. The relevance of this kind of analysis is due to the shortage of on field monitoring data for the case study investigated, and in general for some Sardinian water bodies. Sardinia is an Italian island which is considered as a single River Basin. As below better described, we tested the model giving only main descriptors values (i.e. the length of a chosen river trait, its maximum and average depth, a rough monthly time series of water flows, and the initial conditions of nutrients), and then compared the results with available field data. The biological component has been described based on other Mediterranean studies, due to the lack of information for Sardinia. Another important fact is that AQUATOX considers each modelled section as a mixed tank: it may sound unusual, but various literature reports show that this assumption is compatible with ecological systems, and AQUATOX is one of the suggested models to analyse the River Basin environmental scale (6).

* Corresponding author: luca.salvadori@unica.it

Compared to other available models, such as CATS, CASM, Qual2K, WASP7, EFDC-HEM3D, QEAFdChn, BASS and QSim, AQUATOX is declared as the one that involves more processes and variables, so it was identified as the more suitable in the studied context. Because of the particular interest in nutrients fate study, AQUATOX has to be preferred to CATS, even if the latter shows good prediction capability, due to its better representation of eutrophication and anoxia phenomena (7). Even if QEAFdChn shows better representativeness of bioaccumulation, it does not include an eco-toxicologic component, which is important for future model application in Sardinian case studies.

2 Case study

The Cixerri river has been considered for this study case (Fig. 1). Located in the South of Sardinia, near the city of Cagliari, it is a torrential trait, as many of Sardinians rivers, and it represents a typical multi-stressor environment. In fact, the Cixerri river receives urban wastewater treatment plants discharges from cities and villages (Iglesias, Villamassargia-Musei, Siliqua and Uta), at the same time it is an important water source both for civil and for industrial use, furthermore it flows in a vast agricultural area, making diffuse pressures an additional important pressure to be taken into account. We studied the trait that starts from the monitoring station of Siliqua (blue point in Fig. 2), includes the Cixerri lake and ends in Santa Gilla pond (red point in Fig. 2). Geographic data has been sourced from an open-access dataset from Regione Sardegna (Sardegna Geoportale). Cixerri lake is an artificial lake, and its dam is Genna is Abis. Location, water volume and structure characteristics of this element have been taken from the website of Ente Acque della Sardegna (ENAS). Geographic data has been processed through BASINS 4.1 software, which is a Geographic Information System (GIS) that can be directly linked to AQUATOX to provide morphological and general site accurate information for the simulation. A GIS data framing of Cixerri river is shown in Fig. 3.

A lapse of two years has been considered for the Cixerri simulation, which started on 01/02/2012 and ended on 01/01/2014. This period has been chosen taking into account the data need and availability and the limitations of the computation time.
Water flow was modelled according to the most recent Sardinian River Basin Management Plan as described in Table 1.

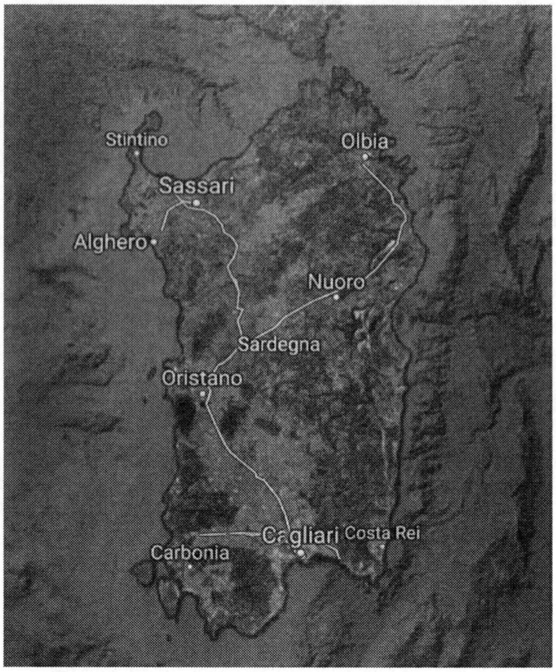

Fig. 1. Satellite framing of the case study.

Fig. 2. Satellite framing of the case study.

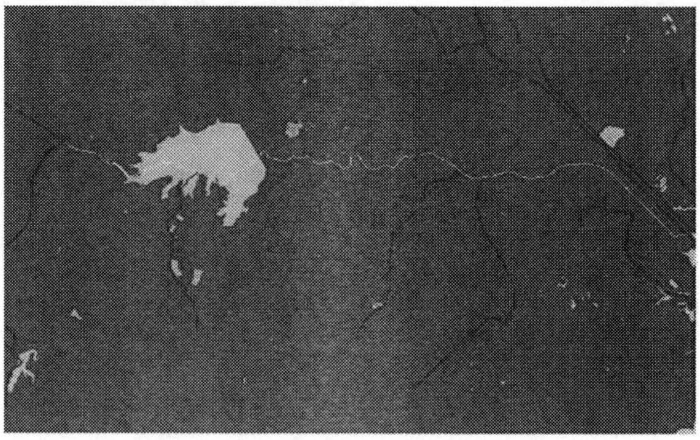
Fig. 3. GIS elements framing of the case study.

Table 1. Monthly water flows in the investigated river. Water flows are expressed in m^3/s.

Month	Water flow
January	5.760
February	7.985
March	5.879
April	3.313
May	1.567
June	0.458
July	0.119
August	0.066
September	0.318
October	2.366
November	3.751
December	5.989

3 Results and discussion

For first simulations, vegetal components and concentrations have been carried as those of a pre-existent simulation available in AQUATOX internal dataset, referred to the Blue Earth River case study simulation. In order to better represent a Mediterranean context, invertebrates have been divided as suggested in (8) and summarized in Fig. 4.

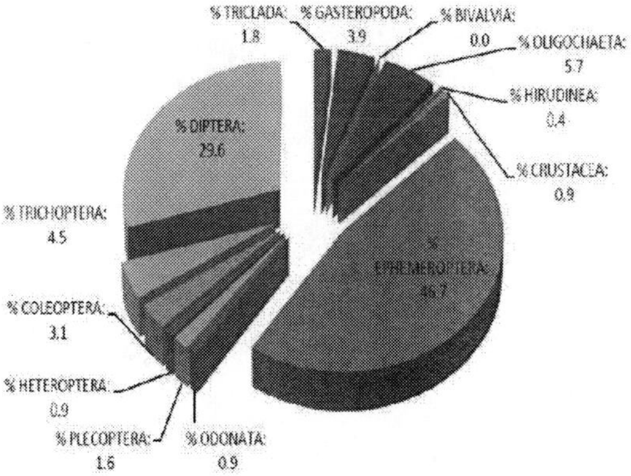

Fig. 4. Pie Chart showing the macrobenthos repartition for Sardinian rivers according to (8).

In Table 2, the complete set of initial conditions for the case study simulation have been reported. It results from the combination of pre-defined data in AQUATOX for Blue Earth River simulation, the implementation of more specific data, as above described, and available monitoring information about examined site. Parameters not included are meant overlooked or were assumed to be zero.

Table 2. Initial conditions for Cixerri simulation.

State Variable Name	Initial Conditions	Units
NH3 & NH4+	0	mg/L
NO3	4	mg/L
Tot. Sol. P	0.2	mg/L
CO2	0	mg/L
Oxygen	9,00	mg/L
Peri Low-Nut Diatom	0.00328494	g/m2 dry
Phyt High-Nut Diatom	0.001009124	mg/L dry
Phyt Low-Nut Diatom	0.000829888	mg/L dry
Phyto, Navicula	0.000842673	mg/L dry
Peri, Green	0.0022381	g/m^2 dry
Phyto, Green	0.000788045	mg/L dry
Cryptomonas	0.000943996	mg/L dry
Tubifex	0.7	g/m^2 dry
Caddisfly, Trichopter	0.58	g/m^2 dry
Mayfly (Baetis)	5	g/m^2 dry
Stonefly	0.23	g/m^2 dry
Gastropod	0.46	g/m^2 dry
Odonata	0.11	g/m^2 dry
Water Vol	9500	m^3
Temp	25	Deg. C
pH	7.5	-

The simulation started with the given configuration. In Table 3, the main results are shown: every pollutant concentration is expressed in mg/l, N indicates total nitrogen as the sum of all its forms and TSS indicates Total Suspended Solids. The values are meant averaged on all simulation period. The comparison between AQUATOX predictions and monitored concentrations shows good results for ammonia and oxygen (respectively, 14,5% and -9,04% of difference), but not so good for the other parameters, where differences are remarkable. In general, trends are not captured by the model, so are not reported.

It has to been noticed that vegetal species and concentrations were assumed as those given by AQUATOX by default for this simulation. Surely, this affects the model capability to represent both of phosphorus and nitrogen (total and its forms) cycles, as a consequence of the uncertain determination of algae concentration.

Table 3. Comparison between AQUATOX results and monitored values for concentrations of some pollutants (every concentration is expressed in mg/l).

	NH_4	N	P	O_2	TSS
AQUATOX	0.03	3.49	0.69	8.43	2.30
Monitoring	0.03	0.78	0.10	9.27	18.21
Difference	0.00	2.71	0.59	-0.84	-15.91

Starting from the above described setting, some response simulations has been conducted, varying different parameters one at the time. Considered animals were *Tubifex tubifex*, Caddisfly, Mayfly (Ephemeroptera), Stonefly (Plecoptera). For plants Diatoms has been chosen. As first, an increment of 20% in nutrients and detritus has been considered. For the whole animal components, percentage variations were found between -10% and +50%; Diatoms duplicate their abundance in average, with peaks of triplication in several periods.

Opposite effects were found decreasing by 20% nutrients and detritus: Mayfly decrease about -100%, and similarly plants. It is worth highlighting that the model was found to be sensitive to initial conditions of nutrients: for both of the aforementioned conditions variations, strong effects were registered in terms of final output of nutrient concentrations (an increment of about 4 times of the output as a consequence of an increment of 20% of the input value for some periods). A variation of 20% of the water flow did not significantly affect final results.

4 Conclusions

As highlighted by the results section, the study of pollutants trends with AQUATOX in conditions of lack of data could be a difficult task. However for some investigated parameters (namely ammonia and dissolved oxygen) a long-term forecast seems to be possible. The importance of this conclusion is related with the aim of this AQUATOX usability study: the need of tools to forecast the water bodies ecosystems response to pressure goes along with the need of the same forecast capability in terms of what-if analysis applied to measurements which are supposed to be taken to improve water quality. These are phenomena which take place in long time horizons, so in this sense a good forecasting model gives an additional support for measures decision and planning management measure. Another important aspect highlighted by the results of this study is the importance of considering ecosystem vegetation for simulations, and hence guide further studies, where a more intensive attention is paid to this aspect.

The consistency of model response to input variations, even with a rough configuration of some parameters, due to limitations in available data, leads to consider this preliminary results on AQUATOX suitability to study water quality in Sardinian context good enough to carry further investigations.

References

1. C.K. Feld, S. Birk, D.C. Bradley, D. Hering, J. Kail, A. Marzin et al., Adv Ecol Res, **44** (2011).
2. R.A.Park, J.S. Clough, *Technical Documentation*, US EPA Office of Water, (2014).
3. R.A. Park, J.S. Clough, M.C. Wellman, Eco Mod, **213**, 1–15 (2008).
4. J. Clough, R. Park, *Aquatox for Windows - a modular fate and effects model for aquatic ecosystems*, (2003).
5. US EPA Office of Water, *Model Validation Reports*, (2000).
6. J. Brils, R. Harris, *Key-findings and recommendations of the RISKBASE project*, (2009).
7. A. Koelmans, A. Van der Heijde, L. Knijff, R. Aalderink, Water Res **35(15)**, (2001).
8. R. Tenchini. *Comunità Macrobentonica nei fiumi della Sardegna: habitat e risposta alle alterazioni morfologiche*, Università Degli Studi Della Tuscia Di Viterbo Dipartimento Di Scienze Ecologiche E Biologiche (2015).

https://doi.org/10.1051/epjconf/202226901051

Using the AQUATOX model to forecast water bodies quality status response to environmental perturbations

Luca Salvadori[1*], *Davide* Moccia[2], *Laura* Melis[1], *Giulia* Folegnani[1], *Antonio* Pusceddu[2], *Alessandra* Carucci[1], and *Simone* Ferrari[1]

[1]DICAAR, Dipartimento di Ingegneria Civile, Ambientale e Architettura, Via Marengo 2, Cagliari, Italia
[2]DiSVA, Dipartimento di Scienze della Vita e dell'Ambiente, Via Fiorelli 1, Cagliari, Italia

Abstract. In the last decades, the scientific community has developed many methods and approaches for the assessment of surface waters quality status and the relative impacts of anthropogenic pressures including, among the others, municipal and industrial wastewater treatment plants, agriculture, animal farming and many other land uses. According to the Water Framework Directive (WFD; 2000/60/EC), each EU Member State has been committed to identifying River Basin Districts, in which, ideally, every water body within European boundaries should have been included. As a contribution to the quality assessment of the Sardinian River Basin District, in the framework of the project "Hydraulic and environmental modelling for the gap analysis and the identification of programmes of measures aimed to satisfy the environmental quality objectives" funded by the Regione Sardegna, we applied a simplified version of the AQUATOX model, developed for investigating its feasibility and validity in case of missing or poor input data and information, on the river Cixerri case study.

1 Introduction

As discussed in previous study on the topic presented, "Validation of a simplified AQUATOX model to assess quality status of a model river in Sardinia", AQUATOX validation can be a difficult task when is applied to the Mediterranean contexts. Such difficulties derived mainly from its large amount of data required, that implies strong differentiations between locations. However, from that study some general conclusions can be drawn, and some similar applications with good results can be found in the literature (1). So, in next sections is shown how including more ecosystem components leads to a more realistic output of AQUATOX model, encouraging its use for water bodies responses to pressures and stressors forecasting.

As a brief description of the main reasons why AQUATOX was chosen, in Table 1, a summary of the comparison with other models is reported, from (2). AQUATOX is perhaps the most complete ecological risk simulation model, and it can be used for different purposes. Initially, it has been created for studying the United States water bodies, but rapidly it has been used also in numerous different countries. For instance, in (3) a study on the Songhua river is presented, related to the accidental discharge of nitrobenzene on November 2005: the model has been used for simulating the nitrobenzene distribution with respect to time, and for determining the potential ecological impact.

Assessing water bodies quality, according to 2000/60/EC directive, means to give a judgement basing on a set of evaluated parameters for that water body, namely biological indicators, morphological status, chemical-physical condition, and chemical pollutants presence.

Table 1. The comparison of AQUATOX and other available models, in terms of variables and processes considered.

State variables and processes	AQUATOX	CATS	CASM	Qual2K	WASP7	EFDC-HEM3D	QEAFdChn	BASS	QSim
Nutrients	x	x	x	x	x	x			x
Sediment diagenesis	x			x	x	x			
Detritus	x	x	x	x	x	x			x
Dissolved oxygen	x		x	x	x	x			x
DO effects on biota	x								x
pH	x			x					x
NH_4 toxicity	x								
Sand/silt/clay	x				x	x			
Sediment effects	x								
Hydraulics						x			x
Heat budget				x	x	x			x
Salinity	x				x	x			
Phytoplankton	x	x	x	x	x	x			x
Periphyton	x	x	x	x	x				x
Macrophytes	x	x	x						x
Zooplankton	x	x	x						x
Zoobenthos	x	x	x						x
Fish	x	x	x					x	x
Bacteria			x						x
Pathogens				x		x			
Organic toxicant fate	x	x			x			x	
Organic toxicants in									
Sediments	x	x			x	x			
Stratified sediments	x				x	x			
Phytoplankton	x	x							
Periphyton	x	x							
Macrophytes	x	x							
Zooplankton	x	x					x		
Zoobenthos	x	x					x		
Fish	x	x					x	x	
Birds or other animals	x	x							
Ecotoxicity	x	x	x					x	
Linked segments	x			x	x	x	x		x

Quality judgement can be expressed basing on indications from D.Lgs. 152/2006, and general method for the ecological state determination, available in several Italian Agenzia Regionale Protezione Ambiente (ARPA) websites (e.g. ARPA Lombardia). Final ecological status depends on the combination of the four aforementioned factors, one of which is called the LIMeco parameter (five levels, Level 1 is the best, Level

* Corresponding author: luca.salvadori@unica.it

https://doi.org/10.1051/epjconf/202226901051

5 is the worst), represents the chemical-physical status of the water body and involves dissolved oxygen, ammonia, nitrate and phosphorus concentration values.

2 Case study

In "Validation of a simplified AQUATOX model to assess quality status of a model river in Sardinia", the Cixerri river was considered as the case study. Here, only a part of the Cixerri river has been investigated: the considered branch has a slope of roughly 23% and a length of 30 km.

Fig. 1. Satellite framing of the investigated trait of Cixerri river with GIS data used.

The satellite framing with the vector data of the Geographic Information System (GIS) used for the study case is shown in Fig. 1 (the free software QGIS was used). Available monitoring data on the Cixerri river were used both to increment the input dataset and to improve the description of the site in order to better compare model results with real observed prospect. The configuration related to the case study of the Blue Earth River has been set as initial in the current model, available in AQUATOX dataset: it is a practice suggested by AQUATOX developers (2), instead of building a simulation starting from zero. This is because the model needs to have every parameter available for the implementation of equations, and for certain cases, rough information is better than no information at all. For the flow rate values, it has been considered the monthly average value, extracted by the most recent public study for the river considered in this case study.

3 Results and discussion

For the model validation, in this study another specification has been included: considering the data of Blue Earth River as described in case study section, algae species presence and the food web, have been modified as suggested in (1) for the Po river. This study case was chosen also since it represents a Mediterranean case study, and it allows to get a more realistic representation of what happens in terms of predator-prey mechanism in a Sardinian river. In Fig. 2, the proposed schematisation is shown.

Table 2 shows the complete set of initial conditions for chemical-physical parameters and some related to algae life-cycle.

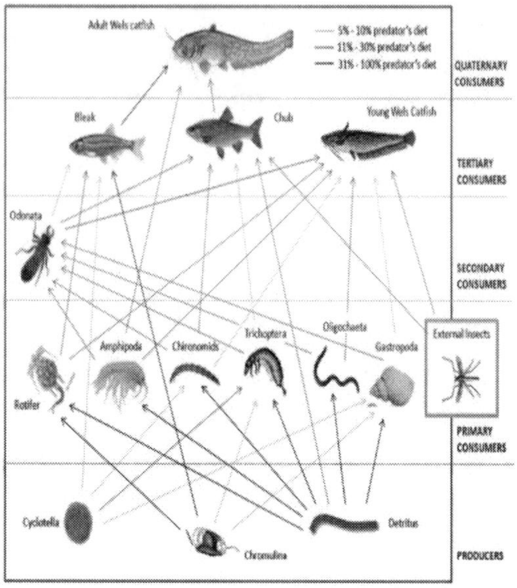

Fig. 2. Food web schematisation for a Mediterranean river (Grechi et al., 2016).

Table 2. Initial conditions for algae (Grechi et al., 2016).

Category	Microalgae	
Default organism/group in the AQUATOX plant library	*Cyclotella*	*Isochryses*
Po river compartment (and abbreviation)	*Cyclotella (Cy)*	*Chromulina (Chro)*
Initial biomass (mg dry/L)	0.12	0.0296
Loading from upstream	Calculated using the AQUATOX option "Enhanced Phytoplankton Retention"	Calculated using the AQUATOX option "Enhanced Phytoplankton Retention"
Saturating light (Ly/d)	22.5	67
P half-saturation (mg/l)	0.05	0.046
N half-saturation (mg/l)	0.117	0.006
Inorg. C half-saturation (mg/l)	0.054	0.054

Category	Microalgae	
Default organism/group in the AQUATOX plant library	*Cyclotella*	*Isochryses*
Po river compartment (and abbreviation)	*Cyclotella (Cy)*	*Chromulina (Chro)*
Temp. response slope	1.8	1.8
Optimum temperature (°C)	23.5	20
Maximum temperature (°C)	30	35
Min. adaptation temperature (°C)	7	2
Max. photosynthetic rate (1/d)	1.87	2
Photorespiration coefficient (1/d)	0.026	0.026
Respiration rate at 20 °C $(g/(g\ d)^{-1})$	0.0752	0.0483
Mortality coefficient $(g/(g\ d)^{-1})$	0.001	0.009
Exponential mortality coefficient $(g/(g\ d)^{-1})$	0.01	0.01
Light extinction	0.02	0.01
Wet to dry mass	5	5
Lipid fraction (kg_{lipid} kg^{-1} wet body)	0.005	0.005

In the following paragraph, a discussion about simulations. As highlighted in Section 3, in this study, the aim was, first, to underline model sensitivity to algae good representation (as the algae species presence and the food web have been modified, as suggested in (1), for the Po river), and secondly to use the model to evaluate the Cixerri water body status. The main morphological data used were:

- River length;
- River mean surface;
- Initial water volume;
- Mean depth;
- Mean slope;
- Input and output water flows;
- Initial total nitrogen concentration;
- Initial total phosphorus concentration;
- Nitrogen and phosphorus discharges.

This type of data is available in the "Piano Stralcio Fasce Fluviali" (PSFF) of Regione Sardegna website. Nitrogen and phosphorus discharges are meant to be both punctual and diffuse pressures, even if in simulations differentiation between these two cases has been made.

Three different types of simulation have been conducted: "Whole", "Parted 1" and "Parted 2". "Whole" simulation is the first, and starts on 01/01/2011, ending on 31/12/2016, this period was chosen based on nutrients data on tap. Initial values for nitrogen and phosphorus concentrations were the average of first year values. "Parted 1" is actually a simulation series, in which the total lapse is divided into sub-periods with one-year duration (five simulations). The first simulation has for N and P input values those observed at the upstream monitoring station in the first part of the monitoring period of 2011. The second receives the output of the first, and so on up to the fifth. It is worth to underline that, even if nitrogen and phosphorus concentrations were taken from precedent simulation output, the other parameters (e.g. algae species and their concentrations) return to initial values once re-starting simulations (an attempt to neglect simulation effects on those parameters, investigating only on nutrients fate). "Parted 2" carries the same scheme as "Parted 1", but with every re-starting of the simulation, input data for N and P is the average of the observed values for the first part of the monitoring period in the respective year. The first simulation received average data registered in the first part of monitoring period in 2011, the second average data registered in the first part of 2012, and so on, up to the fifth simulation.

In following tables the results of Cixerri simulation are reported: Fig. 3, Table 3 and Fig. 4,

Table 4 for total nitrogen and phosphorus concentration, respectively, considering only punctual pressures and neglecting diffuses; Fig. 5, Table 5 and Fig. 6,

Table 6 for nitrogen and phosphorus, respectively, considering all pressures in the water body.

In general, it can be difficult to determine the best method: in most cases, and considering total duration averaged results, "Par 2" method seems to show a better correspondence with observed data, comparing single-year averaged concentrations, but it is not ever true (e.g. for year 2014 of Fig. 5); a decreasing accuracy from 2011 to 2016 can be detected in the "Whole" simulations (for instance in Fig. 3). Considering diffuse pressure improves model performance more in reference to phosphorus rather than to nitrogen.

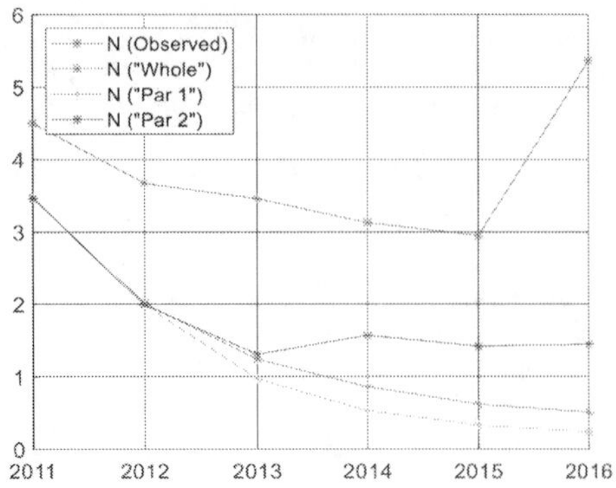

Fig. 3. Comparison between model results and observed concentrations, yearly averaged values (total nitrogen).

Table 3. Total Nitrogen results from model simulation, considering only punctual pressures (total simulation duration average).

N (Observed)	N ("Whole")	N ("Par 1")	N ("Par 2")
3.54	1.47	1.26	1.87

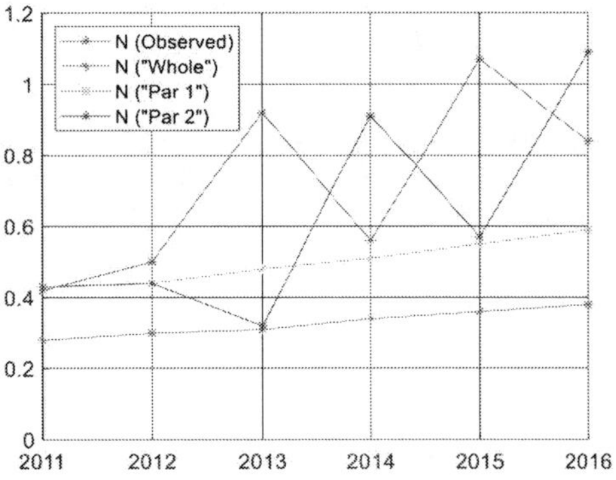

Fig. 4. Comparison between model results and observed concentrations, yearly averaged values (phosphorus).

Table 4. Phosphorus results from model simulation, considering only punctual pressures (total simulation duration average).

Year	P (Observed)	P ("Whole")	P ("Par 1")	P ("Par 2")
Total duration	0.70	0.33	0.50	0.63

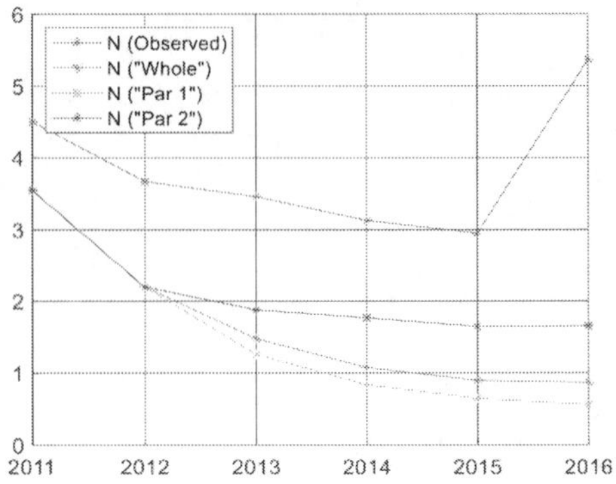

Fig. 5. Comparison between model results and observed concentrations, yearly averaged values (total nitrogen).

Table 5. Total Nitrogen results from model simulation, considering all pressures (total simulation duration average).

Year	N (Observed)	N ("Whole")	N ("Par 1")	N ("Par 2")
Total duration	3.54	1.68	1.51	2.11

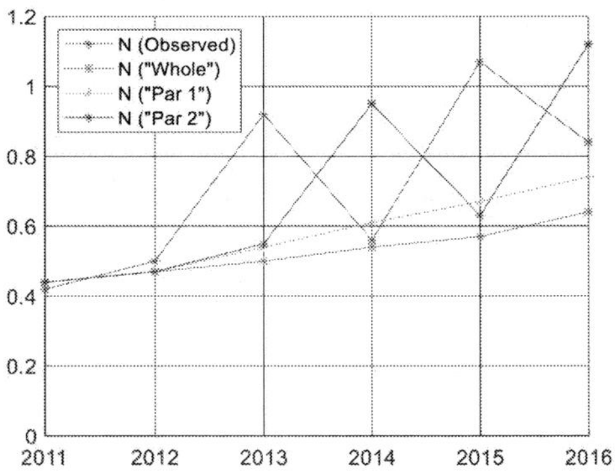

Fig. 6. Comparison between model results and observed concentrations, yearly averaged values (phosphorus).

Table 6. Total Phosphorus results from model simulation, considering all pressures (total simulation duration average).

Year	P (Observed)	P ("Whole")	P ("Par 1")	P ("Par 2")
Total duration	0.70	0.53	0.58	0.70

According to the examined period, considering oxygen data in Cixerri water body, and involving simulation outputs in terms of nutrients concentration, it is possible to estimate the water body environmental quality, in terms of LIMeco, for the considered years, and in relation with different types of simulation.

Table 7. LIMeco levels for the case study, comparison between observations and simulation results.

Year	(Observed)	("Whole")	("Par 1")	("Par 2")
2011	Level 4	Level 4	Level 4	Level 4
2012	Level 4	Level 4	Level 4	Level 4
2013	Level 4	Level 4	Level 4	Level 4
2014	Level 4	Level 3	Level 3	Level 4
2015	Level 4	Level 3	Level 3	Level 4
2016	Level 4	Level 3	Level 2	Level 4

As shown in Table 7, except for the last year of "Par 1" simulation, in which underestimation of concentrations seems to be too relevant, there is a good agreement between monitoring-derived judgement and simulation output-derived judgement. As described above, this is an important fact in view to use forecast modelling to make decisions about measures to take in a certain water body.

4 Conclusions

One of the main results of this study is the obtaining of better accuracy of the model when implementing the ecosystem components: AQUATOX simulates the whole eco-process, and it is not supposed to be used to reproduce nutrients transportation and transformation without a certain accuracy degree in characterizing the ecological component of the river, vegetation and animals. This could be noticed comparing results with other similar studies ("Validation of a simplified AQUATOX model to assess quality status of a model river in Sardinia").

In particular, these tests' results underline that anthropological pressures have an higher impact on the biota community and the ecosystem answer compared to the hydraulic behaviour of the stream (and in general to the total water volume of the water body), even in terms of nutrients transportation (and, even more, transformation).

Even if for the "Whole" simulation a decreasing accuracy in time can be noticed, it still remains a good representativeness in terms of LIMeco determination basing on results and so the possibility to involve AQUATOX and this method in long-term river water quality assessment and forecast.

References

1. L. Grechi , A. Franco, L. Palmeri, A. Pivato, A. Barausse, Eco Mod. **332**, 42–58 (2016).

2. R.A. Park, J.S. Clough, *Technical Documentation*, US EPA Office of Water (2014).

3. B. Lei, S. Huang, M. Qiao, T. Li, Z. Wang, Jour Env Sci, **20**, (2008).

https://doi.org/10.1051/epjconf/202226901052

Experimental investigation of multiphase hydrodynamics of the ocean-atmosphere boundary layer within laboratory modelling

Daniil Sergeev[1,*], *Alexander* Kandaurov[1]

[1]Institute of applied physics RAS, Geophysical Research Division, 603950 Ulyanova st. 46, Nizhny Novgorod, Russia

Abstract. Laboratory modelling of the processes of interaction between the atmosphere and ocean in the boundary layers is one of the most interesting from the point of view of the features of the hydrophysical experiment. The main characteristics that need to be controlled in these experiments include, first of all, the air velocity field over the rough surface waves and the underwater flow, as well as the shape of the free surface. However especial features as spray of droplets, the bubbles in the water and foam generated during the breaking of waves should also be taken into account when modelling extreme weather conditions associated with strong winds. Thus, from the point of view of experimental hydrodynamics, we are dealing with a multiphase turbulent flow with a free boundary. For purpose of experimental study on the wind-wave flume, an integrated approach is required that allows simultaneous measurements of the velocity flow fields, the sizes of droplets and bubbles, the shape of waves and parameters of foam coverage. Contact methods often do not allow for universality. This review study describes developing approaches to the use of optical methods for performing these studies. Presented results were obtained in experiments carried out on several wind-wave flumes. To study the processes of fragmentation of the water surface leading to the formation of droplets an foam, high-speed multi-angle video taking is used in combination with the shadow imaging method.

1 Introduction

The boundary layer between the atmosphere and hydrosphere (first of all the world ocean is meaning) of our planet is characterized by a huge number of various physical processes. From [1] a schematic representation is given, systematizing the numerous physical processes of the interaction of the atmosphere and the hydrosphere (see Fig. 1).

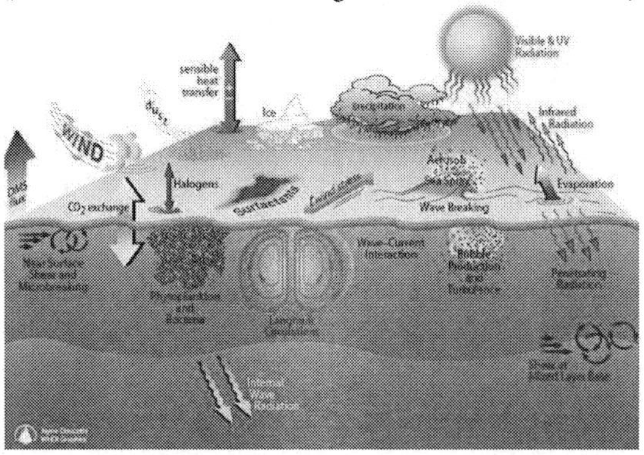

Fig. 1. Picture illustrating air-sea interaction from the website of Center on Air-Sea Interaction and Marine Atmospheric Sciences at Woods Hole Oceanographic Institution (see https://www2.whoi.edu/site/casimas/)

It is clear from it that all physical processes in the atmosphere and hydrophysical processes of the ocean cannot be separated from each other, and they are closely interconnected. For example, the wind in the marine boundary layer of the atmosphere accelerates the waves, while the aerodynamic drag coefficient (characterizing the momentum exchange in the boundary layer) from the sea surface depends on the wave parameters, which vary greatly not only in space but also in time, unlike similar processes over a solid land surface. Depending on the amplitude of the wind speed and the scale of its spatio-temporal variability, characterizing the fetch parameter, regular wave collapse can occur, with the formation of whitecap zones and foam crests, the generation of sprays and the involvement of air in water with the formation of bubbles that float and break the surface, also lead to splashing. Small micron-sized droplets carried away by the wind can levitate for a considerable time in the air, rising to considerable heights and forming a marine aerosol. Large drops, on the other hand, fall back onto the water surface and can "give birth" to small sprays. Spray significantly intensifies heat transfer compared to a smooth water surface, and therefore they are called "fuel" for hurricanes. Foam, also formed during the breaking of waves, also strongly affects the exchange of momentum and heat in the boundary layer. The collapse of waves under the influence of wind leads to the generation of turbulent flows and the intensification of mixing in the surface layer. This is one of the main

* Corresponding author: daniil@ipfran.ru

https://doi.org/10.1051/epjconf/202226901052

mechanisms of the formation of the temperature distribution - stratification in the water column, which in turn determines the temperature regime of the ocean, heat transfer with the atmosphere. Various relationships and chains can still be built quite a lot, but it is obvious that from the point of view of the tasks of experimental research of the processes of interaction between the atmosphere and the hydrosphere, the main problem is the need to simultaneously measure as many physical characteristics as possible with high spatial and temporal resolution, in a wide range of conditions including extreme (storms, hurricanes, etc.). it is quite difficult to solve this problem during in situ measurements. Therefore it is actively resorted to laboratory modeling, creating large-scale experimental facilities - wind-wave channels, see Fig. 2. At these facilities they are trying to realize conditions close to natural. However, in addition to realizing the conditions, it is also necessary to provide measurements of various quantities characterizing physical processes in the air flow, at the interface and processes in the water layer. Now, if we consider this problem from the point of view of a laboratory aero/hydrodynamic experiment, we can conclude that it is necessary to study the turbulent process of wind-wave interaction, dealing with a complex dynamic interface between the liquid (water) and gas (air) phases.

(a)

(b)

(c)

(d)

(e)

Fig. 2. Panorama view on the most known wind-wave flumes: a) SUSTAIN, University of Miami b) Large Air-Sea Interactions Facility (LASIF) Marsei University France c) TSWiWaT, IAP RAS, Russia d) Kyoto University High Speed Wind Wave Facility e) AEOLOTRON - circle wind-wave tank in University Heidelberg.

Among the main problems, it should be highlighted: 1) the need for measurements in air and water close to the interface, which varies significantly in space and time depending on the characteristics of the wind and waves 2) the need for measurements with high spatial and temporal resolution to obtain the necessary ensemble of data for subsequent statistical processing 3) the multiphase nature of the air and water flows due to the formation of sprays and underwater bubbles in strong winds with regular breaking waves. Such studies on wind-wave facilities require significant modernization and adaptation of standard, as well as the development of new methods of laboratory aero/hydrodynamic experiment. Optical methods based on visualization of flows compares favorably point measurements. It allows simultaneous observations in significant volumes of continuous media with high spatial resolution, and also do not introduce disturbances into the medium under study. Another important advantage of optical measurement methods is universality, which consists in the possibility of their application (without significant change) for the study of different processes, the characteristic spatial and temporal scales of which vary over a wide range. The present work is devoted to the history of development and application of these methods and recent results for the comprehensive study of the processes of interaction of the air flow with an rough water surface under laboratory modeling of extreme weather conditions, primarily associated with strong winds. Field measurements in such conditions are very difficult. That is why this study with laboratory modeling focused on the measurements of sprays and foam coverage under which are the features of the boundary layer under severe conditions, the measurements of which are often more complicated than the measurements of the characteristics of flows in the surface water layer. At high wind speeds, regular wave breaking begins with the fragmentation of the water surface, leading to the generation of sprays and the formation of foam on the surface. The first part of the work is devoted to the study of the processes of generation of spray droplets from the rough surface, and their characteristics. The background of research and the current state are described. The second is to study the characteristics of the foam and methods for its artificial creation in laboratory experiments (the results of recent studies are presented).

2 Investigation of spray within laboratory modelling of the interaction between atmosphere and hydrosphere

Sea sprays are typical element of the marine atmospheric boundary layer and important environmental effect. Difficulties of direct measurements in hurricane conditions and insufficient knowledge about the mechanisms of the spume droplet's formation are the main reasons for insufficient knowledge on the SSGF, which characterizes the size distribution of droplets injected from a unit of surface in a unit of time, leading to significant uncertainties in estimations of sea sprays influence on the marine atmospheric boundary layer [2, 3]. For example, the empirical function of spray generation may differ by six orders of magnitude based on different observation reports (see a review of experimental data in [4, 5]). Even the spray-generation mechanisms in extreme winds remained undetermined up to present time. Such difficulties are primarily associated with the problems of obtaining a sufficient amount of empirical data not only in field, but also in laboratory modeling. The development of methods for measuring the characteristics of sprays in laboratory modeling can be considere be considered as competition between point methods and panoramic methods based on visualization and photo/video filming.

Initially, the main source of information about the quantitative characteristics of the spray of droplets generated during fragmentation of the water surface within wind-wave interaction was the measurement results basing on the point methods (see Fig. 3). In [6], the technique of electric wire microgauges was used, similar to hot-wire anemometric sensors, the principle of which was to register an abrupt increase in voltage (decrease in resistance) when droplets hit the sensor. In a similar study [7], a sensor was used based on charge transfer during the contact action of a drop on a wire under high voltage, which causes charge transfer and a voltage drop. In [8], a laser-optical sensor was proposed that works in light when shaded by droplets crossing its beam. It allows to to evaluate not only the size, but also the speed of the droplet in contrast to the previous ones. The electronic and optical sensors listed above had rather low accuracy and allowed measuring only sufficiently large particles (low threshold from 50-100 μm depending on the work), although they outperformed the measurement methods based on photo / video shooting in similar experiments at the same time. The measurements were also carried out for weak to moderate winds (the equivalent wind speed U_{10} at a height of 10 m is less than 15 m/s), when the spray concentration was low. Also the were always performed above the wave crests, i.e. the area of throughs remained unexplored. Significant progress was achieved using the PDA method [9], which was developed at the end of the 80s. Unlike the LDA method (a modification of which is), PDA made it possible to measure size of microparticles in a continuous medium (multiphase flows) in a wide range in addition to measurement velocities. So in [10], droplet sizes were measured in the range of 5–120 μm, with a resolution of only 0.5 μm. Moreover, the wind speed U_{10} was already more than 30 m/s.

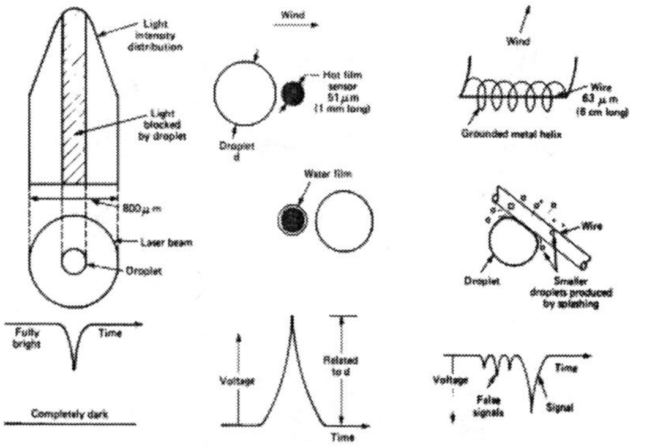

Fig. 3. Optical and electronic sensors for spray measurements: a) from work [8], b) [6], c) [7].

Photography and video began to be used even before contact and non-contact point measurements. In the early 1950s, high-speed imaging was first used to study the generation of droplets by bursting underwater bubbles in a liquid without flows (see Fig. 4a in [11]). And then, in the well-known [12], Toba used photography to study the processes of generation of droplets by bursting bubbles within laboratory modeling on a wind-wave flume. Koga [13] used the MOET method for visualization. This method is based on sequential pulse exposure of the same image with light of a different range (two pulses of the red and blue range were used), see the example in Fig. 4 b. This made it possible to single out a new mechanism for generating sprays, the so-called projection, which consists in breaking a drop directly from the crest of a breaking wave when growing a liquid thread (sometimes they say filament) along the direction of the wind, which ends with a separation of 1-2 large drops. The distribution of projections by size and speed was obtained. Also, an analysis of the displacement between images in different colors makes it possible to find the velocities of droplets (see Fig. 4 c).

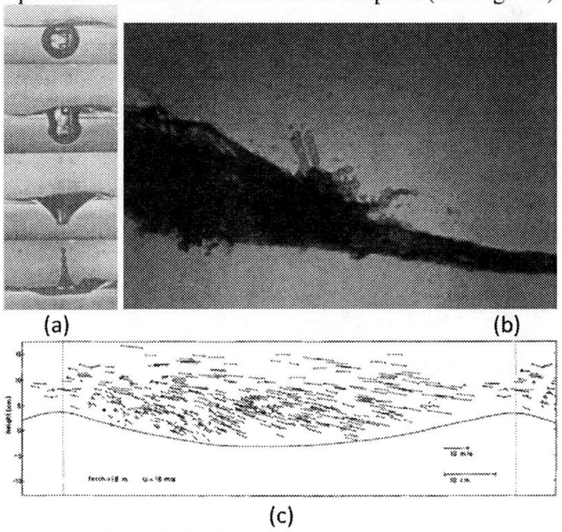

Fig. 4. a) Form [11]. Bursting bubbles produced sprays b) From [13] photographs of projections obtained with MOET technique. Wind direction is from left to right. Successive two images are shown by blue and red colors. One division of the photographed measure (right end) is 1 cm. Fetch F = 16 m, representative wind velocity V=.16 m/s, flushing interval t = 5 ms. c) Movements of directly produced droplets along the representative wave. Each droplet in the figure indicates actual size traced from photographs. Arrows indicate the droplet velocity vector in a coordinate moving with the phase speed of the wave.

It should be noted that despite the high resolution of the film photos used, and the possibility of digitizing the image with subsequent computer processing, which appeared in the 80s, the analysis of such images was very robust. Given the sporadic nature of the occurrence of droplets, with light and moderate winds, the data obtained with filming were often not enough for correct statistical processing in these experiments. Therefore, the information for constructing the spray generation function was taken from the data of point measurements. At the end of the 90s, a great common shift to digital video technology began in a physical experiment. [14] was one of the first works in which a study of generation of sprays was carried out on a wind-wave flume using digital video equipment. Images were obtained with a resolution of 650 μm, which did not greatly exceed 800 μm in the above-mentioned work [13], and allowed only very large droplets (more than 1 mm) to be correctly identified. The shooting speed of 30 frames per second did not allow to restore the drop shift between the frames and determine its velocity. But in this work, an array of data with the image of breaking waves was almost 20 times higher than in the previous work, which significantly increased the accuracy and brought it closer to point measurements. The work [15] can be considered as milestone study on the development and application of methods based on visualization. For this study data were obtained for a droplet with a size from the threshold of 140 microns for equivalent wind speeds of up to 33 m/s, due to the higher spatial resolution. But most importantly, that due to high-speed filming and a large ensemble of implementations, it was possible for the first time to observe a new mechanism of fragmentation of the water surface under the influence of wind, the so-called bag breakup instability mechanism (see Fig. 5) leading to the formation and breaking of sails, which was previously known only in engineering problems associated with the fragmentation of droplets and jets in air flows. Also the shadow method was used for visualization in [15].

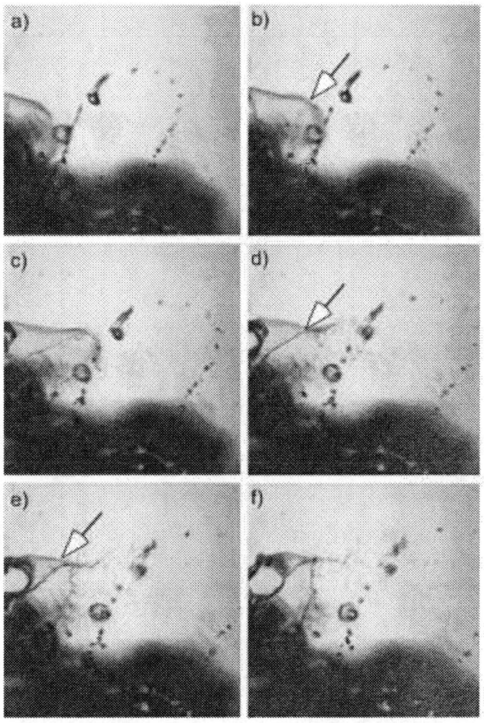

Fig. 5. From [15]. Images of the surface and spray generation taken with the high speed camera at a rate of 1000 fps. Each imageis 6.38*6.38 cm and separated by 1 ms. The arrow shows a liquid sheet inflated by the airflow and its subsequent fractionation that yields to the generation of small water drops which are transported in the airflow. The 10-m equivalent wind speed is $U_{10} = 31.3$ m/s.

The main feature of the studies performed by our group in the last three years ([16-17]) was simultaneous multi-view high speed and resolution filming of the water surface (see Fig. 6 a), also with a shadow visualization method with backlighting by powerful LED spotlights.

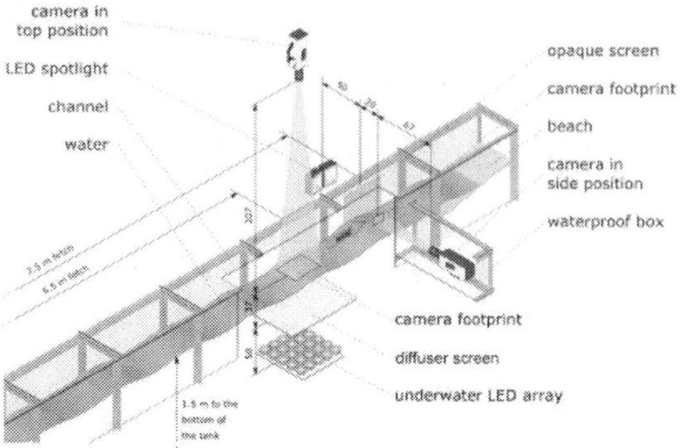

(a)

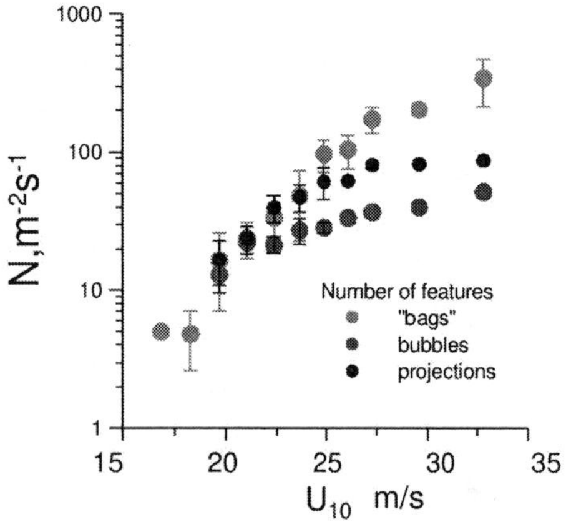

(b)

Fig. 6 a) Schematic diagram of experimental setup on TSWiWaT, from [16], b) Dependence of the specific numbers (per unit time per unit area) of the spray-generating phenomena on the wind speed. Blue circles indicate bursting of floating bubbles, black the liquid filaments, and red the bag breakup events is adapted from [17].

Experiments were carried out on a unique facility of the TSWiWaT. This level is a highest for the present time in experiments on the study of sprays in the conditions of laboratory modeling wind-wave interaction. Basing on the results of [16, 17], the classification of phenomena (projections, bursting bubbles, bagbreakup) leading to the generation of sprays was clarified. It was possible to obtain statistics of precisely the phenomena leading to the generation of splashes due to the additional filming of a top view of the water surface illuminated by underwater lights, which was carried out for the first time. It demonstrated that at high wind speeds the bag breakup mechanism is dominant (see Fig. 6 b).

The obtained statistics data allowed us to offer a fundamentally new approach to obtaining the necessary SSGF. It can be obtained as a combination (convolution) of the statistics of phenomena leading to the generation of sprays and statistics of the generation of sprays in a single phenomenon. At the same time, it is rather difficult to study in detail a single phenomenon under natural conditions of wind-wave interaction, even in laboratory modeling, given its sporadic nature and the difficulty of performing measurements, especially at high wind speeds. Therefore, the idea was proposed to artificially initiate these phenomena under controlled conditions for detailed study. In our last study [18], we succeeded in realizing the artificial initiation of the bag-breakup phenomenon in the modified TSWiWaT. Arterially produced bag-breakup phenomenon turned out to be close to that observed under natural conditions of the wind-wave interaction (see the comparison in Fig. 7a, b).

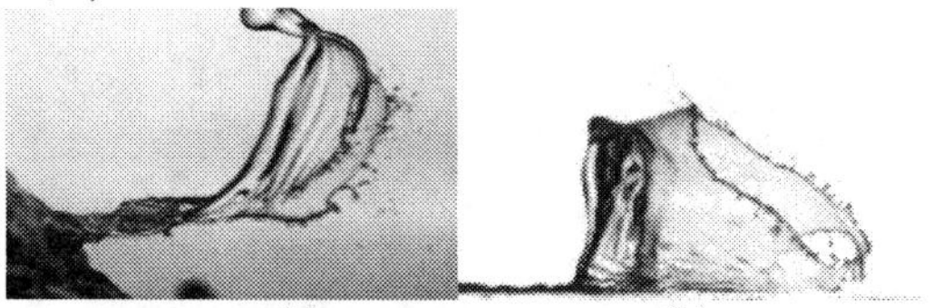

(a) (b)

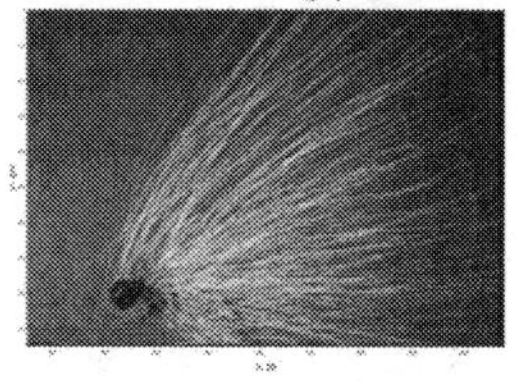

(c)

Fig. 7 a) bag-breakup on the real wavy surface b) artificially induced bag-breakup c) trajectories of the droplets generated by bag-breakup.

New algorithms were developed for identifying droplets, plotting trajectories, and estimating velocities (see Fig. 7 c). However, due to the fact that the phenomenon has a very complex spatial structure, the resolution for observing a drop of size less than 10 μm in the experiments was not enough. Therefore, at present we use the results of previous measurements of droplets induced by bursting bubbles, when they are artificially generated (see [19]), as well as splashes generated by crushing a liquid filament (see [20]) to construct the SSGF. At the same time, the problems of a detailed study of the bag breakup mechanism and refinement of the SSGF based on it are planned to be solved in subsequent works with the artificial initiation of the phenomena leading to the generation of spray of droplets.

3 Methods of laboratory modelling and measurements phenomena of foam coverage over the water surface

Foam (or whitecapping), like spray, is almost always present on the sea surface, with the exception of only very weak waves and wind. In its characteristics, it is very different from both water and air, which determines its special role in the processes of exchange between the atmosphere and the sea surface. The main mechanism for the formation of foam is a strong breaking of the waves (macrobreaking). It leads to intensive air entrainment, the formation of underwater bubbles and their whipping into foam. Foam is also formed due to large-scale near-surface currents, for example, Langmuir circulation. By combining various factors, the relative area of the foam coverage increases sharply in hurricane winds (U10 more than 30 m/s) and exceeds 70% (see the results of field measurements in [21]). In [22], basing on the results of [21], a quantitative model was proposed for describing the contribution of foam to momentum exchange, in which the aerodynamic roughness parameter due to the foam coverage was determined through the size distribution of the bubbles. However, such data obtained in natural conditions are characterized by a large scatter, which complicates their use in the development of theoretical models. Therefore, in the present work, a system for the artificial generation of foam was first developed, as well as a technique for measuring the characteristics of the foam (relative coverage area and bubble sizing) specifically for research in laboratory modeling.

Experiments to study the characteristics of the foam in the framework of laboratory modeling were carried out at the TSWiWaT. The general scheme of the experiments is presented in Fig. 8 a. In conditions of clean water, as well as small accelerations, it is difficult to ensure regular generation of foam. Therefore, to create a stable foam with the necessary density on the surface of the water, a special device (foam generator) was developed (Fig. 8 b). The device consisted of two diffusers (tubes 35 cm long with a diameter of 1 cm with side holes with a diameter of 2 mm with a pitch of 7 mm) connected by side surfaces to each other, wrapped in foam rubber. The device was located horizontally, perpendicular to the channel, with a depth of less than 5 mm. The distance from the beginning of the flume (air flow inlet) is 1.2 m. A solution of a foaming substance enters through one of the diffusers under a pressure of a height of 1.5 m, compressed air (1.5 atm) is supplied through a second diffuser to force the solution of the foaming substance through a layer of foam. As a result, finely dispersed foam forms on the surface of the water. During the experiments, the level of the solution of the substance and the pressure of the compressed air were kept constant, which ensured a constant foam rate. A special series of measurements aimed at assessing the effect of the foam generator on surface waves in the case when only compressed air is supplied to the system showed that changes in the parameters of surface waves compared to a completely disabled foam generation system are insignificant.

The foam characteristics were measured in the seventh section of the channel (6.5 m acceleration) for 6 different values of the equivalent wind speed U10 from 12 to 34 m/s. To visualize the foam on the surface of the water, we used the shadow method of visualizing processes on the rough surface of the water with backlight under the water, which has proven itself in studying the statistics of spray generation phenomena (see earlier), the general scheme is shown in Fig. 8 (a). More detailed description of the experimental setup please find in [23].

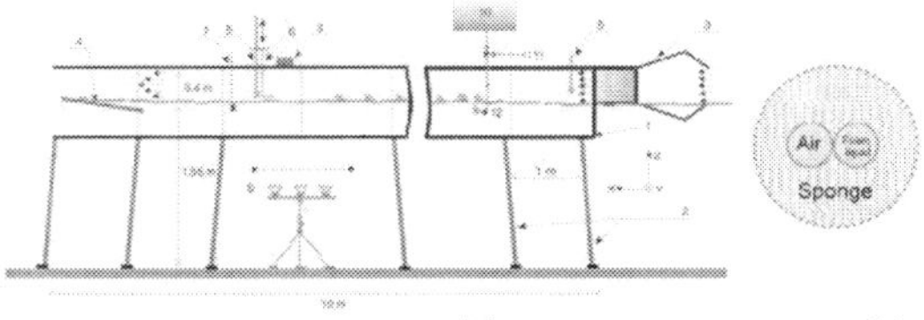

(a) (b)

Fig. 8. Experimental setup: 1, flume; 2, supports; 3, honeycomb; 4, wave-damping beach; 5, hot-film probe; 6, pitot tube; 7, threechannel wave gauge array; 8, high-speed camera; 9, illumination system; 10, SLS solution reservoir; 11, compressed-air supply; 12, foam generator. Sizes are in meters. (b) Enhanced cross section of the foam generator.

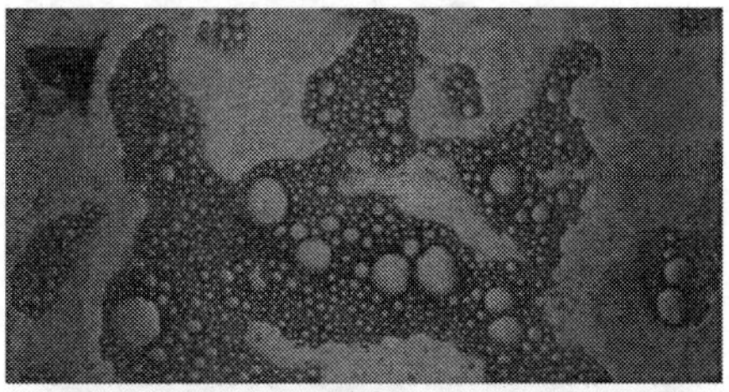

(a)

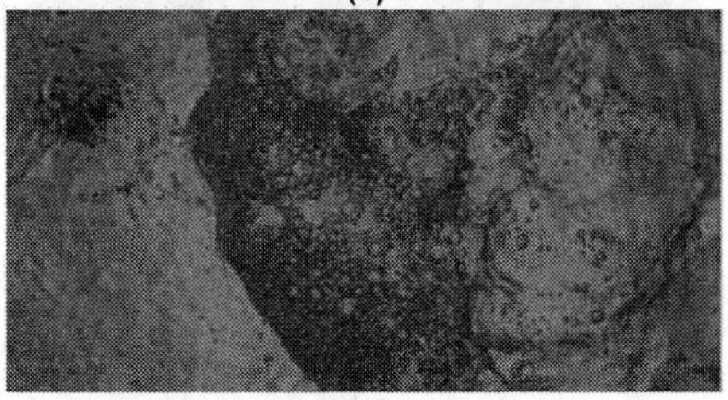

(b)

Fig. 9. a) The photo of the water surface covered with foam for wind speed U_{10}=12 m/s (b) U_{10}=35 m/s

Opaque screen was installed, which is highlighted from below by an array of LED power lamps in the working section of the channel. The size of the shooting area was 231 × 123 mm, the scale was 113 μm/px, and the average frame rate was 50 fps. Such performance was quite enough to study the statistical characteristics of the foam.

The resulting digital images (see examples in Fig. 9 a, b) were processed with specially developed algorithms.

The fraction of the surface area covered with foam was calculated on the basis of the number of pixels darker than a certain threshold. To eliminate the influence of noise and small optical inhomogeneities, the original images were divided into regions, in each of which the average brightness was calculated, ie, the resolution of the frames decreased 50 times on each side.

Using reference records made without wind, one background image was calculated as the median value for each pixel. A background image was subtracted from each frame, then a histogram common to all frames was constructed, i.e. distribution of areas of all images by brightness. For each histogram, one can calculate the fraction of regions with brightness below the threshold, which naturally depended on the choice of the threshold.

To determine the threshold, a direct calculation of the foam coverage area using morphological image processing was performed. It was carried out in several stages shown on Fig 10 a-f. The background was subtracted from the original image and cropping was carried out; the image was binarized with a fixed threshold; the morphological opening with the structural element "disk" of 1 px was performed followed by the morphological closure with a "disk" of 21 px; closed regions of the image were filled with flood-fill operation. The ratio of the number of pixels found to the total number of pixels was taken as the fraction of the surface area covered with foam.

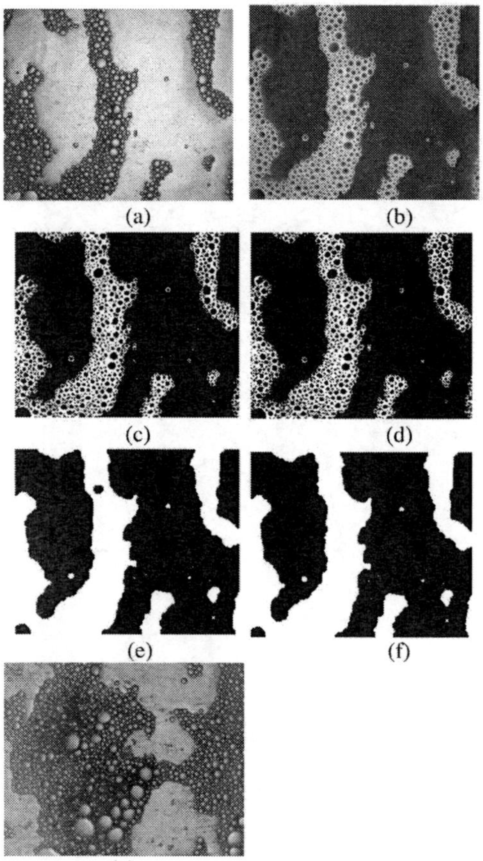

(a) (b) (c) (d) (e) (f) (g)

Fig. 10. Fig. 10 (a-f) sequences of stages of morphological analysis of the images to determine the relative foam coverage (g) an example of bubble determination.

Based on this analysis, a relative coverage was found for sequences of frames from several records in experiments with artificial foam at U10 = 12 m/s. For each record, a relative brightness threshold is selected for analysis by histograms, which gives the corresponding coverage. Then this threshold value averaged over several over all realizations for U_{10} = 12 m/s was used for all cases of wind speeds.

Information about the characteristic size of the foam bubbles was obtained additionally to data on the area of foam coverage. From each record, several frames were selected, on which the foam is clearly visible. Areas were identified on which the bubbles merge (located in several layers) and others unsuitable for automatic determination of the area. For the remaining areas of the images after the preliminary filtering, all circles were found (see Fig. 10 g) by two-stage circular Hough transform (see [24]) and their distribution in radius was calculated.

4 Conclusion

The history of the development of methods for studying the processes of fragmentation of the water surface under wind forcing demonstrates that methods based on visualization of flows and photo/video filming helped us to study microphysics of boundary layer within laboratory modeling for severe conditions. It allow to identify and study in detail the main phenomena that lead to the generation of sprays and obtain information about the characteristics of the foam in contrast to point methods. Multi-angle high speed video filming with a shadow method of visualization, including a top view of the rough surface with backlighting from under water, provided data on statistics of the main phenomena leading to spray generation (projection, bubbles, bag-breakup), as well as data on the area of foam coverage and bubbles size. However, it should be noted that for further advancement in the direction of obtaining information about the SSGF, when taking measurements under conditions of artificial initiation of the phenomena leading to spray generation, it is advisable to use a combined approach that combines visualization methods (high-speed filming with the shadow method) and the most advanced point measurement methods (e.g. PDA). This will provide statistics on the size and velocity of the droplets formed during various types of single phenomena of surface fragmentation and spray generation, which will supplement the previously obtained statistics of the phenomena themselves.

Carrying out experiments themselves was supported by Russian Science Foundation (Agreement No. 19-17-00209); designing of methods of measurements including: optical and visualization scheme, methods of measurements of the air flow and wave field parameters were supported by Russian Foundation of Basic Research (No 21-55-52005, 20-05-00322); the development of the software was supported by the Grant of the President no. MK - 5503.2021.1.5, and the work under processing of the video was supported by Russian Science Foundation (Agreement No.21-19-00755).

Abbreviations

LDA - Laser Doppler Anemometry
MOET - Multi-colored overlapping exposure technique PDA - Phase Doppler Anemometry
SSGF - Sea Spray Generation Function
SUSTAIN - Surge-Structure-Atmosphere Interaction
TSWiWaT - Tthermostratified Wind-Wave Tank

References

1. https://www2.whoi.edu/site/casimas/
2. Andreas, E. L. & Emanuel, K. A. J. Atmos. Sci. 58, 3741, (2001).
3. Bao, J.-W., Fairall, C. W., Michelson, S. A. & Bianco, L. Mon. Weather Rev. 139, 3781, (2011).
4. Andreas, E. L. A review of spray generation function for the open ocean in Atmosphere-Ocean Interactions Volume 1 (ed. Perrie, W.), WIT Press 46, (2002).
5. Veron, F., Annu. Rev. Fluid Mech., 47, 507,(2015).
6. Lai, R. J., and O. H. Shemdin, J. Geophys. Res., 79, 3055,(1974).
7. Wang, C. S., and R. L. Street, Dyn. Oceans Atmos., 2, (1978).
8. Wu, J., Spray in the atmospheric surface layer: Laboratory study, J. Geophys. Res., 78, 511, (1973).
9. Bauckhage, K. Part. Part. Syst. Charact., 5, 16, (1988).
10. M. L. Banner, W. L. Peirson, W. Asher, and R. P. Morison, J. Geophys. Res., 114, C10001, (2009).
11. AH Woodcock, CF Kientzler, AB Arons, DC Blanchard - Nature, 172, (1953).
12. Toba, Y., Drop production by bursting of air bubbles on the sea surface, III, Study by use of a wind flume. Kyoto, Ser. A, 29, 313, (1961).
13. Koga, M. Tellus, 33, 552, (1981).
14. Anguelova, M., R. P. Barber, and J. Wu, J. Phys. Oceanogr. 29, 1156, (1999).
15. Veron, F., Sea spray spume droplet production in high wind speeds. Geophys. Res. Lett., 39, L16602. (2015).
16. Troitskaya, A. Kandaurov, O. Ermakova, D. Kozlov, D. Sergeev, and S. Zilitinkevich, Sci. Rep., 7, 1614, (2017). https://doi.org/10.1038/s41598-017-01673-9
17. Yu. Troitskaya, A. Kandaurov, O. Ermakova, D. Kozlov, D. Sergeev, and S. Zilitinkevich J. Physical. Ocean., 48 (9), 2167 (2018). https://doi.org/10.1175/JPO-D-17-0104.1
18. Troitskaya, Y.I., Ermakova, O.S., Kandaurov, A.A. et al. Dokl. Earth Sc. 477, 1330. (2017) https://doi.org/10.1134/S1028334X17110174
19. Lhuissier H., Villermaux E. // J. Fluid Mech. 696, 5, (2012).
20. Marmottant P., Villermaux E. // J. Fluid Mech. 498. 7 (2004).
21. Holthuijsen L. H., M. D. Powell and J. D. Pietrzak J. Geophys. Res. 117. C09003, (2012).
22. E.Golbraikh and Y. M. Shtemler Dyn. Atmos. Oceans. 73, 1, (2019).
23. Yu Troitskaya, D. Sergeev, A. Kandaurov M. Vdovin, S.S. Zilitinkevich J. Physical. Ocean., 49(4), 959, (2019). https://doi.org/10.1175/JPO-D-18-0168.1
24. H.K Yuen, .J. Princen, J. Illingworth, and J. Kittler. Image and Vision Computing. 8(1), 71,(1990).

PIV-system for flow measurements on the benchmark of promising fast neutron nuclear reactor

Daniil Sergeev[1,*], *Alexander* Kandaurov[1]

[1]Institute of applied physics RAS, Geophysical Research Division, 603950 Ulyanova st. 46, Nizhny Novgorod, Russia

Abstract. Carrying out spatio-temporal flow fields parameters is one of the most important and complicated problem within the modeling of heat and hydrophysical processes in operating regimes of the power plants (including nuclear (NPP)). Particle Image Velocimetry (PIV) method (based on the flow visualization, using laser illumination) fits best to satisfy the demands of such measurements. For the experiments on large-scale models of NPP, there is a number of problems in the application of this method, for example, large scales of the flows of the investigated model objects; a noticeable variability of the flow parameters processes with strong temperature fluctuations, leading to optical distortions; necessity of combining imaging and contact gauges probing etc. In this paper we describe PIV- system specifically designed for the study of heat and hydrophysical processes in large-scale model setup of promising fast neutron reactor. The system allows the PIV-measurements for the conditions of complicated configuration of the reactor model, reflections and distortions section of the laser sheet, blackout, in a closed volume. The illumination system using of two lasers and two video cameras with synchronization, allows to obtain an image of the flow in the whole volume of investigation. The use of filtering techniques and method of masks images enabled us to reduce the number of incorrect measurement of flow velocity vectors by an order. The method of conversion of image coordinates and velocity field in the reference model of the reactor using a virtual 3D simulation targets, without loss of accuracy in comparison with a method of using physical objects in filming area was released. The results of measurements of velocity fields in various modes, both stationary (operational), as well as in non-stationary (emergency). These results are compared with measurements of temperature distribution, which were carried out simultaneously. The obtained data set can be used for verifying CFD codes.

1 Introduction

In engineering practice, models of different scales and complexities are often used to research thermohydraulic processes in Nuclear Power Plants (NPP).

In this case, the research can be subdivided into three basic types:

(1) the simplest type is the simulation of individual, often small nodes, for example, of the flows in pipes and tees [1, 2];

(2) the simulation of thermohydraulics of individual elements of NPS, for example, the investigation of processes in the fuel-element assemblies [3-5];

(3) the most complicated type includes benchmarks - complex large-scale models of NPP [6-8].

The main objective of all investigations on the models is to try to simulate the actual situation as closely as possible, study it under controlled conditions and, if needed, verify the calculation procedures. Recently, Compute Fluid Dynamics methods have been used more often for the numerical simulation of thermohydraulic processes in NPP. The most complete and representative verification of the numerical simulation results can be performed using experimental data about velocity fields. The Particle Image Velocimetry (PIV) is the most modern method for their measurement. There are many variations of this method [9]; its essence lies in the visualization of flows by adding microparticles to them, illumination of required sections with a laser sheet, filming the visualized field of view with one or more digital cameras, and software processing to calculate velocity fields. During the laboratory experiments of the first and the second types, the standard serially produced measuring PIV systems are usually applied with minimal modifications [1-3]. However, in the experiments on large-scale benchmarks, a number of difficulties occur that are associated, for example, with a complex configuration and the large volumes of the investigated model objects, a wide range of changes in the parameters of thermohydraulic processes with strong temperature fluctuations, resulting in optical distortions, with the necessity of combining the visualization and contact measurements, etc.

There is no published data about investigations on the characteristics of using PIV methods on test benches that are comparable in their parameters with the model under consideration here. Therefore, the objective of this

* Corresponding author: daniil@appl.sci-nnov.ru

work is to fill this gap in using PIV methods. A special system and methods of measurements were developed.

2 General description of the reactor model

In order to confirm the effectiveness of the emergency heat-removing system of a proposed fast reactor, the TISEI test bench was developed and manufactured and an experimental study of heatexchange and hydrodynamic processes was performed at the OKBM Afrikantov. The test bench consists of a large-scale model of a reactor and the loop of emergency heat-removing system. The external outlines and elevations of the test-bench equipment and reactor are geometrically similar; the simulation scale was taken equal to 1 : 5. Water is used in the test bench as a coolant unlike in a regular plant, in which liquid sodium is used. The power of the core model in the rated operation mode does not exceed 350 kW. To use visualization-based methods, polyamide particles with a 1.03 g/cm3 density and 20 μm in diameter were added to the water.

The reactor model is a sector with an opening angle of 80 degrees; it includes all of the basic components of a proposed fast reactor (Fig. 1). The test bench was described in more detail in [8].

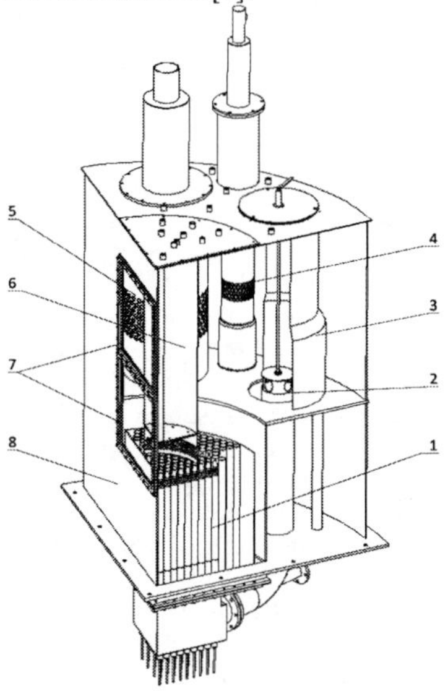

Fig.1. The model of a proposed fast reactor: (1) core with heating elements; (2) valve with manual drive; (3) the displacer of the main circulation pump of the primary coolant circuit (a part of the displacer is shown only); (4) independent heat exchanger of the emergency heat removingsystem; (5) intermediate heat exchanger; (6) the displacer of the central column; (7) transparent windows; and (8).

3 Scheme and hardware of the PIV-system

The internal chamber of the model of the reactor tank sector has a very complex configuration. In addition, thermal probes are installed inside it (these are not shown at Fig. 1), which consist of vertical rods with a diameter of ~7 mm on which the temperature gauges are fastened. The problem was posed to obtain the velocity fields in a maximally large volume area inside the tank, simultaneously using both the top and the bottom windows for measurements. The measurements should be carried out in the sections on the independent and intermediate heat exchangers. The arrangement of laser illumination and filming of the required sections in the presence of numerous shading and reflecting elements is a complicated problem. Taking the features of the construction of the reactor model into account, the configuration of the illumination and video filming was chosen simultaneous using two compact continuous diode-pumped lasers and two cameras (see Fig. 2). The general arrangement of applying the PIV method is similar to that described in [10].

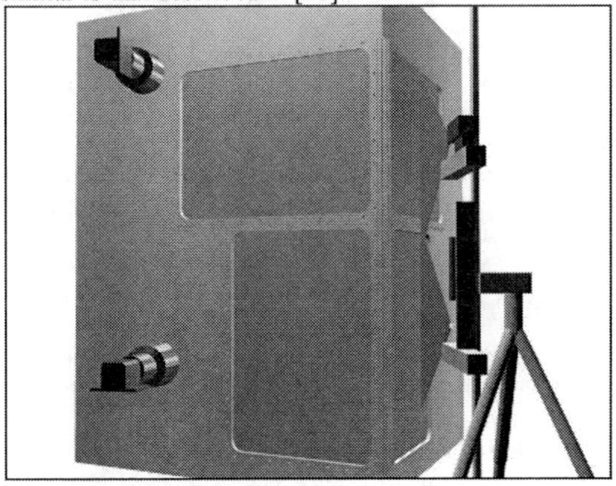

Fig. 2. The scheme of camcorders and lasers disposition (isometric projection).

Solid-state diode-pumped CW lasers in the visible range of the LCS-DTL-413 type were used. The emission wavelength was 527 nm (green light) and the laser power was 1.5 W. Two highly sensitive CygnetRaptor camcorders were used for video filming (the rated resolution was 2048×1088 pixels at a frame rate of 50 frames/s). Both lasers and optical elements were installed on a single tripod (Fig. 2), which greatly facilitated the displacement of the system elements for the replacement of the illumination section.

The required sections of view on the independent and intermediate heat exchangers were chosen taking in to account the structural features of the test bench and thermal probes used regularly to measure the temperature distribution. Fig. 3 illustrates the top view to the cross-section.

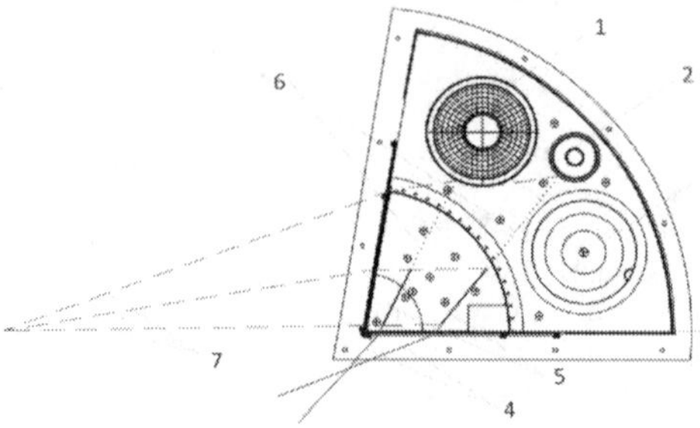

Fig.3. The outline of the sections and filming region inside the test bench (the top view of the test bench): (1) intermediate heatexchanger; (2) independent heat exchanger; (3) the displacer of the main circulation pump of the primary coolant circuit; (4) sections of laser planes before the inlet to the reactor-tank model; (5) filming regions that are under the central rotary column on the images; (6) the remaining part of the image over the entire height; and (7) the contours of boundaries of the optical image. The circles that are crossed out denote the positions of the thermal probes.

The positions of the cameras was chosen based on the necessity to maximize the dimensions of the field of vision. The bottom camcorder was placed at the level of the bottom of the central column and the top camcorder was placed in the middle of the height of the top window of the model of reactor. To reduce the distortion level to the same scale and dimension of the filming area, the camcorders were installed at the maximum possible distance from the reactor sector.

For filming through the bottom window, the camcorder was installed in the usual orientation, since the area was horizontally stretched to provide the ability to capture the area under the central column; for filming through the top window, the cameras was turned by 90 degrees to maximize the vertical dimension. Fig. 3 presents the general arrangement of the cameras and lasers from different angles. The plane of filming the sections on the independent and intermediate heat exchangers was located at an angle to the optical axis of 51 and 62 degrees, respectively. Therefore, in order to obtain a focused image in the entire area, special lenses of the tilt-shift type were applied with the ability for mechanical rotation and displacement of the lens axis relative to the plane of the receiver of the CCD camera.

Visual fields of the bottom and top cameras partly overlap. This is done for the further linking of the measured velocity fields. It was revealed during the preliminary experiments that using a frame rate below the maximum value of 50 frames/s leads to the impossibility of correct calculation of particle displacements as they are moving away from the laser sheet. The cause is both the high value of the velocity component that is normal to the laser sheet and thermal distortion, which can lead to a local displacement of the laser plane. Therefore, the maximum rate was chosen with the conservation of the base resolution for all modes.

4 Image preprocessing and determination of the velocity fields

To calculate the velocity fields from the obtained digital images, the standard for PIV-method digital image processing was used [9]. The images were partitioned with a rectangular grid into fragments (interrogation windows) with a size of 64×64 pixels. For the functions of the intensity distribution, the cross-correlation function (CCF) was then computed on the images of two interrogation windows with the same location on a pair of consecutive frames. The location of its maximum corresponded to the average value of the displacement in the region of this fragment between frames.

By dividing the displacement value the time interval between the frames, one can define the velocity. The field velocity in the investigated flow is characterized by strong heterogeneity (for example, see Section 5). The subpixel adjustment of the location of the CCF maximum at three points that are the closest to the peak was used by means of a 2D Gaussian function (for more detail, see [9]).

To reduce the influence of the features of the background image on the results of the CCF computation, the background was subtracted. A dynamic procedure of subtraction of the background was used. The background image for each frame was calculated on the basis of several neighboring frames: the so-called sliding window method was applied. The window dimension equals 11 frames: the initial frame, 5 previous frames and 5 subsequent frames were used for calculation. The minimum value of the intensity at this point for the entire set of 11 frames in the window is chosen for the intensity level corresponding to the background image. In this way, all bright objects that are displaced during the calculation period are removed from the image: if the particle image is lacking even in one of the frames on this pixel (in the case of the movement of particles), the pixel will take the value of the background image.

Then, when calculating the CCF in each experiment (record), at first, the uniform rectangular grid with a pitch of 32 pixels is formed (50% overlap for the comparison window with a dimension of 64×64 pixels).

However, there are fields in the images in which the images of particles were overlapped by construction elements, including the thermal probes (the image configuration is complicated). Therefore, it is not necessary to carry out calculations at the grid nodes that fall within these fields, as erroneous displacement (velocity) values are derived in any case. To avoid unnecessary calculations at the grid formation, a masking method was developed. The mask is a file obtained from the initial image by binarization: the null values correspond to the masked regions, which must be excluded from the calculations (particles are not visible in these regions); the nonzero values correspond to the regions in which particles are visible and in which their displacement can be calculated. Fig. 4 shows example of such mask.

Before the generation of a grid of interrogation windows for calculating the CCF, a rectangle that includes all of the unmasked points was found in the image for each mask. Hereafter, the interrogation windows to calculate the CCF were chosen only within this rectangle. In this case, the CCF was derived only for interrogation windows that consist of at least half of the unmasked points. To calculate the velocity field, the maximum possible number of frames was used. Successive frames were chosen for comparison and each frame was compared with the following frame. Thus, for a subrecord consisting of 50 frames (1 second of record), 49 velocity fields were computed and such calculations were carried out for all subrecords in the record.

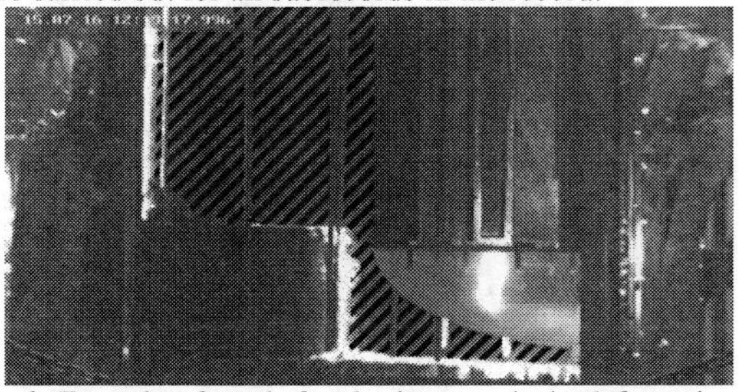

Fig. 4. Example of mask for the images obtained from the bottom camcorder. The shading denotes the fields that are not covered with a mask, i.e., suitable for calculating the displacements of particles.

As a result of the above-described processing procedure, the velocity fields were obtained for each image (except the last one in the subrecord). However, despite the struggle with static noise by subtracting thebackground and using precise masks, correct measurements of the particle displacements were not performed for each interrogation window. The final stage of filtration of the velocity values was performed via their difference from the time-average subrecord value. However, for dynamic modes, such filtering is inadmissible because the true velocity value at the points may change considerably. Therefore, the technique of two-pass median filtering based on the subrecords was developed. Processing was performed for each subrecord for each value of displacement.

The value of the median average was determined at the first pass. All of the values of the displacement whose difference from the average value exceeds the chosen threshold value of five pixels per frame were then rejected in the subrecord. The second stage is similar to the first one, but the average value was calculated from already filtered values and was closer to the true velocity value. The filtering interval at the second pass was chosen to be narrower at two pixels per frame.

After all of the processing stages, the instant velocity fields (in pixels per second) were derived, which were used for comparing the neighboring frames. These contained many missing values (due to errors); thus, the fields derived by averaging over time intervals of the subrecord were used as the main working fields. It should be noted that due to the sequential recording with two cameras, the velocity fields obtained in this way were shifted in relation to each other by a time of 1 second.

5 Rescaling between image and laboratory system

The real dimensions and coordinates of the obtained images in the reference system locked to the laser cross-section are unknown in the reference of reactor model. In addition, it should be noted that during processing by PIV methods, all calculations are performed in the coordinates of the image and dimensions are expressed in pixels. Therefore, the flow displacements (velocities) obtained as the result of the processing are expressed in pixels (pixels per frame).

To recalculate the coordinates of the velocity - field grid nodes (the coordinates of the centers of interrogation windows) and velocities themselves into the coordinate system of the model, one or two (in the case of anisotropy of the camera sensor) scale factors are typically used. However, the features of the filming configuration (at an angle to plane, see Fig. 3) do not make it possible to introduce such a factor, as the scales of different parts of the image are very different. For correct recalculation, the transformation of the coordinates from the coordinate system of the image into the coordinate system of the laser cross-section is required; it is necessary to calculate a special projection matrix.

The simplest solution for determining the projection matrix is the arrangement of real physical marks with known dimensions and coordinates in the reference system of the benchmark in the field of the laser cross-section inside the tank of the benchmark, obtaining the images of such targets, and calculating the coordinate-transformation matrix. In this case, the marks should be used in the working configuration, i.e., they should be located in test bench filled with water (for the proper accounting of refraction) and in the laser plane, since the particles in this plane during the filming are illuminated with the laser and recorded with the camera.

However, installing the real calibrating elements into the model was a necessary procedure. In this case, when changing the laser sheet positions, new holes for new calibration would be required. Therefore, the method of virtual 3D-simulation of marks and their images was developed. The Blender free-access software environment was used (Fig. 5a), which makes it possible to synthesize images derived from virtual camcorders taking a great number of factors into account, including the light refraction. The position of the camcorders in 3D models was chosen close to the measured one. A system of automatic tuning of camcorders based on the attainment of the coincidence of artificial and real images of the motionless elements of the model (Fig. 5b) was developed as well. To determine how the image coordinates are transformed into real coordinates, calibration grids placed on the 3D model in the field in which the laser plane was placed during experiments (Fig. 5a, 5c) were used.

https://doi.org/10.1051/epjconf/202226901053

(a)

(b) (c)

Fig. 5. a) The 3D model with the grid inserted into the region of the laser sheet used for calculations in Blender software; (b)combination of the image derived in the experiment with the top camcorder with the calculated image; (c) an example of rendering the image derived from the top camcorder; the grid is located on the site of the laser plane directed to the intermediate heat exchanger.

The difference of the results of this transformation from the transformations based on the physical marks amounted to 0.2% on average, thus showing the high accuracy of the proposed method. Fig. 6 shows examples of the obtained velocity fields in the coordinates of the reactor benchmark in the laser sheet sections in the intermediate and independent heat exchangers in one of the modes.

6 Conclusion

A system for performing measurements by the PIV method, which is designed to study the thermohydraulic processes in a large-scale model of a proposed fast reactor was developed. The system makes it possible to measure reflections and distortions of the laser sheet section as well as shadowing at the maximum (among such measurements that are known to the authors) spatial scales for a complicated reactor model configuration, as well as in a closed volume.

This work was partial supported by state assignment of IAP RAS (Theme No. 0035-2019-0007).

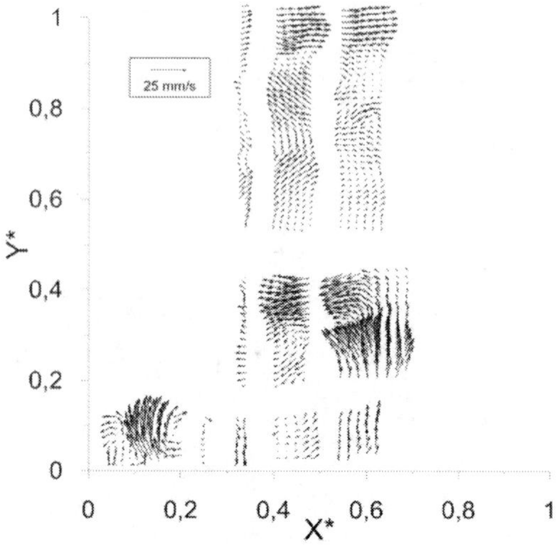

(a)

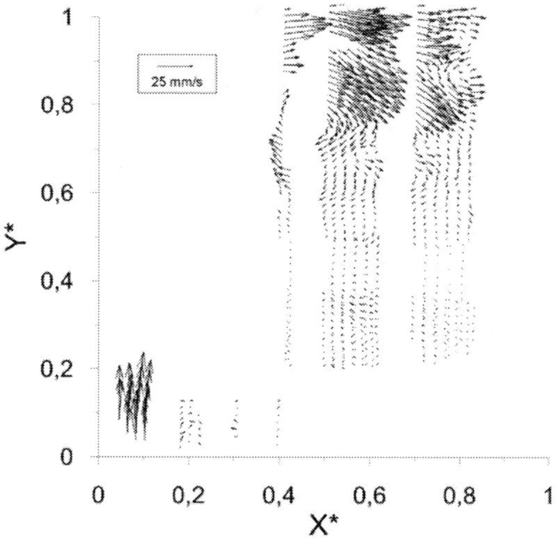

(b)

Fig. 6. Examples of the velocity fields for one of the modes that were derived by averaging over a subrecord for the sections on independent (a) and intermediate (b) heat exchangers. The coordinates are dimensionless: X* = X [mm]/1200, Y* = Y[mm]/1200.

References

1. N. Fujisawa, T. Yamagata, S. Kanno, A. Ito, T. Takano, Nuclear Engineering and Design, **252,** 19 (2012).
2. N. Kimura, H. Ogawa, H. Kamide., Nuclear Engineering and Design **240**, 3055, (2010)
3. T.D. Nguyen, N. Goth, P. Jones, S. Lee, R. Vaghetto, Y.A. Hassan, Int. J. Heat Fluid Flow, **65**, 47 (2017)
4. Dmitriev S.M., Doronkov D.V., Legchanov M.A., Pronin A.N., Solncev D.N., Sorokin V.D.,

Hrobostov A.E. Thermophysics and Aeromechanics **23(3)**, 369, (2016).

5. Dmitriev, S.M., Dobrov, A.A., Legchanov, M.A., Khrobostov, A.E., Journal of Engineering Physics and Thermophysics, **88(5)**, 1297, (2015).
6. Kolesova, Yu.A., Masalov, D.G., Osipov, S.L., Pakholkov, V.V., Rogozhkin, S.A., and Shepelev, S.F., Proc. Sci.-Tech. Conf. "Thermophysics-2012", Obnibsk: FEI, 209, (2012).
7. V. V. Pakholkov, A. A. Kandaurov, A. I. Potseluev, S. A. Rogozhkin, D. A. Sergeev, Yu. I. Troitskaya and S. F. Shepelev Experimental Thermal Engineering, **64**, 496, (2017)
8. U.Biedera J. Maillarda Y. Gorssea D. Guenadoub, Nuclear Engineering and Design 350 **67**, (2019).
9. Raffel M., Willert C.E., Wereley A.T., Kompenhans J. *Particle Image Velocimetry: a Practical Guide, Heidelberg: Springer-Verlag*. 448, (1998)
10. Sergeev, D.A., Instrum. Exp. Tech., 2009, **52**(3), 438, (2009).

Evolution and implosion of cavitation bubbles towards solid surface

Petr Schovanec[1,*], *Darina* Jasikova[1], *Michal* Kotek[1], and *Vaclav* Kopecky[1]

[1]Department of Physical Measurement, Technical University of Liberec, The Institute for Nanomaterials, Advanced Technology and Innovation, Studentska 1402/2, Liberec 461 17, Czech Republic

Abstract. Cavitation bubbles generated via laser-induced breakdown are investigated experimentally. The present work focuses on the direction of the first bubble collapse near a solid surface in distilled water. The solid surface is placed first to the right side in a cuvette filled with distilled water and then placed to the top of the cuvette. In this experiment, it is observed in which direction the cavitation bubble collapses. The cavitation bubble is visualized by a high-speed camera of frequency 68kHz.

1 Introduction

One of the earliest articles that deal of focused laser light in liquids was published in the 1960s by Askar'yan [1]. Then the study of cavitation bubble dynamics generation by laser-induced breakdown was emphasized by Lauterborn [2]. The dependence of the bubble radius and the duration of the laser pulse energy was investigated by Teslenko [3]. In the following years up to now, a great deal of research and application of laser-induced breakdown has been emerging and expanding. Because cavitation belongs not only to industrial applications, but also to biomedical fields such as medicine, biology or pharmacy. The cavitation phenomenon has great potential e.g. in drug delivery systems or cell walls disintegrations.

Cavitation can be defined as a set of effects associated with the formation, activity, and collapse of macroscopic bubbles in a fluid. In nature and in real applications, cavitation bubbles usually form a cloud, yet the bubbles do not separate. Bubbles create structures that work together, but the essential elements of these structures are the individual bubbles [4].

We still cannot produce one controllable bubble according to the cavitation definition, by decreasing the pressure in the volume of the liquid even though the current state of technology is at a very high level [5]. Most methods, such as sparks or laser generated bubbles, are based on evaporating a small volume of liquid, that is closer to the boiling. In our case we use the laser-induced breakdown (LIB) technique, where there is still a lack of information in the experimental part of the bubble cavity investigation [6].

Previously, cavitation research has focused primarily on the behavior of bubbles near solid or flexible boundaries and the behavior of bubbles in various liquids [6, 7]. This present work follows the investigation the collapse of a cavitation bubble towards a solid surface located from different directions.

*e-mail: petr.schovanec@tul.cz

2 Experimental setup

The experiment part leans against the previous work done by Jasikova et al [6]. It is the calculation of the ideal cavitation bubble process and based on the energy losses in the system. The experiment contains laser, a camera, a timer box and a continuous light. The laser serves to generate plasma, followed by a shock wave and cavitation bubble in the liquid using the laser-induced breakdown the method. The continuous light is used to illuminate a cuvette for camera to take photos. The timer box is used to synchronize the time between the laser and the camera.

2.1 Laser Induced Breakdown

Laser-induced breakdown is a method in which laser light in a liquid is focused using appropriately designed optics. This method of LIB in aqueous media and its collateral effects has been described in detail by Kennedy [8]. In addition is followed by Vogel's work [9] on energy distribution during the growth and collapse of laser induced bubble.

In this work we used high power laser with a short pulse (t = 10ns) and wavelength (λ = 532nm). A 10ns short laser pulse was generated using a Q-switched Nd: YAG Quantel EverGreen PIV pulse laser. This laser worked with one cavity to generate a single shot. The camera running in the trigger mode with the laser was synchronized by the Q-switch signal.

To correlate the input laser energy with the bubble size generated by our optical adjustment, we performed a set of calibration and size measurements. The energy of the laser beam was adjusted and measured by the pyroelectric energy sensor Ophir.

Due the losses in the optical path on each of the optical elements, comparable to this one, we had to increase the energy level that enters the entire system. The set laser output energy is taken into account in relation to the bubble diameter. The energies needed to generate the bubbles are shown in Figure 1.

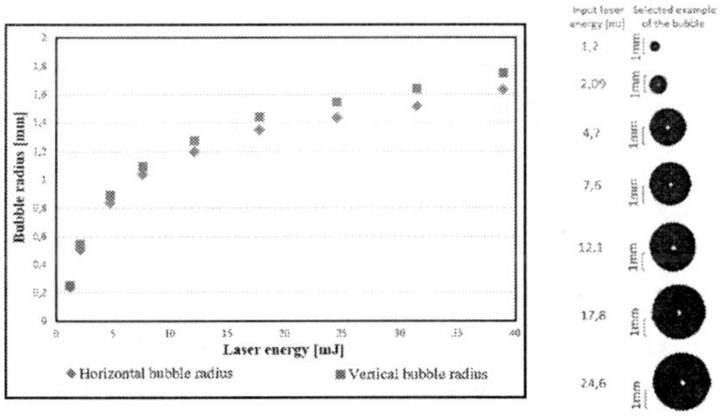

Fig. 1. The graph of maximal bubble diameter.

2.2 Optical setup

Optical setup is based on the direct optical way described by Brujan [10] and the follow-up work by Jasikova et al [7]. The LIB has been set up as an optical direct mode and is shown in Figure 2. The laser beam output diameter was 5mm with Gaussian intensity characteristics. In this arrangement was followed by a concave lens fs = 30mm and two planar convex lenses fs = 75mm and fs = 50mm of 1 inch in diameter. Laser point - probe with diameter <0.1mm was created by focusing the laser beam.

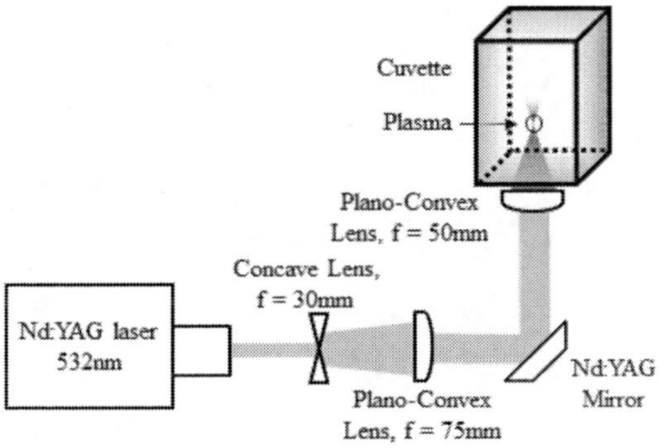

Fig. 2. Optical setup.

2.3 Visualization setup

Visualization setup follow the work [5, 6]. We used here shadowgraphy setup for the bubble visualization. This setup consists of continuous LED matrix daylight lamp Veritas, MiniConstellation 120 - 5000K of illuminance 92klux in 0.5m, set with optical diffuse filter. Opposite to the light source was placed high speed CMOS camera SpeedSense. This camera is working on frequency 68kHz with resolution of (256 x 256)px or lower frequency with higher resolution up to (1280 x 800)px, and the dynamic range 12bit.

The exposure time for the camera was set to $1\mu s$ and subpixel resolution was $20\mu m$. The camera was mounted with optical lens system INFINIPROBE^{TM} TS-160 universal macro/micro imaging system that enables 4x, and 16x magnification. The camera was mounted with edge pass and long pass filter cut-on wavelength 550nm low pass optical filter to reduce the backward laser flashes to the camera and also to eliminate the flash generated while plasmatic breakdown. Visualization setup is shown in Figure 3.

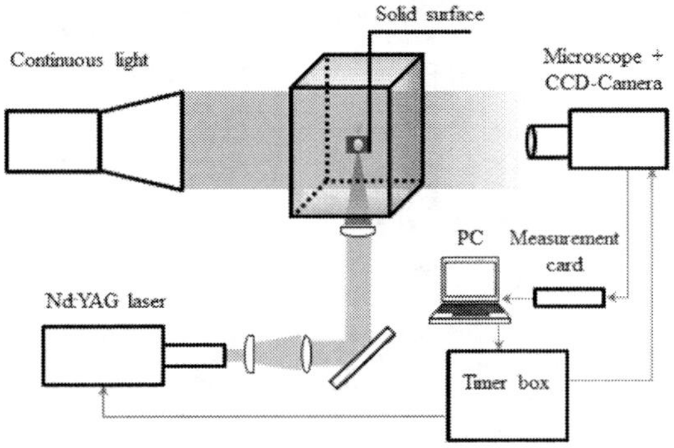

Fig. 3. Experimental setup the visualization system for shadow graphics.

3 Results

Plasma is generated at the location of the laser energy concentration and the plasma is visible for emission. At this plasma point in the liquid, the pressure rises to 10^3bar and the temperature to 10^3K. This leads to plasma expansion at supersonic velocities, and is followed by a cavitation bubble effect. For Figure 4 at time $t_0 = 0\mu s$, it is important that the laser light comes from the bottom side in the pictures.

Figure 4 represents a selected part of a series of images show the formation, evolution, implosion, and behavior of bubbles. In the first column of Figure 4, the solid surface is placed from the right side and in the second column is placed from the up side.

Bubble was generated with the laser energy of 4mJ and 3mJ. The first frames in the time of t_0 show the position of the optical breakdown. The images in the time of t_1 show the enlargement of the cavitation bubble that is very close to the maximum of its diameter.

In the other images in the time of t_2, the direction of collapse of the first phase of the bubble life is seen. With the collapse of the bubble at the second stage of its lifetime in the times of t_3 to t_6, the collapse is attracted by the solid surface in its vicinity. Even when the solid surface is applied from the top, the bubble collapse is pointing upward as seen in the time t_3.

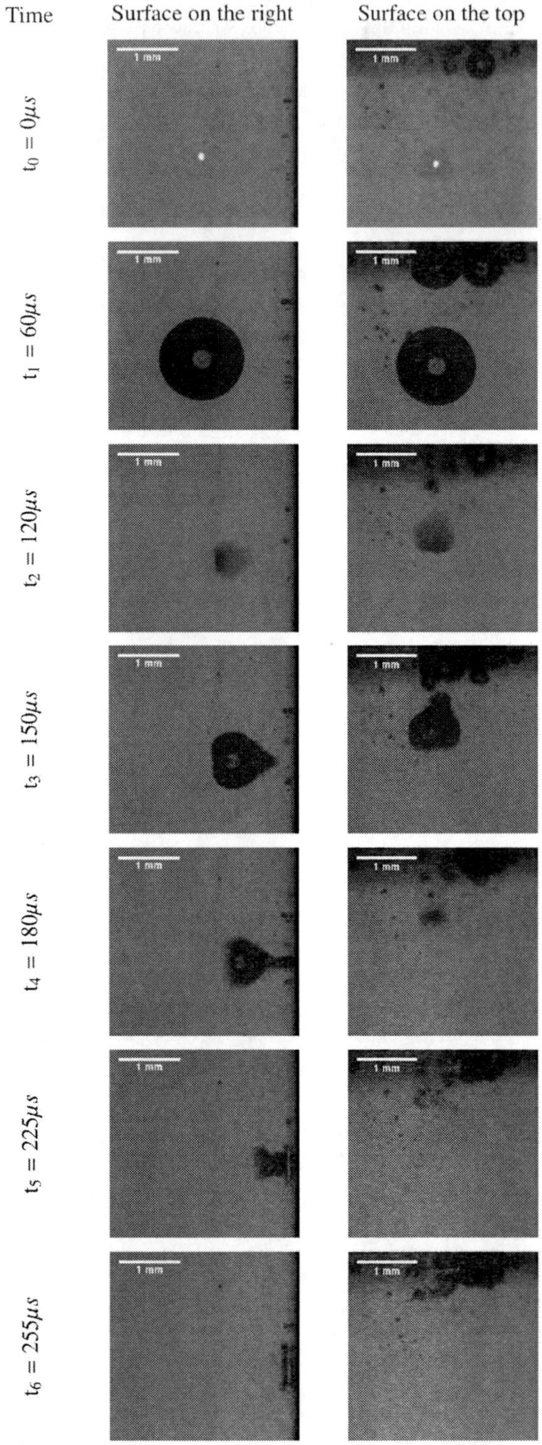

Fig. 4. Evolution of bubbles towards solid surface.

Figure 5 represents a selected part of a series of images show the implosion of bubbles. In the first column of Figure 5, the solid surface is placed from the right side and in the second column is placed from the up side. For the images located in the left column of Figure 5, laser energy of 3mJ and 10mJ was used for the images in the right column. That is reason why the time of bubble implosion was different. The frames at time t_0 show the direction of implosion of the bubble generated.

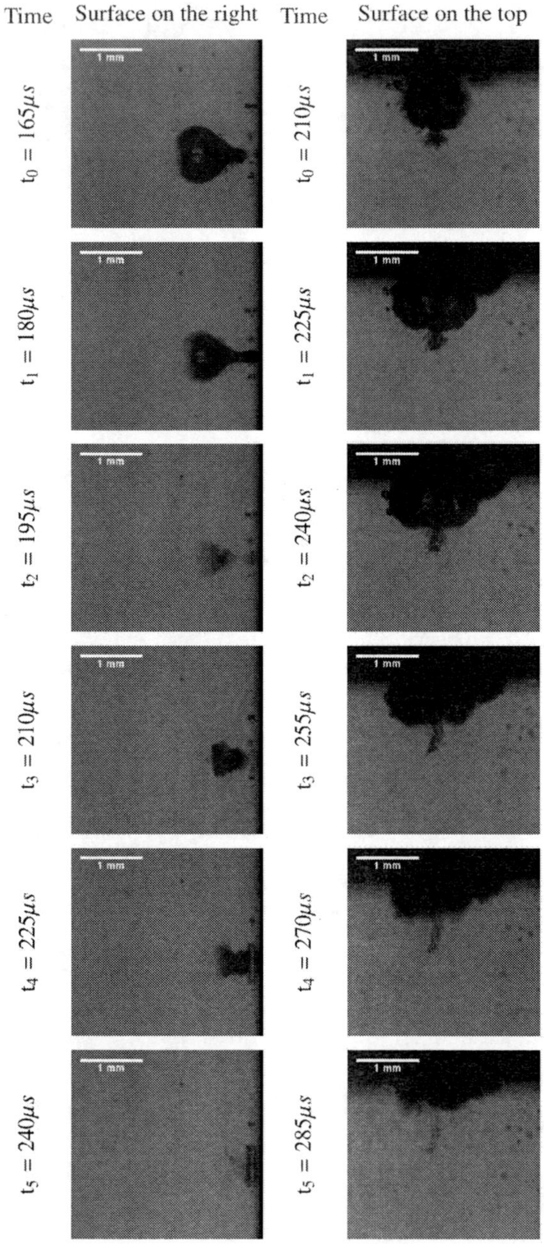

Fig. 5. Detail on micro-jet creation through collapsing cavitation bubble.

This implosion is directed towards a solid surface. In the frames at time t_1 - t_5, the effect of the bubble implosion on a solid surface is seen. Primary demonstration of micro-jet formation through the collapse of cavitation bubbles. The micro-jet was well captured specifically at time t_0 in Figure 5.It is therefore evident from these series of images that the implosion of the bubble is not directed like gravitational force, but towards a solid surface, no matter where it is located.

4 Conclusion

When studying the cavitation process, visualization of cavitation bubbles is a very important and useful tool. We visualized the limits of the bubbles in their growth, size and direction of collapse.

The plasma expansion is followed by a shock wave and during this sequence the liquid evaporates and the cavitation bubble grows. This bubble is filled with steam. The bubble collapse occurs during the second phase of bubble lifespan. The collapse is caused by a decrease in internal pressure and cooling, which is influenced by the external environment. Bubble collapse is attracted towards a solid surface near it.

Acknowledgements.

This work was supported by the Ministry of Education, Youth and Sports of the Czech Republic and the European Union - European Structural and Investment Funds in the frames of Operational Programme Research, Development and Education - project Hybrid Materials for Hierarchical Structures (HyHi, Reg. No. CZ.02.1.01/0.0/0.0/16_019/0000843), and the SGS project no. 21338/115 of the Technical University of Liberec.

References

1. Askar'yan, G.A., Prokhorov, A.M., Chanturiya, I.F., Shipulo, G.P., Sov. Phys. - JETP **17**, 1463 (1963)
2. Lauterborn, W., Applied Physics Letters **21**, 27 (1972)
3. Teslenko, V.S., Soviet Journal of Quantum Electronics **7**, 981 (1977)
4. Schovanec, P., Garen, W., Koch, S., Neu, W., Dancova, P., Jasikova, D., Kotek, M., Kopecky, V., ACC JOURNAL **25**, 58 (2019)
5. Jasikova, D., Schovanec, P., Kotek, M., Muller, M., Kopecky, V., Proceedings of SPIE **10151**, 154 (2016)
6. Jasikova, D., Schovanec, P., Kotek, M., Muller, M., Kopecky, V., EPJ Web Conf. **143**, 02044 (2017)
7. Jasikova, D., Schovanec, P., Kotek, M., Kopecky, V., EPJ Web Conf. **180**, 02038 (2018)
8. Kennedy, P.K., Hammer, D.X., Rockwell, B.A., Progress in Quantum Electronics **21**, 155 (1997)
9. Vogel, A., et. al, Applied Physics B **68**, 271 (1999)
10. Brujan, E.A., Nahen, K., Schmidt, P., Vogel, A., J Fluid Mech **433**, 251 (2001)

A software tool for blade design

Jan Šimák[1,*]

[1]VZLÚ – Czech Aerospace Research Centre, Beranových 130, 199 05 Prague, Czech Republic

Abstract. An interactive software tool for blade design of axial flow machines was created. It is written as an extension module to the open-source software FreeCAD. In its graphical interface, the user can modify the blade profiles, stack them to create the whole blade and generate end walls and other stage features. Or everything can be controlled by a simple Python script. Results can be saved as STEP file or STL mesh and export to a mesh generator and CFD solver. Blade profiles are given by a set of parameters describing Bézier curves, the blade is represented by b-spline surfaces. Up to now, the tool helps the designer to find the right shapes by a 1D method, which evaluates the velocity triangles and the state variables through the machine stage on a mean line (mean radius) from the given design parameters (mass flow, total states in front of the stage, rotational speed etc.) and distribute them along the radius from the hub to the tip. The CFD results of a sample designed stage are included.

1 Introduction

Nowadays, the turbomachinery laboratory of VZLÚ – Czech Aerospace Research Centre provides the aerodynamic research using their turbomachinery test rigs to help the industrial partners increase the efficiency and reduce emissions of turbomachines. Apart from experimental data, it provides also numerical analyses using CFD software.

To do that a software tool which can simply generate the necessary geometry model is very useful. In addition, if design or optimization processes of the flow paths of new turbomachines are incorporated, such a tool is almost a necessity.

In the past decades, many gas turbine design software were written, however, a lot of them is a private property of the manufacturers. There can be found some available, commercial or free, for example *MULTALL* [1] or *T-AXI* [2] or other. They are more or less sophisticated, their predictions based on various methods and empirical relations, but they usually lack the possibility of intuitive geometry modelling.

The presented application has no ambition to compete with professional design tools. It can more likely serve as a teaching tool to get insights into the gas turbine design problem or as a basic geometry tool. Owing to the scripting ability and parameterization it can be integrated into an optimization process.

2 Tool implementation

The tool is written as an extension module to the open-source software FreeCAD [3], which is a multi-platform 3D CAD/CAE parametric modelling application. Using *Open CASCADE* geometry kernel, it allows complex 3D operations on geometry objects. FreeCAD has an intuitive Graphical User Interface (GUI) with a Python console, where all user operations are mirrored. So, the user has a possibility to work with the geometry by his/her mouse or typing commands to the console or directly to a command line interpreter.

The extension module is written partially in C language and in Python. All of its features are accessible through the Python interface, which allows an easy way of control and scripting. Python is an intuitive language and is easy to learn. The most frequent features are also available through the GUI.

The tool helps the designer to create the blade geometry from new or given profiles and check the basic geometric properties such as throat width, centre position or moments of inertia. It is possible to create end walls given by b-splines or piecewise linear curves and prepare a 2D or 3D geometry to do a CFD analysis. The resulting geometry can be saved to STEP, IGS or other exchange formats or as an STL surface mesh.

3 Blade design description

The blade design starts with a profile geometry. The user has to supply one or more profiles of the blade sections. These can be either data files with point coordinates (e.g. from some database) or profiles created by a parameterization.

3.1 Profile parameterization

The profile parameterization is based on the method [4] where the geometry is described by two Bézier curves of fifth degree and an arc on the trailing edge. The total number of 16 parameters can be divided into two groups.

* Corresponding author: simak@vzlu.cz

The first group contains parameters such as chord length c, blade inlet and exit angles α_{ib}, α_{eb}, trailing edge thickness angle ψ, leading and trailing edge radiuses *RLE*, *RTE* and stagger angle γ. The second group consists of length parameters describing the Bézier curves (see Fig. 1.). The connection of the two curves on the leading edge ensures continuity of second order, in other words, the curvature has no jump.

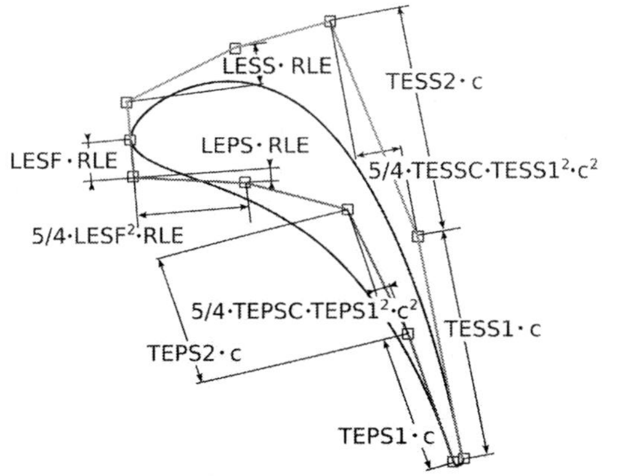

Fig. 1. Length parameters of the profile parameterization.

This way of parameterization ensures a broad set of possible shapes and allows quite straightforward modifications. In the GUI, the user sees an immediate response of a parameter change to the control polygon and therefore to the profile shape, together with throat visualization and basic profile properties.

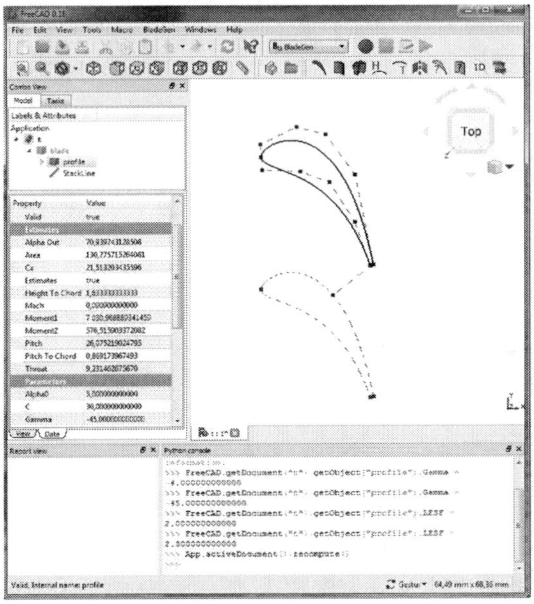

Fig. 2. Design of a profile shape in the GUI.

3.2 Profile analysis

The tool offers also a possibility to check some of the geometry quantities such as area, principal moments of inertia, throat width and its position, gas outlet angle estimate and others.

The real gas outlet angle α_e is unfortunately different from the blade exit angle α_{eb}. The difference, called a deviation angle, depends on the blade shape (especially on the suction side behind the throat), Mach number and boundary layers and so on. If the profile is near end walls then a secondary flow or clearances and seal outlets also affect the angle. Its precise value can be evaluated by time-consuming flow solution which is not what we are interested in. There exist many more or less precise estimates based on empirical data. The one used in this application is based on the work [5]. The outlet gas angle

$$\alpha_e \approx f(\arccos(o/t)) + 4t/e, \quad (1)$$

where function $f(x) = -11.51 + 1.156x$ correlates α_e and $\arccos(o/t)$ for "straight-backed" blades and $e = j^2/8z$ is the mean radius of curvature of the suction side between the throat and the trailing edge (see Fig. 3.). For Mach number below 0.5 the outlet angle is given by eq. (1), for Mach number near unity it closely relates to $\arccos(o/t)$. Angle for Mach between 0.5 and 1.0 is a linear combination of these two values.

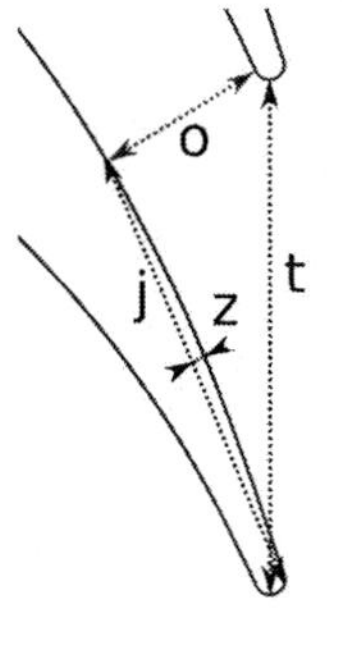

Fig. 3. Mean radius of curvature of the suction side for outflow angle estimate.

3.3 Blade creation

The blade geometry is defined by one or more cross sections at given blade heights. If only one profile is given a prismatic blade is created. Otherwise, the profiles are connected by b-spline surfaces and the blade solid shape is generated from them.

The profiles can be stacked by three different ways: the radius going through leading edges, going through centres of trailing edges or through centres of mass. This stacking line can be further leaned in the axial or tangential direction.

The blade height is given by hub and tip wall radiuses. The end walls can be also added to the model and a 3D geometry of the part of the blade wheel corresponding to one blade could be generated and exported to a mesh generator. FreeCAD offers a lot of basic or advanced operations with shapes such as intersection, fusion, difference, mirroring etc. which can also be used.

The estimate of the number of blades is based on the Zweifel loading coefficient [2].

4 1D model analysis

The design of gas turbine stages usually starts with a specification of requirements and constraints depending on our problem. After the input data are determined, an application of a 1D analysis is common. The outputs of this model are velocity triangles and state values through the stage. The designer gets an instantaneous preview of the stage performance and the blading requirements.

The next steps are usually a 2D through flow solution in a meridional cut of the blade channel to obtain the radial variations, then a 2D blade profile designs on cylindrical surfaces and finally a complete 3D flow through the stage. However, these models are time consuming and cannot be done in real time. At this moment, they are not part of the tool and must be used as external solvers if needed.

4.1 Mean line model

The 1D model assumes a simplified flow on a mean line inside the blade channel on a given radius. The implemented model assumes a mean radius but it is possible to modify the algorithm to accept different radiuses.

There is a lot of possibilities how to choose input data, it can be a combination of pressure, temperature and velocity at given positions *0*, *1* and *2* in the turbine stage together with angles, dimensionless coefficients or other design parameters. The application has up to now a possibility to choose from five variants of input data:

1) P_{0c}, T_{0c}, α_0, α_2, Λ, $\dot{m}$, r_H, r_T, RPM, Y_1, Y_2,
2) P_{0c}, T_{0c}, α_0, Π, Λ, $\dot{m}$, r_H, r_T, RPM, Y_1, Y_2,
3) P_{0c}, T_{0c}, α_0, α_1, β_2, Π, r_H, r_T, RPM, Y_1, Y_2,
4) P_{0c}, T_{0c}, α_0, α_2, Π, Λ, r_H, r_T, RPM, Y_1, Y_2,
5) P_{0c}, T_{0c}, α_0, α_2, Λ, $\dot{m}$, r_{0H}, r_{0T}, RPM, Y_1, Y_2 (*const* C_a)

The first variant means a prescription of input and output absolute flow angles together with a mass flow and a degree of reaction

$$\Lambda = (T_{1is} - T_{2is}) / (T_{0c} - T_{2is}). \quad (2)$$

The pressure drop through the stage depends on the given data. The second variant replaces the outflow angle with the pressure ratio $\Pi = P_{0c} / P_2$. The variant 3 can be useful when the geometry of stator and rotor are given (i.e. angles $\alpha 1$, β_2). The fourth variant evaluates the stage performance from stage inlet and outlet absolute angles, pressure ratio and degree of reaction. All of these variants suppose that the end wall radiuses through the stage are given and so the axial velocity changes. On the other hand, the last variant assumes a constant axial velocity.

The effect of a boundary layer on the blade surface and other loss mechanisms are introduced into the model as pressure loss coefficients Y_i, where

$$Y_1 = (P_{0c} - P_{1c}) / (P_{1c} - P_1), \quad (3)$$
$$Y_2 = (P_{1c,rel} - P_{2c,rel}) / (P_{2c,rel} - P_2). \quad (4)$$

The area of the annulus is evaluated from the radiuses r_H and r_T and the blade velocity U from revolution RPM.

The solution of the problem uses two basic assumptions. The first one is the conservation of mass which implies constant (or given) mass flow through the stage. The second one is the Euler turbine equation which leads to the relation

$$c_p(T_{1c} - T_{2c}) = 2\pi\, RPM/60\, (r_1 C_{1u} - r_2 C_{2u}). \quad (5)$$

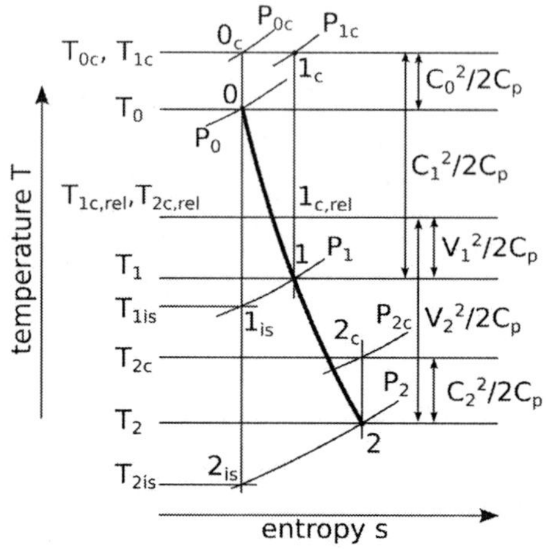

Fig. 4. T-s diagram of a reaction stage.

Only the first and last variants will be shown in detail, the other variants are derived in a similar manner. All computations assume a perfect gas with no heat transfer through walls into the surroundings.

4.1.1 Variant 1

In the first variant of the 1D method, the absolute inlet and exit angles of the turbine stage are given. The blade heights are also given and so the axial velocity can vary. The total temperature can be defined as

$$T_c = T + C^2 / 2c_p \quad (6)$$

and it represents a temperature of a gas which was brought to rest in an isentropic process. In an isentropic flow, pressure and temperature can be related with

$$P = P_c \left(T / T_c\right)^{\gamma/(\gamma-1)}. \quad (7)$$

Since no work is done in the stator, the total temperature should stay the same (see Fig. 4.). To evaluate all unknown parameters involved in the stage design, the temperature after stator T_1 is assumed an unknown variable at this moment. Then the static pressure P_1 is evaluated from eq. (3) and (7)

$$P_1 = P_{0c} \frac{\left(T_1 / T_{0c}\right)^{\gamma/(\gamma-1)}}{1 + Y_1\left(1 - \left(T_1 / T_{0c}\right)^{\gamma/(\gamma-1)}\right)}. \quad (8)$$

Using the relations (2) and (7), the stage exit pressure P_2 is obtained

$$P_2 = P_{0c}\left[\frac{\left(P_1/P_{0c}\right)^{(\gamma-1)/\gamma} - \Lambda}{1-\Lambda}\right]^{\gamma/(\gamma-1)} . \quad (9)$$

The stator outflow angle α_1 is computed from the equation for mass flow

$$\dot{m}_1 = C_1 cos\alpha_1\, A_1 P_1 / RT_1 \quad (10)$$

where $C_1 = \sqrt{2c_p\left(T_{0c} - T_1\right)}$. The components of the relative velocity are $V_a = C_a$, $V_u = C_u - U$, thus

$$T_{1c,rel} = T_1 + \frac{C_{1a}^2 + \left(C_{1u} - U_1\right)^2}{2c_p}, \quad (11)$$

and

$$T_{2c,rel} = T_{1c,rel} - \frac{U_1^2 - U_2^2}{2c_p}. \quad (12)$$

After the remaining values of $P_{1c,rel}$, $P_{2c,rel}$, T_2 and $C_2 = \dot{m}_2 RT_2 / A_2 P_2\, cos\alpha_2$ are evaluated, all the necessary variables are put to the rewritten equation (5)

$$2c_p(T_1 - T_2) + C_1^2 - C_2^2$$
$$- 2(C_1 U_1 \sin\alpha_1 - C_2 U_2 \sin\alpha_2) = 0 \quad (13)$$

and the equation is solved for the unknown T_1. The equation can have more than one solution, so we need some condition how to choose one. There are usually three roots of the eq. (13), the first corresponds to a very low temperature drop and almost no work done, the last root corresponds to a high temperature drop and often high supersonic velocities and so the middle one promises the best results. From this follows that the total-to-static efficiency

$$\eta_{TS} = (T_{0c} - T_{2c}) / (T_{0c} - T_{2is}) \quad (14)$$

seems to be a reasonable criterion.

4.1.2 Variant 5

This is up to now the only variant where the blade heights are part of the solution. The axial velocity is assumed constant through the stage and so the blade height is rising. The geometry is designed to satisfy the given absolute flow stage inlet and outlet angles and the dimensions of the stage inlet (r_{OH} and r_{OT}) are given to get the right scale. This method can be used to design a part of a turbine with repeating stages (that means stages with the same blade geometry).

The stator inlet area A_0 and mean radius are evaluated directly from the input data. The relation between Mach number and mass flow can be derived as

$$\frac{M\sqrt{\gamma/R}cos\alpha}{\left(1+\frac{\gamma-1}{2}M^2\right)^{\frac{\gamma+1}{2(\gamma-1)}}} = \frac{\dot{m}\sqrt{T_C}}{AP_C} \quad (15)$$

and thus the Mach number M_0 can be obtained together with the remaining unknowns before the stator. Now the axial velocity is known and it is possible to continue in the computations. The relations are written with respect to the unknown variable T_1, namely eq. (8), (9), (11) and (12). The velocities are evaluated from the known axial velocity C_{0a} and the angles α_0 and α_2. At the end, we obtain two relations for the temperature behind the rotor,

$$T_2 = T_{2c,rel} - V_2^2 / 2c_p\,, \quad (16)$$
$$T_2 = T_{2c,rel}\,(P_2 / P_{2c,rel})^{(\gamma-1)/\gamma}. \quad (17)$$

Solution of this system is the searched value of T_1. Again, there could be more than one solution so the η_{TS} criterion is used.

4.2 Radial distribution

The above-mentioned model solves the velocity triangles on a simplified streamline that lies on a mean radius. However, the blade velocity rises with increasing radius and the whirl velocity component causes the static pressure to vary across the blade height. From this reason the velocity triangles must also vary with the radius.

It is widely supposed in turbine design methods that any radial flow is done inside a blade row and outside is the radial velocity zero and the flow is axisymmetric. This is called a radial equilibrium flow with satisfies

$$\frac{1}{\rho}\frac{dP}{dr} = \frac{C_u^2}{r}. \quad (18)$$

The change of total specific enthalpy with respect to radius can be expressed as

$$\frac{dh_c}{dr} = \frac{d(h+C^2/2)}{dr} = \frac{dh}{dr} + C_a\frac{dC_a}{dr} + C_u\frac{dC_u}{dr}. \quad (19)$$

Combining this with the thermodynamics relation $Tds = dh - (1/\rho)dP$ results in equation

$$\frac{dh_c}{dr} = T\frac{ds}{dr} + C_a\frac{dC_a}{dr} + C_u\frac{dC_u}{dr} + \frac{C_u^2}{r}. \quad (20)$$

If the total enthalpy (and thus total temperature) together with the entropy are the same along the radius and the axial velocity remains also the same, then the eq. (20) becomes

$$\frac{dC_u}{dr} + \frac{C_u}{r} = 0, \quad (21)$$

which implies

$$C_u r = const. \quad (22)$$

This is known as a free-vortex flow.

5 Comparison with CFD

As follows from the description of the 1D method mentioned above, it cannot deal with any secondary flow or three-dimensional flow structures. The distribution of losses along the radius is not implemented yet. From that reason, a simple turbine stage with constant radiuses and zero clearances at the stator and rotor blade ends is

https://doi.org/10.1051/epjconf/202226901055

assumed. The geometry of the blade was designed according to the given data:

Table 1. Input data.

P_{0c}	120 kPa	Λ	0,5
T_{0c}	300 K	r_H	0,15 m
α_0	0°	r_T	0,2 m
α_2	0°	***RPM***	9000
$\dot{m}$	3 kg/s		

The stator blade according to the velocity triangles is almost prismatic. On the other hand, the rotor blade was chosen twisted to correspond to the expected inlet angles. The resulting geometry is visible in the Fig. 5. It must be said that this stage design is only illustrative. Its handmade shape takes no care about losses and has many shortcomings to be applied in practice. However, its almost random design is useful for the verification purposes.

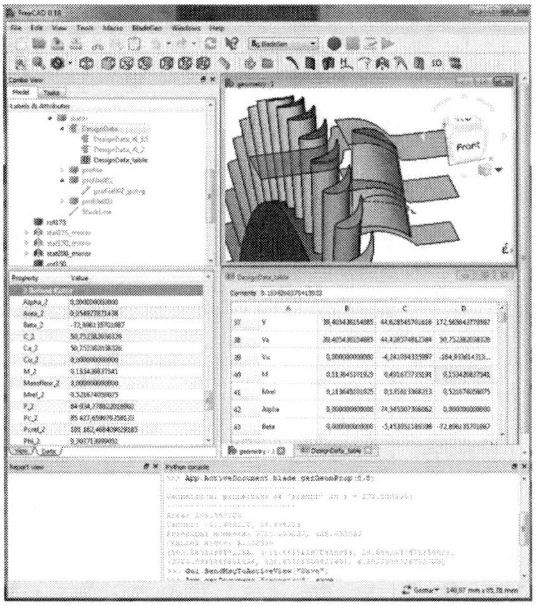

Fig. 5. Geometry of blades with 1D analysis of the stage.

The design was evaluated with the CFD software Edge [6] in 2D blade to blade cylindrical cuts and 3D geometry solution. The stator-rotor interaction was simulated by a mixing plane boundary condition, the flow was fully turbulent. To suppress secondary flow effects, the annulus end walls were treated as a slip (Euler) wall boundary.

The results of the 1D analysis together with 2D and 3D CFD results are summarized in the next tables. Since the analysis needs predictions of pressure losses and also of deviation angles, the whole design task is an iterative process. The presented results are from the second iteration, where the angles and losses estimates were slightly changed according to 2D CFD result on the mean radius.

Table 2. Results on the hub end wall (r = 0,15 m).

r = 0,15	**1D Analysis**	**2D CFD**	**3D CFD**
P_1 *[kPa]*	95,2646	94,6372	96,7134
P_2 *[kPa]*	84,0348	84,0315	84,2259
T_1 *[K]*	281,523	281,506	283,739
T_2 *[K]*	272,333	272,952	275,587
Ca_1 *[m/s]*	44,427	45,398	44,315
Ca_2 *[m/s]*	50,752	49,581	50,580
α_1 *[°]*	76,668	76,344	75,593
α_2 *[°]*	0,000	5,028	-4,091
β_1 *[°]*	46,061	45,096	35,540
β_2 *[°]*	-70,252	-70,133	-70,515

Table 3. Results on mean radius (r = 0,175 m).

r = 0,175	**1D Analysis**	**2D CFD**	**3D CFD**
P_1 *[kPa]*	100,876	100,7256	100,9361
P_2 *[kPa]*	84,0348	84,0422	84,0278
T_1 *[K]*	286,164	286,079	286,107
T_2 *[K]*	272,333	272,399	272,214
Ca_1 *[m/s]*	44,427	45,175	45,061
Ca_2 *[m/s]*	50,752	51,540	50,825
α_1 *[°]*	74,546	74,291	74,155
α_2 *[°]*	0,000	1,222	0,018
β_1 *[°]*	-5,453	-5,459	-7,046
β_2 *[°]*	-72,896	-72,536	-72,653

Table 4. Results on the tip end wall (r = 0,2 m).

r = 0,2	**1D Analysis**	**2D CFD**	**3D CFD**
P_1 *[kPa]*	104,6419	105,7455	104,1285
P_2 *[kPa]*	84,0348	84,0371	83,9642
T_1 *[K]*	289,177	289,840	288,520
T_2 *[K]*	272,333	272,394	275,623
Ca_1 *[m/s]*	44,427	43,868	41,432
Ca_2 *[m/s]*	50,752	51,878	46,168
α_1 *[°]*	72,465	72,002	73,874
α_2 *[°]*	0,000	-4,409	5,523
β_1 *[°]*	-47,148	-50,637	-47,038
β_2 *[°]*	-74,930	-74,931	-75,748

The CFD results of the 2D profile solution, together with the 3D solution of the whole geometry are in good agreement with the preliminary results from the stage analysis. This shows that the analysis together with the radial distribution can give good insight into the problem, especially in simple cases.

6 Conclusions

The presented tool can be useful in preliminary stages of design process. The user can interactively check a prediction of a stage performance and its response to changes in the input data. It is also possible to create and export the geometry and automates the whole process using a simple Python script. The design tool will be available in the near future at www.vzlu.cz.

Acknowledgements: The work was supported by the Long-term Framework Advancement Plan provided by the Ministry of Industry And Trade of the Czech Republic.

Nomenclature

A	area
c	chord length
C	absolute flow velocity
c_p	specific heat capacity at constant pressure
e	mean radius of curvature
h	specific enthalpy
M	Mach number
$\dot{m}$	mass flow
o	throat width
P	pressure
r	radius
R	specific gas constant
RPM	revolutions per minute
s	specific entropy
t	blade pitch
T	temperature
U	blade velocity
V	relative flow velocity
Y	pressure loss coefficient
α	absolute flow angle (measured from the axis)
β	relative flow angle (measured from the axis)
γ	adiabatic constant
η_{TS}	total-static efficiency
η_{TT}	total-total efficiency
Λ	degree of reaction
Π	pressure ratio, $\Pi = P_{0c} / P_2$
ρ	density

Subscripts:

0	stator inlet values
1	rotor inlet values
2	rotor outlet values
a	axial direction
b	blade angles
c	total state
H	hub
i, e	inlet, exit
is	isentropic
rel	relative values (moving system with rotor)
T	tip
u	tangential direction

References

1. J. Denton. *Multall: An Open Source,CFD Based, Turbomachinery Design System. Journal of Turbomachinery*. 139. 10.1115/1.4037819. (2017)
2. M. Turner, A. Merchant, D. Bruna. *A turbomachinery design tool for teaching design concepts for axial-flows fans, compressors, and turbines*. ASME Turbo Expo, Barcelona (2006)
3. www.freecadweb.org
4. Pacák A., Krivánka D., Synáč J., Šimka Z. *Parametrizace ŠKODA AP1 – nový popis profilových mříží*. Sborník Současné trendy při návrhu a výpočtu turbostrojů, Praha (2009)
5. D. G. Ainley, G. C. R. Mathieson. *An Examination of the flow and Pressure Losses in Blade Rows of Axial-Flow Turbines*, London (1955)
6. *Edge - Theoretical Formulation*, FOI dnr 03-2870, Issue 5.1, June 2010

New ventilation schemes of car cabin evaluated by CFD simulations

Jan Sip[1,*], *Jan* Pokorny[1], *Jan* Fiser[1], *Frantisek* Lizal[1], and *Miroslav* Jicha[1]

[1]Brno University of Technology, Faculty of Mechanical Engineering, Energy institute, Technicka 2896/2, Brno 616 69, Czech Republic

Abstract. Computational fluid dynamics (CFD) was used to investigate new ways of air conditioning of a car cabin during summer conditions. Three different schemes of ventilation were considered: classical mixing ventilation (dashboard vents), large-area ceiling vents, and underfloor vents delivering air under the seats. The Star-CCM+ was used to carry out unsteady RANS simulations and resulting temperature and velocity fields were compared. Furthermore, the draft risk and age of air were evaluated for each scheme. For air conditioning of the cabin, mixing ventilation had the best results.

1 Introduction

Nowadays, the most of car's cabin ventilation system using mixing ventilation (dashboard and floor vents). Different situation is in the case of vans, microbuses, which uses ceiling vents. The benefit of this more flexible interior (removable seats etc.). With the developments of autonomous car and flexible interiors the convectional solution of cabin ventilation became restrictive and there are posibilies for new concepts and desings. In the case of the flexible interior, it is assumed that the front seats can be rotated in the opposite direction of travel, so the convectional ventilation system loses its efficiency. From this perspective, the research into new ventilation systems is important.

The new concepts of autonomous vehicle ventilation system has already been investigated experimentally and computationally by Dehne [1]. Dehne has been tested three new modes of vertical ventilation: air intake under the seats, large-area diffusers in the roof of the car, and the combination of both. Inside the cabin there were sitting four manikins, and the cabin was fitted with a number of air velocity, temperature and relative humidity probes. For the CFD simulation the OpenFoam was used, just computational mesh was created in the commercial software Star-CCM+ with base size of cell 4 mm. The turbulence model k-ω SST and. the "buoyantBoussinsq" model were used. From this study vertical ventilation has many advantages with combined air supply achieved the best results. However, the effect of direct solar radiation was not included in the study.

In the presented study we investigated summer test case (including the solar radiation) when three different variant of cabin cooling were simulated. The criteria for each variant was the value of the draft risk and age of air were evaluated with the focus on the driver seat. These simulations help to identify behaviour of the ventilation system in the early phase of mock-up design process before its own construction, which will follow.

2 Materials and methods

As first the concept of cabin mock-up was sketched and than 3D stl geometry created using CAD. For the CFD simulation the Star-CCM+ ver. 10.04 was used. Own simulations were performed on the virtual mock-up using computational cluster at NETME centre.

2.1 Geometry and mesh

Three dimensional geometry of the mock-up (**fig. 1**) was created in terms of Josef Bozek Competence for

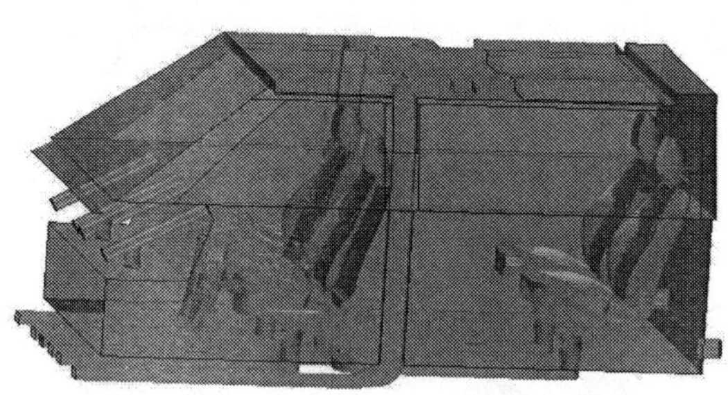

Fig. 1. Geometry.

Automotive Industry The mock-up represents just the cabin without hood and width and height of the cabin corresponds to the typical van with the dimensions $3\times1.4\times1.38$ m (just the length was reduced from 4.5 m to 3 m, we not considered the 3rd row of seats). The total air volume of cabin is 5.7 m^3. In the cabin there are four seats

* Corresponding author: jan.sip@vutbr.cz

and one manikin at the driver's seat. The polyhedral computational mesh was used with base size 0.025 m and with the total number of cells 4.5 million. All parts of the geometry except the supply ducting contained five prism layers. The surface mesh of the manikin and the seats was more refined to Relative Target Size 0.0075 m. Furthermore, the volume mesh was refined locally in the vicinity of active vents.

Table 1 HVAC boundary conditions

Type	Label	Velocity inlet [m/s]	Temperature [°C]
Mix	D1	4.17	12.5
	D2	4.28	13.0
	D3	4.60	11.5
	D4	4.24	11.5
	Def1	0.17	18.8
	Def2	0.33	18.8
Floor	F1 F2 F3 F4	4.14	12.1
Ceiling	C1 C2	8.29	12.1

2.2 Physical models

The simulations were performed as unsteady with timestep 0.001 s, and Courant number of 5. The temporal discretization was performed by the implicit unsteady algorithm with the second-order temporal discretization scheme, and the convection term was discretized using the second-order upwind scheme. For the radiation modelling the view factors were calculated and the surface to surface model was used. The buoyancy, Ideal gass and solar loads in the cabin were taken into account. Turbulence was modelled by the k-ω SST model.

2.3 Boundary condition

Environmental

For summer test case was important to consider Multiband Thermal Radiation model which splits radiation spectrum into shortwave (0,01 - 2,5 μm) and longwave (2.5 - 1000 μm). The value of the solar loads was 800 W/m2 with a 90° elevation angle. The ambient temperature was 30 °C.

Cabin

The material properties of mock up were defined based on the properties of real silver painted car with tinted window, see [2, 3]. The cabin boundary conditions of the wall type were set on the car parts. The surface temperature, the thermal resistance of the material and the heat transfer coefficient were defined. Emissivity and transmissivity of the materials were defined. Transmissivity for both type of radiation was prescribed.

HVAC

The volumetric flow rate for individual ventilations modes was constant (99 l/s). Two boundary conditions of the velocity inlet type are prescribed for ceiling ventilation (CV). The inlet air speed is 8.3 m/s and the air temperature is 12.1 °C. Four identical boundary conditions were prescribed for floor ventilation (FV). The inlet air speed is 4.1 m/s and the air temperature is 12.1 °C. An overview of boundary conditions for mixing ventilation (MV) is in the **table 1**.

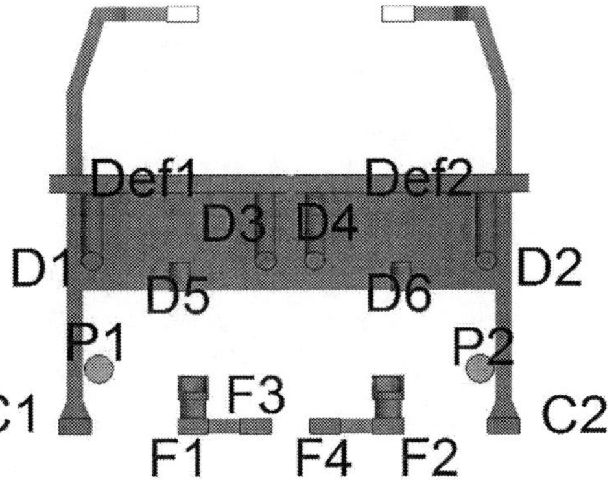

Fig. 2 Inlets.

Manikin

The surface temperature was prescribed 34 °C on non-clother parts of the manikin. The skin temperature was defined as 34 °C and the temperature resistance of the clothing related to the summer period on the parts of the manikin on which the clothing occurs.

2.4. Post-processing and outputs

The comparison was made between flow velocities and temperature of air at the monitoring points at important points (ankle, hand, head). Furthermore, the Draft risk was elavuated by ISO 7730 [3] as:

$$DR = (34 - t_A)(u - 0.05)^{0.62}(0.37u\sqrt{\tfrac{2}{3}k} + 3.14) \quad (1)$$

where t_A is the air temperature (°C), u is the velocity magnitude (m/s) and k is the turbulent kinetic energy (m^2/s^2). The part $\sqrt{\frac{2}{3}k}$ is the turbulence intensity. Criteria for index DR can be split into Category A (DR 10 %), Category B (DR 20 %), Category C (DR 30 %).

Age of air (s) is definited:

$$\tau_i = \int_{t_1}^{t_2} \left[1 - \frac{C_i(t_1)}{C_i(t_2)}\right] dt \quad (2)$$

where C_i is the contaminant concetration (kg_c/kg_s) and t is the time (s).

3 Results

The results of the numerical simulations are presented as mean scalar field Age of air (AA) and Draft risk (DR). The object of interest was mainly the plane going through the centre of the driver.

The average AA fields in **fig. 3** also show that the freshest air near the driver's face is in the case of FV, MV, and the oldest air is in the case of CV.

In terms of DR, the best results were provided by the MV type, where there are no higher DR values near the driver. For type CV, the DR values in the thigh area are high, but the sensitivity to DR is lower thanks to the clothing layer.

In the FV type, there are very high draft risk values in the calf area, which are expected to have a summer case, their high susceptibility to DR.

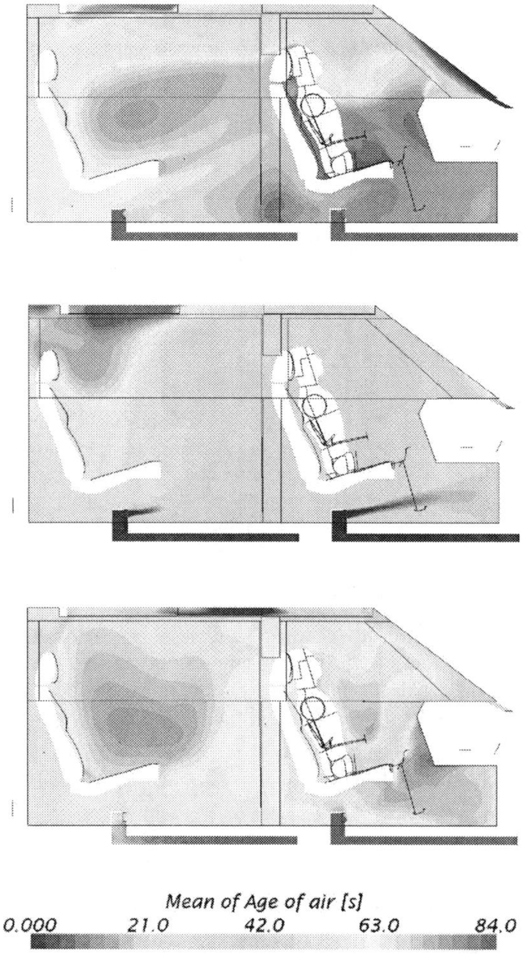

Fig. 3. Mean of Age of air (MV,FV,CV).

Table 2. Time averaged temperature, velocity magnitude, AA and DR in point at driver's nose.

	T [°C]	U [m/s]	AA [s]	DR [%]
Mix	25.45	0.178	52.9	7.8
Floor	22.17	0.148	45.2	10.4
Ceiling	26.85	0.163	65.7	6.1

The **table 2** shows the time averaged values of temperature, velocity magnitude, AA and DR in point near driver's nose. The highest temperature at the monitoring point is in the case CV, while the lowest in the case FV. The speeds for individual types of ventilation are up to 0.2 m/s. The highest AA is in case of CV and the lowest in the case FV. The maximum DR value is for type FV. Other types have DR up to 10 %.

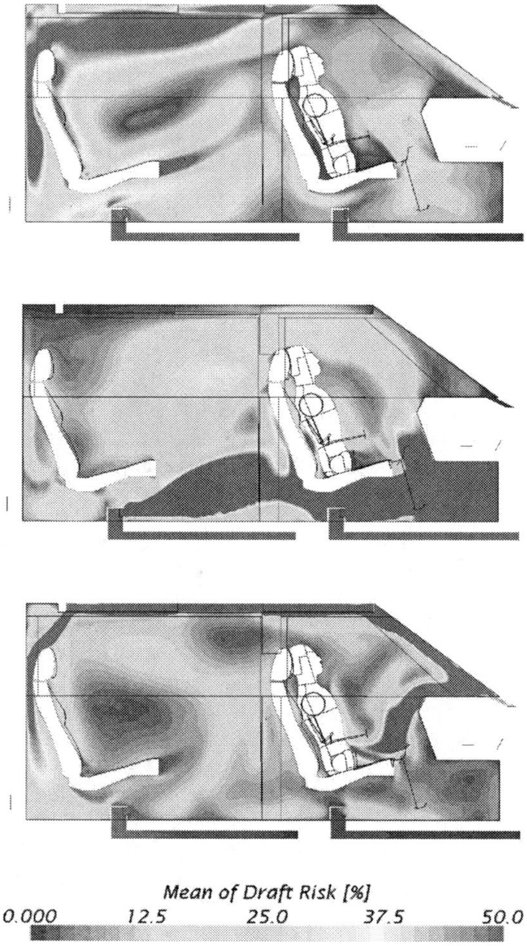

Fig. 4. Mean of Draft risk (MV,FV,CV).

From the results it is also evident that in the type of ventilation MV open ceilings vents in the flowing air, which leaves them in the rear of the car and can have a negative effect on the passenger. In the CV type, the main current core of the large-area vents diffusers adheres to the front and rear window respectively. This is due to the Coanda effect.

4 Discussion

In this first study just the classical interior was investigated.
MV is the best type for driver's comfort in terms of DR index. This is followed by CV and FV.
MV also provides the best results of AA, followed by CV and FV.

The mixing ventilation was evaluated as the optimal, however in case of flexi interior we can expect that the new designs of ventilation system will be more beneficial. To evaluate whole cabin is not presented in this study because should be considered effect of manikins sitting on rear seat

If we wanted to determine AA and DR for other passengers, they would have to be included in the simulation, because the presence of the manikin has a substantial influence on the flow field.
However, from the results (only with a manikin on the driver's seat), a higher DR level in the rear seat behind the driver can be included. In the FV type, it is possible to deduce a higher DR level in the rear seat passenger at the calf area.

5 Conclusion

The virtual mock-up was designed and the CFD simulation of internal air flow were performed with focus on temperature and velocity field Furthermore the index age of air and draft risk were evaluated especially in vicinity of manikin.
From obtained results can be concluded:

- The geometry of the manikin has a great influence on the flow pattern inside vehicles, so there is no point in evaluating the thermal comfort values in the seat where the manikin is not located.
- In mixing ventilation case large-area ceiling vents should be closed to avoid massive flow through ceiling channels allows to In the next step, these vents will be defined as a porous baffle, which should eliminate this effect.

For air conditioning of the cabin in summer conditions the MV had the optimal AA and DR.
The next step is to optimize ratio between flow rates from FV and CV and carried out evaluation of thermal comfort using equivalent homogenous temperature.

Acknowledgement

This research has been realized using the support of Technological Agency, Czech Republic, programme National Competence Centres, project # TN01000026 Josef Bozek National Center of Competence for Surface Vehicles. This support is gratefully acknowledged.

References

1. T. Dehne, P. Lange, A. Volkmanns, Building and environment, **129** (2018)
2. C. Kučera, Mater Diploma Thesis, Brno University of Technology, 2018
3. V. Pavlík, Master Diploma Thesis, Brno University of Technology, 2018

Boundary Layer Transition by Advese Pressure Gradient

Vladislav Skála[1,*], and *Pavel* Antoš[1]

[1] Institute of Thermomechanics of the CAS, v. v. i. Dolejškova 1402/5 182 00 Praha 8, Czech Republic

Abstract. For some cases of diferrent of boundary layer development conditions (roughness of surface, turbulence in main flow) by adverse pressure gradient was determined tranzitiom region. Boundary layer Tranzition region was found based on intermittency development of flow inside the boundary layer. Intermitency was evaluated using modiffied TERA method algorithm.

1 Introduction

Results of data analysis of experimental investigation of the transition of the laminar boundary layer to turbulence on a flat plate into a diffuser channel with and without adverse pressure gradient are presented. Different cases with a combination of some boundary layer development conditions were investigated.

It was studied cases with combination of different roughness of surface by different free stream turbulence intensity and by pressure gradient along boundary layer. This combination of conditions is usually common in real flow tasks.

Better understanding and description of effect of an adverse pressure gradient and another conditions like roughness of surfaces or turbulence of external flow on the boundary layer transition, location and size of the transition zone by transition from laminar boundary layer to turbulence, seems to be important.

The transitional intermittency represents by distribution of intermittency factor γ the boundary layer development. The distributions of an intermittency factor γ and on its based determination of onset and terminating of transition zone are presented in this work.

2 Pressure Gradient

The pressure gradient is often present in the free stream dependently on the changings of the mean velocity along the flow. An adverse (unfavorable) pressure gradient in the external flow is often present in technical tasks, so its a frequent phenomenon in internal and external aerodynamics. It can be identified for example along the diffuser walls or airfoil and blade profiles (tail part on the suction side), that means very serious cases. The unfavorable pressure gradient in the external stream has a strong destabilizing effect on the boundary layer.

3 Experiments

3.1 Experimental setup

The experiments were carried out in the closed type wind tunnel and the investigated boundary layers were developing on the smooth or rough flat plate. The orthogonal coordinate see Fig. 1.

Cross section of the test section of the wind tunnel was 0.9 m x 0.5 m. The length of a flat plate was of 2.65 m. Flat plates with different roughness of surface can be changed (aerodynamically smooth and sandpaper roughness P32 and P60 particles/cm^2). The flow acceleration begins about 1.2 m upstream the flat plate.

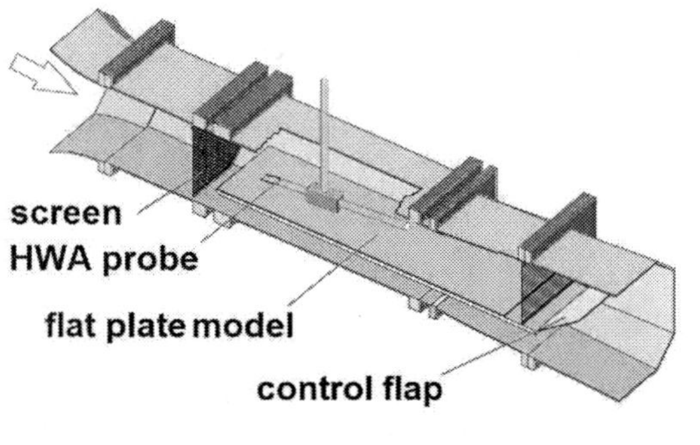

Fig. 1. Experimental setup - flat plate in the windtunnel test section

The flow deceleration starts at the level of the leading edge of the flat plate (x = 0) where the flow velocity is set to U_e = 10 m/s. Cases with adverse pressure gradient are provided by plane diffuser expanding part putted into test section. The diffuser channel was created in the test section for make the adverse pressure gradient in main flow belong flat plate. Length of this diffuser section is 1.4 m and opening angle was 5°. It was checked carefully that there was no flow separation on the deflected wall.

* Corresponding author: skala@it.cas.cz

The free stream turbulence was controlled by square plane mesh (grid) with cylindrical rods and square mesh holes. The free stream turbulence intensity Iu_e at the leading edge was naturally Iu_e = 0.005 or set by the special screen (square grid) to Iu_e = 0.03. The leading edge of the plate had a super-ellipse shape MSE6.

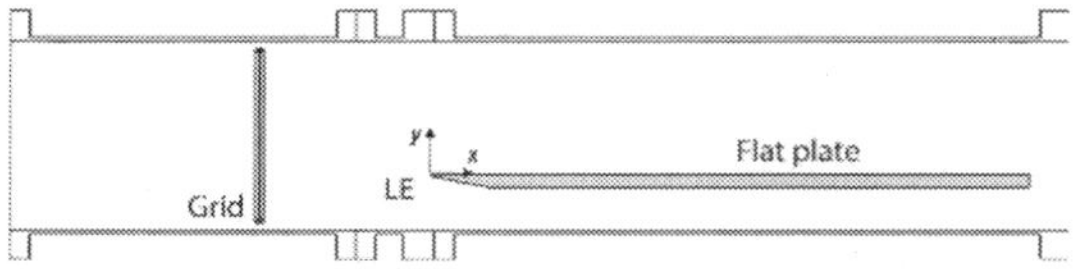

Fig. 2. Experimental setup - flat plate without diffusser part in the windtunnel test section.

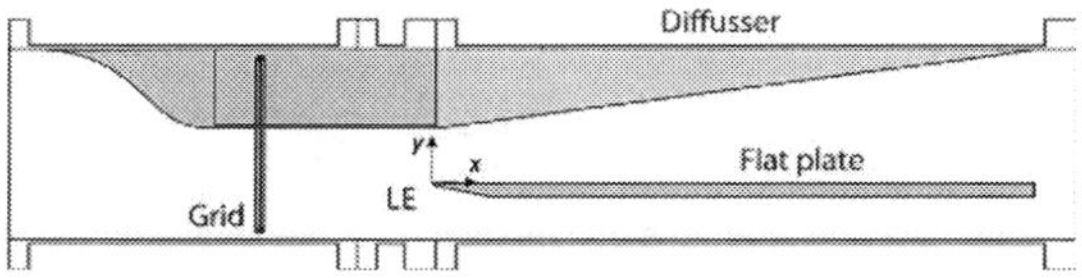

Fig. 3. Experimental setup - flat plate with diffusser along the model of flat plate in the windtunnel test section.

3.2 Measurement techniques

Flow in the boundary layer was studied by constant temperature anemometer technique. It was used a hot wire anemometry equipment CTA system Dantec Streamline.

The hot-wire probe was put into defined positions by the computer controlled traversing device. The distance y from the wall was checked manually by cathetometer with accuracy ±0.02 mm relative to the surface or to peaks of roughness (used fine control plate with area 5mmx20mm below the hot-wire probe).

The reference value of the free stream velocity was measured by a Prandtl pressure probe (diameter 6 mm) at the inlet of test section. Prandtl probe signal and barometric pressure was connected to pressure transducer Druck DPI 145 (range 7 kPa, accuracy ±0.005% FS). The flow temperature was measured by thermometer Pt100. Output voltage from the transducers were read by Data acquisition unit HP 34970A.

A single hot-wire probe was used for HWA measurement. A sensor of the probe was tungsten wire. Diameter of the wire was d_w = 5e-06 m and length of the wire was l_w = 1.25e-03 m. Operating wire temperature during measurement was set T_w=493 K. Hot wire sensor was oriented parallel to leading edge of flat plate. The output anemometer signal was digitalized using the A/D transducer National Instruments data acquisition system and recorded by the PC using the LabVIEW scripts (sampling frequency 75 kHz, 16 bit, time of every measurement was set to 20 s).

A cooling law for heated sensor of Koch and Gartshore (1972), was used for hot-wire measurements.

Data analysis TERA method [2] was performed using the LabVIEW scripts, results were organized using MS EXCEL.

3.3 Measurement uncertainties

The experimental uncertainties Δ are estimated based on calculated root mean square errors of interpolations and observed repeatability. The upper limits of relative errors of evaluated quantities depend mainly on precision of location of the probe (the distance y from surface of flat plate) and accuracy of hot-wire local instantaneous velocity U measurement:

$\Delta U/U = \pm 0.005$; (U_e = 10 m/s);
$\Delta y = 0{,}001\text{m}$;
$\Delta H_{12}/H_{12} = \pm 0.01$;
$\Delta\, \gamma\, /\, \gamma = \pm 0.2$.

3.4. Determination of the intermittency

The transitional intermittency is describing the laminar to turbulence transition process development in the boundary layer along the surface of the flat plate, see Fig.4.

Boundary layer starts from leading edge of the flat plate in stream wise direction and in specific distance from leading edge starts the process of laminar to turbulent flow transition. It relates to the generation and propagation of turbulent spots, turbulent spots are detected using TERA method. The transitional intermittency factor $\gamma(x)$ is one of the key parameters in physically right recognition of the transitional boundary layer state.

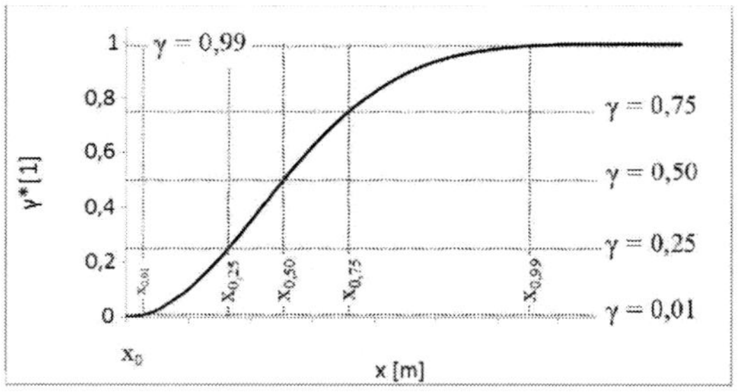

Fig. 4. Boundary layer tranzition – process mdescription using intermittency coeffitient.

Onset of transition corresponds to minimum of its value (ideally $\gamma \approx 0$) and the end of transition region is indicated by1 $\gamma \approx 1$. Increasing value of the intermittency coefficient upstream the transition onset could be explained by disturbances penetrated from outside the boundary layer and interactions with particless of fluid from boundary layer. [3]

The intermittency factor computed by a direct method is defined as time ratio of turbulent flow occurrence time over whole measurement time

$$\gamma(x) = \sum_{i=1}^{N} \frac{I(x,t_i)}{N}, \quad (1)$$

where $I(x,t_i)$ is indicator function in point x and time t_i, derived from digitalized instantneous velocity record captured close to the surface of flat plate in distance x from leading edge of the flat plate, see below and Fig. 4 - Fig. 6.

Determination of onset and terminating of transition zone is usually covenanted. The transition onset is covenanted in $x_{0,99}$, where $x_{0,99j}$ is a point, where $\gamma(x_{0,01}) = 0{,}01$; the transition terminating is covenanted in $x_{0,99}$, where $x_{0,99j}$ is a point, where $\gamma(x_{0,99}) = 0{,}99$.

Direct method TERA of determination of the intermittency factor is very effective in detecting the start and the end of transition with proper choice of parameters for signal processing and analysis. A lot of methods of determination γ is available [1, 2], the TERA (Turbulent Energy Recognition Algorithm) method [2] was chosen.

The TERA method consists of several consecutive steps, with modify raw hot wire signal of instantneous velocity to the criterion function, see [x1, x2].. At the first, the raw instantaneous velocity fluctuations signal from CTA anemometer is filtered to $\tilde{u}$ (t) by Butterworth filter with low pass frequency 1 kHz to eliminate parasitic noise from the signal. Sensor HWA was sensitive to stream wise velocity component and to vertical velocity component.

The second step is derivation of the signal to the detector function $D(t)$. Time derivation of signal makes the differences of the signal time behaviour during turbulent and non-turbulent periods more recognizable. Detector function is defined as

$$D(t) = \left| u \frac{\partial u}{\partial t} \right| \quad (2)$$

where u is fluctuations of the instantaneous velocity. Then the HWA signal (Fig. 2) is filtered (Fig. 3) and derived to the detector function (Fig. 4). Then is detector function smoothed to criterion function (Fig. 5). The threshold value K_P is applied onto criterion function $K(t)$, Fig. 5, and a its result is an indicator function I (t), Fig. 6, details are presented in [x4]). The threshold value K_P is in base TERA method given by

$$K_P(x) = C_2(\tilde{u}(x,t)\mathrm{d}\tilde{u}(x,t)/\mathrm{d}t)_{RMS}), \quad (2).$$

index RMS means the standard deviation evaluated during a time of measuement of one data point. [1, 2]

The disadvantage of TERA algorithm is well known. It is necessary manually find the "right value" of threshold constant C_2 in equation (2).

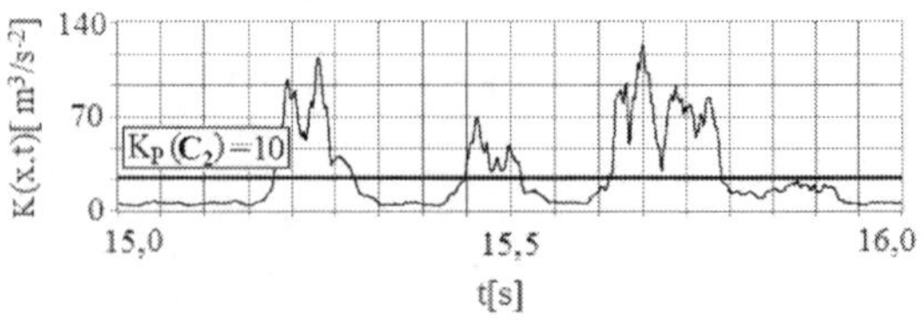

Fig. 5. Criterion function K(t) with threshold value $K_P(C_2)=10$.

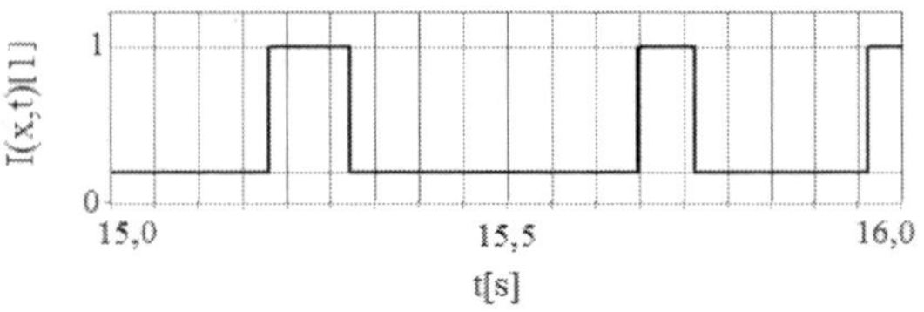

Fig. 6. Indicator function $I(t)$, [3].

It was discussed different ways, how to find an optimal and objective value C_2^* of threshold constant.

The most successful is a criterion using quality of interpolation of evaluated experimentally data points γ with universal intermittency function γ^* [7]

$$\gamma^*(x) = 1 - exp\left(-0{,}412\left(\frac{x - x_{0,01}}{x_{0,75} - x_{0,25}}\right)^2\right). \quad (3)$$

First step is calculation of a group of γ_j^* functions for different values C_{2j} from (empirically recomended) interval

$$C_{2j} \in (0{,}05; 0{,}6). \quad (4)$$

where j = ⟨1, M⟩, M is a number of iterations for evaluation C_2^*.

Universal intermittency function γ_j^* for C_{2j} is expressed

$$\gamma_j^*(x) = 1 - exp\left(-0{,}412\left(\frac{x - x_{0,01j}}{x_{0,75j} - x_{0,25j}}\right)^2\right), \quad (5)$$

where $x_{0,25j}$ is a point, where $\gamma_j(x_{0,25j}) = 0{,}25$; $x_{0,75j}$ is a point, where $\gamma_j(x_{0,75j}) = 0{,}75$, for given C_{2j}.

Values *of* $x_{0,25j}$ and $x_{0,75j}$ (for given C_{2j}) should be found using TERA method from only locally interpolated data points of $\gamma_j(x)$.

The exact value of $x_{0,01j}$ (point of start of the transition proces) is very difficult to find directly from locally interpolated data points of $\gamma_j(x)$, function γ is here very flat and experimental data are usually not enough smooth. Value of $x_{0,01j}$ can be better found using another form of universal intermittency function $\gamma^{**}(x)$, see eq. (6), [8]. This form uses experimentally data points $\gamma_j(x)$ from the middle part of transition zone, function γ is here » *well shaped* ». The value of $x_{0,01j}$ is the last searched value in eq. (5)

$$\gamma_{0,01j}^{**}(x) = 0{,}01 = 1 - exp\left(-0{,}412\left(\frac{x_{0,01j} - x_{0,5j}}{x_{0,75j} - x_{0,25j}} + 1{,}3\right)^2\right), \quad (6)$$

where $x_{0,5j}$ is point, where $\gamma_j = 0{,}5$. Values of $x_{0,25j}$ and $x_{0,25j}$ are known, see before, the value of $x_{0,5j}$ can be found as the local interpolated value from experimental data point of $\gamma_j(x)$ too. This is analytical algorithm, how to find the γ^* or γ^{**} in every "j- iteration".

Another possibility how to find parameters of $\gamma_j(x)$ function consist in direct using of a simple optimization process internally in every "j-iteration" using interpolation of experimentally data points of $\gamma_j(x)$ by γ^* or γ^{**} function.

Experimental data points, which appears to be sure inside the transition zone of the boundary layer, $\gamma^*(x_i, C_{2j}) \in (0,2;0,8)$ or $\gamma^{**}(x_i, C_{2j}) \in (0,2;0,8)$, is necessary take with some higher weight-coefficient into internal optimizing process to supress influence of the data points, which appears to be far outside the transitional zone.

The quality of approximation ϕ_j of (from experimentally data evaluated) intermittent factor γ_j with universal intermittency function γ_j^* can be expressed (for N points of γ and γ_j^*)

$$\phi_j(C_{2j}) = \frac{1}{N}\sum_{i=1}^{N} \sqrt{\gamma(x_i, C_{2j})^2 - \gamma^*(x_i, C_{2j})^2} \,. \qquad (7)$$

Function $\phi_j(C_{2j})$ depends on the value of the threshold value C_{2j}. Optimization criterion to find the optimal and objective value of the treshold value C_2^* can be expressed

$$\min(\phi_j(C_{2j})) \rightarrow C_2^* \,. \qquad (8)$$

This process of experimental data points postprocessing seems be capable and simple improvement of the known TERA method. Now is it possible set the founded optimal value C_2^* into (up to now unknown) constant C_2 in equation (2) and the objective distribution of intermittency factor $\gamma(C_2^*)$ or in the form of the universal intermittency function γ^* can be now clearly expressed.

Another classical and historically older and undirect way musing Emmons hypothesis, how to evaluate the intermittency factor, consit in observing of some integral boundary layer parameter X development along boundary layer, such a boundary layer shape factor, skin friction coeffitient, etc.

$$X = (1-\gamma) X_L + \gamma X_T \qquad (9)$$

where δ_1 is a displacement thickness of boundary layer and δ_2 is a momentum thickness of boundary layer. For some basic cases, such as flat flate boundary layer, are values of X_L and value of X_T for turbullent boundary layer known. For example, shape factor H_{12} can be used

$$H_{12} = \frac{\delta_1}{\delta_2} \,, \qquad (10)$$

4 Results of the experiment

It was carried out a lot of hot wire measurement. It was made fine measurement in whole boundary layer region and its neighbourhood with good space and time resolution for cases of combination different conditions of boundary layer development (Tab. 1).

Table 1. List of the test cases. Combination of the external conditions of the development of the boundary layer along flat plate.

label	α [°]	surface	Iu_e [1]
0-S-0,005	0	smooth	0,005
0-S-0,03	0	smooth	0,03
0-R-0,005	0	rough P32	0,005
0-R-0,03	0	rough P32	0,03
5-S-0,005	5	smooth	0,005
5-S-0,03	5	smooth	0,03
5-R-0,005	5	rough P32	0,005
5-R-0,03	5	rough P32	0,03

Especially in cases with adverse pressure gradient it is not easy find a point of velocity profile (or a distance of the probe from the surface of the flat plate), where to investigate transitional intermittency, see Fig. 7. Position and size of identified transitional zone is in different points different. The best results seems to be for distance about 1,5 mm from the surface or peaks of surface roughness (corresponds to $y^+ \approx 100$), more precise criterion was not found yet. In cases without adverse pressure gradient is possible use signal from the position of probe in the mean velocity profile, where the sign of the skewness parameter $S(u)$ is changing (maximum of turbulent spots generation).

Intermittency factor $\gamma(Re_x)$, shape factor H_{12} and pressure gradient or outer mean velocity decrease $\overline{U_\infty} / \overline{U_{\infty_0}}$ respectively was evaluated. Distributions for different conditions of developing of the boundary layer (Tab. 1) along the flat plate are plotted in Fig. 8 and Fig. 9, dependently on the distance x from the leading edge of the plate or on the local Reynolds number Re_x.

 https://doi.org/10.1051/epjconf/202226901057

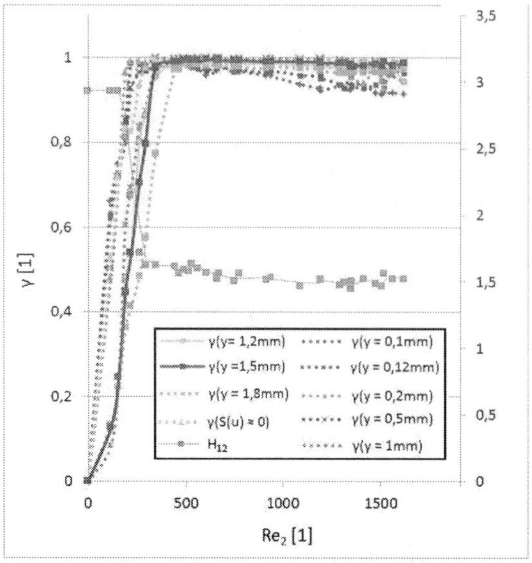

Fig. 7. Intermittency factor and shape factor in different points of mean velocity profiles of boundqary layer, example. Position and size of identified transitional zone seem to be different. Plotted usingn Reynolds number Re_2 related to the momentum thickness δ_2.

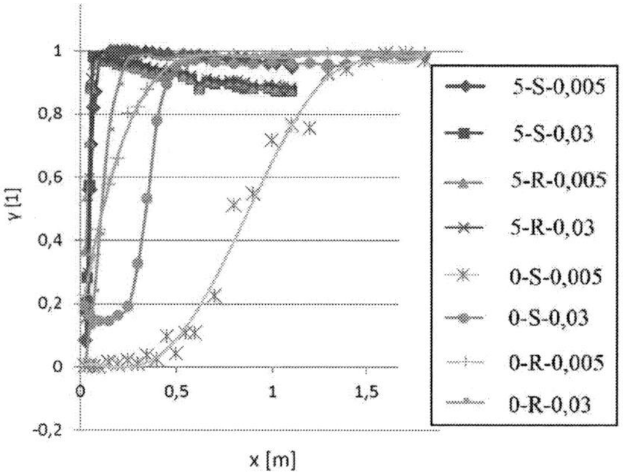

Fig. 8. Universal intermittency function γ^* describes boundary layer transition process, all cases.

5 Conclusions

Adverse pressure gradient accelerates the transition process to bypass transition of boundary layer. Transition starts much closer to the leading edge of the flat plate and the length of the transitional zone is much shorter. Higher free stream turbulence intensity and effect of and roughness of the surface are too recognizable and significant, but the separation of its influence is probably not possible, the process of bypass transition is nonlinear with significant nonlinear interactions in flow field in the boundary layer. Results can be used to numerical models validations.

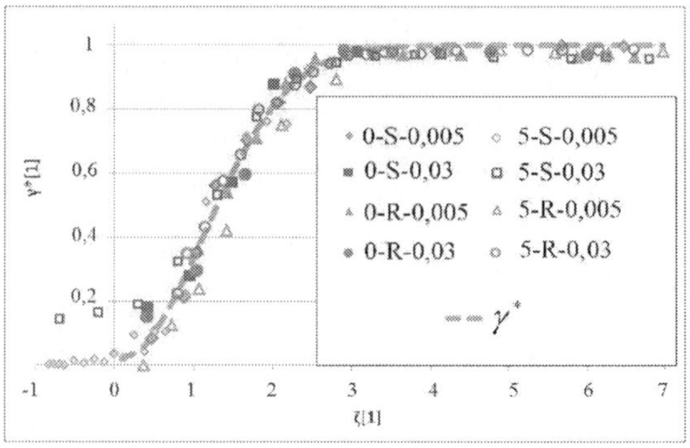

Fig. 9. Universal intermittency functionγ^* describes boundary. layer transition process, mcases without outer pressire gradient – influence of free stream turbulence and roughness of surface of the flat plate. Label explained in Tab. 1.

Acknowledgement:

This work was supported by the Grant Agency of the Czech Republic, project No. 17-01088S.

References

1. Z. Zaric, R.E.Falco, R.F. Blackwelder, Technique for the Detection of Coherent Structures in Wall-Bounded Flows and its Application to the Analysis of MultipleWire Signals, (1984) pp. 8. 1.
2. R. E. Falco, C.P.Gendrich, Turbulence Burst Detection Algorithm of Z. Zaric,. Proc. of the Int. Centre for Heat and Mass Transf.1988, pp. 911-931.
3. R. Narasimha, The Laminar – Turbulent Transition Zone in the Boundary Layer, Prog. Aerospace Sci. Vol. 22, pp 29-80, Pergamon Press, 1985.
4. R. Narasimha, On the Distribution of Intermittency in the Transition Region of a Boundary Layer, J. Aero. Sci. 24, (9), 711-712, 1957.
5. H. W. Emons, The Laminar Turbulent Transition in Boundary Layer, Jour.Aero.Sci.18 (1951) 490-498.
6. V. Skála, P. Antoš, O. Hladík , Intermitence factor evaluation by bypass boundary layer tranzition in flows on rough surface on flat plate, Proc. of EFM 2016, (2016).
7. V. Skála, V. Uruba, P. Antoš., P. Jonáš.: Intermitence Factor Evaluation by Bypass Boundary Layer Transition in Flows on Rough Surface on Flat Plate. In: *Experimental Fluid Mechanics 2018*. Liberec: Technická univerzita v Liberci. Katedra energetických zařízení. 2018.

Overshot water wheel efficiency measurements for low heads and low flowrates

Jaroslav Štigler[1,*]

[1]Brno University of Technology, Facutly of Mechanical engineering, Energy Institute, 616 69, Technická 2, Brno, Czech Republic

Abstract. Utilizing of low renewable energy sources is an actual topic nowadays. This paper is focused on the locations with the low hydro-energy potential. It means locations with a head about 1m and flow rates about tens of litres per second. Hydro turbines for such parameters belong to the pico-turbines. It is not easy to design the hydro-turbine for such parameters with an acceptable efficiency. One of the possibilities is to use a water wheel. It is possible to find in a literature that the efficiency of overshot water wheels achieves 85%. This is the very good efficiency. It is not so easy to find some experimental data which proves this statement. Aims of this work are to design a simple manufactured overshot water wheel, measure the hydraulic efficiency of this wheel for wide range of parameters, finding an effect of the leading edge adjusting and the effect of the channel outflow edge position on the water wheel efficiency.

1 Motivation

A big effort has been expend to utilize renewable energy sources nowadays. The attention is not focussed only on big energy sources but also on a small one. It means the sources with the power of hundreds or tens of watts. These small sources can be very useful in case of remote residences without an electric grid. Moreover, the prices for the electric energy will grow more and more.

This paper is focussed on the water renewably energy. It is not so easy to design a water turbine for very small hydro energy potential with the high efficiency. There are number of different interesting solutions how to transfer the water energy to the electrical energy. One of the very interesting solutions is the bladeless water turbine SETUR [1]. This turbine utilizes very interesting flow phenomena Coanda effect to transfer water energy to mechanical energy but efficiency is very low for low hydro energy potential [2, 3].

Another possibility to utilize this low hydro potential is using a water wheel. Water wheels were used to transfer the water energy into the mechanical energy for many years. Many water wheel types, for different flow conditions, were developed and tuned to obtain maximal efficiency. These wheels have rather complex shape of blades and its production is not easy. But the efficiency is rather high. Well designed water wheels can reach efficiencies 75% (undershot) to 85% (overshot) [4].

The water wheels are processing mainly potential energy of water (Overshot), but some of them are designed to process a kinetic energy (Undershot) [5].

The only problem, to produce electric energy, is low speed of the water wheels. It is necessary to use a gearbox.

Many do-it-yourself men made different constructions of simple water wheels but no one measure its efficiency.

The water wheel, which can be produced easily, was designed in our department. Many parameters, which affect the efficiency, got into the experiment preparation. There was build the tests ring in our laboratory for basic characteristics water wheel measuring.

2 Water wheel design

The basic design idea was to make buckets of the water wheel from the pipe. The quarter of pipe was cut off. These buckets are then squeezed between two plastic circular discs. This construction allows an adjustment of the bucket leading edge positions against the inlet stream. Water wheel and shape of the bucket is in the figure 1.

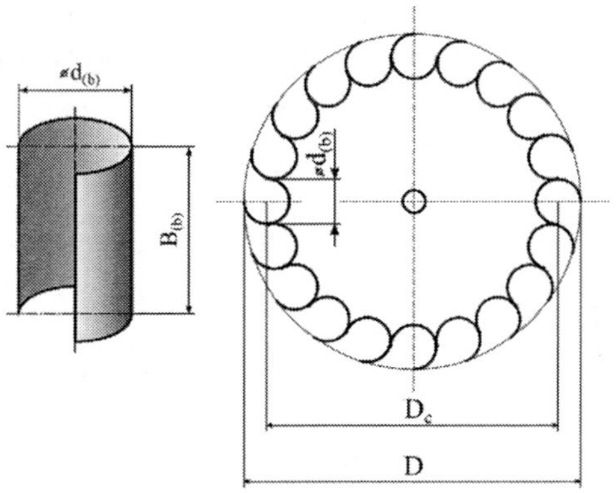

Fig. 1. Bucket and water wheel shape.

* Corresponding author: stigler@fme.vutbr.cz

Beside the easy bucket design, the idea was that this bucket shape allows processing not only potential energy but also the kinetic energy of the water stream. Basic wheel dimensions are in the next table

Table 1. Basic wheel dimensions.

Description	Label	Units	Value
Maximal wheel diameter	D	[mm]	900
Bucket centres diameter	$D_{(c)}$	[mm]	760
Bucket diameter	$d_{(b)}$	[mm]	140
Bucket width	$B_{(b)}$	[mm]	290

3 Experiment layout

Water wheel location and experiment layout is depicted in the figure 2. Water is flowing from a large container through rectangular channel. The end of the channel can be elongated. Than the water stream impacts water wheel buckets.

Some parameters are mentioned in the picture. Total head ($H_{(t)}$) is measured from the wheel bottom up to the water level in the container. If this total head is used for efficiency evaluation then the hydraulic losses in the channel are included into this value. The head H has to be taken for evaluation of the hydraulic water wheel efficiency.

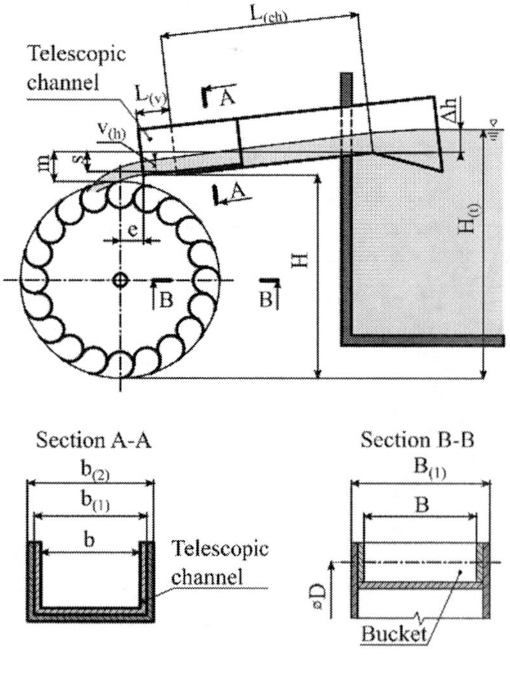

Fig. 2. Experiment layout.

A few words is necessary to say about the channel. The channel consists of two parts. One part is fixed and another part is movable. It allows the total channel length elongation. The length of the channel, it is better to say the location of the channel outflow edge, plays significant role in the water wheel efficiency. Length of the fixed channel part is signed as $L_{(ch)}$ and the channel elongation is signed as $L_{(v)}$. The difference of the inlet and outlet fixed channel edge altitudes is signed as s. The slope of the channel bed can be expressed as $s/L_{(ch)}$.

The difference of the water wheel top altitude and the inlet channel edge altitude is signed as m. The horizontal distance between water wheel axes and outlet channel edge is signed as e. The water depth near the channel outlet is signed as $v_{(h)}$. This is an important parameter for kinetic energy determining. The vertical distance of the water level in the tank from the inlet channel edge is singed as Δh.

4 Basic parameters measuring

Testing circuit has been built in the laboratory of Department of Fluid Engineering. The schema of this circuit is outlined in the Figure 3

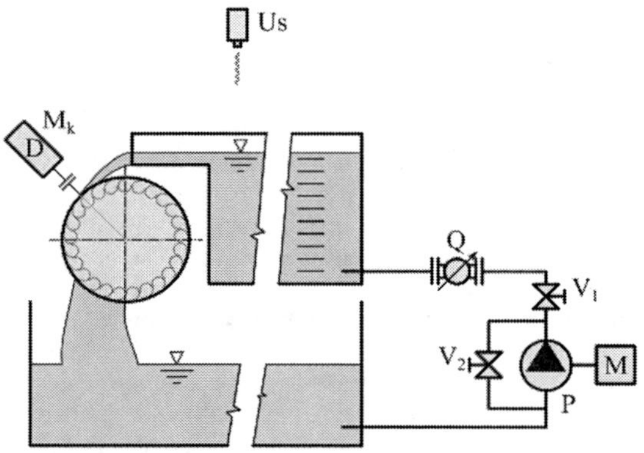

Fig. 3. Testing circle schema.

Feeding pump (P) delivers water from the suction tank to the upper tank. Then the water flows through a slates grid to calm the flow disturbances down. The channel, directing the water on the water wheel, is placed on the opposite side of the upper tank. The water falls down to the suction tank after it leaves the wheel. The fluid flow rate is controlled by the bypass equipped with two valves ($V_{(1)}$, $V_{(2)}$). The flow rate was measured by the magnetic flow meter (Q) DN 80 type MQI series 99 SMART (precision ±0.2% from the measured value). The horizontal water level position was measured by an ultrasonic water level sensor (Us) type 2HMU series 99 Smart (sensor capacity is 2m, precision is 0.8% from the capacity) . The speed of rotation and the torque on the wheel shaft was measured by the dynamometer (D). Torque was measured by torque measuring flange typeT10F HBM. Torque capacity is 50 Nm. Precision is 0.2% of this capacity. Wheel speed is measured by optical incremental sensor which is a part of the torque measuring flange. Precision is 0.1% of the setting capacity (3000 rpm).

All measured data were collected in the computer. All data were measured within 30 s time interval with the frequency 10 Hz. After that the average values were calculated.

Bolt with sharp end above water level was used for water depth measuring in the channel. The measurement was rather rough because of the water level waves. This measurement was done in two locations of the channel span and average value was then calculated.

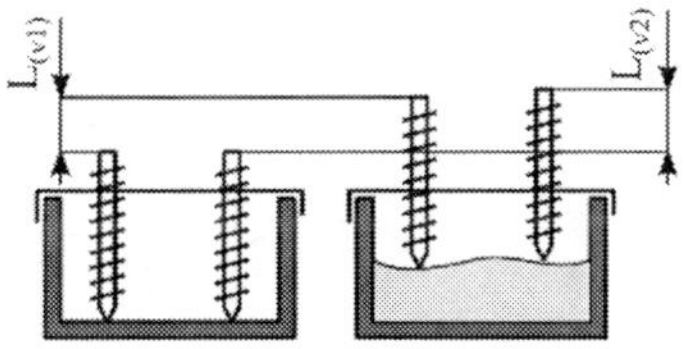

Fig. 4. Measuring of the water depth in the channel.

5 Measuring procedure

There are two adjustable parameters in this water wheel arrangement. First of them is bucket adjusting and second is horizontal position of the outflow channel edge. Bucket position labelling is outlined in the figure. 5. The value means the angel from the vertical axis.

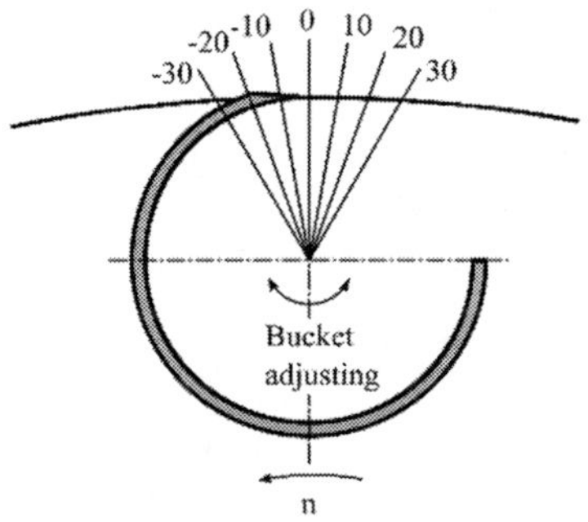

Fig. 5. Labelling of the bucket adjusting.

Two-bucket positions were tested 0 and -20. Labelling of the horizontal position of the outflow channel edge is outlined in the Fig. 6. Position value 0 means that the outflow channel edge is exactly in the water wheel axis. It means that the value of the channel edge position corresponds with the dimension *e* in the figure 2. All six-outflow channel edge positions were tested. Twelve configurations of bucket and outflow channel edge positions were tested. About ten different flow rates, in a range from 5 $l.s^{-1}$ to 25 $l.s^{-1}$, were tested for each configuration. Lower limit is set due to the flow meter precision and upper limit is set due to the safety overflow edge placed in the upper tank. The upper limit of flow rate was lower for some outflow channel edge positions because the water stream flowed behind water wheel. It means that part of the flow rate was not utilized. About eight different wheel speeds was set for each flowrate setting.

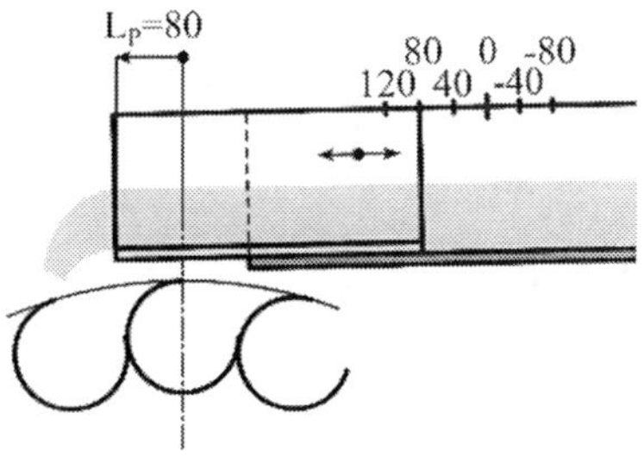

Fig. 6. Labelling of outflow channel edge position.

The wheel speed was set within the range 15 rpm to 65 rpm approximately. Lower limit was about 35 rpm for some high flowrates because the dynamometer torque limit was reached in these cases.

6 Basic expressions

The aim of these tests were to gather all data to be able calculate hydraulic water wheel efficiency and power on the wheel shaft and then find influence of bucket and outflow channel edge positions on them. Available input power can be calculated in two ways. The easiest one is to express it from the total head $H_{(t)}$.

$$P_{(int)} = \rho.Q.g.H_{(t)} \tag{1}$$

The efficiency value is slightly lower than in the next expression because including of channel hydraulic losses in the value. The different head value is taken into consideration in the next input power expression. The head (H) is measured as vertical difference between outflow channel edge position and the wheel bottom in this case. I is necessary to take into account kinetic energy of the stream into the input specific energy. Another expression for input power then is

$$P_{(in)} = \rho.Q.\left(\frac{v^2}{2} + H\right) \tag{2}$$

The kinetic energy of the stream is calculated from the average velocity (*v*) in the channel outflow. The hydraulic losses in the channel are excluded in this approach. The average velocity can be expressed this way.

$$v = \frac{Q}{b.v_{(h)}} \tag{3}$$

The only problem is in height of water in the channel as it is explained at the end of the chapter 4.

The expression (1) will be taken as the input power in this case.

The power on the wheel shaft can be calculated from the expression (4).

$$P_{(out)} = M_{(k)}.2.\pi.n \quad (4)$$

Speed of wheel is signed as n [s^{-1}]. Hydraulic efficiency ($\eta_{(h)}$) can be then calculated this way

$$\eta_{(h)} = \frac{P_{(out)}}{P_{(in)}} = \frac{M_{(k)}.2.\pi.n}{\rho.Q.\left(\frac{v^2}{2}+H\right)} \quad (5)$$

Another, maybe interesting, result is calculation of unit speed ($n_{(11)}$) – unit flow rate ($Q_{(11)}$) diagram. Unit speed can be expressed this way.

$$n_{(11)} = \frac{n.D}{\sqrt{H}} \quad (6)$$

It is a speed of imaginary wheel with diameter 1 m and operating under head 1 m.

Unit flow rate can be expressed similar way like in case of Banki turbine

$$Q_{(11)} = \frac{Q}{D.b.\sqrt{H}} \quad (7)$$

7 Results and discussion

It is impossible to show all obtained results, due to a huge number of measurements, here in this paper. Only some of them will be chosen to show them.

The results of efficiency will be discussed first. The input power was calculated from expression (1). If the expression (2) is taken into consideration then the efficiency is slightly higher.

The efficiency diagram for configuration bucket -20 and channel edge position is in the figure 7. The dots represent values obtained from the measurements. The lines represent polynomial approximation of the function between efficiency and wheel speed.

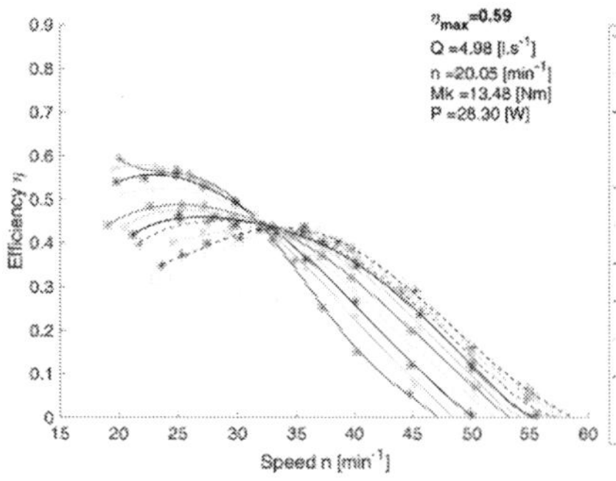

Fig. 7. Efficiency for configuration, bucket -20 and channel edge position -80 for different flow rates.

The number beside the Q (flow rate), in the diagram, means the rounded value of flowrate. Maximal efficiencies are reached for low flow rates and low speeds. The efficiency will be probably higher for lower wheel speeds. But it was difficult to do measuring below wheel speed 15 rpm due to high torque pulsations. And also low wheel speeds are not convenient for electricity production.

The efficiency diagram for configuration with total maximum efficiency of these tests is in the Figure 8. Total maximum efficiency was reached for configuration of the bucket 00 and channel position -40.

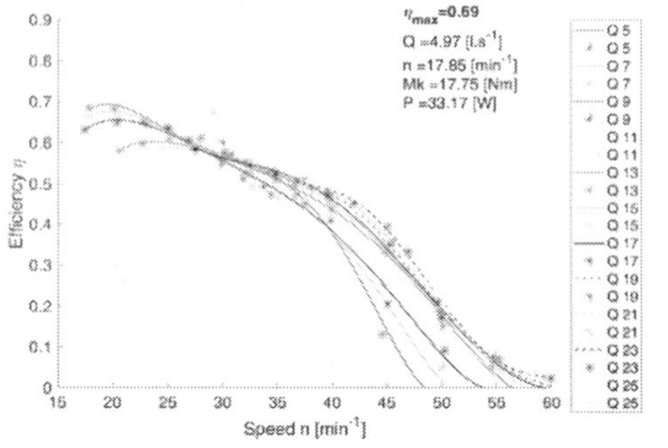

Fig. 8. Efficiency for configuration, bucket 00 and channel edge position -40 for different flow rates.

Basic parameters for maximal efficiency at each configuration are listed in the Table 2. Maximal efficiency for bucket configuration is signed by the grey background. It is apparent that the efficiency is higher for bucket configuration 00 in all cases of channel edge position.

Table 2. Basic parameters for maximal efficiency obtained at each configuration.

Configuration		η_{max}	Q	N	P
Buck	Chan	[-]	[$l.s^{-1}$]	[min^{-1}]	[W]
-20	**-80**	**0.59**	**4.98**	**20.05**	**28.30**
	-40	0.58	5.02	21.31	28.29
	0	0.55	6.84	23.66	36.92
	40	0.53	12.95	35.37	69.47
	80	0.53	9.00	40.03	47.18
	120	0.56	6.96	35.56	38.33
00	-80	0.66	7.00	20.63	45.29
	-40	**0.69**	**4.97**	**17.85**	**33.17**
	0	0.66	4.98	25.15	31.82
	40	0.65	5.02	25.78	31.86
	80	0.66	6.91	24.36	45.25
	120	0.66	6.98	34.76	45.18

The explanation is that the bucket in configuration -20 is more open and water is leaving the wheel prematurely. It is interesting that in the case of bucket configuration -00 the maximal efficiencies are reached under almost the same flow rates for each channel position and the efficiency does not variate so much too.

Different situation is in case of bucket configuration -20. The maximal efficiencies are reached under different flow rates and the difference between them is bigger.

The only problem with this measurements and results is with the dynamometer torque limit for the high flowrates. It is apparent from torque diagram for the total maximum efficiency configuration (bucket 00, channel position -40) in the Figure 9. It was not possible to decrease wheel speed due to torque limit. Probably the efficiency would not be higher but the power could be bigger. It will be discussed later.

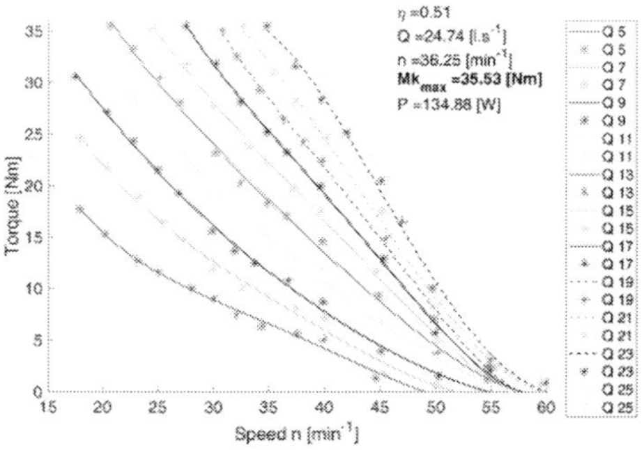

Fig. 9. Torque for configuration, bucket 00 and channel edge position -40 for different flow rates.

Now it will be useful to say a few words about efficiency. Maximum efficiency and maximum power do not correspond for given configuration. From the owner point of view is probably more important the output power then the high efficiency. Output power diagram for configuration bucket -20 and channel edge -40 is in the Figure 10. This is the case of maximal power for bucket configuration -20. The output power grows with the flow rate. Maximum output power is for flow rate 25 $l.s^{-1}$.

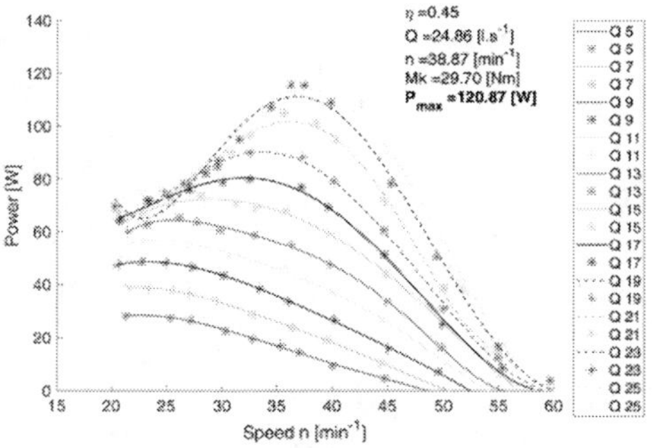

Fig. 10. Output power for configuration, bucket -20 and channel edge position -40 for different flow rates.

The wheel speed related to the maximal output power is higher than the wheel speed related to the maximal efficiency. The wheel speed related to the maximal power is about half of run away speed.

The power diagram for the maximal power for bucket configuration 00 is in the Figure 11. There is apparent influence of torque limit. If it is possible to measure with lower wheel speed then the maximum power will be probable higher. In this case the wheel speed related to the maximum output power will be shifted to the lower values than the half of run-away speed. In this case the wheel speed related to the maximum efficiency and the wheel sped related to the maximum power will be closer than in the case of bucket configuration -20.

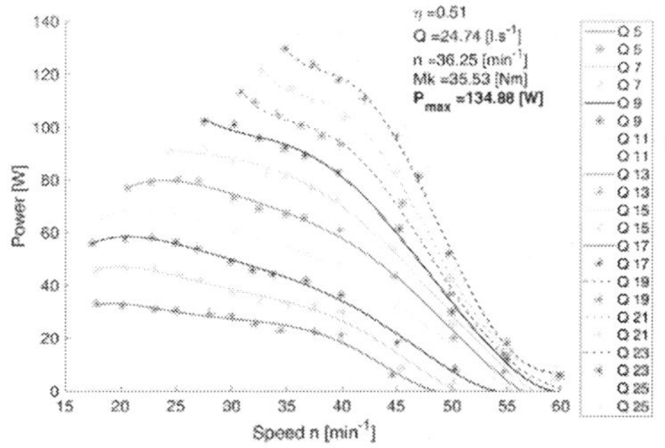

Fig. 11. Output power for configuration, bucket 00 and channel edge position -40 for different flow rates.

Table 3. Basic parameters for maximal output power obtained at each configuration.

Configuration		η	Q	N	P_{max}
Buck	Chan	[-]	[$l.s^{-1}$]	[min^{-1}]	[W]
-20	-80	0,43	24.89	33.79	114.47
	-40	**0,45**	**24.86**	**38.87**	**120.87**
	0	0,49	18,95	39.87	96.95
	40	0,48	14.94	34.85	73.56
	80	0,47	10.96	34.32	52.00
	120	0,56	6.96	35.56	38.33
00	-80	0.50	24.70	35.58	132.33
	-40	**0.51**	**24.74**	**36.25**	**134.88**
	0	0.58	18.93	34.91	113.78
	40	0.62	14.98	34.64	94.43
	80	0.65	11.08	35.18	71.59
	120	0.56	9.01	35.33	49.74

The basic parameters for maximum output power at each bucket and channel edge configuration are written in the Table 3. It is similar like in case of maximum efficiency. One can see that flow rate related to the maximal output power is strongly decreasing from the channel edge configuration 0, in both cases of bucket configuration.

Explanation is that the water stream is partially falling behind the wheel for these channel edge configurations (0, 40, 80, 120). It means that part of the flowrate is not utilized in the wheel. These channel edge configurations cannot be used for higher flow rates.

It is also necessary to say one note to the efficiency and output power. There was one problem with the water wheel and the channel design. The inner channel width is 235 mm and the inner wheel width is 271 mm. It is good that channel width is smaller than the wheel width but this difference is not sufficient. The water stream width is slightly bigger than the wheel width for high flow rates. It means that part of the flow rate is missing in case of high flow rates. If all water fall on the wheel then the power and efficiency could be slightly higher. 0

It is also interesting to draw the diagrams of unit speed (n_{11}) and unit flow rate (Q_{11}). The diagrams for three configurations are shown here. Two of them are for bucket configuration -20 with channel edge configuration -80 (Figure 12) and -40 (Figure 13) and one is for bucket configuration 00 and channel edge configuration -40 (Figure 14). These are chosen because the efficiency diagram or power diagram were aforementioned.

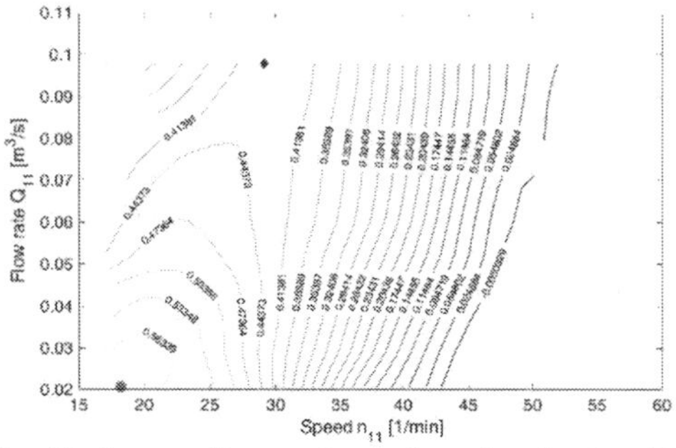

Fig. 12. Q_{11} n_{11} diagram for configuration, bucket -20 and channel edge position -80 for different flow rates.

The lines in the diagrams are the lines of constant efficiency. One can see that shapes of constant efficiency lines are oval. If this diagram is modelled in 3D then it is possible to see a ridge where efficiencies are rather high. The ridge sides are rather steep. It seems to be another local optimum of the efficiency in the diagrams Q_{11}–n_{11}. It rely happens in some other configurations.

It is apparent that some data due to torque limit are missing in the Figure 15. If one look carefully on this diagram then he can see that probably this is the case when the second local efficiency optimum will be there.

There are two dots in these three diagrams. Red circle dot shows point with the maximal efficiency and the blue diamond represents point with the highest output power.

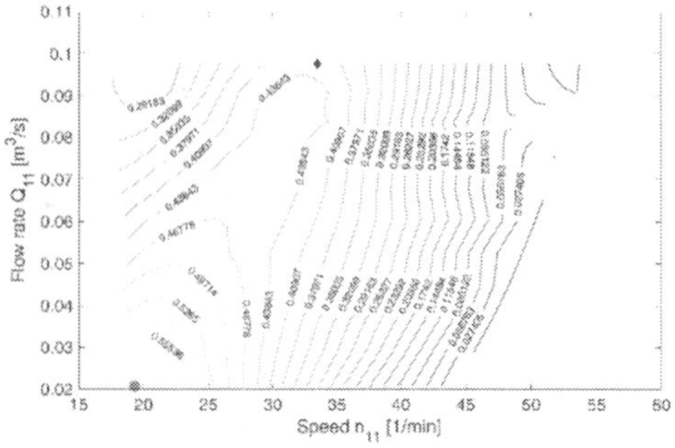

Fig. 13. Q_{11} n_{11} diagram for configuration, bucket -20 and channel edge position -40 for different flow rates.

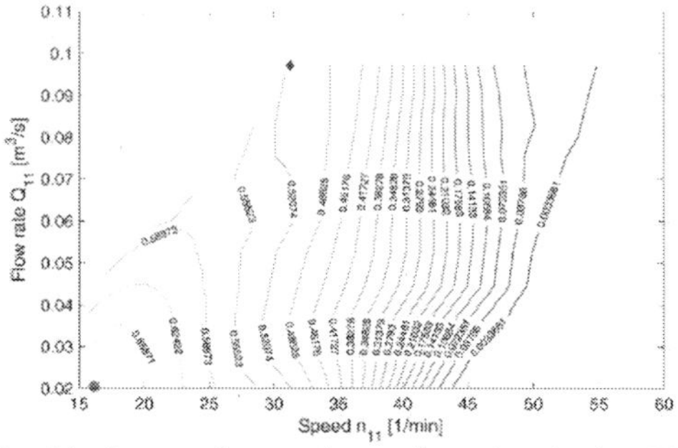

Fig. 14. Q_{11} n_{11} diagram for configuration, bucket 00 and channel edge position -40 for different flow rates.

As it is possible to see and as it was already mentioned the highest efficiencies are obtained for low flowrates and low wheel speeds. This is really interesting information in case of utilizing of hydro power sources with low potential.

8 Conclusion

The overshot water wheel with cylindrical buckets was tested and some results were presented in this paper. Advantages of this wheel type are in its easy production and possibility of bucket adjusting. Tested water wheel is in the figure 15.

Fig. 15. Tested water wheel.

Measurement results for two bucked configurations (-20, 00) and for six horizontal channel edge configurations (-80, -40, 0, 40, 80, 120) were presented here. Basic water wheel parameters were measured for about ten different flow rates, within the range 5-25 $l.s^{-1}$, and for about eight different wheel speeds, within the range 15rmp-run away speed, for each water wheel configuration. It was found that the hydraulic efficiency maximum and the output power maximum were obtained for the bucket configuration 00 and the channel edge configuration -40. The efficiency was calculated as the total efficiency it means that the head was measured as a distance between the wheel bottom and the water level in the upper container. The maximal efficiency value was 0.69, which is not bad for such small water wheel. The maximal output power was 135 W. The output power maximum was influenced by the dynamometer torque limit so it means that it would be probably higher.

There are some reserves in the efficiency maximum and output power maximum due to some volumetric losses. Part of the flow rate flowed beside water wheel. It is shown in the Figure 16.

Fig. 16. Volumetric losses for high flow rates.

Possible improvements are as follows

- minimize losses in the channel
- minimize volumetric losses

How to decrease hydraulic losses in the channel? The flow in the channel is supercritical, it is apparent from the Figure 17. Supercritical flow means that there are high velocities and the self-induced waves are generated. The hydraulic losses in this flow case are rather high. If the subcritical flow will be reached then the losses will be lower.

The subcritical flow can be reached by placing a sluice gate at the channel end, which drives the flow on the water wheel.

What about the volumetric losses decreasing? These losses can be decreased by the channel outflow adjusting or by the wheel width increasing.

What about the future research? The ways of future research are as follows

- Performance of tests with more bucket configuration, mainly positive bucket positions (10, 20 ...)
- Channel with the sluice gate at its end testing.
- Testing water wheel with different bucket diameters, higher and lower diameter.
- Testing the water wheel for bigger vertical distance of the channel outflow edge.

Fig. 17. Supercritical flow in the channel.

It is necessary to continue with the experiments to understand how all configuration parameters affect efficiency and output power. It was found that the bucket position 00 is better than the position -20 generally. Now more closed buckets will be tested, it means position 10 or 20. The water will not be poured out of the buckets easily but it is possible that some water stays in the bucket behind the lowest position and it means that this water will acts against the wheel rotation.

Next way will be testing modified channel with the sluice gate at the channel end. This will bring two advantages first, the hydraulic losses decreasing in the channel and second, the better flow stream regulating. There is one more parameter to be tested at the end. This is a diameter of the buckets under the constant wheel diameter. It is expected that it influences the flow rate which will be related to the maximal efficiency. It means that the wheel with the smaller bucket diameter will have maximal efficiency under lower flow rates. The wheel with bigger bucket diameter will probably have maximal efficiency under higher flow rates. This is only assumption and it is necessary to prove it.

The tests with the vertical channel outflow edge position could be interesting to carry out. In this case the channel outflow edge was placed as close as possible to the water wheel. It means that main energy transfer was from the potential water energy to the mechanical energy. When the channel outflow edge will be placed with bigger vertical distance then the transfer of kinetic energy to mechanical energy will play more dominant role.

Final summary of this work is that the water wheel of this construction is easy to make and the efficiency is rather big for the low heads and low flow rates. It is suitable for the remote residences due to its simplicity and high efficiency.

This research was financially supported under the project "Multiphysical problems of fluid mechanics ", number: FSI-S-17-4615.

References

1. V. Beran, M. Sedláček, F. Maršík. A new bladeless hydraulic turbine. *Applied Energy*. 2013, **104**, 978-983. DOI: 10.1016/j.apenergy.2012.12.016. ISSN 03062619.
2. M.Polák, L. Dlabal, Operating characteristics of a bladeless turbine for irrigation purposes. *Soil and Water Research*. 2016, **10**(No. 4), 278-283. DOI: 10.17221/28/2015-SWR. ISSN 18015395.
3. L. Dlabal, M. Polák, Characteristics of bladeless turbine. *Research in Agricultural Engineering*. 2016, **61**(No. 2), 87-91. DOI: 10.17221/70/2013-RAE. ISSN 12129151.
4. G. Müller, K. Kauppert, Performance characteristics of water wheels. *Journal of Hydraulic Research*. 2010, **42**(5), 451-460. DOI: 10.1080/00221686.2004.9641215. ISSN 0022-1686.
5. G. Helmizar, G. Müller. The effect of using upper shroud on the performance of a breashoot water wheel. *Journal of Physics: Conference Series*. 2019, 1175. DOI: 10.1088/1742-6596/1175/1/012269. ISSN 1742-6588.

Experimental Investigation of Unsteady Static Pressure Field Behind a Circular Cylinder

Jakub Suchý[1,*], *Jiří* Nožička[1], and *Václav* Papež[1]

[1]Department of Fluid Dynamics and Thermodynamics, Czech Technical University in Prague, Faculty of Mechanical Enginnering, Technicka 4, 160 00, Praha 6, Czech Republic

Abstract. In this paper a measurement of a static pressure in the wake behind a circular cylinder for the different Reynolds numbers is presented. For this purpose, a static pressure probe with built-in pressure transducer was used. The main goal is to describe the unsteady pressure field behind the cylinder and determine a frequency of the generated Von Kármán vortex street. The measurement was conducted in a low speed wind tunnel for a cylinder with 110 mm in diameter. The results were compared to an unsteady RANS simulation.

1 Introduction

The wake of a circular cylinder has been investigated by many researchers. The static pressure field behind the circular cylinder can be value source of information about the effects caused by the cylinder in the airflow. It is known, that the periodically shedded vortices called Von Kármán vortex street are responsible of generating tonal noise. [1] This can be considered as one of the best examples of aerodynamically generated sound. For the further research in the tonal sound generated by the cylinder in airflow it is necessary to know a frequency at which the vortices are shedded. This paper evaluates simple approach to determine this frequency by measuring the static pressure fluctuations behind the circular cylinder. To support this approach a CFD simulation using Ansys Fluent was also performed. Validation of this approach is necessary for the future aeroacoustical study of the generated tonal sound.

1.1 Von Kármán vortex street

The relation between Strouhal and Reynolds numbers for a circular cylinder was investigated for example by Lienhard [2]. For the Reynolds number between $3 \cdot 10^2$ and $3 \cdot 10^5$ the Strouhal number is close to 0.2 depending on the cylinder surface and other factors of the airflow. This also means that in this range the frequency of shedding vortices depends on velocity linearly.

For the experimental measurement presented in this paper a cylinder with 110 mm in diameter was chosen. The measurement are conducted in the range of Reynolds numbers approx. 75 000 - 390 000, and Reynolds number is defined as:

$$Re = \frac{vD}{\nu} \quad (1)$$

in which D is the diameter of the circular cylinder, v is the free stream velocity a ν is the kinematic viscosity of air.

Strouhal number is defined by a formula:

$$St = \frac{fD}{v} \quad (2)$$

in which f is the vortex shedding frequency. Strouhal number can be considered as a non-dimensional representation of the frequency.

2 Experimental setup

The experimental measurement was performed in the wind tunnel of Department of Fluid mechanics and Thermodynamics, FME CTU in Prague. This circulation wind tunnel has the test section of dimensions 450 x 950 x 1200 mm and the maximum airflow velocity is 60 ms^{-1}.

The static pressure is measured using a static pressure probe with a built-in pressure transducer Kulite LQ-125, see the figure 1. This probe is designed according to Russo [3] to measure small and fast fluctuations of static pressure of the parallel flow. The diameter of this probe is 6 mm and the diameter of pressure holes is 0.5 mm. The pressure holes are placed 18 mm behind the leading half-sphere of the probe. The pressure transducer itself is placed inside the probe near the pressure hole. The pressure transducer Kulite LQ-125 is a miniature pressure transducer with full Wheatstone bridge sensing principle.

The circular cylinder is placed between two parallel planes and cross to the flow, so 3D effects would be minimal. As a data acquisition device for this measurement National Instruments cDAQ-9174 with an analog input voltage module (NI-9215) is used. The absolute error of this measurement platform is according to the manufacturers' calibrations ±57 Pa. Measurements are taken using National Instruments LabView software and 1000 Hz

*e-mail: j.suchy@fs.cvut.cz

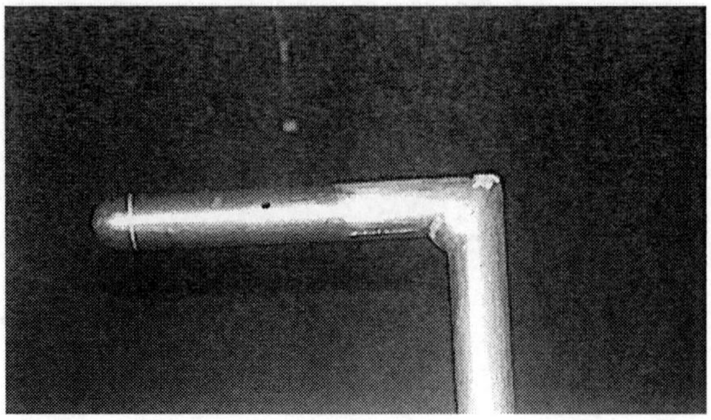

Fig. 1. The static pressure probe.

sampling frequency for 10 seconds. The data is taken separately for each position. The evaluation of the measured data is conducted using MathWorks MATLAB. The main issue of the data evaluation is transformation from the time domain to the frequency domain. For the transformation is used Welch's method to compute the power spectral density and from the power spectral density is then estimated the values of amplitudes. The evaluated pressure data are normalized by dynamic pressure p_d of the free stream.

The static pressure profile i s m easured i n distance $0.75D$ behind the cylinder. The values of the static pressure are obtained in 10 positions spaced equally 10 mm from the centerline of the test section, see the figure 2. The static pressure profile is supposed to be symmetrical for the time averaging values.

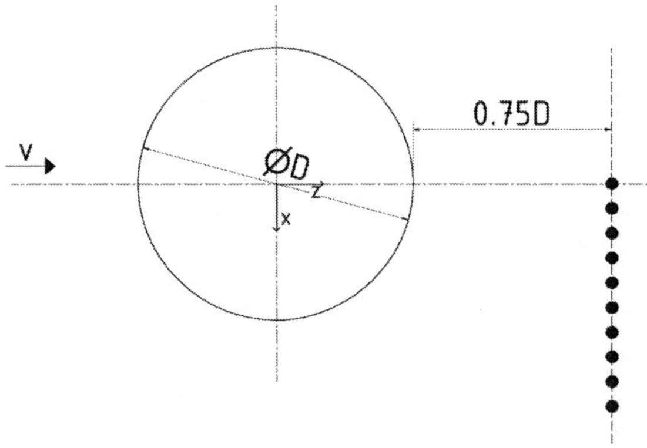

Fig. 2. Scheme of the measurement in the wind tunnel test section.

3 CFD simulation

The Unsteady Reynolds Averaged Navier Stokes (URANS) simulation was conducted using commercial Ansys Fluent Software with non-iterative time advancement method. The computational domain is created with consideration the effects of the closed part of the test section during the experiments. The computation domain is $8.6D$ in width, which is the exactly same ratio as in the wind tunnel test section. For the turbulence model $k - \omega$ SST model is used and the solver is set to use the second order spatial discretization. The boundary condition at the inlet is set to velocity and the outlet is set to the pressure boundary condition. For each simulation the Courant number is below 1 for the most of the cells in the domain.

3.1 CFD prediction

The CFD simulation is calculated for the 9 different inlet velocities, so it can be compared to the experimental results (in the section 4). The vortices are periodically shedded from the two sides of the cylinder. The period of the shedding vortices from one side is the characteristic frequency of the Von Kármán vortex street. The pressure field (in the figure 3) behind the circular cylinder is effected of the shedded vortices from the both sides.

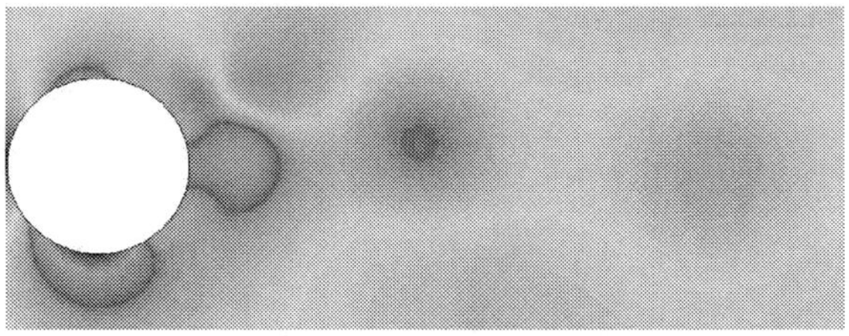

Fig. 3. Prediction of an instant static pressure field behind the circular cylinder for Re = 157 140.

The comparison of the predicted static pressure fluctuations caused by vortex shedding in two positions behind the cylinder is shown in the figure 4. It is evident that the frequency of the static pressure fluctuations in the centerline behind the circular cylinder is double and also that the amplitudes of the fluctuations differ. This is caused by the highest influence of the shedding vortices from the both sides of the cylinder just in the centerline.

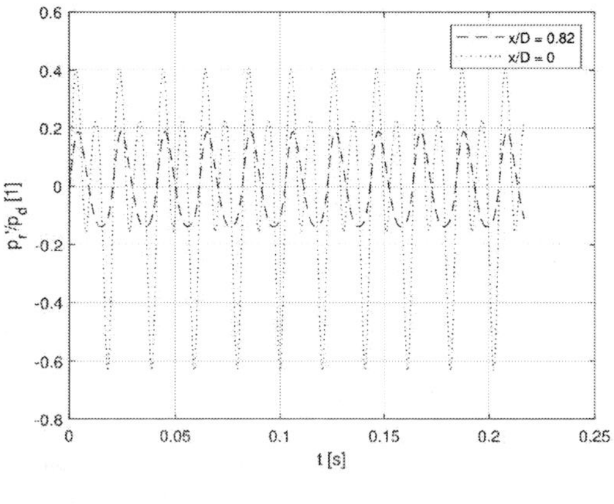

Fig. 4. Prediction of the static pressure fluctuations behind the circular cylinder for Re = 157 140 at two positions.

https://doi.org/10.1051/epjconf/202226901059

4 Experimental results

The measurements are taken for 9 different Reynolds number matching with the CFD prediction. For Reynolds number between 78 000 and 353 000 the characteristic frequency is present in the static pressure spectra at every measured position. For the Reynolds number about 390 000 there is no dominant frequency present at the positions which are the most distant from the centerline, see the figure 5. The dominant frequency is present at the positions between $0.5D$ and the centerline. This phenomenon can be caused by the transition to the turbulent boundary layer on the cylinder [2] and then by the narrower wake behind the circular cylinder.

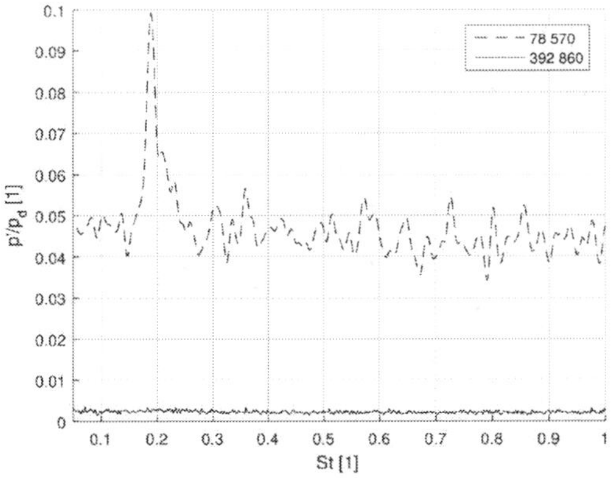

Fig. 5. The measured amplitude spectra normalized with dynamic pressure for the lowest and the highest measured Reynolds numbers at the position $x/D = 0.82$.

The figure 6 shows comparison of normalized amplitudes at the three different measuring positions. The first peak in this spectrum is at St = 0.18 and the second one is at St = 0.37, but this one is perceptible only in positions closer to the centerline. This measurement is in agreement with previously stated theory and predicted periodical values of the static pressure by the CFD simulation, where in the positions closer to the centerline the frequency of the changes in the static pressure is double due to the effect of the vortex at the distant side of the cylinder.

For the measured static pressure for the Reynolds number between 78 000 and 353 000 the dominant Strouhal number was determined. In figure 7 the dominant Strouhal number is presented in the relation to the position x/D, except the centerline, where there is no possibility to undoubtedly determine the dominant frequency because of the effect of the vortex from the opposite side of the cylinder.

The mean values of the Strouhal numbers presented in the figure 7 for different Reynolds numbers are presented in the figure 8. Using the least square method the relation between Strouhal and Reynolds number can be determined as:

$$St = 0.1897 - \frac{13.9}{Re} \tag{3}$$

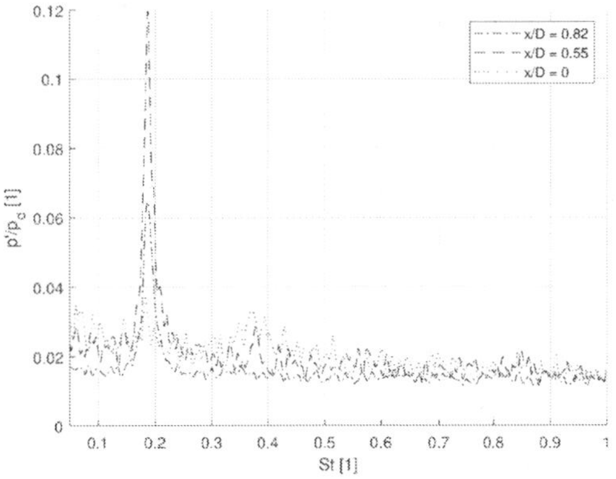

Fig. 6. The measured amplitude spectra normalized with dynamic pressure for Re = 157 140 in the 3 different positions.

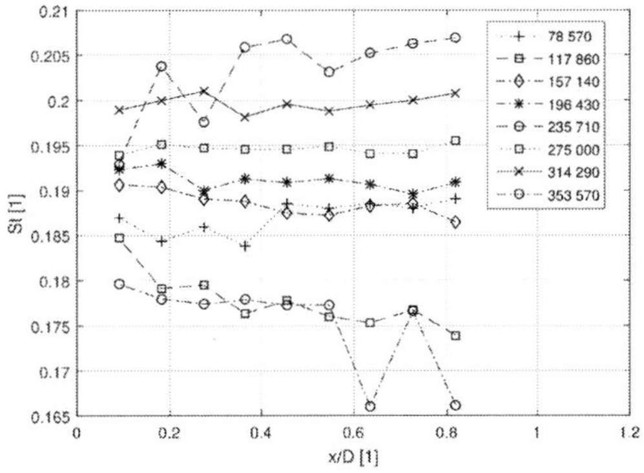

Fig. 7. Dominant Strouhal number of the static pressure fluctuations in relation to position x/D for 8 different Reynolds numbers.

The relation 3 is in a good agreement with the relation proposed by Tyler [4]. This comparison shows that the major fluctuations in the static pressure behind the circular cylinder have the same frequency as Von Kármán vortex street. Also the pressure probe design does not distort significantly the measured frequency.

In the figure 9 there is presented relation between the position of the measurement and the amplitude of the static pressure fluctuations. In general the amplitude of pressure fluctuations due to Von Kármán vortex street is lower in the centerline of the wake of the circular cylinder and the maximum of this amplitude is near the projected edge of the cylinder.

The amplitudes for the Reynolds numbers between 117 860 and 235 710 are close to a constant. This similarity can be also seen in the figure 10 where the mean values of the amplitudes are shown.

Between Reynolds number 100 000 and 250 000 the normalized mean values of the static pressure fluctuations amplitudes seem to be almost constant. This can

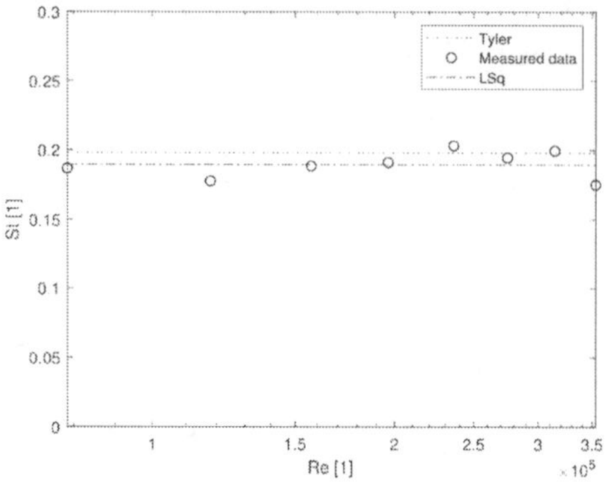

Fig. 8. Relation between Reynolds number and Strouhal number.

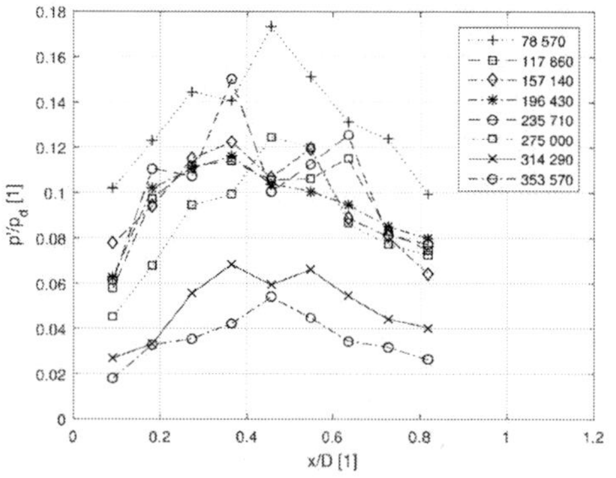

Fig. 9. The amplitude of the vortex shedding frequency in posi-tions x/D for 8 Reynolds numbers.

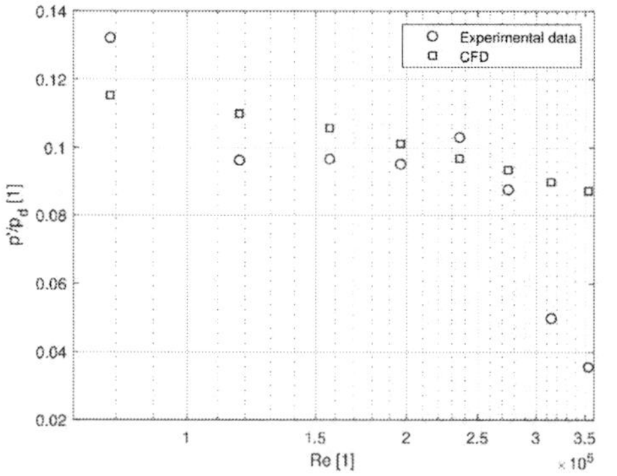

Fig. 10. Relation between Reynolds number and mean static pressure amplitude.

be explained by the almost constant part of Strouhal and Reynolds numbers relation, which was mentioned at the beginning of this paper. In the figure 10 the static pressure fluctuations amplitudes obtained from the CFD simulation is compared to the experimental measurement. The computed normalized static pressure fluctuations amplitudes are for most of the Reynolds number higher than the experimental ones. This can be explained due to lack of noise in the computed results and that the turbulence model is linear. Also in the figure 10 the measured values over Reynolds number $3 \cdot 10^{-6}$ are significantly different, which can be implication of the transition to the turbulent boundary layer on the cylinder.

5 Conclusion

The unsteady static pressure field behind the circular cylinder was measured in 10 different positions for 9 different Reynolds numbers. It was shown that for the most distant measured positions from centerline of the wake behind the cylinder the Strouhal number of the static pressure fluctuations is reliable to determine the frequency of the shedding vortices to the Von Kármán vortex street, however the width of the wake must be considered. It was shown that the wake for Reynolds number over 353 570 is narrower. For this range of Reynolds number between 80 000 and 390 000 the static pressure probe should be placed in the position $0.5D$ from the centerline and $0.75D$ behind the cylinder to be able to measure data for Strouhal number of Von Kármán vortex street evaluation.

Acknowledgements. This work was supported by the Grant Agency of the Czech Technical University in Prague, grant No. SGS18/125/OHK2/2T/12 and the Centre of 3D volumetric anemometry - COLA, reg. no. CZ.2.16/3.1.00/21569 - Operat-ing Program Prague competitiveness.

References

1. Goldstein, S., Modern developments in fluid dynamics: an account of theory and experiment relating to boundary layers, turbulent motion and wakes. Vol. 2. Oxford: Clarendon Press, 1957.
2. Lienhard, J.H., Synopsis of lift, drag, and vortex frequency data for rigid circular cylinders. Bulletin (Washington State University, College of Enginnering, Research Division) 300, Washington State University, 1966.
3. Russo, G.P., Aerodynamic Measurements: From Physical Principles to Turnkey Instrumentation. Cambridge: Woodhead, 2011.
4. Tyler, E., Inst, F., Vortex formation behind obstacles of various sections. The London, Edinburgh, and Dublin Philosophical Magazine and Journal of Science. vol. 11 (72). 1931. p. 849-890

Modelling of multiphase solid-liquid laminar flow with non-Newtonian carrier in pipelines using a layered model

Lukáš Svoboda[1,*], and *Mikoláš* Kesely[1]

[1]Faculty of Civil Engineering CTU in Prague, Department of Hydraulics and Hydrology, Thákurova 7, 160 00 Prague, Czech Republic

Abstract. This paper discusses two-phase laminar flows of mixtures of solids in non-Newtonian carrier consisting of spherical particles in a 50-mm pipe at the Czech Technical University in Prague. Special attention is paid to the frictional head loss and thickness of transport layers. For the prediction of various transport characteristics, a layered model is used. Suitability of the layered model to predict the transport characteristics is investigated. Prediction results are compared to own experimental data.

1 Introduction

Many slurries of industrial interest, such as mixtures of thickened tailing, long distance hauling of coal, minerals, ore and solid commodities or material processing, are mixtures of larger particles carried by a medium that exhibits non-Newtonian behaviour. Due to the high viscosity of the non-Newtonian carrier fluid, laminar flow regime may occur during transport industrial processes. As there is no other mechanism apart from particle collisions, slurries are generally stratified, with sliding bed formed by the coarser particles.

It has been well demonstrated in literature, that coarse particles, which would be under stationary conditions fully suspended in the carrier fluid due to the fact that submerged weight of the particles would not be enough to balance out the yield stress of the carrier fluid, still form the sliding bed during flow conditions, as the local viscosity of the sheared carrier fluid drops, allowing the coarse particles to settle down to the pipe invert (Cooke 2002, Pullum and Graham 2000). [1, 2]

This paper focuses on laminar steady-state flows of mixtures formed by non-Newtonian carrier fluid of Herschel-Bulkley rheological type with coarse particles tending to form stratified flow regime in a horizontal pipeline.

Mixtures of Carbopol Ultrez 10 Polymer with virtually monodisperse glass beads were used to obtain an experimental database containing runs with various rheological properties and various concentrations of solids.

Based on experimental measurements authors evaluate mechanisms that affect the height of the sliding bed layer and present their own empirical equation for layer height prediction. Own experimental measurements are also used to evaluate the laminar two-layer flow model described by Matoušek 2015 [3]. Predicted pressure gradients are compared to measured pressure gradients and discussed.

2 Experimental modelling and set up

In order to evaluate the mechanisms that affect the height of the sliding bed layer, 7 test series with total number of 50 data points in laminar flow regime were carried out in 50 mm pipe loop at the Czech Technical University in Prague. As an analogue to the slurries of industrial interest glass beads (TK15) were used to simulate coarse particles in carrier fluid represented by aqueous solutions of Carbopol polymer (CBP).

A typical experimental run involved measuring of delivered concentration of particles, pressure drop for a set of flow rate, visual observation of the particle flow patterns and set of rheological measurements using carrier fluid samples from before and after experimental run.

2. 1 Carrier fluid

Carbopol Ultrez 10 polymer is a acidic powder of particle size ranging from 2 to 7 microns, which aftrer dispersion in water forms a non-Newtonian visco-plastic fluid of the Herschel-Bulkley (HB) rheological type behaviour. In order to evaluate the viscosity effect on particle flow patterns, diverse solutions of CBP were used for every test run. Rheological properties for test runs are shows below in Table 1. The density of CBP solutions were takes as $\rho_f = 1000$ kg.m^{-3}. The diverse rheological properties of the carrier fluid were achieved by changing the mass concentration of CBP powder in water.

The rheology of the carrier fluid was measured in the rotational viscometer (Haake VT550 with a standard sensor) before and after the test series along with the fluid temperature to make sure that the rheological properties of the carrier fluid remained unchanged during the test series.

* Corresponding author: lukas.svoboda.2@fsv.cvut.cz

Table 1. Rheological properties of CBP solutions.

Test run no.	τ_y	K	n
1	3.76	2.03	0.48
2	1.55	1.16	0.52
3	1.40	0.98	0.55
4	0.42	0.85	0.52
5	0.35	0.61	0.56
6	1.73	0.69	0.59
7	0.18	0.60	0.57

2.2 Particles used in experiments

Coarse glass beads (TK15) were used as an analogue to the coarse particles in thickened tailings handled in the mineral industry. The particles of TK15 are practically mono-disperse with d_{50} = 1,55 mm and the density ρ_s = 2488 kg.m^{-3}.

2.3 Experimental rig

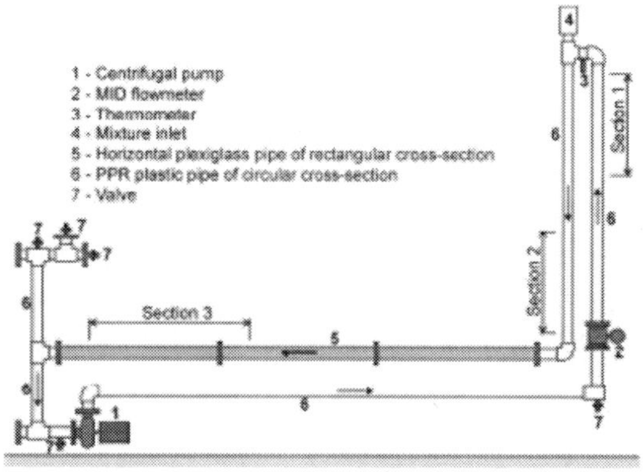

Fig. 1. The experimental rig.

The test series were carried out in Water Engineering Laboratory of the Czech Technical University in Prague (CTU). A pipe loop (Fig. 1) was used to study the slurry flow behaviour. The loop is composed of pieces of a PE pipe (I.D. 51.4 mm, blank pipe in Fig. 1) and a piece of transparent acrylic pipe (I.D. 50.0 mm, grey pipe in Fig. 1). Total length of the loop is 22.96 m and its volume is 45.08 l. The length of the horizontal section is 6.20 m. The pump EBARA 3M 40-200/7,5 kW is driven by an electric motor with a variable frequency converter TECO GD100-011G-4 11 kW. Pump parameters are: power 7,5 kW, impeller diameter 200 mm, maximum flow 11.67 l.s^{-1}, total head from 58 m to 44 m (valid for water for maximum flow). [3]

Differential pressures are measured over vertical Sections 1, 2 (1.3 m long) and the horizontal Section 3 (1 m long) using the differential pressure transducers Fischer Rosemount DP1151 (Sections 1 and 2) and the transducer Siemens Sitrans P DSIII (Section 3). An electromagnetic flow meter Krohne Optiflux 5000 was used to measure the flow rate in the vertical pipe mounted to the discharge outlet of the centrifugal pump. The temperature of the flowing medium was measured in the vertical invert pipe.

2.4 Visual observation of flow patterns

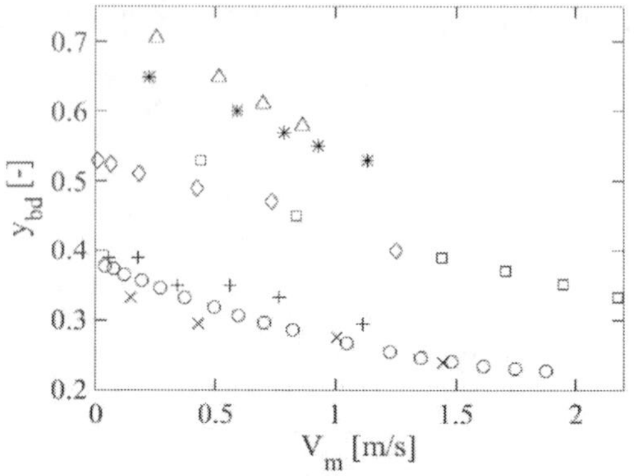

Fig. 2. Measured values of the sliding bed layer thickness y_{bd}. Legend: x – Run no.1; o – Run no.2; + – Run no.3; – Run no.4; □ – Run no.5; * – Run no.6; Δ – Run no.7.

Special attention was paid to the visual observation, as it provided the only means to determine the particle flow patterns at given mean velocity of the laminar flow.

At very low velocities, the thickness of the sliding bed could be determined very accurately. With increasing velocity however, a layer of rolling particles on top of the fully stratified sliding bed started to developed, which with further increase of the mean velocity transformed into shear layer of colliding particles, making the observation of the interfaces between layers much more difficult. The observation of the particle flow patterns ended either at the point of the transition into turbulent flow regime or when a human eye could no longer recognize the sliding bed layer, meaning that only collision shear layer remained under laminar flow conditions.

Camera (NIKON D5100) was used to capture the motion of the particles and millimetre scale, which was fixed on the transparent pipe was used to calculate the actual thickness of the sliding bed layer. It should be noted, that the thickness of the shear layer is not negligible and with increasing velocity, it can even become dominant in the cross section, while the flow is still under laminar flow regime. In Fig. 2 measured values of the sliding bed layer thickness y_{bd} are plotted against mean velocity of the flow V_m.

3 Analysis of sliding bed layer thickness

Thickness of the bed layer along the pipe invert is an important parameter for mathematical modelling of multi-phase slurry flow and for predicting parameters such as hydraulic gradient I_m or mean delivered concentration of solids C_{vd}.

Fig. 2 shows somewhat linear trend in dissipation of the sliding bed layer thickness, that seems not to be affected by the viscosity of the carrier fluid, as the runs no. 1-3 have very different rheological properties of the Carbopol solution, yet the trend seems to be fairly identical. However, the dissolution of the layer thickness seems to be affected by the volumetric concentration of solids in the cross section of the pipeline, as the dissipation of the layer thickness seems to be steeper when the initial thickness of the sliding bed layer is higher.

In order to predict the thickness of the sliding bed layer, following empirical equation 2 was derived based on the assumption, that the layer thickness dissolution in laminar flow regime of Herschel-Bulkley type of non-Newtonian carrier liquid is affected only by the initial bed thickness.

$$y_{bd} = \frac{y_b}{D} \tag{1}$$

Where D is diameter of pipe.

$$y_{bd} = \frac{y_{bd0}}{k} . V_m + y_{bd0} \tag{2}$$

Where k = -4.7 is the found constant and y_{bd0} is initial thickness of the bed layer, which is given by the volumetric concentration of particles in the rig.

Fig. 3 shows experimental data fitted with the predictions by equation 2. The agreement between the experimental data and predictions is very good, even though some of the experimental data runs could be fitted more accurately with 3-parametric curve. Authors however decided to use the linear fitting, as the nonlinearity of the bed thickness dissipation can be caused by the visual observation inaccuracy.

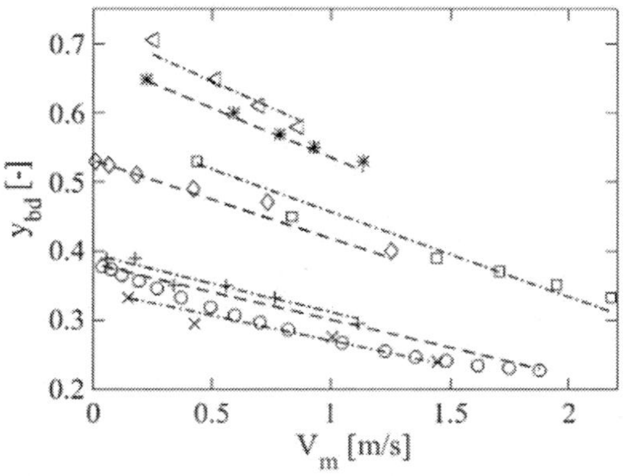

Fig. 3. Sliding bed layer thickness expressed as a function of initial thickness of the bed and mean velocity of the flow.

4. Predictive modelling using two-layer model

Various versions of layered model approaches are available in the literature, when Newtonian fluid is considered as a carrier. However, only a few non-Newtonian carrier based models have been discussed so far. Pullum et al. [4] presented a simple generic two-layer model for laminar and turbulent flow of power-law and visco-plastic carriers based on concept developed by Wilson [5]. This model uses empirical parameters and formulae based on comprehensive experimental database. Later Rojas and Saez [6] introduced and experimentally verified their version of two-layer model. This is a two-layer model to predict laminar and turbulent flows of mixture non-Newtonian fluid (Casson fluid) and dense and fine particles.

In this study two-layer model developed by Matoušek et al. was used, as in authors previous work [8], this model approach was used to calculate the value of deposition limit velocity (the mean velocity of the mixture at which the solid particles at the bottom of a pipeline start to form a stationary deposit) with reasonably accurate results.

Inputs of the predictive model are rheological parameters of carrier fluid τ_y – yield stress, K – consistency index, n – flow index, density of fluid – ρ_f and basic characteristic of particles – ρ_s, diameter of the particles – d, bed concentration – c_b, coefficient of mechanical friction between sliding bed and pipe wall μ and thickness of the sliding bed layer y_{bd}. In this study value of μ was set to the value of 0.38, based on the experimental investigation by Matoušek [3], bed concentration was taken as equal to 0.54 and the values of y_{bd} were predicted using the equation 2, where y_{bd0} was derived from the volumetric concentration of solids in the experimental rig.

The model predicts hydraulic gradient – I_m, delivered concentration of grains – C_{vd}.

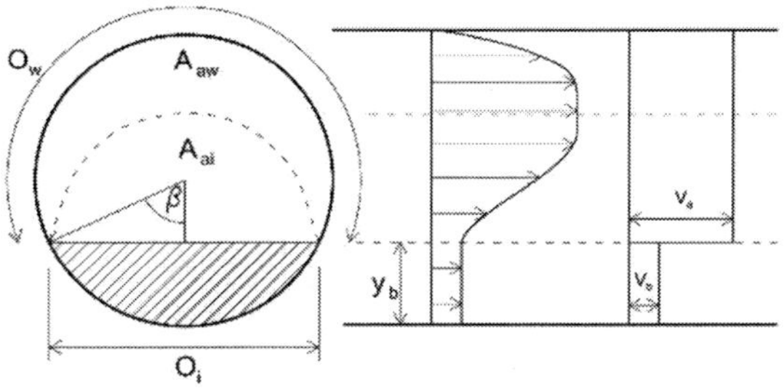

Fig. 4. Schematic of the two-layer model approach.

5. Prediction of the hydraulic gradient using two-layer model

In order to test the accuracy of the Matoušek [3] two layered model, experimental database consisting of 7 test runs with total number of 45 data points were used. 4 of these test runs were used to derive the equation 2, 3 new test runs were added to verify the equation 2 (see Table 2 for test run properties).

Table 2. Rheological properties of CBP solutions used for layered model predictions.

Test run no.	τ_y	K	n	C_{vi}
1	3.76	2.03	0.48	0.08
2	1.55	1.16	0.52	0.08
3	1.40	0.98	0.55	0.1
7	0.18	0.60	0.57	0.27
8	2.99	3.18	0.45	0.11
9	2.82	2.23	0.45	0.26
10	5.06	3.74	0.46	0.07

The results indicate, that the model predictions show good agreement with experimental data, with the highest error no more than 15%. The results also indicate, that for higher velocities near the threshold between laminar and turbulent flow regime the model calculations tend to underpredict the trend shown by experimental measurements (Fig. 5, 6). This can be caused by several reasons. Firstly, the deviation can be caused by a change in rheological parameters due to the rising temperature of the mixture, which was observed during the measurements. Secondly, the model takes in account only the contribution of the sliding bed layer to the resulting value of hydraulic gradient. Experiments however showed occurrence of two layers - sliding bed (bed layer) and a shear layer above it, which gets more dominant with higher flow velocities.

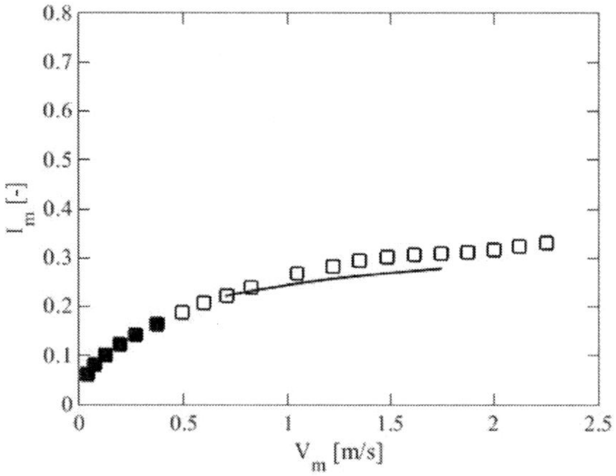

Fig. 5. Figure of hydraulic gradient predictions for run no. 2. Legend: filled squares – measured point in stationary bed, squares – measured points in sliding bed, continuous line – prediction.

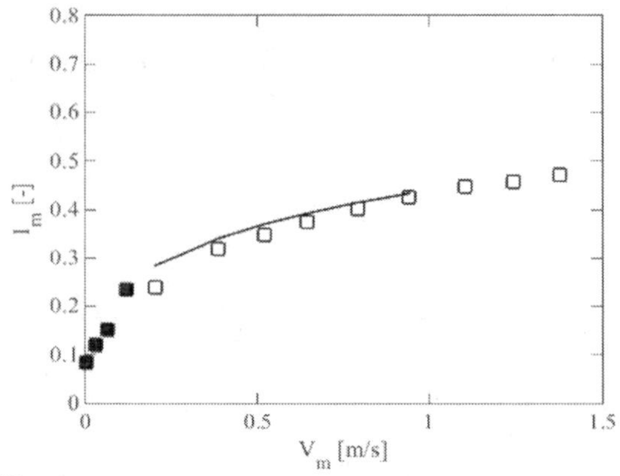

Fig. 6. Figure of hydraulic gradient predictions for run no. 8. Legend: filled squares – measured point in stationary bed, squares – measured points in sliding bed, continuous line – prediction.

Tendency of the model to underpredict the frictional losses for higher mean velocities can be also seen in a Fig. 7, where measured hydraulic gradient is plotted against predicted hydraulic gradient.

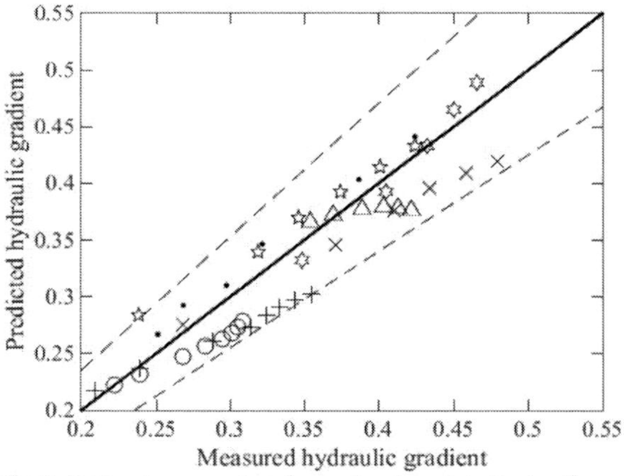

Fig. 7. Parity plot comparing observed and predicted hydraulic gradients. Legend: x – Run no.1; o – Run no.2; + – Run no.3; Δ – Run no.7; five-pointed star – Run no. 8; six-pointed star – Run no. 9; • – Run no. 10; dashed line +- 15% quartile.

6 Conclusions

Experimental investigation of laminar steady-state flows of mixtures formed by non-Newtonian carrier fluid of Herschel-Bulkley rheological type with coarse particles tending to form stratified flow regime in a horizontal pipeline showed somewhat linear trend in dissolution of the sliding bed layer with increasing velocity. Based on experimental measurements, an empirical formulae predicting the thickness of the sliding bed in laminar flow was derived and used in two-layer predictive model to calculate frictional losses in horizontal pipeline.

The comparison of model predictions with experimental data showed good agreement with highest error equal to 15%. At higher mean velocities however the model predictions seem to underpredict the trend given by experimental measurements. These underpredictions may be caused by change in temperature of the carrier fluid and therefore slightly lesser viscosity in later stages of the experiments, but more likely the underpredictions show, that the contribution of the shear layer to the overall frictional losses cannot be neglected in calculations. The implementation of the shear layer contribution will be subject of the future work.

References

1. R. Cooke, Laminar flow settling: The potential for unexpcted problems. *Proc. 15th Int. Conf. on Slurry Handling and Pipeline Transport*, 121–133 (2002)
2. L. Pullum, L. Graham, The use of MRI to probe complex hybrid suspension flows. *Proc. 10th Int.Conf. on Transport and Sedimentation of Solid Particles*, 421–433 (2000)
3. V. Matoušek, V. Pěník, L. Pullum, A. Chryss, Experimental study of bed friction in stratified flow with viscoplastic carried in pipe, *Transport and sedimental of solid particles.* (2015)
4. L. Pullum, L. Graham, P. Slatter, A non-Newtonian two layer model and its application to high density hydrotransport. *Proc. 16th Int. Conf. on Hydrotransport.* (2004)
5. K. Wilson. A unified physically based analysis of solid-liquid pipeline flow. *Proc. 4th Int. Conf. on Hydrotransport of Solids in Pipes,* 1–16. (1976)
6. M. R. Rojas, A. E. Saez. Two-layer model for horizontal pipe flow of Newtonian and non-Newtonian settling dense slurries. *Ind. Eng. Chem. Res.* **51**, 7095-7103 (2012)
7. K. Svoboda, Modelling of coarse slurry flow with non-Newtonian carrier in pipe using two-layer model. (2018)
8. M. Kesely, V. Matoušek, L. Svoboda, Modelling of coarse-grained bed sliding in pipe flow of viscoplastic carrying liquid, *9 th International Conference on Conveying and Handling of Particulate Solids.* (2018)
9. P. K. Swamee, N. Aggarwal. Explicit Equations for Laminar Flow of Herschel-Bulkley Fluids. *The Canadian Journal of Chemical Engineering,* **89:6**, 1426,1433 (2011)
10. R. A. Chilton, R. Stainsby, Pressure loss equations for laminar and turbulent non-Newtonian pipe flow. *J. Hyd. Eng.* **124**, 522-529. (1998)

https://doi.org/10.1051/epjconf/202226901061

Static characteristics of the new artificial pneumatic muscle

Jakub Takosoglu[1,*]

[1]Faculty of Mechatronics and Mechanical Engineering, Department of Manufacturing Engineering and Metrology, Kielce University of Technology, Aleja Tysiaclecia Panstwa Polskiego 7, 25-314 Kielce, Poland

Abstract. The pneumatic artificial muscles are used as driving elements of mobile, anthropomorphic, bionic and humanoid robots as well as rehabilitation and physiotherapeutic manipulators. The PAMs are also increasingly used for the automation of industrial processes. The article presents test stands and methods used to determine the static, isobaric, isotonic and isometric characteristics of the new pneumatic artificial muscles. The muscles have been designed and developed at the Kielce University of Technology. Comparative tests of technical parameters of the designed muscle with the muscles available on the market have been performed.

1 Introduction

The pneumatic artificial muscles belong to a group of pulling single-acting pneumatic actuators. They are characterized by low consumption of compressed air, are completely sealed, can work in contaminated and harmful environments, even underwater [1]. They generate much greater force with the same operating and technical parameters in relation to the standard pneumatic actuators. They are very durable and cheap to operate (no moving parts). There are only two products on the pneumatic industry market: Festo – Fluidic Muscle MAS and DMSP muscles, and Shadow Robot Company – Shadow Air Muscle SAM muscles (Fig, 1).

Fig. 1. Festo – Fluidic Muscle DMSP muscle and Shadow Robot Company – Shadow Air Muscle SAM muscle.

The Festo muscles are more popular due to their marketing campaigns and very good technical support. The book [2] published by Festo presents 99 examples of applications in industrial processes for pneumatic muscles. Festo has also implemented several bionic projects with pneumatic muscles, which were presented at the largest trade fair in the world. Due to the design and the simple structure, the pneumatic artificial muscles are the subject of many scientific works [3–12]. The article [13] presents a broad review of pneumatic muscles developed between 1930 and 2005. Fig. 2 shows the Pneumatic Artificial Muscles PAM-20-330 (under pressure, normal and relaxed) developed at the Kielce University of Technology.

Fig. 2. The pneumatic artificial muscles PAM made at the Kielce University of Technology: a) general view, b) connection.

Table 1 shows the technical parameters of the PAM-20-330, MAS-10-300, S30AM-S-1 pneumatic artificial muscles.

Table 1. Technical parameters of the pneumatic artificial muscles.

Parameter	Value		
	Festo Fluidic Muscle	Shadow Air Muscle	Kielce University of Technology Muscle
Symbol	MAS-10-300	S30AM-S-1	PAM-20-330
Mode of operation	Single-acting, pulling	Single-acting, pulling	Single-acting, pulling
Internal diameter D_n	10 mm	19 mm	20 mm
Internal diameter when fully contracted* D_{min}	8.7 mm	15 mm	33.0 mm
Internal diameter when fully extended* D_{max}	20.4 mm	30 mm	12.7 mm
Nominal length L_n	300 mm	173 mm	330 mm

* Corresponding author: qba@tu.kielce.pl

 https://doi.org/10.1051/epjconf/202226901061

Length of the muscle when fully contracted* L_{min}	252 mm	152 mm	290 mm
Length of the muscle when fully extended* L_{max}	306 mm	213 mm	410 mm
Max. operating pressure p	8 bar	3.5 bar	5 bar
Max. permissible pre-tensioning** ε_{min}	-2 % of L_n	-23.1 % of L_n	-24.2 % of L_n
Max. permissible contraction** ε_{max}	*16 % of L_n*	28.6 % of L_{max}	29.3 % of L_{max}
Lifting force at max. permissible operating pressure F_{max}	630 N	700 N	775 N
Max. additional load, freely suspended	30 kg	-	-
Operating frequency f	35 Hz	210 Hz	-
Max. hysteresis	3% of nominal length	-	-
Max. relaxation	4% of nominal length	-	-
Repetition accuracy	1% of nominal length	-	-
Ambient temperature	-5°C+60°C	0°C+50°C	-
Contraction times	-	0.1-0.5 s	-

The designed and produced muscle is a rubber bladder surrounded by a plastic braided mesh sleeve. The muscle connections are sealed threaded joints with bladder clamped at the ends. Slightly modified standard hydraulic elements sized ½" were used in the design. In order to prevent cutting the muscle during clamping, special plastic inserts were made on 3D printers using the additive manufacturing technology. The inserts were placed between the joint and the bladder. Before the clamping, the bladder was pre-extended to increase the relative contraction and to minimize the hysteresis effect. Thanks to the use of generally available parts, the cost of the muscle production is several times lower than the muscles available on the market from professional companies.

2 Test stand

The technical parameters as well as static and dynamic characteristics of the pneumatic artificial muscles are extremely important for their application. To determine the static characteristics, a test stand was designed, the diagram of which is shown in Fig. 3, while Fig. 4 shows its general view.

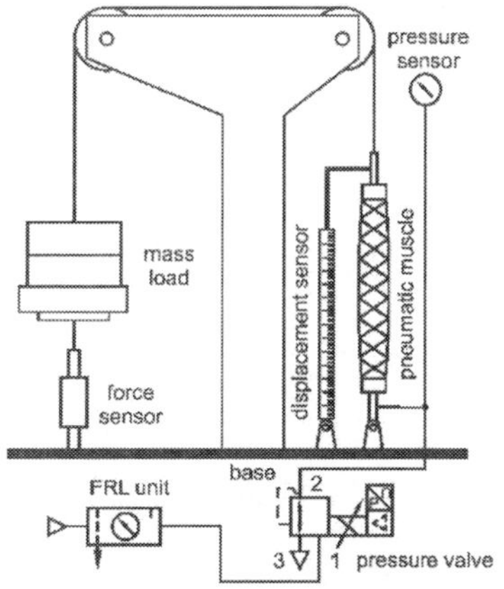

Fig. 3. Diagram of the test stand.

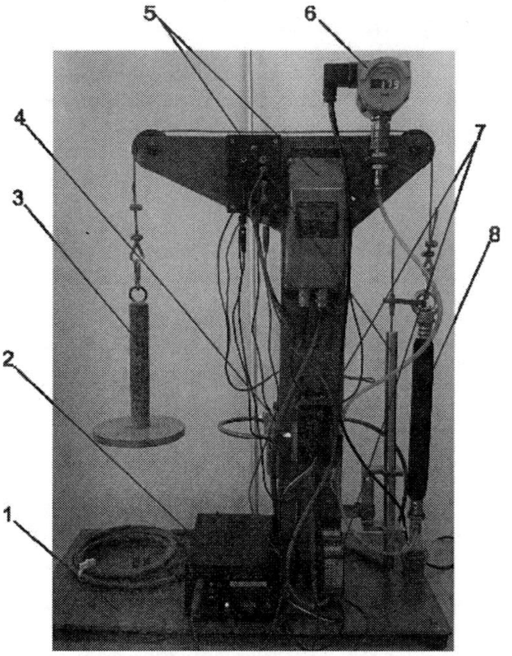

Fig. 4. General view of the test stand: 1 – base, 2 – power supply, 3 – mass load, 4 – FRL (Filter-Regulator-Lubricator) unit, 5 – displacement sensor, 6 – pressure sensor, 7 – proportional pressure valve, 8 – pneumatic artificial muscle.

A non-symmetrical T-shaped arm was mounted on a steel base. Two pulleys, cooperating with a steel wire, were mounted in the arm. One end of the muscle was attached to the base, while the steel wire was attached to the other end. The wire was passed through the pulleys and the other end was connected to the mass load. A displacement sensor was mounted axially to the pneumatic artificial muscle. A proportional pressure valve was used to control the muscle, and a pressure sensor was used to measure the pressure inside the muscle. A force sensor is installed between the mass load in order to determine the pulling force of the muscle. The stand allows to determine the static characteristics: isobaric, isotonic and isometric of the pneumatic artificial muscles. To determine the static characteristics it is necessary to calculate the relative contraction. To calculate the relative contraction, the following relationships were developed [13]:

$$\varepsilon = \frac{L_{max}-L}{L_{max}} \cdot 100\% , \quad (1)$$

$$\varepsilon_{min} = \frac{L_n-L_{max}}{L_n} \cdot 100\% , \quad (2)$$

$$\varepsilon_{max} = \frac{L_{max}-L_{min}}{L_{max}} \cdot 100\% , \quad (3)$$

where:
L – actual length of the muscle,
L_n – nominal length of the muscle,
L_{max} – length of the muscle when fully extended,
L_{min} – length of the muscle when fully contracted.

3 Experimental data

The contraction ε, depending on the pulling force F resulting from the mass load, is measured to determine the isobaric characteristics $F=f(\varepsilon)$ of the pneumatic artificial muscle at constant pressure $(p=const)$ inside the muscle. The principle of determining the isobaric characteristics is shown in Fig. 5.

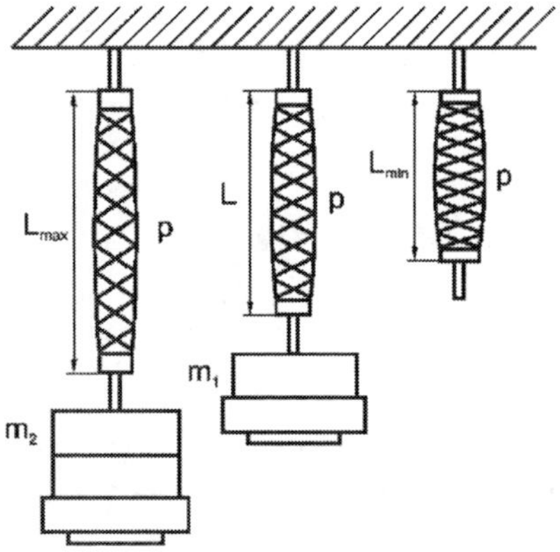

Fig. 5. The principle of determining the isobaric characteristics of the pneumatic artificial muscle.

The isobaric characteristics are limited by a maximum pulling force of 775N at the max. permissible pre-tensioning, at which max. stretching is ε_{min}=-24.2% of L_n n, the max. contraction is ε_{max}=29.3 % of L_{max} and max. working pressure is p_{max}=0.5 MPa. Fig. 6 shows the obtained isobaric characteristics of the pneumatic artificial muscle PAM-20-330.

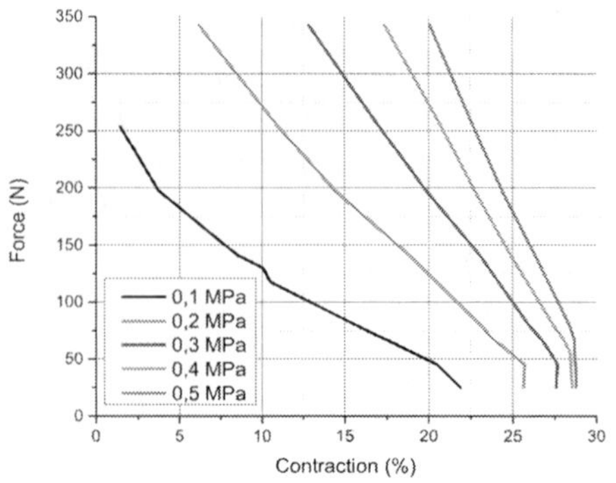

Fig. 6. The isobaric characteristics of the pneumatic artificial muscle PAM-20-330.

The principle of determining the isotonic characteristics $\varepsilon=f(p)$ of the pneumatic muscle is shown in Fig. 7.

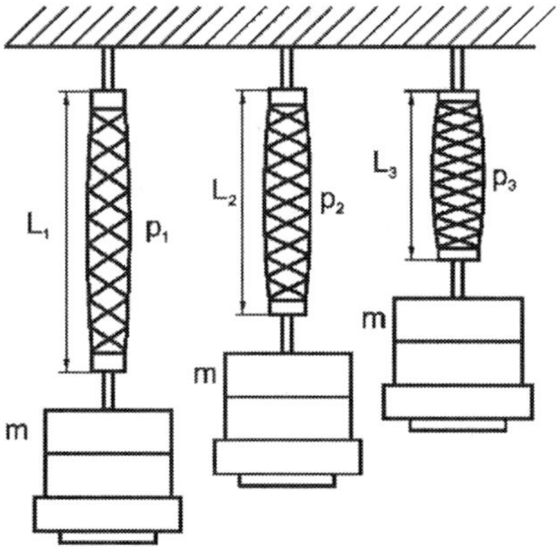

Fig. 7. The principle of determining the isotonic characteristics of the pneumatic artificial muscle.

The relative contraction ε is measured depending on the pressure p inside the muscle and with constant muscle strain $(F=const)$. The isotonic characteristics are limited by the max. permissible relative contraction ε_{max}=29.3% and the max. permissible relative stretching ε_{min}=-24.2%. Fig. 8 shows the obtained isotonic characteristics of the pneumatic artificial muscle PAM-20-330.

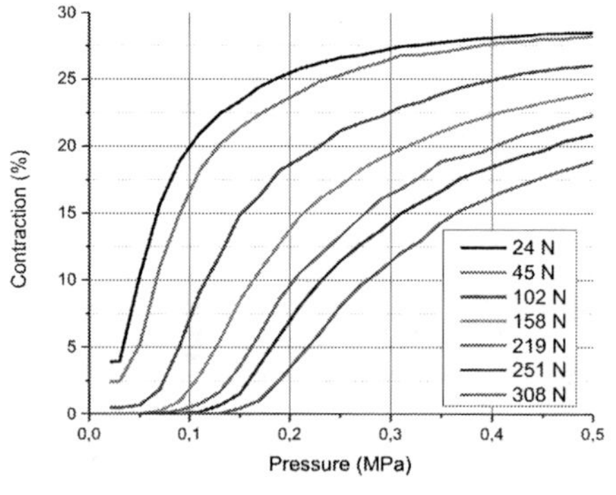

Fig. 8. The isotonic characteristics of the pneumatic artificial muscle PAM-20-330

The principle of determining the isometric characteristics $\Delta F=f(p)$ of the pneumatic muscle is shown in Fig. 9.

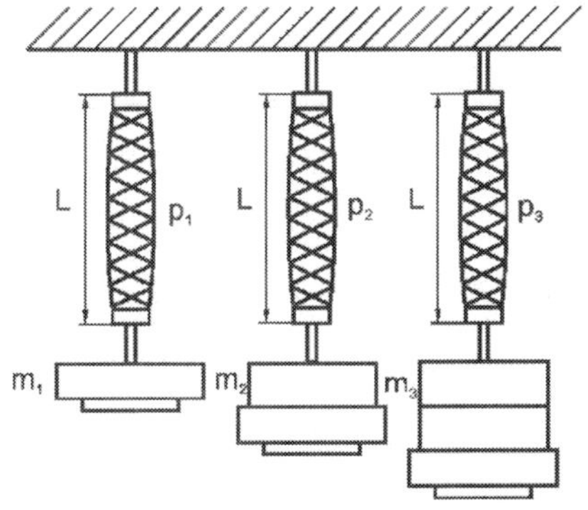

Fig. 9. The principle of determining the isometric characteristics of the pneumatic artificial muscle

The change in pressure p inside the pneumatic muscle and the change in muscle tension ΔF are measured at constant contraction (ε=const). Fig. 10 shows the obtained isometric characteristics of the pneumatic artificial muscle PAM-20-330.

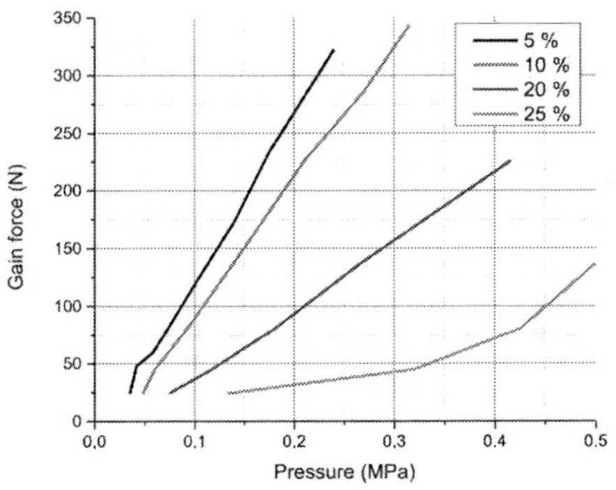

Fig. 10. The isometric characteristics of the pneumatic artificial muscle PAM-20-330.

4 Summary

The work presents the experimental data on the new artificial pneumatic muscle PAM-20-330 designed and developed at the Kielce University of Technology. The test stands and methods used to determine the static, isobaric, isotonic and isometric characteristics are presented here. The PAM-20-330 muscle achieves better technical and performance parameters compared to the muscles offered by the leading manufacturers of the pneumatic industry. The PAM-20-330 has a high pulling force of F=775N at a supply pressure of 0.5MPa. The hysteresis effect was minimized due to the pre-tensioning of the bladder. As in the case of the Shadow Air Muscle by Shadow Robot Company, the pre-stretching is eliminated by the use of a rectilinear bladder. The designed PAM muscle is also characterized by very good static characteristics. The presented characteristics can be successfully compared with the characteristics of the MAS and DMSP type muscles, as a specially developed formula was used to calculate the relative contraction. This type of static characteristics for the bladder-type muscles are presented for the first time and are not found in the available literature.

References

1. J.E. Takosoglu, (Inst. Thermomechanics, Acad. Sci. Czech Republic, Prague, 2017)
2. S. Hesse, (Festo AG & Co., Esslingen, 2001)
3. F. Daerden, D. Lefeber, B. Verrelst, R. Van Ham, (2001)
4. F. Daerden, D. Lefeber, *Eur. J. Mech. Environ. Eng.* **47**, (2000)
5. I. Gaiser, in: R. Wiegand (Ed.), IntechOpen, Rijeka, 2012: p. Ch. 22
6. G.K. Klute, J.M. Czerniecki, B. Hannaford, *Int. J. Rob. Res.* **21**, 4 (2002)
7. D.S. Pietrala, (Inst. Thermomechanics, Acad. Sci. Czech Republic, Prague, 2016)
8. J.E. Takosoglu, (Inst. Thermomechanics, Acad. Sci. Czech Republic, Prague, 2016)
9. B. Verrelst, R. Van Ham, B. Vanderborght, D. Lefeber, F. Daerden, M. Van Damme, *Adv. Robot.* **20**, 7 (2006)
10. N. Yee, G. Coghill, (Auckland, New Zeland, 2002)
11. Z. Zhang, M. Philen, *J. Intell. Mater. Syst. Struct.* **23**, 3 (2011)
12. P.A. Laski, J.E. Takosoglu, S. Blasiak, (Brno Univ. Brno Univ. Technol., Inst. Solid Mechan. Mechatron. & Biomechan., 2014)
13. J.E. Takosoglu, P.A. Laski, S. Blasiak, G. Bracha, D. Pietrala, *Meas. Control.* **49**, 2 (2016)

https://doi.org/10.1051/epjconf/202226901062

Interactive flow around a circular cylinder controlled by a small rod

Takayuki Tsutsui[1,*]

[1] Department of Mechanical Engineering, The National Defense Academy, 1-10-20 Hashirimizu, Yokosuka, Kanagawa 238-8686 Japan

Abstract. In this paper, a unique flow control method using a small rod, called the forced reattachment method, is proposed for the control of the flow around a bluff body. As described in previous papers, the forced reattachment method is a type of separated shear layer control that reduces drag and generates lift, and it can only occur under certain conditions. The forced reattachment phenomenon occurs as a result of an interactive flow between a circular cylinder and a small rod. This phenomenon is characterized by many special features, such as a large stagnant region behind the circular cylinder and the reattachment flow, which adheres to the rear face of the circular cylinder. It is also interesting how it reduces the total drag, including that of the small rod. The aim of this study was to elucidate the details of the stagnant region, the properties of the reattachment flow, and the reduction of the total drag.

1 Introduction

The use of the forced reattachment phenomenon as a unique flow control method has been reported in previous papers [1 – 6]. In this method, which can be categorized as a shear layer control method, a small rod is placed near a circular cylinder, dividing the separated shear layer from the circular cylinder into upper and lower components. The upper shear layer is elongated, and the lower shear layer reattaches and adheres to the rear face of the circular cylinder, forming a large stagnant region behind the circular cylinder. As a result, the drag of the circular cylinder is reduced by 20 % to 30 %, and a lift with a lift coefficient of C_L = 1.0 is generated. Recently, we have reported formulas describing the features of this phenomenon and the optimal conditions of the rod size and position [6].

The forced reattachment phenomenon occurs as a result of an interactive flow between the circular cylinder and the small rod. Figures 1(a) and (b) respectively show a flow visualization photograph and a schematic demonstrating the phenomenon. This phenomenon is characterized by several features for example, a large stagnant region and the adhered flow on the rear face of the cylinder. When this flow control method is used in engineering applications, the total drag reduction, including that of the small rod, must be known.

In this study, we describe several features of the forced reattachment phenomenon and the total drag reduction including that of the small rod.

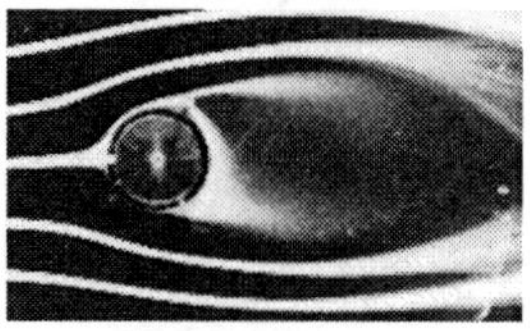

(a) Flow visualization (D = 40 mm, d = 2 mm, G = 6 mm, α = 121°, Re = 1.6×10^4)

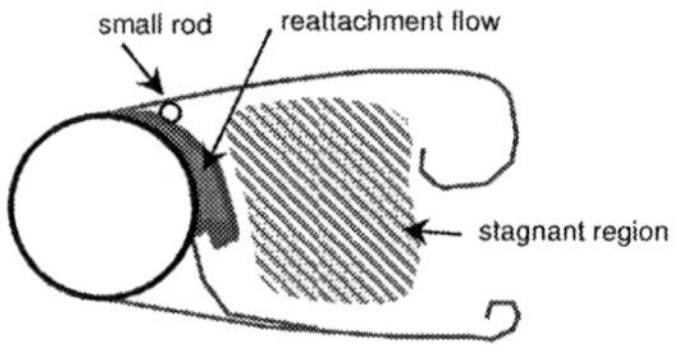

(b) Schematic of the forced reattachment phenomenon

Fig. 1. Forced reattachment. [(5)]

* Corresponding author: tsutsui@nda.ac.jp

2 Experimental device and methods

The coordinate systems and notation used in the present study are shown in Figure 2. The diameter of the circular cylinder was D = 40 mm, and the diameter d of the small rod was 2 mm. The width G of the gap between the circular cylinder and the small rod was 6 mm. The angular position α of the small rod was varied from 90° to 180°.

Experiments were carried out in a low-speed wind tunnel, the working section of which has a height of 1,000 mm, a width of 150 mm, and a length of 1,200 mm. The free-stream velocity U was 16 m / s, and the turbulence intensity was approximately 0.4 % in this velocity. The Reynolds number Re based on D was 4.2×10^4.

In experiment, the surface pressure on the circular cylinder and the small rod were measured, and the drag and lift coefficients were obtained by integrating the pressure distribution. The time-averaged velocity in the wake and the gap were measured with an I-type hot wire anemometer.

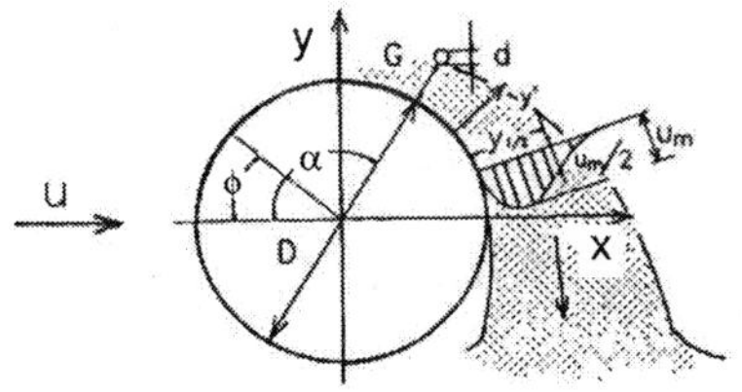

Fig. 2. Coordinate system and notation.

3 Results and discussion

3.1 Forced reattchment

Figure 3(a) and (b) shows the contours of the time-averaged velocity $\bar{u}$ behind the circular cylinder with and without the small rod, respectively. In Fig. 3(a), a wide low speed region ($\bar{u}$ / U ≤ 0.4) formed behind the circular cylinder, and a very low speed region ($\bar{u}$ / U ≤ 0.1) formed just behind the reattachment flow, this latter region is completely stagnant. A high-speed flow appeared in the gap between the cylinder and the small rod. This is the lower component of the divided shear layer, and it reattached and adhered to the rear face of the circular cylinder. The maximum velocity of this reattachment flow was $\bar{u}$ / U = 1.6. The reattachment flow decreased gradually along the rear face of the circular cylinder. A comparison of the forced reattachment case in Fig. 3 (a) with the case with no small rod in Fig. 3 (b) indicates that the stagnant region and the high-speed reattachment flow are characteristic of the forced reattachment phenomenon.

3.2 Total drag reduction

Typical flow patterns at different angular positions α are shown in Figures 4 and 5, which are already described in a previous paper [6].

To assess the total drag, including that of the small rod, the pressure distribution around the small rod was measured for various values of α, and the results are shown in Figure 6.

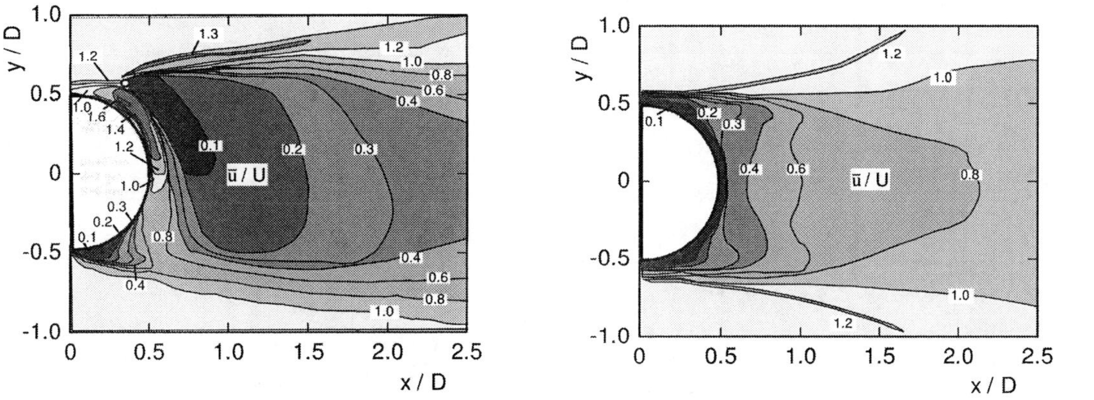

(a) the forced reattachment (D = 40 mm, d = 2 mm, G = 6 mm, α = 121°, Re = 1.6×10^4).

(b) single circular cylinder (D = 40 mm, Re = 1.6×10^4).

Fig. 3. Contours of the time-averaged velocity behind the circular cylinder.

At α = 115 °, the small rod is outside of the shear layer separated from the circular cylinder. The base pressure coefficient C_{pbr} was approximately − 2.5, which is considerably lower than that of the small rod alone, and the pressure coefficient at the front stagnant point was Cp = 1.1, which is greater than 1.0. Because the approaching velocity to the small rod was equal to the velocity of the separated shear layer from the circular cylinder $U_s = U\sqrt{(1 - Cp_b)}$.

At α = 121 °, the forced reattachment phenomenon occurred. The front stagnant point was located at ϕ = 20 °, and C_{pbr} increased to − 0.7. Because the shear layer separated from the circular cylinder impinged on the front face of the small rod, a large stagnant region formed behind the small rod.

At α = 124 °, the small rod is located just inside the separated shear layer. The forced reattachment phenomenon did not occur in this case, but the separated shear layer was elongated by the support of the small rod. The entirety of the small rod was in the stagnant region, causing the pressure coefficient distribution to take a negative value.

At α = 130 °, the small rod is completely buried in the stagnant region. The pressure coefficient had a nearly flat distribution at this position. Cp took a constant value of − 1.1.

Figure 7 shows the drag coefficients of the circular cylinder and small rod and the total drag coefficient. The total drag coefficient C_{DT} is given by

$$C_{DT} = C_D + \frac{d}{D} C_{Dr},$$

where C_{Dr} is the drag coefficient of the small rod and $\frac{d}{D} C_{Dr}$ is the additional drag coefficient. The additional drag coefficient is less than 5 % for 90 ° ≤ α ≤ 121 °, and for 121 ° < α ≤ 180 °, the small rod has no effect on the drag coefficient. The total drag coefficient C_{DT} in the case of the forced reattachment phenomenon is approximately 25 % less than that in the case of the circular cylinder without the small rod.

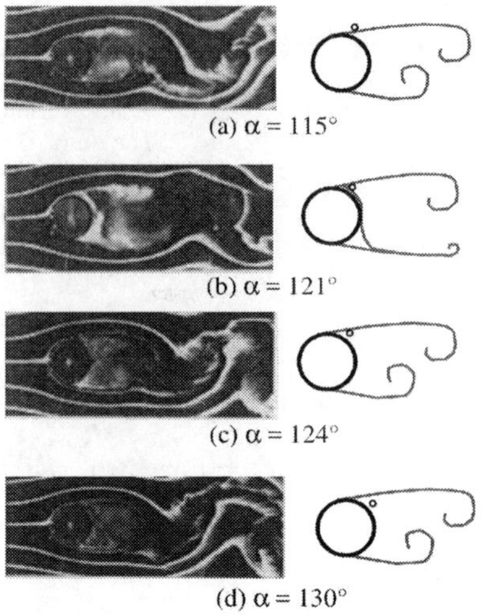

Fig. 4. Flow patterns. [5]

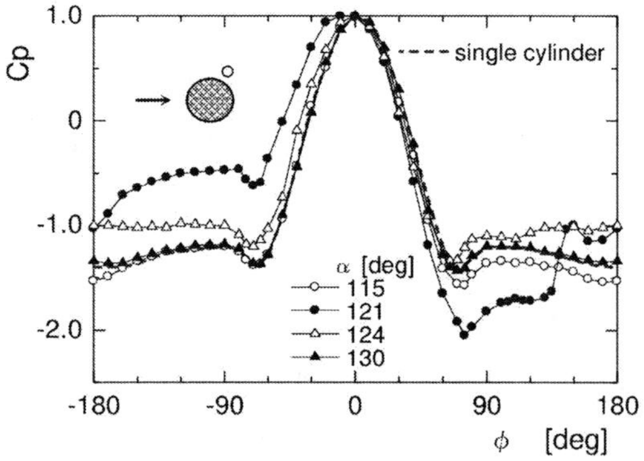

Fig. 5. Pressuer distribution of the circular cylinder for each flow patterns [5] (D = 40 mm, d = 2 mm, G = 6 mm, Re = 4.2 × 10^4)-

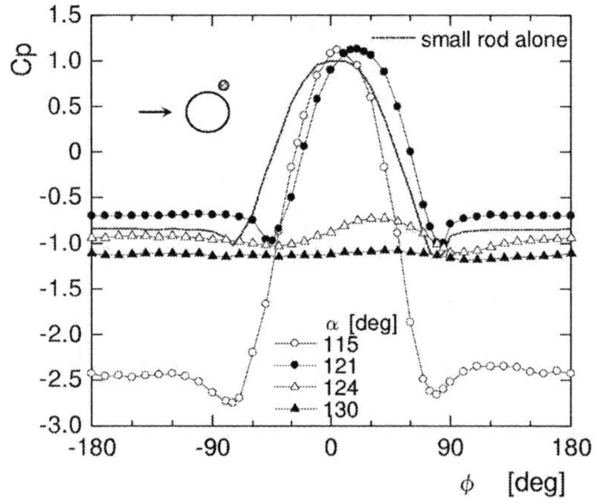

Fig. 6. Pressuer distribution of the small rod for each flow patterns (D = 40 mm, d = 2 mm, G = 6 mm, U = 16 m / s).

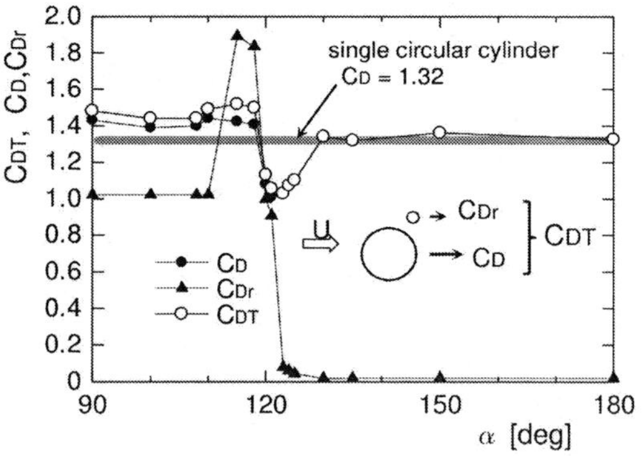

Fig. 7. Drag coefficients of the circular cylinder, the small rod and the total drag coefficient.

 https://doi.org/10.1051/epjconf/202226901062

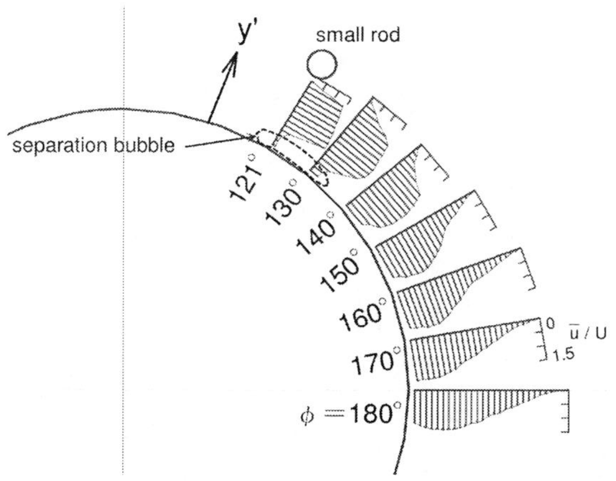

Fig. 8. Time-averaged velocity of the reattachment flow (D = 40 mm, d = 2 mm, G = 6 mm, $\alpha = 121°$, Re = 4.2×10^4).

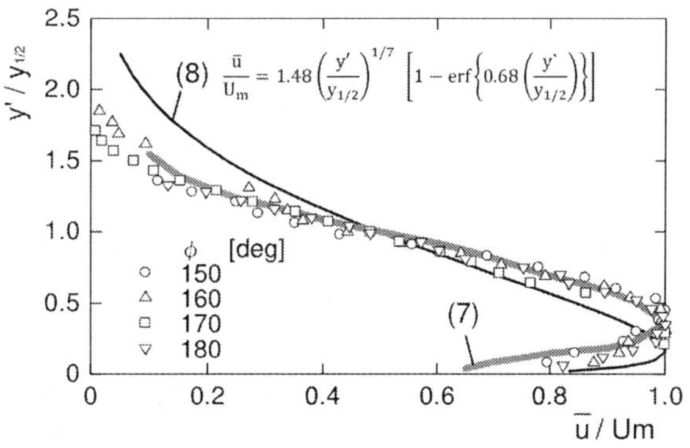

Fig. 9. Comparison between the reattachment flow and turbulent wall jet.

3.3 Reattachment flow

Figure 8 shows the distributions of the time-averaged velocity of the reattachment flow along on the rear face of the circular cylinder at different angular positions ϕ.

The velocity distributions at $\phi = 121°$ and $130°$ exhibit a low velocity near the rear face of the circular cylinder because of the separation bubble [1]. The velocity distributions gradually developed with increasing angle of ϕ.

The velocity profiles for the reattachment flow and a turbulent wall jet are shown in Figure 9 for comparison. The data for the velocity distributions at the $\phi = 121°$, $130°$, and $140°$ sections are omitted because they were not fully developed. The velocity profile of the reattachment flow is plotted along with two other velocity profiles in the figure. Shabayek [7] has reported that the velocity profile was off to up and down from the classical wall jet profile [8] increase in roughness. In this study, the same tendency was observed.

It was thus concluded that the reattachment flow has the same properties as the turbulent wall jet.

4 Conclusions

Flow control by the forced reattachment phenomenon was considered in this study. The features of this phenomenon, for example the details of the stagnant region, the properties of the reattachment flow, and the reduction in the total drag reduction, were discussed.

A wide low speed region ($\bar{u} / U \leq 0.4$) formed behind the circular cylinder, and a very low speed region ($\bar{u} / U \leq 0.1$) formed just behind the reattachment flow.

The total drag coefficient C_{DT} was approximately 25 % less than that of the circular cylinder without the small rod.

The properties of the reattachment flow are similar to those of a turbulent wall jet.

References

1. T. Tsutsui, T. Igarashi, *Flow Control around a Circular cylinder by a Small Cylinder*, Proc. of the 11th Australasian Fluid Mechanics Conference. Hobart, Australia. pp. 14-18, (1992)
2. T. Igarashi ,T. Tsutsui, *Flow Force Acting on a Circular cylinder Controlled by a Small Rod.* Proc. of the 3rd JSME-KSME Fluids Engineering Conference. Sendai, Japan. pp. 571-576, (1994)
3. T. Tsutsui, T. Igarashi and K. Kamemoto, Interactive Flow around Two Circular cylinders of Different Diameters at Close Proximity Experiment and Numerical Analysis by Vortex Method, J. Wind Eng. Ind. Aerodyn., 69-71, pp. 279-291, (1997)
4. T. Tsutsui, T. Igarashi, *Flow Control around a Circular cylinder by a Small Cylinder (Properties of Reattachment Jet).* Proc. of the 3rd ASME-JSME Joint Fluids Eng. Conf., San Francisco, California, FEDSM99-6943, (1999)
5. T. Tsutsui, *Instantaneous fluid force acting on a circular cylinder control by a small rod,* EPJ Web of Conferences, 180, 02110 (2018)
6. T. Tsutsui, Flow control around a circular cylinder by a small rod, ACC journal 2019, vol. 25, Issue 1, DOI:10.15240/tul/004/2019-1-007.
7. S. Shabayek, 2D turbulent jets on rough and smooth boundaries, IOSR journal of Environmental Science, vol. 12, Issue 4, Ver. I, pp. 7-18 (2018), DOI:10.9790/2402-1204010718.
8. N. Rajaratnam, Turbulent jets, Elsevier Publishing Co. Amsterdam, p.217, 1976.

Experimental Research on the Flow in the Steam Turbine Axial Exhaust Hood

David Tupy[1,*], *Vaclav* Slama[1], and *Robert* Kalista[1]

[1]Doosan Skoda Power, Experimental Research, Tylova 1/57, Pilsen, Czech Republic

Abstract. There is the still remaining task to increase the efficiency of power plants and steam turbines. One possibility to increase the thermodynamic efficiency is to apply a suitable exhaust hood. Axial exhaust hoods are mainly designed for small turbines up to 180 MW. They are a very important part of steam turbines and its main purpose is to take away the steam from the last stage outlet to a condenser with minimal loss or even with pressure compression and an increase in the last stage enthalpy drop. This paper concerns experimental measurements on the axial exhaust hood wind tunnel model. The main goal of this paper is to determine how given geometrical and flow parameters affect the resulting pressure recovery coefficient. Main observed parameters are velocity and circumferential angle profiles at the diffuser inlet along with a number, shape and geometrical configuration of internal struts. Several geometrical variants are measured for two types of circumferential angles at the diffuser inlet and for two values of Mach number in the presented work. Multi-hole pneumatic probes and pressure taps are used for flow parameters measurement. The proposed experimental model, along with corresponding CFD (computational fluid dynamics) simulations, will help to optimize the flow parameters in the exhaust hood and allow a decrease in pressure losses in this part.

1 Introduction

One reason to increase the power plant efficiency is to lower the amount of burned fuel and in this way also the operational costs. Another reason is the present trend to reduce harmful emissions. Thermodynamic efficiency can be increased by using a suitable exhaust hood. This paper deals with the experimental research of flow in the axial exhaust hood and optimization of its shape.

Axial exhaust hood are often used as a type of the steam turbine outlet. They are mainly used for turbines of lower outputs up to 180 MW, with the condensers located at the same height level. The axial exhaust hood 3D view is shown in Fig. 1.

Fig. 1. Exhaust hood 3D view.

In comparison with radial hoods, the advantage of axial hoods is a simpler shape and thus also the construction and the related energy losses. Exhaust hoods are construction nodes of considerable dimensions in comparison with other turbine parts, which results in various designs and technological constraints. That is why exhaust hoods are made mainly as weldments with simple shape without much emphasis on aerodynamic properties. Condensing turbines operate at a condensing pressure which is substantially lower than the atmospheric pressure. In the case of axial hoods, among other things, there is a problem with static rigidity of the bearing stand that is surrounded by flowing water steam. For this reason it is necessary to produce hoods with various struts and supports to prevent the casting from collapsing due to external atmospheric pressure. A number of pipelines (oil, seal, regenerative) go very often through the exhaust hood and provide required performance characteristics. The mentioned inner elements form obstacles in the steam flow and generally have an adverse effect on pressure losses. However, these losses can be considerably reduced by a suitable choice of internal struts shape, size and configuration. It is also possible to reach uniformity of velocity field at the hood outlet that reduce unfavourable vibrations of pipe bundles in the condenser, even if in the case of axial exhaust hood the pipe bundle is located rather far from the exhaust hood. A detailed study of the influence of internal support in exhaust hood on its pressure recovery factor is presented in e.g. [1]. The main task of the

* Corresponding author: david.tupy@doosan.com

construction design is to delay or eliminate the occurrence of the boundary layer separation, which has a crucial impact on energy losses. There are also a number of measures that contribute to reducing energy losses, such as a boundary layer suction or, conversely, blowing steam into the boundary layer on the outer side of the diffuser.

The paper is devoted to an experimental verification of the flow on the scale model of the steam turbine axial exhaust hood. The main aim of the work was to define the influence of selected geometric and flow parameters on the c_p (pressure recovery coefficient). Among the observed parameters, there were the velocity profile at the exhaust hood inlet, the profile of circumferential angle α_1 at the exhaust hood inlet and the geometrical configuration of the exhaust hood internal struts. The paper deals with description of the experimental device, experimental verification of the flow, summary of methodology used and results obtained.

2 Experimental device

The experimental device is in fact a scale model of the air axial exhaust hood. The model is attached to the wind tunnel outlet, see Fig. 2. In the real machine it is very difficult to change the body geometry, arrangement of internal struts or to apply various experimental methods for flow research. For this reason a model of axial exhaust hood was made. On the other hand, in experimental conditions it is complicated to set the conditions at least close to the real operation. It is in fact impossible to reach an agreement. However, the results from experimental measurements are necessary for verification of the concurrently running numerical calculations. Using numerical calculations, it is possible to model states that cannot be reached in laboratory conditions, but with many simplifications due to limited computational capacities, see [2] and [3].

Fig. 2. Exhaust hood model with extension pipe.

The exhaust hood geometry represents a scale model corresponding to the real concept that was designed for the turbine with 35 MW output. The aerodynamic design of this scaled air test model of the axial exhaust hood is described in e.g. [4] and [5]. The steam turbine will be operated with the possibility of experimental verification of operational properties. Based on the available information it will be possible to compare data from the real machine and the air model. But it is not a part of this paper.

Mach number (Ma) is one of the key parameters characterizing the flow at the axial exhaust hood inlet. This parameter is considerably limited by the possibilities of the experimental laboratory mainly by the wind tunnel fan output. Experimentally, 0.4 Ma can be reached, while in the real machine the axial velocities reach minimally 0.5Ma. Nevertheless, the main task when designing the experimental model is to guarantee the required circumferential angle (α_1) of flowing medium at the exhaust hood inlet. The circumferential angle that changes along the inlet channel height (along the last rotor blade length) has a significant effect on the interaction between the flow and internal struts. In the real machine there is also a radial outlet angle of the flow that arises during the rotation of rotor blades of the steam turbine last stage. This angle is very hard to adjust on the testing device. For this reason the model was designed to achieve the required angles in the circumferential direction. The fact that the influences on the last stage output can be quite important is discussed in e.g. [6] and [7]. All together two types of a stator blade cascade were designed with circumferentially shaped blades. The first blade profile (Variant 1) is designed so that the required profile of the circumferential angle α_1 along the blade height is in agreement with the referential real machine. In the second case (Variant 2) 1.5 times the circumferential angle is achieved. In order to quantify the influence of the circumferential angle on the exhaust hood behaviour a blade cascade was designed with only the axial component of velocity (Variant 3). The required angle of the circumferential angle α_1 (the so called swirl angle) along the channel height for all three types of the blade cascade and the value of Mach number on the diffuser inlet 0.18 Ma can be seen in Fig. 3. Within this paper, only the results from Variant 1 and Variant 3 are presented.

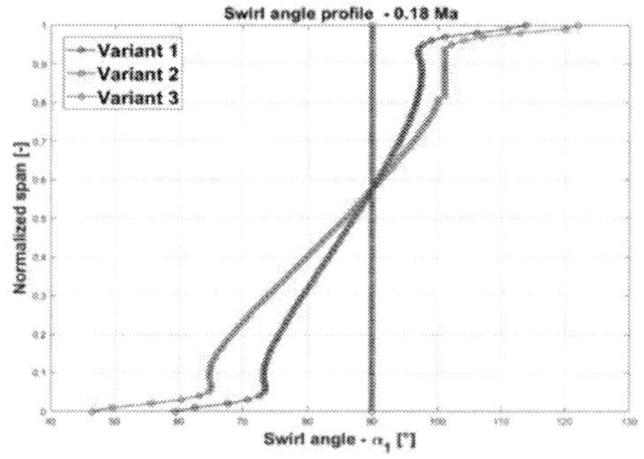

Fig. 3. Profile of circumferential angle α_1.

3 Measurement description

The exhaust hood diffuser consists of an annular and conical part. To the conical part an extension ring pipe is attached. In Fig. 4 a simplified scheme is shown with marked important plane:

https://doi.org/10.1051/epjconf/202226901063

- plane 0 – the plane upstream of the blade cascade,
- plane 1 – axial exhaust hood inlet ,
- plane 2 – axial exhaust hood outlet,
- plane 3 – extension pipe outlet into the open space.

Planes 0 and 1 are divided circumferentially into four measuring segments where it is possible to measure using multi hole pneumatic probes and a Prandtl probe. The segments are marked I to IV in the flow direction. It is possible to measure using pneumatic probes in the plane 1′, 1′′ and 2′. These planes are again divided around the perimeter into four measuring segments. In the plane – a, b, c, d, e static pressure taps are made. Due to the taps it is possible to obtain information about static pressure values in several planes along the exhaust hood length. On the exhaust hood, two temperatures are measured. The first one is taken in the wind tunnel settling chamber and the second one is taken in the plane upstream of the blade cascade (plane 0). For pressure measuring on the exhaust hood model pressure scanners NetScanner 9116 were used. Atmospheric pressure was measured using pressure transmitter Rosemount 3051S performance class ultra and temperatures were taken using resistance thermometers Pt100.

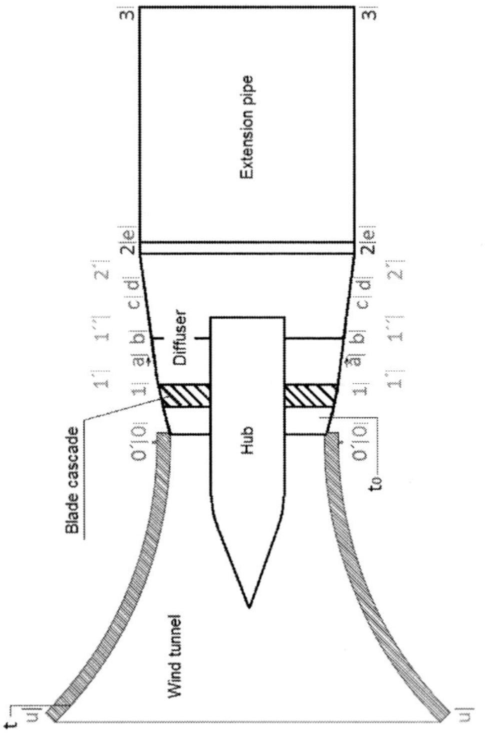

Fig. 4. Scheme with marked planes.

4 Data assessment

The ratio of dynamic pressures upstream and downstream of the blade cascade (p_p) and the loss coefficient of the cascade (ζ_m) are the most important parameters of stator blade cascade characteristics. In order to define these parameters it is necessary to measure in detail the blade cascade using traversing by pneumatic probes. The obtained parameters are then used as constants in equations for the evaluation of the so-called simple measurement when the pneumatic probe is placed in only one characteristic location of the flow field at the blade cascade inlet. Due to measurements in a simple mode the time consuming probe traversing is eliminated. The blade cascade is located at the inlet of the exhaust hood. Data are collected using multi hole pneumatic probes upstream and downstream the blade cascade. In the equations below there are the respective total pressures - p_c, static pressures - p_s and dynamic pressures - p_d. Parameters of the cascade are defined using the following equations:

$$\zeta_m = \frac{p_{c0} - p_{c1}}{p_{d0}}, \tag{1}$$

$$p_p = \frac{p_{d0}}{p_{d1}}. \tag{2}$$

One of the main tasks of the work was to define loss (ζ_D), or pressure recovery coefficient of the diffuser (c_{p_D}). The loss coefficient of the diffuser can be defined for incompressible flow using the following relation:

$$\zeta_D = \left(\frac{p_{c0} - p_{s2}}{p_{d0}} - \zeta_m\right) \cdot \frac{p_{d0}}{p_{d1}} = \\ = (\zeta_{mD} - \zeta_m) \cdot p_p. \tag{3}$$

Pressure recovery coefficient can be then defined using the relation:

$$c_{p_D} = 1 - \zeta_D. \tag{4}$$

Loss coefficient of the diffuser and the extension ring pipe (ζ_C) can be calculated for incompressible flow using the relation:

$$\zeta_C = \left(\frac{p_{c0}}{p_{d0}} - \zeta_m\right) \cdot \frac{p_{d0}}{p_{d1}} = \\ = (\zeta_{mC} - \zeta_m) \cdot p_p. \tag{5}$$

Pressure recovery coefficient can be then defined using relation:

$$c_{p_C} = 1 - \zeta_C. \tag{6}$$

For the calculation of both equations it is necessary to know the mean values of blade cascade parameters. For the purpose of this paper the blade cascade parameters were weighted by mass flow. Then the blade cascade parameters are marked - $\bar{\zeta}_m$ and $\bar{p}_p$.

Using static pressure taps it is possible to calculate the loss coefficient and pressure recovery coefficient also in other planes of the exhaust hood, i.e. in plane: a, b, c, d, e, see Fig. 4. Then it is possible to define the dependence of c_p in individual investigated planes on the relative distance of the planes from the trailing edge of the blades. The pressure recovery coefficient on the trailing edge of the blades is zero and gradually increases towards the outlet.

5 Numerical model

The package of ANSYS software tools was used to carry out required CFD analyses. The model consists of three parts: the inlet part which contains the inlet confusor, the middle part which contains the blade cascade and the outlet part which contains the investigated diffuser. Each section can be changed according to the investigated variant. The whole numerical model is shown in Fig. 5, the detail is in Fig. 6. The model is not symmetric due to different variants of the diffuser internal struts. Therefore, the model has to be complex without using a rotational periodicity.

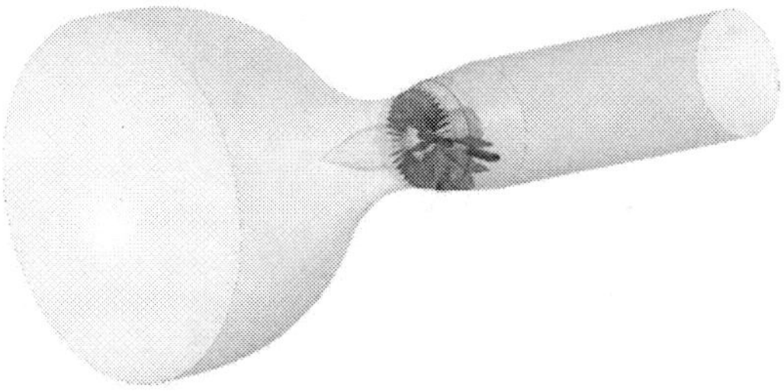

Fig. 5. Numerical model (the inlet part – green, the middle part – pink, the outlet part – blue).

Fig. 6. Detail of the numerical model (the middle part with the blade cascade and the outlet part with diffuser internal struts).

ANSYS ICEM 18.2 and ANSYS Turbogrid 18.2 tools were used to create a fully structured hexahedral mesh for the inlet and middle part and a hex-core mesh for the outlet diffuser part. The picture of the hexahedral mesh is shown in Fig. 7, the hex-core mesh is shown in the cross section of the diffuser part in Fig. 8. The boundary layer with a 0.001 mm thickness of the first cell was created in all mesh parts and all mesh densities in order to keep the value of y+ lower than 5. In order to find the reasonable mesh density, a mesh sensitivity analysis was carried out. For this purpose, the global number of mesh nodes was changed. For the sake of simplicity, it is described by a size factor which is shown in Table 1 along with other details.

Table 1. Mesh sensitivity analysis: different mesh densities presented by mesh nodes (the outlet part is with all diffuser internal struts).

Number of nodes	Coarse	Medium	Fine
Size factor	0.25	0.50	1.00
Inlet part	298 973	936 000	1 872 000
Middle part	4 069 440	16 170 440	46 875 400
Outlet part	1 579 478	4 944 899	9 889 798
Total	5 947 891	22 051 339	58 637 198

Fig. 7. Detail of the hexahedral mesh of the middle section.

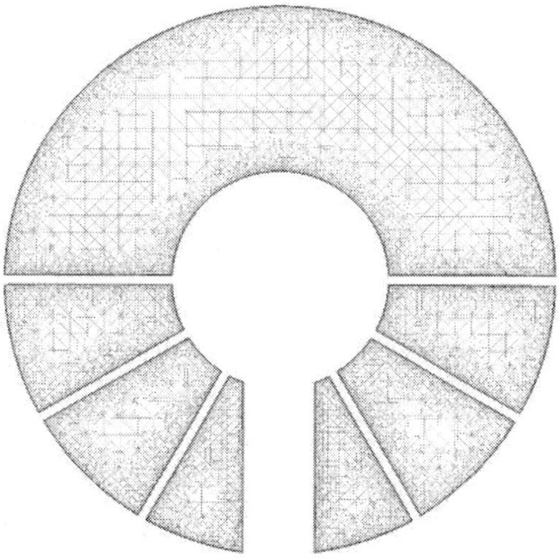

Fig. 8. Detail of the hex-core mesh in the cross-section through the outlet part.

In order to evaluate the mesh density influence, the integral values of c_p were compared for nominal boundary conditions of the test rig. As it can be seen in Table 2, the c_p result when using medium dense mesh is not so different from the fine mesh. As a result, the medium coarse mesh was chosen to be used for all the numerical analyses of different diffuser variants.

Table 2. Static pressure recovery factor depending on mesh densities.

Mesh density	Coarse	Medium	Fine
c_p (at position L = 0.44)	0.5710	0.4434	0.4429
Difference from the medium mesh [%]	28.8	0.0	-0.1

The solver ANSYS CFX 18.2 was used to carry out the numerical computations. This solver provides the steady state RANS (Reynolds-Averaged Navier-Stokes) solution with a compressible turbulent model. The k-ω SST turbulent model whose fidelity was validated in many prior Doosan Skoda Power projects using numerical and experimental results comparison was used. The medium was air which was defined as an ideal gas. A high resolution scheme was used to deal with the advection term and the turbulence numeric. A total pressure and a static temperature were defined at the domain inlet and a static pressure was defined at the outlet according to measurement results. The connections between the domains were defined as interfaces with a general connection.

Residual values and global imbalances were used to quantify simulation convergence criteria. The RMS (root mean square) residual type was chosen and 10^{-4} residual target value was defined. The global imbalances were less than 0.01 %. The values of required c_p were also monitored during each simulation in order to check that its convergence is appropriate.

6 Measured variants and results obtained

6.1 Blade cascade parameters - traversing mode

Loss coefficient of the cascade (ζ_m) and the ratio of dynamic pressures (p_p) upstream and downstream the blade cascade are characteristic parameters of the blade cascade. Within this paper four segments were measured on both variants of stationary blade cascade (Variant 1 and 3) for the value of Mach number 0.2 and 0.4 at the diffuser inlet (plane 1). The circumferential angle of each segment was chosen 27 °. This size of the angle enables to traverse over three blades of the cascade and in this way it is possible to adequately capture the vortex that arise behind the trailing edge of the blades. First, it was necessary to establish an appropriate procedure for data collecting. All together two types of measuring mesh were examined, the fine one and the coarse one. The fine mesh is formed by 324 points. For this type of mesh it was traversed along the channel height from 10 to 115 mm with the step of 10 mm (the last step was only 5 mm) and in the circumferential direction it was traversed by 1 °. In each point 100 readings were taken of 20 Hz frequency. This mesh measuring was very time consuming. For this reason then the coarse mesh was measured. The coarse mesh is made by 112 points. In each point again 100 readings were taken of 20 Hz frequency. For this type of mesh it was traversed along the channel height from 10 to 115 mm with the step of 15 mm and in the circumferential direction by 2 °. The coarse mesh can be seen in Fig. 9.

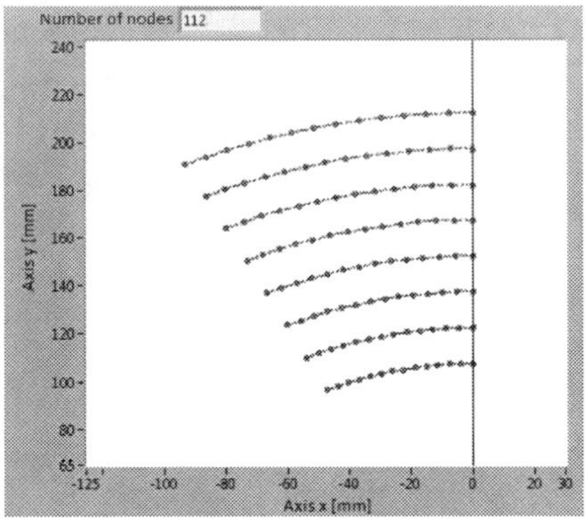

Fig. 9. Coarse mesh.

In order to verify the influence of the extension pipe even the variant without the extension pipe was measured. Measurements were carried out on the right upper segment (in the flow direction) for 1300 rpm, which corresponds with about 0.2 Ma on the diffuser inlet. The obtained results are arranged in Table 3. In the table the loss coefficient value and the ratio of pressures for simple average (Sa) and average weighted by mass flow (Mf) are shown.

Table 3. Mesh comparison – Variant 1 (0.2 Ma).

	Variant 1 – coarse mesh without pipe		Variant 1 – coarse mesh with pipe		Variant 1 – fine mesh with pipe	
	Sa	Mf	Sa	Mf	Sa	Mf
ζ_m	0.036	0.032	0.030	0.026	0.028	0.025
p_p	1.598	1.589	1.588	1.579	1.587	1.579

From the table it is evident that an extension of the hood has certain influence. For this reason, all the segments were measured with ring pipes fitted. The differences between the fine and coarse mesh with the ring extension pipe are practically negligible. Due to the time consumption of the fine mesh measuring the variant with coarse mesh was chosen.

The distribution of loss coefficient for Variant 1 and 0.2 Ma at the inlet can be seen in Fig. 10. From the picture it is evident that the loss coefficient reaches the highest value at the root diameter of the blade. This is where the biggest losses occur. It can be also observed that a significant increase of loss coefficient occurs behind the trailing edge of the blades where also the emerging floods are present. From the picture it is also evident that the loss coefficient value in the right half is slightly higher than the loss coefficient in the left half. Considering this fact it is always suitable to measure several segments distributed over the entire space.

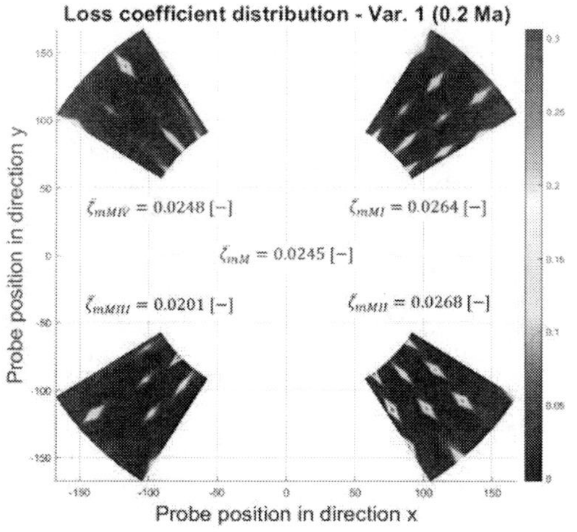

Fig. 10. Loss coefficient distribution.

The values of blade cascade parameters weighted by mass flow are given for individual variants in Table 4. From the table it is evident that the loss coefficient decreases with the increasing Ma value at the inlet. The loss of the blade cascade that models only the axial part of velocity (Variant 3) is approximately the same as for the cascade with circumferential shaped blades (Variant 1).

Table 4. Blade cascade parameters.

	Variant 1		Variant 3	
	0.2 Ma	0.4 Ma	0.2 Ma	0.4 Ma
$\bar{\zeta}_{mM}$	0.025	0.017	0.027	0.016
$\bar{p}_{pM}$	1.574	1.539	1.419	1.407

For illustration purposes, the profile of the swirl angle α_1 was plotted along channel height for several radials. Individual radials correspond with circumferential angle 32 °, 42 °, 50 ° and 58 °. For the sake of a comparison the results from CFD calculations was plotted as well and there is an obvious agreement between the results obtained from CFD calculations and those from the experiment. The blade cascade is for Variant 3 prismatic. In this case the flow does not curve along the blade height and the curvature angle at diffuser inlet reaches $\alpha_1 \cong 90$ °, see Fig. 11 - upper part. In Variant 1 the flow curvature can be seen with the inflection point at about 2/3 of the blade height, see Fig. 11- lower part. The courses are similar for both level of Mach number.

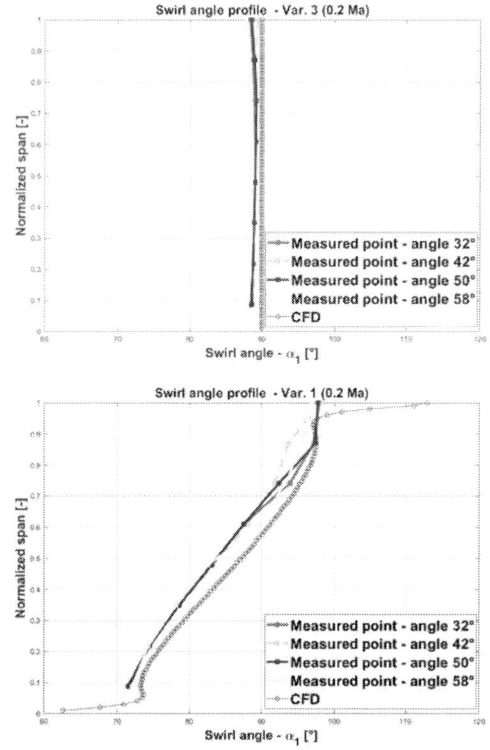

Fig. 11. Profile of swirl angle α_1 along the channel height.

6.2 Simple mode

Based on the knowledge of characteristic parameters of blade cascades of both variants and on verification of the circumferential angle distribution at the diffuser inlet it was possible to perform further measurements in the simple mode. During the measurements in the simple mode the Prandtl probe was located in the corresponding referential position. All together 6 variants were measured for each blade cascade for both Ma levels at the diffuser inlet.

6.2.1 Reference variant

The exhaust hood with the extension pipe without internal struts is considered as a reference variant. Pressure recovery coefficient values are given in Table 5. It is evident that in Variant 3 with increasing Ma value at the inlet the pressure recovery coefficient decreases, while in Variant 1 it in fact remains the same. As mentioned above, the prismatic blade cascade only derives the axial component of velocity and thus it is valid that $\alpha_1 = 90$ °. From the results, it is evident that the circumferential angle of the flow curvature at the exhaust hood inlet has certain influence on the final c_p. However, not as it was expected in a case of the hood with internal struts. The transition of the entering flow kinetic energy into pressure energy is between 67.2 to 73.8 %. The theoretical maximum value of pressure recovery coefficient (c_p) is often stated, which can be defined using the relation:

$$c_{p_{max}} = 1 - \frac{1}{\left(\frac{S_2}{S_1}\right)^2}, \quad (7)$$

where:

S_1 - axial exhaust hood inlet cross section,

S_2 - axial exhaust hood outlet cross section.

The theoretical maximum value of c_p is 0.74 for the given ratio of cross section. Then, the exhaust hood losses for all reference variants are between 0.002 and 0.068, which are very good values.

Table 5. Pressure recovery coefficient values – reference variant.

	Variant 1		Variant 3	
	0.2 Ma	0.4 Ma	0.2 Ma	0.4 Ma
$c_{p,ref}$ [-]	0.674	0.672	0.738	0.699

The dependence of c_p on the length L of the exhaust hood supplemented by the ring extension pipe can be observed in Fig. 12. The black vertical line indicates the end of the exhaust hood. From the picture it is obvious that c_p increases along its entire length. The steepest increase is at the beginning of the hood. This increase corresponds with the sharp increase of static pressure caused by a large initial opening angle. The pressure recovery coefficient in the cone diffuser and ring pipe increases more slightly.

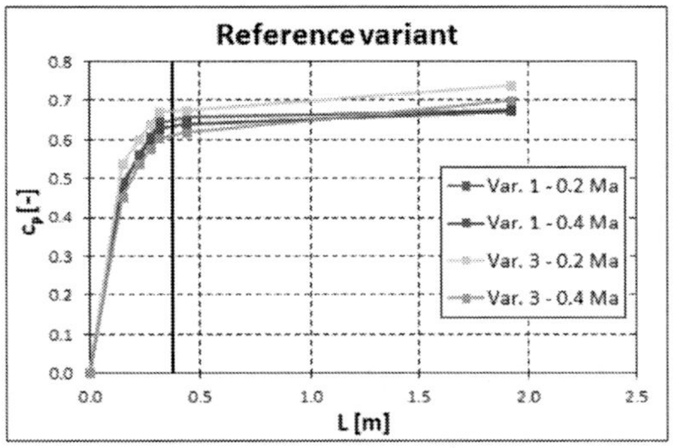

Fig. 12. Dependence c_p on L – reference variant.

Based on the knowledge of boundary conditions it was possible to perform CFD calculations and to compare the obtained results with experimental measurements. In Fig. 13 it is possible to observe the dependence of c_p on the length L for Variant 1 (0.2 Ma). Trends of both curves are similar. However, there is a certain difference between the obtained results.

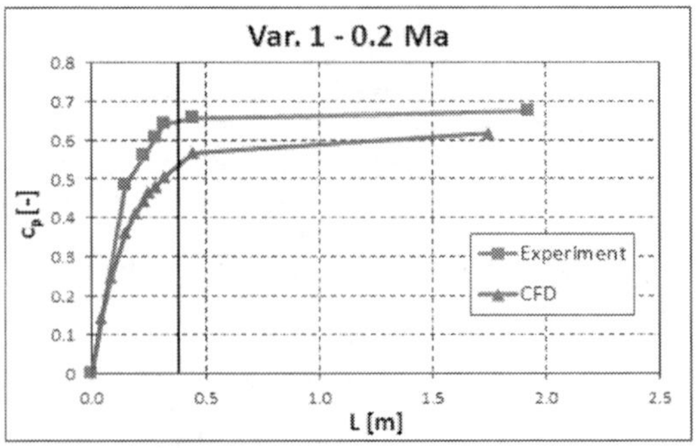

Fig. 13. Comparison of results – reference variant.

This difference is partly caused by the measurement in the simple mode. The traversing mode is a detailed process and the results of this type of measurement are reliable. Due to time constraints it is not possible to verify all variants using the traversing mode, and thus measurement in the simple mode is used. All measured data must be transformed in a very complex process. Many regressions must be used for both the measurement in the centre of gravity of each element of the cascade and for defining the reference position. Subsequently, the probe is placed into an exact position in one chosen segment and it is assumed that the circumferential flow does not change much. Based on these simplifications there will be a certain deviation from the real values and it can make the difference between the results obtained from experimental measurements and the results from CFD calculations.

Pressure recovery coefficient values of the exhaust hood extension pipe obtained from measurements in the simple and traversing mode and the values obtained from CFD calculations are given in Table 5. It is evident that the difference between the traversing mode results and CFD calculations are smaller than the differences between the simple mode results and CFD calculations. For better illustration a bar chart was created that can be seen in Fig. 14.

Table 6. Comparison of pressure recovery coefficient values.

	Variant 1		Variant 3	
	0.2 Ma	0.4 Ma	0.2 Ma	0.4 Ma
Simple mode	0.674	0.672	0.738	0.699
Traversing mode	0.660	0.661	0.678	0.662
CFD	0.618	0.626	0.644	0.646
Simple mode - CFD	**0.056**	**0.046**	**0.094**	**0.053**
Traversing mode - CFD	**0.041**	**0.034**	**0.034**	**0.016**

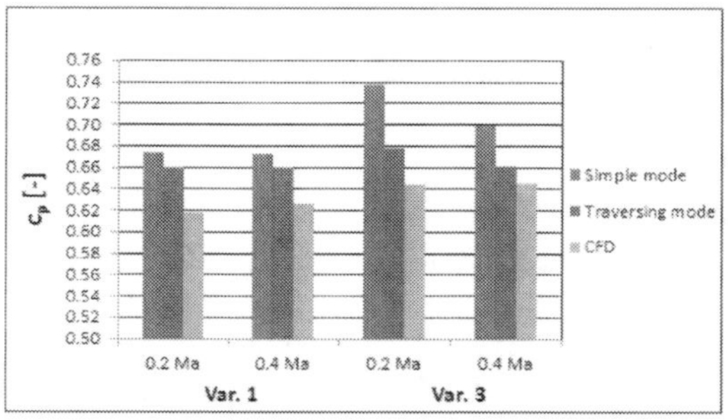

Fig. 14. Comparison of pressure recovery coefficient values.

Fig. 15 shows how the velocity profile develops in dependence on the length L for both measured variants with 0.2 Ma at the inlet. Gradually several radials were measured along the hood length in the plane 1, 1′, 1′′ and 2′ using a three hole pneumatic probe, see the scheme in Fig. 4. Radials were measured in the upper right segment (segment *I*). In the picture the exhaust hood is shown schematically. It is evident that towards the outlet the flow velocity is decreasing. Kinetic energy is transformed into pressure energy. In the last two planes it was traversed from the hood axis to the wall. However, behind the hub a considerable vortex occurs and the flow is here strongly turbulent. For this reason, it was possible to assess only the data taken at a certain distance from the centre of the channel.

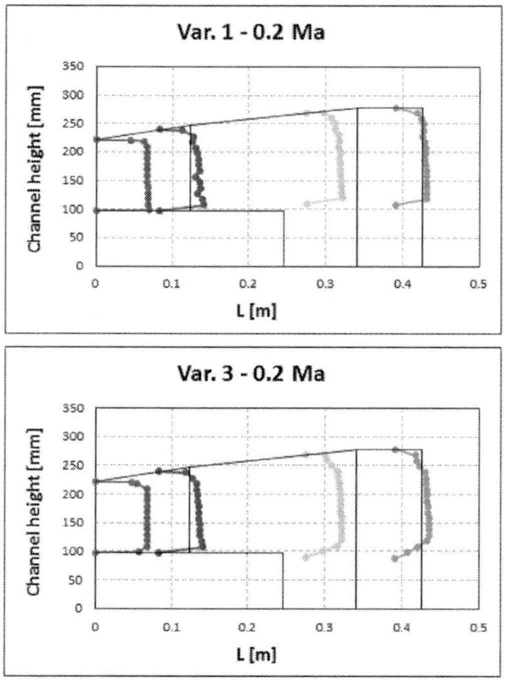

Fig. 15. Velocity profile in dependence on L.

6.2.2 Variant with all struts – Variant e

The location of individual struts can be seen in Fig. 16. All are located in the lower half of the hood. In Table 7 c_p values are given. For comparison also the values corresponding with the reference variant are presented here as well as the drop of c_p value. It is evident that the internal struts significantly influence the increase of loss coefficient values or the decrease of pressure recovery coefficient values. As a result, it is evident that the influence of Mach number on c_p is minimal for both variants of blade cascades.

Fig. 16. Variant with all struts – Variant *e*.

Table 7. Pressure recovery coefficient values – all struts.

	Variant 1		Variant 3	
	0.2 Ma	0.4 Ma	0.2 Ma	0.4 Ma
$c_{p,e}$ [-]	0.616	0.615	0.703	0.672
$c_{p,ref}$ [-]	0.674	0.672	0.738	0.699
Drop c_p [-]	**0.058**	**0.056**	**0.035**	**0.027**

In Fig. 17 the dependence can be seen of c_p on length L for an empty and full hood. (0.2 Ma). A similar trend can be reached also for the value 0.4 Ma at the inlet. The beginning and the ending of all struts is marked with black vertical lines. It can be seen that at the beginning the pressure recovery coefficient increases sharply for all variants. However, in the full hood the static pressure decreases significantly, which is caused by the struts. The struts reduce the cross section of the diffuser, which causes a local increase of flow velocity. The static pressure is restored and rising steeply in the remaining part of the diffuser. The pressure recovery coefficient in the ring pipe increases more slowly, which is given by a slight decrease of flow velocity.

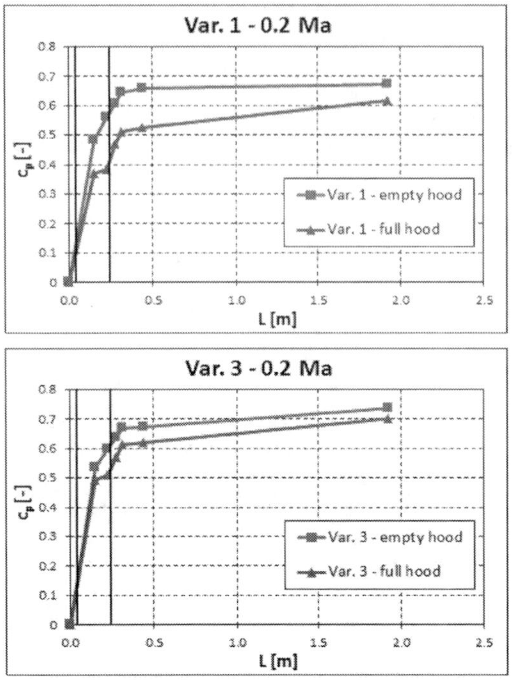

Fig. 17. Comparison of pressure recovery coefficient of the full and empty hood.

It is evident from the results that better results are obtained when the trailing angle of the flow into the strut is axial. For the curved blade cascade the flow entering the diffuser is curved too and thus the flow angle in the strut is not ideal. Due to poor flow around the internal struts larger energy losses occur, which is reflected in the increase of loss coefficient value. In order to reach higher values of c_p it is suitable to turn the struts, if possible, in the flow direction.

For the sake of comparison, the CFD calculations were carried out again. The trend of c_p along the length L for Variant 1 and 0.2 Ma can be seen in Fig. 18. Similar trend is reached for other measured variants.

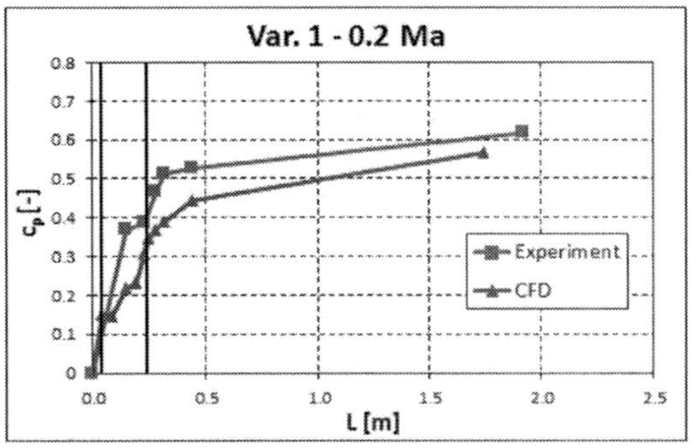

Fig. 18. Result comparison - full hood.

6.2.3 Other measured variants

Individual measured variants, locations of internal struts and variant labelling can be seen in Fig. 19. Measurements were carried out for both types of blade cascade and for both Ma values. As for the strength, only the variant with all struts is suitable, but in order to assess the influence of individual struts also the variants were measured with a lower number of internal struts.

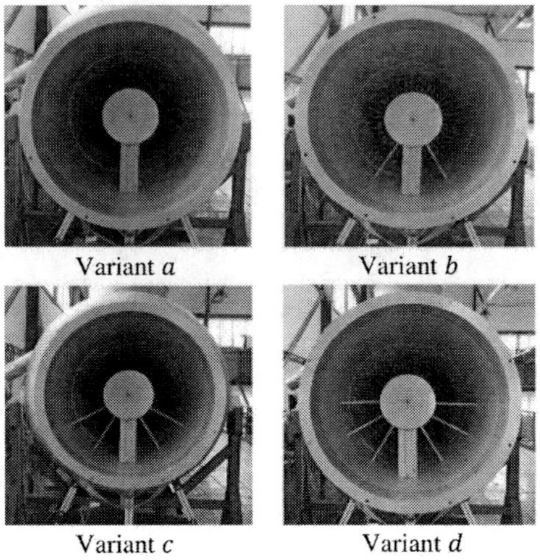

Fig. 19. Measured variants.

In Table 8 all values of c_p are shown for all measured variants. From the table it is evident that by adding each strut (pair of struts) a decrease of c_p arises compared to the previous variant. The only exception is adding the first pair of struts in Variant 3. In Fig. 20 it can be seen how c_p value drops with adding another strut (pair of struts). Based on this bar chart it can be defined which strut has the biggest influence on c_p value drop. For 0.2 Ma it is evident that inserting flat struts in the horizontal plane (Variant d) and traverse struts (Variant e, see Fig. 16) is the most influential. On the other hand for 0.4 Ma inserting bottom struts is most influential (Variant a). Conversely, inserting the first pair of flat struts (Variant b) participates minimally and for Variant 3 even an increase of c_p arises in comparison with the previous variant.

Table 8. Pressure recovery coefficient values.

	Variant 1		Variant 3	
	0.2 Ma	0.4 Ma	0.2 Ma	0.4 Ma
$c_{p,ref}$ [-]	0.6739	0.6716	0.7379	0.6993
$c_{p,a}$ [-]	0.6685	0.6538	0.7363	0.6871
$c_{p,b}$ [-]	0.6614	0.6536	0.7318	0.6889
$c_{p,c}$ [-]	0.6518	0.6445	0.7263	0.6874
$c_{p,d}$ [-]	0.6331	0.6299	0.7143	0.6816
$c_{p,e}$ [-]	0.6164	0.6153	0.7027	0.6718

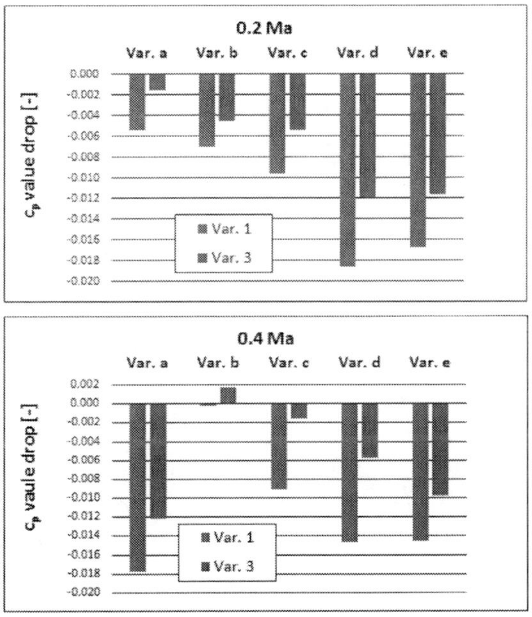

Fig. 20. Influence of individual struts.

For illustration the Variant *c* is compared here again (Variant 1 – 0.2 *Ma*). The dependence of c_p on the exhaust hood length is shown in Fig. 21. From the picture, an agreement is evident between experiment measurement results and CFD results.

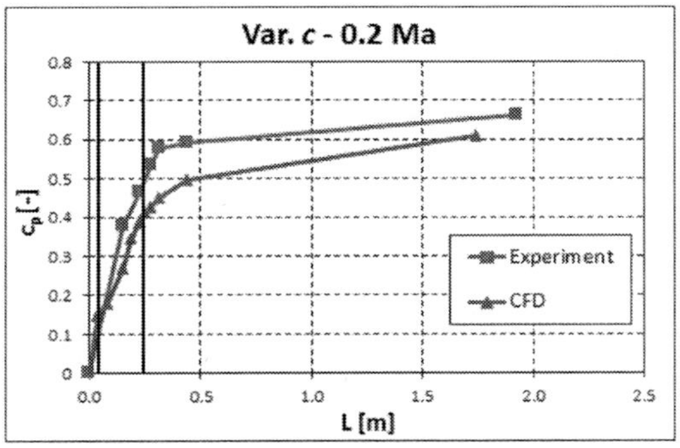

Fig. 21. Result comparison – Variant *c*.

7 Conclusion

The main aim of this paper was to define the pressure recovery coefficient (c_p) increase on the experimental model of the axial exhaust hood, the influence of the circumferential angle (α_1) and the influence of individual struts. The obtained results were compared to CFD calculations based on corresponding boundary conditions.

First of all, measurements were carried out for an empty hood variant. Then, the hood was fitted with various internal struts and their influence on c_p was evaluated. Measurement was carried out for two variants of the circumferential angle at the diffuser inlet and two levels of Mach number at the diffuser inlet, namely 0.2 and 0.4 *Ma*. The required inlet angle was defined by circumferentially shaped stator blades. The first blade profile (Variant 1) was designed so that the required angle was in agreement with a real turbine. The second blade profile (Variant 3) was designed with only the axial velocity component, which made it possible to describe the influence of the flow circumferential angle on the exhaust hood behaviour.

The first important part of the paper was to define characteristic parameters of the blade cascade, namely the cascade loss coefficient (ζ_m) and the ratio of dynamic pressures (p_p). In order to define these parameters it was necessary to measure several segments along the perimeter in the traversing mode. Based on the knowledge of blade cascade parameters it was possible to continue the measurements in the simple mode and in this way to define c_p in an empty hood which ranges for particular variants from 0.672 to 0.738. As a result, it is evident that about 70 % of kinetic energy of the flow is transformed into pressure energy. For the given ratio of cross sections the maximal pressure recovery coefficient is 0.74. Then, the exhaust hood losses for all reference variants (empty exhaust hood) are between 0.002 and 0.068, which are very good values. Consequently, measurements were carried out with individual variants of internal struts. The presence of struts had in all cases a negative impact on c_p compared to the reference variant. In Variant 1, adding the third pair of flat struts had the biggest impact on c_p (Variant *d*). In variant 3, the biggest impact on c_p was measured for bottom strut location (Variant *a*). Due to the strength only the variant with all internal struts is suitable (Variant *e*). Compared to the reference variant, in Variant 1 the decrease was by about 0.058 and in Variant 3 by about 0.03. From this result, it is evident that the circumferential angle negatively contributes to the pressure recovery coefficient for the full hood. It is caused by different behaviour of the flow behind the blade cascade and thus flow running around the struts is different for both cases.

Based on pressure recovery coefficient trend along the exhaust hood length it can be seen that most kinetic energy is transformed into pressure energy at the beginning of the hood. For this reason it is recommended to place the struts as far as possible from the trailing edges of the last rotor blades. From the results the negative impact of internal struts on c_p also evident and therefore it is advantageous to insert as few struts as possible. If possible, it is suitable to turn the struts in the flow direction, see the circumferential angle influence and the flow entering the struts.

The results and knowledge obtained can be transferred to CFD calculations which can be thus refined. The aim is to achieve a certain universal procedure in determining the pressure recovery coefficient of the exhaust hood so that it would not be necessary to perform experiment measurements for each type of the hood.

Acknowledgement

The authors of the paper would like to thank TAČR project Zéta TJ01000048 „Lowering outlet losses in steam turbines of modern construction" and project "LoStr" no. CZ.02.1.01/0.0/0.0/16_026/0008389 for financial support.

Reference

1. R. Kalista at al., *The experimental investigation of the internal support effects on exhaust casing pressure recovery*, ASME IMECE 2017, IMECE2017-70279, Tampa, USA, (November 2017).
2. J. Fu, J. Liu, *Investigation of Influential Factors on the Aerodynamic Performance of a Steam Turbine Exhaust Systém*, In proceedings of ASME Turbo Expo. GT2010-22316, (2010).
3. A. Hirschmann, S. Volkmer, M. Schatz, C. Finzel, M. Casey, *The influence of the Total Pressure Profile on the Performance of Axial Gas Turbine Diffusers*, In Proceedings of ASME Turbo Expo. GT2010-22481, (2010).
4. R. Kalista, L. Mrozek,V. Slama at al., *Experimental measurement on the air test rig of the axial steam turbine exhaust casing*, AIP Conference Proceedings 2047, 020006 (2018).
5. R. Kalista at al., *Experimental and numerical design of axial exhaust diffuser test rig for investigation the aerodynamic performance of steam turbine and its enhancement by shape modifications*, ASME Turbo Expo 2019, GT2019-91313, Phoenix, Arizona USA, (June 2019).
6. M. Hoznedl at al., *The pressure field at the output from a low pressure exhaust hood and condenser neck of the 1090 MW steam turbine – experimental and numerical research*, ASME Turbo Expo 2018, GT2018-75248, Oslo, Norway, (June 2018).
7. M. Hoznedl, et al., *Experimental Investigation on Flow in Diffuser of 1090 MW Steam Turbine*, Proc. XX. International Conference The Application of Experimental and Numerical Methods in Fluid Mechanics and Energy, (2016).

On 3D structure of a circular cylinder wake

Václav Uruba[1,2,*], *Pavel* Procházka[1], and *Vladislav* Skála[1]

[1]Institute of Thermomechanics of the Czech Academy of Sciences, Dolejškova 5, Praha 8, CR
[2]University of West Bohemia, Faculty of Mechanical Engineering, Department of Power System Engineering, Universitní 8, Plzeň, CR

Abstract. The 3D structure of the wake behind a circular cylinder in a cross-flow is studied experimentally using stereo PIV technique, Reynolds number was around 5 thousands. The statistical characteristics along the cylinder axis are effectively constant and thus 2D, however instantaneous flow-field in the wake is fully 3D. Within the wake dynamical flow, the streamwise vortices and spots of streamwise velocity component pulsations are detected.

1 Introduction

Many types of flows in practice could be characterized by 2D boundary conditions very often. Typical example could be cross-flow around a prismatic body. In spite of inherent dynamical nature of the flow-field if turbulent, this case could be considered as a plane, 2D flow, invariant along the body (excluding the ends). Such a case is also mathematically modelled as a 2D flow as a rule, taking into account only a single cross-section, supposing the same flow picture for any other section along the body axis.

The case of circular cylinder in cross-flow is one from the family of so called canonical cases. We could find extensive information source on this case in available literature covering theoretical, experimental and numerical studies respectively. Most of available studies treat the prismatic cylinder as forced 2D case, thus no 3D structures are detected, of course. The 3D structure of the wake behind a body of 2D geometry has been already addressed in several papers, see e.g. in [1]. The paper [2] deals with the d'Alembert paradox (zero forces) for inviscid flow and shows that this paradox is due to forcing 2D stationary flow pattern, which represents one of the solutions of the Naiver-Stokes equations, however instable one. The other, stable solution is represented by 3D and non-stationary flow pattern. This complies with reality. Then, the adequate forces are generated even in the inviscid case and the d'Alembert paradox is thus resolved.

It is known that the real wake behind circular cylinder contains several types of flow patterns: steady, unsteady periodical and chaotic, some of them are 2D character, but most of them are 3D. The control parameter is Reynolds number defined using the incoming velocity, cylinder diameter and fluid kinematic viscosity. Description of the wake topology including its dynamics could be find in any good book on fluid mechanics, more information on the details are in numerous papers, see e.g. [3].

In the presented paper the 3D non-stationary structures in the wake are to be addressed.

2 Experimental setup

The experiments were carried out in the Laboratory of Turbulent Shear Flows in Institute of Thermomechanics of the Czech Academy of Sciences. The blow-down facility was used with open test section, cross-section was 250 x 250 mm^2. Quality of the flow was good, velocity regularity was within 1% of the outlet velocity U_e = 5 m/s, intensity of velocity fluctuation was less than 0.5% of it.

2.1 Physical model

The model of cylinder ϕ d = 15 mm was fixed on the contraction output flange in its axis, perpendicularly to the output flow, 250 mm of the cylinder length was placed in the flow. The situation including measuring device is depicted schematically in Fig. 1.

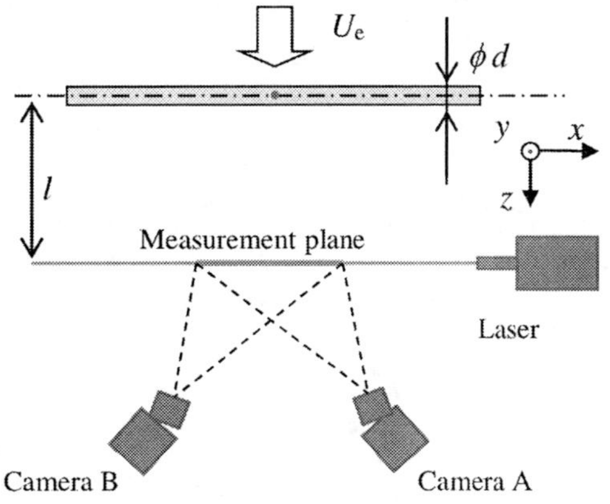

Fig. 1. Experimental setup.

* Corresponding author: uruba@it.cas.cz

The wake is studied in the distance $l = 8\ d$ (120 mm) behind the cylinder, the measurement plane is perpendicular to the incoming flow and parallel to the cylinder axis (shown in green in Fig. 1).

In Fig. 1 the Cartesian coordinate system is introduced with z axis parallel to the flow and x axis parallel to the cylinder axis. The origin is placed to the cylinder axis in the middle of the Field of View (hereinafter FoV) – the red dot in Fig. 1.

2.2 Experimental technique

The Stereo PIV time-resolved measuring technique was used to evaluate all three velocity component in the plane of measurement. The measuring apparatus is by Dantec company.

The laser (see Fig. 1) New Wave Pegasus, Nd:YLF double-head, was used to illuminate the particles. The used light was with wavelength of 527 nm, maximal frequency is 10 kHz, shot energy is 10 mJ (for 1 kHz) and corresponding power is 10 W per head. Two CMOS cameras NanoSense MKIII with Scheinpflug mounting of lenses were used (in Fig. 1 the "Camera A" and "Camera B") to capture the tracing particles distributed in the flow. Each camera has resolution of 1280 x 1024 pixels, maximal corresponding acquisition frequency is 512 double-snaps per second.

The data acquisition was controlled by the Dynamic Studio software. The calibration was performed using standard calibration target 100 x 100 mm. Precise positioning of the calibration target was performed by the stepper motor in five positions (-2, -1, 0, 1, 2 mm). The pinhole camera model was used for the pictures dewarping. The calibration error was approximately 0,2 pixels. 1600 double-snaps have been acquired with frequency 500 Hz, representing 3.2 s of the time record. The schematic view of stereo PIV layout is in Fig. 1.

The Safex generator of particles in the form of oil droplets, the mean diameter of the particles was 1 μm.

2.3 Analysis methods

First, the data is subjected to classical statistical analysis. The methods as calculation of mean values, variances and covariances are used.

To study the dynamical properties of the flow-field the Oscillation Pattern Decomposition method (OPD) was adopted resulting in series of OPD modes. Each OPD mode is characterized by its topology in complex form (consisting of real and imaginary parts), frequency and attenuation of the pseudo-periodic (oscillating) behaviour. Attenuation or amplitude decay is described by so called e-folding time τ_e representing the mean time period of the mode amplitude decay by "e". The other decay characteristic is dimensionless "periodicity" p which expresses the e-folding time in multiples of periods of the mode. The details on OPD method could be found in [4,5].

3 Results

All results are to be shown in dimensionless form. The geometric dimensions are expressed in multiples of the cylinder diameter d and the velocities in multiples of the inlet velocity U_e.

3.1. Mean values

The time-mean values of the acquired spatio-temporal data are to be shown in the form distributions within the plane of measurement. The measurement plane is located in the position $z = 8$, the cylinder axis in the dash-dot line $y = 0$ and projection of the cylinder contours are depicted as dashed lines $y = +-0.5$.

In Fig. 2 the mean velocity distribution is shown, the in-plane velocity vector components are represented by arrows, while the out-of-plane component W (in z direction) is represented by colour. The scale of the vectors is shown in left-bottom corner, showing the size of 0.1. The vector-lines of in-plane velocity vector components are added arbitrarily for better visibility of the vector-field.

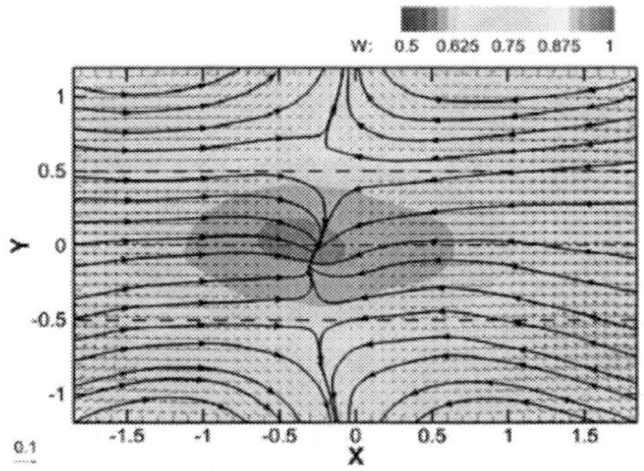

Fig. 2. Mean velocity distribution.

The minimum of the mean velocity is located just in the middle close to the coordinate's origin, forming a sink. In reality no sink could exist in the mean velocity field because of continuity condition. The weak sink in our case is a consequence of a view distortion by the cameras located on sides – see Fig. 1. The mean velocity distribution is close to 2D situation, independent on x. Minimum of the mean streamwise velocity component is located on the cylinder axis x. The spanwise mean velocity components are close to zero.

The Turbulent Kinetic Energy (hereinafter TKE) distribution is in Fig. 3. Maximum of the TKE is within the wake behind the cylinder body, close to the cylinder axis. The distribution is again close to 2D.

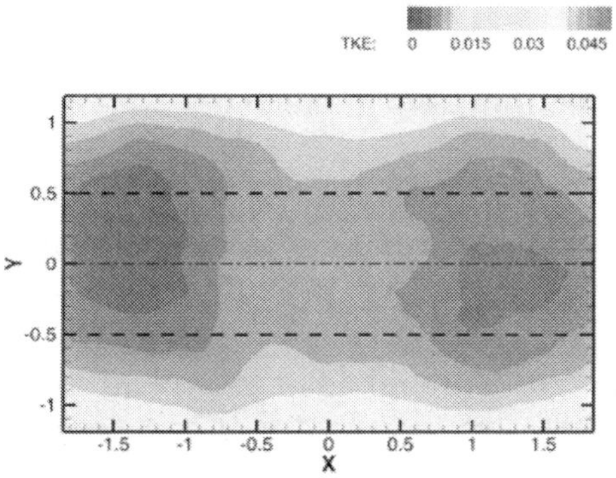

Fig. 3. TKE distribution.

All 3 correlation coefficients between individual velocity components have been evaluated, however only the correlation coefficient (VW) between streamwise and spanwise perpendicular to the cylinder axis velocity components contains important values, see Fig. 4. The rest two coefficients approaches zero in the whole region.

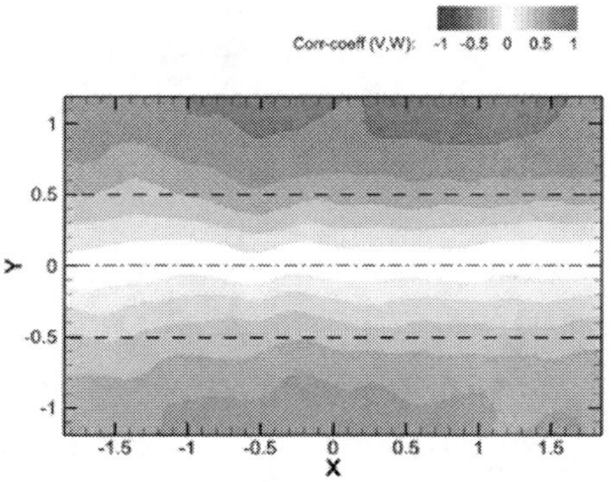

Fig. 4. (VW) correlation coefficient distribution.

This result show turbulence production in shear layers on above and below the cylinder location. Disproportion between the correlation coefficients indicates important anisotropy of the turbulence in the cylinder wake.

3.2 Dynamics

The OPD analysis has been performed on the acquired data. The numerical parameters of all evaluated OPD modes are shown in Table 1.

The OPD modes numbering is made according to their frequency in ascending order. 15 OPD modes have been evaluated.

The OPD modes could be divided into several groups. In Table 1 the dominant OPD mode with high periodicity value is marked by red colour. Blue colour means modes with periodicity about 0,2, green about 0,1 and white (no colour) are mode with very small periodicity values, less than 0,05.

Table 1. OPD modes.

No.	f [Hz]	τ_e [ms]	p [1]
1	0	0.126	0
2	0	0.796	0
3	0	0.506	0
4	0	0.358	0
5	5.14	0.565	0.029
6	5.72	0.283	0.016
7	9.27	0.372	0.035
8	13.17	0.333	0.044
9	17.24	0.516	0.089
10	23.54	0.412	0.097
11	27.12	0.311	0.084
12	35.16	0.583	0.205
13	43.19	0.431	0.186
14	45.36	3.664	1.662
15	62.25	0.461	0.287

Next, the OPD modes are represented graphically in the Frequency-Periodicity plane in Fig. 5 in the form of the spectrum.

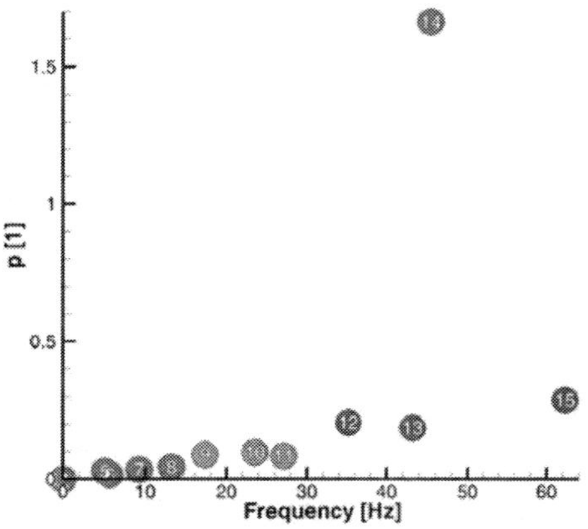

Fig. 5. OPD spectrum.

In Fig. 5 the OPD modes are distinguished by numbers and colours, similarly as in Table 1.

The topology of the OPD modes consists of real and imaginary parts representing the cyclostationary process in different phases shifted by $\pi/2$ from each other.

The dominant OPD mode 14 is characterized by very high periodicity and it is directly related to the vortex shedding process in the wake. It is described by parameters marked by red colour in Table 1. Real and imaginary parts of this OPD mode are shown in Figures 6 and 7 respectively.

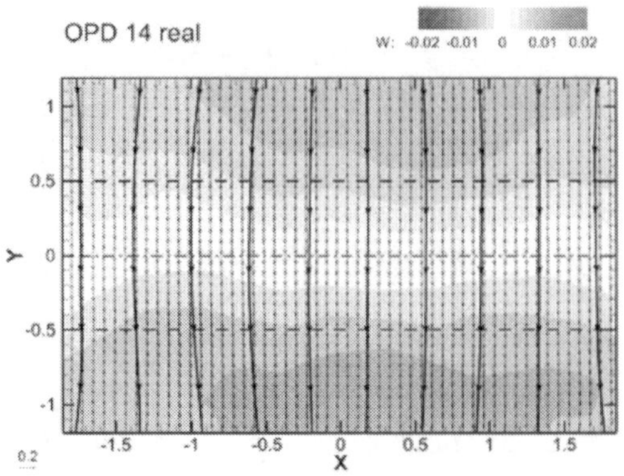

Fig. 6. Real part of the OPD mode 14 topology.

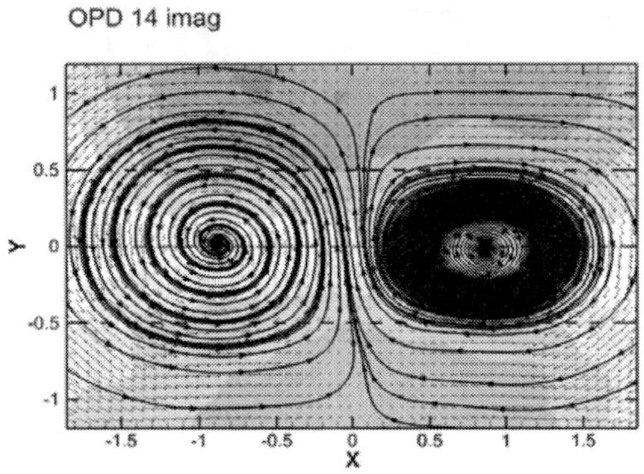

Fig. 7. Imaginary part of the OPD mode 14 topology.

The real part of the OPD mode is shown in Fig. 6 and documents 2D structure of the fluctuation flow-field. On the other hand the imaginary part in Fig. 7 and it contains a contra-rotating pair of streamwise vortices with spacing 2 cylinder diameters. The distribution of the w fluctuation velocity component is similar as for the real part but opposite.

The von Kármán vortex street dynamics have been studied in details in [6], where results of measurement in the plane parallel to the flow and perpendicular to the cylinder axis are shown.

The real part of its topology shows parallel flow in y direction transversal to both flow and cylinder axis. The streamwise velocity component forms shear region with the velocity amplitudes in the top and bottom of the FoV respectively. In the middle of the FoV, close to the cylinder axis, there is fluctuation close to zero.

The topology of the POD mode 14 suggests periodic appearance of the system contra-rotation streamwise vortices during one half of the period, then disappearing and reappearing again. This process is accompanied by pulsations of the flow in y direction and streamwise velocity component shear configuration.

Obviously, the OPD mode 14 is connected with the vortex-shedding process known as von Kármán Vortex Street. These vortices are oriented in x direction shedding alternatively from the cylinder sides with the same frequency as the OPD mode 14 and moving in streamwise direction. Existence of the OPD mode 14 suggests, that the von Kármán vortices are structured along the cylinder axis x containing streamwise vorticity.

Next, we are going to present the 3 OPD modes 12, 13, 15 with periodicity value around 0,2. The real and imaginary parts of the OPD mode 12 are shown in Figures 8 and 9. The mode is represented by alternation of single streamwise vortex and contra-rotating pair of streamwise vortices. The streamwise fluctuating velocity component forms spotty structures in the domain. All the involved structures are smaller and less regular than those in the case of dominant OPD mode 14, confirm Figures 6 and 7.

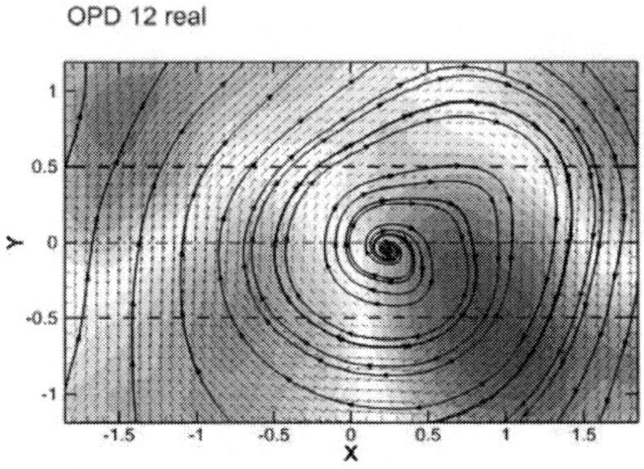

Fig. 8. Real part of the OPD mode 12 topology.

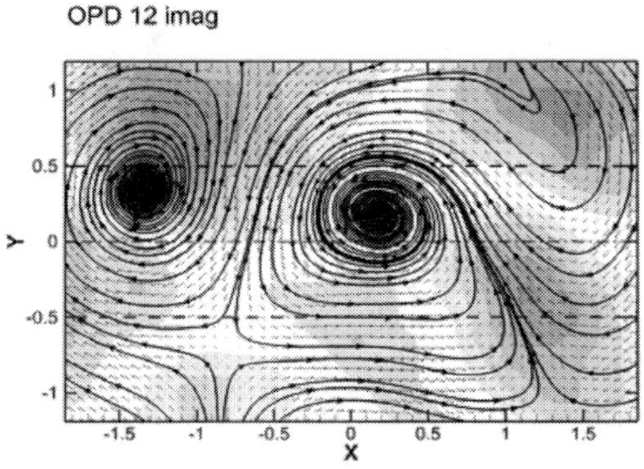

Fig. 9. Imaginary part of the OPD mode 12 topology.

The OPD modes 13 and 14 are quite similar to each other. The dominant structures are big amplitude positive and negative fluctuations of streamwise velocity component along the cylinder axis x. Only real parts of the OPD modes 13 and 14 are shown in Figures 10 and 11 respectively.

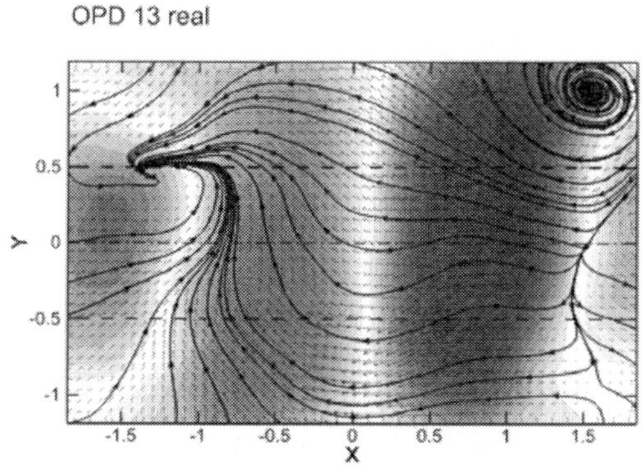

Fig. 10. Real part of the OPD mode 13 topology.

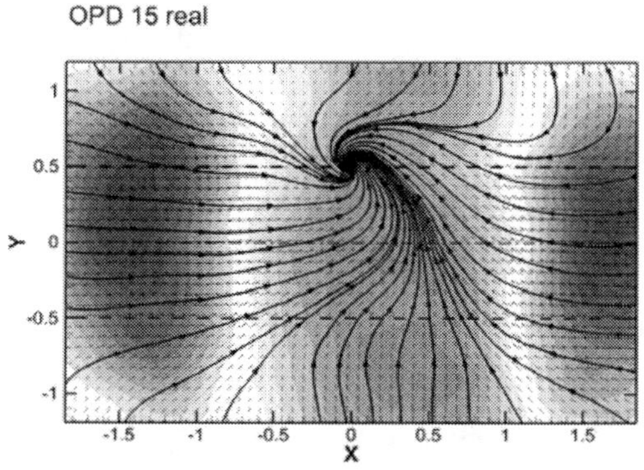

Fig. 11. Real part of the OPD mode 15 topology.

The last group of the OPD modes to be presented here are theses with the periodicity value around 0,1, the modes number 9, 10 and 11. Again, only real parts of the OPD modes are shown, the imaginary counter-parts are qualitatively similar, however the topology is shifted in space. The modes could be characterized by a couple of contra-rotating vortices accompanied with spots of positive and negative fluctuations of streamwise velocity component. The real parts of the OPD modes 9, 10 and 11 are shown in Figures 12, 13 and 14 respectively.

The rest of the OPD modes are characterized by strong decay in time and/or very low frequency. The modes 1-4 are characterized even by zero frequency, this means that those modes are not of cyclostationary nature, but only simply decaying in time, the topology is constant.

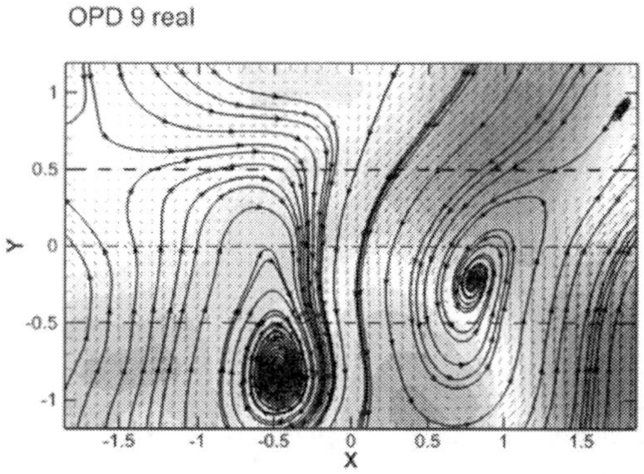

Fig. 12. Real part of the OPD mode 9 topology.

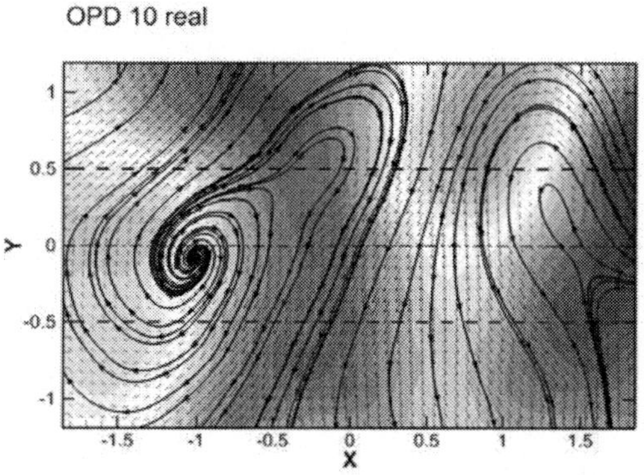

Fig. 13. Real part of the OPD mode 10 topology.

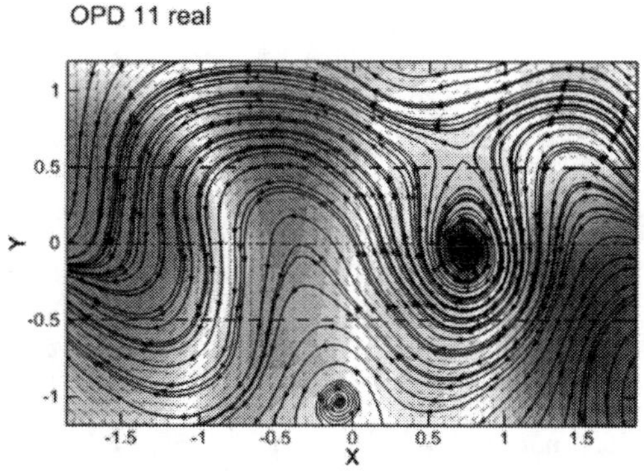

Fig. 14. Real part of the OPD mode 11 topology.

4 Conclusions

The flow in the wake behind prismatic circular cylinder could be considered to be 2D in sense of statistical characteristics, however instantaneous flow patterns are always 3D. The 3D dynamical structure of the wake was studied experimentally using stereo PIV time-resolved technique. The dynamical analysis of the acquired data was performed with help of the OPD method capturing decaying spatial harmonic oscillations, represented by the OPD modes.

The dominant OPD mode with the most stable oscillations characterizes the von Kármán vortex street existing within the wake, which is characterized by the same frequency. The mode topology could be described by oscillating shear layer across the cylinder axis and periodic appearance of the system contra-rotation streamwise vortices during one half of the period, then disappearing and reappearing again.

The remaining dynamical content represented by the other OPD modes is much more random in nature and weaker as for the amplitudes. The topology od modes corresponding to this behaviour could be characterized by system of streamwise vortices in combination with spotty pulsations of the streamwise velocity component.

This work was supported by the Grant Agency of the Czech Republic, project No. 17-01088S.

References

1. V. Uruba, On 3D Instability of Wake behind a Cylinder, AIP Conference Proceedings 1745 (1), 020062, (2016)
2. J. Hoffman, C. Johnson, Resolution of d'Alembert's paradox, *J. Math. Fluid Mech.*, **12** (3), pp.321-334, (2010)
3. C.H.K. Williamson, Vortex Dynamics in the Cylinder Wake, *Annu. Rev. Fluid. Mech.* (1996) **28** pp.477-539
4. V. Uruba, Decomposition methods in turbulent research, EFM11, *EPJ Web of Conferences*, **25** 01095 (2012)
5. V. Uruba, Near Wake Dynamics around a Vibrating Airfoil by Means of PIV and Oscillation Pattern Decomposition at Reynolds Number of 65 000, *Journal of Fluids and Structures*, **55**, pp. 372-383 (2015)
6. P. Procházka and V. Uruba, Reynolds Number Effect on Velocity Field and on Coherent Structures behind a Cylinder, AIP Conf. Proc. **2118**, 030037-1–030037-4; (2019)

Measurement of Pool Boiling Heat Transfer Coefficient with an Insulated Stainless-Steel Block

Viktor Vajc[1,*], and *Martin* Dostál[1]

[1]Department of Process Engineering, Faculty of Mechanical Engineering, Czech Technical University, 166 07 Prague, Czech Republic

Abstract. We have designed an apparatus dedicated for measurement of heat transfer coefficient during saturated nucleate pool boiling. After several performed measurements, we have applied three important modifications of the apparatus motivated by discrepancies between measured and correlated heat transfer coefficients. Firstly, the cubic stainless-steel heated block of the apparatus no longer accommodates heating cartridges. It now stands on a copper carrier of the cartridges, which are vertically oriented. Secondly, we have insulated the heated block with plates made of cellular glass. Thirdly, thermocouples used for measurement of temperatures inside the block have been fixated with a special high-temperature adhesive. These modifications were applied in order to improve the temperature distribution inside the block and to eliminate the impact of surroundings on measured heat transfer coefficients. This contribution presents and discusses experimental results obtained with these modifications.

1 Introduction

We have performed several experiments with our apparatus designed for measurement of heat transfer coefficient during saturated nucleate pool boiling. Obtained results are presented in our article [1], which was written for EFM 2018 conference. The results were compared with five verified correlations.

Eight experimentally found datapoints were published in [1]. Three of them evinced a strong discrepancy for heat flux slightly above 80 kW m^{-2}. Remaining five datapoints evinced a trend which was more or less comparable with trends of correlations, but obtained heat transfer coefficients were slightly higher than correlated values. These eight datapoints were obtained for a very small range of heat fluxes from around 80 up to 100 kW m^{-2}. This article presents and discusses a set of 58 datapoints obtained for heat fluxes from 20 up to 150 kW m^{-2}. Such a set is much more appropriate for making better-founded conclusions.

2 Correlations Used for Comparison

In order to verify our experimental results, we use five correlations briefly described in [1]: Rohsenow [2], Mostinski [3], Stephan and Abdelsalam [4], Nishikawa [5] and Yagov [6]. In this article, Forster and Zuber correlation [7] in the form taken from [8]

$$\alpha = \frac{0.00122\,\Delta T^{0.24}\,\Delta p^{0.75} c_{p,\mathrm{L}}^{0.45}\,\varrho_{\mathrm{L}}^{0.49}\,\lambda_{\mathrm{L}}^{0.79}}{\sigma^{0.5}\,\Delta h_{\mathrm{LG}}^{0.24}\,\mu_{\mathrm{L}}^{0.29}\,\varrho_{\mathrm{G}}^{0.24}} \tag{1}$$

*e-mail: viktor.vajc@fs.cvut.cz

is also included. However, correlation (1) is implicit, because superheat during saturated boiling

$$\Delta T = T_{\mathrm{w}} - T_{\mathrm{sat}} = q/\alpha \tag{2}$$

and difference of saturation pressures

$$\Delta p = p_{\mathrm{sat}}(T_{\mathrm{w}}) - p_{\mathrm{sat}}(T_{\mathrm{sat}})\,. \tag{3}$$

are both functions of heat transfer coefficient α. Therefore, we solved correlation (1) iteratively.

3 Experimental Apparatus

Our apparatus, see Fig. 1, is dedicated for measurement of heat transfer coefficient during saturated nucleate pool boiling under standard atmospheric pressure. It consists of stainless-steel bottom with a square cut. Stainless-steel block, see Fig. 2, is inserted through the cut. The gap between the bottom and the block is sealed with a combination of silicon *Ceresit CS 28* and epoxy *EPO-TEK 302-3M-R*. The block stands on a copper carrier of five vertically oriented heating cartridges, see Fig. 3. The carrier and the block are insulated with five plates made of cellular glass *Foamglas Perinsul S*. Top face of the block with dimensions of 48 × 48 mm is the wall on which boiling occurs. Temperature field inside the block is measured with six sheathed thermocouples *Omega GKMQSS-M100U-150* accommodated inside six holes (D = 1 mm, L = 24 mm) drilled into the block. These holes were filled with alumina-based adhesive *Aremco Ceramabond 670*. Adhesives and other

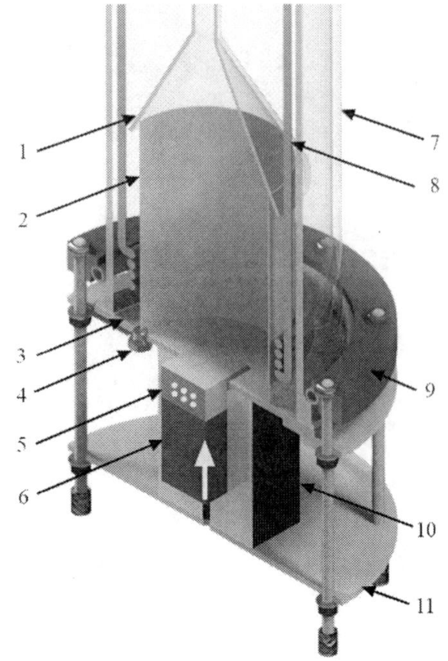

Fig. 1. Experimental apparatus: 1 Funnel roof, 2 Separating tube, 3 Bottom, 4 Plug, 5 Heated stainless-steel block, 6 Copper carrier of heating cartridges, 7 Glass cylinder, 8 Auxiliary heater, 9 Flange, 10 Thermal insulation, 11 Assembly flange. Six holes for thermocouples and orientation of heating cartridges inside the carrier are marked with yellow color.

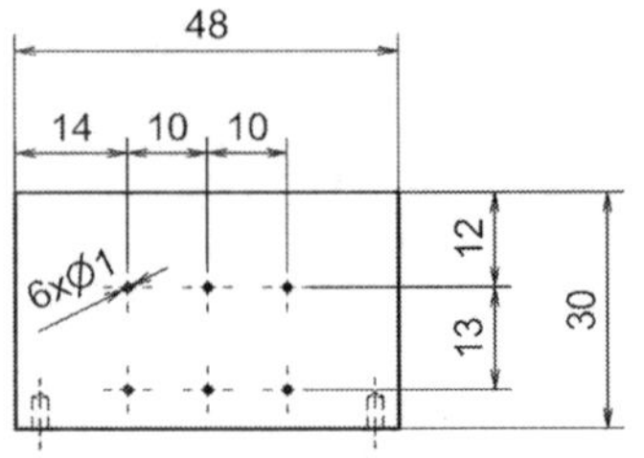

Fig. 2. Heated block.

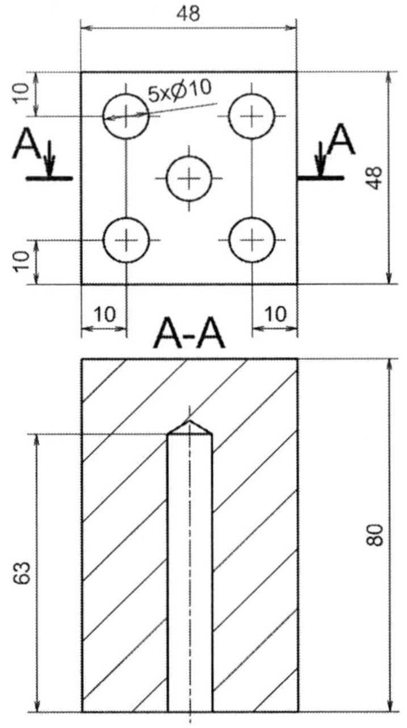

Fig. 3. Carrier of heating cartridges.

methods for attachment of thermocouples to the block are discussed in [9].

On the bottom stands a glass cylinder with inner diameter of 200 mm and length of 500 mm. PTFE gasket is placed between the bottom and the cylinder. The cylinder is opened to the surroundings and serves as a reservoir for the boiling liquid. Temperature of the boiling liquid is measured with thermocouple probes *Omega TJ2-CPSS-M15U-600*. Auxiliary heating spiral is inserted into the cylinder from its top. Power to the heating cartridges and to the heating spiral is supplied and regulated with variacs. Inside the cylinder, on the bottom stands a stainless-steel tube (D = 70 mm, L = 200 mm). Glass funnel was placed on the top of the tube.

Cold junctions of all thermocouples were placed into calibration reference chamber *Omega TRCIII-A* via cold-junction probes *Omega TRP-K-36* and *Omega TRP-T-36*. DAQ module *JanasCard AD24USB* provided A/D conversion between thermocouples and PC.

4 Modifications of the Apparatus

The current experimental apparatus in Fig. 1 was developed from an older version of the apparatus described in [1]. In this section, we mention major modifications and justification for them.

4.1 Elimination of Subcooling

In [1], we have mentioned that we were unable to achieve saturated boiling. Furthemore, temperature of the boiling liquid was not stable when we heated the boiling liquid only with primary cartridge heaters. Therefore, we installed secondary auxiliary heater. Boiling was, however, still subcooled in average by 0.6 °C for datapoints presented in [1]. Because of the subcooling, we were forced to perform experiments only for heat fluxes higher than 80 kW m^{-2}. For lower heat fluxes, the subcooling would have been at least several degrees Celsius. We tried to eliminate this subcooling using insulation of glass cylinder made of glass wool, but the situation did not improve.

After that, we tried to separate and enclose the space above the heating wall. For that, we used a stainless-steel tube, see part no. 2 in Fig. 1, and on the top of the tube we placed a funnel roof, see part no. 1 in Fig. 1. This modification completely eliminated the subcooling. Moreover, we were able to remove insulation from the glass cylinder. Now, we are able to reach saturated boiling for heat fluxes higher than approximately 25 kW m^{-2}. For lower heat fluxes, subcooling of several tenths of °C emerges.

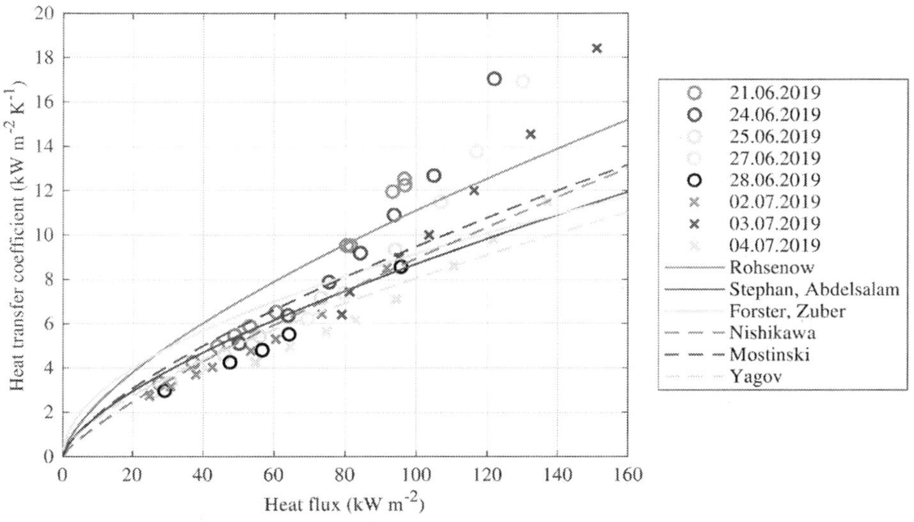

Fig. 4. Results of experimental measurements.

Another positive effect of the funnel roof and stainless-steel tube is that boiling above the heating surface is no longer affected by boiling on the auxiliary heater.

4.2 Thermal Insulation for Heated Block

In the past, we insulated the heated block with glass wool. The disadvantage of glass wool is its pliability which made the installation complicated. Another disadvantage is that glass wool is not suitable into wet environments (and the environment is wet in our case as the apparatus in Fig. 1 is opened to surroundings). Therefore, we did a literary review in order to find a suitable insulation material.

Literature very often mentions insulations made of polymers (PTFE, PEEK, PI, Teflon, Nylon, etc.), wools (fiberglass, mineral, rock, kaolin and other wools) and ceramic in the form of tablets, blankets made of ceramic fibers or custom-made parts. Combined insulation composed from several layers made of these materials is also often used. However, none of these materials seemed to be suitable for our application. Therefore, we had chosen not to follow the literature and used insulation made of cellular glass *Foamglas Perinsul S*, see part no. 10 in Fig. 1. The main advantages of this insulation are that its thermal conductivity is lower than 0.05 W m^{-1} K^{-1}, maximum operational temperature is 430 °C and that the material is water-resistant and impermeable for vapor.

4.3 Two-Piece Heated Block

Since the beginning of our experiments, we have observed horizontal temperature differences inside the heated block. These differences were strongly dependent on axial position of heating cartridges inside holes. In [1], the block was made of a single stainless-steel piece. In order to decrease the differences, we split the block into a copper carrier, see Fig. 3, and the block itself, see Fig. 2. The carrier accommodates five vertically oriented heating cartridges (we used three horizontally oriented cartridges in [1]). Vertical orientation enables us to insulate the carrier and the block more easily. Five holes for these cartridges (D = 10 mm, L = 63 mm) were precisely manufactured in order to reduce the contact thermal resistances between cartridges and the carrier as much as possible. Copper was chosen in order to provide an uniform distribution of temperature and heat flux. We are, of course, fully aware that we created another contact resistance between the carrier and the block by splitting the single-piece block originally used in [1] into two pieces. However, we believe that the contact resistances between cartridges and the carrier are more important with respect to the temperature field inside the block. Moreover, when we want to switch the heated block for another piece, we will not have to deal with precise manufacturing of holes for heating cartridges again.

4.4 Thermocouples

Heat transfer coefficients measured with our apparatus are strongly dependent on temperatures measured inside the heated block. From our experience, even a slight change of position of a thermocouple, which is accommodated inside the heated block, has a strong impact on measured temperature. Therefore, we tried to fixate our thermocouples with a suitable agent.

In article [9], we did another literary review and found out that three methods are commonly used for attachment of thermocouples in pool boiling apparatuses: peening or crimping, melting and fixation with an adhesive. We have chosen to use an adhesive for our *Omega GK-MQSS-M100U-150* thermocouples. However, the most often used adhesives such as epoxies or cyanoacrylate-based glues are not acceptable for our case, because their operational temperature is at most 250 °C. Yet, sometimes, we measured temperatures around 360 °C for the lower line of thermocouples. Therefore, we have chosen an alumina-based adhesive *Aremco Ceramabond 670* with maximum

Table 1. Analysis of influence of horizontal temperature difference of ±1 °C on resulting values for three different datapoints.

Case	ΔT_{up} (°C)	ΔT_{lo} (°C)	T_{up} (°C)	T_{lo} (°C)	T_w (°C)	q (kW m^{-2})	α_i (kW m^{-2} K^{-1})	α_i/α_0 (–)	$\alpha_i - \alpha_0$ (kW m^{-2} K^{-1})
0	0.0	0.0	137.6	169.7	108.0	37.0	4.2	1.000	0.0
1	+1.0	+1.0	138.6	170.7	109.0	37.0	3.8	0.898	−0.4
2	−1.0	−1.0	136.6	168.7	107.0	37.0	4.8	1.129	+0.5
3	+1.0	−1.0	138.6	168.7	110.8	34.7	3.0	0.708	−1.2
4	−1.0	+1.0	136.6	170.7	105.1	39.3	6.6	1.573	+2.4
0	0.0	0.0	172.8	243.3	107.6	81.4	9.7	1.000	0.0
1	+1.0	+1.0	173.8	244.3	108.6	81.4	8.7	0.894	−1.0
2	−1.0	−1.0	171.8	242.3	106.6	81.4	11.0	1.135	+1.3
3	+1.0	−1.0	173.8	242.3	110.5	79.1	7.0	0.726	−2.7
4	−1.0	+1.0	171.8	244.3	104.8	83.7	15.1	1.554	+5.4
0	0.0	0.0	181.7	262.6	107.1	93.3	11.8	1.000	0.0
1	+1.0	+1.0	182.7	263.6	108.1	93.3	10.5	0.888	−1.3
2	−1.0	−1.0	180.7	261.6	106.1	93.3	13.5	1.145	+1.7
3	+1.0	−1.0	182.7	261.6	109.9	91.0	8.5	0.717	−3.3
4	−1.0	+1.0	180.7	263.6	104.3	95.6	18.9	1.602	+7.1

operational temperature of 1 650 °C and dynamic viscosity of 5 Pa s.

The adhesive should have eliminated the unwanted change of position of thermocouple-junctions inside the holes during our experiments. Furthermore, it should have reduced contact thermal resistances between thermocouple-junctions and the heated block in order to reduce the horizontal temperature differences mentioned in section 4.3. While the first objective was accomplished, the second was not. More details are given in section 5.

4.5 Minor Modifications

In [1], we mentioned that we had observed limescale deposits on the heating wall. In this article, all experiments were performed with water demineralized with unit *Aqual 29 XL*.

In [1], we have used ordinary epoxy *Kittfort 1200* together with high-temperature silicone *Versachem Mega Copper*. This combination evinced noticeable degradation after each experiment and has to be reapplied after several conducted experiments. Because of that, we switched it for low-viscous epoxy *EPO-TEK 302-3M-R* and high-temperature silicone *Ceresit CS 28*. This combination has long-term temperature durability up to 260 °C. We were able to obtain all datapoints presented in this article with a single application of the silicone and the epoxy.

5 Experimental Measurements

This section presents results of eight experimental runs with 58 obtained datapoints. All of the datapoints were obtained after reaching steady-state boiling of demineralized water under atmospheric pressure.

5.1 Results

Results of our experimental measurements can be seen in Fig. 4. We can distinguish between two regions in Fig. 4:

1. For heat fluxes lower than approximately 90 kW m^{-2}, measured heat transfer coefficients seems to be slightly lower than correlated values. Also, there seems to be a trend of modest (although not monotone) decrease of heat transfer coefficient with time.
2. For heat fluxes higher than approximately 90 kW m^{-2}, measured heat transfer coefficients outreach correlated values by a significant amount. The discrepancy between measured and correlated values increase with increased heat flux. The decreasing trend of heat transfer coefficient with time seems to be pronounced for these heat fluxes.

From Fig. 4, we can observe that the trend of increased heat transfer coefficients for heat fluxes higher than 90 kW m^{-2} is noticeable for all performed experiments. Also, the trend is reproducible within each measurement, as is obvious from two points obtained on 21 June for heat fluxes of about 80 kW m^{-2} and three points obtained on the same day for heat fluxes slightly below 100 kW m^{-2}.

5.2 Discussion

In article [1], we mention that we have observed unwanted horizontal temperature differences inside the heated block manufactured from single stainless-steel piece. According to the article, the differences were from about 3 up to 4 °C for the upper line and from about 6 up to 8 °C for the lower line of thermocouples. In order to reduce these differences, we modified the apparatus according to section 4. However, we have to admit that implemented modifications did not eliminate these differences. For measurements presented in Fig. 4, we observed horizontal differences of about 10 °C in the lower line and of about 2 °C in the upper line of thermocouples. However, these differences were dependent on heat flux and tended to increase with increased heat flux flowing through the heated block.

We currently believe that the differences are caused by these thermal resistances:

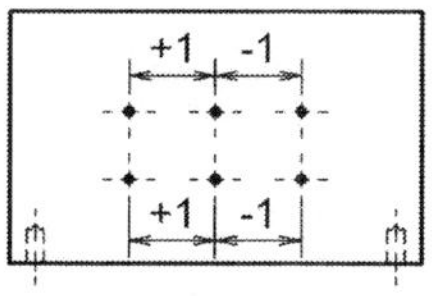

Fig. 5. Symmetrical temperature differences.

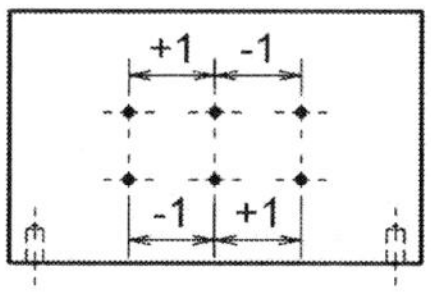

Fig. 6. Crossed temperature differences.

1. Contact resistances between cartridge heaters and the copper carrier
2. Contact resistances between thermocouples and stainless-steel block
3. Contact resistance between the carrier and the block

We also believe that the differences might be responsible for the observed and unexpected trend in Fig. 4. It is certain that horizontal temperature differences have a strong impact on measured heat transfer coefficients.

6 Analysis of Horizontal Differences

In order to understand the impact of temperature differences on resulting values, we analyze effect of artificially induced difference of 1 °C on resulting values in this section. Let us assume, that the temperature of the upper central thermocouple is 172.8 °C, temperature of the lower central thermocouple is 243.3 °C and that water boils at 99.2 °C (these are average temperatures[1] taken from one datapoint measured on 21 June). From these temperatures we can calculate[2] heat flux of 81.4 kW m^{-2} and heat transfer coefficient of 9.7 kW m^{-2} K^{-1}. Now, we fix temperature of the boiling liquid at 99.2 °C and suppose four different cases:

Case 1: There are temperature differences of +1 °C for both thermocouples on the left. That means that for the left side the upper thermocouple measures 173.8 °C and lower thermocouple 244.3 °C. Form these temperatures, we obtain heat flux 81.4 kW m^{-2} and heat transfer coefficient 8.7 kW m^{-2} K^{-1}.

Case 2: There are temperature differences of −1 °C for both thermocouples on the right. For the upper right thermocouple, we then get 171.8 °C and for lower right thermocouple 242.3 °C. Heat flux is then 81.4 kW m^{-2} and heat transfer coefficient 11.0 kW m^{-2} K^{-1}.

Case 3: There is a temperature difference of +1 °C for the upper and −1 °C for the lower thermocouple on the left. This gives us upper temperature of 173.8 °C and lower temperature of 242.3 °C, heat flux of 79.1 kW m^{-2} and heat transfer coefficient 7.0 kW m^{-2} K^{-1}.

Case 4: There is a temperature difference of −1 °C for the upper and +1 °C for the lower thermocouple on the right. Upper temperature is then 171.8 °C, lower temperature 244.3 °C, heat flux 83.7 kW m^{-2} and heat transfer coefficient 15.1 kW m^{-2} K^{-1}.

Results for all cases were put into Tab. 1 together with two other datapoints obtained on 21 June. *Case 0* in Tab. 1 are reference values without any imposed differences. One can make several observations from Tab. 1:

- Equal imposed differences do not affect the heat flux, but heat transfer coefficient is affected due to a change in extrapolated wall temperature T_w.
- When we impose temperature differences of opposite signs, there is a higher difference between resulting heat transfer coefficient α_i and original heat transfer coefficient α_0 compared to differences with the same sign. This difference of heat transfer coefficients unfolds for datapoints obtained under higher heat fluxes.
- Relative fractions of α_i/α_0 are mutually comparable for corresponding temperature differences across all three considered datapoints. However, for datapoints obtained under higher heat fluxes, α_0 is higher. Due to that, the difference in heat transfer coefficients $\alpha_i - \alpha_0$ is higher for these datapoints.

Table 2. Arithmetic means $\bar{\alpha}$ and maximum absolute deviations Δ calculated for datapoints from Tab. 1.

Distribution Fig. 5 and Fig. 6	$\bar{\alpha}$ (kW m^{-2} K^{-1})	Δ (kW m^{-2} K^{-1})
symmetrical	4.3	0.5
crossed	4.6	2.0
symmetrical	9.8	1.2
crossed	10.6	4.4
symmetrical	11.9	1.6
crossed	13.1	5.9

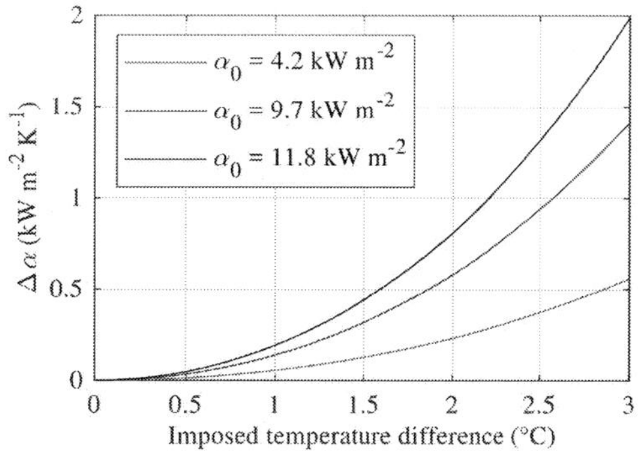

Fig. 7. Maximum absolute deviations of heat transfer coefficients between side and central rows of thermocouples for symmetrical temperature differences.

[1] We do not directly take temperatures measured with thermocouples in the central row, because they might be influenced by temperature differences as well.

[2] We suppose that thermal conductivity of the stainless steel is 15 W m^{-1} K^{-1} and that the heated block block has dimensions drawn in Fig. 2.

- As the extrapolated wall temperature T_w gets closer to the temperature of the boiling liquid T_L, the effect of temperature differences on resulting heat transfer coefficient gets more and more severe.

Let us consider two possible distributions of temperature differences inside the heated block. Fig. 5 shows symmetrical distribution, which is a combination of *Case 1* and *Case 2*. On the other hand, Fig. 6 shows crossed distribution, which is combined from *Case 3* and *Case 4*. From previous considerations, one can conclude that the symmetrically distributed differences should have lower impact on resulting values. To support this conclusion, Tab. 2 presents arithmetic means of heat transfer coefficient $\bar{\alpha}$ and maximum absolute deviations

$$\Delta = \max |\alpha_i - \bar{\alpha}| \ , \tag{4}$$

where i is an integer number and goes from 0 up to 4.

From Tab. 2, it is immediately clear that symmetrically imposed temperature differences results in lower Δ. From Tab. 1 and Tab. 2, it is also noticeable that values of heat transfer coefficient measured above the central row of thermocouples lies very close to the arithmetic mean for symmetrically imposed differences. However, it is a coincidence which occurs due to the fact that imposed differences were small. In order to prove that, Fig. 7 shows value of

$$\Delta\alpha = \frac{\alpha_1 + \alpha_2}{2} - \alpha_0 \tag{5}$$

as a function of imposed temperature difference for symmetrical temperature differences. The fraction in (5) is the arithmetic mean of heat transfer coefficients above the side rows of thermocouples and α_0 is heat transfer coefficient above the central row of thermocouples (without any imposed differences). From Fig. 7, it is obvious that the difference $\Delta\alpha$ non-linearly increases for higher imposed temperature differences for all investigated heat fluxes.

For crossed temperature differences, the situation is quite different compared with symmetrical differences. Arithmetic means $\bar{\alpha}$ are considerably shifted from values calculated above the central row of thermocouples and maximum deviations Δ are significantly higher, as can be seen in Tab. 2.

7 Conclusion

This article discusses modifications of experimental apparatus designed for measurement of heat transfer coefficient during saturated nucleate pool boiling. It presents experimental results obtained with these modifications and mentions problematic horizontal temperature differences which need to be eliminated or at least significantly lowered in order to achieve valuable experimental results. Analysis of effect of artificially induced temperature differences on resulting values is given. The analysis implies that symmetrical temperature differences inside the heated block affect the measurements to a much lower extent compared to crossed non-symmetrical temperature differences.

Our closest plan with the apparatus is to lower the horizontal temperature differences by reducing the contact thermal resistance between the carrier and the block. We want to bolt both parts together. We also intend to use a special graphite sheet as a thermal interface material.

This work was supported by the Ministry of Education, Youth and Sports of the Czech Republic under OP RDE grant number CZ.02.1.01/0.0/0.0/16_019/0000753 "Research centre for low-carbon energy technologies" and by the Grant Agency of the Czech Technical University in Prague, grant number SGS18/129/OHK2/2T/12.

Nomenclature

c_p	isobaric specific heat capacity	($\mathrm{J\,kg^{-1}\,K^{-1}}$)
D	diameter	(m)
L	length	(m)
p	pressure	(Pa)
T	temperature	(°C)
α	heat transfer coefficient	($\mathrm{W\,m^{-2}\,K^{-1}}$)
Δ	maximum absolute deviation	
Δh_{LG}	specific latent heat of vaporization	($\mathrm{J\,kg^{-1}}$)
ΔT	superheat	(°C)
$\Delta\alpha$	difference of α defined in (5)	($\mathrm{W\,m^{-2}\,K^{-1}}$)
λ	thermal conductivity	($\mathrm{W\,m^{-1}\,K^{-1}}$)
μ	dynamic viscosity	(Pa s)
ϱ	density	($\mathrm{kg\,m^{-3}}$)
σ	surface tension	($\mathrm{N\,m^{-1}}$)

Subscripts:

0	reference value
G	related to vapor phase
i	ordinal index
L	related to liquid phase
lo	lower
sat	at saturation state
up	upper
w	related to the heating wall

References

1. V. Vajc, M. Dostál, EPJ Web of Conferences **213** (2019)
2. W.M. Rohsenow, Tech. rep., Div. of Industr. Coop., Heat Transf. Lab., Cambridge, MASS, USA (1952)
3. I.L. Mostkinski, Teploenergetika **10**, 66 (1963)
4. K. Stephan, M. Abdelsalam, Int. J. of Heat Mass Transf. **23**, 73 (1980)
5. K. Nishikawa, Y. Fujita, H. Ohta, S. Hidaka, Proc. 7th Int. Heat Transf. Conf. **4**, 61 (1982)
6. V.V. Yagov, Heat Mass Transf. **45**, 881 (2009)
7. H.K. Forster, N. Zuber, AIChE Journal **1**, 531 (1955)
8. G.F. Hewitt, in *Handbook of Heat Transfer* (McGraw-Hill, New York, NY, USA, 1998), pp. 989–1156, 3rd edn., ISBN 0-07-053555-8
9. V. Vajc, M. Dostál, Proc. 14th Int. Conf. on Heat Transf., Fluid Mech. and Thermodyn., pp. 808–813 (2019)

On the possible origin of a vapour cone occurring during the transonic flight in moist air

Magda Vestfálová[1], *Jindřich* Hála[2,*], *Pavel* Šafařík[3], and *David* Šimurda[2]

[1]Technical University of Liberec, Studentská 1402/2, 461 17, Liberec, Czech Republic
[2]Institute of Thermomechanics of the Czech Academy of Sciences, Dolejškova 1402/5, 182 00, Prague 8, Czech Republic
[3]Czech Technical University in Prague, Faculty of Mechanical Engineering, Technická 4, 166 07, Prague 6, Czech Republic

Abstract. In aerodynamics as in other sciences, there are some phenomena which are widely explained in a way that under a closer look turns out to be based on wrong assumptions and thus not very convincing. One of these is the effect referred to as vapour cone or shock collar which is the cloud of condensed water of the conical shape that forms around objects travelling at transonic speeds through the moist air. This paper aims to shed light on some basic principles that might stand behind this effect and based on the simple physical model provide possible explanation and correct some of the common misconceptions which are repeated ad nauseam not only on the internet sites of aerodynamics enthusiasts but also in some textbooks.

1 Introduction

Certainly, many people have been amazed by a similar photo as is shown in Figure 1 of the low flying F/A-18 Hornet or other fast aircraft surrounded by a strange-shaped cloud called *vapour cone* or *shock collar*. Such cloud is a product of an aircraft travelling at transonic speed in moist and warm air. The air which is in the relative motion with respect to an aircraft expands and eventually reach the supersonic speed - the local supersonic region closed by the strong shock wave is formed downstream of which the relative flow velocity is subsonic again. This paper aims to explain under which conditions this specific cloud might appear and why it disappears at higher supersonic speeds.

Fig. 1. F/A-18 Hornet jet fighter travelling at transonic speed in moist air (U.S. Navy photo by Ensign John Gay).

2 Atmospheric moist air properties

Homogeneous atmospheric air could be considered as an ideal gas. The atmospheric air is a mixture of dry air and humidity. All humidity in the homogeneous moist air is present in the form of water vapour. Based on the thermodynamics of moist air it is possible to determine the parameters on the saturation line beyond which the water condensate and forms water droplets (mist) or deposits to solid particles (snow, frost, rime, etc). Curves of saturated moist air for different values of specific humidity x in the $p-t$ diagram are shown in Figure 2 [1]. In this diagram the quality of the moist air expressed in terms of specific humidity is apparent for an arbitrary point - see Figure 3 and Table 1.

*e-mail: hala@it.cas.cz

3 Simplified aerodynamic model

To describe processes in the flow field around a flying aircraft the one-dimensional model of the compressible fluid flow, representing the stream tube in the three-dimensional flow field is used. The energy balance equation for the referred model includes only two forms of energy - enthalpy and kinetic energy. Assuming ideal gas model the energy equation can be written as

$$c_pT + \frac{v^2}{2} = c_pT_0 = const., \qquad (1)$$

Table 1. Quality of the selected states of the moist air for three different values of the specific humidity x (see Figure 3).

State	$x = 0.001\ \frac{\mathrm{kg_V}}{\mathrm{kg_{DA}}}$	$x = 0.01\ \frac{\mathrm{kg_V}}{\mathrm{kg_{DA}}}$	$x = 0.1\ \frac{\mathrm{kg_V}}{\mathrm{kg_{DA}}}$
A	Heterogeneous: saturated moist air containing solid phase of H_2O	Heterogeneous: saturated moist air containing solid phase of H_2O	Heterogeneous: saturated moist air containing solid phase of H_2O
B	Homogeneous: unsaturated moist air	Heterogeneous: saturated moist air containing solid phase of H_2O	Heterogeneous: saturated moist air containing solid phase of H_2O
C	Homogeneous: unsaturated moist air	Heterogeneous: saturated moist air containing liquid phase of H_2O	Heterogeneous: saturated moist air containing liquid phase of H_2O
D	Homogeneous: unsaturated moist air	Homogeneous: unsaturated moist air	Heterogeneous: saturated moist air containing liquid phase of H_2O
E	Homogeneous: unsaturated moist air	Homogeneous: unsaturated moist air	Homogeneous: unsaturated moist air

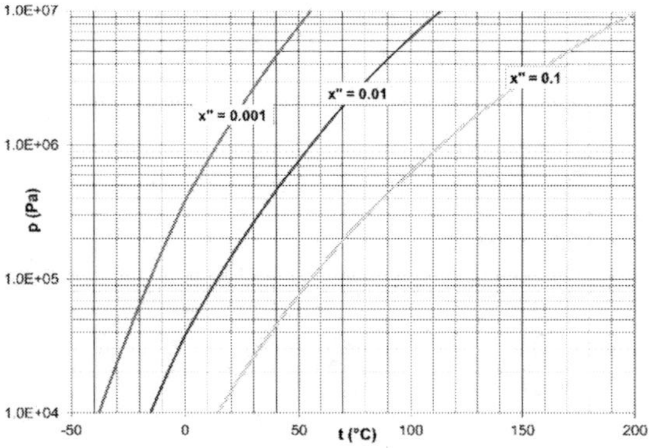

Fig. 2. Saturation lines in $p - t$ diagram for the moist air and different values of specific humidity x [1].

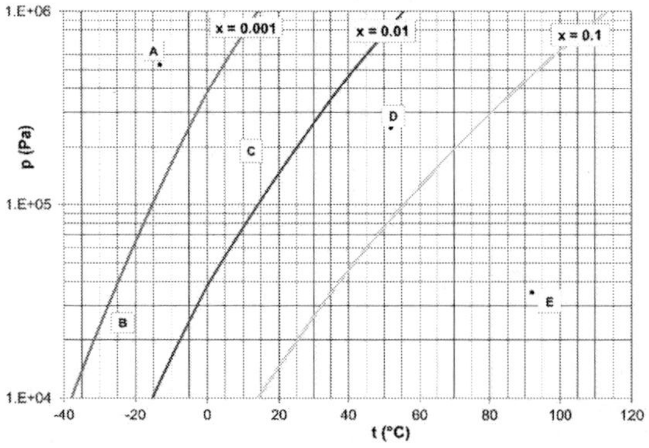

Fig. 3. Saturation lines in $p - t$ diagram for the moist air and different values of specific humidity x. Different states are marked and its quality is described in Table 1.

where c_p ($\mathrm{J \cdot kg^{-1} \cdot K^{-1}}$) is the specific heat capacity at constant pressure, T (K) temperature, T_0 (K) stagnation temperature and v ($\mathrm{m \cdot s^{-1}}$) velocity. It is evident from equation (1) that for higher flow velocities v the temperature T decreases and vice versa.

4 Calculation of the moist air parameters using simplified model

Video records of such events as can be seen in Figure 1 suggests that the favourable conditions for observations usually takes place in a warm maritime climate and when the wind is calm. Parameters of the moist air under these conditions might be as follows: atmospheric pressure p_a = 0.1 MPa, temperature $t_a = 30\,^{\circ}\mathrm{C}$ and relative humidity $\varphi = 70\%$ i.e. specific humidity $x = 0.0019\,\mathrm{kg_V/kg_{DA}}$, speed of sound $a = 350.7\,\mathrm{m \cdot s^{-1}}$, specific heat capacity at constant pressure $c_p = 1020.5\,\mathrm{J \cdot kg^{-1} \cdot K^{-1}}$ and the ratio of specific heats $\gamma = 1.397$.

Assuming an isentropic process, the moist air would saturate at the pressure approximately p_a = 92.1 kPa and the temperature $t = 23\,^{\circ}\mathrm{C}$ (see Figure 4).

For a given Mach number M, the speed of the aircraft is

$$v = Ma, \tag{2}$$

difference of the stagnation and atmospheric temperature

$$\Delta t = \frac{v^2}{2c_p}, \tag{3}$$

and thus the stagnation temperature

$$t_0 = t_a + \Delta t. \tag{4}$$

For an idealised model of isentropic fluid flow the stagnation pressure is given by equation

$$p_0 = p_a \left(\frac{T_0}{T_a} \right)^{\frac{\gamma}{\gamma - 1}}. \tag{5}$$

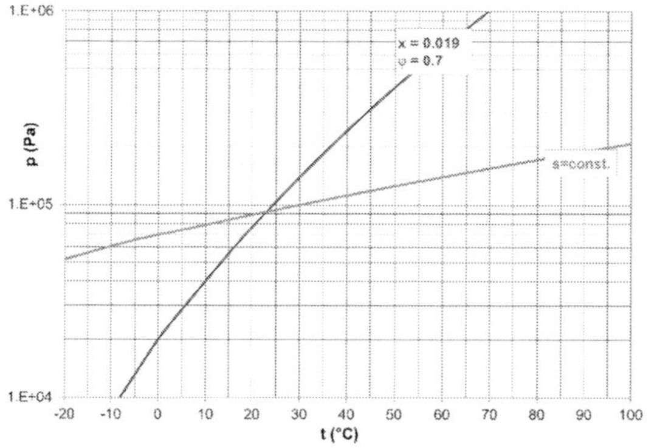

Fig. 4. $p-t$ phase diagram of the moist air with the specific humidity $x = 0.001\,\mathrm{kg_V/kg_{DA}}$ and the isentrope (s = const.) crossing the point of the state $t_a = 30$ °C , $p_a = 0.1$ MPa and $\varphi = 70\%$.

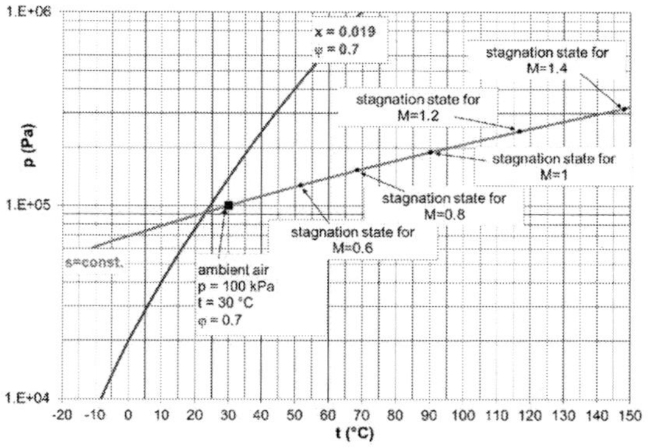

Fig. 5. $p-t$ phase diagram of the moist air with the specific humidity $x = 0.0019\,\mathrm{kg_V/kg_{DA}}$ and the isentrope (s = const.) crossing the point of the state $t_a = 30$ °C, $p_a = 0.1$ MPa and $\varphi = 70\%$ with the marked stagnation states for the selected values of the aircraft Mach number.

Figure 5 pictures the phase diagram of the moist air with marked stagnation states corresponding to different Mach numbers for atmospheric conditions ($p_a = 0.1$ MPa, $t_a = 30$ °C, $\varphi = 70\%$).

Table 2 shows for various Mach numbers of the aircraft the Mach numbers of the flow for which the air in the stream tube would become saturated under the assumption of the isentropic process.

In Figure 6 we can see the dependence of the temperature of the moist air upon the Mach number for the case as described in Table 2. The temperature of the atmospheric air as well as the dew point for pressure and relative humidity considered are indicated.

5 Flow regimes

In the case of the isentropic flow past an aircraft (or an aerofoil) the air might possibly reach the state in which

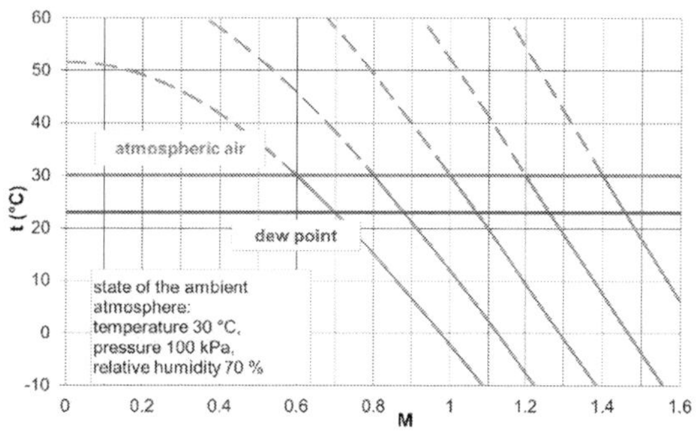

Fig. 6. Dependence of the moist air temperature t upon the aircraft Mach number M assuming isentropic flow conditions (red lines).

the air is saturated or even over saturated as it expands (accelerates) along the body and its conveniently shaped surface features such as canopy and wings. During the subsequent deceleration the air eventually returns back to unsaturated state.

The flow past an aircraft can be isentropic only if the velocity anywhere in the flow field is subsonic. This occurs when the aircraft Mach number is lower than critical Mach number $M_{\mathrm{crit.}}$, which depends on the exact shape of an aircraft and is typically in the range $M_{\mathrm{crit.}} \in (0.8; 0.9)$.

When the speed of an aircraft is above the critical Mach number $M_{\mathrm{crit.}}$ but lower than unity, the region where the flow exceeds sonic speed emerges together with the shock-wave at the rear part of this region. Such flow regimes are referred to as *transonic*. If an aircraft further accelerates the supersonic region will grow in size and the rear shock will travel towards the trailing edge. The shock waves in transonic flow fields might be considered as normal shock waves in one dimensional model of stream tube used in this work. When the flow passes the shock wave the pressure and temperature increases suddenly and the Mach number drops. Downstream the shock wave the flow is isentropic again.

If the speed exceeds certain Mach number over unity, the attached shock wave forms at the leading edge causing decrease in the flow velocity, however, the speed downstream the shock wave might be still supersonic. In this case the flow along the aerofoil is isentropic and expands until the exit shock wave knocks down the speed to the value upstream the aerofoil. Entire flow field is supersonic in this case.

6 Change of flow parameters across the shock wave

The change of pressure and temperature across the shock wave depends on the Mach number at which the shock wave occurs. In the case of normal shock wave the following relations apply

$$\frac{p_2}{p_1} = \frac{2\gamma}{\gamma+1} M_1^2 - \frac{\gamma-1}{\gamma+1}, \tag{6}$$

Table 2. Table view of the results shown in Figure 6 including Mach number at dew point for different values of aircraft Mach number *M*.

$M\,(-)$ aircraft Mach number	0.6	0.8	1	1.2	1.4
$v\,(\mathrm{m \cdot s^{-1}})$ aircraft speed	210.4	280.5	350.7	420.8	491.0
$\Delta t\,(°\mathrm{C})$ difference between total temperature and ambient atmospheric temperature	21.69	38.56	60.25	86.76	118.09
$t_0\,(°\mathrm{C})$ stagnation temperature for given aircraft Mach number and atmospheric temperature	51.69	68.56	90.25	116.76	148.09
$p_0\,(\mathrm{Pa})$ stagnation pressure for given aircraft Mach number and atmospheric pressure	127501.4	152342.2	189141.5	242268.5	317910.2
$M_{\mathrm{DP}}\,(-)$ Mach number at dew point	0.698	0.880	1.069	1.262	1.458

$$\frac{T_2}{T_1} = \frac{\left(1 + \frac{\gamma - 1}{2} M_1^2\right)\left(\frac{2\gamma}{\gamma - 1} M_1^2 - 1\right)}{\frac{(\gamma + 1)^2}{2(\gamma - 1)} M_1^2}. \tag{7}$$

Table 3 shows the flow parameters upstream and downstream of the normal shock wave for two selected supersonic values of aircraft Mach number. Figure 7 depicts the states downstream the normal shock wave in the $p - t$ phase diagram of the moist air. Corresponding stagnation pressure and temperature are also shown.

It is obvious from the foregoing paragraphs that for sufficiently high relative humidity of the air the favourable conditions for condensations might take place even for subsonic flow. However, due to the shock free flow, the condensation is not suddenly ended by the shock wave.

In the case of transonic flow when the aircraft Mach number is above critical Mach number $M_{\mathrm{crit.}}$ but below unity, the condensation most likely takes place in the regions where the expansion is high enough to exceeds the dew point temperature. That is in the supersonic regions which are terminated by the shock waves. Across the shock waves the conditions suddenly change and prevent further condensation.

When the aircraft Mach number is above unity, the shock wave forms upstream the body. Across the shock wave the temperature and pressure rise and the relative humidity drops (see Figure 7). This causes a significant shift of the state downstream the shock wave further from the dew point temperature and thus from the conditions favourable for condensation. It is the reason why the vapour cone is much less likely to appear while the aircraft speed is supersonic.

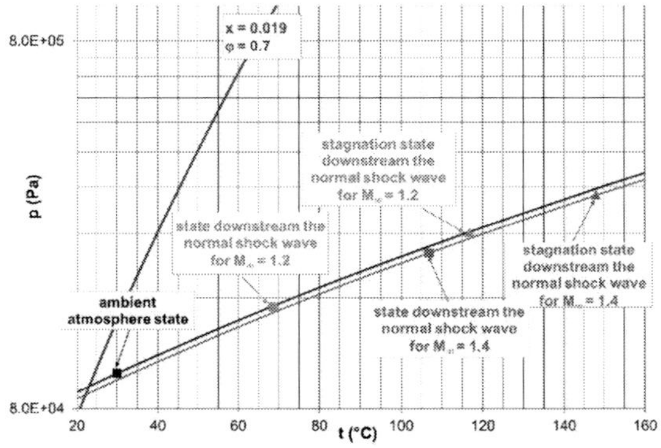

Fig. 7. $p - t$ phase diagram of the moist air with the specific humidity $x = 0.0019\,\mathrm{kg_V/kg_{DA}}$ and the isentrope (s = const.) crossing the point of the state $t_a = 30\,°\mathrm{C}$, $p_a = 0.1\,\mathrm{MPa}$ and $\varphi = 70\%$ with the marked state downstream the normal shock wave and its corresponding stagnation state for aircraft Mach numbers $M_\infty = 1.2$ and 1.4.

7 Amount of water condensed

The amount of condensed water depends upon the atmospheric conditions and the maximal Mach number achieved on the aircraft surface.

The maximal Mach number on the aircraft surface depends upon its exact shape. For the simplicity we might consider the flow of an inviscid fluid about an ellipsoid of revolution [2]. Thick line in Figure 8 shows an incremental velocity as referred in [2] for an ellipsoid of revolution with an aspect ratio $AR = 0.3$. The dashed line shows extrapolated values to higher velocities of the free stream.

For the case of ambient atmospheric conditions considered in foregoing paragraphs (atmospheric pressure $p_a = 0.1\,\mathrm{MPa}$, temperature $t_a = 30\,°\mathrm{C}$ and relative humidity $\varphi = 70\%$) the maximal Mach number achievable

Table 3. Flow properties across the normal shock wave for two selected supersonic upstream Mach numbers.

M (–) Mach number	upstream shock-wave 1.2	downstream shock-wave 0.84	upstream shock-wave 1.4	downstream shock-wave 0.74
v (m·s^{-1}) speed	421	312	491	290
p (MPa) static pressure	0.1	0.152	0.1	0.212
t (°C) static temperature	30	68.59	30	106.79
p_0 (MPa) stagnation pressure	0.242	0.241	0.318	0.305
t_0 (°C) stagnation temperature	116.76	116.76	148.09	148.09

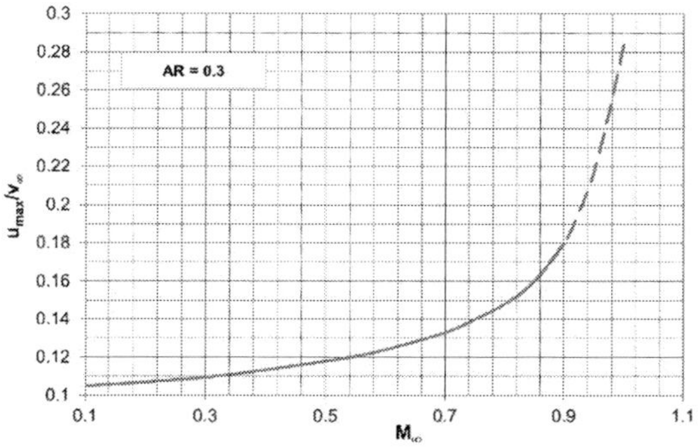

Fig. 8. Graph of the maximum incremental velocity as a function of free stream Mach number M_∞ on an ellipsoid of revolution of the aspect ratio $AR = 0.3$.

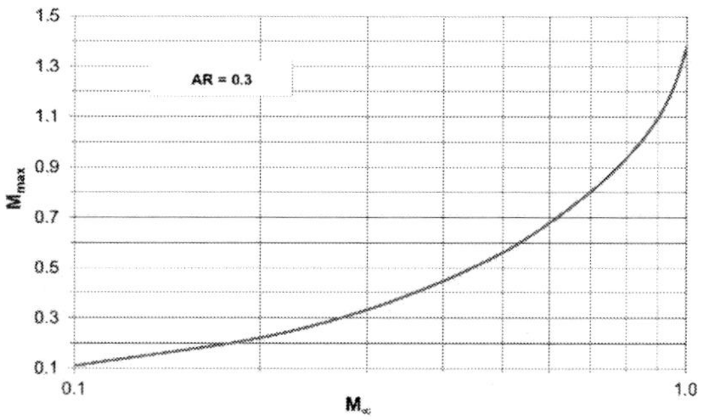

Fig. 9. Graph of the maximum Mach number $M_{\max}$ as a function of free stream Mach number M_∞ on an ellipsoid of revolution of the aspect ratio $AR = 0.3$.

on the ellipsoid surface can be read from the graph in Figure 8. The graph of the maximal Mach number with respect to free stream Mach number is pictured in Figure 9. To each value of maximal Mach number the corresponding value of moist air temperature and pressure can be attributed and thus also the value of maximal specific humidity $x''_{M_{\max}}$. The difference between the maximal specific humidity $x''_{M_{\max}}$ which is in the region of maximal velocity around the object and the specific humidity of the ambient air x'' (in our case $x'' = 0.019\,\mathrm{kg_V/kg_{DA}}$) determines the amount of water which can condensate $x_{\mathrm{cond.}}$ for given conditions of the ambient moist air.

If the Mach number of an aircraft is above unity $M > 1$, the shock wave forms in front of it. Assuming the normal shock wave, the Mach number downstream the shock wave lower than unity can be calculated. Then the flow past the object might be isentropic again and the maximal Mach number can be determined (Figure 9) similarly as in the case described above, however, the decrease in the stagnation pressure across the shock wave must be taken into account. Results of calculations are summarized in Table 4 and Figure 10.

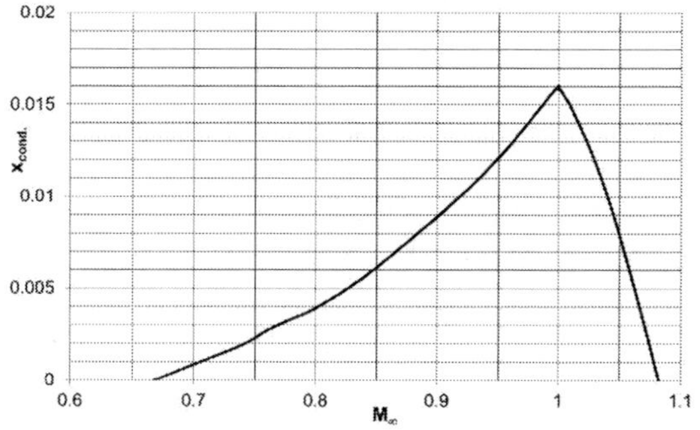

Fig. 10. Graph of the maximum possible amount of water that might condensate in the moist air $x_{\mathrm{cond.}}$ as a function of free stream Mach number M_∞ for the case of the flow past an ellipsoid of revolution of the aspect ratio $AR = 0.3$.

Table 4. The following table summarizes the results of calculations for a simple case of flow past ellipsoid of revolution in the moist air.

M (−) aircraft Mach number	0.6	0.8	1	1.2	1.4
t_0 (°C) stagnation temperature for given aircraft Mach number and atmospheric temperature	51.69	68.56	90.25	116.76	148.09
p_0 (Pa) stagnation pressure for given aircraft Mach number and atmospheric pressure	127501.4	152342.2	189141.5	242268.5	317910.2
p_{02} (Pa) stagnation pressure for given aircraft Mach number and atmospheric pressure	N/A	N/A	189141.5	240522.5	304605.1
M_2 (−) Mach number downstream the normal shock wave	N/A	N/A	1	0.842	0.740
M_{max} (−) maximal Mach number around the ellipsoid of revolution	0.680	0.935	1.370	1.0	0.855
p_{min} (Pa) pressure in the location of M_{max}	93599.25	86738.08	62049.63	127165.4	189062.8
t_{min} (°C) temperature in the location of M_{max}	24.35	17.98	−8.48	52.12	94.65
x'' maximum possible specific humidity in the location of M_{max}	0.020922	0.015128	0.002991	0.075076	0.491829
$x_{\text{cond.}}$ amount of water that might condensate in the location of M_{max}	N/A	0.00393	0.016067	N/A	N/A

8 Conclusions

The objective of this paper is to describe and explain the phenomenon of the phase change of the atmospheric moist air in the vicinity of an object (most frequently a jet fighter) travelling at the transonic speed. The parameters used for the calculations were estimated using the model of the inviscid flow around an ellipsoid of revolution. It was shown that the emergence of the *vapour cone* is limited for the cases of transonic speeds. For supersonic speeds the vapour cloud lapses.

It is necessary to state that the transonic flow is still a relevant research subject and in combination with multiphase flow it represents deeply uncharted territory.

The phenomenon of the phase transition of the expanding moist air takes place for example in turbines with the air as a working fluid o r d uring t he t ests i n high-speed wind tunnels [3].

Acknowledgements: This project has received fund-ing from the Technology Agency of the Czech Republic under the project of the National Center for Power Engineering No. TN01000007.

References

[1] P. Šafařík, M. Vestfálová, *Thermodynamics of Moist Air*, CTU Publishing House, Prague, 2016 (in Czech)

[2] A. H. Shapiro, *Compressible Fluid Flow*, The Roland Press Company, New York, 1993

[3] M. Vestfálová, P. Šafařík, D. Šimurda, *Technique for Determination of Phase Changes in Moist Air Flow in a Blade Cascade*,In: Proceedings of XXIVBiennial Symposium on Measuring Techniques in Turbomachinery, Prague, 2018, pages 19–22

Study of fibre movement in a bifurcation

Jaroslav Volavy[1], *Arpad* Farkas[2], *Frantisek* Lizal[2,*], *Jakub* Elcner[2] and *Miroslav* Jicha[2]

[1]Ademco CZ s r.o., Turanka 106a, 627 00 Brno, Czech Republic
[2]Faculty of Mechanical Engineering, Brno University of Technology, Technicka 2, 616 69 Brno, Czech Republic

Abstract. The nature of fibre motion is much more complex than the motion of spherical particles. A complex model must be used for tracking of the fibre carried by the flow. The model must take into account the orientation of the fibre against the flow and solve the consequent rotation of the fibre. Description of such model can be found in this article. It is then used for simulation and analysis of fibre movement in a single planar bifurcation.

1 Introduction

The prediction of fibre orientation during flow through bifurcating channels such as human airways is required for accurate modelling of the fate of inhaled fibres. However, although there were several studies published in the past years (e.g. [1]), there is still no widely available model applicable to realistic inhalation and whole lungs. Obviously, the motion of fibres is much more complex compared to spherical particles, as fibres rotate around their own axis and around the centre of mass.

This article presents a model which accounts for the translational and rotational motion of fibre experiencing a drag force in a bifurcation channel. The attention is paid to the influence of the velocity gradient on the rotation of the fibre.

2 Model of fibre movement

In this study, fibres were tracked by the Lagrangian approach. Superposition principle was applied for the description of fibre motion. The motion of fibre is considered to consist of translation and rotation around its centre. Equations are provided for both translation and rotation parts of the motion. These equations are based on Newton's second law.

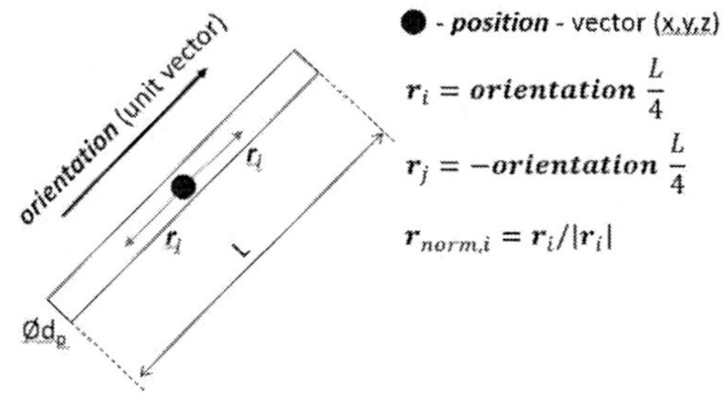

Fig. 1. Fibre description

For the construction of the equations of motion, an approach similar to that presented by Lo Iacono et al. [2] has been applied. Several modifications to the original equations were implemented in this study.

The fibre is divided into discrete segments. In this study, only two segments were defined. The centre of the fibre was characterized by a position vector, while its orientation was given by a unit vector. Vectors $\boldsymbol{r_i}$ and $\boldsymbol{r_j}$ are connections of fibre centre with centres of each segment – see figure 1.

This study focuses on only one force acting on the fibre which is the drag force from the surrounding fluid. At each segment, the drag force $\boldsymbol{F_i}$ is decomposed into two components – the force $\boldsymbol{S_i}$ parallel with fibre's axis and force $\boldsymbol{D_i}$ perpendicular to fibre's axis. The schematic decomposition of the drag force can be seen in figure 2.

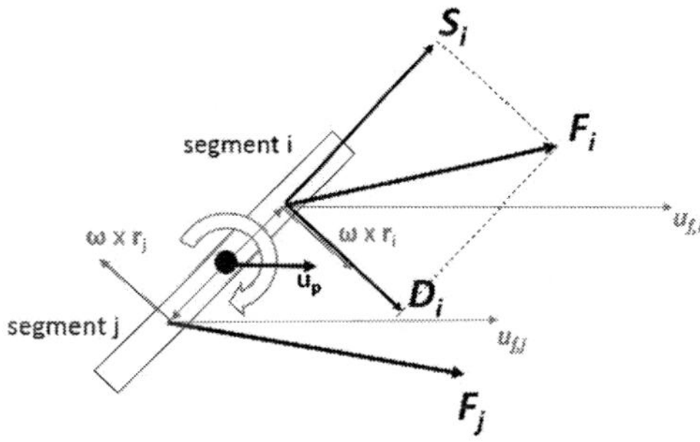

Fig. 2. Forces acting on the fibre during movement in the fluid

The relative velocity of the fibre to the velocity of the surrounding fluid is required for evaluation of the forces $\boldsymbol{S_i}$ and $\boldsymbol{D_i}$. The velocity of segment *i* is the sum of fibre translation velocity $\boldsymbol{u_p}$ and a contribution from rotation $\boldsymbol{\omega} \times \boldsymbol{r}_i$, where $\boldsymbol{\omega}$ is the angular velocity of fibre around its centre. Inclusion of term $\boldsymbol{\omega} \times \boldsymbol{r}_i$ is the main

* Corresponding author: lizal@fme.vutbr.cz

https://doi.org/10.1051/epjconf/202226901067

difference between Lo Iacono et al. [2] and our study. The relative velocity is defined as:

$$\boldsymbol{u}_{rel,i} = \boldsymbol{u}_{f,i} - (\boldsymbol{u}_p + \boldsymbol{\omega} \times \boldsymbol{r}_i), \quad (1)$$

where $\boldsymbol{u}_{f,i}$ is the fluid velocity at the centre of the segment i. The parallel relative velocity is obtained as a projection of the relative velocity to axis direction defined as:

$$\boldsymbol{u}_{\parallel,rel,i} = \boldsymbol{r}_{norm,i}(\boldsymbol{u}_{rel,i} \cdot \boldsymbol{r}_{norm,i}) \quad (2)$$

and perpendicular relative velocity is obtained in a way that the sum of parallel and perpendicular component must result in relative velocity:

$$\boldsymbol{u}_{\perp,rel,i} = \boldsymbol{u}_{rel,i} - \boldsymbol{u}_{\parallel,i}. \quad (3)$$

The parallel component of the force $\mathbf{S}_i$ is modelled as the friction force acting on a flat plate of equivalent surface area to the cylinder:

$$\boldsymbol{S}_i = \frac{2.6}{Re_l^{1/2}} \frac{1}{2} \rho |\boldsymbol{u}_{\parallel,rel,i}| \boldsymbol{u}_{\parallel,rel,i}, \quad (4)$$

where

$$Re_l = \rho L |\boldsymbol{u}_{rel,i}| / \mu. \quad (5)$$

Perpendicular force component $\boldsymbol{D}_i$ is computed according to the classical drag expression for a cylinder of infinite length:

$$\boldsymbol{D}_i = \frac{1}{2} C_D \rho |\boldsymbol{u}_{\perp,rel,i}| \boldsymbol{u}_{\perp,rel,i} d_p |\boldsymbol{r}_i|, \quad (6)$$

where C_D is the drag coefficient as measured by Munson and Wielesberger-Lamb [3].

The overall force acting on segment i is then the sum of parallel and perpendicular component:

$$\boldsymbol{F}_i = \boldsymbol{D}_i + \boldsymbol{S}_i. \quad (7)$$

Only the perpendicular force component $\boldsymbol{D}_i$ generates a torque:

$$\boldsymbol{T}_i = \boldsymbol{r}_i \times \boldsymbol{D}_i \quad (8)$$

With knowledge of force acting on a fibre, the equations of motion can be formed. Second Newton's law of motion is applied for computation of the change of translation velocity of fibre:

$$m_p \frac{\mathrm{d}\boldsymbol{u}_p}{\mathrm{d}t} = \boldsymbol{F}. \quad (9)$$

$\boldsymbol{F}$ is the sum of forces over all segments. The change of angular velocity is expressed as:

$$I_p \frac{\mathrm{d}\boldsymbol{\omega}}{\mathrm{d}t} = \boldsymbol{T} \quad (10)$$

where I_p is the moment of inertia of the cylinder rotating about a perpendicular axis going through its centre of mass.

The solution of the system formed by equations (9) and (10) will give the trajectory of the fibre and orientation of the fibre along its trajectory. The two equations are linked together through the orientation of the fibre. The force acting on the fibre in equation (9) is strongly dependent on the orientation of the fibre.

3 Test case

Performance and abilities of the introduced model of fibre motion were tested on case of simple planar bifurcation. In the previous work, the fibre motion was studied in a simple channel with a parabolic velocity profile. The flow in bifurcation is more complex and the velocity field cannot be described as a mathematical expression. Therefore, the velocity field was calculated in ANSYS Fluent and velocity at fibre position is interpolated from the calculated field. The calculation of fibre trajectory and orientation was done in MATLAB.

The bifurcation consists of an inlet channel with a height of 2H = 0.0056 m. Centre of the coordinate system is in the middle of the channel height at the beginning of the channel. After 8.5 mm, the channels split into two branches under angle of 17.2°. The length of the outer edge of the sub-channel is 9.8 mm and width of the sub-channel is 4.5 mm. The geometry of the bifurcation studied in this work can be seen in figure 3. The flow direction coincides with direction of the x-axis. At the inlet, parabolic velocity profile is prescribed with a centre-line velocity of 1.35 m/s (bulk-velocity 0.675 m/s). The density of the fluid was considered to be 1 kg/m^3, the dynamic viscosity of 1.8e-5 Pa.s. Fibres are from material with a density of 2500 kg/m^3. Other important parameters that specify the fibre are its length and diameter. These parameters are varying in this study. Fibres were released into the channel with zero initial velocity and angular velocity in all presented cases.

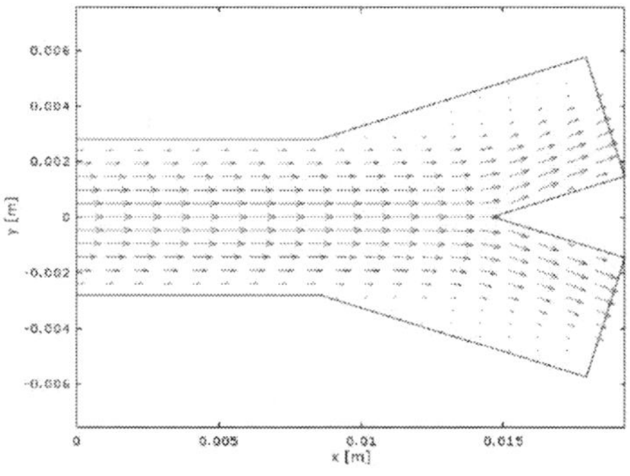

Fig. 3. The geometry of the bifurcation

4 Results

First, fibres with a length of 10 µm and diameter of 1 µm were tracked. Trajectory and orientation of fibre with various initial positions are depicted in figure 4.

https://doi.org/10.1051/epjconf/202226901067

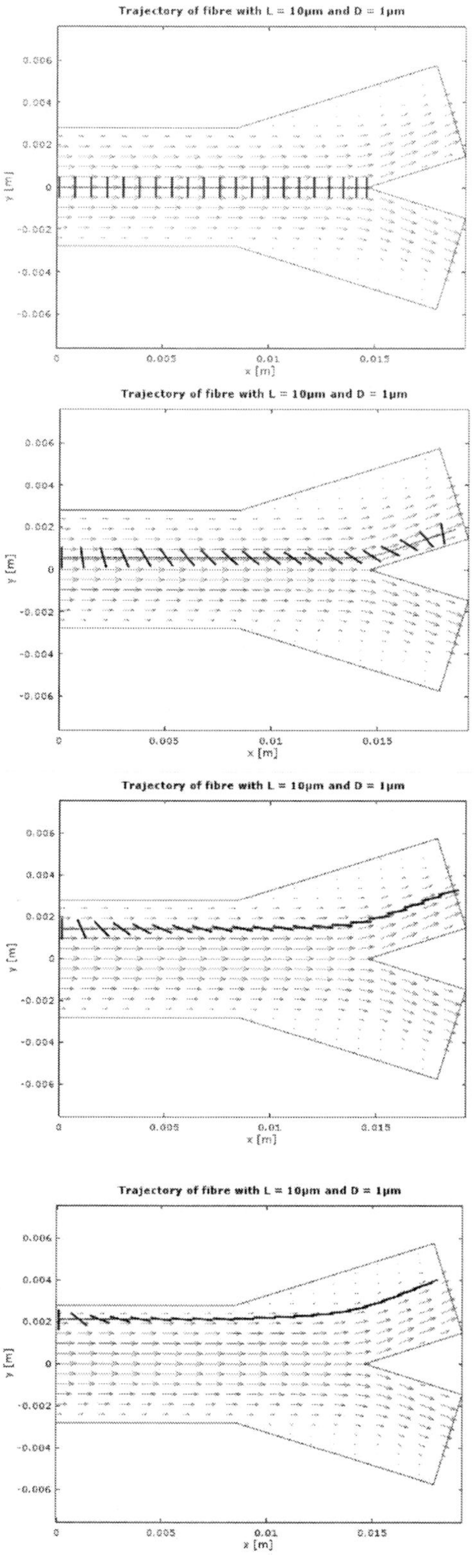

Fig. 4. Trajectory and orientation of fibre with initial positions of y/H = 0, y/H = 0.2, y/H = 0.5 and y/H = 0.75, respectively.

The initial fibre's orientation is traverse to the flow. Length of fibre in figures with trajectories does not reflect the real fibre length. Fibre length in the figure was enlarged for visualisation purposes only. The effect of velocity gradient on fibre orientation is clearly visible in figure 4. Fibre with an initial position at the centre of the channel experiences zero velocity difference at its segments and therefore no torque is generated. Fibre retains its orientation during its motion. Finally, the fibre deposits at the channel junction. Velocity gradient increases with decreasing distance from the wall of the initial position of fibre and this leads to increased torque. Fibre located further from channel axis tends to align with flow direction more and more rapidly. It takes only a small distance for the fibre located near to wall to orient into the flow direction.

All these facts are summarized in figure 5 where the x-components of orientation are plotted as a function of time. Initially, this component is zero because of initial fibre orientation in the wall-normal direction. Then it starts to go toward -1 which indicates that fibre is aligning to the flow direction. The only exception is fibre with an initial position of y/H=0. It retains the same orientation until it deposits. At the end of the inlet channel, the orientation vector of the fibres is almost (-1, 0) and fibre is oriented parallel to the wall. The orientation is a unit vector. The negative sign in x-component of orientation is because the upper end of fibre in initial position is the end of the orientation vector.

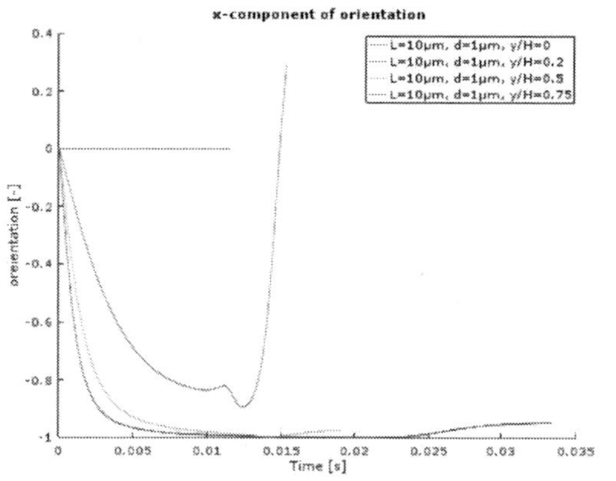

Fig. 5. The x-component of fibre orientation in time

The fibre with an initial position y/H = 0.2 is close to the channel centre where the velocity gradient is relatively low. The fibre gains orientation of the flow more slowly compared with other fibres closer to the wall. In the inlet channel, the fibre rotates counterclockwise, the torque is positive. When it reaches the region where the flow splits, the fibre enters boundary layer of inner bifurcation wall and experiences negative torque. The fibre starts to rotate clockwise. This is represented by the steep rise of x-component of the orientation from time 0.012 s. Fibres with the initial position of y/H = 0.5 and 0.75 behave similarly. Due to the strong velocity gradient, they orient horizontally in the inlet channel and then parallel to the wall after the flow splits.

In figure 6, the x-velocity component is depicted. Fibres accelerate almost immediately to the flow velocity. Fibres with these dimensions have very low particle response time and react very quickly to changes in the velocity of the carrier phase.

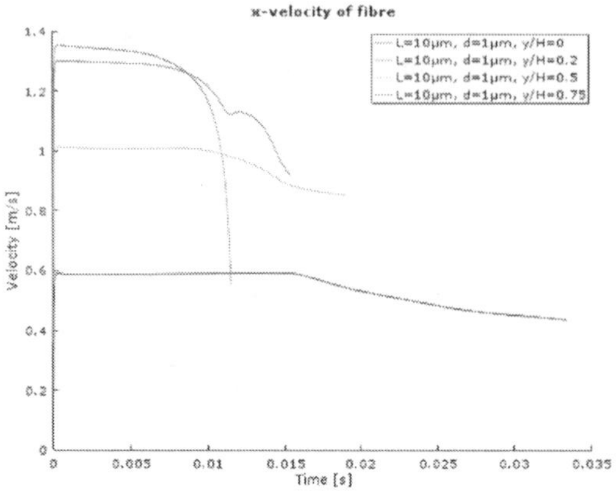

Fig. 6. The x-velocity component of fibre in time.

The x-components of the orientation of various fibres diameters are compared in figure 7. In the region of the inlet channel, all the fibres behave similarly. Changes in fibre diameter have only limited effect on the ability of the fibre to orient into flow direction. This can be explained using the equation (10) for angular velocity change. When the diameter is increased, then higher force **D** is acting on fibre segments and more torque is generated. On the other hand, an increase in diameter will increase the moment of inertia as well. So, the higher torque is partially compensated by higher moment of inertia of fibre and this results in similar behaviour of fibres. Higher differences in fibre behaviour can be observed when fibres get near the bifurcation. A fibre with larger diameter has higher rotation momentum and it takes more time to reverse the character of rotation. A fibre with a diameter of 5 μm leaves domain shortly after it changes the character of rotation.

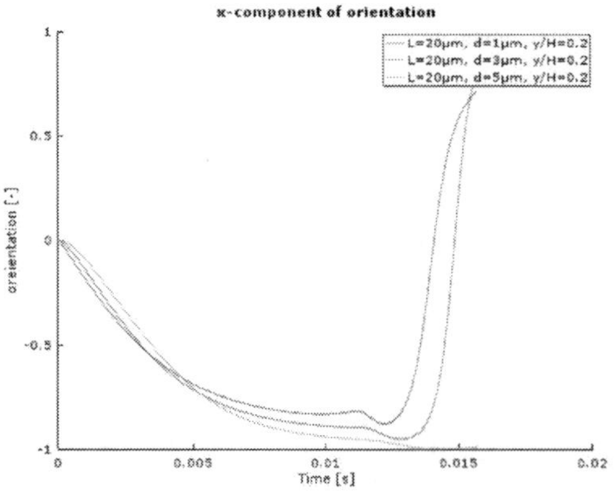

Fig. 7. The streamwise component of the orientation of fibre L=20μm and d_p=1, 3 and 5μm

5 Conclusion

The motion of fibres in a simple planar bifurcation was studied in this work. The character of the fibre motion is more complex than in the simple channel. Fibres with an initial position near the centre of the inlet channel undergo change of rotation direction.

It is important to mention that the results of this model were not confronted by measurements. Such measurements are now being evaluated at Faculty of Mechanical Engineering, BUT. After the confrontation, some modification will probably be necessary to the model.

The most important parameter that affects the fibre rotation in the flow is the velocity gradient at fibre positions. Length and diameter have a lower impact on the fibre motion.

Acknowledgement: This work was supported by the Czech Science Foundation under the grant GA18-25618S and by the project FSI-S-17-4444.

References

1. A. Roshchenko, W. H. Finlay, and P. D. Minev, Aerosol Science and Technology 45 (10), 1260 (2011); L. Tian, G. Ahmadi, Z. Wang, and P. K. Hopke, Journal of Aerosol Science 45, 1 (2012); L. Tian, G. Ahmadi, and J. Y. Tu, Aerosol Science and Technology 50 (5), 474 (2016).
2. G. Lo Iacono, P. G. Tucker, and A. M. Reynolds, International Journal of Heat and Fluid Flow 26 (4), 558 (2005).
3. B. R. Munson, D. F. Young, and T. H. Okiishi, Fundamentals of fluid mechanics, 5th ed. (Wiley, New York, 2006).

Mathematical simulation of the high capacity pumping system for emergencies

Martin Vrábel[1,*], *Sylva* Drábková[1], and *Pavel* Dobeš[2]

[1] VŠB-TU Ostrava, Faculty of Mechanical Engineering, Department of Hydromechanics and Hydraulic Equipment, 17.listopadu 2172/15, Ostrava-Poruba 70800, Czech Republic

[2] VŠB-TU Ostrava, Faculty of Safety Engineering, Department of Civil Protection, Lumírova 630/13, Ostrava-Výškovice 70030, Czech Republic

Abstract. The article deals with the possibilities of mathematical simulation of the high-capacity pumping module used in emergencies, such as floods and fires. These situations can be broad and varied. In order to respond to these events effectively, formulation of methodology followed with simulation of various scenarios can be very helpful. Based on the high-capacity pumping module MČS-25-330-K produced by Sigma company, and the related certified methodology for its application in the field, mathematical simulations of the system were carried out in MATLAB Simulink. The discharge pressure, flow rate and the pressure loss through straight fire hoses were evaluated for various operational conditions of the pump and partly compared with the field measurement.

1 Introduction

The experience of the fire rescue units on domestic and foreign missions shows that in many cases of emergency respond (floods, flash floods, forest fires, fires in cities and in industrial areas, etc.) the long-distance transport of water is necessary. For this purpose, the high-capacity pumping module is required, providing the desired flow-rate and the pressure to cover the pressure losses through the fire hoses.

In the framework of the research project of the Ministry of the Interior of the Czech Republic (No. VG20132015111, "VYMOCERMIS", solved in 2013 - 2015), the prototype of the new high-capacity pumping module MČS-25-330-K, Sigma, was developed and tested in the field. The practical findings on field deployment of high-capacity pumping module for long-distance water transport were included into the following elaborated certified methodology "Operationally tactical instructions for field deployment of high capacity pumps for long-distance transport of water in emergencies"[1], (hereinafter referred to as Methodology). The intended users of this HCPM are primarily the intervening units of professional firefighters, namely the commanders of intervening units in emergencies.

The Methodology defines the basic procedures for the effective and safe deployment of the named high-capacity pumping module (hereinafter referred to as HCPM), as well as the technical and organizational requirements for its setting, start-up, operation and shut down.

Besides basic principles and guidance for the field deployment of HCPM, the methodology presents a tool for the estimation of the pressure loss in the hose line according to available professional sources and field measurements. The prevailing source in a long distance transport is friction and so the pressure loss through a straight hose is proportional to its length. In emergencies, where there is no time for complicated calculations, tables and graphs are used, which provide approximate results. Most often, pressure drop per 100m length of pipe (or hose) is defined, based on the pipe diameter and flow-rate.

For a long-distance water transport, the distance between the water source and the end site and the elevation between both sites will play an important role. These two values will significantly affect the requirements on pump pressure and flow-rate. The discharge pressure from the pump covers several purposes:

- create a water stream that can be a compact or shattered with the desired spray rate, the required minimum pressure in the nozzle conditions the formation of such a stream,
- overcome the frictional losses in the hose line as well as local losses in fittings,
- overcome the geodetic head between the water source and the fire site.

The Methodology [1] contains estimated pressure losses in fire hoses based on specified field conditions. As the data provided with the experimental measurement were limited, the main objective of this work was to extend the range of pressure loss determination for various operational conditions of the high-capacity pumping module (HCPM) using mathematical simulations in MATLAB Simulink.

* Corresponding author: martin.vrabel.st@vsb.cz

2 Field measurement and experimental investigation of pressure losses

Mobile high capacity pump module MČS-25-330 K, produced by Sigma company [3], consists of horizontal spiral pump A-200-KIDR-550, diesel engine Caterpillar C27 ACERT (783 kW, 1800-2100 rpm) and accessories placed on a steel frame, which is inserted into a floating container, adapted to be handled and loaded by a single-arm heavy-duty container carrier according to DIN 30722-1. The weight of the HCPM container with the power unit is about 10.5 t with a diesel tank. The weight of the hose container with 2 km of DN150 dispensing hoses including couplings is approx. 9 t.

The basic performance parameters of HCPM at BEP are as follows [3]:

- n = 1900 rpm, Q = 330 l/s (19 800 l/min), p = 16.5 bar, P = 645 kW.

For normal operation, the pump works at the speed range from 1400 to 1800 rpm, and from affinity laws the corresponding parameters are as follows:

- n = 1800 rpm, Q = 315 l/s (18 900 l/min), p = 14.5 bar, P = 547 kW
- n = 1600 rpm, Q = 280 l/s (16 800 l/min), p = 11.4 bar, P = 385 kW
- n = 1400 rpm, Q = 245 l/s (14 700 l/min), p = 8.8 bar, P = 258 kW

Recommended hoses diameter for the long-distance transport are DN150 (minimum), DN200 or DN250 (optimum).

Field experiment was carried out in 2015 and 2016 in Moravian-Silesian region. Fig. 1 shows simple scheme of HCPM MČS-25-330-K with suction and discharge system. Suction system includes suction humper, two parallel hoses DN250 with the length of 6m. Discharge part of pumping module consists of return valve, flow divider that enables distribution of water to one, two or three straight-line hoses and one backflow hose (for the potential testing of future hydrodynamic booster pump prototype).

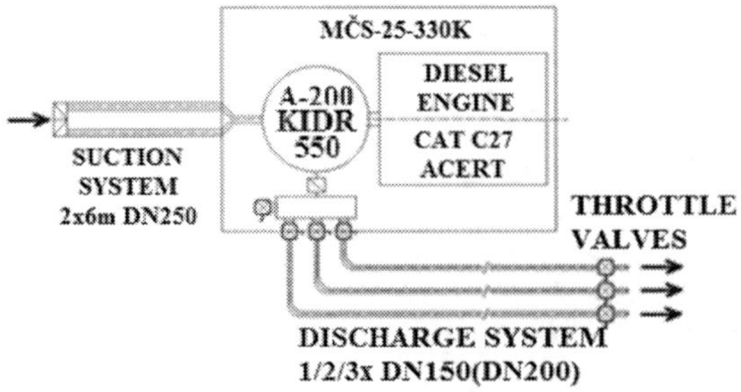

Fig. 1. Scheme of HCPM with suction and discharge system.

The field measurement considered two main scenarios [1]:

- Scenario 1 – long distance water transport for fire extinguishing. Pressure required at the end of the nozzle varies from 4 to 7 bar, single hose DN150 and parallel conduits of two and three hoses DN150 were considered.
- Scenario 2 – long distance water transport for drainage, required end pressure is assumed 0 bar, single hose DN150 and parallel conduits of two and three hoses DN150 were considered.

The main objective of the field experiment was to specify the maximum possible hose length in discharge part of pumping system in the range of defined flow and pressure conditions:

- Max. discharge pressure p_{max} =16 bar
- Pump speed: 1400, 1600, 1800 rev/min
- Pump flow rate: 60, 80, 100, 120, 140, 160, 180, 200, 220 l/s
- Number of pressure hoses in discharge system: 1, 2, 3
- Pressure at the end of hoses: 0, 4, 7 bar

From the results of the field measurement the optimum values of pressure, pump speed and potentially possible lengths of the hose line for the long-distance transport of water were calculated. Subsequently, the values were summarized and put into well-arranged tables and graphs in order to simplify as much as possible the necessary considerations for the optimal setting of the HCPM for possible emergency water scenarios.

3 Mathematical model of high capacity pumping system in MATLAB Simulink

Mathematical simulation of the whole hydraulic system enables to test various combinations according to the upper mentioned condition of field experiments [6,8]. In total, 177 combinations were developed and tested, that were partly compared with the data provided by the Methodology. The scheme of the system is illustrated in Fig.2.

The mathematical model is composed of following elements: centrifugal pump, resistive tubes and fixed orifices, which simulates end regulation valves or fire monitors within the model. Parameters of those elements significantly affect calculation results. Flow-rates and pressures that correspond to the Methodology [1] were determined through the combination of these parameters.

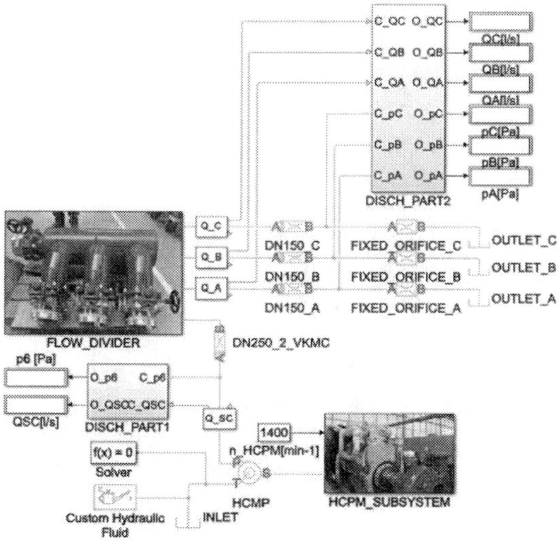

Fig. 2. Example of mathematical model of high-capacity pumping module setup in MATLAB Simulink.

4 Pressure drop calculation

Parameters of the modeled system affect searched maximum hose length. In practice, available power of HCPM is significantly affected by choice of type, length and DN of transportation rubber hoses. In case of DN150 hoses, this specific HCPM could be run only on 1/2 or 2/3 of its maximum available power.

In order to consider the result acceptable, it is necessary to monitor and control the flow rate and the pressure at the end of hoses. The end pressure p_A is given by the difference between the discharge pressure p_6 and pressure loss Δp (1).

$$p_A = p_6 - \Delta p \tag{1}$$

Pressure drop in the hose is given by

$$\Delta p = \rho g h_f = \rho g \left(\lambda \frac{l}{d}\frac{v^2}{2g}\right) \tag{2}$$

The average velocity v is determined from continuity equations for incompressible flow

$$v = \frac{4Q}{\pi d^2} \tag{3}$$

$$\Delta p = \rho g \left(\lambda \frac{8l}{\pi^2 d^5}\right) Q^2 \tag{4}$$

Friction factor λ is calculated based on the regime of water flow in a hose. For turbulent flows we assume $Re >= Re_T$, where minimal value of Reynolds number is $Re_T = 4000$. Friction factor λ is calculated according to Colebrook-White formula in MATLAB Simulink [2].

$$\lambda = \frac{1}{\left[-1{,}8 log\left(\frac{6{,}9}{Re} + \left(\frac{k}{3{,}7d}\right)^{1{,}11}\right)\right]^2} \tag{5}$$

The end pressure in discharge system is determined by the definition of the "fixed orifice". It is possible to set up orifice area A, flow discharge coefficient C_D, laminar transition specification and laminar flow pressure ratio as depicted in Fig.3.

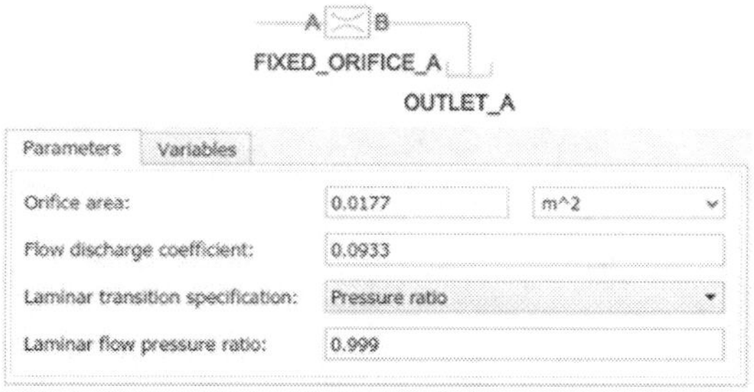

Fig. 3. Fixed orifice setup in MATLAB Simulink [2].

Roughness height on the pipe internal surface was determined as 0.00003m [2]. Flow-rate in fixed orifice was calculated from equation (6),

$$Q = C_D . A \sqrt{\frac{2}{\rho}} \cdot \frac{p}{(p^2 + p_{crit}{}^2)^{1/4}} \tag{6}$$

where

C_D Flow discharge coefficient [-]
Q Flow-rate [m^3/s]
A Orifice passage area [m^2]
ρ Fluid density [kg/m^3]
p_{crit}- Minimum pressure for turbulent flow [Pa]
p Pressure differential p_A-p_B

Minimum pressure for turbulent flow is calculated from equation (7).

$$p_{crit} = \frac{(p_{avg} + p_{atm})}{(1 - B_{lam})} \tag{7}$$

where,

p_{avg} Average pressure between the block terminals $p_{avg} = (p_A + p_B)/2$
p_{atm} Atmospheric pressure 101325 Pa
B_{lam} Pressure ratio at the transition between laminar and turbulent regimes. The default value is 0.999.

5 MATLAB Simulink solution method applied for the case of HCPM

Equations (1)-(7) contain constants and variables that are possible to use on the debugging of pumping system model within the MATLAB Simulink software [2]. The pumping system model was debugged with respect to required pressure and flow values [4]. This process was implemented through the step by step change of the hose length and flow discharge coefficient. The simulation is possible to solve with automatic script or with manual setting. Automatic solution requires fixed and fine iteration step for length and discharge flow coefficient. It leads to a large number of iterations. Figure 4 shows script flowchart for automatic solution.

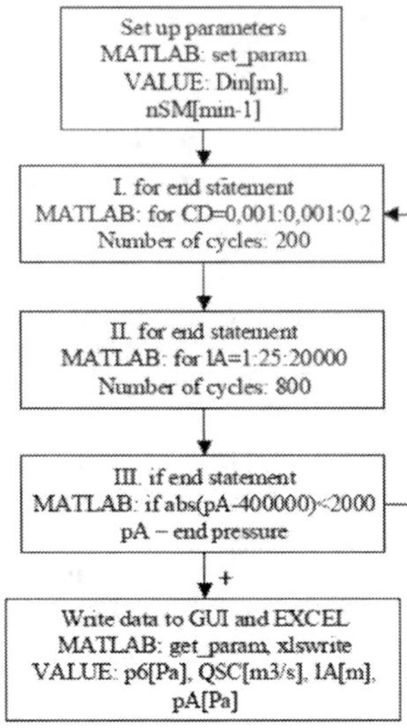

Fig. 4. Script flowchart of automatic solution mathematical model.

It is more advantageous to solve the simulation with manual setting by GUI (graphic user interface). Manual solution allows to skip several combination of C_D and l that are very distant from estimated right solution. Solution accuracy was set to the end pressure ±2000Pa and flow rate ±0.15 l/s. Figure 5 shows graphic user interface for manual setting.

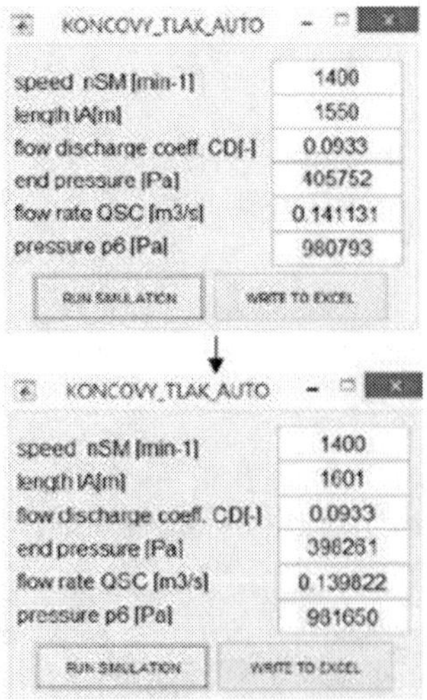

Fig. 5. Manual setting of slx file using MATLAB GUI.

The main goal of the calculation was to define maximal useful length of hoses for defined pump speed, flow-rate and required end pressure. The obtained data were compared with those presented in Methodology and are presented in Tab 1 for one hose DN150.

Table 1. Maximal pipe length determined with Methodology and in MATLAB Simulink.

SPEED n [min-1]	FLOW R Q [l/s]	METH. l1 [m]	MATL. l2 [m]	DIFF. l2-l1 [m]	DEV. [%]
1400	60	1038.4	1040.6	2.2	0.21
1600	60	1726.9	1573.1	-153.8	8.91
1800	60	2302.6	2176.9	-125.7	5.46
1400	80	647.6	592.4	-55.2	8.52
1600	80	971.4	901.4	-70	7.21
1800	80	129.2	1248.1	-47.1	3.64
1400	100	414.5	378.7	-35.7	8.61
1600	100	621.7	578.5	-43.2	6.95
1800	100	828.9	805.4	-23.6	2.85

The result of maximal length from Methodology [1] and mathematical modelling were compared only with each other. Deviation between these two lengths was calculated by equation (8).

$$deviation\ [\%] = \frac{100\%.Abs(l_1 - l_2)}{MAX(l_1;l_2)} \quad (8)$$

The smallest deviation from all simulation was 0.12% and the highest deviation was 27.33%. Maximal length calculated with mathematical modelling are smaller than values from field deployment methodology. Difference is negative in this case.

6 Differences and evaluation between the Methodology and MATLAB Simulink calculation

The maximal hose length is affected with the discharge pressure p_6 behind the pump and pressure loss in the hose system. The discharge pressure p_6 was used to determine the loss head (9).

$$h_f = \frac{p_S}{\rho g} = \frac{p_6 - p_A}{\rho g} \quad (9)$$

In the Methodology [1], the discharge pressure was approximated from the pump curves at various pump speed and was considered constant in specified range of the flow-rate (Fig.5), while in MATLAB [9] the discharge pressure was calculated with higher accuracy from the vector function p-Q (see Tab. 2). Pressure p_A is the end hose pressure fixed to 4 or 7 bar in case of fire extinguishing and zero in case of drainage.

The second difference was in determination of friction losses in the hose line. The field Methodology defines the resistance constant k based on constant value of friction coefficient λ, which does not account for the dependence on Reynolds number.

$$h_f = kQ^2 \tag{10}$$

$$k = \lambda \frac{8l}{\pi^2 d^5} \tag{11}$$

MATLAB Simulink defines friction factor according to Colebrook-White equation (5). Table 2 and 3 shows the comparison of friction factor calculated in methodology and mathematical modelling.

Table 2. Comparison of pressure p_6, loss head for MATLAB and field deployment methodology.

SPEED [min-1]	MATL. p6[bar]	METH. p6[bar]	MATL. hz[m]	METH. hz[m]
1400	10.01	10	62.23	62.09
1600	13.09	13	94.07	93.15
1800	16.58	16	130.18	124.19
1400	9.95	10	61.63	62.09
1600	13.06	13	93.78	93.14
1800	16.55	16	129.85	124.19
1400	9.87	10	60.71	62.09
1600	12.96	13	92.72	93.14
1800	16.47	16	129.09	124.19

Table 3. Comparison of friction factor for MATLAB and field deployment methodology.

-	-	MATL.	METH.
v[m/s]	Re[-]	λ[-]	λ[-]
3.53	509296	0.01527	0.01377
4.71	679061	0.01494	0.01377
5.88	848826	0.01473	0.01377

Figures 6, 7, 8 illustrate the comparison between field deployment methodology and mathematical modelling for one hose DN150 and parallel connection of two and three hoses DN150. It can be observed that for the same value of the loss head the flow-rate predicted by MATLAB Simulink is lower in comparison with the methodology.

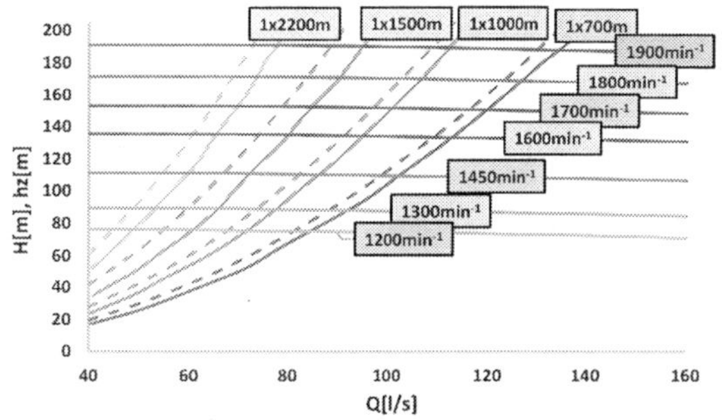

Fig. 6. Pipeline characteristics for 1xDN150 - 700, 1000, 1500, 2200 m (solid line – Methodology, dashed line – MATLAB).

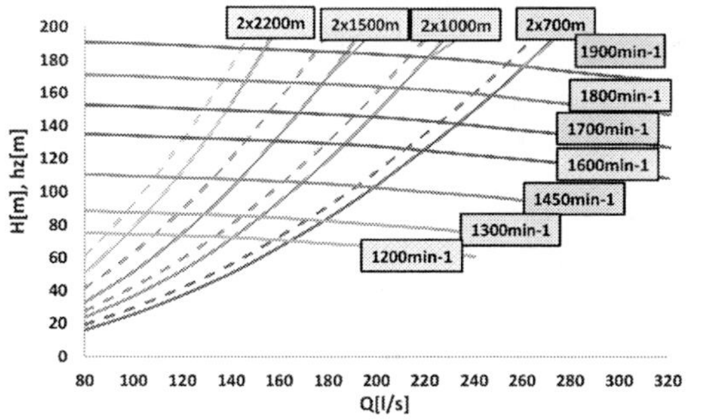

Fig. 7. Pipeline characteristics for 2xDN150 - 700, 1000, 1500, 2200 m (solid line – Methodology, dashed line – MATLAB).

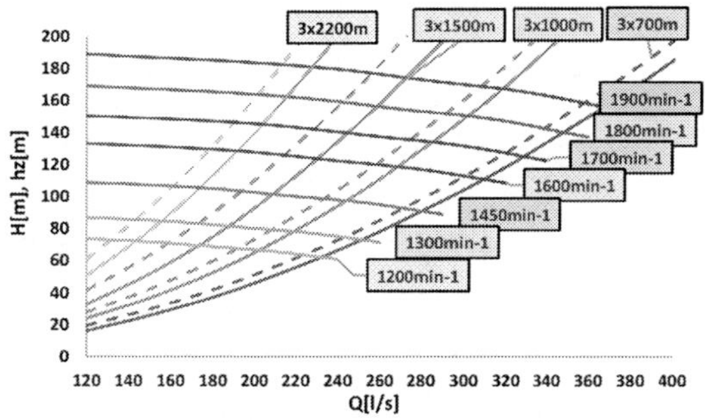

Fig. 8 Pipeline characteristics for 3xDN150 - 700, 1000, 1500, 2200 m (solid line – Methodology, dashed line – MATLAB).

7 Model uncertainties & simplifications

The mathematical model always represents only a partial replacement and simplification of the real model [6, 8, 9]. In practice, there can always be factors that the mathematical model does not take into account. The purpose of creating a mathematical model is to get as close as possible to reality. The field experiments are very difficult in term of time, finance, staff or safety. Therefore, it is very advantageous to solve HCPM problem with mathematical modeling based on the field experiment. There were several simplifications in HCPM model that could affect the final results. First, the suction system with a length of 6m was neglected. This omission has minimal effect on final result. The pressure loss in flow divider and another elements (couplings) in HCPM setup were also neglected with respect to prevailing friction losses. These minor pressure losses could be more significant. If the pressure loss in the flow divider was included in the calculation, the maximum stated length of hoses would be even lower. It would be useful to determine pressure loss in atypical flow divider by measurement or numerical modelling in CFD software.

8 Conclusion

The firefighters, namely the commanders of intervening units in emergencies, are facing many decisions that must be done rapidly, with high responsibility. Prepared scenarios with corresponding instructions can be the useful tool in emergency situations. Field measurement is the most common approach for obtaining the necessary data, but enables to investigate only limited range of conditions, under lower accuracy and high uncertainties. Operation of HCPM is affected by many parameters, the combination of which cannot be experimentally tested in the whole range. For this purpose, mathematical modelling can be applied to create a large volume of useful data as the background for preparing available tables and graphs. Estimation of pressure losses in long distance transport is very important. Differences between the field deployment Methodology and mathematical modelling were evaluated. Resulting hose line characteristics determined in Methodology and mathematical modelling show good agreement, however the pressure losses predicted in MATLAB Simulink are higher than those determined in Methodology.

In some situations, for fully effective utilization of HCPM, it is necessary to overcome the geodetic head between the water source and the fire site. The HCPM pump A-200-KIDR-550 requires the specified minimum suction pressure to avoid the cavitation causing the loss of pressure, efficiency and erosion of pump internals. The booster pump will increases the pressure of water so that the required suction pressure is ensured. For these purpose two booster pumps will be used together with HCPM that are currently under development in Sigma Company. The next work will be focused on mathematical simulation of HCPM operating with boosters.

The article was supported by project No. VI20152019030 "Research and development of a booster pump for high-performance pumping sets" (acronym: PAHDE), supported by the Ministry of the Interior of the Czech Republic for solution within the period 2015 - 2019.

This work has been supported by The Ministry of Education, Youth and Sports of the Czech Republic from the Specific Research Project SP2019/123 (Numerical Modeling of Dynamic Phenomena in Fluid Mechanics with Support of Experimental Research).

References

1. Dobeš, P., Poledňák, P., Nevěřil, J., Šoukal, J., Šputa, O., Komárek, M., Míček, D., Rostek, P., Ondruch, J., Kročová, Š.: *Operationally tactical instructions for field deployment of high capacity pumps for long-distance transport of water in emergencies* (certified methodology). Lutín, 2015. (Certificate of the methodology certification: CERO 3/2016. S. 38. Praha, 2016).
2. https://www.mathworks.com/support/learn-with-matlab-tutorials.html
3. Nevěřil, J., Šoukal, J., Šputa, O., Komárek, M., Dobeš, P., Poledňák, P.: *The prototype of the high capacity pumping module for remote water transport* (type designation: MČS-25-330-K, Sigma), Lutín, 2016.
4. Menon, E. Shashi Menon, Pramila S. (2010). Working Guide to Pumps and Pumping Stations - Calculations and Simulations. Elsevier. ISBN: 978-1-85617-828-0.
5. Vrábel, M., Drábková, S., Kozubková, M., Dobeš, P., 2018. Investigation of pumping system for emergency situations, AIP Conference Proceedings
6. Dobeš, P., Vrábel, M., Šoukal, J., Drábková S., Kozubková M., 2017. Suction Network Modelling Via Matlab SimHydraulics and Matlab Simulink, SGEM GREEN
7. Vrábel, M., Drábková, S., Kozubková, M. 2018. Využitie čerpacej techniky v hydrostatickom systéme pre mimoriadne situácie, Lutín
8. Kozubková M., 2009. Simulace a modelování hydraulických systémů, VŠB-TU Ostrava
9. L. Hružík, A. Bureček, M. Vašina, EPJ Web of Conferences, EFM13, *Non-stationary flow of hydraulic oil in long pipe*, **67**, 02042, (2014), DOI: 10.1051/epjconf/20146702042

Synchronization of the movement for multi-cylinder electro-hydraulic servo driver

Piotr Woś[1,*], *Ryszard* Dindorf[2]

[1] Faculty of Mechatronics and Machine Design, Kielce University of Technology, Kielce, 25-314, Poland
[2] Faculty of Mechatronics and Machine Design, Kielce University of Technology, Kielce, 25-314, Poland

Abstract. The article presents the problem of synchronization of piston rods velocity of the multi-cylinder electro-hydraulic servo drive. This task was presented on the example of a heavily loaded mobile caterpillar platform. For levelling the platform, a four-post electro-hydraulic support mechanism with a proportional valve control system was used. A linear external controller to synchronize the movement of the cylinders was used and a non-linear internal control system for each of the cylinders lifting the platform.

1 Introduction

The synchronous operations of many hydraulic cylinders under load have important performance implications. This issue is most evident in hydraulically controlled multi-point lifting devices. In the case of open loop operation, due to the uneven load and inherent differences in the parameters of the hydraulic circuits, the lifting height between the linear cylinders will be different. There are many approaches to solve the problem of synchronization of many hydraulic cylinders. The simplest approach is to design a flow divider circuit that will maintain the same cylinder speed while maintaining the same flow rate into the cylinders. The efficiency of synchronization depends on the efficiency of the flow divider, as well as the compressibility of the working fluid and the construction of hydraulic components. Another approach is the mechanical connection of hydraulic cylinders using a connecting structure. The disadvantage of mechanical synchronization is the additional complexity of the system structure, as well as the limitation of the scope of equipment operation. Compared with a clean hydraulic and mechanical approach, electro-hydraulic synchronization is a flexible alternative. By using electro-hydraulic synchronization control strategies, they can be designed to compensate for uneven loading and external disturbances associated with the hydraulic system.

Many authors [1-4] deal with the issue of synchronization of unevenly loaded hydraulic cylinders. The results of this work indicate that to achieve synchronization of unevenly loaded cylinders, individual control of each cylinder is required [5].

Achieving minimal errors in the implementation of the given trajectory of drive motion is a basic requirement for modern electro-hydraulic systems [6,7]. The basic causes of errors in the reference trajectory of the actuator are: imperfections in the construction elements, elasticity, friction, phenomena of thermal expansion as well as the imperfection of control systems and lack of synchronization between individual drive mechanisms. In single servo drives, the reduction of trajectory errors is commonly implemented by positioning systems [8]. Electrohydraulic positioning systems are systems in which a change in the position, speed and acceleration of an actuator relates to a changing technological load. Modern electrohydraulic drives are characterized by high motion dynamics with high positioning accuracy, work stability and easy control. These systems are built in conventional, proportional and servo technology.

These techniques successfully use both proportional directional control valves and servo valves. Digital cards and sets of regulators, amplifier cards, setpoint modules as well as sensors and setters are used for control. Unfortunately, in a multi-axis system, a very good motion tracking system for a single axis does not ensure the elimination of the resulting motion effect of the machine effector [9-11].

Reduction of motion trajectory errors can be done by appropriate optimization of controller settings using AC adaptive techniques (AC), fuzzy logic (FLC), forward coupling controllers (FFC), as well as the use of synchronous cross-control (CCC) [12-16]. In conventional control systems, the positioning errors of one of the drive axes are corrected only for itself, while the other axes carry out their previous operation. Lack of knowledge about positioning errors of individual axes causes unacceptable errors of synchronization of the final position of the entire machine.

This work presents the synchronous control system of electrohydraulic servo drives used in the transport unit which is the Mobile Hydraulic Module (MMH). Ultimately, this module is intended for work in the construction industry.

* Corresponding author: wos@tu.kielce.pl

2 Dynamic model of the MMH

Mobile Hydraulic Module (MMH) consists of two support modules. Each module contains two hydraulic cylinders. Figure 2 shows the model of the Mobile Hydraulic Module (MMH). The problem of synchronization of many linear hydraulic cylinders arises in heavy, hydraulically controlled applications such as lifting devices, construction and agricultural machinery. This problem is most often solved by introducing to each parallel actuator branch an additional hydraulic resistance adjusted manually or automatically. In the case of manually adjustable hydraulic resistance, this boils down to the use of a throttle valve, and in the case of automatically adjustable hydraulic resistance, the use of a flow regulator. In both cases, the idea of solving the problem of motion synchronization boils down to creating additional pressure drops in the power supply branches of individual drives so that the pressure in the pump discharge line, determined from the equilibrium pressure equations arranged for the power branches of individual drives, is the same [17-19].

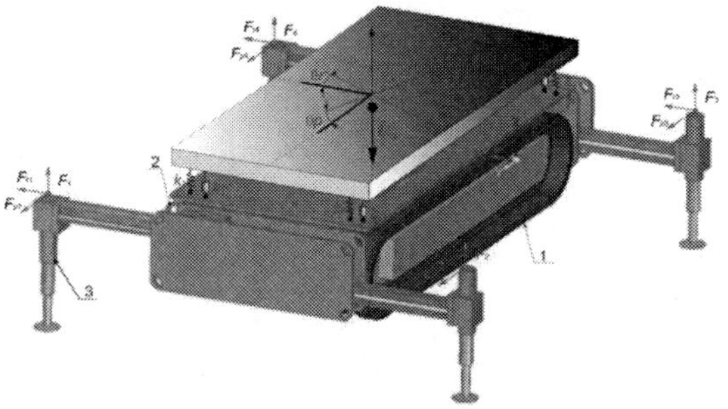

Fig. 1. Mobile Hydraulic Module (MMH).

MMH is a support platform for further development of devices used in cooperating construction works, such as: construction manipulator, material warehouse or lifting platform. It is made of a mobile platform (1), which ensures movement of MMH and two supporting modules (SM) (2). Each of the support modules has retractable arms, at the ends of which there are supports in the form of electro-hydraulic servo drives (3).

Calculation diagram of the single supporting module (SM) consisting of the double-acting cylinder with one-sided piston rod and proportional directional control valve is presented in the Figure 2. Equations containing mechanical and hydraulic variables are described by:

$$\begin{aligned} F_i &= A_{1i} p_{1i} - A_{2i} p_{2i} \\ Q_{1i} &= A_{1i} \dot{z}_i + C_1 \dot{p}_{1i} + G_{lei}(p_{1i} - p_{2i}), \quad i = 1,2,\cdots,n, \quad (1) \\ Q_{2i} &= A_{2i} \dot{z}_i - C_2 \dot{p}_{2i} + G_{lei}(p_{1i} - p_{2i}) \end{aligned}$$

where: Q_{1i} - volumetric flow rate, C_1, C_2 - fluid capacitances in the cylinder chambers, G_{le} - coefficients of leakages in the cylinder, A_{1i}, A_{2i} – effective area of cylinder's piston, p_{1i}, p_{2i} - pressure in individual cylinder's chambers.

The hydraulic forces developed by the actuators are given by:

$$\left[p_{1i} \cdot A_{1i} - p_{2i} \cdot A_{2i}\right] = \left[A_{1i} \cdot p_0 - A_{2i} \cdot p_T - \frac{A^3{}_{1i} + A^3{}_{2i}}{(K_1 \cdot u + K_{0.1})^2} \cdot z_i^{\,2}\right], \quad i = 1,2,\cdots n \quad (2)$$

where: u_i – control inputs, K_1, $K_{0.1}$ – constants factors, p_0 – supply pressure, p_T – pressure in the tank which correspond to the main and leakage valve flow paths.

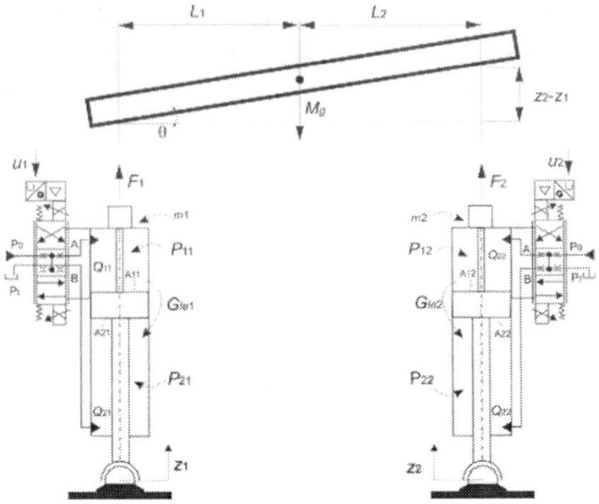

Fig.2. Scheme of the support modules.

The equation of motion of each hydraulic cylinder piston is:

$$m_i \ddot{z} = (P_{1i} - P_{2i}) A_{2i} - b_i \dot{z} - F_i \quad (3)$$

and equations of motion of the support module are:

$$\begin{cases} F_1 + F_2 - Mg = M\ddot{x}_M \\ F_1 L_1 + F_2 L_2 = J\ddot{\theta} \end{cases} \quad (4)$$

where: $x_M = (z_1 L_2 + z_2 L_1)/(L_1 + L_2)$ and $\theta = (z_1 - z_2)/(L_1 + L_2)$

Due to the symmetrical shape of MMH, it can be assumed that the center of gravity of the platform is in the middle, $L_1 = L_2$.

For the module (MMH) (Fig. 1) the following equations of motion along the direction of gravity and rotation around the axis r (roll) and p (pitch) can be obtained:

$$\begin{aligned} M\ddot{z}_M &= \sum_{i=1}^{4} F_i + \sum_{i=1}^{4} F_{ri} \sin(\theta_r) + \sum_{i=1}^{4} F_{pi} \sin(\theta_p) - M_g \\ J_r \ddot{\theta}_r &= \sum_{i=1}^{4} ((F_i + F_{ri} \sin(\theta_r)) L_{ri} \quad (5) \\ J_p \ddot{\theta}_p &= \sum_{i=1}^{4} ((F_i + F_{pi} \sin(\theta_r)) L_{pi} \end{aligned}$$

where: z_M - module center of gravity position, M – total module mass. F_i, F_{ri}, F_{pi} – reaction forces for electrohydraulic servo drives along the z axis and relative to the r (roll) and p (pitch) axes for $i = 1,2,\cdots n$, L_{ri}, L_{pi} - arms length for moments of forces with respect

to the axis r (roll) and p (pitch), J_r, J_p - moments of module inertia about the axis r (roll) and p (pitch), θ_r, θ_p - angles of rotation relative to the axis r(roll) and p (pitch). If we assume, that θ_r, θ_p is small then $\sin(\theta_r) \approx 0$ and $\sin(\theta_p) \approx 0$.
In contact with the hydraulic supports with the ground, the following occurs:

$$G_i = k_i(z_0 - z_i) + b_i(\dot{z}_0 - \dot{z}_i), \quad i = 1,2,\cdots,n \tag{6}$$

where: k_i, b_i - stiffness and damping ratio.
The system can be described by the formulas:

$$\begin{cases} \mathbf{M}_L \ddot{\mathbf{z}}_q = \mathbf{L}_m \mathbf{G} - \mathbf{M}_g \\ \mathbf{G} = \mathbf{K}_i(\mathbf{z}_0 - \mathbf{z}_i) + \mathbf{B}_i(\dot{\mathbf{z}}_0 - \dot{\mathbf{z}}_i), \quad i = 1,2,\cdots,n \\ \mathbf{m}\ddot{\mathbf{z}}_L = \mathbf{F} - \mathbf{G} \end{cases} \tag{7}$$

where:

$\mathbf{G} = [G_1, G_2, \cdots, G_n]^T$, is vector loading forces,

$\mathbf{z}_q = [z_1, z_2, \cdots, z_n]^T$, is platform displacement vector,

$\mathbf{z}_L = [z_{l1}, z_{l2}, \cdots, z_{ln}]^T$, is hydraulic cylinder displacement vector,

$\mathbf{L}_m = \begin{bmatrix} 1 & 1 & \cdots & 1 \\ L_{r1} & L_{r2} & \cdots & L_{rn} \\ L_{p1} & L_{p2} & \cdots & L_{pn} \end{bmatrix}$, is an arm matrix with a fixed moment relative to the axis r (roll) and p (pitch),

$\mathbf{M}_g = \begin{bmatrix} M_g \\ 0 \\ 0 \end{bmatrix}$, is a mass load matrix,

$$\mathbf{F} = \begin{bmatrix} A_{11} \cdot p_{11} - A_{21} \cdot p_{21} - m_1 g \\ A_{12} \cdot p_{12} - A_{22} \cdot p_{22} - m_2 g \\ A_{13} \cdot p_{13} - A_{23} \cdot p_{23} - m_3 g \\ \cdots \\ A_{1n} \cdot p_{1n} - A_{2n} \cdot p_{2n} - m_n g \end{bmatrix},$$

is the matrix of forces generated by hydraulic cylinders.
$\mathbf{M}_L = diag([M \quad J_r \quad J_p])\mathbf{L}_{rp}^{-1}$, is the load inertia matrix,

were: $\mathbf{L}_{rp} = \begin{bmatrix} 1 & L_{r1} & L_{p1} \\ 1 & L_{r2} & L_{p2} \\ \vdots & \vdots & \vdots \\ 1 & L_{rn} & L_{pn} \end{bmatrix}$,

$\mathbf{K}_i = diag([k_1 \quad k_2 \quad \cdots \quad k_n])$, is the contact stiffness matrix, $\mathbf{B}_i = diag([b_1 \quad b_2 \quad \cdots \quad b_n])$, is the contact viscous damping coefficient matrix.

3. Control of MMH

3.1. Control strategy

Design of the control system for the tested MMH was carried out in the following stages: selection of control laws, virtual prototyping of the control system, simulations of the controller in real time, and implementation on the target MMH.
The control analysis is based on the system dynamic and hydraulic model (8).

$$\mathbf{F}_p = \mathbf{M}^+(z)[\ddot{\mathbf{z}}_d + \mathbf{K}_p(\mathbf{z}_d - \mathbf{z}) + \mathbf{K}_v(\dot{\mathbf{z}}_d - \dot{\mathbf{z}})] + \mathbf{G}^+(z) \tag{8}$$

where: $\mathbf{K}_p, \mathbf{K}_v$ - diagonal matrices, which represent the control gains of the system, $\mathbf{Z}_d$ - desired cylinder displacement vector.
Equations of motion are derived,

$$\mathbf{M}^+(z)\ddot{\mathbf{z}} + \mathbf{G}^+(z) = \begin{bmatrix} A_{11} \cdot p_{11} - A_{21} \cdot p_{21} - m_1 g \\ A_{12} \cdot p_{12} - A_{22} \cdot p_{22} - m_2 g \\ A_{13} \cdot p_{13} - A_{23} \cdot p_{23} - m_3 g \\ A_{14} \cdot p_{14} - A_{24} \cdot p_{24} - m_4 g \end{bmatrix} \tag{9}$$

where: $\mathbf{M}^+(\mathbf{x})$, is positive definite mass matrix and $\mathbf{G}^+(\mathbf{x})$ is gravity forces vector.
Figure 3 shows a diagram of a closed system control MMH.

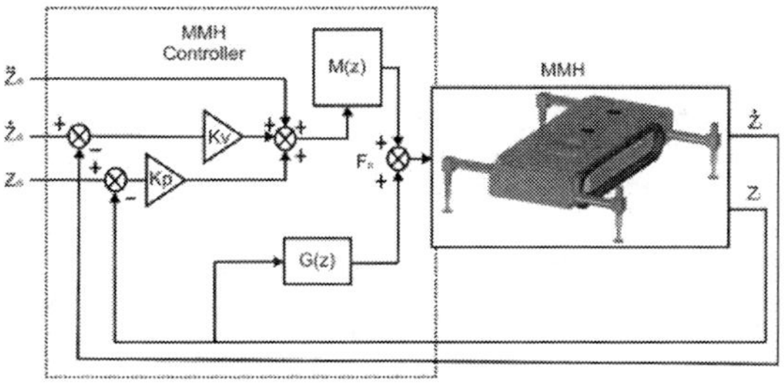

Fig. 3. The block diagram of closed loop control MMH.

Initial axis controller settings were chosen in an experimental way in the process of virtual prototyping. Axis controller gain coefficients $\mathbf{K}_p$, $\mathbf{K}_V$ were chosen in such a way so that the position error of the movement trajectory would be minimized. Only after determining the correct coefficient values for the virtual control system, it was possible to proceed to the phase of control rapid prototyping.

3.2. Synchronization control of MMH

In multi-axis control systems, errors in the position of a single driving axle are transferred to all others. To ensure the required accuracy of the control system, a four-axis cross-coupled control has been proposed for the system. The synchronization control strategy is to stabilize the position of the four hydraulic actuators. In the presented control system there is a set of internal regulators responsible for the correct positioning of individual supports, and an external regulator responsible for the correct synchronization of movements of individual hydraulic supports. A decentralized motion tracking system enables the correction of disturbances in internal loops based on defined synchronization errors for individual drives.

During synchronization, the differential position errors between the hydraulic actuators converge to zero. The position error of i-th hydraulic actuator is defined as follows:

$$e_i(z) = z_i^d(t) - z_i(t), \qquad (10)$$

where $z_i^d(t)$ is the desired generated length and $z_i(t)$ is the actual measured length of i-th hydraulic actuator.

Figure 4 shows the control system diagram for MMH with synchronization.

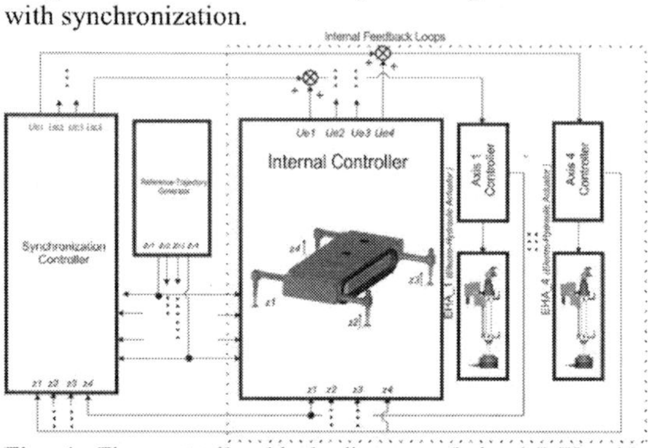

Fig. 4. The controller block diagram of the MMH with synchronization.

The position error vector of three actuators is:

$$\mathbf{e}(t) = [e_1(t), e_2(t), e_3(t), e_4(t)]^T \qquad (11)$$

If the ratio of the actual position of each hydraulic actuator at each sampling time is equal to that of all other actuators, then the MMH moves in a synchronous manner. The synchronization goal is understood to be as follows:

$$e_1(t) = e_2(t) = e_3(t) = e_4(t) \qquad (12)$$

Thus, the synchronization error is defined as:

$$\begin{cases} \varepsilon_1(t) = e_1(t) - e_2(t) \\ \varepsilon_2(t) = e_2(t) - e_3(t) \\ \varepsilon_3(t) = e_3(t) - e_4(t) \\ \varepsilon_4(t) = e_4(t) - e_1(t) \end{cases} \qquad (13)$$

where: $\boldsymbol{\varepsilon}(t) = [\varepsilon_1(t), \varepsilon_2(t), \varepsilon_3(t), \varepsilon_4(t)]^T$

and $\mathbf{T} = \begin{bmatrix} 1 & -1 & 0 & 0 \\ 0 & 1 & -1 & 0 \\ 0 & 0 & 1 & -1 \\ -1 & 0 & 0 & 1 \end{bmatrix}$ is 4x4 synchronization transformation matrix.

The position control system is designed by combining the position error and synchronization error. Herein, in conceiving a cross-coupled controller (CCC), the cross-coupling error c_i includes the position error e_i and the synchronization error ε_i [20]:

$$c_i(t) = e_i(t) + \beta_i \int_0^t \varepsilon_i(\omega) d\omega, \qquad (14)$$

where β_i is positive coupling parameter which determines the weight of the synchronization error, wherein as β_i increases, the synchronization error ε_i decreases, ω is a parameter at a variable time, from $t = 0$ to t.

The position synchronization error looks for the difference between the position errors of many axes and is based on the synchronization function. The block example diagram of the synchronization function designed in Matlab/Simulink is shown in the Figure 5.

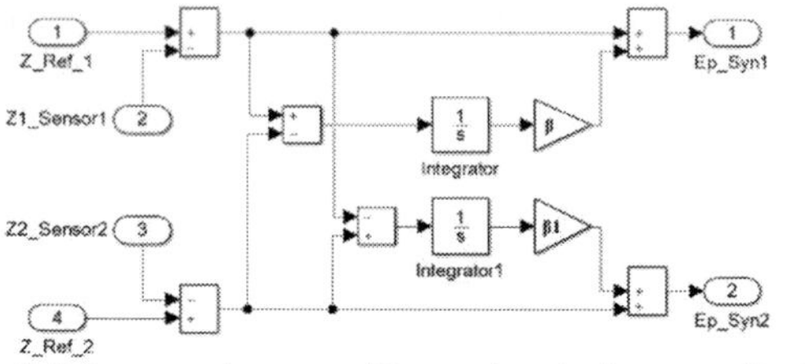

Fig. 5. Scheme a subsystem of the synchronization controller in Matlab/Simulink.

For the simulation purposes constant values of parameters were accepted and implemented (Table 1).

Table 1. Simulation parameters

p_s	16 *MPa*	m_i	8 *kg*
A_{1i}	1.256 x10^{-3}*m*^{2}	*Mg*	1000*kg*
A_{2i}	1.134 x10^{-3}*m*^{2}	*Li*	1.5 *m*

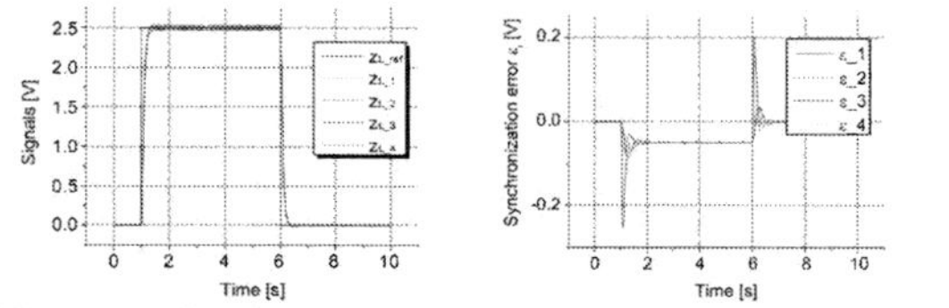

Fig. 6. Synchronization error ε_i for references rectangular input signals u_{iref} with synchronization of the MMHs.

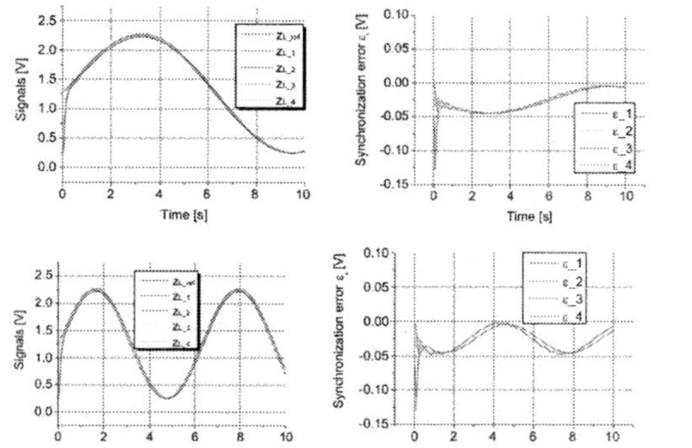

Fig. 7. Synchronization error ε_i for references sinusoidal input signals u_{ref} (0.08 Hz and 0.16 Hz) with synchronization of the MMHs.

4 Summary

The paper presents the results of research on control algorithms for the lifting system of the Mobile Hydraulic Module (MMH). To minimize lag and mapping errors, a cross-coupled controller (CC-Control) was used to control it. Therefore, the purpose of the research was to create a control system that would counteract the position error and synchronization error of each hydraulic actuators. The proposed synchronization controller and parameter estimator employs cross-coupled control (CCC) by feeding back hydraulic actuator position errors and differential position errors.

Errors in the construction elements and technological clearances are the basis for the lack of synchronization of the manipulator drives movement. In addition, such phenomena as non-linear friction, deformations of structural elements due to external forces and generated during operation occur separately for each axis. The effect of these phenomena is the increase of module errors (MMH). The aim of the study was to create such a control system that will counteract these phenomena by taking into account the synchronization errors of individual lifting supports. The presented MMH control system uses feedback from support position errors and synchronization errors. The presented solution can be used in various types and lifting platforms in industries such as mining, engineering and handling.

References

1. D. Sun, R. Lu, J.K. Mills and C. Wang, I.J. of Rob Res., **25**, 1137-1147, (2011).
2. S. Ning and S. Wang, In. Conf. on M. T. and Mech. Aut., 710–713, (2010).
3. S. Kassem, T. S. El-Din, and S. Helduser, Int. J. Fluid Power, **13**, 51–60, (2012).
4. P. Wos and R. Dindorf, Mech.: *Id. for Ind. Appl. edited by J. Awrejcewicz et.al.*, Book, **317**, 269-277, (2015).
5. D. Sun, R. Lu, J. K. Mills, and C. Wang, Int. J. Rob. Res., **25**, no. 11, 1137–1147, (2006).
6. Y. Pi, X. Wang and X. Gu, J. Cent. South Univ. Technol., **18**, 1554–1562, (2011).
7. P. Wos and R. Dindorf, Tech. Tran., **7** 209-217, (2017).
8. R. Dindorf, P. Wos, *Monograph M72*. Kielce University of Technology, (2016).
9. P. Wos and R. Dindorf, A.J. of Con., **15**, 1065-1080, (2013).
10. T. Kheowree, S. Kuntanapreeda, A. J. of Con., **17**, 855–867, (2015).
11. R. Dindorf and P. Wos, Ad. Mat. Res., **874**, 57–62, (2014).
12. J. Yao, X. Cao, Y. Zhang, and Y. Li, J. Mech. Sci. Technol., **32**, 5, 2261–2271, (2018).
13. S. Mengfei, L. Jianxiong, and F. Yiming, 34th Ch. Cont. Conf., 3361–3366, (2015).
14. D. Sun, Aut., **39**, 997–1005, (2003).
15. D. Sun, IFAC Proced., **35,** 313-318, (2002).
16. W. Lu, K. Ji, H. Dong, J. Zhang, Q. Wang, and L. GUO, Chinese J. Mech. Eng., **30**, 2,272–285, (2017).
17. Z. Liu, Q. Gao, C. Yu, X. Li, W. Guan, and G. Deng, Adv. Mech. Eng., **2014**, 1–18, (2014).
18. S. Blasiak, J. E. Takosoglu, and P. A. Laski, Proc. of 20th In. Conf. on Eng. Mech., pp.,**2014**, 96–99 (2014).
19. W. Depczyński, L. Nowakowski, P. Hepner, and E. Miko, Metal., **56**, pp. 364–366, (2017).
20. K. Ishizaki, B. Sencer, and E. Shamoto, Int. J. Autom. Technol., **7**, 514–522, (2013).

The aspect of the flow around a circular cylinder in a free surface water flow

Yoshifumi Yokoi[1,*]

[1]National Defense Academy of Japan, Department of Mechanical Engineering, 1-10-20 Hashirimizu, Yokosuka 239-8686, Japan

Abstract. The purpose of this study is in collection of the basic data for development of the simple flow velocity measurement equipment. A 10mm diameter circular cylinder was inserted into the free surface stream, and investigation of the water surface height and the relationship of the flow velocity which creep up the circular cylinder surface were performed. The experiment performed in the closed circuit water channel apparatus (the range of stream velocity is 0.07 - 0.98m/s) and carried out image-processing measurement using the video recording apparatus. In the insertion depth of various circular cylinders, observation of the height which creeps up the circular cylinder when varying the flow velocity was carried out. As a result, it became clear that the height of the water surface which creeps up by the increase in the flow velocity increases. And it also became clear for the phenomenon not to be based on the insertion depth. It was obtained that measured rising height and the flow velocity is required from Bernouilli's equation using the value. In the measurement, it became clear that 5 or more times of the cylinder diameter are required for the required insertion depth.

1 Introduction

In our research group, flow visualization observation of the around of a body has been carried out using water tank experimental apparatus (a closed circuit water channel and a towing water tank). And the aspect of various flows has been shown in comparatively low Reynolds number [1][2]. These reports of research were carried out at stream velocity and towing velocity at which the water surface of the around of a body is not choppy in a water tank with a free surface. Generally, if stream velocity exceeds 23cm/s, it is known that a surface wave will occur on the water surface. Moreover, if stream velocity and towing velocity become fast, a necklace vortex will be formed so that the bluff cylinder may be surrounded the front of the bluff cylinder inserted in the flow. The phenomenon in which the water surface which is formed by the bow of a ship by the bluff cylinder front part of the jumping phenomenon of the water accompanying formation of this necklace vortex rises is seen. About the phenomenon produced at such a flow field, it is treated in fields, such as civil engineering and marine engineering, for many years, and many knowledge is accumulated [3][4]. However, it seems that there is no example which applied the phenomenon in which this water surface rose to the flow velocity measurement in an author's knowledge. Although rising of the water surface can be seen in a bridge pier at the time of rise of water of a river, and a flood, the report which determined for the height of this rising and the relationship of stream velocity is not found. The flow velocity measurement method of a river differs in the method by the flow velocity. When the flow velocity is slow, on foot observation, observation by a ship, and observation from a bridge are carried out. On the other hand, when the flow velocity is fast, it is carried out by stretching a wire and locating sensors, such as an electromagnetism current meter and a price current meter, among both banks, at predetermined depth of water. Anyway, advance preparations and large-scale equipment are required. In the site at the time of rise of water of a river, and a flood, it seems that it is important to get to know stream velocity immediately also when planning subsequent measures. So, an idea which gets to know stream velocity simply with easy equipment is desired.

In this study, for development of the simple flow velocity measurement equipment, the circular cylinder was inserted in the stream which has a free surface, the water surface height and the relationship of the flow velocity which creep up the circular cylinder surface were investigated, and some obtained knowledge was shown.

2 Experimental apparatus and method

2.1 Outline of experimental apparatus

The experimental apparatus consists of a small closed circuit water channel, a test circular cylinder unit, and a video recording apparatus. The closed circuit water channel apparatus is a small water tank with a water capacity of $0.4m^3$, and the size of test section of water

* Corresponding author: yokoi@nda.ac.jp

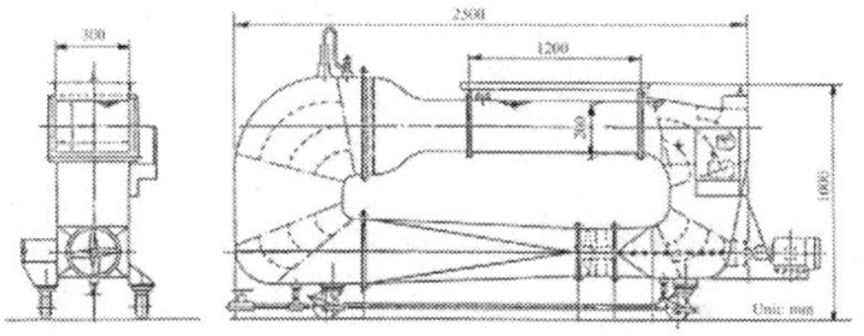

Fig. 1. Schematic diagram of closed circuit water channel.

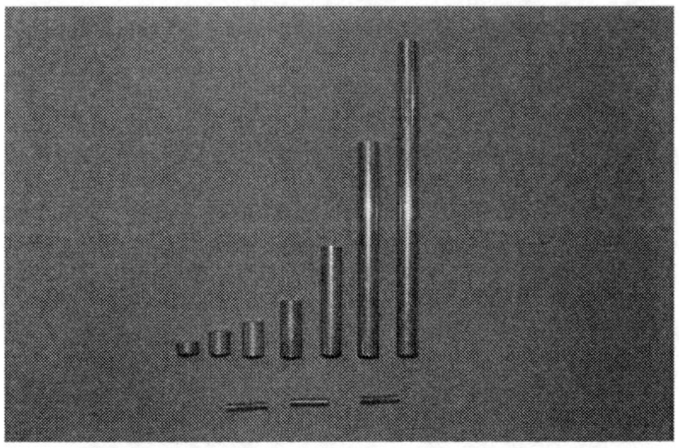

Fig. 2. The elements of underwater section, from left hand side; 5mm, 10mm, 15mm, 25mm, 50mm, 100mm and 150mm; They are attached with the screw shown below.

tank are length of 1.2m and cross-section of 0.3m × 0.3m square meters. The test section is an open channel. The board for circular cylinder attachment is prepared in the open section of the test section. At the time of an experiment, the water level was set as 0.2m. The sketch of the closed circuit water channel is shown in **Fig. 1.** At the time of the experiment, the water level was set as 0.2m. The stream is generated by carrying out input operation of the operation frequency f Hz in a digital inverter. The stream velocity v is obtained by v=0.0327× f(Hz).

A test circular cylinder is 10mm in diameter, and it is blocked by the water upper part (100mm) and the underwater section (5-200mm) so that the length under the water surface can be adjusted. The block of an underwater section was prepared seven kinds and length was 5mm, 10mm, 15mm, 25mm, 50mm, 100mm, and 150mm, respectively. They are shown in **Fig. 2.** Adjustment of the length of the underwater section was carried out by combining these seven blocks.

The video recording apparatus consists of a camera for photography, and a personal computer for image processing. The camera for photography used the digital camera (Nikon D40) and the digital camcorder (WAT-250D2) properly by the use. Those cameras fixed to the tripod mount were set in the manometer front. The personal computer (NEC LaVie) commercial note type was used for image-processing measurement. Measurement of the length of picture was performed using "Microsoft Paint" [5] installed as application software. The video signal converter (Alterplus AX-200) was used for dealing with the image signal from the digital camcorder with the personal computer.

2.2 Experimental procedure

The experimental procedure is the following. Water is filled in a closed circuit water channel, and the water level of the test section is adjusted to 200mm. Warming up is performed. A test circular cylinder is set in the test section. Since the record camera is set beside the test section, the aspect of the flow around a circular cylinder seen from the side is photographed. The focus and zoom of the record camera are adjusted. Those conditions are held after completing this adjustment until a series of recording is completed.

The flow velocity for which it wishes operates the inverter for the flow velocity setup, and is set up. Here, the digital input of the frequency for operation f (Hz) is carried out to an inverter, and the flow velocity v (m/s) is obtained from the authorization formula. If the flow velocity for which it wishes is obtained, the aspect of a flow will be recorded with a digital camera, and the aspect of a flow is observed about another flow velocity.

The obtained picture is downloaded to the computer; the image-processing measurement is performed.

2.3 Measurement procedure of the rising water-head length with the aid of image processing

The "Paint" is started and the target picture file is opened. A reference position is decided in the displayed picture. Here, a reference position uses the same position over a series of measurement. A square is chosen from the "figure" in the section of a menu display of the software screen upper part. The displayed square is drawn on the target point. It meets with the reference position which determined one of the sides of the square. Another side is met with the rising water-head position. For example, the upper side will be met with the rising water-head position if the base of the square is met with the reference position. Here, the reference point is the water surface at the time of flow rest. Since the information on the length of two sides of the square (the number of horizontal pixels, x, the number of vertical pixels) is displayed on the software screen bottom, the information on the number of vertical pixels is read. The actual dimension per pixel is obtained by authorization. For example, it is investigated by the box function whether an actual length of 10mm shown on the screen is displayed with how much number of pixels. The thickness of the drawing line sets up so that it may be easy to operate it.

2.4 Experimental parameters

The experimental parameters are the flow velocity and underwater circular cylinder length. The flow velocity is the range of 0.07 - 0.98m/s (the inverter input frequency f is 2Hz unit to 2-30Hz) in 15 kinds. The underwater circular cylinder length is 5mm, 10mm, 15mm, 20mm, 25mm, 30mm, 40mm, 50mm, 100mm, 150mm, and

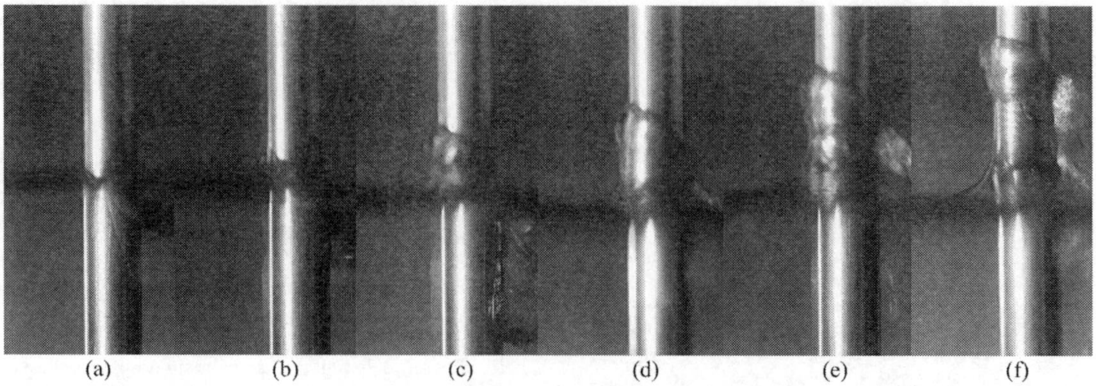

Fig. 3. The aspect of rising of the water surface, submerge depth is 200mm; (a) 0.33m/s, (b) 0.46m/s, (c) 0.59m/s, (d) 0.72m/s, (e) 0.85m/s, (f) 0.98m/s.

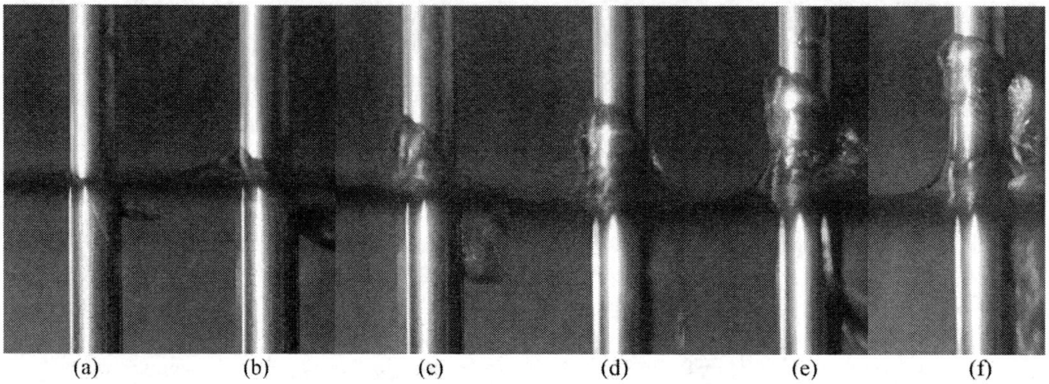

Fig. 4. The aspect of rising of the water surface, submerge depth is 50mm; (a) 0.33m/s, (b) 0.46m/s, (c) 0.59m/s, (d) 0.72m/s, (e) 0.85m/s, (f) 0.98m/s.

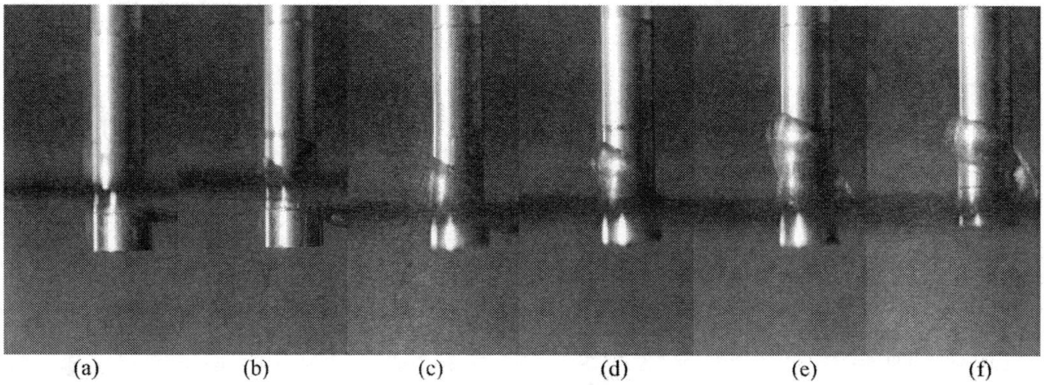

Fig. 5. The aspect of rising of the water surface, submerge depth is 5mm; (a) 0.33m/s, (b) 0.46m/s, (c) 0.59m/s, (d) 0.72m/s, (e) 0.85m/s, (f) 0.98m/s.

200mm in 11 kinds, respectively. Here, in the case of 200mm length, the circular cylinder has arrived at the bottom of test section.

3 Experimental results and discussion

The height of the water which creeps up the circular cylinder corresponding to the flow velocity was

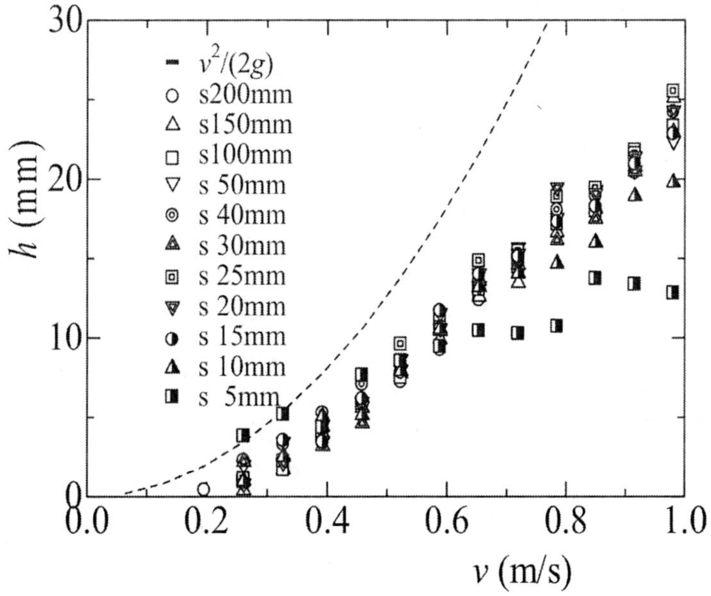

Fig. 6. The relationship between the flow velocity and the height of water surface.

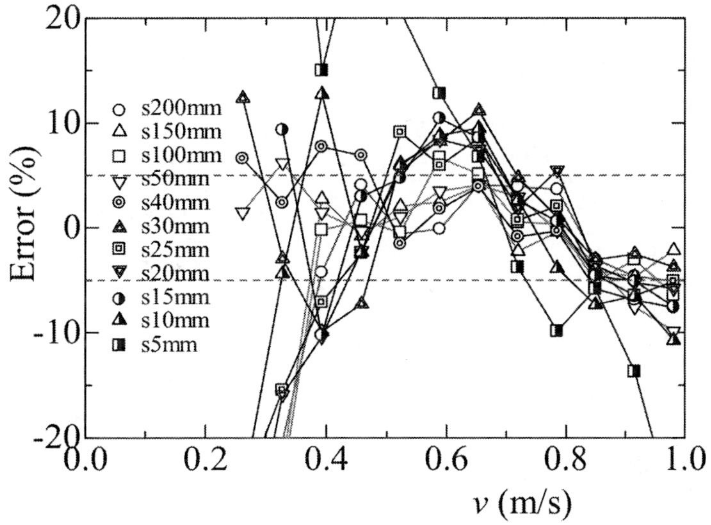

Fig. 7. The error of the guess flow velocity for every submerge depth.

observed for every underwater circular cylinder length. An example of the aspect of the water surface which creeps up the circular cylinder is shown in **Figs. 3 - 5.** It became clear that the height which creeps up by the increase in the flow velocity is increasing, and that it is not dependent on underwater circular cylinder length. **Figure 3** is the case where underwater circular cylinder length is 200mm, and the lower end of the circular cylinder is in contact with the test section bottom. In this case, the flow which passes over the circular cylinder will go to the direction of cylinder side and the direction of water surface. **Figure 4** is the case where underwater circular cylinder length is 50mm, and the lower end of the circular cylinder exists underwater. In this case, the flow which passes over a circular cylinder will go to the circular cylinder side, the water surface, and the circular cylinder end. Although underwater flow around a circular cylinders differ in both, it is shown that the height of the water which creeps up the circular cylinder is almost same. **Figure 5** is the case where underwater circular cylinder length is 5mm, and although there is a lower end of the circular cylinder underwater, it is located in a very shallow position. Therefore, it seems that the height of the water which creeps up the circular cylinder as compared with the case where underwater circular cylinder length is 50mm differs. The tendency for change of the aspect of the water surface which creeps up the circular cylinder which is not based on underwater circular cylinder length, but is observed to be the same was obtained. While the height of the water surface which creeps up is comparatively low, there is also no turbulence in the surface, and rising is smoothly shown as it, but if the height which the flow velocity becomes fast and creeps up becomes high, a knot will occur in the top part. **Figure 6** shows change of the height by the flow velocity which creeps up for every underwater circular cylinder length. An abscissa is the flow velocity and an ordinate is the height. The dashed line in the figure is the height ($h=v^2/(2g)$) corresponding to the flow velocity required from Bernouilli's equation. If the case where underwater circular cylinder length is 5mm and 10mm is removed, it is shown that the height which is not based on underwater circular cylinder length, but creeps up becomes the almost same quantity. This supports a possibility that the flow velocity can be known, in a flow with the free surface of a river etc. by inserting a bar thing like a circular cylinder from the surface of river. Accordingly, the flow velocity can be guessed using Bernouilli's equation, applying a coefficient to the actually measured rising height. Here, it becomes important to get to know how much length of an underwater circular cylinder is required. It is classified when it is in the case where the situation of the end is in contact with the bottom when inserting a bluff cylinder, and underwater. Furthermore, it is divided into whether the position of the underwater circular cylinder end is close to the water surface, or it is separated. Actually, the case where the end is underwater is assumed in on-site situation. The error of the flow velocity and the real flow velocity which were guessed is defined like an equation (1). Generally, the flow velocity measurement requires few errors and little scatter. It is necessary to estimate an acceptable value. In this study, the range of permissible error was made into ±5% in view of being rough measurement.

(Guess flow velocity - Real flow velocity)
/ (Real flow velocity) x100(%) (1)

The result of having calculated the error over the real flow velocity obtained from the equation (1) for every length of an underwater circular cylinder is shown in **Fig. 7**. An abscissa is the flow velocity and an ordinate is an error. The dashed line in the figure shows the range (±5%) of with error. If the underwater circular cylinder length is the case of 50mm or more, fitting in the scatter error (±5%) can be seen. Accordingly, it will be said that the circular cylinder diameter d (10mm) is required for underwater circular cylinder length 5 or more times. And also the flow velocity ranges from which an error becomes ±5% were 0.39 - 0.85m/s (an underwater circular cylinder length of 50mm), and 0.39 - 0.98m/s (an underwater circular cylinder length of 100mm, 150mm). Here, in an underwater circular cylinder length of 100mm, and 150mm, the flow velocity range which satisfies ±5% of scatter errors was the same as the state where the circular cylinder edge was attached to the test section bottom. This showed that enough measurement accuracy was obtained, when setting underwater circular cylinder length to $5d$ or more.

4 Conclusions

The circular cylinder was inserted into the stream with the free surface generated with a closed circuit water channel, and image-processing observation of the height of the water surface which creeps up the circular cylinder surface was carried out. The following conclusions were obtained.

(1) The height in which the water surface creeps up the circular cylinder by the increase in the flow velocity increases.

(2) It is not dependent on underwater circular cylinder length, and the phenomenon occurs.

(3) The flow velocity can be guessed if the height of the water surface which creeps up the circular cylinder is got to know.

(4) Required underwater circular cylinder length is 5 or more times of the circular cylinder diameter.

References

1. Y Yokoi, AIP Conf. Proceedings, **1745**, 020066-1-020066-6 (2016)
2. Y.Yokoi, R. Vitkovičová, EPJ Web of Conf. **143**, 02146-1-02146-6(2017)
3. E.g., K. Nakayama, et al., J. Hydraulic Engineering, JSCE, **43**, 365-370(1999)
4. E.g., T. Suzuki, et al., J. Visualization Soc. Japan **13**-51, 37-42(1993)
5. https://en.wikipedia.org/wiki/Microsoft_Paint

Comparison of fibre-based phase Doppler analysers

Matous Zaremba[1,*], *Milan* Maly[2], *Vojtech* Mraz[1] and *Jan* Jedelsky[2]

[1] WTtech.CZ s.r.o., Lhotecká 214, 290 01 Sokoleč, Czech Republic
[2] Brno University of Technology, Faculty of Mechanical Engineering, Energy Institute, Technicka 2896/2 Brno, 616 60, Czech Republic

Abstract. Laser diagnostics techniques are widely used in experimental fluid mechanics. By far the most widely used systems for getting spatial velocity fields and turbulence data are laser Doppler anemometers (LDA). Further, in the cases of two-phase flows phase Doppler analysers (PDA) are typically chosen to measure the size and velocity of the droplets or bubbles. The PDA system is non-intrusive laser technique with high spatial and temporal resolution. Moreover, the PDA system does not require additional calibration as, for example, hot-wire anemometers. However, with a growing number of PDA users, there is a need for verification of the results among the workplaces and systems themselves. The current paper deals with the comparison of two fibre-based PDA systems. The main scope of the investigation is an evaluation of the system's age and the influence of lasers type. One of the systems is older, operated with Argon-Ion laser and the second one is brand new equipped with Diode-pumped solid-state lasers. Both have the same optics and were manufactured by the same company. Various properties of the PDA system are tested to show particular influence on the quality of results when measuring in a spray generated by a small air-blast atomizer.

1 Introduction

Optical and electronic technologies are, in general, well established in fluid mechanics. Nowadays trend increases demands on flow examination with a good spatial and temporal resolution. Laser Doppler anemometry (LDA) has been one of the most used non-intrusive techniques in the experimental fluid mechanics thanks to their small measurement volumes and fast processing enabling detailed flow examination in space and time [1]. The LDA enables point measurement of particles velocity in up to three directions. In the case of two-phase flows, a phase Doppler (PDA) analysers are frequently used to obtain information about a size of droplets or bubbles in the flow. The PDA system includes the LDA for velocity measurement and with an extension of measuring a phase difference in light signals hitting photomultipliers, it estimates a sphericity of a droplet or a bubble [2].

The LDA and PDA techniques underwent significant development during last few decades and play an important role in various research applications, such as combustion of liquid hydrocarbon fuels in automotive and aircraft engines or in power plants, in spray drying processes in inkjet print devices and others [1]. Advantages of LDA and PDA systems are good spatial and temporal resolution, directional response, sensitivity and no need for additional calibration in comparison with, for example, hot-wire anemometers.

Together with a rapid increase of experimental studies aimed at a flow investigation, there is also a need for verification of the experiments and the systems themselves. Moreover, the PDA technique involves numerous complex physical processes which require broad interdisciplinary understanding. There are many ways the system can be set which influences the results significantly. So in every literature research, there should be a question of what was the uncertainty of the measurement and the exact experimental setup and used technique of the previous results.

In a view of evaluation of LDA systems, the researchers focused on a determination of uncertainty of velocity measurements, such as [3] and [4]. In the case of PDA, a frequent aim was to evaluate mass flux [5] or particle concentration [6]. In general, the uncertainty of the droplet size measurements is assumed relatively good, typically in a range from 5 to 10% after amplitude and phase validations depending on the droplet size and optical configuration [7]. The situation is a bit more complicated in the case of real dense spray where more variables influencing the uncertainty of flux measurements are presented such as: obscuration effect, unwanted reflections, trajectory or Gaussian beam effects, slit effect etc. [2].

A number of studies were aimed to compare various sizing techniques e.g. [8–12]. The studies showed that the comparison is possible with relatively good agreement among the instruments. But it was shown that harsh conditions like an unknown refractive index of liquid or dense spray regions make the quality of results dependent on a given experimental setup and technique.

A comparison of various laser Doppler system is relatively rare. For example, [13] used two different PDA processors made by Dantec: Enhanced 58N50 and BSA P80 to measure LDA and PDA data simultaneously to improve system performance. Another study by [5] focused on an

* Corresponding author: matous.zaremba@wttech.cz

evaluation of measured mass fluxes by classic PDA and Dual PDA and Qiu and Sommerfeld (QS) on two different atomizers, i.e. the pressure swirl and an air-blast. They pointed out that the improved configurations of the PDA, Dual and QS systems, achieve a better estimation of flux measurements. However, due to a large number of influencing parameters, they made no universal explanation from the obtained results. Other researchers [14, 15] compared the PDA with optical patternator and showed relatively good absorptance agreement. Nevertheless, they also pointed out that the agreement depends on a drop arrival rate which influences the calculated mean diameter used for flux determination.

A recent study focused on a comparison of various configurations of laser systems, a fibre based PDA with an older-classic PDA construction [16]. This showed that there was an only minor deviation in velocities. But the different processors, Dantec P80 and older Dantec Model 58N50 used for data collection had a demonstrable effect on the measured drop size distribution. The main issue here was detectability of droplets with minimal and maximal detectable size. Especially the largest droplets have a significant influence on the spray properties such as flux or Sauter mean diameter, *D32,* used in an evaluation of sprays in combustion applications.

When comparing such a complex system like PDA, there are many variables which play a role in the results quality. There was a series of studies aimed at an evaluation of the measured mass flux within the spray. Especially, measurement of droplet size is critical in many applications. The recent study showed that even two similar PDA systems might measure demonstrable different droplet size spectrum. Thus, our current approach is to simplify the situation to the necessary minimum number of variables which might influence the results. For that purpose, we present a comparison of two identical fibre-based PDA systems which differ in age, used lasers and processors. The rest of components is identical for both systems. For the sake of simplicity, only 1D measurements were performed.

2 Materials and Methods

Experiments were performed at the cold test bench at still ambient conditions and room temperature at two workstations: the Spray laboratory at Brno University of Technology (system A) and at WTtech.cz company (system B).

2.1 Atomizer

The experiments were conducted using a standard twin-fluid air-blast atomizer (air-brush) which is normally used for paint spraying. This atomizer generates stable spray with small droplets, typically smaller than 50 microns. The spray is generated by suction of water into the high-speed air stream which disrupts the liquid into small droplets. This atomizing principle is basically the same as for the air-blast atomizers [17].

The tested liquid was distilled water, experiments were performed at room temperature of 23 °C and atmospheric ambient pressure. The air supply line was equipped with a pressure sensor (BD sensors 331). The pressurized air was taken from a small mobile compressor ABAC pro air compressor 50 Hp3. An air filter and valve were placed into the line behind the compressor. This system (nozzle, sensor, filter and compressor) was used in both experiments i.e. in systems A and B. Therefore, the identical operating conditions were achieved and number of variables which might play a role in the results quality was decreased. The pressure was kept at a given value of 190 ± 5 kPa.

The nozzle was mounted to 3D positioning system. Positioning error of the mount is less than 0.5 mm.

2.2 Principles of PDA technique

The laser Doppler techniques (LDA and PDA) are point-wise optical methods that capture a Eulerian description of the flow using droplet and bubbles or seeding particles. In principle, a continuous-wave monochromatic laser beam is split into two coherent beams that intersect symmetrically in a focal point of transmitting optics. The crossing beams forms the measurement volume, where interference generate series of parallel planes of higher and lower light intensities, so-called fringes. A particle presented in the flow can generate a signal when passing through the fringes in the measurement volume. In order to identify the particle direction, a frequency of one beam is typically shifted using Bragg cell. This principle solves the directional ambiguity and enables to measure the zero velocity. The generated light signal is then collected in the receiving probe using photodetectors which produces a "Doppler burst" signal with a Doppler frequency f_D. This frequency is proportional to the velocity component u_n of the measured particle according to the following relation [2, 18]:

$$f_D = f_B + \frac{u_n}{s} = f_B + \frac{2sin\left(\frac{\theta}{2}\right)}{\lambda} u_n \qquad (1)$$

where λ is the laser light wavelength. The signal is then processed to yield the velocity and arrival time of each measured and validated particle.

The PDA is extension of above described LDA technique and provide an estimation of the particle velocity and size simultaneously. The PDA transmitting optics is the same as for the LDA. However, the PDA uses more complex receiving probe which contains typically three photodetectors. One detector is located in the observation plane while the other two are elevated at an angle with respect to that plane. The signals collected at different angles show a phase difference with respect to each other. This phase difference is proportional to the particle diameter, when assuming spherical object, and it is also dependent on a scattering direction and difference in elevation angles between the detectors. Moreover, it might depend on a particle/fluid refractive index. Using the fringe model, it gives a simple relationship between the diameter and the measured phase shift $\Delta\varphi$:

$$\Delta\varphi = \frac{2\pi D_p n_p}{\lambda}\phi \qquad (2)$$

The parameter ϕ depends on the scattering mode and the PDA configuration. A detailed description can be found elsewhere [2].

2.3 PDA systems

A Two fibre-based PDA systems are described in this section. Generally, the PDA system consists of the laser, transmitting optics, receiving optics, processor and a computer. Schematic layout of the PDA system is in Fig. 1.

In the past, majority of the laser Doppler systems were operated using the gas lasers, namely the Argon-Ion or Helium-Neon. These lasers have the advantage of sufficient laser power output needed for PDA measurements and very good beam quality. Nowadays trend, however, is to use a diode and diode-pumped solid-state lasers which experienced a rapid development in recent years. New diode lasers with improved properties such as high-power and fundamental-mode emission enable usage in laser Doppler techniques [19], but with much higher energy efficiency compared to gas lasers.

Both systems used in this paper consist of the same type of the optical parts (transmitting and receiving optics and optical fibres) but they differ in the used lasers and processors. The older system (system A) has Argon-Ion multicolour laser and BSA P80 processor. In contrast, the newer system (system B) which is equipped with diode laser and processor BSA P800-2D. The laser beam is directed from the laser head into, so called, transmitter box which in the case of system A divides a multi-line laser beam into individual colours and provides a frequency shift to one of the laser beams in a pair. The system B uses a monochromatic laser and the transmitting box only splits the beam and provide the frequency shift. The frequency shift is the same for both systems and it is 40 MHz. The transmitter box also focuses the laser beams into the optical fibres which outcomes in the transmitting optics. From transmitting optics there is only one pair of laser beams coming out and intersecting in measurement volume, see Fig. 1. A detailed description of the individual components of the examined PDA system is in Table 1 below.

The scattering angle, the angle between X-axis and receiving optics, was set to 70° due to good Brewster condition. It means that the first order of refraction is dominant and other modes of light scattering on a droplet are minimised, a similar setup was used by [13].

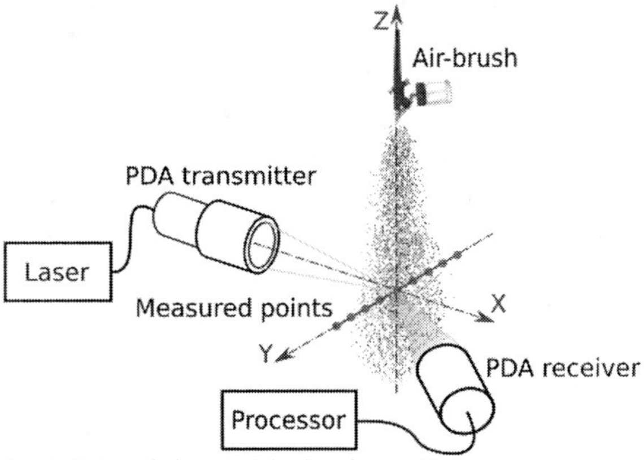

Fig. 1. Schematic layout of the fibre based PDA system.

Table 1. Parameters of the PDA systems.

Component	System A	System B
Laser type	Spectra physics Stabilite 2017 Argon laser	Quantel ELBA-M-546 nm DPSS laser
Wavelength	514.5 nm	546 nm
Processor	BSA P80	BSA P800-2D
Transmitter box	60X41 Transmitter, which splits the beam into its individual colour components, provides a frequency shift and directs the beams into fibres.	
Bragg cell	A cell is implemented inside the transmitter box which gives frequency shift 40 MHz to each beam from given pair.	
Transmitting optics	60X81 2D mm transmitting optics, the focal length of the final lens of 310 mm.	
Receiving optics	57X50 112 mm diameter fiber PDA receiver optics with a spatial filter and focal length of the lens of 500 mm. Slit aperture set to 200 μm.	

2.4 Parameters of the measurement volume

Two intersecting laser beams coming out from the transmitting optics create in the intersecting point so-called measuring volume (m.v.). This ellipsoid consists of layers of high and low intensities of light i.e. fringes. Dimensions of the m.v. depends on a current optical configuration which is the same for both examined laser systems, i.e. systems A and B. The optical setup results in a fringe spacing of 2.2 μm, dimensions of m.v. 0.05×0.05×0.4 mm and a number of fringes equals to 22. The effective volume of m.v. has a shape of an ellipsoid truncated by a slit shaped by a spatial filter placed in the receiver optics [2] as shown in the Fig. 2. We assume that m.v. is symmetric along the main axis. The effective length of the m.v. is determined by the slit used in the receiving optics. In both systems, the slit was set to 200 μm. The PDA receiver enables the usage of different aperture plates inside the PDA receiver. These apertures, so-called masks, are interchangeable and determine the droplet size range which can be measured. In both systems mask B was used.

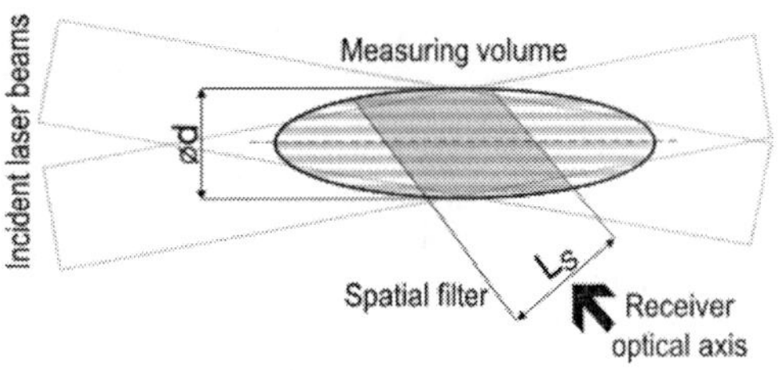

Fig. 2. Measurement volume geometry.

2.5 PDA setup

The scope of the present experiment is to map an influence of laser power of photomultipliers sensitivity on quality of results. The signal gain was set to zero value to limit noise which might be generated by the amplification. The only two variables were changed in the experiment: laser power in range from 25 mW up to 100 mW for each laser beam, and sensitivity of photomultipliers in range from 600 V to 1200 V. The rest of the variables which can be set in the software was left at a default value.

To control PDA measurement a PC with a given software is used. In addition to the experimental setup, there is a series of variables which might be sent via software and optimise the measurement conditions. In both systems, a BSA flow software version 6.5 by Dantec Dynamics was used.

As mentioned in the introduction, the main variables which are normally withdrawn from PDA data are droplets mean diameter, *D10*, and, so-called, Sauter mean diameter, *D32*. The *D10* is calculated as a mean value of the measured diameter of all recorded droplets in one position. The *D32* is the diameter of the drop whose ratio of volume to surface area is the same as that of the entire spray [17]. It is used for mass transfer, combustion and reaction application due to its sensitivity to the large droplets. The *D32* is calculated according to the following relation:

$$D32 = \frac{\sum N_i D_i^3}{\sum N_i D_i^2} \quad (3)$$

where *N* is a number of drops and *D* is a drop diameter.

3 Results and discussion

In this section the results are discussed to obtain a general information about the spray and to determine the best possible region for comparative measurements. Further, the comparison between the two system is made and concluding remarks are withdrawn at the end.

3.1. Preliminary results - spray examination

The preliminary measurements were performed to map the spray. Five axial distances downstream from the nozzle were examined; 10, 20, 30, 40 and 50 mm. Sample results are shown in the Fig. 3. Images: a), b), c) shows contours of the data rate in three planes. It is clear that in the plane closest to the nozzle exit, the highest frequencies, up to 8.5 kHz, are reached close to a centre of the spray. The data rate then decreases with growing axial distance. A plane at 50 mm from the nozzle then reaches a maximum of 900 Hz which points on a rapid decrease in the droplet concentration. This could be caused by the evaporation of droplets and by entrainment of air into the spray and consequent spray-air interaction [20].

A similar trend was observed in the case of mean axial velocities, Fig. 3 images: d) e) f), where the maximum values were measured closest to the nozzle, up to 75 m/s. The highest velocity is reached in a centre of the spray. The velocity then decreases with the axial distance and reaches values of 30 m/s at 50 mm downstream from the nozzle. Moreover, the velocity field gets more uniform which supports the assumption of the spray-air interaction mentioned above.

From the results above it is deduced that the best possible region for comparative measurements is close to the nozzle due to the high data rate and wide range of different conditions as results of high and low concentration of the droplets inside the spray, i.e. centre of the spray and the edges respectively. It is also evident that the spray is not perfectly symmetrical. Therefore, the results from both systems can be affected by positioning error of the nozzle with respect to m.v.

3.2 Preliminary results - repeatability

A series of nine measurements were performed to get the repeatability of the velocity and size measurements in the PDA system. One radial profile was measured from -5 to +5 mm in the Y-axis direction with a step of 1 mm. 20,000 samples were collected in each measured point or the measurement was finished after five seconds to save the measurement time.

Mean measured values of basic spray properties are shown in the Table 2. below. Repeatability of each property is calculated as a deviation from the mean value. The value of repeatability in the table is a mean value calculated from the measured points in a given profile. Average repeatability of the *D10* value is 0.27 %, the *D32* is 0.58 % and for the velocity is the average repeatability equal to 0.96 %. The repeatability is assumed to be within acceptable range. Thus, further we focus on testing selected parameters of the PDA.

3.3 Sensitivity

The effect of sensitivity was investigated by changing the voltage of photomultipliers from 600 V to 1200 V but with constant laser power of 25 mW. The velocity profiles in top part of Fig. 4 show that the lowest measured sensitivities (600 V and 700 V) yield demonstrably lover values than the rest of the regimes. This behaviour is the most obvious in the centre of the spray. The outer regions tend to show opposite trend i.e. the higher velocity is achieved at lower sensitivity and vice versa. These differences are linked with different measurable particle sizes, as discussed below. In the spray centre, where a dense spray region is located, the small particles are not measured using low sensitivity, see middle part of Fig. 4. The measured large particles have no time to accelerate enough to reach the velocity of the air flow. On the spray boundary, the effect is opposite. The velocity is smaller there and also the smallest droplets are rapidly decelerated by the surrounding stationary air. Since only the large droplets are measured in the case of 600 and 700 V, the velocity is higher there compare to the other sensitivities. The regimes with sensitivity above 1000 V show a small deviation from the main trend in the data.

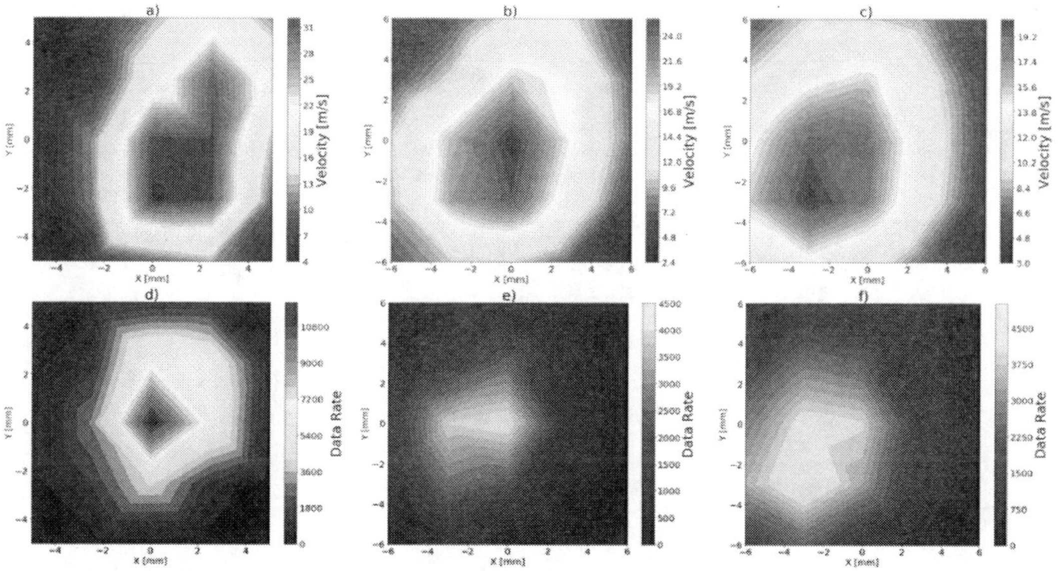

Fig. 3. Contours of the data rate in three axial distances from the nozzle: a) 10 mm, b) 20 mm, c) 30 mm. Bottom: Contours of mean axial velocity in three distances from the nozzle: a) 10 mm, b) 20 mm, c) 30 mm.

Table 2. System B, statistics made from 9 measurements.

Radial Position	*D10*	Repeatability of *D10*	*D32*	Repeatability deviation of *D32*	Mean velocity	Repeatability of velocity
[mm]	[μm]	[%]	[μm]	[%]	[m/s]	[%]
-5	9.25	0.23	21.4	0.66	16.40	0.69
-4	8.28	0.16	19.3	0.51	27.43	0.98
-3	7.77	0.11	18.9	0.27	43.12	0.84
-2	7.77	0.10	19.5	0.55	48.64	1.24
-1	8.16	0.15	21.9	0.54	43.29	0.91
0	8.47	0.13	23.8	0.50	34.69	0.90
1	8.73	0.20	23.9	0.70	27.82	1.29
2	9.20	0.20	24,0	0.72	22.13	1.17
3	10.33	0.30	24.9	0.59	16.35	0.82
4	11.78	0.50	27.1	0.61	13.17	0.78
5	14.00	0.91	30.2	0.75	10.83	0.98

The influence of the sensitivity on the diameter statistics is shown in the Fig. 4 below. The mean diameter, *D10,* shows a clear trend that the *D10* decreases with an increase in sensitivity. This is expected results because at low sensitivity only larger particles reflect sufficiently strong light signal which can be detected. When the sensitivity reach value of 1100 V then, even for higher sensitivities, the results are very similar. This means that for the value of 1100 V system reaches its limit for detection of the smallest droplets. Similar behaviour can be observed for the *D32* profiles, see bottom part of Fig. 4. The highest *D32* values are reached for the lowest sensitivity and vice versa. The sensitivity strongly influences the visibility of the smallest droplets. The results seems to be relatively stable when the sensitivity reaches value of 1100 V. When comparing the systems A and B it can be seen that only the results from 600 V differs demonstrably. System A reaches values from 34.4 to 40.5 μm and the System B reaches values from 39.6 to 57.4 μm.

3.4 Laser power

The influence of the laser power was investigated changing the laser power output from 25 mW to 100 mW using constant sensitivity of 1000 V. The velocity profiles are shown in a) and b) of **Chyba! Nenalezen zdroj odkazů.**. There is only a small influence of the laser power on the velocity profiles. This means that even the lowest laser power used in this experiment seems to be sufficient for detection of the smallest droplets in the spray and further increase does not change the results. Note here, that the laser power and sensitivity should act in the same way. The lowest laser power regime investigate was found sufficient in the previous chapter and further increase in laser power should result in even stronger signal.

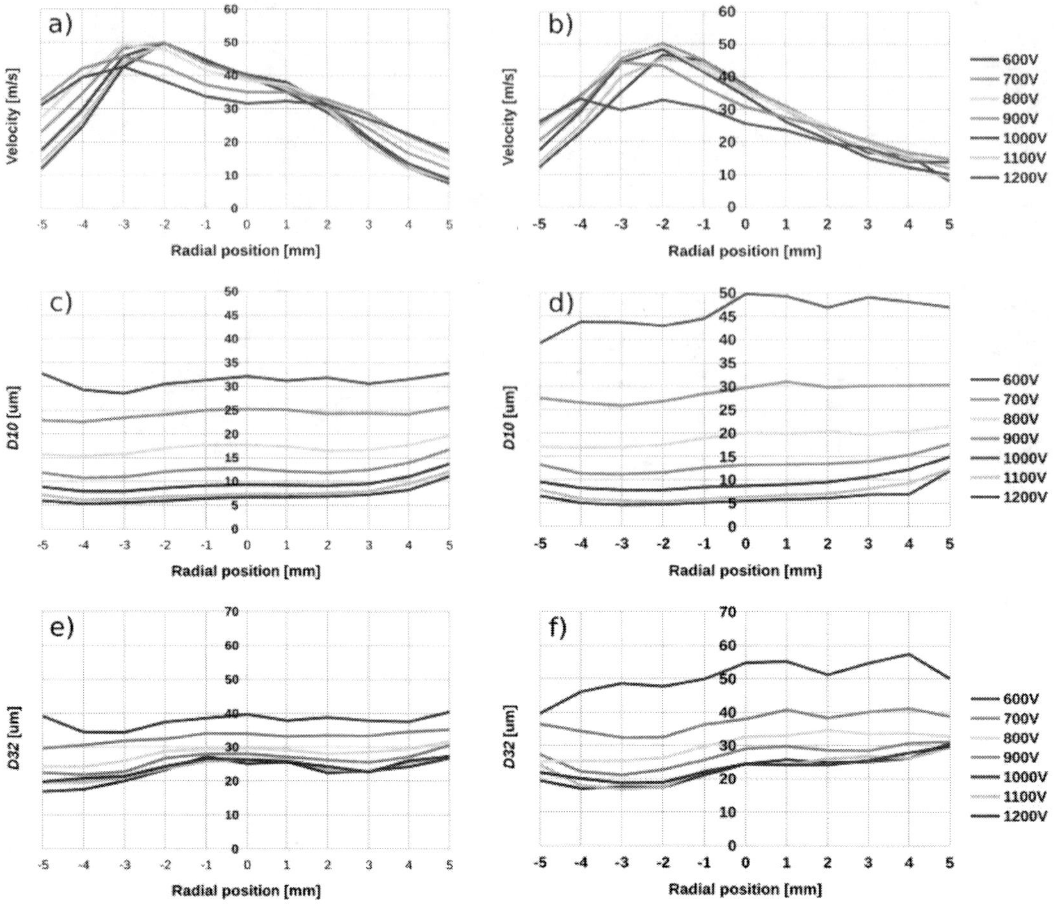

Fig. 4. Effect of the photomultiplier sensitivity: a), b) radial profiles of mean axial velocity in the axial distance of 10 mm from the nozzle, c), d) radial profiles of *D10* and e), f) radial profiles of *D32*. Left column a), c), e): system A. Right column b), d), f): system B.

A clear trend can be seen in the *D10* profiles c) and d) of Fig. 5. The value of *D10* decreases as the laser power increases for both systems. However, the *D32* profiles seems to be relatively unaffected by the laser power as shown on e) and f) of Fig. 5. There are only small variations in the *D32* profiles. Majority of the variations are within a range of 10 µm without any distinguishable trend. In contrast, the laser power has crucial influence on the *D10* profiles whereas the *D32* are relatively unaffected. This feature points on a fact that when the laser power is increased the mean diameter and also that the system is able to detect the smaller droplets. The portion of the largest droplets seems to be unaffected as illustrated by the *D32* profiles. This means that the laser power determines the mean value of diameter to which the system is focused but the number of largest particles remains relatively unchanged. This is due to the fact that the largest particles are more likely to be detected by the system than the small ones. It is also a manifestation of the effective measurement volume which is dependent on the droplet size [2].

Note here, that too strong signal oversaturate the photomultipliers and the signal is therefore rejected. This effect should tend to reduce *D32*, but it is not evident here.

The system A differs slightly in the regime of the lowest laser power (25 mW). It reaches higher values of *D10* comparing the system B. The values are in a range of 9 to 11.8 um.

3.5 Diameter statistics

For a global evaluation of results an integral value of D32 is used so called Integral Sauter mean diameter (ID32) [21]. This value is used for global evaluation of spray especially for combustion applications when analysing PDA data. It can be calculated from the following relation:

$$ID32 = \frac{\sum_{i=2}^{n}(r_i \cdot D_{30,i}^3 \cdot f_i)}{\sum_{i=2}^{n}(r_i \cdot D_{20,i}^2 \cdot f_i)}$$

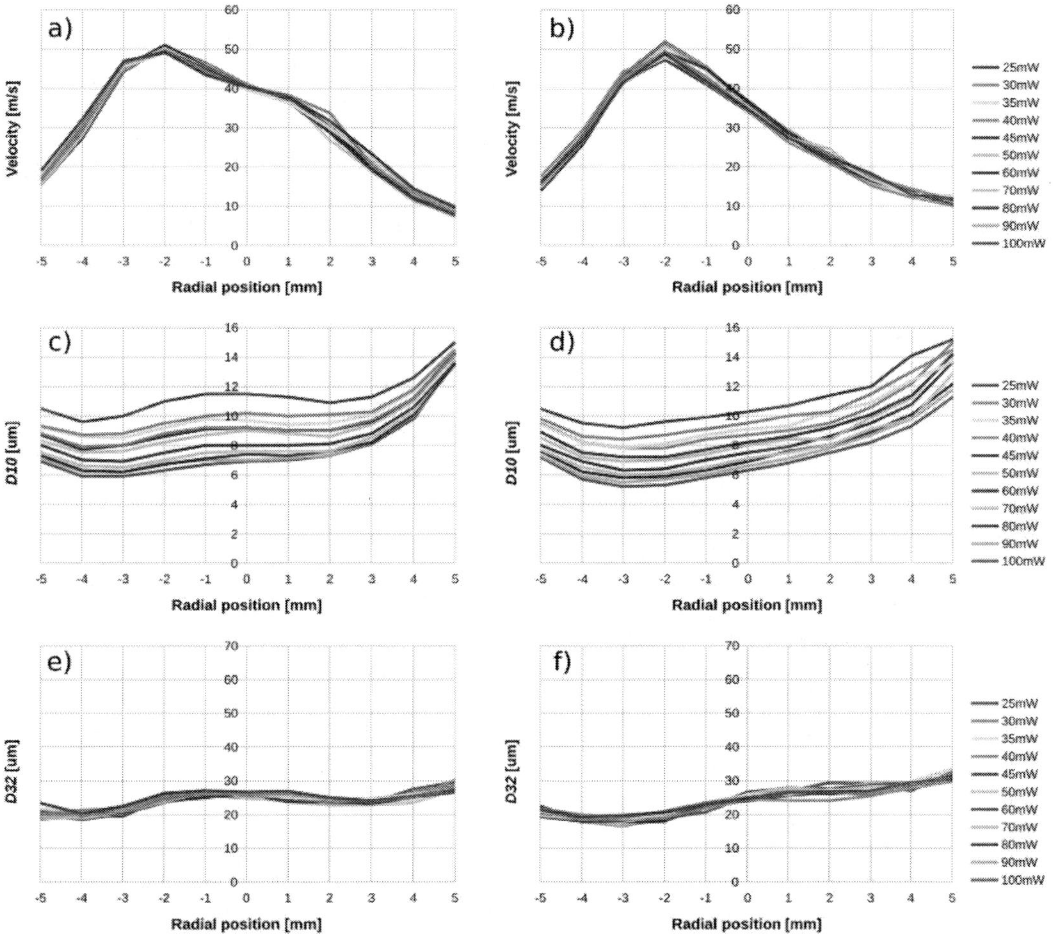

Fig. 5. Effect of laser power. From top: radial profiles of mean axial velocity in the axial distance of 10 mm from the nozzle, radial profiles of *D10* and radial profiles of *D32*. Left: system A. Right: system B.

where r is the radial distance from the centre of the spray, D is a droplet diameter and f is a mean data rate (number of measured droplets per second) at a given point.

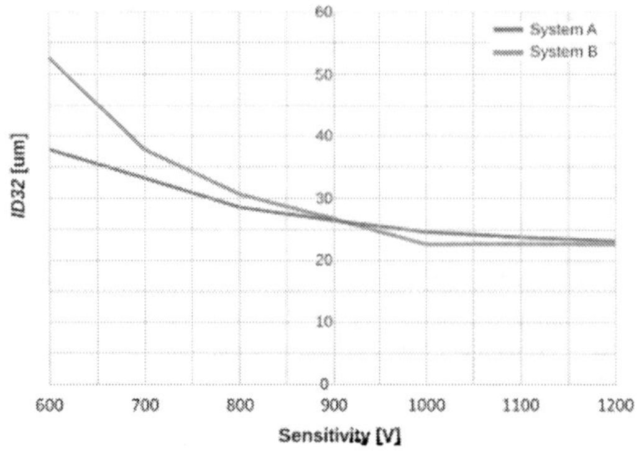

Fig. 6. Systems compared by ID32 varying sensitivity of photomultipliers.

A clear trend can be seen when comparing ID32 and sensitivity of both systems, see Fig. 6. The ID32 grows when sensitivity decreases. This corresponds to findings mentioned in the sections above. Low sensitivity means that large particles are detected and the small ones do not reflect sufficiently strong signal to be measured. Interestingly, the ID32 values are nearly the same for both systems. The largest difference in ID32 is 14.7 which corresponds to 5.8 % from the measurement range. This difference is at the lowest sensitivity, 600 V. The difference between the systems then decreases with growth of sensitivity. At the sensitivity of 800 V and higher the difference is very small, below 1 % from the measurement range. Thus, from this threshold of sensitivity the systems show nearly the same results in quantitative evaluation of droplet size.

4 Conclusion

Two similar PDA systems were compared under various laser power and photomultiplier voltage - sensitivity. The older system used Argon-Ion laser and Dantec BSA processor P80 while the newer one used a diode-pumped solid state laser and P800 processor.

The repeatability error of both system was found to be less than 1 %.

Low sensitivity setups were unable to measure the smallest droplets and therefore the droplet mean diameter, *D10*, and Sauter mean diameter, *D32*, were overestimated. This effect diminishes for sensitive higher than 1000 V. Note here, that this results is valid for 25 mW of laser beam power.

Increasing the laser power above this value generate stronger signal, but its effect on *D32* and velocity profiles is negligible. Only *D10* was affected as it decreases with increasing laser power output because stronger signal allows the smallest droplets become measurable.

When both PDA systems compared, differences are less than 5 % for sensitivities > 900 V and all the laser power investigated.

The proper PDA setup should be tailored for a given application. Higher sensitivity and laser power allows to measure very small droplets, but their effect on *D32* and velocity is usually small. If the spray contains wide distribution of droplet sizes, too high laser power may lead to oversaturation of photomultipliers and underestimation of *D32*.

The authors acknowledge the financial support from the project No. 18-15839S funded by the Czech Science Foundation and the project "Computer Simulations for Effective Low-Emission Energy Engineering" funded as project No. CZ.02.1.01/0.0/0.0/16_026/0008392 by Operational Programme Research, Development and Education, Priority axis 1: Strengthening capacity for high-quality research.

References

1. P. Buchhave, L. Lading, and G. Wigley, *Optica Diagnostics for Flow Processes* (1994)
2. H. E. Albrecht, N. Damaschke, M. Borys, and C. Tropea, *Laser Doppler and Phase Doppler Measurement Techniques* (2013)
3. J. W. Czarske, Optics Letters, 21, 5 (1996)
4. M. Annoni, Journal of the International Measurement Confederation, **45**, 11 (2012)
5. K. Dullenkopf, M. Willmann, S. Wittig, F. Schone, M. Stieglmeier, C. Tropea, and C. Mundo, Particle & Particle Systems Characterization, **15**, 7 (1998)
6. E. Calvo, J. A. Garcia, I. Garcia, and L. A. Aisa, Experiments in Fluids, **47**, 19 (2009)
7. Y. Hardalupas and C. H. Liu, Progress in Energy and Combustion Science, **23**, 23 (1997)
8. L. G. Dodge, Appl. Opt. **26**, 13 (1987)
9. L. G. Dodge, D. J. Rhodes, and R. D. Reitz, Appl. Opt. **26**, 9 (1987)
10. T. Berg, J. Deppe, D. Michaelis, H. Voges, and S. Wissel, ICLASS 2006, (2006)
11. N. Fdida, J. B. Blaisot, A. Floch, and D. Dechaume, Atomization and Sprays, **20,** 22 (2010)
12. N. Fdida and J. B. Blaisot, Measurement Science and Technology, **21** (2010)
13. G. Wigley, M. Goodwin, G. Pitcher, and D. Blondel, Experiments in Fluids, **36**, 10 (2004)
14. A. R. Muliadi, P. E. Sojka, Y. R. Sivathanu, and J. Lim, IMECE, **6**, 12 (2007)
15. A. R. Muliadi, P. E. Sojka, Y. R. Sivathanu, and J. Lim, Journal of Fluids Engineering, **132** (2010).
16. M. Zaremba, M. Maly, J. Jedelsky, and M. Jicha, EFM (2016)
17. A. H. Lefebvre, *Atomization and sprays* (1988)
18. F. Lizal, J. Jedelsky, K. Morgan, K. Bauer, J. Llop, U. Cossio, S. Kassinos, S. Verbanck, J. Ruiz-Cabello, A. Santos, E. Koch, and C. Schnabel, Eur. J. Pharm. Sci., **113**, 37 (2018)
19. J. W. Czarske, Measurement Science and Technology, **17**, 21 (2006)
20. S. D. Sovani, P. E. Sojka, and A. H. Lefebvre, Progress in Energy and Combustion Science, **27**, 39 (2001)
21. Jedelsky, J. , Jicha, M. , Slama, J. , Otahal, J. , 2009. Development of an effervescent at- omizer for industrial burners. Energy Fuels 23 (Dec), 6121–6130 .

Flow inside impeller sidewall gaps and disk friction of radial centrifugal pump

Lucie Zemanová[1,*], *Pavel* Rudolf[1], *Andrey V.* Naumov[2], and *Alexander V.* Volkov[2]

[1]Victor Kaplan Dept. of Fluid Eng., Faculty of Mech. Eng., Brno University of Technology, Technická 2, 616 69 Brno, Czech Republic [2]National Research University Moscow Power Engineering Institute, Moscow 111250, Russia

Abstract. Disk friction losses can significantly influence the overall pump efficiency. It is most pronounced in the case of machines with low or moderate specific speed, which has to be kept in mind during a design process of a pump. However, analyses using computational fluid dynamics (CFD) often completely neglect the sidewall gaps geometry. Correctly solved flow in the sidewall gaps can have a great impact on the whole flow pattern inside the pump and therefore it can considerably influence the calculated efficiency. The paper is focused on the precise simulation of the flow in the spaces between impeller shrouds and casing. CFD study using embedded Large Eddy Simulation (LES) approach in order to achieve accurate insight into the flow in sidewall gaps and reasonable computational time was performed. Obtained results are compared with data from experimental measurements.

1 Introduction

The disk friction losses are generated on a hub and a shroud of an impeller rotating in fluid inside a casing. It is caused by shear stress appearing on the surfaces. It can have a major impact on hydraulic efficiency drop especially in case of low and moderate specific speed pumps. According to [1], at specific speed $n_s = 10$, the disk friction losses consume typically to about 50 % of the useful power.

The problem of the disk friction was first studied on simplified cases, see Fig. 1 left. Plain disk rotating in cylindrical casing was extensively experimentally studied by Daily and Nece [2]. The friction moment can be expressed as:

$$M = c_m \frac{1}{2} \rho \omega^2 D^5 \qquad (1)$$

where M is the friction moment, c_m is dimensionless moment coefficient dependent on flow regime, ρ is density of the fluid, ω is angular velocity of the disk, D is the outer diameter of the disk. Considering Reynolds number (Re), the gap width h and the disk diameter D, four flow regimes can occur:

- Regime I: laminar, merged boundary layers
- Regime II: laminar, separated boundary layers
- Regime III: turbulent, merged boundary layers
- Regime IV: turbulent, separated boundary layers

For each of the regimes, moment coefficient determining friction can be expressed by different equation [2]. Later, the study was expanded to include also an influence of surface roughness of the disk and the casing [3].

It was found out, that a presence of throughflow can change the friction in a significant way. Coren et al. [4] demonstrated it on a set of experiments with open cavity. For low level of throughflow, the flow structure inside the cavity is dominated by rotation. With increasing flow rate Q, the tangential velocity decreases, boundary layers become thinner leading to increase of the disk friction. Hu et al. [5, 6] dealt with the influence of the direction of the throughflow and correlated it to particular four flow regimes.

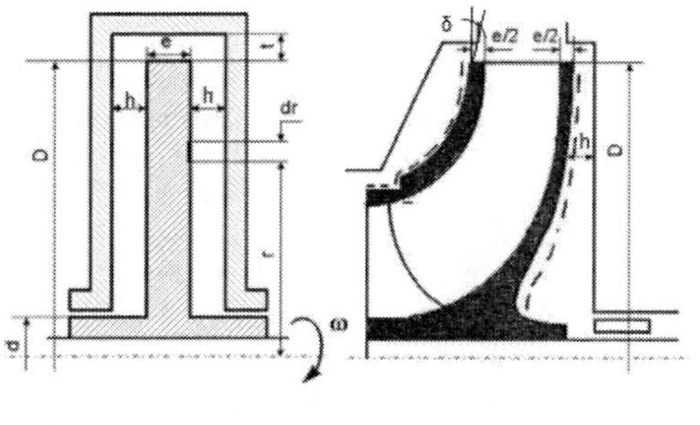

Fig. 1. Geometry of the simplified case and real impeller [8].

Although the governing parameters for disk friction losses were identified, the derived equations are valid only for simple disk geometry and their usability in turboma-

*e-mail: lucie.zemanova1@vut.cz

chinery applications is limited. Güllich [1] described variables on impeller, which plays decisive role in disk friction losses in real pumps, as is shown in Fig. 1 right. Effect of Reynolds number, surface roughness and axial gap in connection with realistic shapes used in turbomachinery were studied, e. g. in [7–9]. Beside the mentioned parameters, disk friction is influenced by the shape of the stator and sidewall gaps, leakage flow, partload recirculation and exchange of momentum, which can be altogether hardly expressed by a single formula [1].

However, in these days, we are able to use methods, which can overcome the problem and all mentioned phenomena can be computationally captured using CFD. Not only friction losses (efficiency) calculations, but detailed insight into flow patterns in the whole pump domain can be obtained by CFD. Considering the main pump fluid regions, the methodology of such analyses is described in many studies, e.g. [9–13].

In the vast majority of the simulations of flow inside the pumps, models based on Reynolds-Averaged Navier-Stokes equation (RANS) are used due to its optimal ratio of sufficient accuracy and computational resources. SST k-ω and k-ε Realizable are the most common and in many cases justified turbulence models. The most of the CFD studies neglects the influence of sidewall gaps on the global flow and the regions of front and back side disk gaps are not involved into the simulation at all, as can be seen in [9–13]. The studies, where it is included [14, 15], do not show any special approach in order to resolve the flow in those regions. Moreover, the mesh in the narrow gaps has often low quality and no attention is paid to these domains, although it was shown, that it can significantly change the performance of the pump [15].

The flow in the sidewall gaps can be very complex. Dependent on the dimensions of the gap (height h, diameter D), and operating conditions (angular velocity ω, flow rate Q), four mentioned flow regimes can occur. Dimensionless numbers used for description of the regimes are aspect ratio G:

$$G = \frac{h}{D} \tag{2}$$

and rotational Reynolds number Re:

$$Re = \frac{\omega D^2}{\nu} \tag{3}$$

Additionally, several types of vortex structures can emerge in boundary layers during transitions from one regime to another. Complex overview of it including conditions where it appears published Launder et al. [16]. It can take a form of spirals rotating in the same or opposite direction than the disk, concentric circular rolls or turbulent spots, as describes [17, 18] and shows Fig. 2.

It has to be pointed out, that modelling of the flow between stationary and rotating disks is quite challenging task for CFD. Due to the complexity of the flow patterns, achieving a realistic description is not a trivial task. There were many approaches to calculate velocity components inside the rotating cavity. Chew [19] used k-ε turbulence model with wall functions, Poncet at al. [20] solved the flow by SST k -ω. Promising results described Elena and Schiestel [21] using RSM. Soghe at al. discussed the differences between results obtained by low- and high- Reynolds approach of modelling near wall regions. More detailed studies were based on direct numerical simulation (DNS), e. g. [22, 23], or large eddy simulation (LES) [24, 25].

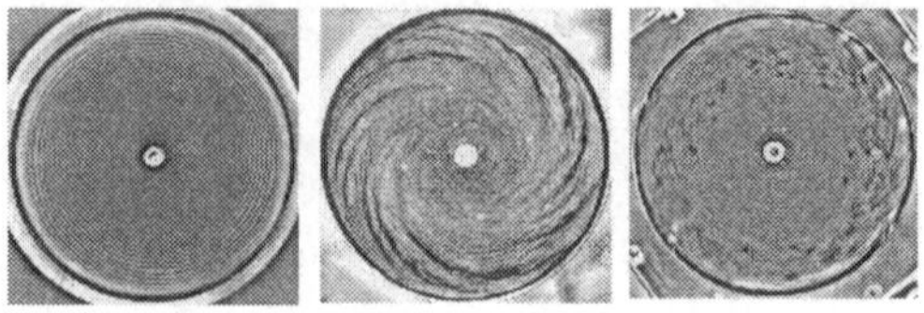

Fig. 2. Different types of unstable structures [17].

Capabilities of particular turbulence models to describe the flow between stationary and rotating disk were published in a previous study [26]. It was found out, that simpler and less computationally demanding RANS models can provide a reasonable solution for base flow, however, they are not able to capture instabilities in boundary layers. Even low-Reynolds approach and hybrid models (Detached Eddy Simulation, Scale Adaptive Simulation) were not able to resolve the boundary layers correctly. Comparison of radial and tangential velocities revealed, that all models are compliant in the core of the fluid. Nevertheless, significant discrepancies can be observed in the near wall regions. These areas are important for the onset and spreading of instabilities. Above that, velocity profiles in boundary layers are crucial for precise determination of the dissipation on the shroud and the hub (disk friction losses).

Following study demonstrates CFD analysis of centrifugal pump with emphasis on the flow inside impeller sidewall gaps. Based on previous knowledge, LES approach is necessary to resolve the boundary layers in those regions correctly. Using LES in the whole domain of the pump would require extreme computational resources. In order to obtain accurate results in reasonable time, embedded LES approach combining Large Eddy simulation in the sidewall gaps and RANS model in the rest of the fluid was used.

2 Numerical simulation - simplified rotating cavity

Before the calculation of the whole pump, flow inside simplified disk cavity with througflow was solved to verify embedded LES approach and its capability to detect instabilities on simpler geometry. The findings were then applied on the complex geometry of the radial pump. All simulations were done using ANSYS 19.1 software, meshes were created by ANSYS Meshing.

2.1 Geometry and mesh

The geometry of the test case was based on the same geometry of rotating cavity as in experimental study [27] and preliminary numerical investigation [26], where instabilities were observed. The main dimensions were kept, inlet and outlet non-rotating domains were added, as shows Fig. 3. Due to the symmetry of boundary conditions and the resulting flow, it was possible to reduce the domain to the fraction to save computational resources [27], therefore, 1/8 of the geometry was taken into account.

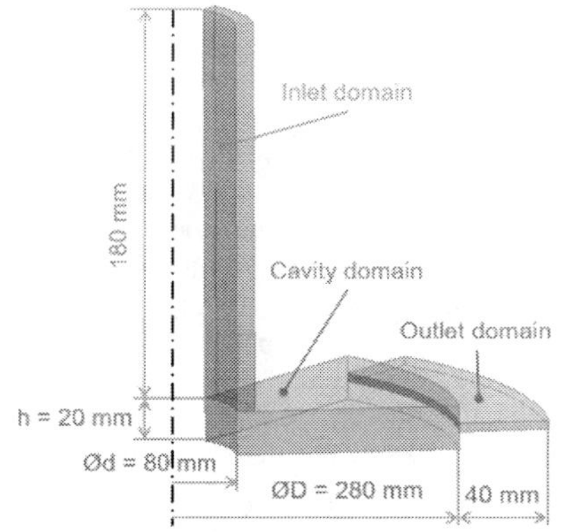

Fig. 3. Geometry and dimensions.

The mesh meeting requirements for LES was generated in the cavity domain. The spatial resolution was (200 × 80 × 120) elements in radial, tangential and axial direction with refinements near walls to comply $y^+ \simeq 1$. For inlet and outlet domain, high-Reynolds turbulence model was applied, thus the requirements on element sizing were not so strict. Refinements in near wall regions were applied to achieve $y^+ \simeq 30$. The whole mesh was conformal, made up of 1 960 000 hexa elements.

2.2 Boundary conditions

Boundary conditions are schematically shown in Fig. 4. There were 3 cell zones, as mentioned before.

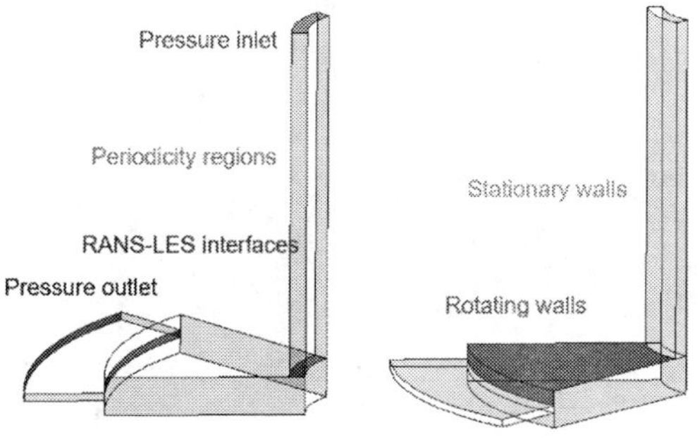

Fig. 4. Boundary conditions.

Inlet and outlet domain were RANS zones, while cavity domain was LES zone. Therefore, RANS-LES interfaces needed to be set up to connect them. These interfaces are conformal and are marked in red in Fig. 4. Grey regions represent periodic boundary condition. Pressure inlet and pressure outlet with gauge pressure = 0 Pa were applied on the borders of the domain. No-slip boundary condition was applied to all of the walls of the cavity. Upper part (violet) was rotating, whilst the rest of the walls was stationary. To achieve unstable structures in the boundary layer, for the cavity with given dimensions (aspect ratio $G = 5$), Reynolds number has to be $Re = 4 \times 10^5$. It corresponds to angular velocity $\omega = 20.4082$ rad.s^{-1}.

2.3 Computational details and solver settings

The default model for liquid water from ANSYS library was used. It was considered incompressible. The calculation was in the first step run as steady-state with k-ε Realizable turbulence model in the whole domain. For pressure-velocity coupling, SIMPLE scheme was used with the first orders of accuracy for advection terms in all transport equation. Non-equilibrium wall functions were used to resolve boundary layer regions. After achieving the convergence, simulation was switched to transient. Embedded LES approach was applied by changing the cavity domain to LES zone, while the rest of the domain was kept RANS. Embedded subgrid-scale model WALE was used with bounded central differencing momentum spatial discretization scheme. The accuracy was switched to second orders for all transport equations and pressure-velocity coupling algorithm was changed to PISO according to recommendation for unsteady cases. This procedure was tested and proved in previous study [26]. With respect to physical nature of the unstable structures and previously reported numerical results [26, 27], time step was set to $\Delta t = 1 \times 10^{-5}$ s.

2.4 Results

The ability to capture coherent vortices in boundary layers of rotating cavity by embedded LES approach was confirmed. Q-criterion was used for visualisation. This metric is commonly used, e. g. in [24, 25], to track unstable flow structures and it was also successfully incorporated in previous study [26]. As shows Fig. 5, so-called Type II instability emerging in form of spirals can be clearly observed and therefore it was proved, the approach can be used for precise modelling of the flow in sidewall gaps of centrifugal pump.

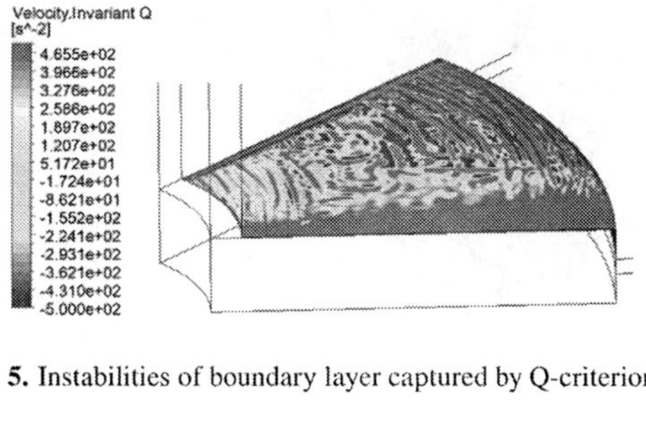

Fig. 5. Instabilities of boundary layer captured by Q-criterion.

https://doi.org/10.1051/epjconf/202226901072

3 Numerical simulation - radial centrifugal pump

The flow in centrifugal radial pump was defined by parameters in Tab. 1. It belongs to the best efficiency point (BEP).

Table 1. Parameters of the pump at BEP.

Number of blades z (–)	6
Average value of blade angle β_2 (°)	60
Impeller diameter D_2 (m)	0,174
Flow rate Q ($l \cdot s^{-1}$)	31
Specific energy Y ($J \cdot kg^{-1}$)	329
Rotational speed n (s^{-1})	48,3
Specific speed n_s (s^{-1})	128

3.1 Geometry and mesh

The inner geometry of the pump was divided into 5 domains, which are shown in Fig. 6. They were meshed separately and then merged within preprocessing in Fluent.

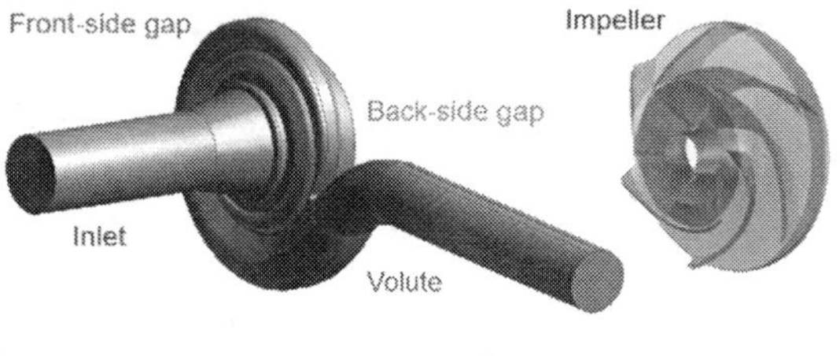

Fig. 6. Instabilities of boundary layer captured by Q-criterion.

Inlet domain was meshed using ANSYS Meshing with fully hexa mesh, wall refinements were set in order to use RANS turbulent model there. Impeller geometry was made using BladeGen and meshed by TurboGrid meshing tool. Volute consisted of tetra mesh with prismatic boundary layer and was created in ANSYS Meshing. Sidewall gaps were meshed in ANSYS Meshing as well with hexa elements and special impact on quality of the mesh, since $y^+ \simeq 1$ was necessary to obtain in order to use LES in those regions. The whole mesh was merged by 5 non-conformal interfaces and consists of approximately 14 500 000 cells. Some details can be seen in Fig. 7.

3.2 Boundary conditions

No-slip boundary condition was applied to the walls. The fluid enters the domain in mass flow inlet which is shown in Fig. 8, where for BEP mass flow rate corresponds to 31 $kg \cdot s^{-1}$. The domain of impeller rotates, while the others are stationary and only walls adjacent to the impeller are treated with rotational wall boundary condition. The rotational speed of the impeller and surrounding walls was set to n = 2900 rpm. There are two pressure outlets according to Fig. 8. The first belongs to the volute, where p = 434 793 Pa obtained from measurement was prescribed. The second pressure outlet with atmospheric pressure is in the shaft region of the backside gap domain. The individual domains are coupled together by interfaces. Sliding mesh type of interface was used in inlet - impeller and impeller - volute regions, while at volute - frontside gap, volute - backside gap and frontside gap - inlet, RANS-LES interfaces were established.

3.3 Computational details and solver settings

Material properties of liquid water were used from ANSYS Fluent library as well as in previous case, it was treated as incompressible fluid. In RANS domains, k-ε Realizable turbulence model was set, front- and back-side gaps of special interest were computed by LES. WALE subgrid-scale model recommended for embedded LES approach was used. Common practice in CFD in turbomachinery is to set such a time step, which ensures rotation of max 2° per Δt. The time step was finally even lower($\Delta t = 1 \times 10^{-6}$ s) in order to fullfil the Courant-Friedrichs-Lewy (CFL) condition. The transient analysis was initialized from converged steady results. First orders of accuracy were first set and after achieving convergence for transient case it was switched to the second orders.

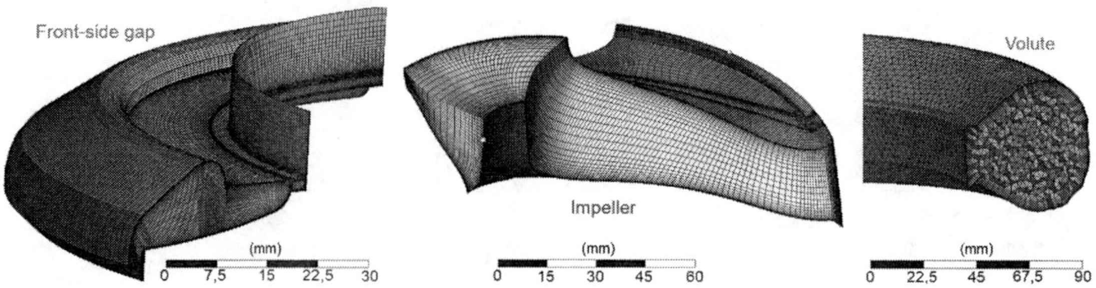

Fig. 7. Details of mesh.

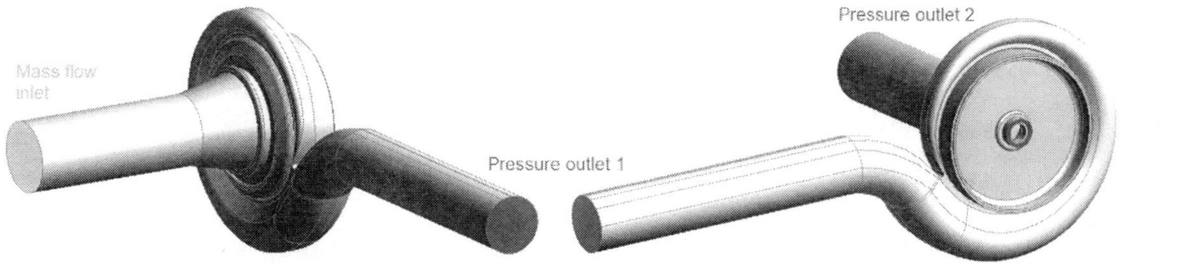

Fig. 8. Boundary conditions.

3.4 Results

The flow inside the pump is very complex and moreover, changes in time with rotation of the impeller. Pressure and velocity in particular point of front-side or back-side gap domain depend on current position of blades. Typical example of overall velocity field in one meridional section shows Fig. 9.

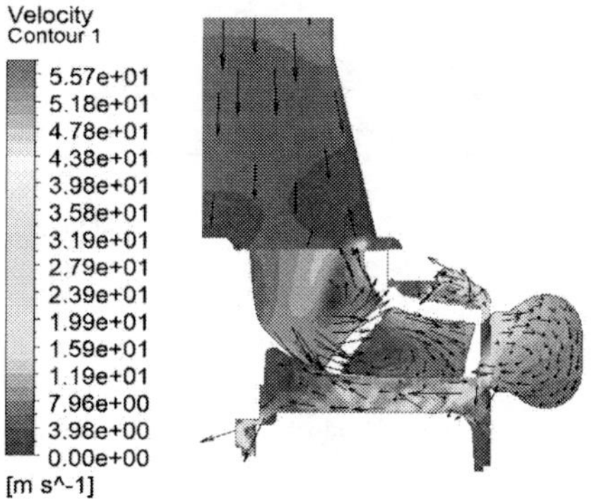

Fig. 9. Vectors of velocity in meridional section.

As a consequence of shape of pump spiral casing, blades passing the volute cutwater produce pressure pulsations. They are well observable on pressure contours on the front-side gap wall. Another effect is that in some circumferential locations, the water flows into the sidewall gaps, whereas in other positions flow of opposite direction occurs. It also causes very diverse velocity field at the entrance to the sidewall gap. Fig. 10 demonstrates it on the the velocity vectors and pressure distribution at the interface between the impeller domain and front-side gap domain.

As the impeller rotates, the direction of the flow into and out of the sidewall gaps changes periodically with passing blades. Fig. 11 shows the changes during 8 time steps corresponding to 1/6 of impeller revolution (1 blade passes the volute cutwater).

The uneven velocity around the entrance into the front-sidewall gap results in very complex flow field in this region. Strong vortical structures of various shapes emerge in the whole domain, as shows Fig. 12.

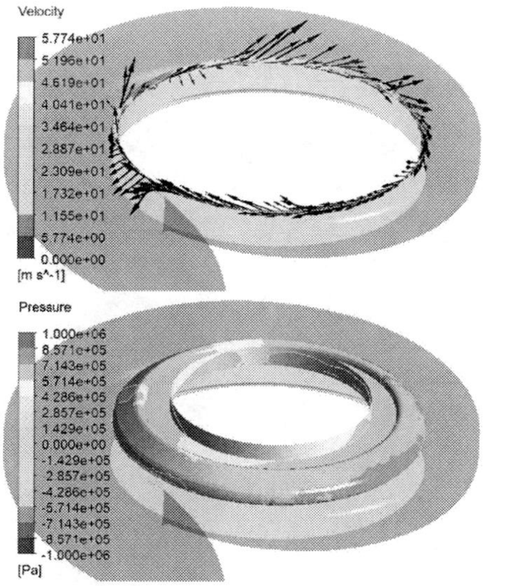

Fig. 10. Vectors of velocity at the interface and pressure distri-bution in front-side gap.

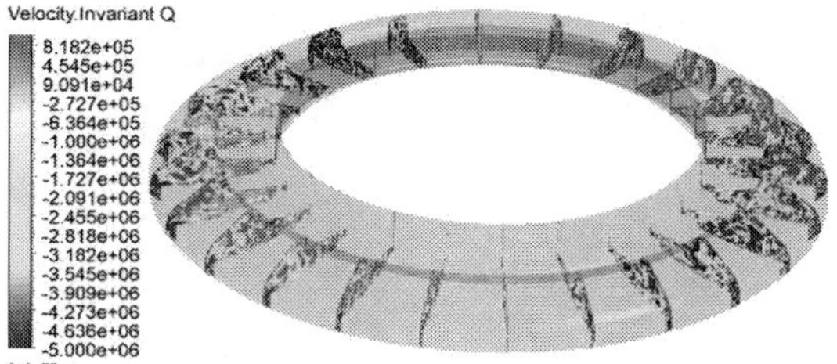

Fig. 12. Q-criterion in front-side gap.

Furthermore, the flow field evolves in time with rotation of the impeller. Fig. 13 shows evolution of radial and tangential components of velocity measured in in the middle of the cross section of the front-side gap. Typical S-shaped radial and tangential velocity profiles known from simplified disk in cylindrical cases are highly distorted due to the complicated shape of the spiral casing.

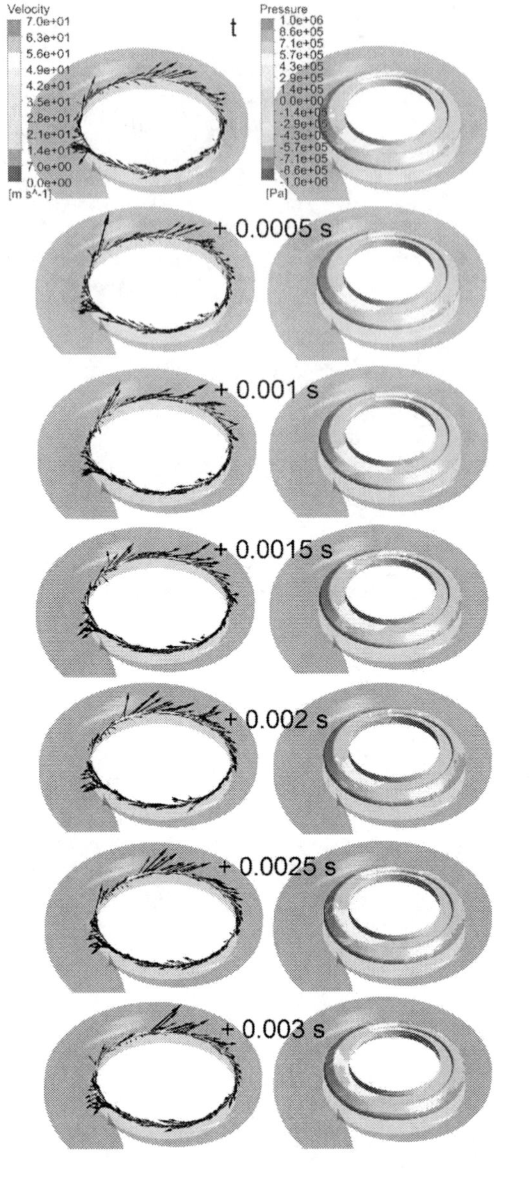

Fig. 11. Velocity and pressure evolution during 1 blade passage.

Analogical situation appears in back-side gap. Again, the flow field is uneven and is strongly influenced by position of blades and volute cutwater. Depending on the circumferential position, the water either flows in or out of the gap, as can be seen in Fig. 14 top. The rotation of the impeller is also accompanied by non-constant pressure distribution and pressure pulsations in time, as shows Fig. 14 bottom.

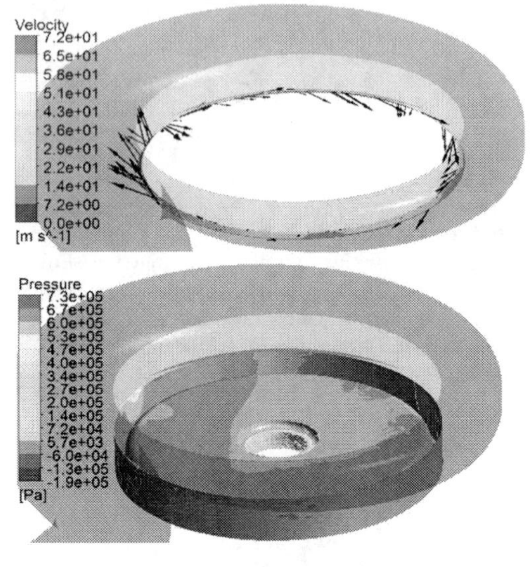

Fig. 14. Vectors of velocity at the interface and pressure distribution in back-side gap.

Such operating conditions result in strong whirls in back-side gap, as documents Q-criterion visualisation in Fig. 15. Taking into consideration only outer diameter, gap width and angular velocity of the impeller, according to stability diagram by Daily and Necce [2], flow regime IV with turbulent separated boundary layers should occur in both sidewall gaps, transitions or vortices should not be visible. However, the results suggests, that this approach would be unjustified simplification for the case of real impeller rotating in spiral casing. The changes of velocity in circumferential direction due to shape of casing and in time as a consequence of impeller rotation lead to more complicated evolution of the flow field in sidewall gaps.

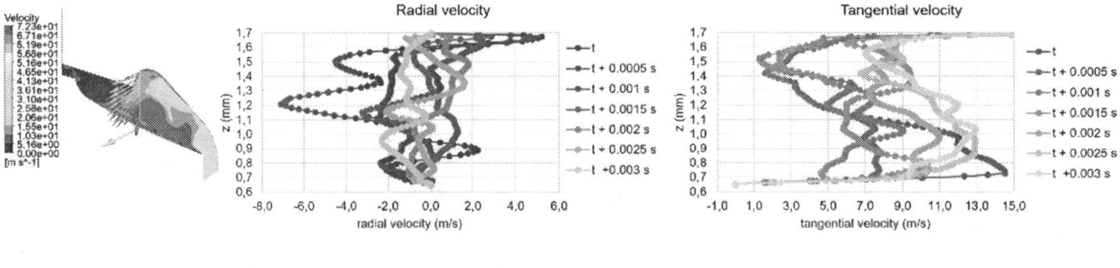

Fig. 13. Radial and tangential velocity in time during 1 blade passage.

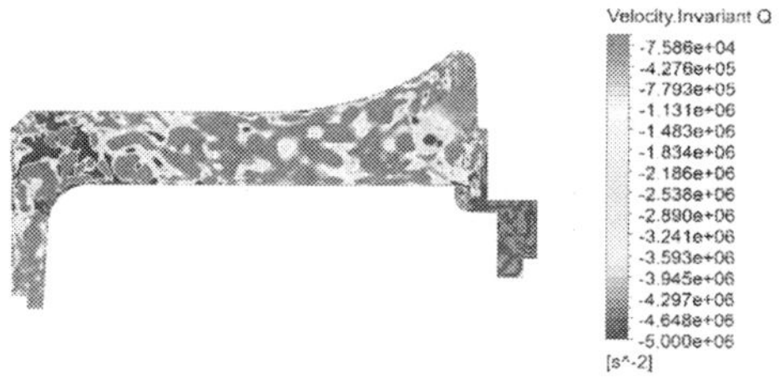

Fig. 15. Q-criterion in back-side gap.

During experiment, pressure in 3 radial positions was measured using pressure probes. Their locations are shown in Fig. 16, where can be seen comparison with CFD results as well. Measured and calculated values summarizes Tab. 2. Quite large discrepancy is apparent. It is probably caused by different procedure of evaluation experimental and calculated data. Experimental data were collected for a period of time (several impeller turns) and average value was taken into account, while due to lack of numerical data for such a time interval, average value from circle of given radius was considered. The trend is similar, however absolute values differ significantly.

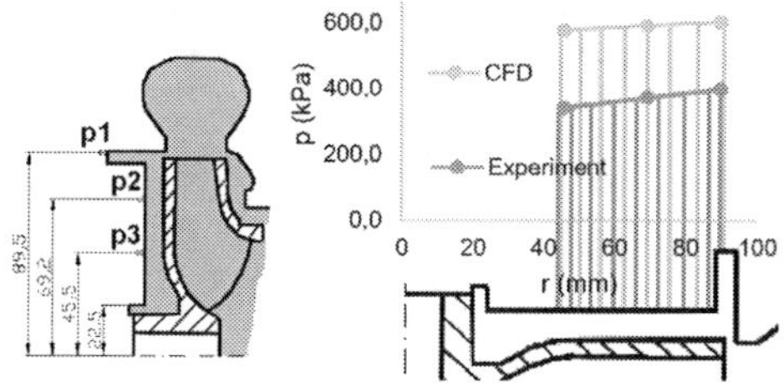

Fig. 16. Pressure acting on the back sidewall gap.

Table 2. Comparison of calculation with experiment.

	Experimental (time averaged)	**Simulation** (domain averaged)
p1 (kPa)	397.9	602
p2 (kPa)	372.8	591.7
p3 (kPa)	340.3	578

The insight into complex flow inside front and back-side wall cavities can be used with benefits for precise determination of disk friction. Comparison of friction moments on hub and shroud calculated using empirical equations and CFD simulation shows Tab. 3. The CFD simulation was performed without consideration of surface roughness, thus smooth surface was taken into account also in empirical calculations in order to compare the results.

Table 3. Comparison of friction moments.

	Empirical	**Simulation**
M shroud ($N \cdot m$)	0.493	0.5
M hub ($N \cdot m$)	0.422	0.235
M total ($N \cdot m$)	0.915	0.735

Very good agreement is achieved in terms of disk friction on shroud. It probably results from relative geometry simplicity of back-side gap, which is close to the model example of disk against flat casing. The gap width is constant in the major part of the domain. The cross-section of front-side gap is more complicated, the gap width is not constant and the shape is not very similar to the geometry for which the empirical relationships were derived. Therefore, the resulting discrepancy is larger. It is then pronounced into the total friction moment.

4 Conclusions

Using of CFD for precise determination of flow field in sidewall gaps was demonstrated on practical application of radial pump. It was proved, that Embedded LES approach is capable of capturing swirling flow in narrow gaps between rotor and stator not only for simplified cases, but also when considering the real pump geometry.

The results suggest, that the flow in sidewall gaps is quite complex and unstable. Fully transient analysis is required in order to describe the flow field sufficiently. The velocity and pressure changes in time with rotation of the impeller and also in space, which excludes the use of periodic boundary condition for subsequent analysis. To obtain proper results, non-axisymmetry of the volute is necessary to take into account.

It was shown, that as the impeller rotates, the direction of flow into and out of the both sidewall gaps periodically changes with passages of blades through area of volute cutwater. It also causes pressure pulsations, which are also well observable in both sidewall gaps domains.

For better comparison with experimental data, more calculated time steps (corresponding to few turns of the impeller) would be necessary.

Experimental data in order to compare friction moments on disks are not available, thus approximate empirical equations which are commonly used in turbomachinery were considered. They are in good agreement with CFD in case of shroud resulting from relative simplicity of the back-side gap geometry which is close to the theoretical cylindrical cavity. In estimation of the disk friction loss on the hub, simplified equations are not very precise. The cross-section is more complicated and the error propagates further to the total friction moment. In such complex geometries, CFD simulation is more appropriate tool for precise estimation of disk friction losses.

Research was supported by project of Ministry of Education, Youth and Sports of the Czech Republic LTARF18019 and by project of BUT FME FSI-S-17-4615.

References

1. J. F. Güllich, Forschung im Ingenieurwesen, **68**, 87–95 (2003)
2. J. W. Daily and R. E Nece, J. of Basic Eng. **82**, 217–230 (1960)
3. R. E. Nece and J. W. Daily, J. of Basic Eng. **82**, 553–560 (1960)
4. D. Coren et al., Proc. of the Institution of Mech. Eng., **223**, 873–888, (2009)
5. B. Hu et al., Int. J. of Turbomachinery, Propulsion and Power, **2**, (2017)
6. B. Hu et al., Int. J. of Turbomachinery, Propulsion and Power, **3**, (2018)
7. S. Mikhail et al., ICFDP7, (2001)
8. A. Nemdili and D. H. Hellmann, The 6th Int. Conf. on Hydraulic Machinery and Hydrodynamics, 235–240, (2004)
9. L. Cho et al., Transactions of the Japan Society for Aeronautical and Space Sci., **55**, 150–156, (2012)
10. K. Matsumara et al., IOP Conf. Ser. Earth Environ. Sci., **240**, (2019)
11. E. Dick et al., Task quarterly, **5**, 579–594, (2001)
12. S. Deniz et al., IOP Conf. Ser. Earth Environ. Sci., **240**, (2019)
13. S. R. Shah et al., Proc. Eng., **51**, 715–720, (2013)
14. S. Yang, et al., Renewable Energy, **48**, 507–513, (2012)
15. A. Ayad et al., Aerospace Sc. and Tech., **47**, 247–255, (2015)
16. B. Launder et al., Annual Rev. of Fluid Mech., **42**, 229–248, (2010)
17. A. Cross and P. Le-Gal, Physics of Fluids, **14**, 3755–3765, (2002)
18. L. Schouveiler et al., Experiments in Fluids, **26**, 179–187, (1999)
19. J. W. Chew, Int. J. of Turbo and Jet Engines, **2**, 253–262, (1985)
20. S. Poncet et al., ECCOMAS CFD 2010, (2010).
21. L. Elena and R. Schiestel, Int. J. of Heat and Fluid Flow, **17**, 283–289, (1996)
22. E. Serre et al., J. of Fluid Mech., **434**, 65–100, (2001)
23. E. Tuliska-Sznitko at al., Comptes Rendus Mécanique, **330**, 91–99, (2002)
24. R. Pasquetti et al., Theoretical and Computational Fluid Dynamics, **22**, 261–273, (2008)
25. E. Séverac and E. Serre, J. of Computational Physics, **226**, 1234–1255, (2007)
26. L. Zemanová and P. Rudolf, Proc. of the Int. conf. Experimental Fluid Mechanics 2018, 705–712, (2018)
27. S. Viazzo et al., Flow, Turbulence and Combustion, **88**, 63–75, (2012)

Design of air nozzle for air curtains

Nikola Žižkovsky[1,*], and *Armand* Drábek[1]

[1]VZLÚ - Czech Aerospace Research Centre, Beranových 130, 19905 Praha 9 - Letňany, Czech republic

Abstract. The procedure, which is used in the design of the air nozzle intended to be used on air curtain, is described in this work. The present work is focused on the isolated flow of one air nozzle, while next work will continue with air nozzles constellation. Nozzle reach in the flow field is monitored with and without tangential flow simulating non-isothermal flow. Several basic shapes of nozzles with the same cross-section are evaluated. After searching the design space of the parametric CAD model simulated in isothermal flow – without tangential flow, the best configuration with the longest range were selected. From this smaller set of nozzles was selected one nozzle with the highest directional stability in non-isothermal flow field. The shape of the nozzle with minimal directional deviation in the non-isothermal flow is subsequently optimized by the adjoint method. This method makes it possible to further reduce the pressure loss in the distribution element by deforming the shape according to the local sensitivity of cost function on shape deformation. The final step of design process is to verify effect of modified nozzle on both isothermal and non-isothermal flow filed.

1 Introduction

The development of all products should lead to an improvement in product properties. In the field of transport technology, on which this project is focused, it can be a wide list of features ranging from passenger's comfort to the energy consumption of vehicles. Recently, there is an emphasis on reducing energy consuption, or emissions of harmful substances during operation. In the case of vehicles, there are several areas suitable for flow optimization. While for high speed transport vehicles is optimal to improve external shape in order to minimize drag forces and in the same time comply, for example in rail vehicles, with respect to the slip stream velocities in the wake, for slower transportation vehicles, where the energy consumption for movement is associated generally with overcoming passive and inertial effects, savings in aerodynamics associated with vehicle air conditioning and ventilation are possible. The common denominator in the energy performance of air conditioning and vehicle ventilation is pressure losses and operational conditions. The pressure losses in the air channels depends on its shape, which is different for each vehicle. Even the operating conditions of individual vehicles may vary, but there is a difference between public transport and individual transport. While in individual transport the vehicle and passengers are moved to the same location, in public transport, passenger rotation is more frequent, leading, in particular in climatically more demanding periods, to higher energy demands for heating or cooling. Depending on the nature of the operation conditions, it is possible to reduce the energy loss caused by the opened entrance door using an air curtain. The benefit of the air curtain depends on the operating cycle, with the increasing ratio of travel time to open door time the importance of the air curtain decreases. Air curtains are beneficial for public transportation with frequent stops. In previous projects, see [1], the dependence of heat loss on the air curtain settings was found.

*e-mail: zizkovsky@vzlu.cz

2 Problem

The basic problem of air curtains is their range, which is shortened due to non-isothermal flow, which mixtures and also bending the stream behind nozzle. For air curtains, in respect of proximity to passanger's heads in the boarding place, it is advisable to limit maximum velocity behind the nozzle, the parameter which affect the air stream range mostly. The air curtain range as parameter is difficult to define, because the velocity profile develops with the distance from the nozzle as well as with the distance from the air stream axis, see [2]. In the case of a non-isothermal flow, the air flow is deformed. For this reason, the air flow behind nozzle is evaluated first for isothermal flow filed, where the nozzle and flow parameters are monitored and then the most promissing configurations for futher optimization are selected. Futher simulations runs with non-isothermal flow field, from which the configurations with minimal deviation of maximum velocity at reference distance from nozzle is selected for optimization.

3 Objectives

The aim of this work is to describe the procedure used in the development of shapes of air nozzles for longer reach of the air curtain. For this purpose several nozzle configuration with simple parameterization are processed. The basic shapes are shown in figure 1. The basic shape of the considered nozzles is the same and only the dimensions and internal parts differ. The oultet area of all nozzles, resp. the sum of the annulus and circles are identical. The basic configuration T0t has the smallest diameter, the variant T1t has the outer diameter larger and has a solid shaft in the nozzle's axis. The T2t variant is similar to the T1t variant, but the solid shaft is replaced by a tube. The diameter of the nozzle T0 is 9 mm, while the outside diameter of the outlet T1 and T2 is 12 mm.

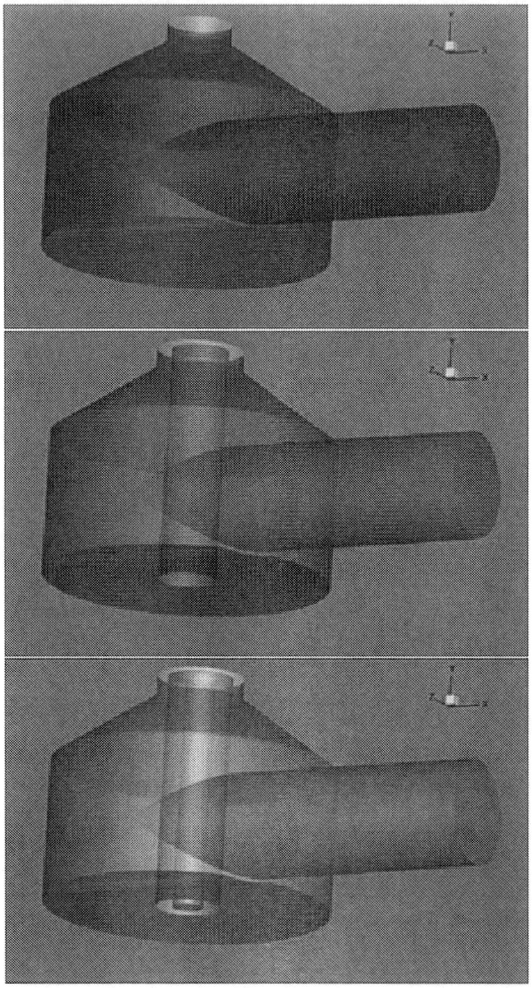

Fig. 1. Geometry from the top: T0t, T1t and T2t for zero angle of tangential input.

4 Methods

4.1 Basic variants

To initially assess the suitability of the air nozzle topology, all three basic variants were simulated in a free isothermal flow field with tangential inlet angles varying from 0 ° (perpendicular to the nozzle's axis) to 90 ° (parallel to the nozzle'saxis). From the analysis of the axial velocities, one to two variants with the highest axial velocity at a distance of 0.4 m from the diffuser were selected from each variant T0, T1 and T2. The automated selection was supplemented by configurations with favorable axis velocities. The simulations was carried on a computational domain with hemispherical shape with radius of 2 m, in which the surface of the hemisphere represented a free stream and the flat plane represented a viscous wall, in which center the nozzle was located. The nozzle is also modeled as a viscous wall. The diffuser has one or two inputs according to the configuration. Selected configurations were subjected to analysis in non-isothermal stream in own solver. The simulations were considered as compressible, viscous and time steady. A coupled solver RANS with W & J EARSM + Hellsten turbulence model was used for k-ω. Wall functions were not used. The computational grid was created in PointWise. The volume is represented by tetrahedrals cells with prismatic cells at the wall surfaces. The computational grid is refined in a distance up to 1 m from the air nozzle. The computational mesh defined in depicted way contains approximately $21 \cdot 10^6$ cells, depending on the configuration.

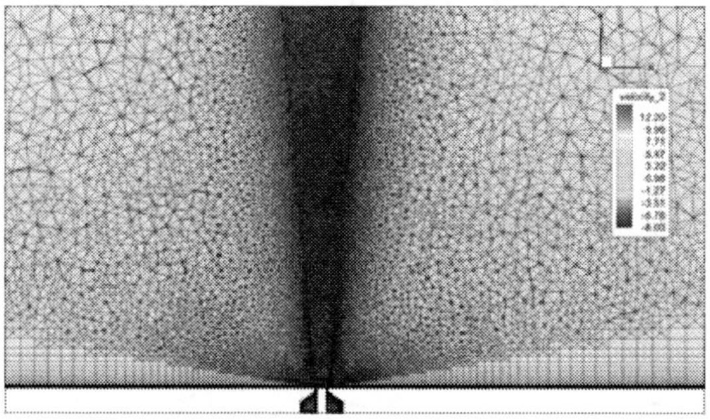

Fig. 2. Computational mesh - T1 variant.

4.2 Non-Isothermal flow

Simulations selected as the most promising in isothermal flow field are simulated in non-isothermal flow field, which is in this case simulated using a constant velocity approaching in the direction perpendicular to the axis of the diffuser. The actual implementation of static buoyancy forces due to different temperatures in the gravitational field leads to physically accurate results, but at the same time the solution imposes numerical instability. The magnitude of the tangential velocity, $1\ m \cdot s^{-1}$, was chosen based on the speeds read from other simulations of the non-isothermal flow on the target air curtains application.

From these variants, three configuration were selected for further processing, of which only the T2t45 variant will be listed here.

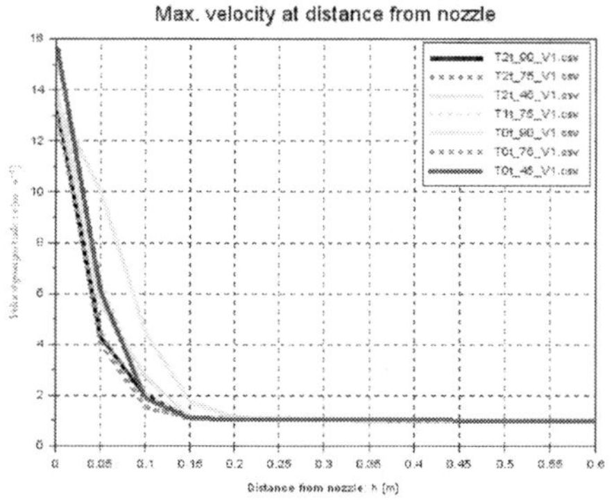

Fig. 3. Distance of maximum velocity behind nozzle.

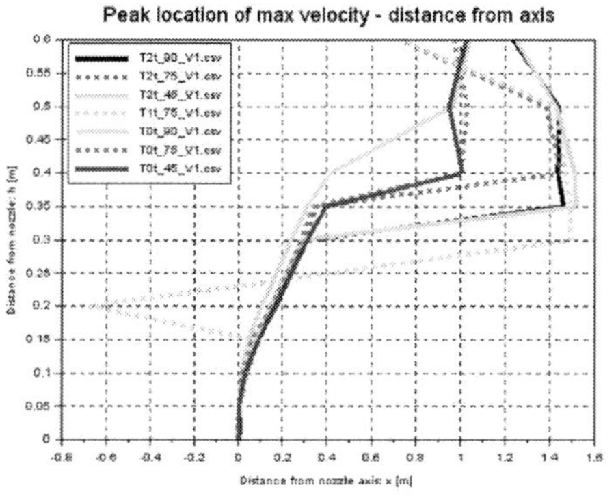

Fig. 4. Maximum velocity peak distance from axis.

4.3 Optimization

The final step of the air nozzle design process is the shape optimization of the considered geometry. Gradient method using Adjoint solver implemented into Star-ccm+ solver is used for optimization. In this case, the optimized geometry is limited only by the interior of the diffuser, see figure 1, without the hemispherical computational domain. The outlet of the diffuser is considered as atmospheric output (*Pressure Outlet*). The boundary condition describing the total state (*Stagnation Inlet*) is used as input, with pressure values corresponding to the total pressure calculated in the previous calculation steps. The computational mesh consists of polyhedral cells in volume, supplemented by prismatic cells on the viscous walls. Generated computational mesh contains abou 921000 cells depending on the configuration. Time steady, compressible, viscous model was used with a turbulence model k-ω SST and a coupled solver. The aim of this optimization process is to minimize the pressure drop across the volume of the diffuser, maximize the uniformity of the outlet speeds and maximize the outlet speed. Based on these requirements, a cost function has been set up, which is maximized by the used solver.

Table 1. Design comparison - monitored parameters.

Parameter	Input geo.	Optimised
Velocity magnitude $[m \cdot s^{-}1]$	11.6074	12.2405
Axial velocity $[m \cdot s^{-}1]$	9.75486	11.2103
Flow uniformity [1]	0.865	0.929
Pressure drop $[Pa]$	43.67	29.11
Cost function	0.2059	0.3765

$$F = 1 - \frac{\sum_f \mid v_f - \bar{v} \mid \cdot A_f}{2 \cdot \mid \bar{v} \mid \cdot \sum_f A_f} \cdot \frac{\sum_f A_f}{\sum_f v_f \cdot A_f} \cdot \Delta p_T^{-1} \quad (1)$$

where v is the speed of the output boundary condition (mean value with overline), A_f is the partial area of the nozzle's boundary condition, p_T is the total pressure, respectively difference in inlet and outlet total pressure.

Figures 5 and 6 shows the comparison of input and optimized geometry.

Numerical comparison of input and final geometry parameters is given in the table 1.

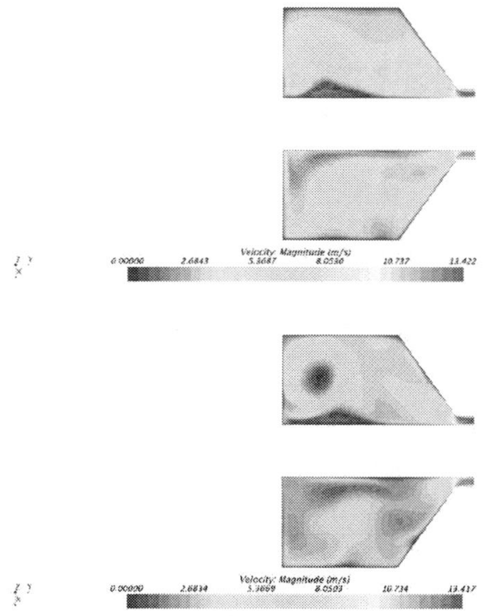

Fig. 5. Nozzle original (upper) and optimized shape - velocity flow field.

https://doi.org/10.1051/epjconf/202226901073

5 Conclusion

The text describes the procedure used in the selection of topologies suitable for the application of circular nozzles as air curtains, which are subsequently optimized by the gradient method with the adjoint solver. Unfortunately, the entire design process, selection and optimization is not fully automated. From whole process, only last mentioned part was automated - the optimisation with predefined cost function. The last step is to verify the proposed shape of the diffuser in the non-isothermal flow. The result of this work is a nozzle with a better cost function value of 83 %, a lower pressure drop of 33 % and a better flow field uniformity of 7 %. The target function also resulted in a 15% increase in the output velocities in the axial direction while the velocity magnitude increased only by 5 %. The target function aligns the original rotating flow.

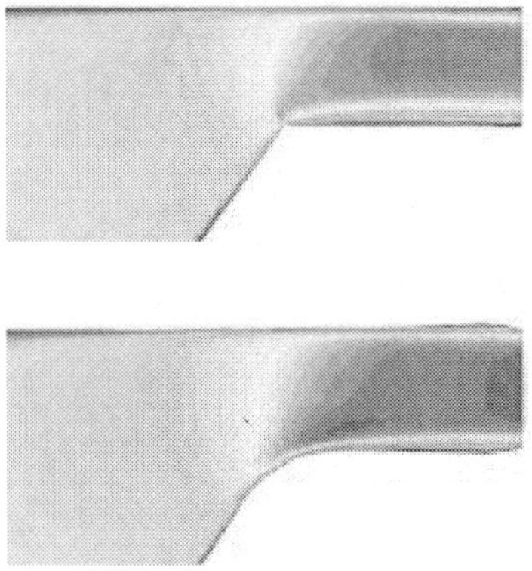

Fig. 6. Nozzle original (upper) and optimized shape - velocity flow field - detail.

In the next part of the project the aim is to create a set of nozzles with lower overall energy consumption, which could replace the slot nozzles used on air curtains.

Acknowledgements. This result was achieved within the institutional support of the Ministry of industry and trade for the development of the research organization.

References

1. N. Zizkovsky, 24th International Conference Current Problems in Rail Vehicles **Volume II.**, 365-371 (2019)

2. E. Janotkova, *Technika prostredi - 1. část* (Vysoké Učení Technické v Brně-Fakulta strojního inženýrství-Odbor termomechaniky a techniky prostˇredí EÚ, Brno, 2011)58

Experimental determination of spatial distribution of granular fractions in flow of bimodal mixture

Štěpán Zrostlík[1,*], *Tomáš* Picek [1], *Lukáš* Svoboda[1], and *Václav* Matoušek[1]

[1]Czech Technical University in Prague, Faculty of Civil Engineering, 166 29 Prague 6, Czech Republic

Abstract. The contribution deals with the determination of the ratio of volumes of particles of individual fractions in flows transporting bimodal granular mixtures. The spatial distribution of the ratio is obtained by analyzing images collected by video recording. The measuring method and post-processing analysis are demonstrated on experiments in our new recirculating fluidization cell. The tested bimodal mixture is composed of two granular fractions of different colors – black and white. The analysis is based on a local identification of particles and their background in the collected images. More precisely, the identification of individual particles is based on the grayscale screening of the images. The method allows to identify both the instantaneous and time-averaged spacious distribution of the grain volume ratio. The experimental results reveal that the measuring method is successful if constant light condition is secured and the medium carrying particles is sufficiently transparent. The method is applicable to laboratory flows with sediment transport no matter whether sediment particles are transported as bed load or as suspended load.

1 Introduction

Transport of solid particles by flowing carrier fluid is common in many industries including mining, dredging, chemistry or food industry. Usually, the transported solids contain particles of different sizes and various investigations have been conducted to look at the effect of the particle size on the transport phenomenon. For instance, Berstein (1962) used the x-ray technique to evaluate the influence of the particle size on solids transport. When it comes to bimodal mixtures, it is preferable to identify the ratio of the individual fractions in the transported bimodal mixture directly during transport. However, the biggest problem associated with such monitoring that the used method should be non-invasive and mostly continuous.

For example, video analysis is a non-invasive method. However, it requires a transparent section of a pipe conveying the mixture, as used by e.g., Mena et al (2005) for an identification of the shape and size of air bubbles. The video was used to identify the flow through the particles by Simon (2005) and Surý (2006). Video can be used for an identification of the total volumetric concentration of solids in flowing mono-size mixture as showed by Spinewine et al (2011) and Krupička et al (2018).

Hlom (2018) used video recordings to determine the ratio of volumes of particles of individual fractions in flows of bimodal granular mixtures composed of a black fraction and a white fraction in open channel flow. The thesis by Hlom summarizes advantages and disadvantages of this method and points out the necessity of constant light conditions.

In this paper, we investigate and further analyse video recordings for a bimodal mixture. Contrary to previous works, the method is tested for the condition of turbulent suspension of particles in which it has not been used yet.

2 Methods

The bimodal suspension experiments were carried out in our new fluidization cell in the Water Engineering Laboratory of the Czech Technical University (CTU) in Prague.

2.1 Materials

Two different solids fractions were used in the tested mixture. Both fractions were plastic grains of similar density and they differed in size and shape. They also differed in colour to make visual observations of the individual fractions in the bimodal mixture more recognisable. Properties of the fractions and the bimodal sediment are summarized in Table 1. Plastic grains of the finer fraction (Hostaform, HSF3) are white, its shape is ellipsoidal, the equivalent sphere diameter d_e = 3.18 mm and the density ρ_s = 1359 kg.m^{-3}. Plastic grains of the coarser fraction (Tiulit, TLT50) are black and of a cylindrical shape with the equivalent diameter d_e = 5.41 mm and the density ρ_s = 1307 kg.m^{-3}.

* Corresponding author: stepan.zrostlik@fsv.cvut.cz

Table 1. Properties of sediments.

Properties	HSF3	TLT50
d_e (mm)	3.18	5.41
v_t (m.s^{-1})	0.131	0.149
ρ_s (kg . m^{-3})	1359	1307
color	white	black

2.2 Fluidization cell

An experiment in a fluidization cell is a suitable way to test a method for the measurement of solids concentration under controlled conditions (Spinewine et al., 2011, Krupička et al., 2017). Recently, we constructed a new fluidization cell based on our previous experience with other cell used previously. The main improvement is a possibility of the recirculation of particles or keeping particles in suspension without recirculation. The fluidization cell has two parts of the same height of 1.4 m. The first part is built from a circular pipe and a propeller was placed into the pipe. The propeller is driven by an electric drilling machine. The second part of the cell has a rectangular cross-section of 20 cm x 11 cm and has transparent walls (Fig. 1). Water flows downstream through the circular part and then upstream through the rectangular part. During our experiments, mixture of water and suspended particles recirculated in the system.

The flow was observed in the upstream part of the setup by a camera Photron Fastcam 1024PCI. This camera has colour resolution. The constant lightness condition was guaranteed by the couple of led lights using daylight shading. The locations of the camera and lights are shown in figure 1. A system of shutters close to the bottom of the rectangular part operated inflow of water and helped to ensure the uniform distribution of flow velocity. Circulation of the mixture through the setup is driven by the propeller.

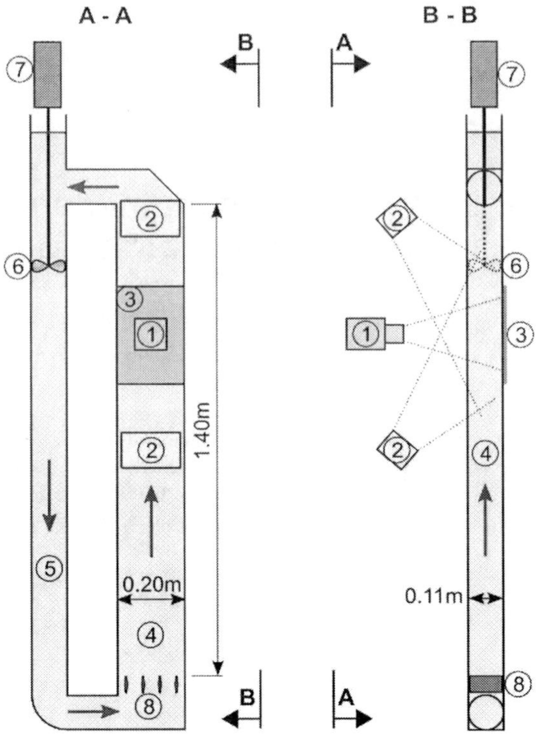

Fig. 1. Schematic sketch of fluidization cell; 1) camera, 2) light, 3) background 4) transparent part of fluidization, 5) pipe part, 6) propeller, 7) electric drilling machine, 8) shutter.

2.3 Experimental work

The experiments were carried out in two series of experiment. The first series started with filling of the cell with black plastic particles only and their recirculation in the mixture driven by the propeller. The total volumetric concentration c was approximately 4%. After some time spent for mixture homogenization, video recordings were carried out. The recorded area was 123x70 mm with resolution 768x432 pixels applied to this area. The high-speed camera was used in a slow regime of 50 fps for the recordings. Each 5th frame was selected for further processing. The reason for this procedure was the guaranties of different picture by its moving. The total record time was more than 30s and the number of analysed frames was 600.

The next step was to change the tested conditions by adding the white-particle fraction to the mixture and to record the new flow conditions. This was recorded with three different modes of shading of background (experiments 2, 3, 4). The middle shade was selected for further experiments.

The addition of white particles and recording of videos continued until the equilibrium between the volumetric concentration of white particles c_w and concentration of black particles c_b was reached (experiments 4-11). At the end of the series all particles were removed from the cell.

https://doi.org/10.1051/epjconf/202226901074

The second series was carried out in a similar way, but only white particles were introduced to the cell at the beginning of the experiment 12. Then black particles were gradually added (12 – 15). Figure 2 shows the order of experiments and the proportion of the concentration of the white particles to the total concentration.

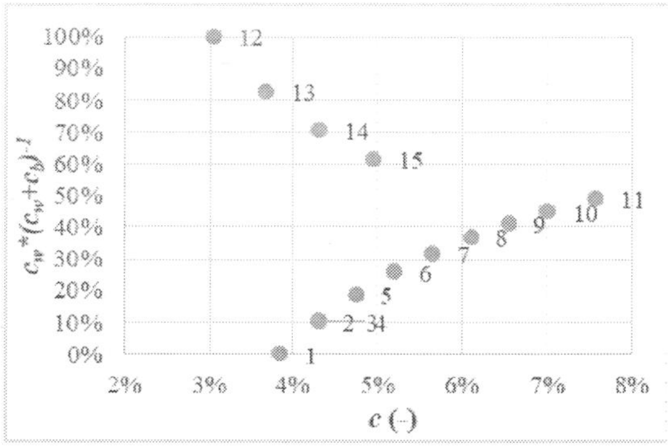

Fig. 2. Order of experiments and proportion of concentration of white particles and total concentrations, numbers in description are order of measurement.

Table 2 summarizes all experiments and gives the volumetric concentration of each particle fraction, the total concentration and the relative concentration of white particles.

Table 2. Cell experiments.

ID	c_b	c_w	$c_w.(c_b+c_w)^{-1}$	c
1	3.85%	0.00%	0.00%	3.85%
2	3.85%	0.45%	10.47%	4.30%
3	3.85%	0.45%	10.47%	4.30%
4	3.85%	0.45%	10.47%	4.30%
5	3.85%	0.90%	18.96%	4.75%
6	3.85%	1.35%	25.97%	5.20%
7	3.85%	1.80%	31.87%	5.65%
8	3.85%	2.25%	36.90%	6.10%
9	3.85%	2.70%	41.24%	6.55%
10	3.85%	3.15%	45.02%	6.99%
11	3.85%	3.70%	49.06%	7.55%
12	0.00%	3.04%	100.00%	3.04%
13	0.63%	3.04%	82.81%	3.67%
14	1.26%	3.04%	70.66%	4.31%
15	1.90%	3.04%	61.62%	4.94%

3 Analysis

The main task of the analysis of video images is to determine the ratio of particle occurrence. As mentioned previously, colour images were taken by the camera. The first step in processing the images was to convert the images to the grey scale. Each pixel has a value between 0 and 255. This number represents the value between white and black. The total white colour is 0 and black is 255.

The analysis consists in identifying the particles and filtering the background in the image. The analysis itself consists of two steps. The first is to find a grayscale boundary where white particles are captured in the image, when black is visible, and when the background of the cell is seen through the mixture. The boundary between the black particles and the background was named the boundary of black b_b and the boundary between the background and the white particles was the boundary of white b_w. The boundaries are illustrated in figure 3.

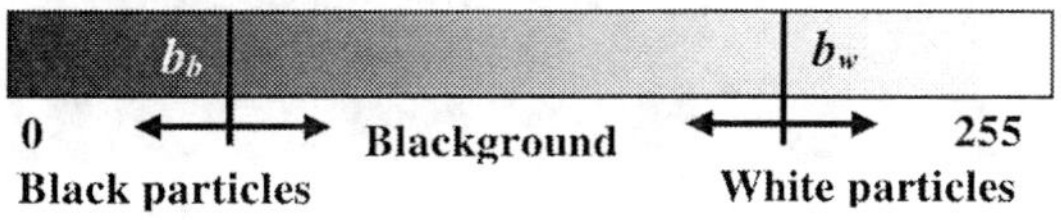

Fig. 3. Range of shade in image and boundary of white and black particles.

An example of the results of the analysis is shown in figure 4. The part a) shows the cut out from raw images for an analysis. The part b) shows identified particles. In the figure, the particles are coloured for a better illustration. White particles are shown as yellow, black particles stay black and background is coloured to red.

The part c) quantifies the relation of white particles to total concentration obtained using the equation

$$c_{relative} = c_w.(c_b+c_w)^{-1} \quad (1)$$

The relation is computed for every column of the analyzed area. In this way, the actual ratio of particles through the flow profile can be determined. The relation for the whole picture is computed as the mean value of values for individual columns. By averaging the values for individual images, we get the value for that experiment.

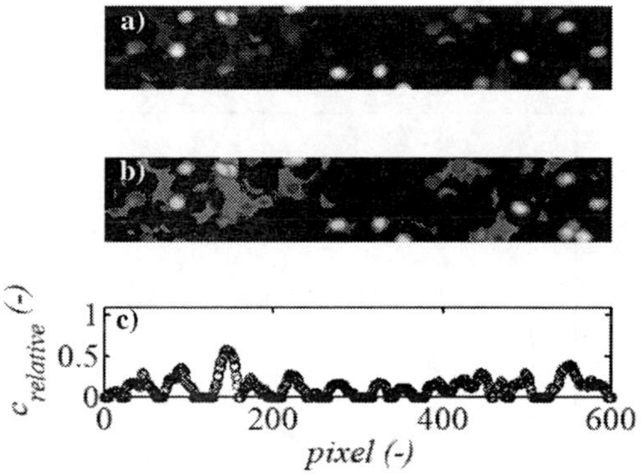

Fig. 4. Example of analyse part of image; a) raw image, b) recognizing area, c) relation of white particles to total concentration in each column of picture.

Pictures have resolution 768x432 pixels. The analysis was carried out at a little smaller area as shown in Figure 5. The analysis itself was preceded by the optimization of boundaries. This was done on an even smaller area, which is also shown in the figure. This area follows from results by Hlom (2017). The minimum size is 6 particle sizes in the direction of flow and more than two thirds of the picture (100x400) was used in the perpendicular direction. The reason for the selected smaller area for optimization was the speed of calculation. Details of optimization are described in the next chapter.

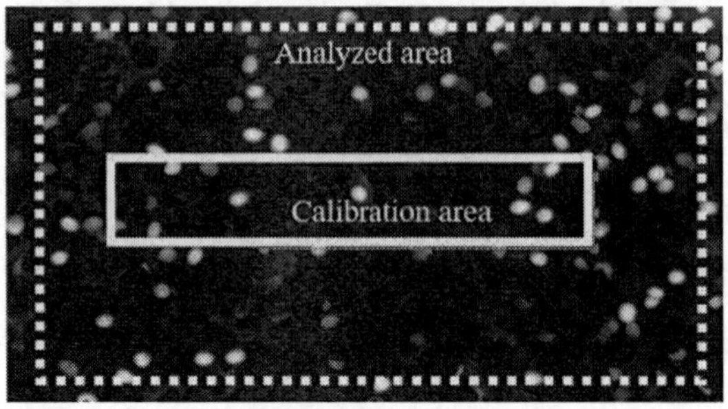

Fig. 5. Example of observed area.

3.1. Optimization of trashed value of boundaries

Two boundaries had to be optimized: the boundary of black and the boundary of white. The reason why the boundaries need to be optimized is the diminishing intensity of light penetrating through the mixture and at the same time the small differences in shades of particles of the same material.

The optimization starts by handling the analysis of darkness of pixels representing each area. It was found that the boundary of black has a smaller scatter on the shade scale. It is in an agreement with the conclusion by Hlom (2017). Based on this finding, the boundary of black to was set to 20.

The range of shade for white particles was wider and the border had to be examined more carefully. The limit value was gradually chosen and the concentration of individual fractions was calculated. Figure 6 shows the result of an optimization for the experiment No. 4. The optimal boundary of the white particles is determined by the point of intersection of the two lines in the plot.

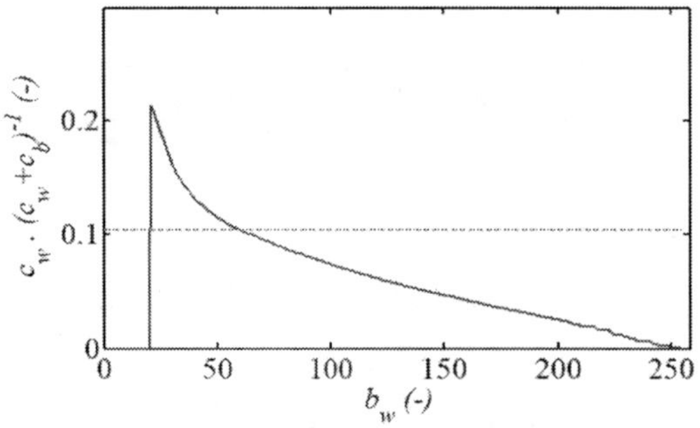

Fig. 6. Result of optimization for experiment No. 4. Blue line is the number of white particles during optimization, red doted line is the measured number of white.

This analysis was done for each mixture experiment and the result of the optimization is plotted in figure 7. It shows how the boundary of white varies with the total concentration. Results for the experiments No. 12 to 14 are missing because no optimum value between value 21 and 255 was found. The solution to this problem is to optimize the boundary of black particles instead of the boundary of white.

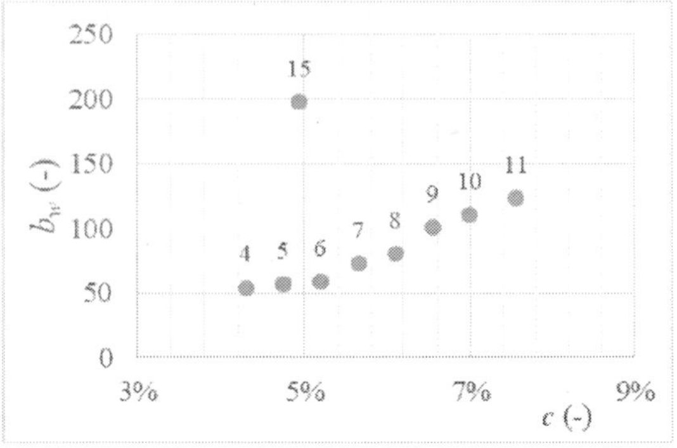

Fig. 7. Optimized boundary of white dependent on concentration.

4 Result and discussion

Since the results of the optimization show that it is not possible to use only one white border value, we discuss how the use of the only one border for white particles influences an evaluation of the other experiments for which the value was not optimized.

For clarity, figure 8 shows the result of relative concentration of white particles for three border values of white. Experiments Nos. 2 and 3 were processed with different colour of background. The effect of the background is shown on the parts of the curves where the experimental values deviate considerably from the curves.

The middle part shows a relatively good match always with better agreement in the area where the limit value was optimized. In general, the method works well in areas with a larger number of black particles when the black border value is not optimized.

The problem with non-optimized values of the boundary of black are at the end of curves where the curves start to deviate from the real values.

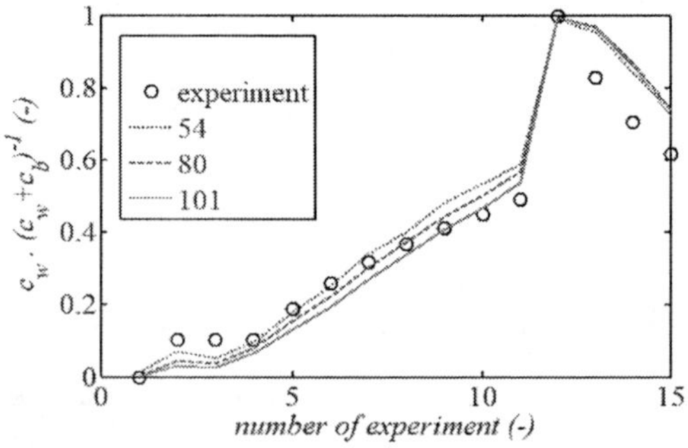

Fig. 8. The result of relative concentration of white particles for three values of boundary white and invariable value for boundary of black.

In order to show the accuracy of the method of determining the volumetric ratio in the mixture, the following figure 9 show parity plot between real value of relative concentration from measurement and evaluate concentration from video. From plot is seen the accuracy of method is with deviation o 10% in area with higher contain of white particles. The maximum deviation is to 20 % in area where there are more black particles than white.

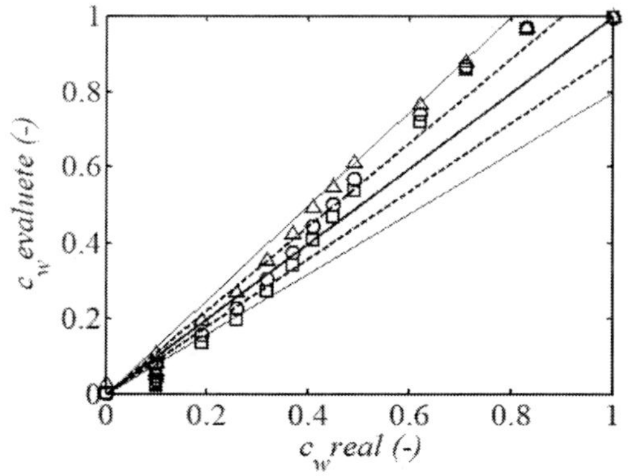

Fig. 9. Parity plot of relative concentration by experiment and by evaluation. Legend: square – evaluation for b_w 101, circle – evaluation for b_w 80, triangle – evaluation for b_w 54, thick line – perfect fit, dash line – deviation 10%, thin line – deviation 20%.

5 Conclusions

A series of experiments was performed to verify the possibility of using video analysis for an identification of the ratio of the local volume concentrations of individual particle fractions suspended in a bimodal mixture.

Previous work suggested that the black border may not be optimized. However, the results of the here presented work indicates that even this boundary must be given due attention.

In general, the method works well in areas with a larger number of black particles when the black border value is not optimized.

The method of determining the ratio has been shown to be suitable for determining the actual or average ratio of particles across flow profile or over time. Assuming a smaller variation in the particle ratio, it can be said that the maximum deviation can be a maximum of 10%.

Acknowledgments

The research has been supported by the Czech Science Foundation through the grant project No. 19-18411S.

References

1. F. Berstein, Adv. X-Ray Anal, **6**, 436-446,(1962)
2. P.C. Menaa, M.N. Ponsb, J.A. Teixeirac, F.A. Rochaa, Chem Eng Sci 60, 5144 – 5150, (2005)
3. M. Simon, *Master Thesis UCT Praha* , (2005)
4. A. Surý, *Master Thesis UCT Praha* , (2006)
5. B. Spinewine, H. Capart, L. Fraccarollo, M. Larcer, Exp. Fluids **50**, 6 (2011)
6. J. Krupička, V. Matoušek, T. Picek, Š. Zrostlík, *EPJ Web of Conferences* 180 (2018)
7. J. Krupička, T. Picek, Š. Zrostlík, ACTA POLYTECH CTU PRAHA, **50**, 3 (2018)
8. J. Hlom, *Master Thesis CTU in Prague* (2018)

AUTHOR INDEX

AUTHOR INDEX